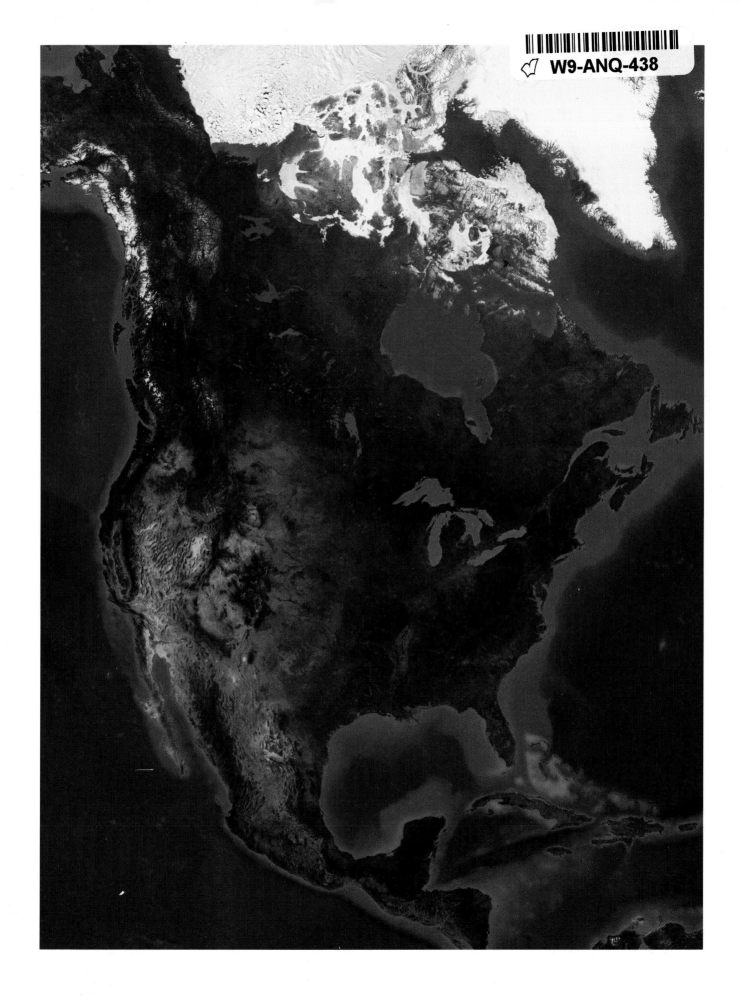

BOOKS IN THE WADSWORTH EARTH SCIENCE AND ASTRONOMY SERIES

Contact your local Wadsworth representative for a review copy of any of the above titles or call Faculty Support at 1-800-423-0563.

EARTH SCIENCE

EARTH SCIENCE

Steven I. Dutch
University of Wisconsin-Green Bay

James S. Monroe
Central Michigan University

Joseph M. Moran
University of Wisconsin-Green Bay

WEST/WADSWORTH
I(T)P   An International Thomson Publishing Company

Belmont, CA • Albany, NY • Bonn • Boston • Cincinnati • Detroit • Johannesburg
London • Los Angeles • Madrid • Melbourne • Mexico City • Minneapolis/St. Paul
New York • Paris • Singapore • Tokyo • Toronto • Washington

Earth Science Editor	Kim Leistner
Senior Media Editor	Pat Waldo
Project Development Editor	Jori Finkel
Editorial Assistant	Elena Bailey
Marketing Manager	Halee Dinsey
Project Editor	Dianne Jensis
Print Buyer	Stacey Weinberger
Permissions Editor	Peggy Meehan
Production Editor	Andrea Cava
Designer	Diane M. Beasley
Photo Researchers	Tobi Zausner, Kathy Ringrose
Copyeditor	Patricia Lewis
Illustrator	Precision Graphics
Cover Designer	Diane Beasley
Cover Image	Fred Hirschmann © Tony Stone Images
Compositor	Carlisle Communications
Printer	R.R. Donnelley & Sons

Page i credit: Face of the Earth, ARC Science Simulations copyright 1997.

For more information, contact Wadsworth Publishing Company, 10 Davis Drive, Belmont, CA 94002, or electronically at http://www.thomson.com/wadsworth.html.

International Thomson Publishing Europe
Berkshire House 168-173
High Holborn
London, WC1V 7AA, England

Thomas Nelson Australia
102 Dodds Street
South Melbourne 3205
Victoria, Australia

Nelson Canada
1120 Birchmount Road
Scarborough, Ontario
Canada M1K 5G4

International Thomson Publishing GmbH
Königswinterer Strasse 418
53227 Bonn, Germany

International Thomson Editores
Campos Eliseos 385, Piso 7
Col. Polanco
11560 México D.F. México

International Thomson Publishing Asia
221 Henderson Road
#05-10 Henderson Building
Singapore 0315

International Thomson Publishing Japan
Hirakawacho Kyowa Building, 3F
2-2-1 Hirakawacho
Chiyoda-ku, Tokyo 102, Japan

International Thomson Publishing Southern Africa
Building 18, Constantia Park
240 Old Pretoria Road
Halfway House, 1685 South Africa

LIBRARY OF CONGRESS CATALOGING-IN-PUBLICATION DATA

Dutch, Steven I.
    Earth science / Steven I. Dutch, James S. Monroe, Joseph M. Moran.
      p.   cm.—(Wadsworth earth science and astronomy series)
    Includes bibliographical references and index.
    ISBN 0-314-20111-4 (pbk. : alk. paper)
    1. Earth sciences.   I. Monroe, James S. (James Stewart)
    II. Moran, Joseph M.   III. Title.   IV. Series.
QE28.D87   1997
550—dc21   97-14327

ABOUT THE AUTHORS

STEVEN I. DUTCH is professor of earth science at the University of Wisconsin-Green Bay, where he has taught since 1976. He received his B.A. in geology from the University of California at Berkeley in 1969 and his Ph.D. from Columbia University in 1976. His research interests are in Precambrian geology and computer applications in geology, and he has done field research in Ontario, Antarctica, and several areas of the United States. For someone who attended Berkeley in the 1960s, he probably seems an unlikely candidate to have done three tours of military duty. Nevertheless, he served in the U.S. Army in 1970–72, including a year in Turkey, and was activated as a Reservist for the Persian Gulf War and for relief efforts in Kurdistan in 1991 and for peacekeeping operations in Bosnia in 1996.

JAMES S. MONROE is professor emeritus of geology at Central Michigan University, Mount Pleasant, Michigan, where he has taught since 1975. He earned his B.A. in geology from California State University at Chico in 1970 and his Ph.D. in geology from the University of Montana in Missoula in 1976. His research interests are in Cenozoic geology and geologic education. He and his colleague Reed Wicander, also of Central Michigan University, have co-authored the market-leading textbooks *Physical Geology, Essentials of Geology, Historical Geology,* and *The Changing Earth,* all published by Wadsworth.

JOSEPH M. MORAN is Barbara Hauxhurst Cofrin Professor in Earth science at the University of Wisconsin-Green Bay, where he has taught since 1969. He received his B.S. in geology in 1965, his M.S. in geophysics in 1967 from Boston College, and his Ph.D. in meteorology from the University of Wisconsin-Madison in 1972. He is a widely published author whose publications include successful textbooks on introductory meteorology and, along with James H. Wiersma and Michael D. Morgan, one of the pioneering textbooks on environmental science. His research interests include analysis of historical weather records in the Great Lakes region. In 1994 he received two signal honors: a Regents Award for excellence in teaching from the University of Wisconsin System, and the Outstanding Alumni Award in science from Boston College.

BRIEF CONTENTS

CONTENTS

CHAPTER 14 — The Atmosphere: Basic Properties 316

CHAPTER 15 — Solar and Terrestrial Radiation 340

CHAPTER 17

Atmospheric Circulation 390

 PROLOGUE 391

CHAPTER 16

Humidity, Clouds, and Precipitation 364

 PROLOGUE 365

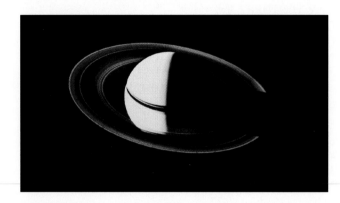

A NOTE TO THE INSTRUCTOR

The goal of this text is to provide undergraduate nonscience students with a contemporary view of the Earth sciences by emphasizing Earth systems and their interconnectedness, the role of unifying processes that link often quite different phenomena, and incorporating the most up-to-date concepts and findings. In addition, we have strived to write this book in a way that makes it as accessible to nonscientists as possible.

One of the most common questions from students is how to study. Students face the problem of finding a proper balance between the details and the processes and interrelationships. Some students attempt to memorize great numbers of facts. Like Gilbert and Sullivan's *Modern Major General,* they have "information vegetable animal and mineral" but fail to understand why the facts are significant or how they relate to other concepts. For example, students might memorize that a drumlin is a streamlined hill, but not realize that drumlins help Earth scientists map ice-flow directions.

Others may learn broad unifying themes but not know enough supporting details to understand the evidence for an idea, or to apply the concept in practice. They might know that glaciers once covered much of North America but not be able to explain how we know that or what it implies for the history of the Great Lakes or the Mississippi River. An understanding of science requires a balance between facts and higher-order concepts.

We have endeavored throughout this book to ask "why should students know this fact?" There seem to be four general justifications for including facts:

1. *Facts are likely to be interesting to students or be of great human significance.* Gemstones, for instance, occupy a place in student interests far out of proportion to their actual geologic or economic significance. Soils are not very glamorous, but they feed the world.
2. *The fact is a necessary preliminary to a more significant fact.* For example, some discussion of energy and simple physics is necessary before attempting to describe the behavior of the atmosphere. Some facts are important because they are supporting evidence for something bigger. Glacial erratics might be mostly curiosities if not for showing that glaciers once covered large areas where they are now absent.
3. *The fact leads to some general principle, scientific law, or fundamental insight into how the Earth works.* Magnetic stripe patterns on the sea floor are significant because they furnish some of the strongest evidence for sea-floor spreading.
4. *The fact helps illustrate how science works, both in terms of actual working techniques, and in terms of how scientists draw conclusions.*

For instance, P- and S-waves do comparatively little damage during earthquakes, but analysis of their travel times gives us most of our knowledge of the Earth's interior. Walter Alvarez's discovery of the evidence for a large impact at the end of the Cretaceous Period is not only a significant contribution to Earth history but a marvelous illustration of the way science often advances through unexpected and accidental discoveries.

Our approach has been to keep terminology to the necessary minimum, employing terms only if they are needed to understand other concepts in Earth science, or if they support some significant finding. We keep boldfaced terms limited to about twenty per chapter, and restrict them to the most basic and essential vocabulary, without which we simply cannot talk about Earth science. We use italics for terms that are significant but not quite as essential. For example, hardness, density, and cleavage are highly diagnostic properties of minerals that are related directly to their atomic structures. Fracture is a property that is considerably more variable and hence not as essential at the introductory level. Often, terms like *precious gem* are italicized to let students know, in passing, what the proper term for some concept is.

Structure and Organization

Two of us, Joseph M. Moran and Steven I. Dutch, began work on this text in 1992. James S. Monroe was later invited to join the project. Dutch wrote the chapters on rocks and minerals, Earth dynamics, and astronomy; Moran the chapters on meteorology; and Monroe those on surface processes and oceanography.

The balance of topics in an Earth science course varies greatly from one instructor to the next. Some attempt to cover all topics briefly, others select topics of special interest and omit others, still others prefer to allot equal time to each of the four subject areas. In trying to meet the needs of a diverse clientele, we have tried to provide adequate coverage of all the most important topics in Earth science, recognizing that we have provided more material than some instructors will use. The solid Earth is inherently far more complex than the atmosphere or the oceans in terms of the number of substances that it comprises and the range of phenomena it exhibits. Thus, we believe that a text that covers geology, meteorology, and oceanography must necessarily spend more time on geology, as we have done. An equally thorough coverage of astronomy would result in an astronomy text, so we have chosen to present only a brief overview of astronomy, with emphasis on the Earth-Moon system, meteoroid impact, and planetary science.

Science textbooks tend to follow one of two basic strategies. The first might be termed "outside in," that is, begin with a broad overview and fundamental concepts and work toward more detailed and specific topics. The other strategy, sometimes termed "inside out," begins with relatively familiar topics and moves to less familiar concepts and topics as students gain familiarity with the subject. Although it is usually necessary to begin with some overview to give students a sense of the nature of the course and to define essential terms and concepts, we believe the outside-in strategy plunges Earth science students headlong into unfamiliar terms and concepts.

We have thus chosen a modified inside-out approach. We begin with an overview of Earth systems (Chapter 1) and Earth materials (Chapters 2 and 3), discuss the solid Earth (Chapters 4 through 11), move on to the ocean (Chapters 12 and 13) and atmosphere (Chapters 14 through 18), and then to the universe beyond the Earth (Chapters 19 through 21). In describing solid Earth systems, we deal first with relatively familiar processes at the surface before delving into the less familiar environment and processes of the deep earth. In the astronomy section, similarly, we begin with the Earth-Moon system, progress to the other planets in the solar system, then to the realm of the stars and galaxies.

Chapter Features

The chapters in our text follow a common organization. We begin with a prologue that seeks to engage student interest by illustrating the practical application or human impact of each chapter topic. Within the chapters, we present additional supplementary material in two ways. There are longer essays called Perspectives. In many chapters, one of the Perspectives deals with an environmental issue and is subtitled Focus on the Environment. Shorter items are presented as small boxes titled Earthfact. Each chapter concludes with a chapter summary, a list of twenty or so key terms, multiple choice and short answer questions, as well as a few questions that demand a bit more analysis. These questions are presented under the heading Points to Ponder. Finally, we include a list of recent additional readings and a short listing of World Wide Web sites. In these, we present a few key Web sites with simple suggestions for accessing some of the more interesting activities. For current updates and more exercises, log on to the Wadsworth Earth Science Resource Center http://www.wadsworth.com/geo.

Author Team

This text is based on the collaboration of three authors each with 20 or more years experience in teaching introductory Earth science classes. In addition, Joseph M. Moran and James S. Monroe are successful authors of other textbooks. The sequencing and balance of topics reflects an approach that we have found successful in the classroom. Within the chapters, the treatment of specific points is often based on approaches that we have found effective in teaching our own classes, and we have made a conscientious effort to anticipate student questions and address them. Such questions as "how do deeply buried metamorphic rocks get to the surface?" or "why did Pangaea break into so many pieces?" are examples. Some of these questions we have not seen addressed in other textbooks.

Qualitative Descriptions

This text is aimed at a nonscientific audience that may include students with little mathematical background. Accordingly, our treatment is mostly qualitative. The few formulas used in the text are mostly to illustrate the way one quantity depends on another rather than for computation. For example, the discussion of meteor impact includes the formula for kinetic energy primarily to show that the energy of an impacting meteor depends on its mass and velocity. We believe that students should have at least an order-of-magnitude appreciation for rates of tectonic uplift, plate movement, erosion and deposition, and climate change. They should also have a similar appreciation for the heights of mountains, the depths of the oceans, and how these quantities compare with the overall size of the Earth. Finally, they should especially appreciate the enormity of geologic time and the distance scales of the universe.

Acknowledgments

Among the many people who assisted us in this project were Joy Phillips, Sharon Dhuey, Pat Bodart, and Barbara Maenpaa; present or former members of the secretarial staff in Natural and Applied Sciences at the University of Wisconsin-Green Bay; and student Pattie M. Dimmer, all of whom assisted in typing at some point or other. This text has a complex history in that it went into production while its publisher was acquired by another company. We salute Jerry Westby of West Publishing, who agreed to the project and gave us the highest level of assistance at every turn, and West staffers Dean DeChambeau, Jana Otto Hiller, and Michelle McAnelly who were deeply immersed in the production process when the turnover occurred. Kim Leistner, our editor at Wadsworth, project development editor Jori Finkel, production editors Dianne Jensis and Andrea Cava, and photo researcher Kathy Ringrose took up the task and saw the project to completion with dedication and enthusiasm.

STEVEN I. DUTCH
JAMES S. MONROE
JOSEPH M. MORAN

REVIEWER LIST

James R. Albanese
State University of New York-Oneonta

Charles Breitsprecher
American River College

Debra Brooks
Rancho Santiago College

Thomas L. Burrus
Miami Dade Community College

Frederick M. Chester
St. Louis University

George S. Clark
University of Manitoba

Emily CoBabe
Indiana University

J. Warner Cribb
Middle Tennessee State University

Ernest L. Kern
Southeast Missouri State University

Roberto Garza
San Antonio College

Marvin Glasser
University of Nebraska-Kearney

Donald Greene
Baylor University

Karen Havholm
University of Wisconsin-Eau Claire

Anne Henry
Sinclair Community College

Greg Holden
Colorado School of Mines

Kenneth Johnson
Wayland Baptist University

Betsy Julian
University of Texas-El Paso

Alan Kafka
Boston College

Michael J. Keables
University of Denver

George W. Kipphut
Murray State University

Kristine Larsen
Central Connecticut State University

Robert D. Lawrence
Oregon State University

Keenan Lee
Colorado College of Mines

Rita Leafgren
University of Northern Colorado

Judy Ann Lowmann
Chaffey College

John A. Madsen
University of Delaware

Craig Manning
University of California-Los Angeles

Neil J. Moloney
California State University-Fullerton

Archie C. Moore
Southeastern Louisiana University

Jonathan Mutter
Lamont Doherty Earth Observatory

Usha Natarajan
University of North Carolina-Greensboro

Marlon A. Nance
California State University-Sacramento

Carl F. Ojala
Eastern Michigan University

Raymond David Perkins
Linn-Benton Community College

Tad Pfeffer
University of Colorado

John E. Poling
California Polytechnic State University

George A. Randall
Gloucester County College

Paul T. Ryberg
Clarion University

Ronda Ryder
Tallahassee Community College

Rich Schweickert
University of Nevada-Reno

Jonathan Scott
State University of New York-Albany

Parvinder S. Sethi
Northeastern Illinois University

Lynn Shelby
Murray State University

Douglas Sherman
College of Lake County

John Silva
University of Massachusetts-Dartmouth

Jonathan Southam
University of Massachusetts-Dartmouth

Morris L. Sotonoff
Chicago State University

James Steart
Chicago State University

J. Robert Thompson
Glendale Community College

Brooke L. Towery
Pensacola Junior College

Gene Walton
Tallahassee Community College

Felicia E. West
University of North Florida

Edmund J. Williams
Ricks College

Carole Ziegler
University of San Diego

◈ SUPPLEMENT LIST

To accompany *Earth Science,* we're pleased to offer a full suite of text and multimedia products. You'll find the product lists below, divided for student and professor use. In each case the supplements specifically designed for this textbook are listed first; general geology supplements follow. In the future we'll also be putting many of our print resources on the Internet for easy access. For current updates and exercises for *Earth Science,* log on to the World Wide Web: http://www.wadsworth.com/geo.

FOR INSTRUCTORS

Instructor's Resource Manual and Test Bank

Steven I. Dutch and Joseph M. Moran (University of Wisconsin-Green Bay)

A full teaching guide that contains chapter outlines and overviews, common misconceptions, critical thinking questions, enrichment topics, learning objectives, lecture suggestions, key term lists, and a bibliography with additional print, multimedia, and Internet resources for each chapter. The authors have also written approximately 75 test questions for each chapter in a variety of formats—multiple choice, true/false, and short answer.
0-314-20769-4

Computerized Testing for IBM and Macintosh

Approximately 2,000 questions in the formats multiple choice, true/false, and short answer are presented.
0-314-21238-8 IBM, 0-314-21234-5 Macintosh

Author's Full-Color Slide Set

The slide set comprises a total of 150 slides—100 photos from the book and 50 additional photos selected by authors.
0-534-53283-7

Wadsworth's Geology Slide Set B

An anthology of over 400 slides featuring key geologic landmarks and materials, both panoramic and zoom views. Geology professors throughout the country have contributed their own photos to make this slide set as comprehensive as possible.
0-314-01213-3

Transparencies

The package contains 150 selections of line art from the text, chosen by the authors.
0-314-20998-0

GeoLink: Wadsworth's Electronic Resource Library for Geology Instructors

GeoLink is a CD-ROM with multitiered indexing, search capabilities, graphs, tables, charts, illustrations, photographs, maps, animations, text, equations, and a glossary. Instructors can search resources to find materials for a classroom presentation or test or to post to an Internet site. It includes a printed user's guide.
0-314-09424-5

Wadsworth's Great Ideas for Teaching Geology

This handbook contains teaching ideas, demonstrations, and analogies gathered from classrooms around the world.
0-314-00394-0

Wadsworth's Earth Science Video Library

The library offers 21 videotapes on topics in geology, oceanography, and meteorology. Qualified adopters can select and exchange from this list. Please order from Wadsworth's marketing department.

FOR STUDENTS

Study Guide

Gustavo Morales and Steven Hardesty (Valencia Community College)

This engaging guide walks students through the 21 chapters at their own pace. The selection of aids includes learning objectives, key word lists, vocabulary and math quizzes, and chapter overviews.
0-314-20888-7

Current Perspectives in Geology, 1998 Edition

Editors Michael McKinney, Parri Shariff, and Robert Tolliver (University of Tennessee-Knoxville)

This book of 50 current readings is designed to supplement any geology or Earth science course and is ideal for instructors who include a writing component in their course. The articles are culled from a number of popular science magazines (such as *American Scientist, National Wildlife, The Economist, Discover, Science, New Scientist,* and *Nature*). It is available for sale to students or shrink-wrapped with any of our Earth science texts.
0-314-20617-5

Geology Workbook for the Web

Bruce Blackerby (California State University-Fresno)

Over 20 chapters to encourage students to explore geology-related sites on the Internet. Chapter exercises range in topics from minerals to plate tectonics. The workbook comes three-hole punched with perforated pages so that worksheets can be handed in as assignments.
0-314-21072-5

Earth Online: An Internet Guide for Earth Sciences

Michael Ritter (University of Wisconsin-Stevens Point)

The first Internet guide written exclusively for the Earth sciences, it can be used as a study guide or Internet resource. It will help you get the most out of the Internet for the following areas: physical geology, geography, meteorology, oceanography, environmental science, astronomy, and global change. It contains material both for the Internet novice who wishes to get on-line and for the more experienced user who would benefit from challenging exercises.
0-534-51707-2

In-TERRA-Active: Wadsworth's Physical Geology Interactive CD-ROM for students

Philip E. Brown (University of Wisconsin, Madison) and Jeremy Dunning (Indiana University)

Interactive CD-ROM lets students explore and simulate key geologic events, such as earthquakes. Students can manipulate variables and data to experience for themselves the cause-and-effect chain of an event. Over 35 minutes of full-motion video and animation to clarify hard-to-illustrate concepts are included. The CD-ROM may be used in the classroom for multimedia lectures or as an assigned study guide. Cross-platformed for Macintosh and Windows.
0-314-04731-x

Formatted for Windows '95
0-534-54297-2

The Wadsworth Earth Science Resource Center

An encyclopedia's worth of links to geology and geography resources, grant information, a map collection, and much more are offered on one level. On another level, you'll find hypercontents, critical thinking questions, and self-quizzes specifically created to accompany *Earth Science*. Log on at http://www.wadsworth.com/geo.

EARTH SCIENCE

CHAPTER

1

Introduction to Earth Science

The Earth as seen from the Moon by the Apollo 8 astronauts in 1968. These were the first human beings ever to see the Earth in its entirety. Even from this distance we can see the white clouds of Earth's dynamic atmosphere and the blue of the hydrosphere, unique in the solar system.

Prologue

As asteroids go, it was only average in size, about 5 km (3 mi) in diameter. After the asteroid had made several close approaches to Jupiter, its orbit changed so that it passed much closer to the Sun and crossed the orbits of Mars and Earth. The new orbit meant that, sooner or later, the asteroid and one of these planets would be at the same place at the same time. One day, 66 million years ago, it happened to be the Earth (□ Fig. 1.1). The asteroid bore down on a small area that would one day be the Yucatán Peninsula of Mexico. At a speed of 30 km/sec (19 mi/sec), the asteroid took about five seconds to penetrate the atmosphere. Nothing under the asteroid lived to see it strike as the atmosphere ahead of it heated to incandescence and compressed to hundreds of times its normal density. All life at the impact site was obliterated in the last hundredths of a second before the asteroid hit.

In a tenth of a second, the asteroid burrowed 3 km (2 mi) deep where its front came to a stop, but the rear of the asteroid, still moving at 30 km/sec, kept on going. The asteroid flattened and spread out, creating a bowl-shaped *transient crater.* Jets of asteroid material and terrestrial rock shot across the floor of the crater and up the sides. Some of the debris was moving almost as fast as the asteroid itself had been. The energy of the asteroid's impact was equivalent to *300 million* Hiroshima-type atomic bombs. Many cubic kilometers of rock were melted and vaporized, and

□ FIGURE 1.1 An artist's conception of the meteorite impact that might have caused the extinction of the dinosaurs

the atmosphere near the impact was heated to thousands of degrees and expanded outward as an incandescent fireball. Any exposed surface within sight of the fireball burned from its radiant heat. As it expanded, the fireball pushed air out of the way, creating a shock wave. For hundreds of kilometers, the shock wave leveled everything in its path; creatures not killed outright by flying debris died of internal injuries caused by the sudden great changes in air pressure.

As the shock wave passed through the Earth, the compressed rock in the floor of the transient crater rebounded to create a central peak, and the walls of the crater collapsed. Within a few minutes of the impact, a crater 200 km (124 mi) across marked the impact site. Vast sea waves hundreds of meters high, triggered by the impact, crashed ashore on nearby coasts. The impact fireball, lighter than the surrounding air, rose at a speed of 200 km/hr to more than 50 km (31 mi). Dust thrown up by the impact began to spread around the world in the upper atmosphere, blocking out the Sun.

Some of the debris ejected from the crater was moving fast enough to escape the Earth and fly off into space. Most of the debris, molten to begin with, heated to incandescence when it reentered the atmosphere; for hours it rained down as brilliant meteors, some falling thousands of kilometers from the impact site. The light and heat radiated by the incoming debris eventually surpassed that of the Sun. Small animals ran for shelter; large dinosaurs may have found shelter or been protected by their thick skins, but plants had no protection. Vegetation thousands of kilometers from the impact dried out and caught fire. The smoke and dust from the impact hid the Sun.

For perhaps a year, the Earth's surface was dark and cold. The heat of the impact had caused nitrogen and oxygen in the air to react, creating nitric acid, which fell as acid rain. The acid rain killed plants and animals, as did the cold, but the real killer was the darkness. Deprived of light, plants died and so did the animals that fed on them.

The dinosaurs may have taken thousands of years to die out after the impact. Different species of dinosaurs likely suffered different fates. So many members of some species may have died as a result of the impact that their numbers dwindled beyond recovery. Others may have starved because critical food plants or prey had been killed off. Others may have struggled on for a time, but faced with competing species that had recovered faster, they were unable to resume their former place in the ecosystem. The dinosaurs who lived in the polar regions were already accustomed to prolonged winter darkness, and they may have been the last survivors.

INTRODUCTION

One of the great mysteries of Earth history is why the dinosaurs suddenly became extinct after flourishing for about 150 million years. The evidence for the impact described in the Prologue is a widespread layer of meteoritic debris in rocks dating from the time of the extinction (Fig. 1.2). Furthermore, a possible crater has been identified, although it is now buried under a kilometer or more of younger rock and sediments on the north coast of Mexico's Yucatán Peninsula. Although the exact size and speed of the asteroid are unknown, the figures used in the Prologue are realistic. Soot deposits found in a number of localities indicate that the incandescent debris ignited widespread fires.

The *impact-extinction hypothesis* for the extinction of the dinosaurs is a good introduction to Earth science. Not only is it an intriguing concept, but the impact affected all of the Earth's major systems (○ Table 1.1). A **system** is a collection of related parts that interact in some organized way (Fig. 1.3). Materials, energy, and information that enter a system from outside are *inputs,* and materials, energy, and information that leave the system are *outputs.* A system can include

■ FIGURE 1.2 The thin white clay layer at about the level of the knee of the person in the photograph represents meteorite debris.

TABLE 1.1 Earth's Systems and the Impact-Extinction Hypothesis

Extraterrestrial: Effects of other celestial bodies on Earth
Orbital interactions between asteroid and other planets that led to collision. Disruption of solar energy to Earth.

Atmosphere: Earth's gaseous envelope
Shock wave, impact fireball, transport of impact dust, heating of impact debris on reentry, reaction of nitrogen and oxygen due to impact heat. Changes in heat trapping and radiation due to suspended particles in upper atmosphere.

Hydrosphere: Seas, lakes, rivers, underground water
Sea waves generated by impact, source of water for acid precipitation.

Cryosphere: Glaciers, permanent ice and frozen ground
Minor effects: there was little or no glaciation on Earth at the time. Cooling may have contributed to glaciation in polar or high-altitude regions, but was not long or severe enough to trigger long-lasting climate change.

Biosphere: Domain of life on earth
Direct death of organisms from impact, heat from reentering debris, acid precipitation, cooling, and lack of light. Destruction of food chains and habitats. Mass extinction of dinosaurs from direct and indirect effects of impact.

Solid Earth: Earth and its interior
Crater formation, shock wave effects within Earth, melting and recrystallization of rocks due to impact.

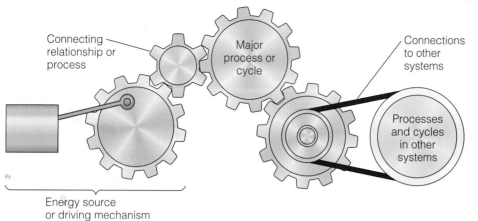

Connecting relationship or process

Major process or cycle

Connections to other systems

Processes and cycles in other systems

Energy source or driving mechanism

FIGURE 1.3 A series of gears can be used to illustrate how some of Earth's systems and processes interact. Pistons and driving rods represent energy sources or driving mechanisms, large gears represent important processes or cycles, and small gears represent connecting processes or relationships. Pulleys are used to show relationships to other systems. Although gears are a useful way to represent systems diagrammatically, remember that real Earth systems are far more complex.

smaller systems, or subsystems, and can also be part of a larger system. An automobile is a familiar example of a system. It consists of an engine system, a brake system, and a fuel system, to name only a few. Each of these smaller systems is connected to one or more other systems, and a change in any one system affects the others. The principal input into the auto is gasoline, and its outputs include motion, heat, and pollutants. At the same time, the automobile is also a component in the traffic flow of a city and thus is part of a still larger system. Like an automobile, the Earth is a system consisting of many related subsystems, all of which are themselves composed of smaller systems. The atmosphere, for example, consists of hundreds of smaller weather systems that constantly form, grow, and break up. Even the solid Earth actually consists of several interconnected systems. Furthermore, Earth itself is part of the solar system and the Milky Way Galaxy, which are larger systems within the universe.

 EARTH SCIENCE

Earth science includes the sciences of geology, meteorology, and oceanography, each of which focuses on one or more of the Earth's major systems (Fig. 1.4). Although the discipline of **geology** is so broad that it is subdivided into

FIGURE 1.4 A schematic diagram of Earth's major systems. The dimensions of the different components are not to scale.

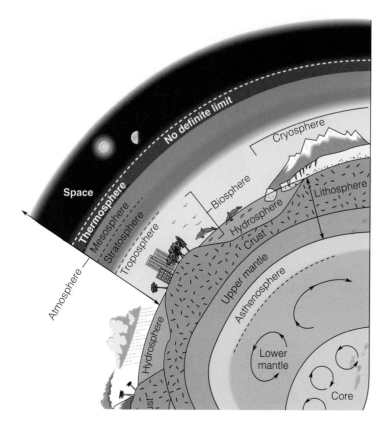

No definite limit

Cryosphere

Space

Thermosphere

Mesosphere

Stratosphere

Troposphere

Atmosphere

Biosphere

Hydrosphere

Lithosphere

Crust

Upper mantle

Asthenosphere

Lower mantle

Core

Hydrosphere

Crust

FIGURE 1.5 Although volcanism is related to Earth's internal processes, specifically melting of rock, it is manifested at the surface. This lava fountain is in Hawaii.

have a significant impact even on land areas that are far from the seashore. By transferring tremendous quantities of heat from the equator toward the polar regions, ocean currents have a profound effect on climates. In addition, vertical circulation in the oceans has important economic consequences because cold, nutrient-rich waters rising from the depths commonly support large fish populations, and some geological resources formed as a result of this same phenomenon.

Because Earth is a planet, affected by other bodies in the solar system, Earth science also includes **astronomy,** the study of the universe beyond the Earth, from nearby objects such as the Moon to the most distant galaxies (Fig. 1.8). Although the Earth is usually not included in astronomy's domain, one must understand geological process on Earth in order to study other planets. Indeed, scientists study extraterrestrial bodies and phenomena by applying laws of physics formulated from experiments and observations conducted mostly on Earth. In a similar manner, geologists, meteorologists, and oceanographers employ concepts and principles from other sciences to study their particular area of investigation.

FIGURE 1.6 Meteorologists study atmospheric phenomena such as this tornado.

many specialties, two general areas of study are recognized—*physical geology* and *historical geology.* Physical geology has three primary concerns: Earth materials, which include minerals and rock; Earth's surface processes, which include such diverse phenomena as running water, glaciers, waves, and wind; and Earth's internal processes, such as earthquake activity, volcanism, and the magnetic field (Fig. 1.5). Historical geology, as the name implies, examines the history of the Earth and its life-forms; the latter are studied through fossils, which are the remains or traces of organisms preserved in rocks.

Meteorology is concerned with the composition, physical properties, and circulation of the atmosphere as well as all other atmosphere-related phenomena (Fig. 1.6). Small- and large-scale atmospheric disturbances including thunderstorms, hurricanes, and various other weather events are of vital concern to meteorologists. Given the impact the weather has on everyone's life, the importance of understanding the atmosphere is clear.

Oceanography involves the study of the physical, chemical, and biological aspects of the oceans as well as the oceans' relationships with the solid Earth and the atmosphere (Fig. 1.7). Through these relationships, the oceans can

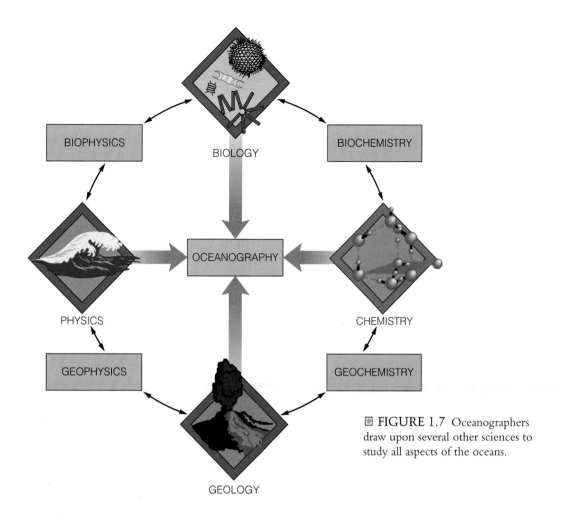

FIGURE 1.7 Oceanographers draw upon several other sciences to study all aspects of the oceans.

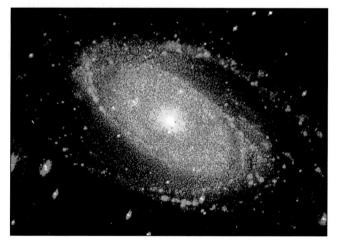

FIGURE 1.8 Among other things, astronomers are interested in the origin and evolution of the universe. After the universe originated, matter condensed to form the stars in this galaxy.

Earth science encompasses such a broad range of sciences because Earth is such a complex system and includes so many interrelated subsystems that no single discipline can do justice to all of them. The next section presents a brief survey of the main features of Earth's major systems.

EARTH AND ITS SYSTEMS

Earth is part of the solar system, which consists of the Sun and its orbiting planets (Fig. 1.9); the solar system in turn is part of the Milky Way Galaxy, which is one of billions of galaxies in the universe (Chapter 21). Accordingly, we can recognize an **Earth–universe system.** Apart from small amounts of natural and artificial nuclear energy, the Sun is the source of all Earth's energy. The amount of energy Earth receives varies, however, as it revolves around the Sun. These variations in the amount of solar energy received, coupled with the tilt of Earth's axis, result in what we call seasons. Earth's distance from the Sun also varies by about 5 million km (3 million mi) over the course of a year, but this variation has only a minor effect on the solar energy received. Earth is farthest from the Sun in July (Northern Hemisphere summer) and closest to the Sun in January (Northern Hemisphere winter). The Moon also affects the Earth, most notably through the tides, the fluctuations in sea level that occur once or twice daily; the Sun also influences the tides but to a lesser degree than does the Moon.

As the impact-extinction hypothesis shows, extraterrestrial influences on Earth can be drastic and unpredictable. Objects a few meters across strike Earth several times a year,

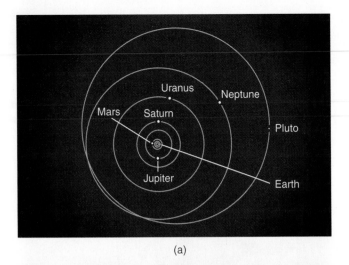

(a)

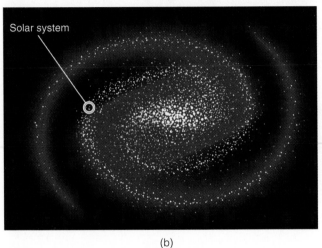

(b)

■ FIGURE 1.9 (a) Earth's place in the solar system. At this scale, Earth is too small to see. If Earth were drawn 1 cm (0.39 in) in diameter, this diagram would be 9 km (5.6 mi) across. (b) Earth's place in the Milky Way Galaxy. The solar system is invisible at this scale. To make the solar system 1 cm in diameter, the diagram would have to be 1,000 km (600 mi) across. The closest star to the solar system would be more than 40 m (130 ft) away.

and meteorite impacts capable of causing damage comparable to that of a large nuclear explosion occur every few thousand years. Indeed, several large explosions occur high in the atmosphere every year. In 1994, a piece of a meteorite that exploded over central Michigan hit and slightly damaged a house.

A second Earth system is the **atmosphere** (Chapters 14–18), which consists of about 79% nitrogen, 20% oxygen, 1% argon, and trace amounts of other gases such as carbon dioxide. The atmosphere regulates the flow of solar energy to and from Earth's surface and transports heat and water vapor from one region to another. The atmosphere also shields Earth's surface from high levels of ultraviolet radiation, charged particles

from the Sun, small meteors, and cosmic rays. The mixing of air masses of different temperatures and humidity drives Earth's weather.

Earth is unique among the planets in the solar system in that much of its surface is covered with liquid water. This water layer of the Earth is the **hydrosphere.** The vast majority of Earth's water is in the oceans (Chapters 12–13), although small but important quantities are in the atmosphere, groundwater, rivers, lakes, and swamps; about 2% of all freshwater on Earth is frozen in glaciers on land (■ Fig. 1.10). The oceans are the great heat reservoir of Earth and the principal source of water vapor for precipitation. Running water and glacial ice, both fed by precipitation, are the principal agents that sculpt the surface of the Earth, although wind can be important locally. Wind-driven waves are also important agents that continually modify seacoasts and the shorelines of large lakes.

The hydrosphere and atmosphere are intimately linked. Water is in constant motion from the liquid state in the hydrosphere to the gaseous state in the atmosphere. Water evaporates from the oceans and is transported as water vapor over the continents, where it falls as rain or snow. Much of the precipitation returns to the oceans in stream and river channels, transporting loose material as it flows downhill and sculpting the landscape (Chapter 5). Some water seeps into the ground, where it chemically attacks the rocks (Chapter 4). This subterranean groundwater may reemerge within hours, or it may infiltrate deep into the rocks, not to emerge for thousands of years. Some water on land is frozen in glaciers where it may remain for thousands of years. Plants and animals use a great deal of precipitation, and some is intercepted for human domestic, industrial, and agricultural use. Eventually, though, all water that falls on the continents returns either to the atmosphere through evaporation or to the sea via rivers and streams. Thus, water is continuously recycled from the oceans to the land and back to the oceans through what is known as the *hydrologic cycle* (Chapter 6).

Another Earth system is the **biosphere,** which includes all areas on Earth occupied by organisms. The biosphere extends from the deepest oceans to about 10 km (6 mi) above sea level. Thus, organisms live upon and within the solid Earth, at all oceanic depths, and in the lower atmosphere (▢ Perspective 1.1). The biosphere plays a profound role in shaping the surface of the Earth. Soil microorganisms assist in the chemical breakdown of rocks to sediment and soil (Chapter 4). Vegetation cover modifies erosion processes. Forests darken the land, absorb solar energy, and help warm Earth's surface. Plants absorb carbon dioxide from the atmosphere and release oxygen, creating the atmosphere as we know it.

The **solid Earth** can also be considered a system. We can characterize Earth as a rocky planet, meaning that it is composed of naturally formed solids consisting of basic constituents known as minerals (▢ Fig. 1.11). The only exception is that part of the Earth's core, representing less than 16% of its total volume, appears to be molten.

☐ FIGURE 1.10 The water frozen in this Alaskan glacier was derived from the oceans and will eventually return to the oceans. This water along with all other water on land is continually recycled through the hydrologic cycle.

(a)

☐ FIGURE 1.11 (*a*) Various minerals are the basic components of rocks. This photograph shows minerals as seen in a thin slice of rock under a microscope. (*b*) The solid Earth is made up of rocks such as these exposed metamorphic rocks in Canada.

(b)

Lost Worlds

Millions of species have lived on Earth and become extinct. If you could travel back in time 50 million years or so, you would find many unfamiliar species and few familiar ones. Yet you would also see forests, grasslands, and coral reefs that resemble those of the present. The particular species of trees, shrubs, and animals occupying the forest would be unfamiliar, for example, but the overall environment would be clearly recognizable as a forest. Have completely unfamiliar environments ever existed on Earth—environments that do not exist today?

About 150 million years ago, both Alaska and Australia lay in polar latitudes. Fossils from this time indicate that the polar environment was temperate enough to support forests, even though these regions would have experienced cold temperatures and weeks or months of darkness each winter. Such *temperate polar environments* do not exist on Earth today, and exactly how plants survived prolonged darkness is still poorly understood. Perhaps they simply became dormant during the winter. Dinosaurs in Alaska could have migrated to warmer climates, but Australia was surrounded by water so dinosaurs there must have been able to survive cold, dark winters. The existence of dinosaurs in this polar temperate environment poses a problem for the impact-extinction hypothesis; if these dinosaurs could survive cold and darkness, why did they become extinct?

Another "extinct" environment existed much more recently. The present environment of northern Alaska and Siberia is *tundra*. On these vast treeless plains, ice near the surface melts during the brief summer, but ice a few centimeters or meters below the surface remains frozen. Meltwater cannot soak into the Earth, but remains on the surface creating wet, marshy conditions. A few thousands of years ago, however, during the ice ages, these areas were dry, cold, polar grasslands with a radically different assemblage of plants and animals.

Dry, cold, polar grasslands do not exist today. Failure to realize how different this environment was has led to some bizarre interpretations of fossils. A number of mammoths, close relatives of present-day elephants, have been found in Siberia and Alaska with vegetation in their stomachs that is not typical of present tundra vegetation. Hoofed animals not well suited to marshy terrain are also preserved as fossils in these areas. Assuming that a grassland habitat must have been warm, some writers have speculated that these fossils mean that the animals were killed suddenly by some mysterious catastrophe. The writers failed to realize that there might once have been environments that do not exist on Earth at present.

Scientists generally depict the Earth as consisting of three concentric layers. The outermost layer, or crust, is a very thin skin, below which lie the mantle and the core (Fig. 1.12). The crust beneath the oceans, or *oceanic crust,* is much thinner but more dense than the crust beneath the continents, or *continental crust.* Together the crust and the upper part of the underlying mantle form the **lithosphere,** which is fragmented into a mosaic of individual segments called plates. These lithospheric plates move with respect to one another, separating in some areas, colliding in others, and simply sliding past one another in others. This concept of moving plates is the basis of *plate tectonic theory,* or what is popularly known as continental drift (Chapter 7).

Lying below the crust at a depth ranging from 5 to 90 km (3 to 55 mi) is the mantle, which constitutes more than 80% of the volume of the entire planet. The mantle is denser than either oceanic or continental crust and is solid except for isolated pockets of molten rock, or magma. At a depth of about 2,900 km (1,800 mi), a rather distinct boundary separates the mantle from the core. Various studies indicate that the core is composed mostly of iron and nickel and consists of a liquid outer part and a solid inner part (see Fig. 1.12).

Perhaps such features as the mantle and core seem so remote as to have little or no effect on other Earth systems. Yet circulating currents in the mantle are probably responsible for the movements of lithospheric plates, and Earth's internal heat is responsible for processes such as volcanism and earthquakes, which can be expressed at the surface even though they are internal processes. The eruption of Mount Pinatubo in the Philippines in 1991 and the Northridge, California earthquake of 1994 are but two examples. Furthermore, Earth's early atmosphere and its surface waters probably formed from volcanic gases derived from within the Earth. Earth's magnetic field is generated in the outer core. It makes compasses work, and ancient terrestrial magnetism preserved in rocks provides clues to the motions of continents in the geologic past.

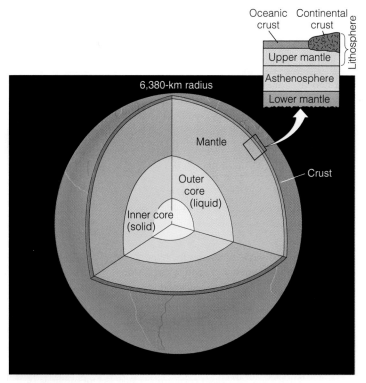

FIGURE 1.12 A cross section of the Earth illustrating the core, mantle, and crust. The enlarged portion shows the relationship between the lithosphere and the underlying asthenosphere.

We have already alluded to the fact that systems do not function in isolation but interact with each other. Earth systems interact in many ways, exchanging matter and energy and recycling them into different forms (○ Table 1.2). Examples of these interactions are numerous. For example, materials of the solid Earth—minerals and rocks—are decomposed and disintegrated by the actions of water, atmospheric gases, and organisms. The resulting materials are commonly transported by water, wind, or glaciers and deposited to form layers that may become soil or a new generation of rocks. Oceanic circulation patterns and their modifying effects on climate are partly controlled by solid Earth processes, particularly moving plates and the origin and evolution of mountains. Present-day plate positions also have a profound effect on the geographic distribution of organisms. Although still controversial, the impact-extinction hypothesis provides a vivid example of how a single event can affect numerous Earth systems (○ Table 1.3).

 EARTH SCIENCE AND THE SCIENTIFIC METHOD

In colloquial usage, the term *theory* means a speculative view of something—hence the widespread belief that scientific theories are little more than unsubstantiated guesses. In scientific usage, though, a **theory** is a coherent explanation for some natural phenomenon that is supported by a large body of objective evidence. Predictions that can be tested by experiment and/or observation are derived from the theory so that its validity can be assessed.

Theories are formulated through the process known as the **scientific method.** This method is an orderly, logical approach that involves gathering and analyzing the facts or data about the problem under consideration. Tentative explanations or **hypotheses** are then proposed to explain the observed phenomena. Next, the hypotheses are tested to see if what they predicted actually occurs in a given situation. Finally, if after repeated tests, a hypothesis is found to be a valid explanation, then it is proposed as a theory. A theory is simply better supported by evidence than a hypothesis is. One should remember, however, that even well-substantiated theories are always questioned and refined as new data become available. In some cases, a new theory may do a better job and replace a previously widely accepted theory.

Each scientific discipline has certain theories that are of particular importance. Biologists and paleontologists rely on the theory of organic evolution to explain certain observations of living and fossil organisms, and Newton's laws of motion are critical to astronomers. *Plate tectonic theory* is particularly important in the study of the solid Earth. It holds that the outer part of the Earth consists of a mosaic of plates that separate, collide, and slide past one another. These plate movements account for several seemingly unrelated phenomena, such as earthquake activity, volcanism, and the origin and evolution of mountain ranges at plate margins.

The impact-extinction concept discussed in the Prologue is a good example of a hypothesis. Researchers have amassed considerable evidence for a huge meteorite impact 66 million years ago at the time the dinosaurs became extinct. This evidence has convinced many scientists that an impact did in fact occur, but not all scientists agree that such an impact necessarily led to extinctions. Furthermore, other hypotheses have been advanced to explain the extinctions. According to one of these, extensive volcanism in what is now India spewed volcanic gases into the atmosphere, bringing about changes that led to the dinosaurs' demise. Perhaps neither of these hypotheses is correct, but both provide a framework for further investigation.

While neither dinosaur extinction hypothesis has been elevated to the status of a theory, many theories are so well supported by evidence that they are firmly established even though they are always subject to question and refinement. No one doubts the validity of the law of universal gravitation or the Copernican theory holding that the Earth revolves around the Sun, rather than the Sun around the Earth. When first proposed, plate tectonic theory was highly controversial and met with skepticism, but evidence amassed during the past several decades has now convinced nearly everyone of its validity.

TABLE 1.2   Interactions of Some Major Earth Systems

Only a few of the most important interactions are listed here. Note that systems can interact with themselves. Also note that many interactions work both ways. For example, the atmosphere provides gases for respiration by organisms, and the gases given off by organisms affect the atmosphere. Chains of interactions also occur: for example, solar heating warms the land, unequal heating of land and water helps drive the winds, which in turn drive ocean currents.

SOURCE SYSTEM	EFFECTS ON				
	Earth-Universe System	Atmosphere	Hydrosphere	Biosphere	Solid Earth
Earth-universe system	Gravitational interaction; recycling of stellar materials.	Solar energy input; creation of ozone layer.	Heating by solar energy; tides.	Solar energy for photosynthesis; daily/seasonal rhythms	Heating by solar energy; meteor impacts.
Atmosphere	Affects escape of heat radiation to space.	Interaction of air masses.	Winds drive surface currents; evaporation.	Gases for respiration; transport of seeds and spores.	Wind erosion; transport of water vapor for precipitation.
Hydrosphere	Tidal friction affects Moon's orbit.	Input of water vapor and stored solar heat.	Mixing of oceans; deep-ocean circulation; hydrologic cycle.	Water for cell fluids; medium for aquatic organisms; transport of organisms.	Precipitation; water and glacial erosion; solution of minerals.
Biosphere	Biogenic gases affect escape of heat radiation to space.	Gases from respiration.	Removal of dissolved materials by organisms.	Predator-prey interactions; food cycles.	Modification of weathering and erosion processes; formation of soil.
Solid Earth	Gravity of Earth affects other bodies.	Input of stored solar heat; diversion of air movements by mountains.	Source of sediment and dissolved materials.	Source of mineral nutrients; modification of habitats by crustal movements.	Plate tectonics, crustal movements.

TABLE 1.3   Earth Systems and the Impact-Extinction Hypothesis

- **Earth-universe system.** Orbital interactions between asteroid and other planets led to collision; disruption of solar energy to Earth.
- **Atmosphere.** Shock wave, impact fireball, transport of impact dust, heating of impact debris on reentry, reaction of nitrogen and oxygen to form acid rain; changes in amount of heat and radiation trapped in atmosphere due to suspended particles in upper atmosphere.
- **Hydrosphere.** Sea waves generated by impact; source of water for acid rain.
- **Biosphere.** Death of organisms from direct or indirect results of impact; destruction of food chains and habitats.
- **Solid Earth.** Crater formation, effects of shock waves within Earth, melting of rocks due to impact.

  TIME AND THE EARTH

The time spans involved in geology and astronomy are almost incomprehensibly vast. Earth scientists speak of millions or billions of years just as routinely as astronomers reckon distance in light-years (■ Fig. 1.13). To gain some conception of the magnitude of **geologic time,** consider that a million seconds is 11.6 days, yet a million years is not very long by geologic standards, especially when one considers that the Earth is 4.6 billion years old and that this 4.6 billion years is only about one-third the age of the universe! Time of such duration is simply beyond human comprehension because we view events from the limited perspective of our own existence.

The dinosaurs are often viewed as failures because they eventually died out. Indeed, the term *dinosaur* is commonly used for something obsolete or inefficient, yet dinosaurs survived for about 150 million years. By contrast, written human history began only 5,000 years ago. For each year of

Accidental Discoveries

When Walter Alvarez made the discovery that led to the impact-extinction hypothesis, he was looking for evidence to unravel the complex crustal movements of Italy. In a thick sequence of limestones near Gubbio, in northern Italy, he noticed a single layer of clay that was deposited at the time the dinosaurs and many other organisms became extinct. In an attempt to estimate how long it took for the clay to be deposited, he and his father, physicist Luis Alvarez, looked for traces of meteorite debris, which falls on Earth at a known rate. They selected the rare metal iridium as an indicator of meteor debris because iridium is much more abundant in meteorites than in most surface rocks. To their surprise, the iridium content of the clay was much greater than expected. This iridium-enriched layer, thought to be far-flung impact debris, has since been found at many other widely separated places (see Fig. 1.2).

The process of accidental discovery is called *serendipity,* after an old tale in which three princes of Ceylon (called Serendip by medieval Europeans) set off on a quest, only to have so many unexpected adventures along the way that they never achieved their original goal. Serendipity accounts for many discoveries in science, and the discovery of the evidence for the impact-extinction hypothesis is one of the best examples.

■ FIGURE 1.13 Rocks provide our record of geologic time. These rocks in the Grand Canyon of Arizona record about 2 billion years of Earth history.

recorded human history, dinosaurs had 30,000 years of existence; for every day, they had 86 years.

Because of the enormous time and space scales involved, Earth scientists often deal with slow processes and infrequent events—at least they seem slow and infrequent from the human perspective. For example, only two volcanoes have erupted in the continental United States during this century, Mount Lassen in California and Mount St. Helens in Washington, both of which are in the Cascade Range of the Pacific Northwest. Certainly, from the human perspective, these eruptions appear to be unusual events, but from the perspective of geologic time, volcanoes are erupting quite regularly in this range. If humans had life spans of thousands of years, our perspective on the frequency of eruptions would be completely different.

CHAPTER SUMMARY

1. Earth science includes the sciences of geology, meteorology, oceanography, and astronomy, all of which rely heavily on other sciences for their concepts and principles.

2. Earth scientists think of the Earth as a system, meaning that it consists of a number of related parts that interact in some way.

3. The Earth's systems include the Earth-universe system, the atmosphere, the hydrosphere, the biosphere, and the solid Earth.

None of these systems operates in isolation; all interact to some degree.

4. Earth is composed of oceanic and continental crust, a mantle comprising most of the planet, and a core. The crust and part of the upper mantle constitute the lithosphere, which is fragmented into several plates that separate, collide, or slide past one another.

5. The scientific method is a logical, systematic approach to solving problems by gathering data and proposing tentative hypotheses.

6. Scientists propose testable hypotheses to explain observed phenomena. If after repeated testing a hypothesis appears to be valid, it may be proposed as a theory. Plate tectonic theory is a good example of a well-established theory.

7. Some aspects of Earth science involve time scales and distances that cannot be readily comprehended by humans. We have no frame of reference for understanding geologic time or distance measured in light-years.

IMPORTANT TERMS

astronomy	Earth-universe system	hypothesis	scientific method
atmosphere	geologic time	lithosphere	solid Earth
biosphere	geology	meteorology	system
Earth science	hydrosphere	oceanography	theory

REVIEW QUESTIONS

1. A scientific explanation needing additional testing to be confirmed is a(n):
   a. _____ fact; b. _____ system; c. _____ hypothesis; d. _____ law; e. _____ principle.

2. Earth's gaseous envelope is its:
   a. _____ hydrosphere; b. _____ biosphere; c. _____ lithosphere; d. _____ atmosphere; e. _____ asthenosphere.

3. According to _____, segments of lithosphere move with respect to one another.
   a. _____ Newton's laws of motion; b. _____ the universal law of gravitation; c. _____ the laws of thermodynamics; d. _____ plate tectonic theory; e. _____ the concept of geologic time.

4. Which part of the Earth makes up most of its volume?
   a. _____ core; b. _____ oceanic crust; c. _____ lithosphere; d. _____ continental crust; e. _____ mantle.

5. Which one of the following is part of the hydrosphere?
   a. _____ cosmic dust; b. _____ water in glaciers; c. _____ the upper atmosphere; d. _____ animals and plants; e. _____ minerals and rocks.

6. Habitats in the biosphere are found in all of the following except:
   a. _____ the lower atmosphere; b. _____ deep oceans; c. _____ Earth's mantle; d. _____ rivers and lakes; e. _____ deserts.

7. According to scientists, the Earth is _____ billion years old.
   a. _____ 1.5; b. _____ 2.0; c. _____ 4.6; d. _____ 14; e. _____ 25.

8. Describe the scientific method, and explain how it may lead to a scientific theory.

9. Name the major layers of the Earth, and describe their general composition.

10. What are the main concerns of astronomy, geology, meteorology, and oceanography?

11. List the major systems discussed in this chapter. Give an example of how systems interact.

12. Why is the impact-extinction concept considered to be a scientific hypothesis?

POINTS TO PONDER

1. Explain why the proposal that a comet or meteorite impact caused dinosaur extinctions qualifies as a scientific hypothesis.

2. Summarize Earth's major systems and give examples of how interactions take place among these systems.

3. Why is the concept of geologic time important in earth science?

Courtillot, V. E. 1990. A volcanic eruption (What caused the mass extinction 65 million years ago?). *Scientific American* 263: 85–92.

Gould, S. J. 1994. The evolution of life on the Earth. *Scientific American* 271: 84–91.

Hartmann, W. K. and R. Miller. 1991. *The History of Earth.* New York: Workman Publishing.

McKinney, M. L., and R. M. Schoch. 1996. *Environmental Science: Systems and Solutions.* St. Paul, Minn.: West Publishing Co.

WORLD WIDE WEB SITES

For current updates and exercises, log on to http://www.wadsworth.com/geo.

AMERICAN GEOPHYSICAL UNION—SCIENCE FOR EVERYONE

http://www.agu.org/sci_soc/everyone.html

This site offers a collection of popular-level articles on many earth science topics of public interest, including several relating to the Chicxulub impact.

UNITED STATES GEOLOGICAL SURVEY—ASK-A-GEOLOGIST

http://walrus.wr.usgs.gov/docs/ask-a-ge.html

An on-line public inquiry forum, this site does not have a Frequently-Asked-Questions (FAQ) page. Before submitting an inquiry, check geology news groups like sci.geo.geology. But beware—responses to inquiries over the Web may or may not be reliable. Be sure your source is knowledgeable, or get responses from several different sources.

CHAPTER

2

Minerals and Mineral Resources

"Steamboat"—red and green tourmaline and colorless quartz crystals. From the Tourmaline King mine, near Pala, San Diego County, California. The specimen is about 28 cm high.

Prologue

Gemstones of one kind or another have been sought for thousands of years. Archaeological evidence indicates that 75,000 years ago people in Spain and France were carving objects from bone, ivory, horn, and various stones. The ancient Egyptians mined turquoise more than 5,000 years ago, and by 3400 B.C. they were using rock crystal (colorless quartz), amethyst (purple quartz), lapis lazuli (a rock composed of a variety of minerals), and several other stones to make ornaments. Turquoise remains a popular gemstone in many cultures, including those of the Native Americans of the southwestern United States.

Beauty and rarity govern the value of gems, along with some practical considerations: the stone cannot be too soft or break too easily, or it will not be durable. Fluorite is commonly transparent and beautifully colored, but it is too soft and breaks too easily to make a good gem. Valuable gems, such as diamonds, also sparkle because they bend or *refract* light strongly. Because fluorite refracts light more weakly than most minerals, it also fails as a gem on this account.

The most valued gemstones today are the *precious gems,* usually considered to include diamond, ruby, and sapphire (gem varieties of corundum); emerald (green beryl); and precious opal. Alexandrite, a variety of the mineral chrysoberyl, is also considered a precious gem because of its color variations; it appears green in sunlight but red under artificial light. Other valued stones, called *semiprecious,* include topaz, aquamarine (blue-green beryl), amethyst (violet quartz), tourmaline, turquoise, and garnet. A number of other minerals are used as *ornamental stones* for decoration but have only minor commercial value. Some minerals are prized for their utility in carving, including alabaster (a variety of gypsum), onyx (banded calcite), rock crystal (transparent quartz), and jade. Low-quality jade is common, but delicately colored jade can be extremely valuable.

A few gems are not true minerals because they are organic in origin. Pearls form around sand grains in marine mollusks, and coral is also produced by marine organisms. Amber, which is fossilized tree resin, often with insects preserved in it, is not widely used in the United States but is popular in Europe. Fossil insects preserved in amber played a key role in the novel and movie *Jurassic Park*. Ivory comes from the tusks of elephants and a few other animals.

The practice of cutting gems probably began when early people found that they could improve the appearance of stones by rubbing off surface imperfections. Over the years, gem cutting has become more systematic. Opaque and translucent gems are commonly cut as *cabochons,* round or oval dome-shaped stones (☐ Fig. 2.1a). Transparent gems are usually cut, or faceted, to yield small plane surfaces (Fig. 2.1b). Originally, faceting

☐ FIGURE 2.1 (*a*) Opaque gems, like this turquoise, are often polished in rounded shapes called cabochons (the stones set in silver). (*b*) Transparent gems, like these diamonds, are faceted to enhance their luster.

(a)

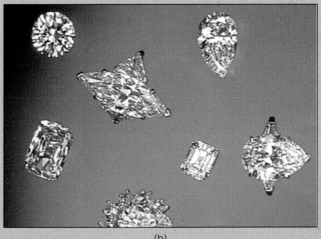

(b)

was merely decorative (Fig. 2.2), but now it is carefully planned to take maximum advantage of the gem's properties. For example, the top facets on a gem are designed to capture as much light as possible, while the rear facets are designed to reflect as much light as possible back out through the front of the gem. Gem cutters may spend months studying a stone before cutting it. Spending thousands of dollars in planning time is worth the cost when a mistake could ruin a stone worth millions.

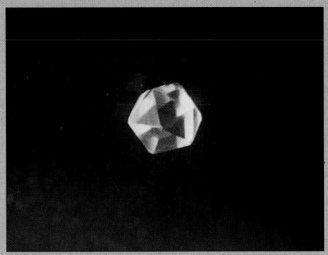

☐ **FIGURE 2.2** This piece of quartz from the Roman city of Ephesus (now in Turkey) was cut 1,800 years ago, making it one of the oldest faceted stones in existence.

 ## INTRODUCTION

The term *mineral* commonly brings to mind dietary substances such as calcium and iron. Actually, however, these are chemical elements, not minerals in the geologic sense. In geology, a **mineral** is defined as a naturally occurring, crystalline, chemical compound. Crystalline means that the compound has an ordered internal arrangement of atoms. All materials are made up of *atoms,* which when joined to other atoms in definite patterns and proportions make up *chemical compounds.* Minerals are simply the chemical compounds that make up rock, which is the main component of the Earth.

In addition to their importance as constituents of rocks, minerals and the rocks they comprise are essential resources upon which all industrialized societies depend. For instance, iron, copper, gold, coal, oil, natural gas, building stone, and many other resources are found in rocks. Many gemstones such as diamonds, topaz, and rubies are minerals that for various reasons we find attractive and valuable (see the Prologue).

 ## WHAT IS THE EARTH MADE OF?

From studies of meteorites (Chapter 19) and materials brought to the surface by crustal movements and volcanic eruptions, scientists have been able to estimate the chemical composition of the Earth. The Earth's interior consists of a *mantle* of rock rich in silicon, oxygen, magnesium, and iron and a *core* of iron, nickel, and probably sulfur (Chapter 8). The Earth as a whole is made mostly of oxygen (35%), iron (24%), silicon (17%), magnesium (14%), and sulfur (6%), with aluminum and calcium each amounting to about 1% (☐ Fig. 2.3a).

The outer part of the Earth beneath the continents, or what is called continental crust, has a different composition than the interior. Notice in Figure 2.3b that iron constitutes only 5.5% of the continental crust, but is the second most abundant element in the Earth as a whole. However, most iron is in the Earth's core. Oxygen (47%) and silicon (27%) make up nearly three-fourths of the continental crust, and together with aluminum, iron, calcium, sodium, potassium, and magnesium, they account for 99% of the crust beneath the continents (Fig. 2.3b). Titanium, hydrogen, phosphorus, and manganese account for most of the remaining 1%.

This list is surprising. Some important resources, such as aluminum and iron, are listed in Figure 2.3b, but copper, lead, zinc, gold, silver, platinum, and uranium are not. These and many other metals make up only a tiny part of the Earth. We can mine them in some localities only because geologic processes have concentrated them to form *ore deposits.*

 ## WHAT IS A MINERAL?

As we noted earlier, a *mineral* is defined as a naturally occurring, inorganic (or nonbiological), crystalline solid. In addition, minerals are characterized by a specific chemical composition and certain physical properties such as hardness, density, and color. About 3,500 minerals have been described and named, and more are being found all the time. Some are very common, but others are so rare they have been found only at one locality.

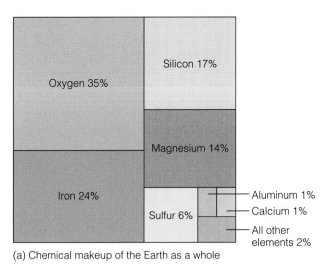

(a) Chemical makeup of the Earth as a whole

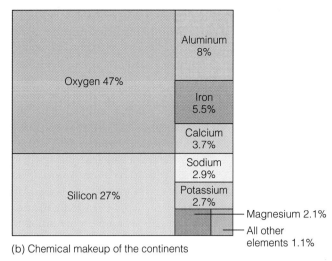

(b) Chemical makeup of the continents

FIGURE 2.3 (*a*) Chemical composition of the Earth as a whole. (*b*) Chemical composition of continental crust. Note that iron comprises 24% of the Earth as a whole but only 5.5% of the continental crust. Most iron is contained in Earth's core. The other elements that together make up 1.1% of the continents are mostly titanium, hydrogen, phosphorus, and manganese. Carbon, an essential element in all organisms, constitutes only 0.03%.

TABLE 2.1   Mineral Names

- *Color:* azurite (azure), olivine (olive green), rhodonite (Greek *rhodos* = red), albite (Latin *albus* = white).
- *Places:* labradorite (Labrador), andesine (the Andes), muscovite (Muscovy, an old name for Moscow), turquoise (Turkey).
- *Uses:* fluorite (Latin *fluere* = to flow; because it makes ores melt more easily), graphite (Greek *graphos* = writing; for its use in pencils).
- *Chemical composition:* cuprite (copper), siderite (Greek *sideros* = iron), uraninite (uranium), calcite (calcium).
- *Properties:* magnetite (from its magnetism), barite (Greek *barys* = heavy).
- *People:* biotite (Jean Biot, a French physicist), sillimanite (Benjamin Silliman, an American chemist).

Minerals are named in a variety of ways. Some have been known since ancient times and have inherited the common names given them by early miners, such as *quartz* and *feldspar.* Some other sources of mineral names are listed in ○ Table 2.1. Minerals are usually named by their discoverer, most often for the place where the mineral occurs, its chemical makeup or properties, or some person the discoverer wishes to honor.

ATOMS AND THEIR STRUCTURE

The fundamental units that make up all familiar materials are small particles known as **atoms** (▣ Fig. 2.4). At the center of an atom is a tiny **nucleus,** which is about 1/100,000 the diameter of the atom, yet contains virtually all the atom's

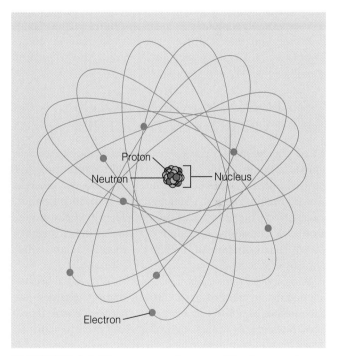

FIGURE 2.4 The structure of an atom. The dense nucleus of protons and neutrons is surrounded by a cloud of electrons arranged in electron shells.

mass. Making up the nucleus are **protons** with positive electrical charges and chargeless **neutrons** with nearly the same mass as protons. Orbiting the nucleus is a swarm of very small, negatively charged **electrons,** which are grouped into distinct **electron shells** (Fig. 2.4). The electrons govern the atom's interaction with other atoms, but the nucleus determines how many electrons each atom has, because the positive protons attract and hold the negative electrons in their orbits.

The identity of an atom is determined by the number of protons in its nucleus, or its **atomic number.** An atom with 1 proton in its nucleus is the chemical element hydrogen, one with 6 protons is carbon, one with 26 is iron, and so on (Appendix I). Thus, each *chemical element,* or simply **element,** is composed of a specific type of atoms. All atoms of an element have the same atomic number, but can have varying numbers of neutrons. The sum of all neutrons and protons in an atom determines its **atomic weight.** For example, oxygen always has 8 protons in its nucleus, but oxygen atoms with 8, 9, or 10 neutrons occur in nature. Atoms of an element with different numbers of neutrons are *isotopes* of that element. Isotopes of an element are chemically identical but may differ in other ways; for example, some isotopes may be radioactive whereas others are stable. Radioactive isotopes play a key role in determining the ages of rocks (Chapter 11).

Each element has a distinctive name and a one- or two-letter symbol (Appendix I). For many elements, such as carbon (C), silicon (Si), and uranium (U), the symbol is simply an abbreviation of the name. The symbols of other elements, such as iron (Fe) and gold (Au), have less obvious origins. These elements were known in ancient times and have symbols based on their Latin names. Iron's symbol Fe comes from *ferrum,* for example, and gold's symbol Au derives from *aurum.*

BONDING: WHAT HOLDS EVERYTHING TOGETHER

Large objects such as planets and stars are held together by gravity, whereas smaller objects are held together by forces between their constituent atoms. The process whereby atoms join into larger structures is called **bonding.** Bonding results from the interaction of the electron shells around atoms. When atoms of different elements bond together, they form a **compound,** such as the mineral quartz, which consists of silicon (Si) and oxygen (O) atoms.

Most matter in everyday life is electrically neutral because an electrically charged object attracts opposite charges until the number of positive and negative charges is equal. If possible, however, atoms can gain or lose electrons to achieve a more stable arrangement of electrons. The most stable arrangement is for an atom to have eight electrons in its outermost shell. Some elements, such as argon and krypton, have that arrangement naturally and rarely combine with other atoms. In other elements, however, the atoms have fewer than eight electrons in their outermost shell and can interact with other atoms to gain or lose electrons to achieve this more stable arrangement. Such atoms, which no longer have equal numbers of negative electrons and positive protons, are electrically charged and are called **ions.**

A simple type of bonding occurs when one atom loses electrons and becomes a positively charged ion, and another atom gains the electrons to become a negatively charged ion. For example, sodium (Na) has a lone electron in its outermost shell, and chlorine (Cl) has seven electrons in its outer shell (Fig. 2.5a). If sodium loses its electron and chlorine gains it, both atoms have eight electrons in their outer shells, but now the sodium atom has a positive charge, and the chlorine has a negative charge. The force of attraction between the oppositely charged ions causes the sodium and chlorine atoms to bond to each other to form sodium chloride (NaCl), the mineral halite, which is better known as table salt (Fig. 2.5b). This sort of bonding, called **ionic bonding,** is the most common kind of bonding in minerals.

Atoms can also bond by sharing outer electrons. The shared electrons orbit both atoms and help fill the electron shells of both. This sort of bonding is called *covalent bonding.* Most carbon-bearing compounds (including most of the compounds in the human body) have covalent bonds. Essentially, people and other living things are held together by covalent bonds.

In metals, electrons wander freely in the space between atoms. In effect, the material consists of atoms held together by *metallic bonds.* The free electrons can move about easily, making metals good conductors of heat and electricity. Layers of atoms in the metal are not bonded tightly together, but can readily slip over one another, making metals easily deformable. The ability to shape metals, which is the basis of our entire industrial civilization, is related directly to the metallic bonding in metals.

In addition to containing ions of a single element, minerals also very commonly contain tightly bonded charged groups called *radicals* that are composed of atoms of different elements but behave as a single unit within the mineral. A common example is the *carbonate radical* (Fig. 2.6), in which three oxygen atoms bond to a carbon atom to form a unit with the formula CO_3, which behaves as an ion with

EARTHFACT

The Not-So-Solid Earth

As a rule, negatively charged ions are larger than positively charged ions. Positively charged ions are usually small because they have given up some of their outermost electrons.

This fact has some astonishing consequences. Oxygen makes up 46 percent of the crust by weight, but because its negatively charged ions tend to be two to three times as large in diameter as the most common positively charged ions, oxygen ions occupy 8 to 27 times more volume and account for about 95 percent of the total volume of all the atoms in the crust. Even more astonishing is another result: space cannot be eliminated no matter how tightly spheres are packed, so about half the volume of most minerals turns out to be interatomic space. By volume, then, the makeup of the apparently solid crust of the Earth is: interatomic space 50 percent, oxygen 48 percent, everything else 2 percent.

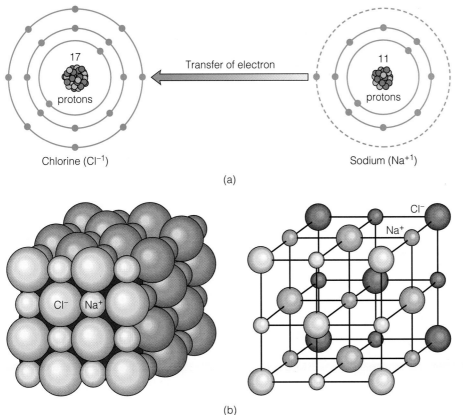

(a)

(b)

■ FIGURE 2.5 Many atoms gain or lose electrons to achieve a more stable arrangement of 8 electrons in their outermost shell. These atoms become electrically charged ions. (*a*) A sodium atom has 11 protons in its nucleus and 11 orbiting electrons. It easily loses its outermost electron and then becomes an ion with a positive charge of 1. A chlorine atom has 17 protons and 17 electrons. It gains an electron to fill its outermost shell and is then a negatively charged ion. The opposite charges of the sodium and chlorine ions attract one another to form an ionic bond. The compound formed by this bond is sodium chloride (NaCl), the mineral halite. (*b*) The crystal structure of halite. The diagram on the left shows the relative sizes of the sodium and chlorine ions, and the diagram on the right shows the location of the ions in the crystal structure.

a -2 charge. The formula CO_3 is a simple **chemical formula** (see ▢ Perspective 2.1). Other common radicals in minerals, with their formulas and charges, are *sulfate* (SO_4, -2), *hydroxyl* (OH, -1), and *silicate* (SiO_4, -4).

 IDENTIFYING MINERALS

Identifying minerals actually makes use of a skill we all use every day: identifying an unknown substance from its physical properties. We have no problem recognizing plastic in all its countless forms or recognizing metal whether it is shiny or dull, clean or tarnished. To identify minerals, we need to be familiar with certain distinctive mineral properties. The most reliable properties of minerals are those directly related to their composition and atomic structure.

The Appearance of Minerals

Color is the most obvious property of any material, but it is of limited use for identification. In everyday life, for example, some materials have distinctive colors, but others come in every imaginable color. Similarly, some minerals have distinctive colors and others do not (▢ Fig. 2.7a). Copper minerals, for example, are commonly bright green or blue, and the colors of metals tend to be fairly constant, but quartz, which is colorless when pure, can be tinted any color by small amounts of impurities. Often the apparent color of a mineral is only a surface coating, and the true color becomes

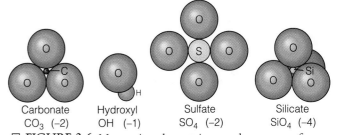

■ FIGURE 2.6 Many minerals contain complex groups of atoms known as radicals that behave as single units.

apparent only when the mineral is broken. Just as many everyday materials change color after long exposure to the sun, rain, and air, minerals can be similarly affected. Therefore we must be certain that any mineral sample we try to identify is unaltered.

By far the most important coloring material in minerals and rocks is iron. *Ferric* iron (iron that has lost three electrons and has a $+3$ electric charge) produces earth tones: red, brown, or yellow. *Ferrous* iron (iron that has lost two electrons and has a $+2$ electric charge) is responsible for the dark green to black colors of many silicate minerals and rocks composed of these minerals. Manganese (pink), chromium (pink, green), and copper (blue, green) are other common elements that sometimes act as coloring agents in minerals.

Understanding Chemical Formulas

The chemical composition of any substance is expressed by a *chemical formula*. For simple substances, the formula consists of the symbol for each chemical element present, followed by a subscript indicating the number of atoms of that element. For example, a water molecule, H_2O, consists of two hydrogen atoms (H) and one oxygen atom (O). A few simple formulas of this sort have already appeared in this chapter. For minerals, whose atoms are bonded in all directions to form a three-dimensional structure, the chemical formula represents the proportions of the atoms present. Thus, halite (NaCl) contains one sodium atom (Na) for every chlorine (Cl), but we cannot say any particular Cl atom belongs to any particular Na atom.

In many minerals, elements can substitute for one another. For example, iron and magnesium often substitute for each other, as in the mineral olivine. We indicate substitution by listing the possible elements in parentheses, separated by a comma. Thus, the chemical formula of olivine is $(Mg,Fe)_2SiO_4$; that is, for each silicon atom, there are four oxygen atoms and two other atoms, which may be either magnesium or iron. Any proportion of iron and magnesium is possible.

Many minerals contain radicals or distinct molecules such as carbonate (CO_3) or sulfate (SO_4). When the molecule is an essential component of the mineral, as in the case of a radical, it is enclosed in parentheses. If the mineral contains more than one such unit, the number of units appears after the

parentheses as a subscript. A very common mineral with a formula of this sort is dolomite, which is $CaMg(CO_3)_2$. In other cases, a molecule is present in a mineral as an "extra." A very common example is water, which may be present in some minerals if the spaces between atoms are large enough. This water is not tightly bonded to the mineral and may be lost if the mineral is heated, so the water molecule is separate from the rest of the formula. Gypsum ($CaSO_4 \cdot 2H_2O$) is one such mineral.

Variable compositions are so common in minerals that it is convenient to have family names that cover an entire series of compositions as well as names for specific chemical compositions. For example, olivine is a general name for a combination of the two minerals *forsterite* (Mg_2SiO_4) and *fayalite* (Fe_2SiO_4). The most widespread series of this type is the *plagioclase feldspar* series, which is a combination of the minerals *albite* ($NaAlSi_3O_8$) and *anorthite* ($CaAl_2Si_2O_8$). Such combinations of two solids are called *solid solutions*.

Finally, some mixtures of closely related minerals are so complex and the amounts of the components are so small and so intermingled that it is impractical, in most cases, to distinguish the individual minerals. Accordingly, geologists give a name to the mixture, while understanding that it is not an exact description. Two of the best-known examples are *limonite,* a mixture of iron oxides and hydroxides, and *bauxite,* a mixture of aluminum oxides and hydroxides that is the principal source of aluminum.

Some of the color variations of minerals can be evened out by crushing the mineral to a fine powder. A convenient way to make such a test is to use a *streak plate,* an unglazed ceramic tile. Rubbing the mineral on the streak plate leaves a trail of fine powder. Hematite has a highly variable appearance—it may be red and earthy, or gray and metallic—but it always leaves a reddish brown streak (Fig. 2.7b).

Luster refers to the way a mineral reflects light. Most minerals can be characterized by one of two broad types of luster: *metallic* and *nonmetallic* (Fig. 2.7c). The distinction between metallic and nonmetallic luster is a fairly reliable property because it is directly related to the type of atomic bonding in the mineral. Metallic luster comes in many forms. For example, aluminum and lead are both recognizable as metals, even though they have very different lusters. Minerals with metallic luster are almost always opaque, meaning that they are not clear enough to transmit light. In contrast, nonmetallic minerals may be opaque, transparent, or translucent, meaning that some light passes through them.

A variety of terms are used to describe the luster of nonmetallic minerals: glassy, waxy, resinous, silky, and so on. These terms are mostly self-explanatory. The term *vitreous* is sometimes used for glassy luster.

Other Properties of Minerals

Density is a reliable property of minerals because it is directly related to the weight of the atoms in a mineral and how closely they are arranged. When we describe lead as "heavier" than aluminum, we really mean "denser." Density, or mass per unit of volume, is usually expressed in grams per cubic centimeter (g/cm^3) or in terms of **specific gravity,** which expresses the density of a material compared to an equal volume of water. Most common minerals have specific gravities in the range of 2.5 to 3.5, which means they are 2.5 to 3.5 times heavier than water. Metallic minerals generally have higher specific gravities than nonmetallics. For instance, the value for galena, the ore of lead, is 7.58, and for pure gold it is 19.3. With practice, it is easy to estimate the "heft" of a mineral with fair accuracy.

Another reliable property of minerals is **hardness,** or resistance to scratching. Although hardness may seem like a fairly trivial matter, it is a direct measure of how tightly the atoms in a mineral are bonded together. On the atomic scale, a nail scratching a mineral is like a snowplow ripping through a drift (▣ Fig. 2.8). The nail scratches the mineral only because the atoms in the nail are bonded more tightly than the atoms in the mineral; if the nail were softer, it would be worn away instead.

The German mineralogist Friedrich Mohs devised a hardness scale by arranging 10 common minerals in order of relative hardness (○ Table 2.2). Mohs placed talc, which is very soft, at 1 on his scale, and diamond, the hardest material known, at 10. Minerals with a high Mohs hardness scratch minerals lower on the scale. Orthoclase (6), for example, scratches fluorite (4), but not the reverse. It is possible to have intermediate hardnesses. A mineral that scratches calcite (3) but not fluorite (4) is said to have a hardness of 3.5.

▣ FIGURE 2.7 (*a*) These specimens of fluorite show that the same mineral can occur in different colors. (*b*) Streak is the color of a powdered mineral. Note that these two varieties of hematite yield similar streaks although the samples are different colors. (*c*) Galena (*left*) shows a metallic luster, whereas orthoclase is nonmetallic.

▣ FIGURE 2.8 Why the simple Mohs hardness scale is so useful. (*a*) A hard material, with the strength of its atomic bonds shown schematically as thicker lines, can scratch a softer (more weakly bonded) material. (*b*) A soft material cannot scratch a hard material but is worn away instead. The Mohs scale is a direct measure of the strength of atomic bonds in minerals.

(a)

(b)

(c)

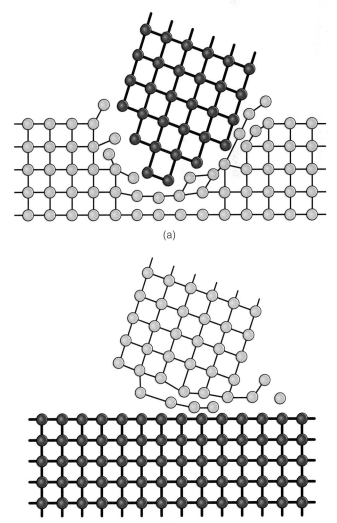

(a)

(b)

TABLE 2.2   Mohs Hardness Scale

HARDNESS	MINERAL	HARDNESS OF SOME COMMON OBJECTS
10	Diamond	
9	Corundum	
8	Topaz	
7	Quartz	
		Steel file (6.5)
6	Orthoclase	
		Glass (5.5–6)
5	Apatite	
4	Fluorite	
3	Calcite	Copper penny (3)
		Fingernail (2.5)
2	Gypsum	
1	Talc	

■ FIGURE 2.9 Quartz showing conchoidal fracture. Fracture is random breakage unrelated to atomic structure.

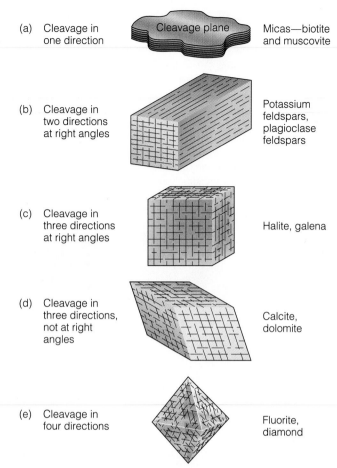

(a) Cleavage in one direction — Cleavage plane — Micas—biotite and muscovite

(b) Cleavage in two directions at right angles — Potassium feldspars, plagioclase feldspars

(c) Cleavage in three directions at right angles — Halite, galena

(d) Cleavage in three directions, not at right angles — Calcite, dolomite

(e) Cleavage in four directions — Fluorite, diamond

■ FIGURE 2.10 Several types of mineral cleavage: (*a*) one direction; (*b*) two directions at right angles; (*c*) three directions at right angles; (*d*) three directions not at right angles; and (*e*) four directions.

To perform a Mohs hardness test, find a large unaltered area on the sample and try scratching the sample with different materials. Also, see if the unknown sample scratches materials of known hardness. Rub the scratch to be sure the mark is a true scratch and not just a streak of powdered mineral. A useful Mohs test can be done with common objects. A fingernail has a hardness of about 2.5, a copper penny about 3, most nails about 5.5, and window glass about 6. The belief that one can identify a diamond by seeing if it scratches glass is a misconception. A diamond will scratch glass, but so will several other minerals such as quartz.

Many minerals break or fracture in distinctive ways. *Fracture* refers to the way a mineral breaks on an uneven surface. Glass, quartz, and many other hard materials break in a smooth, curving manner yielding a *conchoidal* fracture (■ Fig. 2.9). Minerals that consist of fine, parallel crystals sometimes exhibit *fibrous* fracture. Most other fracture terms such as *uneven* or *splintery* are self-explanatory.

A much more useful property than fracture is **cleavage,** a tendency for minerals to break along smooth planes. Cleavage occurs along planes of weakness between atoms and is one of the most consistent and distinctive properties of minerals. Minerals known as micas cleave into thin sheets, amphiboles and pyroxenes into long splinters, halite and galena into cubes, and calcite into skewed boxlike shapes (■ Fig. 2.10). Ironically, diamond, the hardest mineral, has excellent cleavage. In fact, part of the process of diamond cutting is actually accomplished by cleaving specimens along their planes of weakness. Quartz is an example of a common mineral that lacks cleavage.

Crystals, which are solids in which atoms are arranged in regular three-dimensional frameworks, are another distinctive feature of minerals (see □ Perspective 2.2). Well-formed crystals are uncommon, however, and the interpretation of crystal forms is complicated. Nevertheless, crystals of some minerals are common; cubic crystals of pyrite and 12-sided garnet crystals are examples (■ Fig. 2.11).

Some minerals have unusual properties. Magnetite is magnetic, talc has a distinctive feel, graphite writes on paper, halite tastes salty, and calcite fizzes in dilute hydrochloric acid.

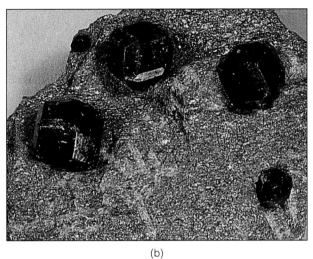

(a)   (b)

■ FIGURE 2.11 (*a*) Crystals of pyrite from Spain—NMNH specimen #R18657. (*b*) Garnet crystals from Alaska.

 NONSILICATE MINERALS

Minerals can be divided into two broad categories: *silicates,* which contain silicon, and *nonsilicates,* which do not. The most common nonsilicate minerals have simpler chemical formulas and crystal structures than most silicate minerals, and we therefore examine them first (◯ Table 2.3).

A few of the most chemically inert elements, including gold and silver, occur uncombined with other elements and

TABLE 2.3   Important Nonsilicate Minerals

NAME	FORMULA	USES
Native Elements: Chemical Elements Not Combined with Other Elements		
Gold	Au	Coins, jewelry, dentistry.
Silver	Ag	Coins, photography, dentistry.
Sulfur	S	Sulfuric acid.
Diamond	C	Jewelry, abrasives.
Graphite	C	Pencil leads, lubricants.
Sulfides: Element Combined with Sulfur (-2)		
Pyrite	FeS_2	"Fool's gold"; minor source of iron; principal contributor to acid runoff and precipitation.
Chalcopyrite	$CuFeS_2$	Principal ore of copper.
Sphalerite	ZnS	Principal ore of zinc.
Galena	PbS_2	Principal ore of lead.
Halides: Element Combined with Chlorine or Fluorine (-1)		
Halite	NaCl	Table salt.
Fluorite	CaF_2	Metallurgy; source of fluorine.
Oxides: Element Combined with Oxygen (-2)		
Hematite	Fe_2O_3	Principal ore of iron. Common rust has the same chemical composition.
Corundum	Al_2O_3	Abrasives. Ruby and sapphire are gem varieties.
Hydroxides: Element Combined with OH (-1)		
Limonite	See Perspective 2.1.	Minor ore of iron; principal coloring agent in soils.
Bauxite		Principal ore of aluminum.
Sulfates: Element Combined with SO_4 (-2)		
Gypsum	$CaSO_4 \cdot 2H_2O$	Principal ingredient of plaster.
Barite	$BaSO_4$	Used in drilling oil wells.
Carbonates: Element Combined with CO_3 (-2)		
Calcite	$CaCO_3$	Main component of limestone; principal ingredient of cement.
Dolomite	$CaMg(CO_3)_2$	Main component of dolostone.

Perspective 2.2

Crystals

The study of mineral crystals, *crystallography*, is one of the most beautiful subjects in the Earth sciences, but also one of the most puzzling to beginners. At first glance, there seems to be no pattern to the way crystallographers classify crystals. In Figure 1, for example, the crystals in the top row look much the same, but crystallographers assign them to different classes. The crystals in the bottom row all look very different, yet crystallographers put them in the same class. The shapes of crystals are not as important as the geometric rules that govern the shapes. These rules, called *symmetry*, also describe how atoms are arranged within the crystal. For example, halite cleaves into cubic fragments because the atoms within a halite crystal have a cubic arrangement. Thus, crystals are a clue to the atomic structures of minerals.

All crystals are made of repeating units of atoms. These repeating units can be pictured as located in imaginary boxes, or *unit cells*. A unit cell is the smallest unit of a crystal that has all the chemical properties of the crystal. Calcite, for example, cleaves into skewed box-shaped fragments in which the angles between the flat surfaces, or crystal faces, are always the same and opposite faces are always parallel. This shape reflects the shape of the unit cell of a calcite crystal. We can account

for all the crystal forms of calcite by stacking these blocks in the right way. Furthermore, the rules for making each shape are always fairly simple, and the simpler the rule, the more likely a shape is to occur as a crystal.

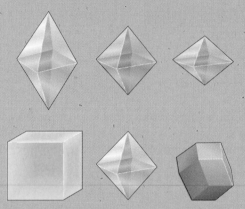

□ FIGURE 1 The crystals in the top row look similar, but crystallographers place them in different classes. The crystal shapes in the bottom row are all quite different, yet they are all in the same class of crystals. It is not the shape of a crystal that is important but the symmetry rules responsible for that shape.

form minerals known as **native elements.** Diamond and graphite are also native elements. Both diamond and graphite are composed of carbon (C), but they are separate minerals because they have different atomic structures. Minerals known as *sulfides* have a positively charged ion combined with sulfur. The sulfides are of great environmental importance because the refining of sulfide ore minerals and the burning of coal that contains the sulfide mineral *pyrite* (FeS_2) are two of the principal causes of *acid precipitation* (Chapter 14). Other important minerals include the *halides,* in which positively charged ions combine with chlorine or fluorine, and the *oxides,* in which positively charged ions combine with oxygen. The *sulfates* contain the complex SO_4 radical, and the *carbonates* contain the CO_3 radical. The common carbonate mineral calcite ($CaCO_3$) is the main constituent of the sedimentary rock limestone. Although most of these minerals are not nearly as common as the silicates, several of them are very important geologic resources.

 ## SILICATE MINERALS

By far the most abundant minerals are the **silicates,** which are composed of silicon and oxygen, which together make up nearly 75% of the Earth's crust by weight. So important is the combination of silicon and oxygen that the two elements in combination are usually referred to as **silica.** Not only is silica abundant, but it is capable of forming a great variety of atomic structures. In silicates, a silicon atom bonds to four oxygen atoms to form a three-sided pyramid, or **silica tetrahedron,** with an oxygen atom at each corner and the silicon atom in the center (□ Fig. 2.12). By sharing oxygen atoms, silica tetrahedra can link together to form rings, chains, sheets, and three-dimensional structures (□ Fig. 2.13 on page 28).

Two subgroups of silicates are generally recognized. One group, known as ferromagnesian silicates, consists of minerals containing iron and magnesium, whereas the other, the nonferromagnesian silicates, lacks these elements. Ferromagnesian silicates are generally darker in color and more dense than nonferromagnesian silicates (□ Fig. 2.14). Several of the most common silicate minerals, including

It is possible to figure out the shape of a mineral's unit cell from its crystal shape. A common analogy that illustrates the process is shown in Figure 2. Stores frequently construct pyramid displays of cans. The slope of the pyramid provides a clue to the shape of the cans, and the shape of the cans is often a clue to the product inside them. Similarly, the crystallographer can reason out the shape of the unit cell from the external shape of the crystal. Several rules aid the crystallographer: the stacking arrangements of unit cells are usually simple, and different faces with the same stacking arrangement often grow at the same rate, so these faces have similar sizes and shapes.

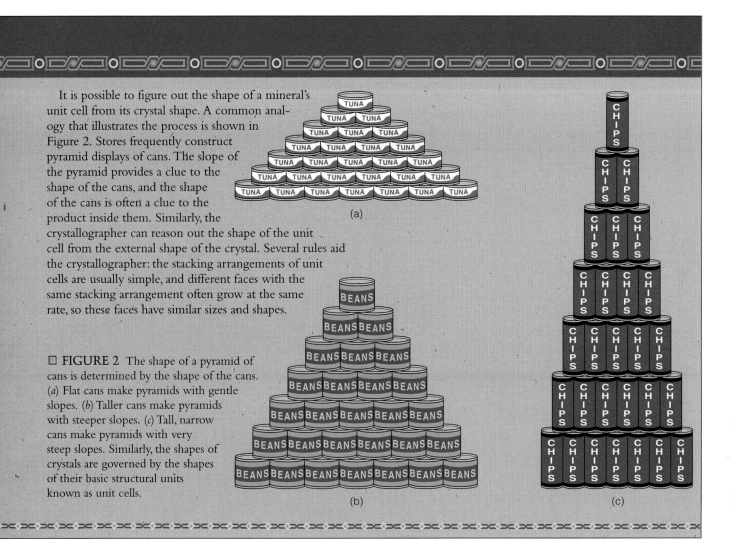

□ **FIGURE 2** The shape of a pyramid of cans is determined by the shape of the cans. (*a*) Flat cans make pyramids with gentle slopes. (*b*) Taller cans make pyramids with steeper slopes. (*c*) Tall, narrow cans make pyramids with very steep slopes. Similarly, the shapes of crystals are governed by the shapes of their basic structural units known as unit cells.

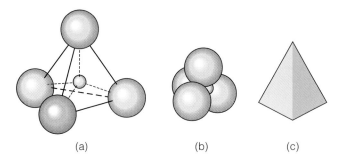

■ **FIGURE 2.12** The silica tetrahedron. (*a*) Expanded view showing oxygen atoms at the corners of a tetrahedron and a small silicon atom at the center. (*b*) View of the silica tetrahedron as it really exists with the oxygen atoms touching. (*c*) Diagrammatic representation of the silica tetrahedron; the oxygen atoms are at the four points of the tetrahedron.

quartz, plagioclase feldspars, and biotite, are known as *rock-forming minerals* to emphasize their importance as constituents of various rocks. Indeed, most rocks are composed largely or entirely of silicate minerals; the only nonsilicate minerals that are particularly common in rocks are the carbonates calcite ($CaCO_3$) and dolomite [$CaMg(CO_3)_2$].

GEOLOGIC RESOURCES

We extract materials from the Earth for purposes as varied as personal adornment to heating our homes. The terms **geologic resource** and *mineral resource* are used to describe useful Earth materials in general, even though many of them are rocks rather than minerals, and some, such as coal and oil, are organic in origin. Geologic resources fall into several broad categories.

1. *Construction materials.* Sand, gravel, limestone (for cement), building stone, and crushed stone. These humble materials account for as much dollar value each year in the United States as all metals combined.
2. *Nonmetallic resources.* Gypsum (for plaster), feldspar (for ceramics), fluorite (for steel making), diamond and corundum (for abrasives), gemstones, and many others.
3. *Metallic resources.* Often subdivided into the *ferrous metals* (iron and steel), the *precious metals* (gold, silver, and platinum), and the *base metals* (copper, zinc, lead, and many others).
4. *Energy resources.* Coal, petroleum, natural gas, uranium, and geothermal energy sources.

FIGURE 2.13 Structures of some of the common silicate minerals shown by various arrangements of silica tetrahedra: (a) isolated tetrahedra; (b) continuous chains; (c) continuous sheets; and (d) networks.

			Formula of negatively charged ion group	Silicon to oxygen ratio	Example
(a)	Isolated tetrahedra (nesosilicates)		$(SiO_4)^{-4}$	1:4	Olivine
(b)	Continuous chains of tetrahedra (inosilicates)	Single chain	$(SiO_3)^{-2}$	1:3	Pyroxene group
		Double chain	$(Si_4O_{11})^{-6}$	4:11	Amphibole group
(c)	Continuous sheets (phyllosilicates)		$(Si_4O_{10})^{-4}$	2:5	Micas
(d)	Three-dimensional networks (tectosilicates)	Too complex to be shown by a simple two-dimensional drawing	$(SiO_2)^0$	1:2	Quartz

Geologic resources are produced by a wide variety of processes. Some are found in volcanic rocks or igneous intrusions (Chapter 3). Aluminum, nickel, and some other metals are concentrated by weathering (Chapter 4). Many building materials, as well as fossil fuels, are recovered from sediments and sedimentary rocks, and other mineral deposits are created by metamorphism, especially by mineral-bearing solutions that alter the composition of rocks (Chapter 3). Deformation of the Earth's crust also creates traps where petroleum and natural gas accumulate. Plate tectonic processes (Chapter 7) govern many of the geologic processes that produce useful resources, and searches for new resources are frequently shaped by an understanding of plate tectonics. Manganese, diamonds, and other resources are known to occur on the ocean floor (Chapter 12), and the technology for extracting these resources is still being developed. The ownership of sea-floor resources is an emerging controversy.

When most people think of mineral exploration, they often think of oil, gold, or precious minerals. In reality, our technological world depends critically on many nonglamorous resources. If all the gem diamonds in the world were to vanish, life would go on largely unaffected; if all the industrial abrasive diamonds were to vanish, our technology would grind to a halt very quickly. Silver is valuable for jewelry and coins, but it is irreplaceable for photography. Without photographic silver, many advanced technologies would be impossible.

EARTHFACT

Asbestos

A silicate mineral with important implications for human health is *asbestos*, which occurs as long, thin, flexible fibers. The properties of the two recognized types of asbestos are directly related to its atomic structure. One type, known as amphibole asbestos, has a fibrous texture. The other variety, called *serpentine,* is a sheet silicate in which the silica sheets curl into a spiral form that is visible under high magnification. Serpentine asbestos is more flexible than amphibole asbestos. Asbestos was originally used for its fire-resistant properties. In the days before fire-resistant construction and smoke detectors, asbestos was considered a lifesaver rather than a danger. After the discovery that asbestos causes cancer in humans, a great effort was made to eliminate asbestos wherever it was found. In many cases, however, the removal process spread more asbestos through the environment than would have occurred if it had simply been left in place; now asbestos is often left undisturbed as long as there is no danger that people will come into contact with it. Some studies indicate that serpentine asbestos dissolves rapidly in the body and poses no threat and that amphibole asbestos is the dangerous variety. However, the danger from any variety of asbestos is still highly controversial.

(a) (b)

◼ FIGURE 2.14 (*a*) Common ferromagnesian silicates: olivine (top left,) augite (top right), hornblende (bottom left), and biotite mica (bottom right); (*b*) some of the common nonferromagnesian silicates: quartz (top left), the potassium feldspar orthoclase (top right), plagioclase feldspar (bottom left), and muscovite mica (bottom right). Nonferromagnesian silicates not only are lighter colored but are also generally less dense than ferromagnesian silicates.

Whether a geologic resource is worth extracting depends on the value of the resource, its quantity, the cost of extracting it, and the distance to the user. For example, most people, given a choice between owning a gravel pit or a gold mine, would pick the gold mine, yet a good sand or gravel pit can be as valuable per unit area as a gold deposit. Gold is valuable, but rare and expensive to extract. Sand and gravel are cheap, but they are plentiful, and virtually everything extracted is salable. On the other hand, gold can be mined profitably anywhere in the world, but sand and gravel must be extracted close to the market to be profitable.

These cost factors explain why the United States imports many resources even though it is rich in minerals. Some resources, like platinum, diamonds, and bauxite (aluminum ore), are simply not abundant in the United States. Others, such as chromium, are abundant, but the ores are not rich enough to extract and refine profitably, so it is cheaper to import chromium from elsewhere. Copper is abundant in the United States but mining costs (labor, taxes, compliance with regulations) are lower elsewhere, so much of our copper is imported as well.

Demonstrated economic resources are termed *reserves*. At any given time, reserves of most materials amount to only a few years' supply, for two reasons. First, mapping reserves is time-consuming and expensive, so most companies do only what is necessary to plan operations for a few years in advance. Second, and more important, in many places reserves are treated as property and are subject to property taxes. For these reasons, many companies map out just enough reserves to plan ahead for a few years at a time.

For the near future, the world's mineral picture is fairly optimistic. Recycling and the use of alternative materials have reduced the demand for new minerals, allowing our known supplies to last longer. The value of recycled metals in the United States each year now exceeds the value of newly mined metals. At the same time, though the decline in demand is good news for manufacturers and consumers,

it has caused a sharp decline in mining in recent years, and there is no immediate prospect of a revival.

EARTHFACT

Precious Metals

Gold, silver, and platinum are all rare, valuable metals used in jewelry, yet gold sets the standard for international monetary exchange. The reason, ironically, is that gold has few other uses. Apart from jewelry, dentistry, and some use in electronics, gold is used almost exclusively for exchange. Silver and platinum are unsuitable for this purpose precisely because they have important practical uses. Silver was used in U.S. coins until 1964, when the rising price of silver for photography made it profitable to melt down coins for their silver content. Platinum sometimes is equal to gold in price, but it is valuable for its resistance to chemical attack and its ability to trigger, or *catalyze,* chemical reactions. When silver and platinum are in short supply, their prices can rise sharply, making them too unstable to serve as standards of monetary value.

At one time in recent history, copper was more valuable than silver. During World War II (1941–1945), copper was essential for artillery shell casings. To conserve copper, some of the facilities engaged in constructing the atomic bomb actually used wiring made from silver furnished by the Treasury Department. The last of this wiring was not replaced until the 1970s. In 1943, because of the copper shortage, pennies were minted from steel. By 1944 the copper shortage had eased, and used artillery shell casings were melted down to make pennies. Collectors refer to these as "shell case" cents.

1. Minerals are naturally occurring, inorganic, crystalline, chemical compounds.

2. Like all materials, minerals are made up of atoms. Each atom consists of a central nucleus of protons and neutrons surrounded by an orbiting swarm of electrons.

3. Each chemical element is made up of a distinct type of atom. All atoms of the same element have the same atomic number, but because the number of neutrons in the nucleus may vary, atoms of the same element can have different atomic weights.

4. The electrons that are important in bonding are those in the outermost shell.

5. In minerals, it is very common for atoms to gain or lose electrons to form electrically charged ions. The attraction between positively and negatively charged ions results in an ionic bond.

6. The physical properties of minerals relate directly to their composition and atomic structure. Hardness, density, luster, and cleavage are properties that are directly governed by the atomic properties of the mineral and are very useful for identification.

7. Color, though obvious, is not a reliable guide to identifying many minerals because many factors can affect color. Color is more consistent for metallic minerals than for nonmetallics.

8. A few minerals, including gold, diamond, and sulfur, consist of a single chemical element and are known as native elements. However, most minerals, including the sulfides, oxides, and silicates, are compounds composed of two or more elements.

9. Many minerals contain groups of atoms, or radicals, that behave as single units. These minerals include the carbonates, which have the CO_3 radical, and the silicates, which have the SiO_4 ion.

10. The silicates are the largest and most varied mineral group. The basic units in silicate minerals are silica tetrahedra, which can link together in many ways. Ferromagnesian silicates contain iron and magnesium, which are lacking in nonferromagnesian silicates.

11. Mineral crystals are geometrically regular forms displayed by minerals. The outward form of a crystal reflects the geometrical arrangement of its atoms.

12. Geologic resources include all minerals and rocks or materials contained within them that are extracted and used. All industrialized societies depend on such resources.

IMPORTANT TERMS

atom	crystal	ion	nucleus
atomic number	density	ionic bonding	proton
atomic weight	electron	luster	silica
bonding	electron shell	mineral	silica tetrahedron
chemical formula	element	native element	silicate
cleavage	geologic resource	neutron	specific gravity
compound	hardness		

REVIEW QUESTIONS

1. The atomic number of an element is determined by the:
   a. ___ number of electrons in its outermost electron shell; b. ___ number of protons in its nucleus; c. ___ diameter of its most common isotope; d. ___ number of neutrons plus electrons in its nucleus; e. ___ total number of neutrons orbiting the nucleus.

2. To which of the following groups do most minerals in Earth's crust belong?
   a. ___ oxides; b. ___ carbonates; c. ___ sulfates; d. ___ halides; e. ___ silicates.

3. When an atom loses or gains electrons, it becomes a(n):
   a. ___ isotope; b. ___ proton; c. ___ ion; d. ___ neutron; e. ___ native element.

4. The two most abundant elements in Earth's crust are:
   a. ___ iron and magnesium; b. ___ carbon and potassium; c. ___ sodium and nitrogen; d. ___ silicon and oxygen; e. ___ calcium and chlorine.

5. Many minerals break along smooth planes and are said to possess:
   a. ___ density; b. ___ cleavage; c. ___ covalent bonds; d. ___ fracture; e. ___ metallic luster.

6. The basic building block of all silicate minerals is the:
   a. ___ silicon sheet; b. ___ oxygen-silicon cube; c. ___ silica tetrahedron; d. ___ silicate double chain; e. ___ silica framework.

7. Minerals are solids possessing an orderly internal arrangement of atoms, meaning that they are:
   a. ___ denser than water; b. ___ crystalline; c. ___ composed of at least three different elements; d. ___ ionic compounds; e. ___ made up of a single element.
8. Calcite (CaCO₃) is a(n):
   a. ___ oxide mineral of great value; b. ___ ferromagnesian silicate; c. ___ common carbonate mineral; d. ___ mineral used in the manufacture of pencil leads; e. ___ important energy resource.
9. Silica is a mixture of:
   a. ___ aluminum and iron; b. ___ silicon and oxygen; c. ___ minerals and rocks; d. ___ quartz and calcite; e. ___ protons and neutrons.
10. Ferromagnesian silicate minerals are rich in:
    a. ___ sodium and calcium; b. ___ potassium and aluminum; c. ___ iron and magnesium; d. ___ sulfur and oxygen; e. ___ chlorine and silicon.
11. The most common type of bonding in minerals is:
    a. ___ covalent; b. ___ metallic; c. ___ hematitic; d. ___ oxide; e. ___ ionic.
12. Native elements are:
    a. ___ composed of at least two elements; b. ___ the most common types of minerals; c. ___ made up of only one element; d. ___ commonly used as energy resources; e. ___ not crystalline.
13. What does it mean to say minerals are crystalline solids?
14. How do compounds and native elements differ?
15. What is a silicate mineral? How do the two subgroups of silicate minerals differ from one another?
16. Describe the mineral property of cleavage, and explain what controls cleavage.
17. What does hardness mean in reference to minerals? What controls mineral hardness?
18. Why is color commonly an unreliable property for identifying minerals?
19. What are some of the factors that determine whether a mineral deposit will be worth mining?
20. What factors determine whether a mineral is also a gemstone?
21. List some geologic resources characterized as construction materials, nonmetallic resources, and energy resources.
22. If diamonds and graphite are both composed of carbon atoms, why are they different minerals?
23. Why is metallic bonding so important to industrialized societies?

POINTS TO PONDER

1. How would the color and density of a rock composed mostly of nonferromagnesian silicate minerals differ from one made up mostly of ferromagnesian silicates?
2. Why must the United States, a nation rich in geologic resources, import a large part of the resources it needs? What are some of the problems created by dependence on imports?
3. Explain how the composition and structure of minerals control such mineral properties as hardness, cleavage, color, and density.

ADDITIONAL READINGS

Cepeda, J. C. 1994. *Introduction to minerals and rocks.* New York: Macmillan.

Dietrich, R. V., and B. J. Skinner. 1990. *Gems, granites, and gravels: Knowing and using rocks and minerals.* New York: Cambridge University Press.

Hochleitner, R. 1994. *Minerals: Identifying, classifying, and collecting them.* Hauppauge, N.Y.: Barron's Educational Series.

Klein, C., and C. S. Hurlbut. 1985. *Manual of mineralogy.* New York: John Wiley & Sons.

Schumann, W. 1993. *Handbook of rocks, minerals, and gemstones.* New York: Houghton Mifflin.

Zoltai, T., and J. H. Stout. 1985. *Mineralogy: Concepts and principles.* Minneapolis: Burgess.

WORLD WIDE WEB SITES

For current updates and exercises, log on to http://www.wadsworth.com/geo.

SMITHSONIAN INSTITUTION
http://galaxy.einet.net/images/gems/gems-icons/html

At this site is a collection of images with brief descriptions of some of the Smithsonian's more spectacular gem and mineral specimens.

UNIVERSITY OF WISCONSIN
http://geology.wisc.edu/online.html

A number of online courses about minerals, crystals, and gems are here, ranging in difficulty from elementary to advanced.

CHAPTER

3

Rocks

Rocks deformed by folding and intruded by small granite dikes, Georgian Bay, Ontario, Canada.

Prologue

At one time, stone was widely used as a building material, but with advances in concrete technology and the proliferation of glass skyscrapers, the building stone industry went into decline. In the United States, it nearly became extinct. Then architects "rediscovered" stone and began to use marble and granite once again in building. As a result, the U.S. stone industry has rebounded dramatically.

Limestone and sandstone are the sedimentary rocks most commonly used for building, granite is the preferred igneous rock, and marble and slate are the most used metamorphic rocks. In the United States, some of the finest marble has come from the Yule marble quarry★ in Colorado. This stone, which rivals the finest European marble, was used in the Lincoln Memorial. Yule marble is also well known to scientists who have used it in many experimental studies on the behavior of rock under high pressure.

Another famous landmark made of Yule marble is the Tomb of the Unknowns (formerly the Tomb of the Unknown Soldier) in Arlington National Cemetery in Virginia. When the stone for the tomb was excavated in 1931, it was the largest block of marble quarried up to that time. The stone was so large and heavy that it taxed the ingenuity of quarry workers and transport facilities to the limit. It took a year to quarry the first block, which weighed 51 metric tons. When the block was finally freed, it turned out to have concealed flaws and was discarded. Quarrying a second stone, along with a third for insurance, took another year. The second stone was acceptable, and the third was simply set aside.

Over the next years, however, the market for marble declined until the Yule quarry was forced to close in 1941. It remained closed until 1990, when the revival of the stone market made it profitable to reopen. Former quarry workers, some long since retired, returned to oversee the operations. An Arizona sculptor, upon learning the quarry was again active, asked what had happened to the third block once intended for the Tomb of the Unknowns. Finally, a retired quarry worker was found who recalled the fate of the stone. It had been pushed aside and forgotten. Over the years, excavations had lowered the quarry floor, leaving the stone stranded in a niche 18 m (60 ft) above the floor. There it sits to this day, too heavy and too inaccessible to lift easily. Nobody can figure out a practical and economical way to get the block out without destroying it.

★A quarry is a surface excavation, usually for building stone.

INTRODUCTION

The Earth is an active planet inside and out. Liquid water on the surface and plate tectonics combine to rework Earth materials endlessly. Not only do erosion and weathering modify the Earth's surface, but igneous activity brings new material from its interior, and movements of the crust fold and fracture rocks. Pressure and high temperatures within the crust often change rocks so thoroughly that their original form is hard to recognize. Movements of the crust also cause some rocks to be deeply buried, while others are brought to the surface to be attacked by air, water, acids, and the activities of organisms.

ROCKS AND THE ROCK CYCLE

Rocks and the minerals composing them are the fundamental materials of the Earth. Indeed, rocks are simply aggregates of one or more minerals. Some rocks are made up of a single mineral. Limestone, for instance, is composed of the mineral calcite. Most rocks, however, are mixtures of several minerals. Granite, made up mostly of feldspars and quartz, also usually contains other minerals in smaller amounts.

Three great families of rocks are recognized, each of which is defined by the processes that form the rocks. Rocks that solidify from the molten state are **igneous rocks** (Latin *ignis* = fire). Igneous rocks may form from molten material that cools within the crust or at the surface, or they may result from consolidation of fragmental material ejected from volcanoes during explosive eruptions. **Sedimentary rocks**

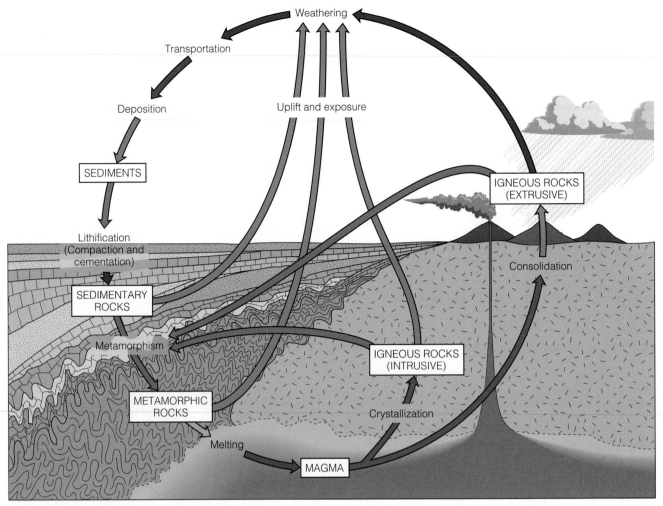

FIGURE 3.1 The rock cycle shows the relationships among the three families of rocks. The arrows inside the circle show possible interruptions in the ideal cycle.

are made up of the particles and chemicals yielded by the breakdown of older rocks. The broken-down residue is transported by water, wind, or glaciers. After deposition, the residue, called *sediment,* is compacted and cemented together to form a new generation of rocks. Rocks altered into new forms by heat, pressure, and chemical fluids in the Earth are termed **metamorphic.**

The **rock cycle** is a useful way to summarize the relationships among the three great families of rocks (Fig. 3.1). As one example, the rock cycle might begin with molten rock from the Earth's interior invading the crust and hardening to form igneous rocks. The igneous rocks in turn are eventually exposed at the surface by erosion and attacked by weathering. The weathered rock fragments are carried by streams and rivers to a lake or the sea and deposited in layers, which harden to form sedimentary rocks. These sedimentary rocks may later be buried deeply enough that they are changed by heat and pressure into metamorphic rocks. The heating may actually melt part of the rock, creating a new generation of igneous rocks and starting the cycle anew.

As shown by the arrows within the circle in Figure 3.1, any possible change can occur in the rock cycle. For example, sedimentary rocks need not be metamorphosed, but in-

stead might be uplifted and exposed at the surface where they are weathered; the particles then are transported and become sedimentary rocks once again. Igneous, sedimentary, and metamorphic rocks can all be attacked by weathering and used as the raw material for new sedimentary rocks. Any of the three rock types can also be changed by heat and pressure into metamorphic rocks or melted to form a new generation of igneous rocks. In short, any type of rock can be transformed into any other type.

MAGMA, LAVA, AND IGNEOUS ROCKS

Igneous rocks form from **magma,** molten rock within the Earth, or from **lava,** magma that reaches the Earth's surface. Contrary to popular misconception, magma does not come from the center of the Earth, nor is the Earth's interior entirely molten. Instead, magma exists in small pockets, or *magma chambers,* a few kilometers in diameter, usually 50 km (30 mi) or less below the surface. During explosive eruptions, particles collectively known as **pyroclastic materials** are discharged. Much of this material is quite small and is designated as *ash,* but much larger *volcanic bombs* are also erupted.

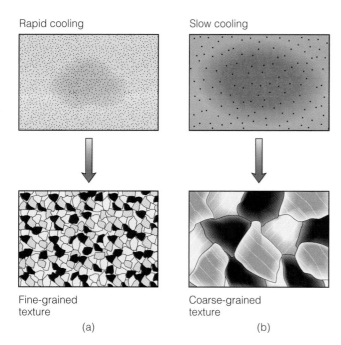

Rapid cooling

Slow cooling

Fine-grained texture
(a)

Coarse-grained texture
(b)

■ FIGURE 3.2 The effect of the cooling rate of a magma on the size of mineral crystals. (*a*) Rapid cooling results in many small crystals and a fine-grained texture. (*b*) Slow cooling allows mineral crystals to grow larger and thus yields a coarse-grained texture.

■ FIGURE 3.3 Igneous rock with a porphyritic texture consisting of large mineral grains in a background of much smaller crystals.

Two categories of igneous rocks are recognized: **volcanic rocks,** which solidify from lava, and **plutonic rocks,** which harden beneath the surface. As a general rule, the faster the molten rock material cools, the smaller the mineral crystals in the resulting rocks (■ Fig. 3.2a). Lava tends to cool quickly so volcanic rocks are generally *fine-grained,* meaning that it is hard to see the constituent minerals with the unaided eye. Volcanic rocks also often contain cavities known as *vesicles* because they cool on the surface where pressure is low and gas bubbles can expand as lava rises to the surface. Plutonic rocks commonly have constituent minerals large enough to see with the unaided eye because they cool slowly and crystals have time to grow to large size. Hence, these rocks are *coarse-grained* (Fig. 3.2b). Plutonic rocks rarely contain vesicles because they cool deep in the crust where pressure is high and gas bubbles cannot expand easily. Masses of igneous rock that have hardened within the Earth are called *intrusions* (Chapter 9).

Melting and Crystallization

Making frozen juice bars in a home freezer is a good way to learn about melting and crystallization. One of the most obvious facts is that the whole mixture does not freeze at once, nor does it freeze uniformly. A half-frozen juice bar consists of fairly pure ice crystals and a concentrated liquid; the mixture has to get colder than the freezing point of water to harden entirely. Mixtures of substances, whether fruit juices or magmas, do not melt or solidify all at once. The melting point of the mixture is usually lower than the melting point of the individual substances in the mixture. In everyday life, we make use of the same principle when we use salt to melt ice or put antifreeze in a car; we create a mixture whose melting point is lower than that of pure water. Similarly, magmas melt at lower temperatures than the individual minerals that comprise them. The freezing of an ice cube and the solidification of a magma are essentially the same process; they differ only in the temperature at which the freezing occurs.

Usually, magma is not much hotter than its melting point and consists of a mush of solid crystals mixed with liquid. The earliest-formed crystals have a head start in growth and tend to be the largest crystals in the rock. If the remainder of the magma cools quickly, the resulting rock has a *porphyritic texture* with large crystals embedded in finer-grained material (■ Fig. 3.3).

Classification of Igneous Rocks

Of the many kinds of igneous rocks, only about a dozen are common. Igneous rocks are classified according to their chemical composition using three criteria:

1. The amount of *silica* (silicon plus oxygen) that is present compared to everything else
2. The kinds of feldspars that are present
3. The other minerals that are present

Igneous rocks rich in silica (>65%) are **felsic,** those poor in silica (45–52%) are **mafic,** and those in between (53–65% silica) are called, logically enough, **intermediate.** Igneous rocks with very low silica (<45%) are *ultramafic* (○ Table 3.1). The commonest felsic rocks are the plutonic rock *granite* and the volcanic rock *rhyolite,* both of which are comprised mostly of silicates such as quartz and various feldspar minerals. These minerals, which lack iron and magnesium, are termed *nonferromagnesian.* The most abundant mafic rocks are the volcanic rock *basalt* and the plutonic rock *gabbro* (■ Fig. 3.4). Unlike felsic rocks, these rocks contain ap-

TABLE 3.1 Classification of Igneous Rocks

TYPE	SILICA CONTENT	MINERAL COMPOSITION	VOLCANIC ROCK NAME	PLUTONIC ROCK NAME
Ultramafic	<45%	Olivine	—	Dunite
		Olivine, pyroxene	Komatiite (rare)	Many types
Mafic	45–52%	Olivine, pyroxene, calcium plagioclase	Basalt	Gabbro
Intermediate	53–65%	Pyroxene, amphibole, calcium plagioclase, sodium plagioclase	Andesite	Diorite
Felsic	>65%	Amphibole, biotite, sodium plagioclase, potassium feldspar, quartz	Rhyolite	Granite

◻ **FIGURE 3.4** (*a*) Basalt and (*b*) gabbro are mafic rocks because they have a low silica content and contain considerable ferromagnesian silicate minerals. Basalt is volcanic and fine-grained, whereas gabbro is plutonic and coarse-grained. (*c*) Granite and (*d*) rhyolite are felsic rocks and thus have high silica contents. Granite is plutonic and rhyolite is volcanic.

(a)

(b)

(c)

(d)

(c)

(a)

(b)

■ **FIGURE 3.5** These igneous rocks are classified according to their texture rather than their composition. (*a*) Obsidian and (*b*) pumice are both varieties of volcanic glass. (*c*) Tuff is composed of pyroclastic materials known as ash.

preciable quantities of iron- and magnesium-bearing silicates (or *ferromagnesian silicates*) and as a result are generally dark colored. Among the intermediate rocks, *diorite,* a plutonic rock, and *andesite,* its volcanic equivalent, are most common (Table 3.1). Generally, these rocks are difficult to distinguish from gabbro and basalt without chemical analysis or microscopic examination. *Ultramafic rocks,* made up mostly of ferromagnesian silicates, are rare at the Earth's surface.

Silica content is also the most important control on a magma's or lava's *viscosity,* or its resistance to flow. Because of its high silica content, felsic lava flows as a thick pasty mass. In contrast, mafic lava with its low silica content has low viscosity and flows easily and much more rapidly than felsic lava. Explosive volcanoes most commonly erupt felsic magma, whereas eruptions of mafic magmas tend to be much less violent (Chapter 9).

Some igneous rocks are given special names because of their distinctive textures. Two of the most common are *obsidian* and *pumice* (■ Fig. 3.5). Obsidian forms when felsic lavas are so viscous and slow moving that atoms cannot move enough to arrange themselves into crystals and instead form a natural glass. Because of its conchoidal fracture, hardness, and uniformity, obsidian was prized by ancient people around the world for making arrowheads and other edged tools. Pumice forms when felsic lava has such a high gas content that it forms a porous froth as it hardens. Pumice has a texture very much like plastic foam and is so light and

porous that it actually floats on water. The volcanic rock *tuff* is composed of volcanic ash (Fig. 3.5c).

Bowen's Series

The geologist N. L. Bowen observed that minerals in most igneous rocks tend to form in a specific sequence. Among the ferromagnesian silicate minerals, olivine tends to form before pyroxene, pyroxene before amphibole, and amphibole before biotite. Among the feldspars, calcium-rich plagioclase grades into sodium-rich plagioclase, and both types of plagioclase tend to form earlier than the potassium feldspars (Fig. 3.6). Notice in Figure 3.6 that Bowen's series has two branches along which ferromagnesian silicates and plagioclase feldspars crystallize simultaneously. If muscovite mica and quartz occur, they usually form last. It is common to find rocks in which a later mineral forms a rim around an earlier-formed mineral or in which some minerals seem to have been dissolved by the very magma that formed them. Bowen found he could arrange all these observations into a composite series of minerals, which is now called **Bowen's series** (Fig. 3.6).

Bowen's series summarizes the processes that occur when magma melts or solidifies. As minerals solidify in a magma, some chemical components leave the liquid magma and go into the solid minerals. The composition of the remaining

FIGURE 3.6 Bowen's series shows the order of crystallization of minerals common in igneous rocks.

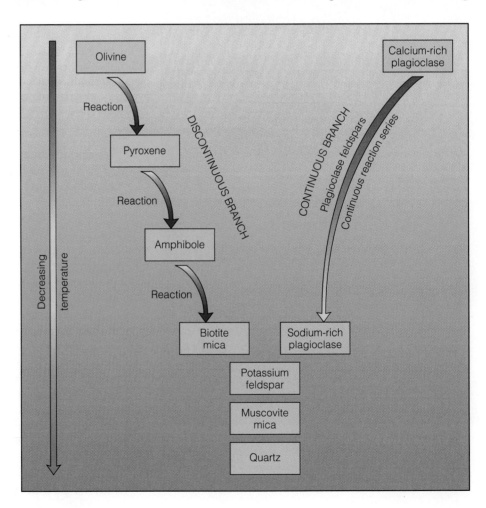

liquid changes, and other minerals begin to form in turn, until the magma has completely solidified. The ferromagnesian minerals and calcium-rich plagioclase form first, and the remaining magma becomes progressively poorer in iron, magnesium, and calcium as it solidifies. The earliest-formed minerals are silica-poor, but as other elements are removed from the magma, the remaining magma becomes more silica-rich.

Several important properties of igneous rocks correlate with Bowen's series (Fig. 3.7). Felsic rocks tend to have more minerals with shared silica tetrahedra than do mafic rocks. Because the silicon-oxygen bonds among the silica tetrahedra are strong, rocks from the felsic end of Bowen's series decompose (weather) less rapidly than mafic rocks (Chapter 4). Rocks at the felsic end of the series also melt at lower temperatures than those from the mafic end. Bowen's series is also related to several volcanic phenomena (Chapter 9).

SEDIMENTARY ROCKS

Various processes collectively known as weathering attack rocks at or near the surface, dissolving some materials and disintegrating solid rock into fragments. When transported, this fragmentary material is termed *sediment;* it is the raw material for *sedimentary rocks* (Fig. 3.1). Running water is a particularly important agent of sediment transport, and glaciers, wind, waves, and ocean currents also transport sediment effectively, although wind can transport only small particles. Dissolved substances derived by weathering are also transported to lakes or the ocean where they may become concentrated and eventually precipitate as minerals. Any area of sediment deposition is a **depositional environment.** Earth scientists recognize three major deposition settings: continental, marginal marine, and marine. Each includes several specific depositional environments (Fig. 3.8).

(a)

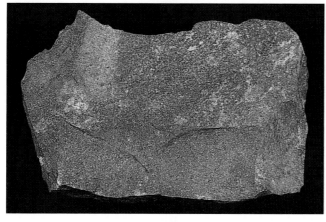

(b)

 FIGURE 3.7 Bowen's series and igneous rocks. (*a*) Granite is composed mostly of minerals that form late in the series, such as quartz (left) and the potassium feldspar orthoclase (center). Granite also contains a variety of other minerals in small quantities including biotite (right). (*b*) Basalt consists of minerals that form early in Bowen's series, including small amounts of olivine (left), and considerable pyroxene (center), and calcium-rich plagioclase (right).

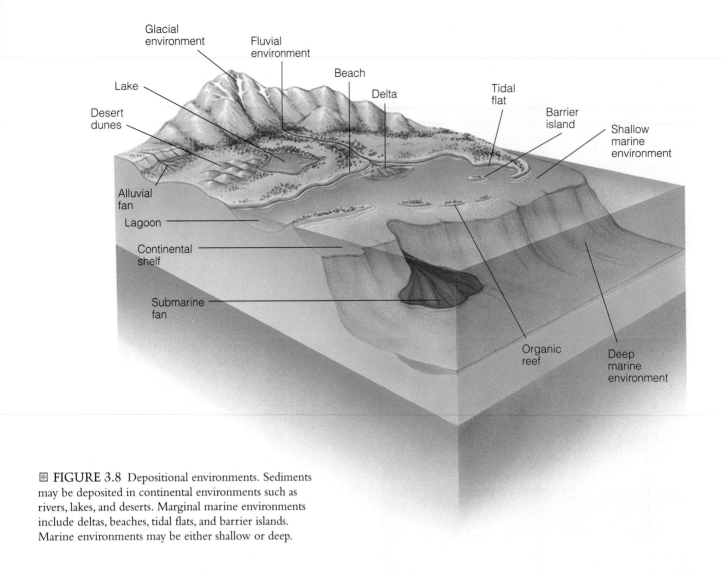

■ FIGURE 3.8 Depositional environments. Sediments may be deposited in continental environments such as rivers, lakes, and deserts. Marginal marine environments include deltas, beaches, tidal flats, and barrier islands. Marine environments may be either shallow or deep.

Lithification: Sediment to Sedimentary Rock

Sedimentary deposits consist of solid particles and pore spaces, which are the voids between particles. The processes of compaction and cementation, collectively referred to as **lithification,** convert sediment into sedimentary rocks. For example, compaction resulting from the pressure exerted by the weight of overlying sediments reduces the amount of pore space and thus the volume of a deposit (■ Fig. 3.9). A mud deposit may be 80% water-filled pore space, but when the deposit is compacted, the water is squeezed out and the deposit's volume may be reduced by up to 40%.

Compaction alone is generally sufficient to lithify mud, but for larger particles such as sand and gravel, cementation is necessary. Cementation results when dissolved substances such as calcium carbonate ($CaCO_3$) and silica (SiO_2) precipitate in the pore spaces of sediment, thereby effectively binding the particles together (Fig. 3.9). The colorful yellow, brown, and red sedimentary rocks in the southwestern United States derive their color mostly from iron cement.

Earth scientists estimate that Earth's crust consists of about 95% igneous and metamorphic rocks, but sedimentary rocks are by far the most common at or near the surface. Indeed, approximately 75% of surface exposures on continents are sediments or sedimentary rocks, and they cover most of the sea floor. Sedimentary rocks are generally classified as *detrital* or *chemical.*

Detrital Sedimentary Rocks

Detrital sedimentary rocks are composed of detritus or the solid particles of preexisting rocks (Fig. 3.9). Each of the several varieties recognized is characterized by the size of its constituent particles. Size is a significant factor in classification because the size of particle transported is a direct indicator of the energy of the transport agent. For instance, detrital particles measuring more than 2 mm in diameter, designated as gravel, cannot be transported by wind or slow-moving water and indicate a high-energy depositional environment. Rocks made of such particles are termed *conglomerate* and *sedimentary breccia* (■ Fig. 3.10a and b). The only difference between these two rocks is that the gravel in sedimentary breccia is angular, meaning that the particles have

FIGURE 3.9 Lithification of sediments by compaction and cementation forms various types of detrital sedimentary rocks.

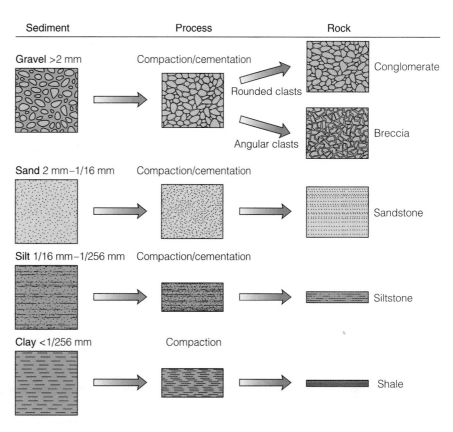

Sediment	Process	Rock
Gravel >2 mm	Compaction/cementation	Conglomerate (Rounded clasts) / Breccia (Angular clasts)
Sand 2 mm–1/16 mm	Compaction/cementation	Sandstone
Silt 1/16 mm–1/256 mm	Compaction/cementation	Siltstone
Clay <1/256 mm	Compaction	Shale

sharp edges and corners, whereas the gravel in conglomerate is rounded.

Sand is simply a size designation referring to detrital particles measuring ⅟₁₆ to 2 mm in diameter; a rock composed of sand is a *sandstone* (Fig. 3.10c). Most sand is quartz (SiO_2) because it is so common and durable, but a number of other minerals are also generally present, although usually in small quantities. Accordingly, quartz sandstone is the most common variety, but sandstone known as *arkose*, which contains abundant feldspar minerals, is also fairly common.

FIGURE 3.10 Detrital sedimentary rocks: (*a*) conglomerate; (*b*) sedimentary breccia; (*c*) sandstone; and (*d*) the mudrock shale.

(a)

(b)

(c)

(d)

Mudrocks include all detrital sedimentary rocks composed of silt-sized ($\frac{1}{256}$ to $\frac{1}{16}$ mm) and clay-sized (<$\frac{1}{256}$ mm) particles; *siltstone, mudstone,* and *claystone* are examples. The term *clay* is used both as a size description and as a name for a group of silicate minerals. In practice, there is little confusion since most clay-sized particles are clay minerals as well. Some mudstones and claystones are designated as *shale* if they break along closely spaced parallel planes (Fig. 3.10d). Mudrocks comprise about 40% of all detrital sedimentary rocks, making them the most common of these rocks.

Chemical Sedimentary Rocks

Chemical sedimentary rocks originate from the materials dissolved during weathering. These dissolved materials are transported to lakes and the oceans where they can be extracted from water to form minerals by either inorganic chemical reactions or the chemical activities of organisms. Rocks formed by this latter process are referred to as *biochemical sedimentary rocks* to emphasize the importance of organisms in their origin.

Limestone, composed of calcite ($CaCO_3$), and *dolostone,* composed of dolomite [$CaMg(CO_3)_2$], are carbonate rocks, meaning they are made up of minerals with the carbonate (CO_3) radical (Fig. 3.11). Some limestone forms as an inorganic chemical precipitate, but most of it results from organic activity and is thus biochemical. For instance, broken seashells are the main components of some limestones, and chalk is made up largely of the microscopic shells of organisms (Fig. 3.11b). Dolostone forms when magnesium replaces some of the calcium in limestone. More than one process may account for the conversion of limestone to dolostone, but most probably forms when magnesium-rich water permeates limestone layers.

Rock salt and *rock gypsum* are known as evaporites because they form by inorganic chemical precipitation of minerals from evaporating water (Fig. 3.12a and b). When seawater evaporates, the amount of dissolved mineral matter increases in proportion to the volume of water and eventually reaches the saturation limit, the point at which precipitation must occur. Rock salt, composed of the mineral halite (NaCl), is simply sodium chloride precipitated from seawater or, more rarely, lake water. Rock gypsum ($CaSO_4 \cdot 2H_2O$) is the most common evaporite rock. A number of other evaporite minerals and rocks are known, but most of these are rare.

Chert is a hard rock composed of microscopic crystals of quartz (SiO_2) (Fig. 3.12c). Some of the color varieties of chert are flint (black), jasper (brown or red), novaculite (white), and agate (banded). Because chert can be shaped to form sharp cutting edges, many cultures have used it for tools, spear points, and arrowheads. Most chert probably forms from accumulations of silica shells, but some results from inorganic chemical precipitation.

The compressed, altered remains of organisms, especially woody land plants, make up *coal,* which, strictly speaking, is not composed of minerals but is nevertheless considered to be a biochemical sedimentary rock (Fig. 3.12d). It forms in swamps and bogs where the water is deficient in oxygen and plant remains fail to decompose completely and form an organic muck. When buried, this muck becomes *peat,* which looks rather like coarse pipe tobacco. When deeply buried and compressed and heated, peat is altered to coal.

Sedimentary Structures and Fossils

Sediments contain a variety of features known as **sedimentary structures** that formed as a result of physical, chemical, and biological processes operating in the depositional

FIGURE 3.11 Two types of limestone: (*a*) limestone containing numerous fossil shells; (*b*) chalk cliffs in Denmark. Chalk is a type of limestone made up of microscopic shells.

(a)

(b)

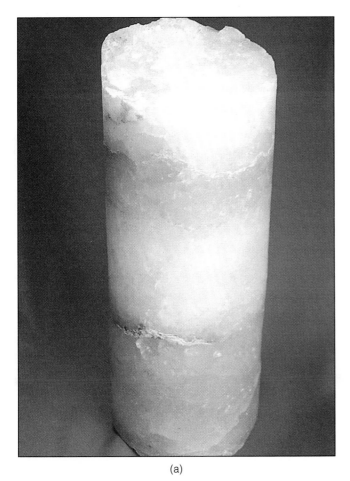

(a)

(b)

(c)

(d)

▣ **FIGURE 3.12** (*a*) Core of rock salt from an oil well in Michigan. (*b*) Rock gypsum. (*c*) Chert. (*d*) Coal.

environment. Among the most common of these features are distinct layers known as *strata* or *beds* (▣ Fig. 3.13). Most sedimentary rocks have some kind of layering, with individual layers showing differences in composition, particle size, color, or a combination of features. Many sedimentary rocks are *cross-bedded* (Fig. 3.13), which means that layers within a layer are inclined. Cross-bedding is common in steam channel and sand dune deposits, and the cross-beds are inclined downward, or dip, in the direction of flow. Thus, one can determine the direction of the current at the time of deposition from cross-beds in ancient rocks.

Ripple marks are another common sedimentary structure. Some result from the to-and-fro motion of waves; these ripple marks tend to be symmetrical and typically develop in sand deposits near lakeshores and seashores (▣ Fig. 3.14a). Ripples formed by currents moving in one direction, as in a stream, are asymmetrical, with a gentle upstream slope and a steep downstream slope (Fig. 3.14b). Like cross-bedding, asymmetrical ripples are good indicators of ancient current directions. Other sedimentary structures include the *mud-cracks* that form when clay-rich sediments shrink and crack as they dry out (Fig. 3.14c).

The remains or traces of prehistoric organisms, known as **fossils,** are most common in sedimentary rocks (▣ Fig. 3.15)

(see ▢ Perspective 3.1 on page 46). Especially common fossils are the shells of marine animals such as clams, oysters, corals, and many others, some of which are extinct. In some varieties of limestone, fossils are the main components of the rock (see Fig. 3.11b). Fossils are useful for several purposes. For example, fossils are the best way of demonstrating the age equivalence of sedimentary rocks in different areas, and the also constitute our only record of prehistoric life (see Ch 11). In addition, fossils are useful in determining the tional environment of sedimentary rocks.

■ FIGURE 3.13 These sandstone layers in Arizona show both horizontal bedding and cross-bedding. Cross-beds indicate ancient current directions by the way they are inclined, to the right in this case.

Environment of Deposition

Because sedimentary rocks commonly contain fossils and various structures indicating how they formed, they provide much of our knowledge of Earth history. Furthermore, it is important to understand how sediments were deposited for economic reasons; most petroleum, natural gas, and coal come from sedimentary rocks. In short, we want to know if a sandstone, for example, was deposited in stream channels, in desert sand dunes, or on a beach.

To interpret the environment of deposition, Earth scientists examine those features of sedimentary rocks that formed as a result of processes operating in the depositional environment. *Sorting* in detrital sedimentary rocks results from processes that selectively transport and deposit sediment. The sand in wind-blown dunes, for instance, is characterized as well sorted because it consists of grains that are all about the same size, whereas glacial deposits tend to be poorly sorted and consist of mixtures of gravel, sand, and mud. Fossils such as corals and sea lilies seem to have lived only in marine environments, and their remains are clear ev-

idence of .... tion. Other fossils, including those .... s, seem almost never to have been .... indicate continental or freshwater

.... s are particularly useful. Although .... s form in a variety of settings, they .... in steam channel deposits. Wind

also generates asymmetrical ripples, but they are usually easily distinguished from those of stream channels. Symmetrical ripple marks typically develop in sand deposited near lakeshores and seashores. Sand dunes may possess large cross-beds, and mudcracks may form in mud exposed near a seashore at low tide.

The following example illustrates how Earth scientists use various features of sedimentary rocks to determine the depositional environment. The Navajo Sandstone (■ Fig. 3.16 on page 47) of the southwestern United States is composed mostly of well-sorted sand grains measuring 0.2 to 0.5 mm in diameter. Tracks of dinosaurs and other land-dwelling animals indicate the original deposit accumulated on land rather than in the sea. These features along with cross-beds up to 30 m (98 ft) high and ripple marks, both of which appear to have formed in sand dunes, lead to the conclusion that the sandstone represents an ancient desert dune deposit. The cross-beds are inclined downward generally to the southwest, indicating that the wind blew mostly from the northeast.

METAMORPHISM AND METAMORPHIC ROCKS

Most metamorphic rocks (Fig. 3.1) form in one of two ways. **Contact metamorphism** takes place when rocks are heated by an adjacent mass of hot rock, such as an intrusion

(a)

(b)

(c)

■ FIGURE 3.14 (*a*) Symmetrical wave-formed ripple marks in ancient rocks in Montana. (*b*) Asymmetrical ripple marks in a streambed in California. (*c*) Mudcracks formed in clay-rich sediments that have dried and shrunk.

■ FIGURE 3.15 Fossils. (*a*) Horse teeth are preserved in this rock. (*b*) Shells of marine animals known as brachiopods.

(a)

(b)

Focus on the Environment:
The Green River Formation

Perspective
3.1

About 50 million years ago, two large lakes existed in what are now parts of Wyoming, Utah, and Colorado. Sand, mud, and dissolved minerals were carried into these lakes where they were deposited and became layers of sedimentary rocks. These rocks, called the Green River Formation, contain the fossilized remains of millions of fish, plants, and insects and are a potential source of large quantities of oil, combustible gases, and other useful substances.

Thousands of fossilized fish skeletons are found on single surfaces within the Green River Formation, indicating numerous mass mortality events (□ Fig. 1). The cause of these events is not known with certainty, but some scientists have suggested that blooms of blue-green algae produced toxic substances that killed the fish. Others propose that rapidly changing water temperatures or excessive salinity at times of increased evaporation was responsible. Whatever the cause, the fish died by the thousands and settled to the lake bottom where their decomposition was inhibited because the water contained little or no oxygen.

The Green River Formation is also well known for its huge deposits of oil shale (□ Fig. 2). Oil shale consists of small clay particles and an organic substance known as *kerogen*. When the appropriate extraction processes are used, liquid oil and combustible gases (hydrocarbons) can be produced from the kerogen of oil shale. To be designated as a true oil shale, the rock must yield at least 10 gallons of oil per ton of rock.

Oil can be produced from oil shale by heating the rock to nearly 500°C (932°F) in the absence of oxygen, vaporizing the hydrocarbons, which are then recovered by condensation. During this process, 25 to 75% of the organic matter of oil shale can be converted to oil and combustible gases. The Green River Formation oil shales yield from 10 to 240 gallons of oil per ton of rock processed, and the total amount of oil recoverable with present technology is estimated at 80 billion barrels. Currently, however, no oil is produced from oil shale in the United States because producing oil by conventional drilling and pumping is less expensive. Nevertheless, the Green River oil shale constitutes one of the largest untapped sources of oil in the world. If more effective processes are developed, it could eventually yield even more than the currently estimated 80 billion barrels.

One should realize that at the current and expected consumption rates of oil in the United States, oil production from oil shale will not solve all of our energy needs. Furthermore, large-scale mining of oil shale would have considerable environmental impact. What would be done with the billions of tons of processed rock? Can such large-scale mining be conducted with minimal disruption of wildlife habitats and groundwater systems? Where will the huge volumes of water necessary for processing come from—especially in an area where water is already in short supply? These and other questions are currently being considered by scientists and industry. Perhaps at some future time, the Green River Formation will provide some of our energy needs.

□ FIGURE 1 Fossil fish from the Green River Formation of Wyoming.

□ FIGURE 2 Layers of oil shale of the Green River Formation are exposed along these hillsides.

■ FIGURE 3.16 The Navajo Sandstone in Zion National Park, Utah. The large cross-beds and other features of these rocks indicate deposition in a desert dune environment.

(■ Fig. 3.17). Contact metamorphic effects rarely extend more than a kilometer (0.6 mile) from the source of heat and diminish in intensity with distance from the source. Most of Earth's metamorphic rocks formed by **regional metamorphism,** resulting from intense pressure and large-scale heating of the crust, usually related to plate tectonic processes (Chapter 7). Regional metamorphism extends over vast areas and is most often accompanied by deformation of the rocks. Another type of metamorphism, sometimes called *metasomatism,* takes place when water rich in minerals permeates rocks and adds or removes chemical compounds. Thus, the agents of metamorphism include pressure, heat, and chemical fluids.

The changes that occur during metamorphism depend not only on the agents of metamorphism but also on the type of rock that is affected. Pure limestone and sandstone are made mostly of single minerals: limestone is made of calcite, and most sandstone is made of quartz. Neither mineral is

■ FIGURE 3.17 Contact metamorphism takes places when rocks are altered by the heat of an adjacent intrusion.

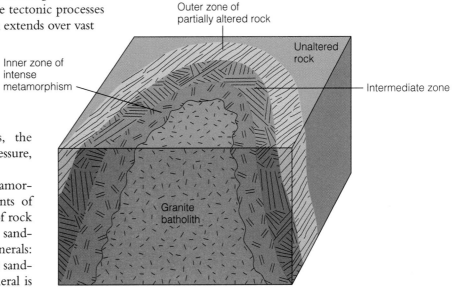

Outer zone of
partially altered rock

Unaltered rock

Inner zone of intense metamorphism

Intermediate zone

Granite batholith

(a)

(b)

▣ FIGURE 3.18 (a) Marble is metamorphosed limestone, whereas quartzite (b) results from the metamorphism of quartz sandstone.

much affected by metamorphism, so the only change that takes place in pure limestone or sandstone during metamorphism is a change in grain structure. Limestone metamorphoses by recrystallizing to *marble,* and sandstone metamorphoses to *quartzite* (▣ Fig. 3.18).

Coal metamorphoses at low temperature into *anthracite,* a hard, dark gray, metallic-looking coal. At somewhat higher temperature, anthracite transforms into *graphite.* Contrary to popular misconception, the final stage in the metamorphism of coal is not diamond. No coal is ever buried deeply enough to reach the temperature and pressure (corresponding to depths of more than 100 km [62 mi]) at which diamond forms.

Mudstone, the most abundant sedimentary rock, undergoes a series of changes (▣ Fig. 3.19). At low temperature and pressure, the platy clay particles in the mudstone reorient themselves at right angles to the greatest pressure, and the mudstone takes on a grain, called *foliation,* and becomes

▣ FIGURE 3.19 These metamorphic rocks possess the property of foliation: (a) slate; (b) schist; (c) gneiss.

(a)

(b)

(c)

still higher pressure and temperature, large crystals of garnet and other minerals may form in schist.

When basalt and andesite metamorphose, the first new minerals to form impart a greenish appearance to the rock, so the rock is called *greenstone*. During more intense metamorphism, abundant black amphibole minerals form, and the rock becomes a black, coarsely crystalline rock known as *amphibolite*. Schist, amphibolite, and other rock types can be metamorphosed at very high temperature and pressure to form a rock in which different minerals separate into bands or streaks, which is yet another type of foliation. This rock is *gneiss* (pronounced "nice"; Fig. 3.19c). Gneiss can also form when granite is deformed and metamorphosed. The various types of metamorphic rocks are summarized in ○ Table 3.2.

Heat and Pressure in the Earth

Most of the heat in the Earth results from the decay of radioactive elements, most importantly, isotopes of uranium, thorium, and potassium. When these isotopes decay, they emit energetic particles and gamma rays; the surrounding rock absorbs this energy and is heated by it. Rock is such a poor conductor of heat that it does not take much radioactivity to produce appreciable heat, given enough time. Observations from deep mines and drill holes indicate that Earth's temperature increases with depth. This temperature increase with depth, or *geothermal* gradient, is about 25°C/km near the surface.

harder as it is transformed to *slate* (Fig. 3.19a). As temperature and pressure increase, the platy mica minerals form and grow big enough to be visible to the unaided eye. This rock is called *schist,* which also possesses foliation (Fig. 3.19b). At

TABLE 3.2   Classification of Common Metamorphic Rocks

TEXTURE	METAMORPHIC ROCK	TYPICAL MINERALS	METAMORPHIC GRADE	CHARACTERISTICS OF ROCKS	PARENT ROCK
Foliated	Slate	Clays, micas, chlorite	Low	Fine-grained, splits easily into flat pieces	Mudrocks, claystones, volcanic ash
	Schist	Micas, chlorite, quartz, talc, hornblende, garnet, staurolite, graphite	Low to high	Distinct foliation, minerals visible	Mudrocks, carbonates, mafic igneous rocks
	Gneiss	Quartz, feldspars, hornblende, micas	High	Segregated light and dark bands visible	Mudrocks, sandstones, felsic igneous rocks
	Amphibolite	Hornblende, plagioclase	Medium to high	Dark-colored, weakly foliated	Mafic igneous rocks
Nonfoliated	Marble	Calcite, dolomite	Low to high	Interlocking grains of calcite or dolomite, reacts with HCL	Limestone or dolostone
	Quartzite	Quartz	Medium to high	Interlocking quartz grains, hard, dense	Quartz sandstone
	Greenstone	Chlorite, epidote, hornblende	Low to high	Fine-grained, green color	Mafic igneous rocks
	Anthracite	Carbon	High	Black, lustrous	Coal

FIGURE 3.20 Pressure is applied equally in all directions in the Earth's crust due to the weight of the overlying rocks. A similar situation occurs when 200 ml styrofoam cups are lowered to ocean depths of approximately 750 m and 1,500 m. Increased pressure is exerted equally in all directions on the cups, and they consequently decrease in volume, while still maintaining their general shape.

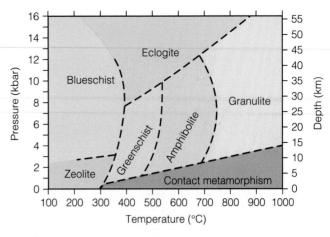

FIGURE 3.21 Metamorphic facies are characterized by particular minerals that form under the same broad temperature-pressure conditions. Each facies is named for its most characteristic rock or mineral.

When rocks are buried, they are subjected to pressure from the weight of the overlying rocks, and with increasing depth, pressure increases. Pressure of this type is applied equally in all directions, much as it is on an object immersed in water (Fig. 3.20). In a similar fashion, deeply buried rocks are subjected to increasing pressure with depth, and under these conditions the minerals become more closely packed together. In addition to the pressure from deep burial, rocks may also be subjected to *differential pressure,* which is not applied equally from all sides, and the rock consequently is distorted.

It is possible to duplicate the conditions of metamorphism in the laboratory. In contrast to the slowness of many geologic processes, the chemical reactions in metamorphic rocks often take place within a few hours or days. Once large hydraulic presses were used to generate the pressures needed for such studies, but recently, a revolutionary device has appeared that surpasses the performance of large presses weighing many tons, yet literally fits in the palm of the hand. The *diamond-anvil pressure cell* consists of opposite jaws, each equipped with a gem-quality diamond window. By tightening the jaws with a simple screw, it is possible to create pressures between the jaws approaching those in the Earth's core. Because the windows are transparent, the sample can be observed through a microscope while it is being tested.

Metamorphic Minerals and Facies

The specific minerals formed during metamorphism depend on the composition of the original rock and the temperatures and pressures involved. The degree of change produced by metamorphism is called the *metamorphic grade.* Slate, for example, retains many characteristics of the shale from which it forms; it is therefore referred to as low grade. However, shale can also be altered to schist, which is consid-

erably different in texture from its source material and accordingly represents a higher grade of metamorphism.

The range of rock types and minerals that form under the same broad conditions of temperature and pressure represents a **metamorphic facies** (Fig. 3.21). Each facies is named for its most characteristic rock or mineral. For example, the green metamorphic mineral chlorite, which forms under fairly low temperature and pressure, is found in rocks belonging to the *greenschist facies.* The blue-gray mineral *glaucophane,* on the other hand, occurs in rocks of the *blueschist facies,* representing metamorphic rocks formed under conditions of low temperature and high pressure. Minerals called amphiboles are common in rocks formed at high temperatures and pressures, the so-called *amphibolite facies.* Accordingly, by examining minerals in metamorphic rocks, scientists can determine the temperature-pressure conditions responsible for the rocks' metamorphism.

ROCKS AND GEOLOGIC RESOURCES

In a sense, most geologic resources are ultimately derived from igneous rocks, because igneous activity is the mechanism by which materials from the Earth's deep interior reach the crust. Some igneous rocks, such as granite for building stone, are resources in their own right, whereas other igneous rocks are sources of important minerals. *Magmatic ores,* for example, are ore minerals that form as constituents of igneous rocks. In some cases, the ore minerals were much denser than the magma they crystallized from and sank rapidly to the floor of the magma chamber where they became concentrated. This process is responsible for many chromite and platinum deposits. Other ores occur as minerals disseminated throughout igneous intrusions. The copper minerals found in small granitic intrusions are among the most important of these.

■ FIGURE 3.22 A quarry for the excavation of pyroclastic materials known as cinders.

In some granitic intrusions, the very last part of the granite to solidify is rich in water and elements that do not fit easily into the major rock-forming minerals. This "leftover" material hardens to form *pegmatite,* which generally has a composition similar to that of granite. Pegmatite is a very coarse-grained rock, and its crystals sometimes grow to a very large size, several meters across in some cases. Pegmatites commonly include rare minerals that contain unusual ingredients derived from the leftover magma. As a result, pegmatites are major sources of gem minerals and ores of some rare elements such as beryllium, as well as more common materials including feldspars for porcelain and micas for electrical insulation.

Volcanic rocks are also sources of some geologic resources. Pumice, for example, is crushed and used as an abrasive, and pyroclastic materials resembling cinders are used for construction, decorative stones, and roadbeds (■ Fig. 3.22). Some metal deposits, notably copper, formed when mineral-bearing solutions precipitated ore minerals within and between lava flows.

Sediments and sedimentary rocks are also the sources of many resources. Sand and gravel are essential to the construction industry, pure clay deposits are needed for ceramics and the manufacture of paper, and limestone is used for cement and in blast furnaces where iron ore is refined to make steel. Rock salt is a source of common table salt and a number of chemical compounds, and rock gypsum is used to manufacture wallboard.

Because of their greater density, gold, platinum, tin, and diamonds are commonly separated from other minerals during sediment transport and form local concentrations called *placer deposits* in stream channels and on beaches (■ Fig. 3.23). Much of the gold recovered during the gold rushes to California and the Yukon Territory of Canada came from placers. Currently, South Africa is the leading producer of gold with the United States a distant second. Most U.S. gold production comes from mines in Nevada, California, South Dakota, and Utah.

Most of the uranium used in nuclear reactors in North America comes from uranium-bearing minerals in sedimentary rocks. *Banded iron formations,* which are chemical sedimentary rocks with alternating thin layers of chert and iron minerals, are a source of iron ores. North America has vast iron ore deposits in the Lake Superior region and in eastern Canada.

Coal, oil, and natural gas are collectively termed *fossil fuels;* they are literally the fossil chemical remains of ancient organisms. Much of the energy needed to generate electricity is produced by burning coal, and coke, a purified carbon fuel made from coal, is used to fire blast furnaces where steel is produced. Most petroleum and natural gas are recovered from porous sedimentary rocks in which they were trapped in sufficient quantities for profitable extraction. Other sources of oil and natural gas that will probably become increasingly important include *oil shale* (Perspective 3.1) and *tar sands.* The United States has about two-thirds of

(a)

world's oil shale deposits, and the most extensive deposits of tar sands are in Canada.

Metamorphism is responsible for a number of geologic resources. In addition to metamorphic building and ornamental stones such as marble and slate, a number of metamorphic minerals are valuable. These include garnet (abrasives and jewelry), talc (talcum powder), kyanite (porcelain), and graphite (lubrication and electrical products). The most important metamorphic resources are those created by hot fluids derived from igneous intrusions that circulate through rocks, changing them chemically and precipitating ore minerals. In some cases, the minerals form in or very near the intrusion. Other deposits form from fluids that have migrated far from their source. For example, many gold deposits are found in quartz veins in metamorphic rocks. Some fluids may travel hundreds of kilometers and deposit their minerals at temperatures below the normal range of metamorphic temperatures. Lead and zinc deposits in the central United States, for instance, were deposited by fluids at only about 200°C (392°F).

□ FIGURE 3.23 Because of its great density, gold is sometimes concentrated by stream action. Recovery methods also make use of gold's great density. (a) Gold panning is used on a small scale to recover gold. As stream gravel and water are agitated in a pan, the gold sinks to the bottom and lighter material is washed away. (b) Hydraulic mining uses high-pressure water to excavate gold-bearing sediment. The sediment washes through a series of traps. The heavy gold is caught by the traps while lighter minerals are washed on through.

(b)

CHAPTER SUMMARY

1. Igneous rocks originate when magma or lava cools and crystallizes or when pyroclastic materials are consolidated. The two categories of igneous rocks are plutonic rocks, which form within the Earth's [...] ic rocks, which form at the surface.

   [...] cks are classified according to their chemical [...] ecially their silica content. Mafic igneous [...] silica, whereas felsic igneous rocks are rich in silica. Those having a silica content between mafic and felsic rocks are characterized as intermediate.

3. A few igneous rocks, including obsidian, pumice, and tuff, are classified on the basis of their texture.

4. According to Bowen's series, minerals crystallize from a magma in a specific sequence. Some properties of igneous rocks can be related to Bowen's series.

5. Detrital sedimentary rocks consist of solid particles yielded during weathering, whereas chemical sedimentary rocks are made up of minerals extracted from solution by inorganic chemical processes and the activities of organisms.

6. Lithification is the process of converting sediment into sedimentary rock through compaction and cementation. Silica, calcium carbonate, and iron are common cements in sedimentary rocks.

7. Sedimentary structures such as cross-bedding, ripple marks, and mudcracks form when sediments are deposited. These structures, along with fossils, aid Earth scientists in interpreting ancient depositional environments.

8. Metamorphism involves the chemical and physical alteration of rocks by heat, pressure, and chemical fluids within the Earth.

9. Metamorphic processes include chemical reactions, changes in rock texture, and the addition or removal of material by fluids.

Metamorphism is usually not reversible because water is lost during the process.

10. The Earth's heat results primarily from radioactive breakdown of isotopes of potassium, uranium, and thorium. Pressure in the Earth is caused by the weight of overlying rocks.

11. Contact metamorphism takes place adjacent to igneous intrusions, but it has limited effects. Regional metamorphism resulting from intense pressure and heat accounts for most metamorphic rocks. Chemical fluids migrating through rocks cause some metasomatism.

12. The presence of certain minerals in metamorphic rocks indicates the broad conditions of pressure and temperature that caused metamorphism.

13. Many igneous, sedimentary, and metamorphic rocks or materials contained within them are valuable resources.

IMPORTANT TERMS

Bowen's series
chemical sedimentary rock
contact metamorphism
depositional environment
detrital sedimentary rock
felsic

fossil
igneous rock
intermediate
lava
lithification
mafic

magma
metamorphic facies
metamorphic rock
plutonic rock
pyroclastic materials

regional metamorphism
rock cycle
sedimentary rock
sedimentary structure
volcanic rock

REVIEW QUESTIONS

1. Volcanic rocks can usually be distinguished from plutonic rocks by:
   a. ___ color; b. ___ composition; c. ___ iron-magnesium content; d. ___ size of mineral crystals; e. ___ specific gravity.

2. Igneous rocks composed largely of ferromagnesian silicates are characterized as:
   a. ___ pyroclastic; b. ___ obsidian; c. ___ intermediate; d. ___ felsic; e. ___ mafic.

3. In which of the following pairs of igneous rocks do both rocks have the same mineral composition?
   a. ___ granite-tuff; b. ___ andesite-rhyolite; c. ___ pumice-diorite; d. ___ basalt-gabbro; e. ___ pumice-pegmatite.

4. The metamorphic rock formed from limestone is:
   a. ___ quartzite; b. ___ marble; c. ___ greenstone; d. ___ slate; e. ___ gneiss.

5. Which type of metamorphism produces the majority of metamorphic rocks?
   a. ___ regional; b. ___ hornblende; c. ___ contact; d. ___ metasomatism; e. ___ lithostatic.

6. Which of the following metamorphic rocks possesses a foliated texture?
   a. ___ marble; b. ___ quartzite; c. ___ greenstone; d. ___ schist; e. ___ sandstone.

7. Which of the following is detrital sediment?
   a. ___ seashells; b. ___ dissolved minerals; c. ___ quartz sand; d. ___ cross-bedding; e. ___ fossil bones.

8. The process whereby dissolved minerals are precipitated in pore spaces of sediment and bind it together is:
   a. ___ rounding; b. ___ sorting; c. ___ layering; d. ___ cementation; e. ___ evaporation.

9. Most limestones have a large component of calcite ($CaCO_3$) that was extracted from seawater by:
   a. ___ an inorganic chemical process; b. ___ weathering and compaction; c. ___ activities of organisms; d. ___ evaporation; e. ___ lithification.

10. The broad conditions of pressure and temperature under which metamorphic rocks form are called a:
    a. ___ metamorphic facies; b. ___ depositional environment; c. ___ mafic rock; d. ___ foliated texture; e. ___ geothermal gradient.

11. The most important control on a magma's or lava's viscosity is:
    a. ___ temperature; b. ___ specific gravity; c. ___ altitude; d. ___ pressure; e. ___ silica content.

12. A volcanic rock composed of pyroclastic materials known as ash is:
    a. ___ gabbro; b. ___ tuff; c. ___ pegmatite; d. ___ andesite; e. ___ pumice.

13. The gravel in conglomerate is _____, whereas the gravel in sedimentary breccia is _____.
   a. ____ quartz/feldspars; b. ____ dense/dark colored; c. ____ rounded/angular; d. ____ shiny/dull; e. ____ felsic/mafic.

14. Metamorphism occurring adjacent to an igneous intrusion is:
   a. ____ dynamic; b. ____ volcanic; c. ____ directed; d. ____ pressure induced; e. ____ contact.

15. Why are felsic lava flows so much more viscous than mafic lava flows?

16. Explain why limestone and quartz sandstone undergo little chemical change during metamorphism.

17. What is lithification and how does it occur?

18. How can scientists tell from metamorphic rocks what the temperature-pressure conditions were when the rocks were metamorphosed?

19. What is meant by low-grade and high-grade metamorphism? Give an example of a metamorphic rock representing each grade.

20. Explain how scientists can determine the directions of ancient currents.

21. Why is most limestone considered to be biochemical sedimentary rock? How is limestone changed to dolostone?

22. What is meant by porphyritic texture and how does such a texture originate?

23. How do the sedimentary rocks known as evaporites form? Name two common evaporite rocks.

POINTS TO PONDER

1. Why would foliated metamorphic rocks make a poor foundation for a dam? What types of metamorphic rocks would make good foundations?

2. How can the color and size of minerals in igneous rocks give clues about the composition and cooling history of a magma?

3. In the United States, about 860 million metric tons of coal from a total reserve of 243 billion metric tons are used annually. Assuming that all of this coal can be mined, how long will it last at the current rate of consumption? Why is it improbable that all of this reserve can be mined?

ADDITIONAL READINGS

Best, M. G. 1982. *Igneous and metamorphic petrology.* San Francisco: W. H. Freeman & Co.

Boggs, S., Jr. 1995. *Principles of sedimentology and stratigraphy,* 2d ed. Columbus, Ohio: Merrill Publishing Co.

Brimhall, G. 1991. The genesis of ores. *Scientific American* 264: 84–91.

Collinson, J. D., and D. B. Thompson. 1989. *Sedimentary structures,* 2d ed. London: Allen & Unwin.

Friedman, G. M., J. E. Sanders, and D. C. Kopaska-Merkel. 1992. *Principles of sedimentary deposits.* New York: Macmillan.

Jayaraman, A. 1984. The high-pressure diamond anvil cell. *Scientific American* 250: 54–62.

LaPorte, L. F. 1979. *Ancient environments,* 2d ed. Englewood Cliffs, N.J.: Prentice-Hall.

Peck, D. L., T. L. Wright, and R. W. Decker. 1979. The lava lakes of Kilauea. *Scientific American* 241, no. 4: 114–29.

Roberts, D. 1990. When the quarry is marble. *Smithsonian* 22, no. 10: 98–107.

Stewart, D. 1988. The quarries hum as architects enter a new stone age. *Smithsonian* 19, no. 7: 86–93.

For current updates and exercises, log on to
http://www.wadsworth.com/geo.

UNIVERSITY OF WISCONSIN GEOLOGY MUSEUM
http://geology.wisc.edu/~museum/

This site gives information about the rocks, minerals, and fossils of Wisconsin. Wisconsin rocks include all three groups: igneous, metamorphic, and sedimentary. Among the fossils on display is a complete mastodon skeleton.

ILLINOIS GEOLOGY SURVEY—MAZON CREEK FOSSILS
http://museum.state.il.us/exhibits/mazon_creek/index.html

The Mazon Creek area in east-central Illinois is one of the most celebrated fossil localities in the world. About 250 million years ago it was an ancient shoreline, and superbly preserved fossils of both terrestrial and marine plants and animals have been found.

CHAPTER

4

Weathering, Soil, and Mass Wasting

This tree near Crystal Falls, Michigan, is contributing to the disintegration of a rock by enlarging a crack.

Prologue

On May 31, 1970, high in the Andes of Peru, an earthquake dislodged a huge mass of snow, ice, and rock from the north peak of Nevado Huascarán, setting in motion an enormous avalanche of debris. Hurtling down one of the mountain's steep valleys at up to 320 km (200 mi) per hour, the avalanche, consisting of more than 50 million m³ of debris, flowed over ridges 140 m (460 ft) high, obliterating everything in its path.

About 3 km (2 mi) east of the town of Yungay, where the valley makes a sharp bend, part of the avalanche flowed over the valley wall and within seconds buried the town, killing more than 20,000 of its residents. The main mass of the avalanche continued down the valley, devastating several villages and burying about 5,000 more people. In a span of roughly four minutes from the onset of the initial shaking, about 25,000 people died, and most of the area's transportation, power, and communications networks had been destroyed.

Ironically, the only part of Yungay not buried was Cemetery Hill, where 92 people survived by running to its top (□ Fig. 4.1). A Peruvian scientist who was giving a French couple a tour of Yungay provided a vivid eyewitness account of the disaster:

> As we drove past the cemetery the car began to shake. It was not until I had stopped the car that I realized that we were experiencing an earthquake. We immediately got out of the car and observed the effects of the earthquake around us. I saw several homes as well as a small bridge . . . collapse. It was, I suppose, after about one-half to three-quarters of a minute when the earthquake shaking began to subside. At that time I heard a great roar coming from Huascarán. Looking up, I saw what appeared to be a cloud of dust and it looked as though a large mass of rock and ice was breaking loose from the north peak. My immediate reaction was to run for the high ground of Cemetery Hill, situated about 50 to 200 m away. I began running and noticed that there were many others in Yungay who were also running toward Cemetery Hill. About half to three-quarters of the way up the hill, the wife of my friend stumbled and fell and I turned to help her back to her feet.

> The crest of the wave had a curl, like a huge breaker coming in from the ocean. I estimated the wave to be at least 80 m high. I observed hundreds of people in Yungay running in all directions and many of them toward Cemetery Hill. All the while, there was a continuous loud roar and rumble. I reached the upper level of the cemetery near the top just as the [avalanche] struck the base of the hill and I was probably only 10 seconds ahead of it.

> At about the same time, I saw a man just a few meters down hill who was carrying two small children toward the hilltop. The [avalanche] caught him and he threw the two children toward the hilltop, out the path of the flow, to safety, although the [avalanche] swept him down the valley, never

□ **FIGURE 4.1** Cemetery Hill was the only part of Yungay, Peru, to escape the 1970 avalanche that destroyed the town. Only 92 people in the town of more than 20,000 survived the destruction by running to the top of the hill.

to be seen again. I also remember two women who were no more than a few meters behind me and I never did see them again. Looking around, I counted 92 persons who had also saved themselves by running to the top of the hill. It was the most horrible thing I have ever experienced and I will never forget it.★

★B. A. Bolt et al., *Geological Hazards* (New York: Springer-Verlag, 1977), pp. 37–39.

This devastating avalanche was not the first to sweep down this valley. In January 1962, a large avalanche had buried several villages and killed about 4,000 people.

 WEATHERING

Many rocks form within the crust where little or no water or oxygen is present and where temperatures, pressures, or both are high. At or near the surface the rocks are exposed to low temperatures and pressures and are attacked by atmospheric gases, water, acids, and organisms. This phenomenon, known as **weathering,** involves the physical breakdown (disintegration) and chemical alteration (decomposition) of rocks and minerals. Thus, physical and chemical weathering alter rocks and minerals so that they are more nearly in equilibrium with a new set of environmental conditions.

Earth scientists are interested in weathering because it is an essential part of the rock cycle (see Fig. 3.1). The **parent material,** or rock being weathered, breaks down into smaller pieces, and some of its constituent minerals dissolve or are altered and removed from the weathering site, a phenomenon known as **erosion.** Running water, wind, or glaciers commonly transport weathered materials elsewhere where they are deposited as *sediment,* which may become sedimentary rock. Whether eroded or not, weathered rock

materials may undergo further modification and form soil. Thus, weathering provides the raw materials for both sedimentary rocks and soils. Weathering is also important in the formation of some mineral resources such as aluminum ores, and it is responsible for the enrichment of other important mineral deposits.

Two types of weathering are recognized, *mechanical* and *chemical.* Both occur simultaneously at the weathering site, during erosion and transport, and even in the environments where weathered materials are deposited.

 MECHANICAL WEATHERING

Mechanical weathering involves forces breaking rocks into smaller pieces without changing their chemical composition. A good example is the mechanical weathering of granite, which yields smaller pieces of granite or individual mineral grains (Fig. 4.2). The processes responsible for mechanical weathering include *frost wedging, salt crystal growth, pressure release, thermal expansion and contraction,* and the *activities of organisms.* Some Earth scientists also consider the size reduction of particles by *abrasion* to be a form of me-

■ FIGURE 4.2 Mechanically weathered granite. The sandy material consists of small pieces of granite (rock fragments) and minerals such as quartz and feldspars liberated from the parent material.

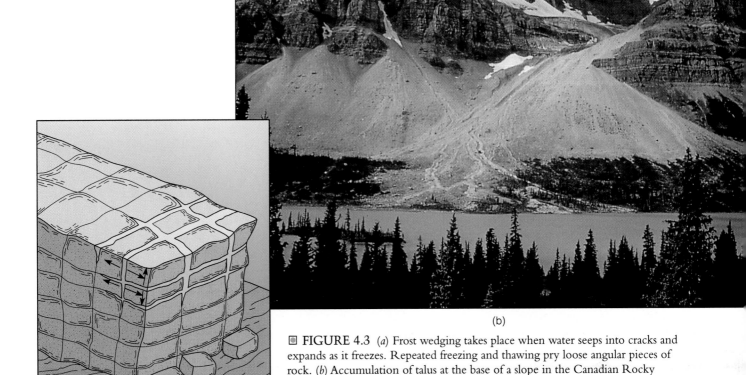

(b)

FIGURE 4.3 (*a*) Frost wedging takes place when water seeps into cracks and expands as it freezes. Repeated freezing and thawing pry loose angular pieces of rock. (*b*) Accumulation of talus at the base of a slope in the Canadian Rocky Mountains.

chanical weathering. Abrasion is most effective on particles during sediment transport as they impact and grind against one another.

Frost Wedging

When water seeps into a crack and freezes, it expands by about 9% and exerts great force on the walls of the crack, thereby widening and extending it by **frost wedging.** As a result of repeated freezing and thawing, pieces of rock are eventually detached from the parent material (Fig. 4.3a). The debris produced by frost wedging in mountains commonly accumulates as large cones of *talus* lying at the bases of slopes (Fig. 4.3b). Frost action is most effective in areas where temperatures commonly fluctuate above and below freezing.

Salt Crystal Growth

Under some circumstances, salt crystals forming from solution can cause rocks to disaggregate. Growing crystals exert enough force to widen cracks and crevices, dislodge particles, and pry loose individual mineral grains. To the extent that **salt crystal growth** produces forces that expand openings in rocks, it is similar to frost wedging. Most salt crystal growth takes place in hot arid areas, although it probably also affects rocks in some coastal regions.

Pressure Release

Pressure release is especially evident in rocks that formed as deeply buried intrusive bodies such as batholiths, but it takes place in other types of rocks as well. When a batholith, forms, the molten rock (magma) crystallizes under tremendous pressure (the weight of the overlying rock) and is stable under these pressure conditions. When the batholith is uplifted and exposed at the surface as the overlying rock is eroded, the pressure is reduced. But the rock contains energy that is released by expansion and by the formation of *sheet joints,* large fractures that more or less parallel the rock surface (Fig. 4.4a). Slabs of rock bounded by sheet joints may slip or slide off the host rock, forming large rounded rock masses known as **exfoliation domes** (Fig. 4.4b).

(b)

(a)

■ FIGURE 4.4 (*a*) Sheet joint formed by expansion in the Mount Airy Granite, North Carolina. (*b*) Exfoliation domes in Yosemite National Park, California.

Thermal Expansion and Contraction

During **thermal expansion and contraction,** the volume of a rock changes as it is alternately heated and cooled. In a desert, where the temperature may vary as much as 30°C (86°F) in one day, rocks expand when heated and contract as they cool. A rock is a poor conductor of heat, so its outside heats up more than its inside, and the surface expands more than the interior. Furthermore, dark minerals absorb heat faster than light-colored minerals, so differential expansion occurs even between the mineral grains of some rocks.

Although thermal expansion and contraction can generate forces in rocks, the magnitude of these forces is generally insufficient to overcome the rocks' internal strength.

Experiments in which rocks are heated and cooled repeatedly to simulate years of thermal expansion and contraction indicate that it is not a very important agent of mechanical weathering.

Activities of Organisms

Animals, plants, and bacteria all participate in the mechanical and chemical alteration of rocks. Burrowing animals constantly mix soil and sediment particles and bring material from depth to the surface where further weathering may occur. The roots of trees and bushes wedge themselves into cracks in rocks and further widen them (see the chapter-opening photo).

CHEMICAL WEATHERING

Chemical weathering involves the decomposition of rocks by chemical alteration of the parent material. Water, acids, atmospheric gases, especially carbon dioxide and oxygen, and organisms play important roles in chemical weathering. Rocks with lichens (composite organisms consisting of fungi and algae) on their surfaces are chemically altered more rapidly than lichen-free rocks. Plants affect the chemistry of soil water, and their roots release organic acids.

Solution is a chemical weathering process during which the ions in a substance become separated from one another, and the solid substance dissolves. When a soluble solid such as the mineral halite (NaCl) is in water, it dissolves as the sodium ions and chloride ions are liberated from the crystal structure (Fig. 4.5).

Most minerals are not very soluble in pure water, but some dissolve quickly if a small amount of acid is present. One way to make water acidic is by the separation or dissociation of the ions in carbonic acid as follows:

$$H_2O \;+\; CO_2 \;\rightleftharpoons\; H_2CO_3 \;\rightleftharpoons\; H^+ \;+\; HCO_3^-$$

| water | carbon dioxide | carbonic acid | hydrogen ion | bicarbonate ion |

According to this chemical reaction, water and carbon dioxide combine to form carbonic acid, and the ions in the carbonic acid become separated to yield hydrogen ions and bicarbonate ions. The concentration of hydrogen ions determines the acidity of a solution; the more hydrogen ions present, the stronger the acid. Once an acid solution is present, rocks such as limestone and marble dissolve rapidly.

Another important chemical weathering process is *oxidation*, which involves the reaction of oxygen with other substances to form oxides. For instance, the iron in a number of ferromagnesian silicates such as olivine and pyroxenes combines with oxygen to form the iron oxide hematite (Fe_2O_3). Acid soils and waters in coal-mining areas result from the oxidation of the mineral pyrite (FeS_2), which yields sulfuric acid and iron oxide.

In the chemical weathering process known as *hydrolysis,* hydrogen (H^+) ions replace positively charged ions in minerals. Particularly susceptible to hydrolysis are the feldspars, which can be altered to form clay minerals that differ both chemically and structurally from the parent material. In fact, a number of important clay minerals originate by chemical weathering of various rocks.

FACTORS CONTROLLING THE RATE OF CHEMICAL WEATHERING

Chemical weathering processes operate on the surfaces of particles. Because chemical alteration occurs from the outside inward, the greater the surface area, the more effective is the weathering. Notice in Figure 4.6 that a block measuring 1 m on a side has a total surface area of 6 m^2; when the block is broken into smaller particles, the total surface area increases, yet the volume remains the same. The fact is that small objects have proportionately more surface area compared to volume than do large objects. Accordingly, me-

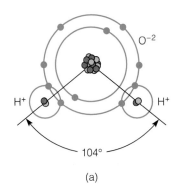

(a)

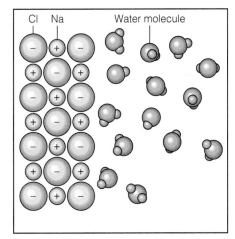

◻ FIGURE 4.5 (*a*) The structure of a water molecule. The asymmetric arrangement of the hydrogen atoms causes the molecule to have a slight positive electrical charge at its hydrogen end and a slight negative charge at its oxygen end. (*b*) Halite (NaCl) dissolves as its sodium ions are attracted to the negative end of the water molecule and its chloride ions are attracted to the positive end.

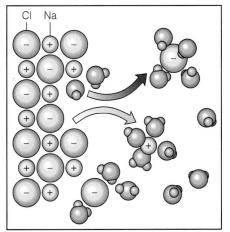

(b)

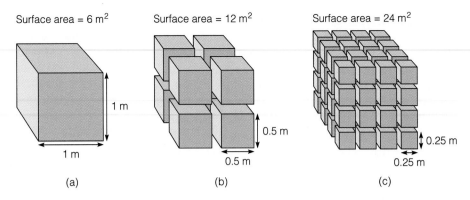

Surface area = 6 m² Surface area = 12 m² Surface area = 24 m²

1 m
1 m
0.5 m
0.5 m
0.25 m
0.25 m

(a) (b) (c)

■ FIGURE 4.6 Particle size and chemical weathering. As a rock is reduced into smaller and smaller pieces, the total surface area increases, but its volume remains the same. In (a) the surface area is 6 m², in (b) it is 12 m², and in (c) 24 m², but the volume remains the same at 1 m³.

chanical weathering, which reduces the size of particles, contributes to chemical weathering by exposing more surface area.

Chemical weathering is most effective in tropical regions, because chemical processes occur more rapidly at high temperatures and in the presence of water. Vegetation and animal life are much more abundant in the tropics than in arid or arctic regions, and they also contribute to more vigorous chemical weathering. Consequently, the effects of weathering extend to depths of several tens of meters in the tropics, but extend only centimeters to a few meters deep in deserts and cold areas.

Some rocks are chemically more stable than others. The metamorphic rock quartzite, for instance, is extremely stable and alters very slowly compared to most other rock types. In contrast, rocks containing abundant ferromagnesian silicates are chemically unstable and decompose rapidly.

One manifestation of chemical weathering is **spheroidal weathering** (■ Fig. 4.7). In spheroidal weathering, a stone, even one that is rectangular to begin with, becomes more

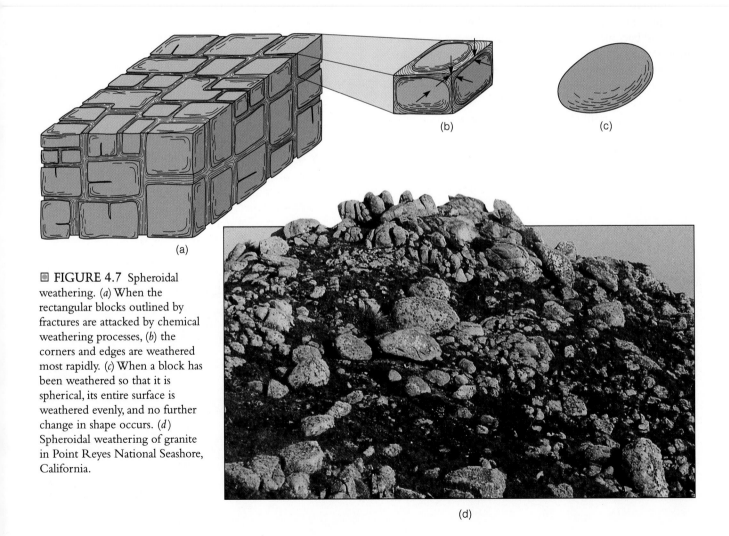

(a)
(b)
(c)
(d)

■ FIGURE 4.7 Spheroidal weathering. (a) When the rectangular blocks outlined by fractures are attacked by chemical weathering processes, (b) the corners and edges are weathered most rapidly. (c) When a block has been weathered so that it is spherical, its entire surface is weathered evenly, and no further change in shape occurs. (d) Spheroidal weathering of granite in Point Reyes National Seashore, California.

spherical as it weathers. On a rectangular stone, the corners are attacked by weathering processes from three directions, and the edges are attacked from two directions, but the flat surfaces are weathered rather uniformly from one direction (Fig. 4.7). As a result, the corners and edges alter more rapidly, the material sloughs off them, and a more spherical shape develops.

SOIL

In most places the land surface is covered by a layer of unconsolidated rock and mineral fragments called **regolith.** Regolith that supports plant growth is recognized as **soil.** A good, fertile soil for gardening or farming is about 45% weathered rock material including sand, silt, and clay. Another essential constituent of productive soils is *humus,* which is derived from bacterial decay of organic matter. Humus contains more carbon and less nitrogen than the original material and is resistant to further bacterial decay.

Some weathered materials in soils are simply sand- and silt-sized mineral grains, especially quartz, but other minerals and rock fragments may be present as well. These solid particles are important because they hold soil particles apart,

allowing oxygen and water to circulate more freely. Clay minerals are also important in soils and aid in the retention of water as well as supplying nutrients to plants.

THE SOIL PROFILE

A soil consists of distinct layers or **soil horizons** that differ from one another in texture, structure, composition, and color (Fig. 4.8a). Starting from the top, soil horizons are designated O, A, B, and C, but the boundaries between horizons are transitional rather than sharp.

The O horizon, composed of organic matter, is rarely more than a few centimeters thick and is not present in all soils. The remains of plants are clearly recognizable in the upper part of the O horizon, but in the lower part only humus is present.

Horizon A, called *top soil,* contains more organic matter than the horizons below. It is also characterized by intense biological activity because plant roots, bacteria, fungi, and animals such as worms are abundant. In soils developed over a long time, horizon A consists mostly of clays and chemically stable minerals such as quartz. Water percolating down through horizon A dissolves soluble minerals and carries them away or downward to lower levels in the soil by a process called *leaching.*

Horizon B, or *subsoil,* contains fewer organisms and less organic matter than horizon A (Fig. 4.8a). It is also known as the *zone of accumulation,* because soluble minerals leached from above accumulate in it as irregular masses. If horizon A is stripped away by erosion leaving horizon B exposed, plants do not grow as well, and if horizon B is clayey, it is

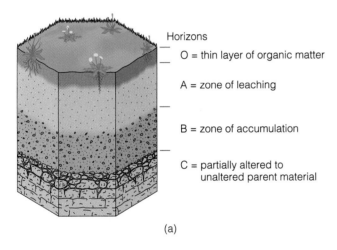

Horizons

O = thin layer of organic matter

A = zone of leaching

B = zone of accumulation

C = partially altered to unaltered parent material

(a)

FIGURE 4.8 (*a*) The soil horizons in a fully developed or mature soil. (*b*) Soil developed directly on bedrock near Denver, Colorado. (*c*) Laterite, shown here in Madagascar, is a deep, red soil formed by intense chemical weathering in the tropics.

(b)

(c)

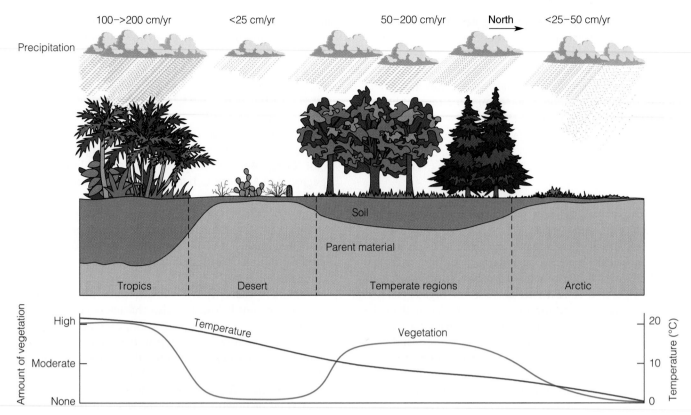

Precipitation

100->200 cm/yr   <25 cm/yr   50–200 cm/yr   **North**   <25–50 cm/yr

Soil

Parent material

Tropics   Desert   Temperate regions   Arctic

Temperature

Vegetation

FIGURE 4.9 Schematic representation showing soil formation as a function of the relationships between climate and vegetation, which alter parent material over time. Soil-forming processes operate most vigorously where precipitation and temperatures are high.

harder when dry and stickier when wet than other soil horizons.

Horizon C consists of partially altered parent material with little organic matter (Fig. 4.8a). In horizons A and B, the composition and texture of the parent material have been so thoroughly altered that they are no longer recognizable. In contrast, rock fragments and mineral grains of the parent material retain their identity in horizon C.

Factors Controlling Soil Formation

It has long been acknowledged that climate is the single most important factor influencing soil type and depth. Intense chemical weathering in the tropics generally yields deep soils from which most of the soluble minerals have been removed by leaching (Fig. 4.9). In arctic and desert climates, on the other hand, soils tend to be thin, contain significant quantities of soluble minerals, and consist mostly of materials derived by mechanical weathering.

A very general classification recognizes three major soil types characteristic of different climatic settings. Soils that develop in humid regions such as the eastern United States and much of Canada are **pedalfers,** a name derived from the Greek word *pedon,* meaning soil, and from the chemical symbols for aluminum (Al) and iron (Fe). Because these soils form in areas with abundant moisture, most of the soluble

minerals have been leached from horizon A. Although it may be gray, horizon A is commonly dark because of abundant organic matter, and aluminum-rich clays and iron oxides tend to accumulate in horizon B.

Pedocals are the soils of arid and semiarid regions as in much of the western United States, especially the southwest. Pedocal derives its name in part from the first three letters of calcite. These soils contain less organic matter than pedalfers, so horizon A is generally lighter colored and contains more unstable minerals because of less intense chemical weathering. As soil water evaporates, calcium carbonate leached from above commonly precipitates in horizons B and C where it forms irregular masses of *caliche.* Precipitation of sodium salts in some desert areas where soil water evaporation is intense yields *alkali soils* that cannot support plants.

Laterite (from the Latin word for brick) is a soil formed in the tropics where chemical weathering is intense and leaching of soluble minerals is complete (Fig. 4.8c). Such soils are red, commonly extend to depths of several tens of meters, and are composed largely of aluminum hydroxides, iron oxides, and clay minerals.

Although laterites support lush vegetation, they are not very fertile. The native vegetation is sustained by nutrients derived mostly from the surface layer of organic matter. Consequently, when societies practicing slash-and-burn agriculture clear these soils, they can raise crops for only a few years. Then the soil is depleted of plant nutrients, the clay-rich laterite bakes brick hard in the tropical sun, and the farmers move on to another area where the process is repeated.

One aspect of laterites is of great economic importance. If the parent material is aluminum rich, aluminum hydroxides may accumulate in horizon B as *bauxite,* the ore of aluminum. Because such intense chemical weathering currently does not occur in North America, the United States and Canada are dependent on foreign sources for aluminum ores.

Although climate is more important than parent material in determining the type of soil that develops, rock type does exert some control. The metamorphic rock quartzite, for instance, will have only a thin soil over it because it is chemically stable, whereas an adjacent body of granite will have a much deeper soil. Also, weathering of pure quartz sandstone will yield no clay, whereas weathering of clay will yield no sand.

Soils not only depend on organisms for their fertility, but also provide a suitable habitat for organisms. Earthworms—as many as one million per acre—ants, sowbugs, termites, centipedes, and millipedes, along with various types of fungi, algae, and single-celled animals, make their homes in the soil. Along with plants, these organisms contribute to the formation of soils and provide humus when they die and decompose. Burrowing animals constantly churn and mix soils, and their burrows provide avenues for gases and water. Some types of soil bacteria are extremely important in changing atmospheric nitrogen into a form of soil nitrogen suitable for use by plants.

Because climate changes with elevation, *relief,* the difference in elevation between high and low points in a region, affects soil-forming processes. *Slope* affects soils in two ways. One is simply slope angle: the steeper the slope, the less opportunity for soil development because weathered material is eroded faster than soil-forming processes can work. The other slope control is the direction a slope faces. In the Northern Hemisphere, north-facing slopes receive less sunlight than south-facing slopes. Consequently, north-facing slopes have soils with cooler internal temperatures, may support different vegetation, and, if in a cold climate, remain frozen or snow covered longer.

Climate and organisms altering parent material through time determine soil properties; the longer these processes have operated, the more fully developed the soil will be (Fig.

4.9). How much time is needed to develop a centimeter (0.4 in) of soil or a fully developed soil a meter (3.3 ft) or so deep? No definitive answer can be given because weathering proceeds at vastly different rates depending on climate and parent material, but an overall average might be about 2.5 cm (1 in) per century. However, a lava flow a few centuries old in Hawaii may have a well-developed soil on it, whereas a flow of the same age in Iceland will have considerably less soil. And given the same climatic conditions, soil will develop faster on sediment than it will on rock.

 SOIL DEGRADATION

Any decrease in soil productivity or loss of soil to erosion is referred to as **soil degradation.** Three types of soil degradation are generally recognized: erosion, chemical deterioration, and physical deterioration. Most soil erosion results from the action of wind and water. When the natural vegetation is removed and a soil is pulverized by plowing, the fine particles are easily blown away (see ☐ Perspective 4.1). Falling rain also disrupts soil particles and carries soil with it when it runs off at the surface (▣ Fig. 4.10). This is particularly devastating on steep slopes from which the vegetation has been removed by overgrazing, deforestation, fire, or construction.

If soil losses to erosion are minimal, soil-forming processes can keep pace, and the soil remains productive. Should the loss rate exceed the formation rate, however, the

▣ FIGURE 4.10 A small channel called a rill was eroded in this field during a rainstorm, but it was later plowed over. If such channels are more than about 30 cm (12 in) deep, they are called gullies, and they cannot be eliminated by plowing. If gullying is extensive, croplands must be abandoned.

The Dust Bowl

The stock market crash of 1929 ushered in the Great Depression, a time of diminished economic activity and rising rates of unemployment. In the decades prior to the 1930s, farmers had enjoyed a degree of success unparalleled in U.S. history. During World War I (1914–1918), the price of wheat soared, and after the war when Europe was recovering, the government subsidized high wheat prices. High prices and mechanized farming resulted in more and more land being tilled. Even the weather cooperated, and land in the western United States that was only marginally productive was plowed.

Beginning about 1930, most of the country, especially the southern Great Plains, experienced several years of drought. Some rain fell but it was insufficient to maintain productivity. And since the land, even marginally productive land, had been plowed, the native prairie grasses were no longer available to bind the soil together. Nothing kept the soil from blowing away. And blow away it did—in huge quantities.

A large region in the southern Great Plains that was particularly hard hit by drought, dust storms, and soil erosion came to be known as the Dust Bowl. Its boundaries were not well defined, but it included parts of Kansas, Colorado, New Mexico, Oklahoma, and Texas. The Dust Bowl and its less affected fringe area covered more than 400,000 km^2 (154,000 mi^2)!

Dust storms were common during the 1930s, and some reached phenomenal sizes (□ Fig. 1). In 1934, one of the largest covered more than 3.5 million km^2 (1.35 million mi^2), lifted dust nearly 5 km (3 mi) into the air, obscured the sky over large parts of six states, and blew hundreds of millions of tons of soil eastward where it settled on east coast cities and on ships some 480 km (300 mi) out in the Atlantic Ocean. Dust storms of regional extent were recorded on 140 occasions during 1936 and 1937. Dust was everywhere. It seeped into houses,

suffocated wild animals and livestock, and adversely affected human health.

The dust was, of course, the top soil of the tilled lands. Blowing dust was not the only problem, however; wind-blown sand piled up along fences, drifted against houses and farm machinery, and covered what otherwise might have been productive soils. Agricultural production fell precipitously, farmers could not meet their mortgage payments, and by 1935 tens of thousands were leaving. Many went west to California and became the migrant farm workers immortalized in John Steinbeck's novel *The Grapes of Wrath*.

The Dust Bowl was an economic disaster of great magnitude. Droughts had stricken the southern Great Plains before, and have done so since, from August 1995 well into the summer of 1996, for instance, but the drought during the 1930s was especially severe. In addition to the drought, political and economic factors contributed to the disaster. Wheat prices were artificially inflated and many farmers were deeply in debt—mostly because they had purchased farm machinery so they could benefit from the higher wheat prices.

If the Dust Bowl has a bright side, it is that soil is no longer taken for granted or regarded as a substance that needs no nurturing. In addition, the disaster stimulated the development of a number of soil conservation methods that have now become standard in agriculture.

□ **FIGURE 1** A huge dust storm strikes Hugoton, Kansas, on April 14, 1935, also known as Black Sunday.

most productive upper layer of soil, horizon A, is removed, exposing horizon B, which is much less productive. The Soil Conservation Service of the U.S. Department of Agriculture estimates that 25% of the cropland in the United States is eroding faster than soil-forming processes can replace it.

Problems experienced during the past, particularly during the 1930s, have stimulated the development of methods to minimize soil erosion on agricultural lands. Crop rotation, contour plowing, and the construction of terraces have all proved helpful (Fig. 4.11). So has no-till planting in which the residue from the harvested crop is left on the ground to protect the surface from the ravages of wind and water.

A soil undergoes chemical deterioration when its nutrients are depleted and its productivity decreases. Loss of soil nutrients is most notable in countries where soils are overused in an attempt to maintain high levels of agricultural productivity. Other causes include insufficient use of chemical fertilizers and clearing soils of their natural vegetation. Chemical deterioration of soils takes place on all continents, but it is most prevalent in South America where it accounts for 29% of all soil degradation.

Other types of chemical deterioration are pollution and salinization, which occurs when salts become concentrated in a soil making it unfit for agriculture. Pollution can be caused by improper disposal of domestic and industrial wastes, oil and chemical spills, and the concentration of insecticides and pesticides in soils.

Physical deterioration of soils results when soil particles are compacted under the weight of heavy machinery and livestock, especially cattle. Once compacted, soils are more costly to plow, and plants have a more difficult time emerging. Furthermore, water does not readily infiltrate compacted soils, so more runoff occurs, which in turn accelerates the rate of water erosion.

WEATHERING AND MINERAL RESOURCES

In a preceding section, we discussed intense chemical weathering in the tropics and the origin of *bauxite,* the ore of aluminum. Such an accumulation of valuable minerals formed by the selective removal of soluble substances is a *residual concentration.* Other important residual concentrations include ore deposits of iron, manganese, clays, nickel, phosphate, tin, diamonds, and gold. Some of the sedimentary iron deposits of the Lake Superior region were enriched by chemical weathering when soluble constituents were dissolved and carried away.

Some deposits of the clay mineral kaolinite in the southern United States were formed by chemical weathering of feldspars in pegmatites and of clay-bearing limestones and dolostones. Kaolinite is used in the manufacture of paper and ceramics.

A gossan is a yellow to reddish deposit composed largely of iron hydroxides that formed by the alteration of iron- and sulfur-bearing minerals such as pyrite (FeS_2). The dissolution of these minerals forms sulfuric acid, which causes other metallic minerals to dissolve and be carried down toward the water table. Gossans have been used occasionally as sources of iron, but they are far more important as indicators of underlying ore deposits. One of the oldest known underground mines exploited such ores about 3,400 years ago in what is now southern Israel.

FIGURE 4.11 One soil conservation practice is contour plowing, which involves plowing parallel to the contours of the land. The furrows and ridges are perpendicular to the direction that water would otherwise flow downhill.

MASS WASTING

Mass wasting (also called *mass movement*) is defined as the downslope movement of material under the influence of gravity. While water can play an important role, the relentless pull of gravity is the major force in all types of mass wasting. Even though most people associate mass wasting with rapid movements on steep slopes, it can also occur on near-level land. Furthermore, the rapid types of mass wasting, such as landslides and mudflows, typically get the most publicity, but the much slower types usually do the greatest amount of property damage.

FACTORS INFLUENCING MASS WASTING

All slopes are in a state of dynamic equilibrium, meaning that they constantly adjust to new conditions. Whenever a building or road is constructed on a hillside or the base of a slope is eroded, the equilibrium of that slope is affected, and the slope then adjusts to this new set of conditions. When the gravitational force acting on a slope exceeds its resisting force, slope failure (mass wasting) takes place. The resisting forces

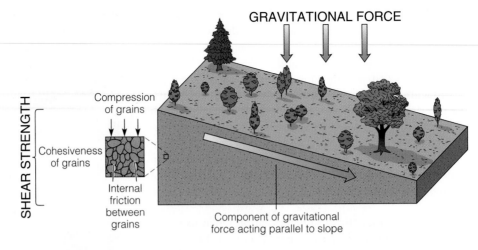

GRAVITATIONAL FORCE

SHEAR STRENGTH

Cohesiveness of grains

Compression of grains

Internal friction between grains

Component of gravitational force acting parallel to slope

◨ FIGURE 4.12 A slope's shear strength depends on the slope material's strength and cohesiveness and on the amount of internal friction between grains. These factors promote slope stability. The force of gravity operates vertically but has a component acting parallel to the slope. When this force, which promotes instability, exceeds a slope's shear strength, slope failure occurs by one or more mass wasting processes.

helping to maintain slopes include the slope material's strength and cohesion, the amount of internal friction between grains, and vegetation. These factors collectively define a slope's **shear strength** (◨ Fig. 4.12). Opposing a slope's shear strength is that part of the force of gravity acting parallel to the slope, thereby causing instability.

The greater a slope's angle, the greater the component of gravitational force acting parallel to the slope, and the greater the chance for mass wasting. The steepest angle that a slope can maintain without collapsing is its angle of repose. At this angle, the shear strength of the slope's material exactly counterbalances the force of gravity. For unconsolidated material, the angle of repose normally ranges from 25° to 40°. Slopes steeper than 40° usually consist of unweathered rock.

A number of processes can account for oversteepening of slopes and the resulting mass wasting. Undercutting by a stream or waves and excavations for roads and hillside building sites remove a slope's base, increase the slope angle, and thereby increase the gravitational force acting parallel to the slope (◨ Fig. 4.13).

Mass wasting is more likely to take place in soil and loose or poorly consolidated sediment than in solid rock. As soon as rocks are exposed at the surface, weathering begins to disintegrate and decompose them, thereby reducing their shear strength and increasing their susceptibility to mass wasting. The deeper the weathering zone extends, as in the tropics, the greater the likelihood of some type of mass movement.

◨ FIGURE 4.13 (a) Stream erosion at the base of a slope increases the slope angle and (b) can lead to slope failure. (c) Excavation at the base of a slope also increases the slope angle and decreases slope stability.

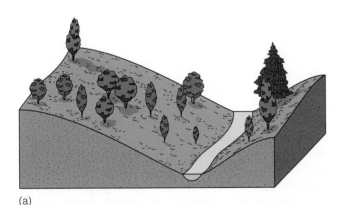

(a)

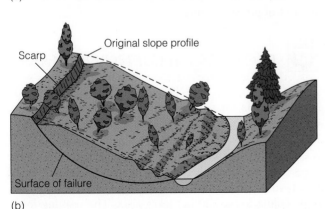

Scarp

Original slope profile

Surface of failure

(b)

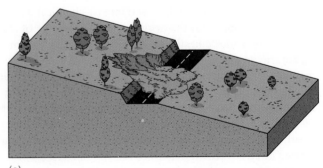

(c)

■ FIGURE 4.14 An aerial view of the slope failure at Aberfan, Wales, that resulted in 144 fatalities, most of them children in a school.

Large quantities of water from melting snow or rainstorms greatly increase the likelihood of slope failure. The additional weight water adds to slope materials can be enough to cause mass movement, and water percolating through slope materials helps to decrease friction between particles, contributing to a loss of cohesion. Slopes composed of dry clay are usually quite stable, but they become an unstable slurry when wet because clay consists of platy particles that easily slide over each other.

Vegetation affects slope stability in several ways. It absorbs water that would otherwise saturate a slope's material, a condition that frequently leads to mass wasting. In addition, the root systems of plants help stabilize slopes by binding soil particles together. The removal of vegetation by either natural causes or human activity is a major cause of many mass movements. Summer brush and forest fires in southern California frequently leave the hillsides bare; when the autumn rains come, they saturate the ground, causing damaging mudslides.

Overloading of slopes is almost always the result of human activity and typically results from dumping, filling, or piling up material on hillsides. The additional weight created by overloading increases the water pressure within the material, which in turn decreases its shear strength, thereby weakening the slope material. A tragic example of slope failure caused by overloading occurred at Aberfan, Wales, in 1966 when a large pile of coal mining debris collapsed, flowed downhill, and killed 144 people, including 116 children in a school (■ Fig. 4.14).

If the rocks underlying a slope are inclined downward or dip in the same direction as the slope, mass wasting is more likely to occur than if the rocks are horizontal or dip in the opposite direction (■ Fig. 4.15). When the rocks dip in the same direction as the slope, water can percolate along the various planes separating the layers and decrease the cohesiveness and friction between adjacent rock layers. This is particularly true where clay layers are present because clay becomes very slippery when wet. Fractures in rocks may also dip in the same direction as the slope. Water migrating through fractures weathers the rock and expands these openings until the weight of the overlying rock causes it to fall.

Although all these factors contribute to slope instability, most—though not all—rapid mass movements are triggered by a force that temporarily disturbs slope equilibrium. The most common triggering mechanisms are strong vibrations from earthquakes (see the Prologue) and excessive amounts of water from a winter snowmelt or heavy rains. Leakage from broken water pipes and swimming pools can also contribute to mass movements. Volcanic eruptions, explosions, and even loud claps of thunder may be enough to trigger a landslide if a slope is sufficiently unstable.

TYPES OF MASS WASTING

Mass movements are generally classified on the basis of three major criteria: (1) rate of movement (rapid or slow); (2) type of movement (falling, sliding, or flowing); and (3) type of material involved (rock, soil, or debris) (○ Table 4.1). *Rapid mass movements* usually occur quite suddenly and involve a visible movement of material. Slow mass movements advance at an imperceptible rate and are usually only detectable by the effects of their movement such as tilted trees and power poles or cracked foundations. Although rapid mass movements are more dramatic and more dangerous, slow mass movements are responsible for the downslope transport of much more material and considerably more property damage.

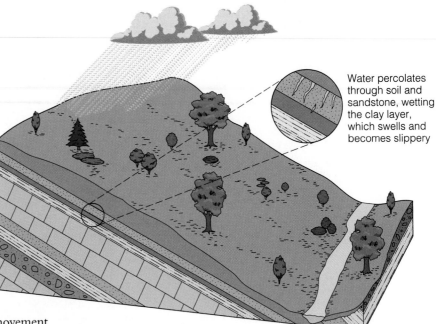

Water percolates through soil and sandstone, wetting the clay layer, which swells and becomes slippery

■ FIGURE 4.15 Rocks inclined downward (dipping) in the same direction as a hill's slope are particularly susceptible to mass wasting. The slope becomes more unstable when its base is undercut by erosion or by the construction of a road.

TABLE 4.1   Classification of Mass Movements and Their Characteristics

TYPE OF MOVEMENT	SUBDIVISION	CHARACTERISTICS	RATE OF MOVEMENT
Falls	Rockfall	Rocks of any size fall through the air from steep cliffs, canyons, and road cuts	Extremely rapid
Slides	Slump	Movement occurs along a curved surface of rupture; most commonly involves unconsolidated or weakly consolidated material	Extremely slow to moderate
	Rockslide	Movement occurs along a generally planar surface	Rapid to very rapid
Flows	Mudflow	Consists of at least 50% silt- and clay-sized particles and up to 30% water	Very rapid
	Debris flow	Contains larger particles and less water than mudflows	Rapid to very rapid
	Earthflow	Thick, viscous, tongue-shaped mass of wet regolith	Slow to moderate
	Quick clays	Composed of fine silt and clay particles saturated with water; when disturbed by a sudden shock, lose their cohesiveness and flow like a liquid	Rapid to very rapid
	Solifluction	Water-saturated surface sediment	Slow
	Creep	Downslope movement of soil and rock	Extremely slow
Complex movements		Combination of different movement types	Slow to extremely rapid

⊡ FIGURE 4.16 Wire mesh has been draped over this steep slope in Hawaii to prevent rocks from falling to the road below.

Falls

A **rockfall** is a common type of extremely rapid mass movement in which rocks of any size fall through the air. Rockfalls result from failure along fractures or planes between rock layers and are commonly triggered by natural or human undercutting of slopes or by earthquakes. Anyone who has driven through mountainous or hilly areas is familiar with the "Watch for Falling Rocks" signs posted to warn drivers of the danger. Slopes particularly prone to rockfalls are sometimes covered with wire mesh in an effort to prevent dislodged rocks from falling to the road below (⊡ Fig. 4.16). Another tactic is to put up wire mesh fences along the base of the slope to catch or slow down bounding or rolling rocks.

Slides

A slide involves slow to rapid movement of material along one or more surfaces of failure (Table 4.1). The type of material may be soil, rock, or a combination of the two, and the slide mass may break apart during movement or remain intact.

A **slump** is a type of slide involving the downward movement of material along a curved surface and is characterized by the backward rotation of the slump block (⊡ Fig. 4.17). Unconsolidated or weakly consolidated sediment or soil is

most susceptible to slumps, which may be small individual sets, such as those along stream banks, or massive, multiple sets that affect large areas and cause considerable damage. Slumps are commonly initiated by stream or wave erosion along the base of a slope, but may also be caused by slope oversteepening resulting from construction of highways and housing developments.

Many slumps are merely a nuisance, but large-scale slumps involving populated areas and highways can cause extensive damage. Many areas along the southern California coast are underlain by weakly consolidated layers of silt, sand, gravel, and clay, and slumping has been a constant problem. In addition, southern California is geologically active, so

FIGURE 4.17 (*a*) In a slump, material moves downward along a curved surface of failure, causing the slump block to rotate backward. Most slumps involve unconsolidated or weakly consolidated material and are typically caused by erosion along a slope's base. (*b*) Undercutting of steep sea cliffs by waves resulted in massive slumping in the Pacific Palisades area of southern California in 1958. Note the heavy earthmoving equipment for scale.

Fractures

Scarp

Slump block

Surface of rupture

(a)

Pacific Palisades
Santa Monica
Los Angeles
Palos Verdes
Long Beach
Pacific Ocean

(b)

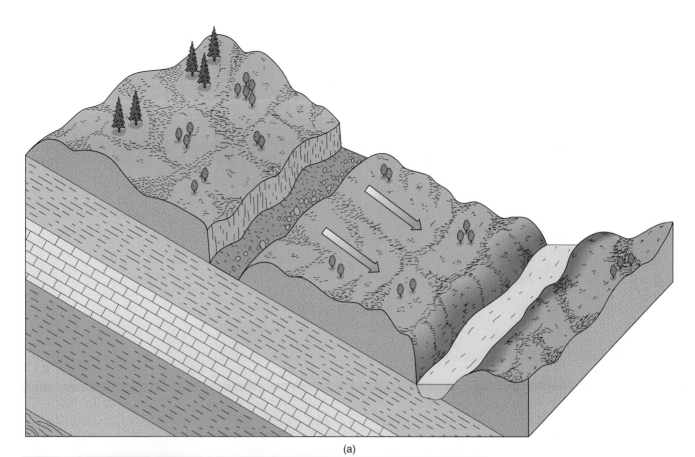

(a)

(b)

■ FIGURE 4.18 (*a*) A rockslide occurs when material moves downslope along a planar surface. (*b*) A 1978 rockslide at Laguna Beach, California, destroyed numerous homes and cars.

many of these deposits are cut by fractures, which allow the infrequent rains to percolate downward rapidly, wetting and lubricating the clay layers.

In a **rockslide,** rocks move downslope along a more or less planar surface. Rock layers dipping in the same direction as local slopes are responsible for most rockslides (Fig. 4.18a), although they can also occur along fractures parallel to a slope. Startled residents of Laguna Beach, California, watched

as a rockslide destroyed or damaged 50 homes on October 2, 1978 (■ Fig. 4.18b). The rocks at Laguna Beach dip about 25° in the same direction as the slope of the canyon walls and contain clay beds that "lubricate" the overlying rock layers, allowing the rocks to slide. During a rockslide at Frank, Alberta, Canada, on April 29, 1903, nearly 40 million m³ of rock slid down Turtle Mountain along fracture surfaces, killing 70 people and partially burying the town.

FIGURE 4.19 A small earthflow near Baraga, Michigan. Earthflows occur most commonly in humid areas on grassy soil-covered slopes.

Flows

Flows are mass movements in which material flows as a viscous fluid or displays plastic movement; their rate of movement ranges from extremely slow to extremely fast (Table 4.1). **Mudflows,** the most fluid of the major mass movement types, consist of at least 50% mud combined with up to 30% water. They are common in arid and semiarid environments where they are triggered by heavy rain that quickly saturates the regolith, turning it into a raging flow of mud that engulfs everything in its path. Because mudflows are so fluid, they generally follow preexisting channels.

Debris flows are composed of larger particles than those in mudflows and do not contain as much water. Consequently, they are usually more viscous, typically do not move as rapidly, and rarely are confined to preexisting channels. Debris flows can be just as damaging as mudflows, however, because they can transport large objects.

Earthflows move more slowly than either mudflows or debris flows. An earthflow slumps from the upper part of a hillside, leaving a scarp, and flows slowly downslope as a thick, viscous, tongue-shaped mass of wet regolith (Fig. 4.19).

Solifluction, the slow downslope movement of water-saturated soil or sediment, can take place in any climate where the ground becomes saturated with water, but it is most common in cold climates where *permafrost* (permanently frozen ground) is present. During the warmer season when the upper portion of the permafrost thaws, water and surface sediment form a soggy mass that flows and produces a characteristic series of lobes (Fig. 4.20).

Many problems are associated with construction in areas of permafrost. When an uninsulated building is constructed directly on permafrost, heat escapes through the floor, thaws

FIGURE 4.20 Solifluction flows near Suslositna Creek, Alaska, show the typical lobate nature of these flows.

the ground below, and turns it into a soggy, unstable mush. Because the ground is no longer solid, the building settles unevenly into the ground, and numerous structural problems result.

FIGURE 4.21 (*a*) Some evidence of creep: (A) curved tree trunks; (B) displaced monuments; (C) power poles tilted downhill; (D) displaced and tilted fence; (E) roadway moved out of alignment; and (F) hummocky surface. (*b*) These fence posts near Clare, Michigan, slant downhill as a result of creep. (*c*) Creep has bent these sandstone and shale beds near Marathon, Texas.

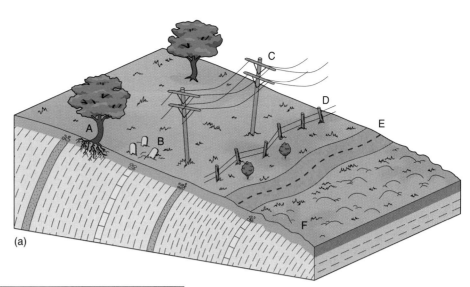

(a)

(b)

(c)

Creep, the slowest type of flow, involves extremely slow downhill movement of soil or rock. Although it can occur anywhere and in any climate, creep is most effective in humid climates. In fact, it is the most common form of mass wasting in the southeastern United States and southern Appalachian Mountains. Creep usually involves entire hillsides and probably occurs, to some extent, on any weathered or soil-covered, sloping surface. Because the rate of movement is imperceptible, we are frequently unaware of creep until we notice its effects: tilted trees and power poles, broken streets and sidewalks, and cracked retaining walls or foundations (Fig. 4.21).

Complex Movements

When one type of mass wasting is dominant, it can be classified as one of the types described thus far. If several types are involved, however, it is called a **complex movement.** The most common type of complex movement is the slide-flow involving well-defined slumping at the head, followed

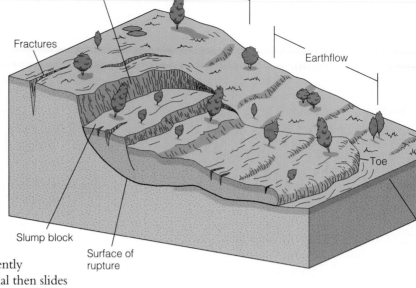

FIGURE 4.22 A complex movement involving slumping at the upper end or head followed by an earthflow.

Labels in figure: Scarp, Slumping, Fractures, Earthflow, Toe, Soil, Slump block, Surface of rupture

by a debris flow or earthflow farther along its course (Fig. 4.22).

A **debris avalanche** is a type of complex movement that typically starts out as a rockfall when large quantities of rock, ice, and snow are dislodged from a mountainside, frequently as a result of an earthquake. The material then slides or flows downslope, picking up additional surface material and increasing in speed as it progresses (see the Prologue).

RECOGNIZING AND MINIMIZING THE EFFECTS OF MASS MOVEMENTS

The most important factor in eliminating or minimizing the damaging effects of mass wasting is a thorough investigation of the region in question. Identifying areas with a high potential for slope failure is important in any hazard assessment study; such studies include identifying former landslides as well as sites of potential mass movement (see Perspective 4.2).

Soil and bedrock samples are also studied in both the field and the laboratory to assess characteristics such as composition, susceptibility to weathering, cohesiveness, shear strength, and ability to transmit fluids. These studies help Earth scientists and engineers predict slope stability under a variety of conditions.

The information derived from a hazard assessment study can be used to produce *slope stability maps* of the area. These maps allow planners and developers to make decisions about where to site roads, utility lines, and housing or industrial developments based on the relative slope stability or instability at a particular location.

Although most large mass movements usually cannot be prevented, various methods can be employed to minimize the danger and damage resulting from them. One of the most

effective and inexpensive ways to reduce the potential for slope failure is through surface and subsurface drainage of hillsides. Drainage serves two purposes. It reduces the weight of the material likely to slide and increases the shear strength of the slope material by lowering the water pressure within the mass.

Surface waters can be drained or diverted by ditches, gutters, or culverts designed to direct water away from slopes. Drainpipes perforated along one surface can be driven into a hillside to help remove subsurface water (Fig. 4.23). Finally, planting vegetation on hillsides helps stabilize slopes by holding the soil together and reducing the amount of water in the soil.

FIGURE 4.23 Drainpipes can remove some subsurface water and thus help stabilize a hillside.

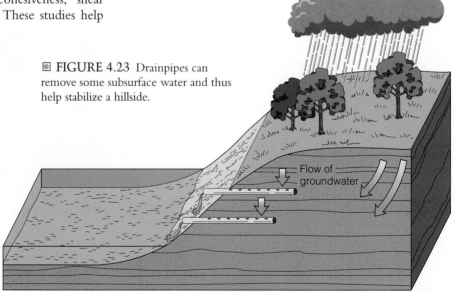

Labels in figure: Flow of groundwater

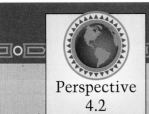
The Vaiont Dam Disaster

On October 9, 1963, more than 240 million m³ of rock and soil slid into the Vaiont Reservoir in Italy, triggering a destructive flood that killed nearly 3,000 people (☐ Fig. 1). Within 15 to 30 seconds, much of the reservoir was filled with debris that rose up to 175 m (575 ft) above the level of the reservoir. The water displaced in the reservoir overtopped the dam by 100 m (330 ft), rushed down the valley, and destroyed everything in its path. It is a tribute to the dam's designer and construction engineer that the dam itself survived the disaster.

Among the geological causes for the slide were the rocks themselves, which were weak to begin with and dipped in the same direction as the valley walls. Active solution of the limestones by slightly acid groundwater further weakened them by developing and expanding an extensive network of cracks, fissures, and other openings.

During the two weeks before the slide, heavy rains saturated the ground, adding extra weight and reducing the shear strength of the rocks. Water from the reservoir also infiltrated the rocks of the lower valley walls, further reducing their strength. It is thought that the actual slide was triggered by an increase in water pressure that facilitated slippage along wet clay layers.

Soon after the dam was completed, a relatively small slide of one million m³ occurred. Following this slide, it was decided to limit the amount of water in the reservoir and to install monitoring devices throughout the potential slide area. Between 1960 and 1963, the eventual slide area moved an average of about 1 cm (0.4 in) per week. Beginning on September 18, 1963, however, movement had increased to about 1 cm per day.

By October 8, the creep rate had increased to almost 39 cm (15 in) per day, and engineers realized that the entire slide area was moving, and quickly began lowering the reservoir level. On October 9, the rate of movement increased even further, in some locations up to 80 cm (31 in) per day, and there were reports that the reservoir level was actually rising. Finally, at 10:41 P.M. that night, during yet another rainstorm, the south bank of the Vaiont valley slid into the reservoir.

The lesson to be learned from this disaster is that a complete and systematic appraisal of an area must be conducted before major construction begins. Such a study should examine the geology of the area, identify past mass movements, assess their potential for recurrence, and evaluate the effects that the project will have on the rocks, including how it will alter their shear strength. Without these precautions, similar disasters will occur and lives will needlessly be lost.

☐ FIGURE 1 An aerial view of the Vaiont Dam in Italy. Displacement of water in the reservoir by a huge slide caused a wave to overtop the dam by 100 m (330 ft).

Another way to help stabilize hillsides is to reduce their slope, thereby decreasing the potential for slope failure. In some situations, retaining walls can be constructed to provide support for the base of a slope (Fig. 4.24a). These are usually anchored into solid rock, backfilled with crushed rock, and provided with drain holes to prevent the buildup of water pressure in the hillside. *Rock bolts,* similar to those employed in tunneling and mining, have been used to fasten potentially unstable rock masses to the underlying stable rock (Fig. 4.24b).

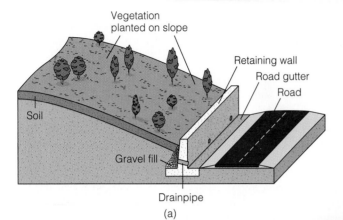

(a)

 FIGURE 4.24 (*a*) A retaining wall with a drainpipe can be used to stabilize a slope. (*b*) Rock bolts are used to help secure rock above the outlet of the west diversion tunnel of the Glen Canyon Dam. As can be seen, however, some portions of rock still broke away.

(b)

CHAPTER SUMMARY

1. Mechanical and chemical weathering result in disintegration and decomposition of parent material so that it is more nearly in equilibrium with new physical and chemical conditions.
2. The solid residue of weathering can be further modified to form soil, or it can be eroded, transported, and deposited as sediment, which may become sedimentary rock.
3. Mechanical weathering includes such processes as frost wedging, salt crystal growth, pressure release, thermal expansion and contraction, and the activities of organisms.
4. Chemical weathering includes processes such as solution, oxidation, and hydrolysis during which the chemical composition of parent material is changed. Chemical weathering gen-

erally proceeds most rapidly in hot, wet environments. It is aided by mechanical weathering processes that break the parent material into smaller pieces, thereby exposing more surface area.

5. In spheroidal weathering, the corners and edges of a rock are altered more rapidly so that even if it is rectangular to begin with, it becomes more spherical as it weathers.
6. Both mechanical and chemical weathering produce regolith. Regolith that contains humus and can support plants is soil.
7. Soils are characterized by horizons that are designated in descending order as O, A, B, and C, each differing from the others in texture, structure, composition, and color.

8. Soils recognized as pedalfers develop in humid regions, whereas pedocals are characteristic of more arid regions. Laterite is a deep, red soil resulting from intense chemical weathering in the tropics.

9. Soil degradation caused by water and wind erosion, chemical deterioration, and physical deterioration is a problem in many areas. Human activities such as construction, agriculture, and deforestation can accelerate soil degradation.

10. Intense chemical weathering is responsible for the origin of residual concentrations, many of which contain valuable minerals such as iron, lead, copper, and clay. Bauxite, the ore of aluminum, is formed by chemical weathering of aluminum-rich parent material in the tropics.

11. Mass wasting involves the downslope movement of soil, sediment, or rock under the influence of gravity.

12. Many factors contribute to mass wasting including slope angle, weathering, water content of slope materials, overloading of slopes, and removal of vegetation.

13. Mass movements are generally classified on the basis of their rate of movement (slow versus rapid), type of movement (falls, slides, flows), and type of material (soil, sediment, rock).

14. Rapid mass movements can cause considerable property damage and numerous deaths, but slow movements, such as creep, cause more damage to buildings, roads, and other structures.

15. Several measures can be taken to reduce or minimize damage from mass wasting. Slopes can be stabilized by building retaining walls, draining excess water, regrading slopes, planting vegetation, and using rock bolts.

IMPORTANT TERMS

chemical weathering
complex movement
creep
debris avalanche
debris flow
earthflow
erosion
exfoliation dome

frost wedging
laterite
mass wasting
mechanical weathering
mudflow
parent material
pedalfers
pedocal

pressure release
regolith
rockfall
rockslide
salt crystal growth
shear strength
slump
soil

soil degradation
soil horizon
solifluction
spheroidal weathering
thermal expansion and
    contraction
weathering

REVIEW QUESTIONS

1. Which mechanical weathering process is involved in the origin of exfoliation domes?
   a. _____ heating and cooling; b. _____ expansion and contraction; c. _____ the activities of organisms; d. _____ oxidation and reduction; e. _____ pressure release.

2. When the ions in a substance separate or dissociate in a fluid, the substance has been:
   a. _____ mechanically weathered; b. _____ altered to clay; c. _____ dissolved; d. _____ oxidized; e. _____ converted to soil.

3. The process whereby hydrogen ions replace ions in a mineral is:
   a. _____ enrichment; b. _____ oxidation; c. _____ creep; d. _____ hydrolysis; e. _____ laterization.

4. Given the same climatic conditions, on which one of the following will soil form most rapidly?
   a. _____ granite; b. _____ quartzite; c. _____ basalt; d. _____ sediment; e. _____ caliche.

5. The soil and unconsolidated rock material covering most of the Earth's surface is:
   a. _____ regolith; b. _____ laterite; c. _____ humus; d. _____ parent material; e. _____ talus.

6. A soil's B horizon is also known as the:

   a. _____ top soil; b. _____ humus layer; c. _____ alkali soil; d. _____ zone of accumulation; e. _____ organic-rich layer.

7. Any loss of soil to erosion or a decrease in a soil's productivity is:
   a. _____ gullying; b. _____ soil degradation; c. _____ weathering; d. _____ leaching; e. _____ exfoliation.

8. The shear strength of slope materials includes:
   a. _____ the strength and cohesion of the material; b. _____ the amount of internal friction between grains; c. _____ gravity; d. _____ all of these; e. _____ answers (a) and (b).

9. Which one of the following factors can actually enhance slope stability?
   a. _____ water content; b. _____ vegetation; c. _____ overloading; d. _____ rocks dipping in the same direction as the slope; e. _____ none of these.

10. A type of mass wasting common in mountainous regions where talus accumulates is:
    a. _____ creep; b. _____ solifluction; c. _____ rockfalls; d. _____ slides; e. _____ mudflows.

11. Downslope movement along an essentially planar surface is a(n):
    a. _____ slump; b. _____ rockfall; c. _____ earthflow; d. _____ solution; e. _____ rockslide.

12. Which one of the following is the most fluid type of mass movement?
    a. ____ earthflow; b. ____ debris flow; c. ____ mudflow; d. ____ solifluction; e. ____ slump.
13. The most widespread and costly type of mass wasting in terms of total material moved and monetary damage is:
    a. ____ creep;   b. ____ solifluction;   c. ____ mudflow; d. ____ debris flow; e. ____ slump.
14. Which one of the following features indicates a hillside is undergoing creep?
    a. ____ fence posts tilted downhill; b. ____ movement along a curved surface of failure; c. ____ little or no vegetation; d. ____ accumulation of talus; e. ____ numerous animal burrows.
15. How does mechanical weathering differ from and contribute to chemical weathering?
16. Describe the process whereby soluble minerals such as halite (NaCl) are dissolved.
17. What is an acid solution, and why are acid solutions important in chemical weathering?
18. Explain why particle size is an important factor in chemical weathering.
19. What is the significance of climate and parent material in the development of soil?
20. What are the forces that help to maintain slope stability?
21. What roles do climate and weathering play in mass wasting?
22. How does vegetation affect slope stability?
23. What is the difference between a slump and a rockslide? Why are slumps particularly common where hillsides have been excavated for roads?
24. What precautions must be taken when building in permafrost areas?

POINTS TO PONDER

1. Consider the following: A soil is 1.5 m thick, new soil forms at the rate of 2.5 cm per century, and the erosion rate is 4 mm per year. How much soil will be left after 100 years?
2. What kinds of weathering and mass wasting would you expect to occur on the Moon? On Venus? Explain your answer.

3. How can you recognize areas that are susceptible to mass wasting? What measures would you recommend to stabilize a slope failing by slump and soil creep?
4. What types of human activities contribute to soil degradation, and how can their impact be minimized?

ADDITIONAL READINGS

Bear, F. E. 1986. *Earth: The stuff of life.* 2d rev. ed. Norman, Okla.: University of Oklahoma Press.

Birkeland, P. W. 1984. *Soils and geomorphology.* New York: Oxford University Press.

Brabb, E. E., and B. L. Harrod, eds. 1989. *Landslides: Extent and economic significance.* Brookfield, Va.: A. A. Balkema.

Buol, S. W., F. D. Hole, and R. J. McCracken. 1980. *Soil genesis and classification.* Ames, Iowa: Iowa State University Press.

Carroll, D. 1970. *Rock weathering.* New York: Plenum Press.

Courtney, F. M., and S. T. Trudgill. 1984. *The soil: An introduction to soil study.* 2d ed. London: Arnold.

Crozier, M. J. 1989. *Landslides: Causes, consequences, and environment.* Dover, N.H.: Croom Helm.

Fleming, R. W., and F. A. Taylor. 1980. Estimating the cost of landslide damage in the United States. *U.S. Geological Survey Circular 832.*

Gibbons, B. 1984. Do we treat our soil like dirt? *National Geographic* 166, no. 3: 350–89.

Kiersch, F. A. 1964. Vaiont reservoir disaster. *Civil Engineering* 34: 32–39.

Ollier, C. 1969. *Weathering.* New York: Elsevier.

Parfit, M. 1989. The dust bowl. *Smithsonian* 20, no. 3: 44–54, 56–57.

Zaruba, Q., and V. Mencl. 1982. *Landslides and their control.* 2d ed. Amsterdam: Elsevier.

WORLD WIDE WEB SITES

For current updates and exercises, log on to http://www.wadsworth.com/geo.

DEVILS MARBLES
http://www.synaptic.bc.ca/gallery/devilmar.htm

This web site is maintained by Patrick Jennings and contains various images of Devils Marbles from Australia. Devils Marbles are examples of spheroidal weathering.

U.S. GEOLOGICAL SURVEY EARTH SCIENCE IN THE PUBLIC SERVICE GEOLOGIC HAZARDS
http://gldage.cr.usgs.gov/

The home page of this site contains links to earthquakes, landslides, geomagnetism, and other geologic hazards.

JCP GEOLOGISTS, INC.
http://www.car.org/rex/jcp/index.htm

This site is maintained by a company that provides disclosure reports for the real estate profession in California. It does contain interesting and useful geological information and links to other geological sites.

CHAPTER

5

Running Water and Groundwater

Cumberland Falls on the Big South Fork River in Cumberland Falls State Park, Kentucky. At 38 m
(125 ft) wide and 18 m (59 ft) high, it is the second largest waterfall east of the Rocky Mountains
and one of the most impressive.

Prologue

From late June to August 1993, flooding in the midwestern United States left 50 people dead and 70,000 homeless and caused at least $15 billion in property damage. More than 200 counties were declared disaster areas in Iowa, South Dakota, North Dakota, Minnesota, Wisconsin, Nebraska, Missouri, Illinois, and Kansas. This flood, known as the Flood of '93, was one of the most costly natural disasters in U.S. history.

Despite the heroic efforts of thousands of volunteers and the National Guard to stabilize levees by sandbagging them, levees failed everywhere or were simply overtopped by rising flood waters. By the end of the first week in July, at least 10 million acres of farmland had been flooded with more rain expected, and by August the total area flooded had risen to 23 million acres. Grafton, Illinois, at the juncture of the Mississippi and Illinois Rivers was 80% underwater. And so much of St. Charles County, Missouri, near the confluence of the Mississippi and Missouri Rivers was flooded that

8,000 people were evacuated from this county alone (☐ Fig. 5.1).

By mid-July flooding was extensive and still more rain was predicted. On July 16, an additional 17.8 cm (7 in) of rain fell in North Dakota and Minnesota, and 15.2 cm (6 in) more fell in South Dakota, further swelling already flooding rivers. Obviously, the direct cause of the flooding was too much water for the Mississippi and Missouri Rivers and their tributaries to handle. Thunderstorms that are usually distributed over a much larger area simply dumped their precipitation in a smaller region, providing as much as 1½ to 2 times the normal amount of rain.

One important lesson learned from the Flood of '93 is that despite our best efforts at flood control, some floods will occur anyway. The 29 dams on the Mississippi and 26 reservoirs on upstream tributaries, as well as about 5,800 km (3,600 mi) of levees, did little to hold the flood waters in check. No doubt debate will now focus on the utility of levees in flood control. Levees are effective in protecting many areas during floods, yet in some cases they actually exacerbate the problem by restricting the flow that formerly would have spread over a floodplain. They are expensive to build and maintain, and their overall effectiveness has been and will continue to be questioned.

No one doubts that some dams and levees have successfully contained floods, at least within the limits of their design. But critics charge that flood control projects make the problem of flooding worse, particularly since flood-prone areas commonly experience more development after flood control projects are completed, and nothing can be done to prevent some floods. As a consequence, flooding continues to be the most destructive and most costly type of natural disaster.

☐ **FIGURE 5.1** Portage des Sioux in St. Charles County, Missouri, on July 16, 1993. Most of this small community near the confluence of the Missouri and Mississippi Rivers was flooded during the Flood of '93.

THE HYDROLOGIC CYCLE

With fully 71% of its surface covered by water, Earth is unique among the terrestrial planets (the inner planets—Mercury, Venus, Earth, and Mars) in having abundant liquid water. The volume of water on Earth is 1.36 billion km^3, most of which (97.2%) is in the oceans. About 2.15% is frozen in glaciers, and the remaining 0.65% constitutes all water in streams, lakes, swamps, groundwater, and the atmosphere (○ Table 5.1).

Although the quantity of water in streams and the groundwater system is small at any one time, during the course of a year very large volumes of water are cycled from the oceans, through the atmosphere, to the continents, and back to the oceans. This **hydrologic cycle,** as it is called, is powered by heat from the Sun and is possible because water at the surface is easily evaporated (◙ Fig. 5.2). About 85% of all water vapor entering the atmosphere is derived from the oceans, and the remaining 15% evaporates from water on land. Water vapor rises into the atmosphere where the complex processes of condensation and cloud formation occur. About 80% of all precipitation falls directly back into the oceans, so for this water the hydrologic cycle is limited to a three-step process of evaporation, condensation, and precipitation. For the 20% of all precipitation falling on land as rain and snow, the hydrologic cycle involves more steps: evaporation, condensation, movement of water vapor from the oceans to the continents, precipitation, runoff, and infiltration into the groundwater system (Fig. 5.2).

◙ FIGURE 5.2 The hydrologic cycle. Water evaporates from the oceans and rises as water vapor to form clouds that release their precipitation either over the oceans or over land. Much of the precipitation falling on land returns to the oceans by runoff, but a considerable amount seeps below the surface where it is stored as groundwater.

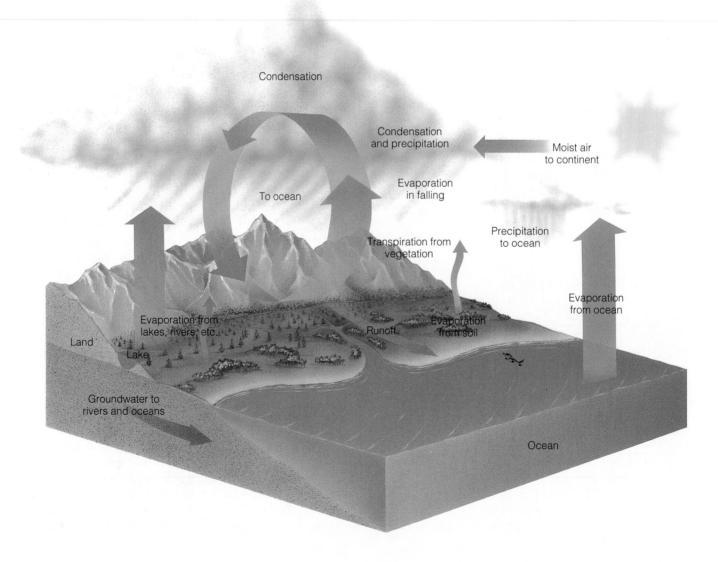

TABLE 5.1 Water on Earth

LOCATION	VOLUME (km³)	PERCENTAGE OF TOTAL
Oceans	1,327,500,000	97.20
Ice caps and glaciers	29,315,000	2.15
Groundwater	8,442,580	0.625
Freshwater and saline lakes and inland seas	230,325	0.017
Atmosphere at sea level	12,982	0.001
Average in stream channels	1,255	0.0001

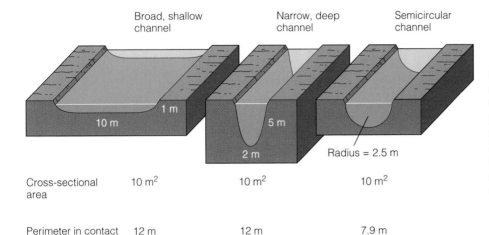

■ FIGURE 5.3 All three of these channels have the same cross-sectional area, but each has a different shape. The semicircular channel has the least perimeter in contact with the water and thus the least frictional resistance to flow. If other variables, such as channel roughness, are the same in all three channels, flow velocity will be greatest in the semicircular channel.

Much of the precipitation falling on land returns to the oceans by *runoff,* the surface flow of streams, but a considerable amount seeps down below the surface where it is temporarily stored as *groundwater.* Some water is also temporarily stored in lakes, snow fields, and glaciers, but it along with groundwater eventually makes its way back to the oceans. The comparatively small quantity of water returning to the oceans as runoff is responsible for many surface features, and groundwater, in addition to being an important resource, effectively weathers and erodes rocks below the surface in some areas.

 RUNNING WATER

The amount of runoff in any area during a rainstorm depends on the *infiltration capacity,* the maximum rate that soil and other surface materials can absorb water. If rain is absorbed as fast as it falls, no runoff occurs, but when the infiltration capacity is exceeded, or surface materials become saturated, excess water collects on the surface and moves downhill if a slope exists. As water moves downslope, it accelerates and may move by *sheet flow,* a more-or-less continuous film of water flowing over the surface. Sheet flow is not confined to depressions, and it accounts for *sheet erosion,* a particular problem on some agricultural lands.

In *channel flow,* runoff is confined to long, troughlike depressions. Channels range from tiny rills containing a trickling stream of water to large rivers such as the Amazon of

South America. Stream and river channels have been, and continue to be, important avenues of commerce; much of the interior of North America was explored by following rivers such as the St. Lawrence, Mississippi, and Missouri.

Streams flow downhill from a source area to a lower elevation where they empty into another stream, a lake, or the sea. The slope a stream flows over is its *gradient,* which is the vertical drop in a given horizontal distance; it is expressed in meters per kilometer (m/km) or feet per mile (ft/mi). Generally, gradients are steeper in the upper reaches of streams where they may be tens of meters per kilometer.

The *velocity* of running water is a measure of the downstream distance traveled per unit of time expressed in meters per second (m/sec) or feet per second (ft/sec). Because of friction, flow velocity is slower near the bed and banks of a stream than it is farther from these boundaries. And water flows more rapidly in semicircular channels than in broad, shallow channels and narrow, deep channels because it encounters less frictional resistance (■ Fig. 5.3). Frictional resistance to flow is also greater in a channel containing large boulders than in one where the banks and bed are composed of sand or clay.

The most obvious control on velocity is gradient, and one might think that the steeper the gradient, the greater the flow velocity. In fact, the average velocity generally increases in a downstream direction, even though the gradient decreases in the same direction. Three factors contribute to this: First, velocity increases continuously in response to the acceleration of gravity unless other factors retard flow.

EARTHFACT

The Amazon River of South America

Although slightly shorter than the Nile River of Africa, the Amazon dwarfs the rest of the world's rivers in all other aspects. It is 6,450 km (4,000 mi) long and at one point is up to 2.4 km (1.5 mi) wide and 91 m (300 ft) deep. In fact, oceangoing vessels can navigate about 3,700 km (2,295 mi) upstream. Even more impressive is the fact that its annual discharge is more than 6.3 trillion m^3, or about 16% of all water discharged into the oceans each year!

Second, streams commonly have boulder-strewn, broad, shallow channels in their upper reaches, so flow resistance is high and velocity is correspondingly slower. Downstream, channels generally become more semicircular, and the bed and banks are usually composed of finer materials, thus reducing the effects of friction. Third, the number of tributary streams joining a larger stream increases in a downstream direction, so the total volume of water (discharge) increases, and increasing discharge results in increased velocity.

The total volume of water in a stream moving past a particular point in a given period of time is its *discharge.* To determine discharge, multiply a stream's cross-sectional area by its flow velocity. The average discharge of the Mississippi River is 18,000 m^3/sec, and for the Amazon in South America it is 200,000 m^3/sec! Discharge varies considerably, being greatest during floods and lowest during long dry spells.

STREAM EROSION AND SEDIMENT TRANSPORT

Erosion involves the removal of dissolved substances and loose particles of soil and rock from a source area. Thus, sediment transported in a stream consists of both dissolved materials, the *dissolved load,* and solid particles, which are transported as suspended load or bed load. *Suspended load* consists of silt and clay, which are kept suspended by fluid turbulence, whereas *bed load* is made up of sand and gravel. Some of the bed load can be temporarily suspended as when sand bounces along a streambed. Particles too large to be suspended even temporarily are transported by rolling or sliding.

Much of the sediment carried in streams is eroded by the power of running water, or what is called *hydraulic action.* Streams also erode by *abrasion* in which exposed rock is worn and scraped by the impact of solid particles. One manifestation

of abrasion is circular to oval *potholes* that form where eddying currents containing sand and gravel swirl around and erode depressions into the streambed. (▣ Fig. 5.4).

The maximum-sized particles a stream carries define its *competence,* a factor related to flow velocity, whereas *capacity* is a measure of the total load a stream carries, which varies as a function of discharge. Capacity and competence may seem quite similar, but they are actually related to different aspects of stream transport. A small, swiftly flowing stream, for instance, may have the competence to move gravel but be unable to transport a large volume of sediment; thus, it has a low capacity. A large slow-flowing stream, on the other hand, has a low competence, but may have a very large suspended load, and hence a large capacity.

STREAM DEPOSITION

Streams do most of their erosion, sediment transport, and deposition when they flood. Consequently, stream deposits result from the periodic, large-scale events of sedimentation associated with flooding, rather than from continuous day-to-day activity.

Braided streams possess an intricate network of dividing and rejoining channels (▣ Fig. 5.5). Braiding develops when a stream is supplied with excessive sediment, which over time is deposited as sand and gravel bars within its channel. Braided streams have broad, shallow channels and are generally characterized as bed load transport streams because of the large amounts of sand and gravel they transport. Their deposits are mostly sheets of sand and gravel (Fig. 5.5).

Meandering streams have a single, sinuous channel with broadly looping curves called *meanders* (▣ Fig. 5.6).

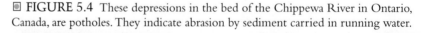

▣ FIGURE 5.4 These depressions in the bed of the Chippewa River in Ontario, Canada, are potholes. They indicate abrasion by sediment carried in running water.

◉ FIGURE 5.5 A braided stream near Santa Fe, New Mexico. The deposits of this stream are composed entirely of sand.

◉ FIGURE 5.6 An aerial view of a meandering stream. The broad, flat area adjacent to the stream channel is its floodplain. Notice the crescent-shaped lakes—these are cutoff meanders, which are known as oxbow lakes.

Meanders are markedly asymmetric, being deepest near the outer bank, or cut bank, because greater flow velocity and turbulence on that side of the channel cause erosion (☐ Fig. 5.7). In contrast, flow velocity is at a minimum on the inner bank, which slopes gently into the channel. As a result of this unequal distribution of flow velocity across meanders, the cut bank is eroded, and a sand body called a **point bar** is deposited on the opposite side of the channel (Fig. 5.7).

The valley floors of meandering streams are commonly marked by crescent-shaped lakes known as **oxbow lakes** (Figs. 5.6 and ☐ 5.8). These form when meanders become so sinuous that the thin neck of land separating adjacent me-

☐ FIGURE 5.7 (*a*) In a meandering channel, flow velocity is greatest near the outer bank. The dashed line follows the path of maximum flow velocity. Because the flow velocity varies across the channel, the outer or cut bank is eroded, and a point bar is deposited on the inner side of the meander. (*b*) Two small point bars in a meandering stream.

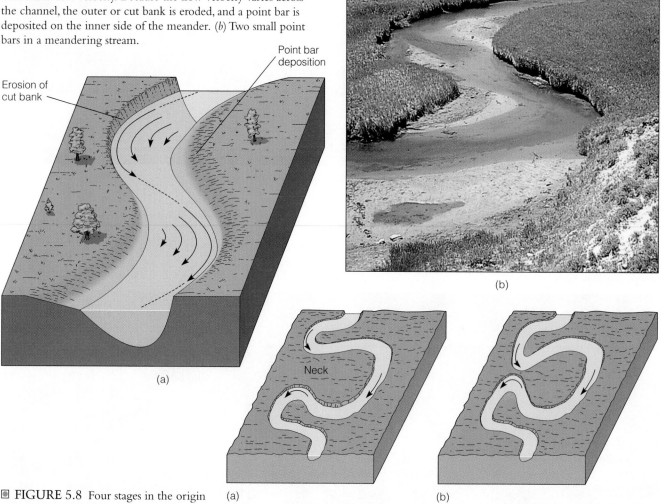

(a)

(b)

☐ FIGURE 5.8 Four stages in the origin of an oxbow lake. In (*a*) and (*b*), the meander neck becomes narrower. (*c*) The meander neck is cut off, and part of the channel is abandoned. (*d*) When it is isolated from the main channel, the abandoned meander is an oxbow lake.

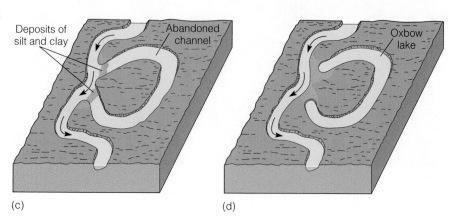

(a)

(b)

(c)

(d)

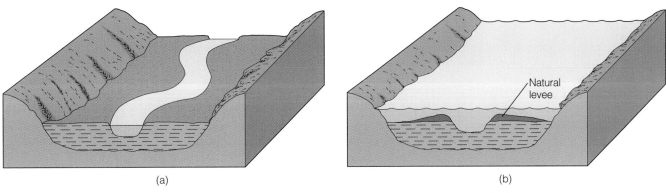

■ FIGURE 5.9 Two stages in the formation of floodplain deposits. (*a*) Stream at low-water stage. (*b*) Flooding stream and deposition of natural levees adjacent to the channel and mud on the floodplain. These deposits build upward during repeated floods.

anders is cut off during a flood. Oxbow lakes eventually fill with fine-grained sediment carried by floods and organic matter.

Most streams periodically receive more water than their channels can carry, so they spread across low-lying, relatively flat areas known as **floodplains** adjacent to their channels (Fig. 5.6) (see ▢ Perspective 5.1). When a stream floods, the velocity of the water spilling onto the floodplain diminishes rapidly because of greater frictional resistance to flow as the water spreads out as a broad, shallow sheet. As a result, ridges of sand called *natural levees* are deposited along the margins of the channel, and mud is deposited on the floodplain (▢ Fig. 5.9).

When a stream flows into another body of water, its flow velocity decreases rapidly, resulting in sediment deposition and the origin of a **delta** (▢ Fig. 5.10). The simplest deltas exhibit a characteristic sequence of beds (Fig. 5.10a). The finest sediments are carried some distance beyond the river mouth, where they settle to form *bottomset beds*. Nearer the river mouth, *foreset beds* form as sand and silt are deposited in gently inclined layers. *Topset beds* consist of coarser sediments deposited in a network of channels traversing the top of the delta.

Many small deltas in lakes have the three-part division just described, but large marine deltas are usually much more complex. The Mississippi River delta, for instance, consists of long finger-like sand bodies, each deposited in a channel built seaward across the delta (Fig. 5.10b). Deltas of this type are called *bird's-foot deltas* because they resemble the toes of a bird.

Stream deposition can also result in the formation of **alluvial fans.** These lobate deposits on land form best on lowlands adjacent to highlands in arid and semiarid regions where little or no veg-

■ FIGURE 5.10 (*a*) Internal structure of the simplest type of delta. (*b*) The bird's-foot delta of the Mississippi River on the U.S. Gulf Coast.

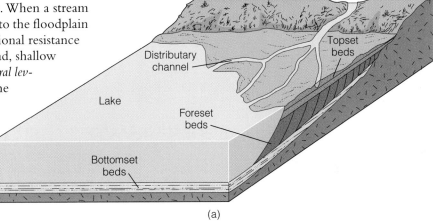

(a)

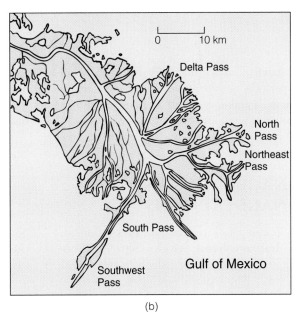

(b)

Predicting and Controlling Floods

To monitor stream behavior, the U.S. Geological Survey maintains more than 11,000 stream-gauging stations, and various state agencies also monitor streams. Stream-gauge data are commonly used to construct *flood-* *frequency curves* (Fig. 1). To construct such a curve, the peak discharges are first arranged in order of discharge; the flood with the greatest discharge has a magnitude rank of 1, the second largest is 2, and so on (Table 1). The *recurrence in-*

TABLE 1   Some of the Data and Recurrence Intervals for the Rio Grande Near Lobatos, Colorado

YEAR	DISCHARGE (m³/sec)	RANK	RECURRENCE INTERVAL
1900	133	23	3.35
1901	103	35	2.20
1902	16	69	1.12
1903	362	2	38.50
1904	22	66	1.17
1905	371	1	77.00
1906	234	10	7.70
1907	249	7	11.00
1908	61	45	1.71
1909	211	13	5.92

The greatest yearly discharge is given a magnitude rank (m) ranging from 1 to N ($N = 76$ in this example), and the recurrence interval (R) is calculated by the equation $R = (N + 1)/m$.

YEAR	DISCHARGE (m³/sec)	RANK	RECURRENCE INTERVAL
1974	22	64	1.20
1975	68	43	1.79

SOURCE: U.S. Geological Survey Open-File Report 79–681.

etation exists to stabilize surface materials (Fig. 5.11 on page 92). When periodic rainstorms occur, surface materials are quickly saturated, and the runoff is funneled into a canyon leading to an adjacent lowland. When the stream discharges onto the lowland, it spreads out, its velocity diminishes, and deposition ensues.

DRAINAGE BASINS AND DRAINAGE PATTERNS

A stream and all its tributaries carry surface runoff from an area known as a **drainage basin,** which is separated from adjacent drainage basins by topographically high areas called **divides** (Fig. 5.12). The regional arrangement of channels in a drainage basin is the basis for defining several **drainage patterns.**

Dendritic drainage, consisting of a network of channels resembling tree branches, is the most common pattern (Fig. 5.13a on page 93). It develops on gently sloping land where surface materials respond more or less homogeneously to erosion.

Channels with right angle bends and tributaries joining larger streams at right angles characterize *rectangular drainage* (Fig. 5.13b). The positions of the channels are strongly controlled by geologic structures, particularly regional fractures intersecting at right angles.

In some areas erosion of folded sedimentary rocks develops a landscape of alternating parallel ridges and valleys. The ridges consist of resistant rocks, such as sandstone, while the valleys overlie less resistant rocks such as shale. Main streams follow the trends of the valleys, and short tributaries flowing from the adjacent ridges join the main stream at nearly right angles, hence, the name *trellis drainage* (Fig. 5.13c).

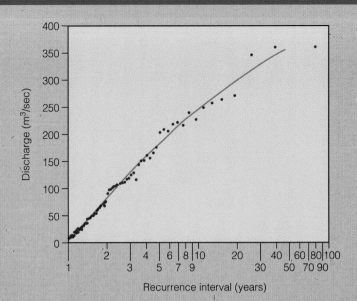

 FIGURE 1 Flood-frequency curve for the Rio Grande near Lobatos, Colorado. The curve was constructed from the data in Table 1.

terval—that is, the time period during which a flood of a given magnitude or larger can be expected over an average of many years—is determined by the equation

$$R = (N + 1)/m$$

where R is the recurrence interval in years, N is the number of years of record, 76 in this case, and m is the magnitude rank. Applying this equation to the data in Table 1 indicates floods with magnitude ranks of 1 and 23 have recurrence intervals of 77.00 and 3.35 years, respectively. Once the recurrence interval has been calculated, it is plotted against discharge, and a line is drawn through the data points (Fig. 1).

According to Figure 1, the 10-year flood for the Rio Grande near Lobatos, Colorado, has a discharge of 245 m³/sec. This means that, on average, we can expect one flood of this size or greater within a 10-year interval. One cannot predict that such a flood will occur in any particular year, however, only that it has a probability of 1 in 10 (1/10) of occurring in any year. Furthermore, 10-year floods are not necessarily separated by 10 years. Two such events might take place in the same year or in successive years, but over a period of centuries, they occur on average once every 10 years.

Unfortunately, stream-gauge data in the United States and Canada have generally been collected for only a few decades, and rarely for more than a century. Accordingly, we have a good idea of stream behavior over short periods, the 2-year and 5-year floods, for example, but our knowledge of long-term behavior is limited by the short period of record keeping. Thus, predictions of 50-year or 100-year floods from Figure 1 are unreliable. In fact, the largest magnitude flood shown in Figure 1 may have been a unique event for this river that will never be repeated. On the other hand, it may actually turn out to be a magnitude 2 or 3 flood when data for a longer time are available.

Although flood-frequency curves have limited applicability, they are nevertheless helpful in making decisions regarding flood control. Careful mapping of floodplains can identify areas at risk for floods of given magnitude, and planners must decide what magnitude of flood to protect against.

In *radial drainage,* streams flow outward in all directions from a central high area (Fig. 5.13d). Radial drainage develops on large, isolated volcanic mountains and in areas where Earth's crust has been arched up by intrusive igneous bodies.

In parts of Minnesota, Wisconsin, Michigan, and Canada that were glaciated until only a few thousand years ago, the previously established drainage systems were obliterated by glacial ice. Following the final retreat of the glaciers, drainage systems became established, but have not yet become fully organized. Consequently, streams flow into and out of swamps and lakes with irregular flow directions, thus illustrating *deranged drainage* (Fig. 5.13e).

BASE LEVEL AND THE GRADED STREAM

The lower limit to which a stream can erode is its **base level.** Theoretically, a stream could erode its entire valley to very near sea level, so sea level is commonly referred to as *ultimate base level* (☐ Fig. 5.14a). In addition to ultimate base level, streams have *local* or *temporary base levels;* a lake or another stream can serve as a local base level for the upstream segment of a stream (☐ Fig. 5.14a). Likewise, where a stream flows across particularly resistant rock, a waterfall may develop, forming a local base level. The escarpment Niagara Falls plunges over is a good example of a local or temporary base level.

Changes in base level can be caused by a rise or fall in sea level with respect to the land or by the uplift or subsidence

◙ FIGURE 5.11 Alluvial fans adjacent to the Panamint Range along the margin of Death Valley, California.

◙ FIGURE 5.12 Drainage basin of the Wabash River, which is one of the tributaries of the Ohio River. This drainage basin covers about 85,500 km² (32,900 mi²), mostly in Indiana. All of the streams within the drainage basin, such as the Vermillion River, have their own smaller drainage basins. Divides are shown by brown lines.

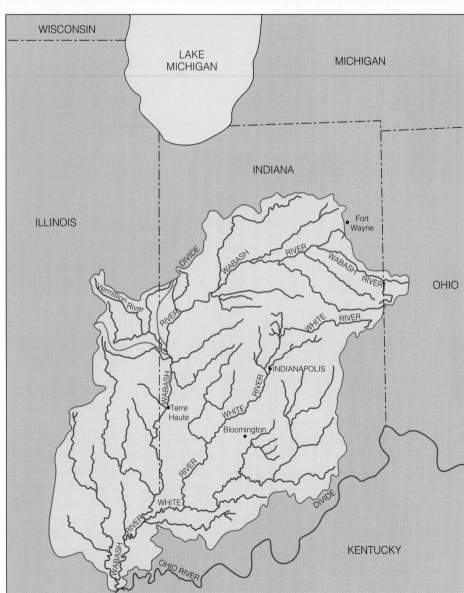

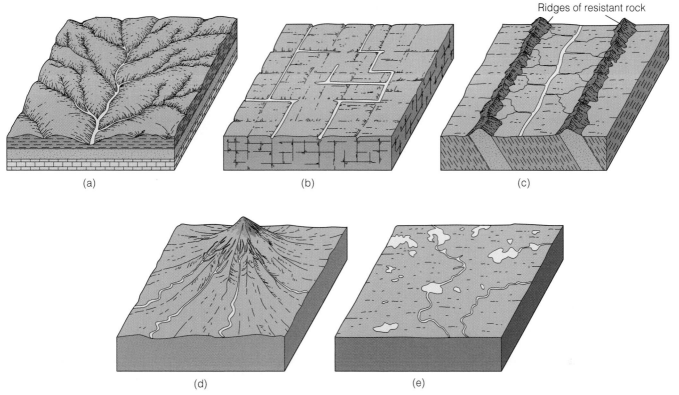

■ FIGURE 5.13 Examples of drainage patterns: (*a*) dendritic drainage; (*b*) rectangular drainage; (*c*) trellis drainage; (*d*) radial drainage; and (*e*) deranged drainage.

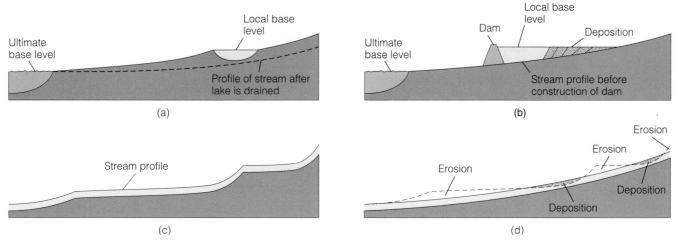

■ FIGURE 5.14 (*a*) Ultimate and local base levels. If the lake were drained, the stream would adjust to the lower base level as shown by the dashed line. (*b*) When a dam is constructed and water is impounded in a reservoir, a local base level is created. A stream deposits much of its sediment load where it flows into the reservoir. (*c*) An ungraded stream has irregularities—local base levels—in its longitudinal profile. (*d*) Erosion and deposition along the course of a stream eliminate irregularities, and the stream tends to develop the smooth, concave profile typical of a graded stream.

of the land a stream flows over. During the Pleistocene when extensive glaciers were present, sea level was about 130 m (425 ft) lower than at present. Accordingly, streams adjusted to the new, lower base level by deepening their valleys. Rising sea level at the end of the Pleistocene caused base level to rise, and the lower ends of stream valleys were flooded.

Scientists are well aware that the process of building a dam and impounding a reservoir creates a local base level (Fig. 5.14b). Where a stream enters a reservoir, its flow velocity diminishes rapidly, and its sediment is deposited, so reservoirs eventually fill with sediment unless they are dredged. Conversely, draining a lake lowers the base level for the part of the stream above the lake, and the stream will very likely respond by rapidly eroding downward (Fig. 5.14a).

A stream's *longitudinal profile* shows the elevations of a channel along its length as viewed in cross section (Fig. 5.14c). The longitudinal profiles of many streams show a number of irregularities such as lakes and waterfalls, which are local base levels. Over time streams tend to eliminate irregularities and develop a smooth, concave longitudinal profile of equilibrium, meaning that all parts of the system are dynamically adjusting to one another (Fig. 5.14d).

A stream possessing an equilibrium profile is said to be a **graded stream;** that is, a delicate balance exists between gradient, discharge, flow velocity, channel characteristics, and sediment load so that neither significant erosion nor deposition occurs within the channel. Even though the concept of a graded stream is an ideal, we can generally anticipate the response of a graded stream to changes altering its equilibrium. A change in base level, for instance, would cause a stream to adjust as previously discussed. Increased rainfall in a stream's drainage basin would result in greater discharge and flow velocity, and the stream may respond by eroding its valley deeper, thereby reducing its gradient until it is once again graded.

 # DEVELOPMENT OF STREAM VALLEYS

Valleys are common landforms, and with few exceptions they form and evolve mostly as a consequence of stream erosion. The shapes and sizes of valleys vary considerably: some are small, steep-sided *gullies,* others are broad and have gently sloping walls, and some such as the Grand Canyon are steep-walled and deep. Several processes, in particular, downcutting, lateral erosion, and mass wasting, contribute to the origin and evolution of valleys.

When a stream possesses more energy than it needs to transport its sediment load, some of its excess energy is expended by *downcutting* and deepening its valley. If downcutting were the only process operating, valleys would be narrow and steep-sided, but in most cases a valley widens as its walls are undercut by the stream. Such undermining, termed *lateral erosion,* creates unstable conditions so that part of a bank or valley wall moves downslope by mass wasting processes.

According to one concept, stream erosion of an area uplifted above sea level yields a distinctive series of landscapes. When erosion begins, streams erode downward; their valleys are deep, narrow, and V-shaped; and they have a number of irregularities in their profiles (■ Fig. 5.15a). As streams cease downcutting, they start eroding laterally, thereby establishing a meandering pattern and a broad floodplain (Fig. 5.15b). Finally, with continued erosion, a vast, rather featureless plain develops (Fig. 5.15c).

Many streams do indeed show an association of features typical of these stages. For instance, the Colorado River flows through the Grand Canyon and closely matches the features in Figure 5.15a. Streams in numerous areas approx-

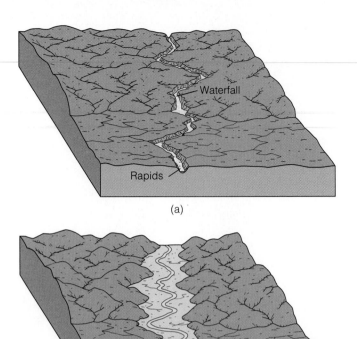

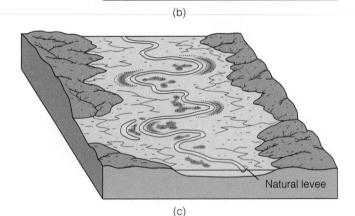

■ FIGURE 5.15 Idealized stages in the development of a stream and its associated landforms.

imate the second stage of development, and certainly the lower Mississippi closely resembles the last stage (Fig. 5.15c).

Nevertheless, the idea of a sequential development of stream-eroded landscapes has been largely abandoned because there is no reason to think that streams necessarily follow this idealized cycle. Indeed, a stream on a gently sloping surface near sea level could develop features of the last stage very early in its history. Furthermore, as long as the rate of uplift exceeds the rate of downcutting, a stream will continue to erode downward and be confined to a narrow canyon.

Some streams are restricted to deep, meandering canyons cut into bedrock, where they form features known as *incised meanders* (■ Fig. 5.16). Because lateral erosion is inhibited once downcutting begins, the meandering course must have

FIGURE 5.16 The Goose Necks of the San Juan River in Utah are incised meanders.

been established when the stream flowed across a gently sloping, sediment-covered surface. Suppose that a stream near base level has a meandering pattern. If the land it flows over is uplifted, the stream is *rejuvenated,* meaning that it possesses more energy than it had previously, and it begins downcutting, thereby incising its meanders into the underlying bedrock.

Adjacent to many streams are erosional remnants of floodplains that formed when the streams were flowing at a higher level. These *stream terraces* have a fairly flat upper surface and a steep slope descending to the level of the lower, present-day floodplain (◫ Fig. 5.17). Although all stream terraces result from erosion, they are preceded by an episode of floodplain formation and deposition of sediment. Subsequent erosion causes the stream to erode downward until it establishes a new floodplain at a lower level. Several such episodes account for the multiple terrace levels seen adjacent to some streams (Fig. 5.17).

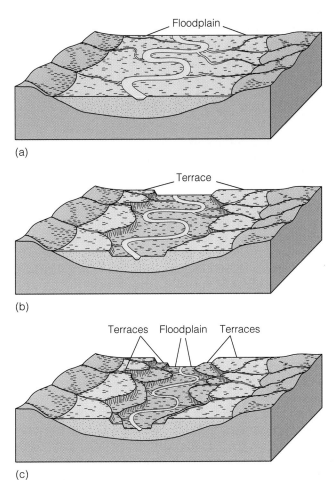

FIGURE 5.17 Origin of stream terraces. (*a*) A stream has a broad floodplain adjacent to its channel. (*b*) The stream erodes downward and establishes a new floodplain at a lower level. Remnants of its old floodplain are stream terraces. (*c*) Another level of terraces forms as the stream erodes downward again.

Development of Stream Valleys   95

GROUNDWATER

The study of **groundwater,** the water stored in the open spaces within underground rocks and unconsolidated material, has become increasingly important as the demand for fresh water by agriculture, industry, and domestic users has reached an all-time high. Demands for groundwater have severely depleted the supply in many areas and led to such problems as ground subsidence and saltwater contamination. In other areas, pollution from landfills, toxic waste, and agriculture has rendered the groundwater supply unsafe.

Groundwater and the Hydrologic Cycle

Groundwater, one reservoir in the hydrologic cycle, represents about 22% of the world's supply of fresh water (Table 5.1), or about 36 times more than the total of all the water in streams and lakes. If all groundwater were spread evenly over Earth's surface, it would be about 10 m (33 ft) deep.

The major source of groundwater is precipitation that infiltrates the ground and moves through the pore spaces of soil and rocks. Other sources include water infiltrating from lakes and streams, recharge ponds, irrigation, and wastewater treatment systems. As groundwater moves through soil, sediment, and rocks, many of its impurities, such as disease-causing microorganisms, are filtered out. Not all soils and rocks are good filters, though, and some serious pollutants are not removed. Groundwater is eventually discharged into streams, lakes, or swamps, or it flows directly back into the oceans, thus completing the hydrologic cycle (Fig. 5.2).

 POROSITY AND PERMEABILITY

Water soaks into the ground because Earth materials have open spaces or pores; the percentage of a material's total volume that is pore space is its **porosity.** Most porosity consists of the spaces between particles in soil, sediments, and sedimentary rocks, but some occurs as cracks, fractures, and vesicles in volcanic rocks (Fig. 5.18). Most igneous and metamorphic rocks as well as many limestones and dolostones are composed of tightly interlocking crystals and thus have very low porosity. Their porosity increases, however, if they have been fractured or weathered by groundwater.

Detrital sedimentary rocks with well-sorted grains have very high porosity because any two grains touch only at a single point, leaving large open spaces between the grains (Fig. 5.18a). Poorly sorted sedimentary rocks typically have low porosity because smaller grains fill in the spaces between the larger grains, reducing the porosity (Fig. 5.18b). Cement between grains also decreases porosity.

Although porosity determines the amount of groundwater Earth materials can hold, it does not guarantee that the water can be extracted. The capacity of a material for transmitting fluids is its **permeability,** which is dependent not only on porosity, but also on the size of the pores or fractures

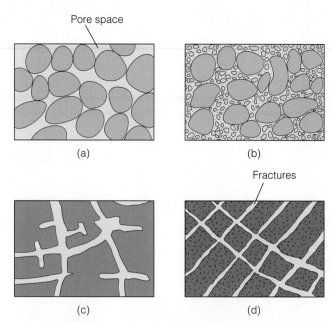

FIGURE 5.18 Porosity in various types of materials. (*a*) A well-sorted sediment has high porosity while (*b*) a poorly sorted one has low porosity. If chemical cement is present, it also reduces porosity. (*c*) In soluble rocks such as limestone, porosity can be increased by solution, while (*d*) igneous and metamorphic rocks can be rendered porous by fracturing.

and their interconnections. Clay has high porosity (Table 5.2) but low permeability because its pores are very small, and the molecular attraction between the clay and the water is great, thereby inhibiting movement of the water. In contrast, the pore spaces between grains in sand and gravel are much larger, and the molecular attraction on the water is therefore low.

MATERIAL	PERCENTAGE OF PORE SPACE
TABLE 5.2 Porosity Values for Different Materials	
Unconsolidated sediment	
Soil	55
Gravel	20–40
Sand	25–50
Silt	35–50
Clay	50–70
Rocks	
Sandstone	5–30
Shale	0–10
Solution activity in limestone, dolostone	10–30
Fractured basalt	5–40
Fractured granite	10

SOURCE: U.S. Geological Survey, Water Supply Paper 2220 (1983) and others.

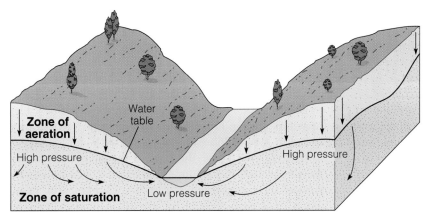

■ FIGURE 5.19 The water table is the surface between the zones of aeration and saturation. Some groundwater moves along the slope of the water table, and the rest moves through the zone of saturation from areas of high pressure toward areas of low pressure.

A porous and permeable layer that transports groundwater is called an **aquifer.** The most effective aquifers are deposits of well-sorted sand and gravel, but limestones with fractures and bedding planes enlarged by solution are also good aquifers. Impermeable rocks and any other materials that prevent groundwater movement are **aquicludes.**

THE WATER TABLE

When precipitation falling on land seeps down from the surface, some of it adheres to the material it is moving through and halts its downward progress. This water is in the **zone of aeration** where pore spaces contain both water and air (■ Fig. 5.19). Beneath the zone of aeration lies the **zone of saturation** where all pore spaces are filled with groundwater. The surface between the zone of aeration and the underlying zone of saturation is the **water table** (Fig. 5.19). In

general, the water table is a subdued replica of the overlying land surface; that is, its highest elevations are beneath hills and its lowest elevations are in valleys. In most arid and semiarid regions, though, the water table is quite flat and is below the level of the river valleys.

When water entering the ground reaches the water table, it continues to move through the zone of saturation from areas where the water table is high toward areas where it is lower, such as at streams, lakes, or swamps (Fig. 5.19). Only some of the water follows the direct route along the slope of the water table; most of it takes longer curving paths downward and then enters a stream, lake, or swamp from below. Water in the zone of saturation also moves from areas of high pressure toward areas of lower pressure. Groundwater beneath a hill is under greater pressure than water at the same elevation beneath a valley; thus, the water moves toward the valley.

SPRINGS, WATER WELLS, AND ARTESIAN SYSTEMS

When percolating groundwater reaches the water table or an impermeable layer, it flows laterally, and if this flow intersects the surface, water discharges as a **spring** (■ Fig. 5.20). Water moving laterally along a perched water table, formed where a local aquiclude lies within a larger aquifer, may also intersect the surface to produce a spring.

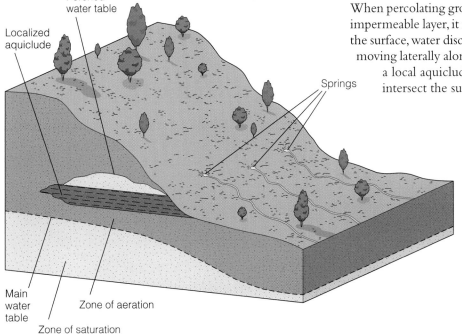

■ FIGURE 5.20 Most springs form when percolating water reaches an impermeable layer and migrates laterally until it seeps out at the surface. If a localized aquiclude, such as a clay layer, is within an aquifer, a perched water table results with springs forming where the perched water table intersects the surface.

Springs have always fascinated people because the water flows out of the ground for no apparent reason and from no readily identifiable source. It is not surprising that springs have long been regarded with superstition and revered for their supposed medicinal value and healing powers.

A *water well* is made by digging or drilling into the zone of saturation where water flows into the well and fills it to the level of the water table. When a well is pumped, the water table in the area around the well is lowered because water is removed from the aquifer faster than it can be replenished. A local lowering of the water table or a **cone of depression** forms around the well, varying in size according to the rate and amount of water being withdrawn (▣ Fig. 5.21). If water is pumped out of a well faster than it is replaced, the cone of depression grows until the well goes dry. Withdrawing tremendous amounts of water for indus-

try and irrigation may create a large cone of depression that lowers the water table sufficiently to cause shallow wells in the immediate area to go dry (Fig. 5.21).

Any system in which groundwater is confined above and below by aquicludes and builds up high hydrostatic (fluid) pressure is called an **artesian system.** The word *artesian* comes from the French town and province of Artois (called Artesium during Roman times), where an artesian well drilled in A.D. 1126 is still flowing today. Water in an artesian system rises above the level of the aquifer if a well is drilled through the confining layer, thereby reducing the pressure and forcing the water upward (▣ Fig. 5.22). In addition to artesian wells, many artesian springs exist where fractures allow water to rise above a confined aquifer. Oases in deserts are commonly artesian springs.

▣ FIGURE 5.21 A cone of depression forms when water is withdrawn from a well. If water is withdrawn faster than it can be replenished, the cone of depression grows in depth and circumference, lowering the water table and causing nearby shallow wells to go dry.

▣ FIGURE 5.22 An artesian system must have an aquifer confined by aquicludes, the aquifer must be exposed at the surface, and there must be sufficient precipitation in the recharge area to keep the aquifer filled. The sloping dashed line shows the highest level to which well water can rise above the aquifer.

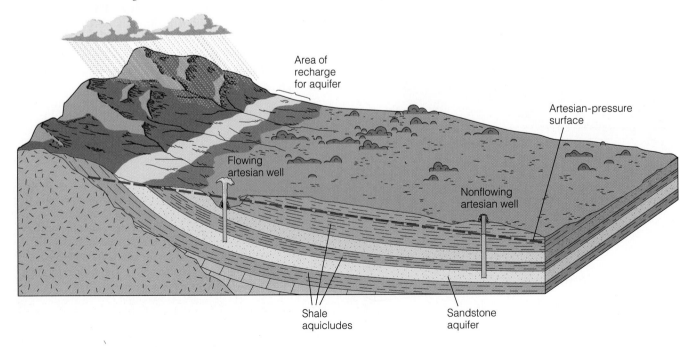

GROUNDWATER EROSION AND DEPOSITION

Limestone is a common sedimentary rock that underlies large areas of Earth's surface. Although limestone is nearly insoluble in pure water, it readily dissolves if a small amount of acid is present. Because the atmosphere contains a small amount of carbon dioxide, and decay of organic matter produces carbon dioxide in soil, most groundwater is slightly acidic (see Chapter 4). When slightly acidic groundwater percolates through openings in limestone, it dissolves the rock and carries it away in solution.

In regions underlain by limestone, the ground surface may be pitted with numerous circular to oval depressions with gently sloping sides known as **sinkholes** (☐ Fig. 5.23). Sinkholes form when soluble rock is dissolved and natural openings are enlarged, leaving shallow depressions. They also form when a cave's roof collapses, usually producing a steep-sided crater. When adjacent sinkholes merge, they form a network of larger irregular depressions called *solution valleys* (☐ Fig. 5.24).

☐ FIGURE 5.23 Sinkholes near Bowling Green, Kentucky.

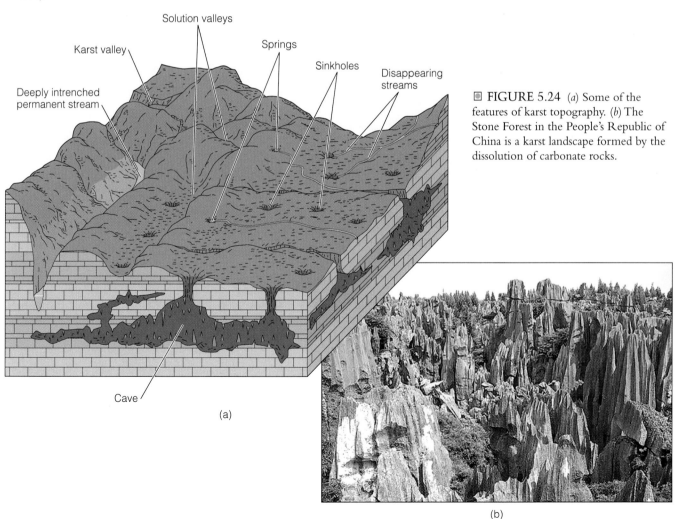

Solution valleys

Karst valley

Springs

Sinkholes

Disappearing streams

Deeply intrenched permanent stream

Cave

(a)

☐ FIGURE 5.24 (*a*) Some of the features of karst topography. (*b*) The Stone Forest in the People's Republic of China is a karst landscape formed by the dissolution of carbonate rocks.

(b)

Caves are defined as naturally formed openings below the surface large enough for a person to enter; a very large cave or system of interconnected caves is a *cavern*. Caves form as weakly acidic groundwater percolates through the zone of aeration and slowly dissolves soluble rocks, yielding a system of horizontal passageways (▣ Fig. 5.25). As the surface streams erode deeper valleys, the water table drops in response to the lower elevation of the streams. The water that flowed through the system of passageways now percolates down to the lower water table where a new system of passageways begins to form.

Most of the several types of cave deposits form in essentially the same manner. As water seeps through a cave, some of the dissolved carbon dioxide in the water escapes, and a small amount of calcite is precipitated, forming various cave deposits. *Stalactites,* icicle-shaped structures hanging from cave ceilings, form as a result of precipitation from dripping water (▣ Fig. 5.26). The water that drips from a cave's ceiling also precipitates a small amount of calcite when it hits the floor, forming an upward-growing projection called a *stalagmite* (Fig. 5.26). If a stalactite and stalagmite meet, they form a *column*.

▣ FIGURE 5.25 Formation of a cave. (*a*) As groundwater percolates through and dissolves soluble rocks, a system of passageways forms. (*b*) and (*c*) As the surface streams erode deeper valleys, the water table drops, and the abandoned channelways form an interconnecting cave system.

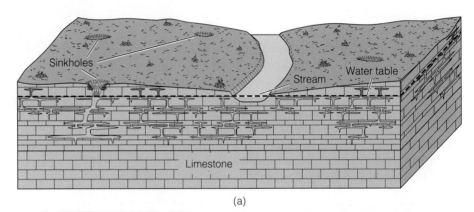

(a)

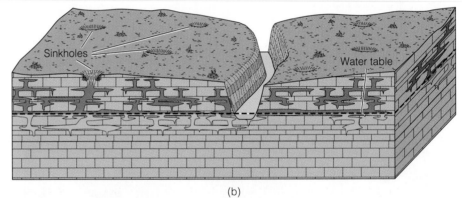

(b)

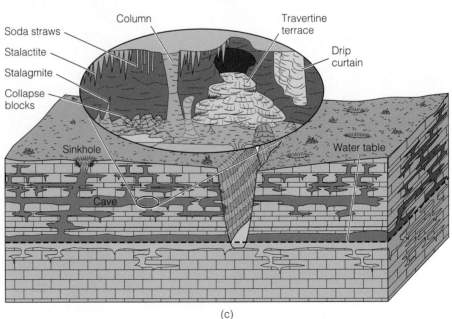

(c)

■ FIGURE 5.26 The icicle-shaped structures hanging from the cave ceiling are stalactites, while the upward-pointing structures on the cave floor are stalagmites. Several columns formed by the joining of stalactites and stalagmites are visible in this chamber of Luray Caves, Virginia.

Karst topography, which develops largely by groundwater erosion, is characterized by numerous caves, sinkholes, solution valleys, and disappearing streams (Fig. 5.24). *Disappearing streams* are so called because they flow only a short distance at the surface and then disappear into a sinkhole. The name *karst* is derived from an area near the borders of Slovenia, Croatia, and northeastern Italy where this type of topography is well developed. In the United States, karst topography is found in parts of Illinois, Indiana, Kentucky, Tennessee, Missouri, Alabama, and Florida.

EARTHFACT

The Great Cave War of Kentucky

During the 1920s, many caves in western Kentucky were developed as tourist attractions to help supplement meager farm earnings. The largest and best known is Mammoth Cave with its underground rivers and dramatic cave deposits. As Mammoth Cave drew more and more tourists, rival cave owners became increasingly bold in attempting to lure visitors to their caves and curio shops. Signs pointing the way to Mammoth Cave frequently disappeared, while "official" cave information booths redirected unsuspecting tourists away from Mammoth Cave. Nevertheless, rival caves attracted few tourists, and in 1926 Mammoth Cave was designated as a national park.

 ## MODIFICATIONS OF THE GROUNDWATER SYSTEM

Currently, about 20% of all water used in the United States is groundwater, but this percentage is increasing and unless it is used more wisely, sufficient amounts of clean groundwater will not be available in the future. Modifications of the groundwater system include (1) lowering of the water table, which causes wells to dry up; (2) loss of hydrostatic pressure, which causes once free-flowing wells to require pumping; (3) saltwater incursion; (4) subsidence; and (5) contamination.

The High Plains aquifer, underlying a large area in the western United States, accounts for about 30% of the groundwater used for irrigation in the entire country. Irrigation from this aquifer is largely responsible for the high agricultural productivity of this region. In some parts of the High Plains aquifer, from 2 to 100 times more water is pumped annually than is recharged, causing the water table to drop significantly in many areas. If long-term withdrawal of water continues, the aquifer will no longer supply the quantities of water needed for irrigation. Suggested solutions to this problem range from going back to farming without irrigation to diverting water from other regions such as the Great Lakes.

Saltwater incursion results from excessive groundwater pumping in coastal areas. Fresh groundwater, being less dense than seawater, forms a lens-shaped body above the underlying salt water (■ Fig. 5.27a). The weight of the fresh water exerts pressure on the underlying salt water, so if the rate of recharge

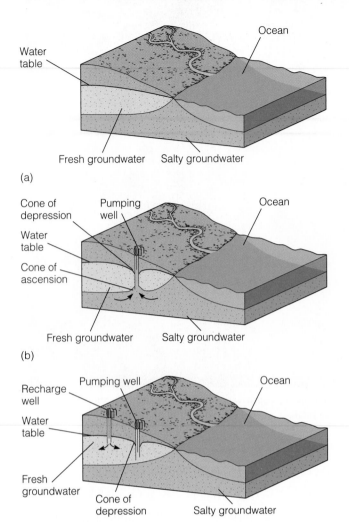

(a)

(b)

(c)

■ FIGURE 5.28 The dates on this power pole dramatically illustrate the amount of subsidence in the San Joaquin Valley, California. The ground surface subsided almost 9 m (29.3 ft) between 1925 and 1975 as groundwater was withdrawn for agriculture.

■ FIGURE 5.27 Saltwater incursion. (*a*) Fresh water is less dense than salt water, so it forms a lens-shaped body above the underlying salt water. (*b*) Excessive pumping causes a cone of depression to form in the fresh groundwater, and a cone of ascension forms in the underlying saltwater, resulting in saltwater contamination of the well. (*c*) Pumping water back into the groundwater system through recharge wells can help counteract the effects of saltwater incursion.

equals the rate of withdrawal, the boundary between the fresh groundwater and the seawater remains stationary. If excessive pumping takes place, however, a deep cone of depression forms in the fresh groundwater, salt water migrates upward, and wells become contaminated (Fig. 5.27b). To counteract the effects of saltwater incursion, water is pumped back into the groundwater system at recharge wells (Fig. 5.27c) or is allowed to infiltrate at recharge ponds.

As excessive amounts of groundwater are withdrawn from poorly consolidated sediments and sedimentary rocks, the water pressure between grains is reduced, and the weight of the overlying materials causes the grains to pack closer together, resulting in subsidence of the ground. Subsidence is becoming a major hazard in many areas and can cause damage to buildings, water lines, utility lines, and roads (■ Fig.

5.28). New Orleans, Louisiana, and Houston, Texas, have both subsided more than 2 m (6.6 ft), and Las Vegas, Nevada, has subsided 8.5 m (27.9 ft). Pumping groundwater from lake sediments beneath Mexico City has caused compaction, and the city is slowly and unevenly subsiding.

The most common sources of groundwater contamination are sewage, landfills, toxic waste disposal sites (see □ Perspective 5.2), and agriculture. Once pollutants get into the groundwater system, they spread wherever groundwater travels, making containment of the contamination difficult. Furthermore, because groundwater moves very slowly, it takes a very long time to cleanse a groundwater reservoir once it has become contaminated.

In many areas, septic tanks are the most common way of disposing of sewage. A septic tank slowly releases sewage into the ground where it is decomposed and filtered as it percolates through sediment in the zone of aeration. In most situations, by the time the water from the sewage reaches the

Focus on the Environment: Radioactive Waste Disposal

One of the problems of the nuclear age is finding safe storage sites for the radioactive waste from nuclear power plants, the manufacture of nuclear weapons, and the radioactive by-products of nuclear medicine. Radioactive waste can be grouped into two categories: low-level and high-level waste. Low-level wastes, when properly handled, do not pose a significant environmental threat. Most low-level wastes can be safely buried in controlled dump sites.

High-level radioactive waste, such as the spent uranium fuel assemblies used in nuclear reactors, is extremely dangerous because of high amounts of radioactivity. Currently, more than 15,000 metric tons of spent uranium fuel are awaiting disposal, and the Department of Energy (DOE) estimates that by the year 2000 the nation will have produced almost 50,000 metric tons of highly radioactive waste.

In 1987, Congress authorized the DOE to study the feasibility of using Yucca Mountain in southern Nevada as the nation's first high-level radioactive waste dump (□ Fig. 1). Such a facility must be able to isolate high-level waste from the environment for at least 10,000 years, which is the minimum time the waste will remain dangerous. The planned capacity of the Yucca Mountain site is 70,000 metric tons of waste. It will not be completely filled until about 2030, at which time its entrance shafts will be backfilled and sealed.

The radioactive waste at the Yucca Mountain repository will be buried in volcanic tuff at a depth of about 300 m (985 ft) in canisters designed to remain leakproof for at least 300 years. The water table in the area is an additional 200 to 420 m (656 to 1,380 ft) below the proposed dump site, so the canisters will be stored in the zone of aeration. Only about 15 cm (6 in) of rain fall in this area per year, and only a small amount percolates into the ground. Thus, the rocks at the depth of the repository are very dry.

One of the concerns of some scientists is that the climate might change during the next 10,000 years. If the region should become more humid, more water will percolate through the zone of aeration, increasing the corrosion rate of the canisters. Furthermore, the water table could rise, thereby decreasing the travel time between the repository and the zone of saturation. This area was much more humid between 1.6 million and 10,000 years ago.

While it appears that Yucca Mountain meets all the requirements for a safe high-level radioactive waste dump, the site is still controversial, and further studies must be conducted to ensure that the groundwater supply in this area is not rendered unusable by nuclear waste.

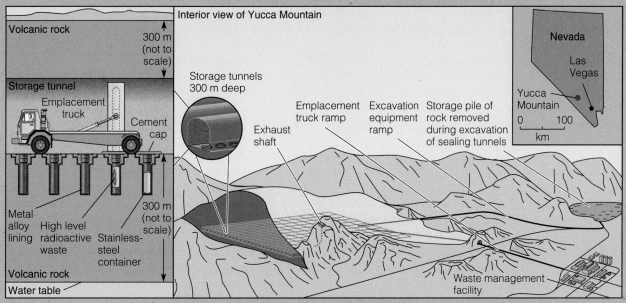

□ FIGURE 1 The location of Nevada's Yucca Mountain and a schematic diagram of the proposed high-level radioactive waste dump.

 FIGURE 5.29 Old Faithful Geyser in Yellowstone National Park, Wyoming, erupts every 30 to 90 minutes.

zone of saturation, it has been cleansed of impurities and is safe to use. If the water table is very close to the surface or if the rocks are very permeable, water entering the zone of saturation may still be contaminated and unfit to use.

Landfills, toxic waste disposal sites, and chemicals from fertilizers, pesticides, and herbicides are also potential sources of groundwater contamination. Unless a landfill is carefully designed and lined below by an impermeable layer such as clay, many toxic and cancer-causing compounds will find their way into the groundwater system. The United States alone must dispose of several thousand metric tons of hazardous chemical waste per year. Unfortunately, much of this waste has been, and still is being, improperly dumped and is contaminating the surface water, soil, and groundwater.

HOT SPRINGS AND GEYSERS

Groundwater percolating through rocks in regions of recent volcanic activity is heated and, if returned to the surface, forms *hot springs* and *geysers*. A **hot spring** is a spring in which the water temperature is warmer than the temperature of the human body; in some hot springs, the temperature reaches the boiling point. Of the approximately 1,100 known hot springs in the United States, more than 1,000 are in the Far West, while the rest are in South Dakota, Arkansas, Georgia, and the Appalachian region.

Recent and continuing igneous activity in the western United States accounts for the large number of hot springs

in that region. Many of these hot springs emit sulfur gases, especially hydrogen sulfide, which gives them their characteristic rotten-egg odor. The water in some hot springs is circulated deep into Earth, where it is warmed by the normal increase in temperature, the geothermal gradient. The spring water at Warm Springs, Georgia, is heated in this manner.

Hot springs that intermittently eject hot water and steam with tremendous force are **geysers.** One of the most famous geysers is Old Faithful in Yellowstone National Park, Wyoming (Fig. 5.29). With a thunderous roar, it erupts a column of hot water and steam every 30 to 90 minutes.

Geysers are found where groundwater percolates down into a network of fractures and is heated by hot igneous rocks. Since the water near the bottom of the fracture system is under greater pressure than that near the top, it must be heated to a higher temperature before it boils. When the deeper water is heated to very near the boiling point, a slight rise in temperature or a drop in pressure causes it to change instantly to steam. The expanding steam quickly pushes the water above it out of the ground and into the air, thereby producing a geyser eruption (Fig. 5.29).

Hot spring and geyser water contains large quantities of dissolved minerals because most minerals dissolve more rapidly in warm water than in cold water. Because of this high mineral content, the waters of many hot springs are believed by some to have medicinal properties. Numerous spas and bathhouses have been built throughout the world at hot springs to take advantage of these supposed healing properties.

When the highly mineralized water of hot springs or geysers cools at the surface, some of the minerals are precipitated, forming various types of deposits. If the groundwater contains dissolved calcium carbonate ($CaCO_3$), then *travertine* or *calcareous tufa* (both of which are varieties of limestone) is precipitated. Upon reaching the surface, groundwater containing dissolved silica will precipitate a soft, white mineral called *siliceous sinter* or *geyserite*.

 GEOTHERMAL ENERGY

Geothermal energy from steam and hot water trapped within Earth's crust could meet about 1 to 2% of the world's current energy needs. In areas where it is plentiful, as in Iceland where it has been used since 1928, geothermal energy can supply most, if not all, of the needed energy, sometimes at a fraction of the cost of other types of energy.

A geothermal electrical generating plant was built in 1960 at The Geysers, about 120 km (75 mi) north of San Francisco, California. Wells drilled into numerous near-vertical fractures reduce the pressure on the groundwater, which changes to steam that is piped directly to electricity-generating turbines. The present electrical generating capacity at The Geysers is about 2,000 megawatts, which is enough to supply about two-thirds of the electrical needs of the San Francisco Bay area.

While geothermally generated electricity is a generally clean source of power, it can also be expensive because most geothermal waters are acidic and very corrosive. Consequently, turbines must either be built of expensive corrosion-resistant alloy metals or frequently replaced. Furthermore, the steam and hot water removed for geothermal power cannot be easily replaced, and over time pressure in the wells drops until the geothermal field must be abandoned.

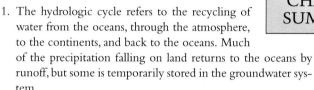

CHAPTER SUMMARY

1. The hydrologic cycle refers to the recycling of water from the oceans, through the atmosphere, to the continents, and back to the oceans. Much of the precipitation falling on land returns to the oceans by runoff, but some is temporarily stored in the groundwater system.

2. Streams erode by hydraulic action, abrasion, and dissolution of soluble rocks. They transport sand and gravel as bed load and silt and clay as suspended load. A dissolved load consists of substances in solution.

3. A braided stream has a complex of dividing and rejoining channels, whereas a meandering stream has a single, sinuous channel with broad looping curves. A stream occupies the floodplain paralleling its channels when too much water is supplied to the main channel.

4. Braided streams deposit sheets of sand and gravel, point bars form in meandering streams, and mostly mud accumulates on floodplains.

5. Many small deltas in lakes conform to the three-part division of bottomset, foreset, and topset beds, but large marine deltas are more complex. Alluvial fans are lobate deposits on land that typically form in arid and semiarid regions.

6. Sea level is ultimate base level, the lowest level to which streams can erode. Streams also commonly have local base levels such as lakes, other streams, or points where they flow across particularly resistant rocks.

7. If a stream has irregularities in its profile, it tends to eliminate them and develop a smooth, concave profile of equilibrium. Such streams are graded, meaning they have a balance between gradient, discharge, flow velocity, channel characteristics, and sediment load so that little or no deposition or erosion occurs within their channels.

8. Stream valleys develop by a combination of processes including downcutting, lateral erosion, and mass wasting. Renewed downcutting by a stream might result in the formation of stream terraces or incised meanders.

9. Groundwater is the water occupying the pores and fractures in soil, sediment, and rock; the water table is the surface separating the zones of aeration and saturation.

10. Both the movement of groundwater and its recovery depend on porosity and permeability. Porous and permeable materials that groundwater moves through are aquifers, whereas materials that prevent groundwater flow are aquicludes.

11. Groundwater moves very slowly to outlets such as streams, lakes, and swamps, thus eventually completing the hydrologic cycle by returning to the oceans. It may be discharged at the surface when a spring forms where the water table intersects the ground surface.

12. When water is pumped from a well, a cone of depression forms. If water is pumped faster than it is replenished, the cone of depression deepens until the well goes dry temporarily.

13. In artesian systems, groundwater rises above the level of an aquifer confined by aquicludes.

14. Karst topography, which is characterized by caves, sinkholes, solution valleys, and disappearing streams, results from groundwater weathering and erosion. Stalactites, stalagmites, and other cave deposits result from the precipitation of calcite in caves.

15. Modifications of the groundwater system, such as excessive withdrawal, can result in dry wells, loss of hydrostatic pressure, saltwater incursion, and ground subsidence. Groundwater contamination from sewage, landfills, toxic waste, and agriculture is becoming a serious problem.

16. Most hot springs and geysers are found where groundwater is heated by hot subsurface volcanic rocks, but some result from deep circulation of groundwater that is heated by Earth's temperature increase with depth.

17. Geothermal energy comes from the steam and hot water trapped within the crust. It is a relatively nonpolluting form of energy that is used to generate electricity.

IMPORTANT TERMS

alluvial fan
aquiclude
aquifer
artesian system
base level
braided stream
cave
cone of depression

delta
divide
drainage basin
drainage pattern
floodplain
geyser
graded stream
groundwater

hot spring
hydrologic cycle
karst topography
meandering stream
oxbow lake
permeability
point bar

porosity
sinkhole
spring
water table
zone of aeration
zone of saturation

REVIEW QUESTIONS

1. The erosive force of running water is:
   a. _____ bed load; b. _____ saltation; c. _____ hydraulic action; d. _____ meander cutoff; e. _____ base level.

2. The vertical drop of a stream in a given horizontal distance is its:
   a. _____ discharge; b. _____ gradient; c. _____ velocity; d. _____ base level; e. _____ drainage pattern.

3. A meandering stream has:
   a. _____ numerous sand and gravel bars in its channel; b. _____ a single, sinuous channel; c. _____ a broad, shallow channel; d. _____ a deep, narrow valley; e. _____ long, straight reaches and waterfalls.

4. Erosional remnants of a floodplain that are higher than the current level of a stream are:
   a. _____ oxbow lakes; b. _____ cut banks;. c. _____ stream terraces; d. _____ incised meanders; e. _____ natural bridges.

5. Infiltration capacity is the:
   a. _____ rate at which a stream erodes; b. _____ distance a stream flows from its source to the ocean; c. _____ maximum rate that surface materials can absorb water; d. _____ vertical distance a stream can erode below sea level; e. _____ variation in flow velocity across a stream channel.

6. A stream with a cross-sectional area of 250 m² and an average flow velocity of 1.5 m/sec has a discharge of _____ m³/sec:
   a. _____ 500; b. _____ 125; c. _____ 375; d. _____ 1,000; e. _____ 200.

7. A drainage pattern in which streams flow in and out of swamps and lakes with irregular flow directions is:
   a. _____ radial; b. _____ longitudinal; c. _____ deranged; d. _____ rectangular; e. _____ graded.

8. What percentage of the world's supply of fresh water is represented by groundwater?
   a. _____ 5; b. _____ 18; c. _____ 22; d. _____ 43; e. _____ 50.

9. The capacity of a material to transmit fluids is:

10. The water table is a surface separating the:
    a. _____ zone of porosity from the underlying zone of permeability; b. _____ aquiclude from the underlying zone of aeration; c. _____ point bar from the underlying zone of saturation; d. _____ zone of aeration from the underlying zone of saturation; e. _____ zone of saturation from the underlying zone of aeration.

11. A perched water table:
    a. _____ occurs wherever there is a localized aquiclude within an aquifer; b. _____ is frequently the site of springs; c. _____ lacks a zone of aeration; d. _____ answers (a) and (b); e. _____ answers (b) and (c).

12. An artesian system is one in which:
    a. _____ water is confined; b. _____ water can rise above the level of the aquifer when a well is drilled; c. _____ water must be pumped; d. _____ answers (a) and (c); e. _____ answers (a) and (b).

13. Rapid withdrawal of groundwater can result in:
    a. _____ a cone of depression; b. _____ ground subsidence; c. _____ saltwater incursion; d. _____ loss of hydrostatic pressure; e. _____ all of these.

14. Which one of the following is a cave deposit?
    a. _____ stalagmite; b. _____ geyser; c. _____ sinkhole; d. _____ disappearing stream; e. _____ cone of depression.

15. Explain how the hydrologic cycle works.

16. Explain what infiltration capacity is and why it is important in considering runoff.

17. How do channel shape and roughness control flow velocity?

18. How is it possible for a meandering stream to erode laterally yet maintain a more or less constant channel width?

19. Sea level is ultimate base level for most streams. If sea level drops with respect to the land, how would a stream respond?

20. What is a graded stream, and why are streams rarely graded except temporarily?

a. _____ porosity; b. _____ permeability; c. _____ solubility; d. _____ aeration quotient; e. _____ saturation.

21. Why is the water table a subdued replica of the surface topography?
22. Why does groundwater move so much slower than surface water?
23. Why are some artesian wells free-flowing while others must be pumped?

24. Discuss the various effects that excessive groundwater removal may have on a region. Give some examples.
25. Discuss the various ways that a groundwater system may become contaminated.

POINTS TO PONDER

1. A stream 2,000 m (6,560 ft) above sea level at its source flows 1,500 km (930 mi) to the sea. What is its gradient? Do you think the gradient you calculated will be accurate for all segments of this stream? Explain.
2. What long-term changes may occur in the hydrologic cycle? How might human activities bring about such changes?
3. How and why would the discharge of a stream vary as it flows through an area of karst topography?
4. What kinds of considerations are important in selecting the site of a sanitary landfill?
5. According to one estimate, 10.76 km³ of sediment are eroded from the continents each year, most of it by running water. Given that the volume of the continents above sea level is 92,832,194 km³, they should be eroded to sea level in only 8,627,527 years. Although the calculation is correct, there is something seriously wrong with the line of reasoning. What is it?

ADDITIONAL READINGS

Bevin, K., and P. Carling, eds. 1989. *Floods.* New York: John Wiley & Sons.

Courbon, P., C. Chabert, P. Bosted, and K. Lindslay. 1989. *Atlas of the great caves of the world.* St. Louis: Cave Books.

Dietrich, R. V. 1993. How are caves formed? *Rocks and Minerals* 68, no. 4: 264–68.

Fetter, C. W. 1988. *Applied hydrogeology.* 2d ed. Columbus, Ohio: Merrill Publishing Co.

Frater, A., ed. 1984. *Great rivers of the world.* Boston: Little, Brown.

Grossman, D., and S. Shulman. 1994. Verdict at Yucca Mountain. *Earth* 3, no. 2: 54–63.

Jennings, J. N. 1983. Karst landforms. *American Scientist* 71, no. 6: 578–86.

———. 1985. *Karst geomorphology.* 2d ed. Oxford: Basil Blackwell.

Knighton, D. 1984. *Fluvial forms and processes.* London: Edward Arnold.

National Geographic. 1993. Water: The power, promise and turmoil of North America's water. Special edition, October.

Petts, G., and I. Foster. 1985. *Rivers and landscapes.* London: Edward Arnold.

WORLD WIDE WEB SITES

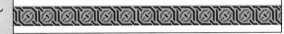

For current updates and exercises, log on to http://www.wadsworth.com/geo.

U.S. GEOLOGICAL SURVEY WATER RESOURCES OF THE UNITED STATES
http://h2o.er.usgs.gov/

This web site contains information about streams, flooding, flood forecasting, flood warnings, and groundwater.

LINKS TO INFORMATION SOURCES FOR HYDROGEOLOGY, HYDROLOGY, AND ENVIRONMENTAL SCIENCES
http://www.us.net/adept/links.html

This site contains links to information sources for hydrogeology, hydrology, and environmental sciences. The links are listed alphabetically. Check out several sites for information about rivers, flooding, surface water usage, both in the United States and elsewhere in the world.

THE VIRTUAL CAVE
http://www.goodearth.com/virtcave.html

This web site contains images from around the world on various cave features. In addition, it contains links to other cave sites and information on caves elsewhere in the world.

YUCCA MOUNTAIN PROJECT STUDIES
http://www.ymp.gov/

This website is devoted to the Yucca Mountain Project which is the only current site being considered for the sto level radioactive waste. It includes informat

CHAPTER
6

Glaciers, Deserts, and the Work of Wind

Glacier on Mount Foresta, Alaska.

Prologue

Following the Great Ice Age, which ended about 10,000 years ago, Earth experienced a general warming trend that was periodically interrupted by short, relatively cool periods. During one cool period, small glaciers in mountain valleys expanded, and sea ice at high latitudes persisted for longer periods than it had previously. This interval, from A.D. 1500 to the mid to late 1800s, is known as the *Little Ice Age.*

During the Little Ice Age, many of the small glaciers in Europe and Iceland expanded and moved far down their valleys, reaching their greatest historic extent during the early 1800s (□ Fig. 6.1). A small ice cap formed in Iceland where none had existed previously, and glaciers in Alaska and the mountains of the western United States and Canada also expanded to their greatest limits during historic time. Although glaciers caused some problems in Europe where they advanced across roadways and pastures, destroying some villages in Scandinavia and threatening villages elsewhere, the impact of glaciers on humans was minimal. Far more important from the human perspective was that dur-

ing much of the Little Ice Age the summers in northern latitudes were cooler and wetter.

Although worldwide temperatures were a little lower during this time, the change in summer conditions rather than cold winters or glaciers caused most of the problems. Particularly hard hit were Iceland and the Scandinavian countries, but at times much of northern Europe was affected. Growing seasons were shorter during many years, resulting in food shortages and a number of famines. Iceland's population declined from its high of 80,000 in 1200 to about 40,000 by 1700. Many times between 1610 and 1870, sea ice was observed near Iceland for up to three months a year, and each time sea ice persisted for long periods, poor growing seasons and food shortages followed.

Exactly when the Little Ice Age ended is debatable. Some authorities put the end at 1880, whereas others think it ended as early as 1850. In any case, during the late 1800s, the sea ice was retreating northward, glaciers were retreating back up their valleys, and summer weather became more moderate.

□ FIGURE 6.1 During the Little Ice Age, many glaciers in Europe, such as this one in Switzerland, extended much farther down their valleys than they do at present. This painting, called *The Unterer Grindelwald,* was painted in 1826 by Samuel Birmann (1793–1847).

Most people have some idea of what a glacier is, but many confuse glaciers with other masses of snow and ice. A **glacier** is a mass of ice composed of compacted and recrystallized snow that flows under its own weight on land. Accordingly, sea ice is not glacial ice, nor are drifting icebergs glaciers even though they may have derived from glaciers that flowed into the sea. Snow fields in high mountains may persist in protected areas for years, but these are not glaciers either because they are not actively moving.

At present, glaciers cover nearly 15 million km² (5.8 million mi²), or about one-tenth of Earth's land surface (○ Table 6.1). Numerous glaciers exist in the mountains of the western United States, especially Alaska, western Canada, the Andes in South America, the Alps in Europe, the Himalayas of Asia, and other high mountains. Even Mount Kilimanjaro in Africa, although near the equator, is high enough to support glaciers. In fact, Australia is the only continent lacking glaciers. By far the largest existing glaciers are in Greenland and Antarctica; both areas are nearly covered by glacial ice (Table 6.1).

About 2.15% of the world's water is contained in glaciers, which constitute one reservoir in the hydrologic cycle (see Fig. 5.2). However, many glaciers at high latitudes, as in Alaska, flow directly into the sea where they melt, or where icebergs break off by a process called *calving* and drift out to sea where they eventually melt. At lower latitudes where they can exist only at high elevations, glaciers flow to lower

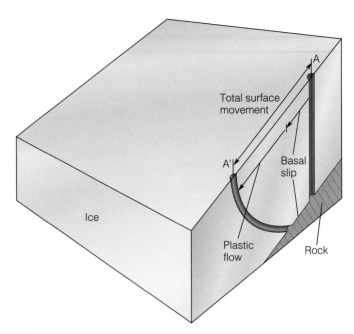

■ **FIGURE 6.2** Movement of a glacier by a combination of plastic flow and basal slip. If a glacier is solidly frozen to the underlying surface, it moves only by plastic flow.

elevations where they melt, and the water returns to the oceans by surface runoff or infiltrates into the groundwater system.

THE ORIGIN OF GLACIAL ICE

Ice is a mineral in every sense of the word; it has a crystalline structure and possesses characteristic physical and chemical properties. Accordingly, glacial ice is a type of rock, although it is easily deformed rock. Glacial ice forms in a fairly straightforward manner; when an area receives more winter snow than can melt during the spring and summer season, a net accumulation occurs. Freshly fallen snow consists of about 80% air and 20% solids, but it compacts as it accumulates, and is finally converted to *glacial ice* with only about 10% air.

When ice accumulates to a critical thickness of about 40 m (130 ft), pressure on the ice at depth is great enough for **plastic flow** to begin. Plastic flow, which is the primary way glaciers move, involves a permanent change in shape of a solid without fracturing. Glaciers may also move by **basal slip,** which takes place when a glacier slides over its underlying surface (■ Fig. 6.2). Basal slip is facilitated by the presence of meltwater that reduces frictional resistance between the glacier and the underlying surface.

TABLE 6.1 Present-Day Ice-Covered Areas	
Antarctica	12,653,000 km²
Greenland	1,802,600
Northeast Canada	153,200
Central Asian ranges	124,500
Spitsbergen group	58,000
Other Arctic islands	54,000
Alaska	51,500
South American ranges	25,000
West Canadian ranges	24,900
Iceland	11,800
Scandinavia	5,000
Alps	3,600
Caucasus	2,000
New Zealand	1,000
USA (other than Alaska)	650
Others	about 800
	14,971,550

Total volume of present ice: 28 to 35 million km³

SOURCE: C. Embleton and C. A. King, *Glacial Geomorphology* (New York: Halsted Press, 1975).

TYPES OF GLACIERS

Two basic types of glaciers are generally recognized: *valley* and *continental*. A **valley glacier,** as its name implies, is confined to a mountain valley or an interconnected system of mountain valleys (Fig. 6.3). Large valley glaciers commonly have several smaller tributary glaciers, much as rivers have tributaries. Valley glaciers flow from higher to lower elevations and are invariably small in comparison to continental glaciers.

Continental glaciers, also called *ice sheets,* cover at least 50,000 km^2 (19,245 mi^2) and are unconfined by topography. In contrast to valley glaciers, which flow downhill within the confines of a valley, continental glaciers flow outward in all directions from a central area of accumulation. Currently, continental glaciers exist only in two areas, Greenland and Antarctica. These glaciers are more than 3,000 m (9,840 ft) thick in their central areas, become thinner toward their margins, and cover all but the highest mountains.

Although valley and continental glaciers are easily differentiated by size and location, an intermediate variety called an *ice cap* is also recognized. Ice caps are similar to, but smaller than, continental glaciers. Some ice caps form when valley glaciers grow and overtop the divides and passes between adjacent valleys and coalesce to form a continuous ice cap. They also form on fairly flat terrain as in Iceland and some of the Canadian Arctic islands.

 THE GLACIAL BUDGET

Just as a savings account grows and shrinks as funds are deposited and withdrawn, glaciers expand and contract in response to accumulation and wastage. Their behavior can be described in terms of a **glacial budget,** which is essentially a balance sheet of accumulation and wastage. The upper part of a valley glacier is a *zone of accumulation* where additions exceed losses, and the glacier's surface is perennially covered by snow. In contrast, the lower part of the same glacier is a *zone of wastage,* where losses from melting and calving of icebergs exceed the rate of accumulation (Fig. 6.4).

At the end of winter, a glacier's surface is usually covered with the accumulated seasonal snowfall. During spring and summer, the snow begins to melt, first at lower elevations and then progressively higher up the glacier. The elevation to which snow recedes during a wastage season is the *firn limit* (Fig. 6.4). One can easily identify the zones of accumulation and wastage by noting the position of the firn limit.

FIGURE 6.3 A large valley glacier in Alaska. Notice the tributaries that join to form the large glacier.

Observations of a single glacier reveal that the position of the firn limit usually changes from year to year. If it does not change or shows only minor fluctuations, the glacier is said to have a balanced budget; that is, additions in the zone of accumulation are balanced by losses in the zone of wastage, and the distal end or *terminus* of the glacier remains stationary. When the firn limit moves down the glacier, the glacier has a positive budget; its additions exceed its losses, its volume increases, and its terminus advances (Fig. 6.4b). If the budget is negative, the glacier's volume decreases and it recedes—its terminus retreats up the glacial valley (Fig. 6.4c). But even though a glacier's terminus may be receding, the glacial ice continues to move toward the terminus by plastic flow and basal slip. If a negative budget persists long enough, though, a glacier recedes and thins until it no longer flows, thus becoming a *stagnant glacier.*

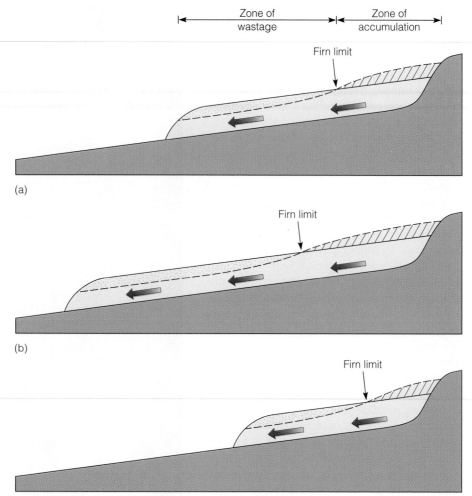

Zone of wastage | Zone of accumulation

Firn limit

(a)

Firn limit

(b)

Firn limit

(c)

 FIGURE 6.4 Response of a hypothetical glacier to changes in its budget. (*a*) If losses in the zone of wastage, shown by stippling, equal additions in the zone of accumulation, shown by crosshatching, the terminus of the glacier remains stationary. (*b*) Gains exceed losses, the volume increases, and the glacier's terminus advances. (*c*) Losses exceed gains, the volume decreases, and the glacier's terminus retreats. Even though the terminus is retreating, the glacier continues to move in the direction indicated by the arrows.

Although we have used a valley glacier as an example, the same budget considerations control the flow of continental glaciers. The entire Antarctic ice sheet is in the zone of accumulation, but it flows into the ocean where wastage occurs.

RATES OF GLACIAL MOVEMENT

In general, valley glaciers move more rapidly than continental glaciers, but the rates for both vary, ranging from centimeters to tens of meters per day. Valley glaciers moving down steep slopes flow more rapidly than glaciers of comparable size on gentle slopes, assuming all other variables are the same. Temperature exerts a seasonal control on valley glaciers because, although plastic flow remains rather constant year-round, basal slip is more important during warmer months when meltwater beneath the ice is more abundant.

Flow rates also vary within the ice itself. Valley glaciers are similar to streams in that the valley walls and floor cause frictional resistance to flow. So the ice in contact with the glacier's boundaries moves more slowly than the ice some distance away, and the most rapid movement takes place near the center and top of the glacier. The upper 40 m (130 ft) or so of a glacier constitute its outer rigid part; there large cracks known as *crevasses* develop that extend down to the zone of plastic flow.

One reason continental glaciers move comparatively slowly is that they exist at higher latitudes and are frozen to the underlying surface most of the time, which limits the amount of basal slip. Nevertheless, some parts of continental glaciers manage to achieve extremely high flow rates. Near the margins of the Greenland ice sheet, for example, the ice is forced between mountains where flow velocities exceed 100 m (328 ft) per day.

GLACIAL EROSION AND TRANSPORT

Glaciers erode by bulldozing, plucking, and abrasion and transport huge quantities of materials, especially sediment and soil. Bulldozing, although not a formal term, is fairly self-explanatory: a glacier simply shoves or pushes materials in its path. *Plucking,* also called *quarrying,* results when glacial ice freezes in the cracks and crevices of a bedrock projection and eventually pulls it loose.

Sediment-laden glacial ice can effectively erode by *abrasion,* or the grinding and scouring of exposed rock surfaces. Abraded bedrock commonly develops a *glacial polish,* a smooth surface that glistens in reflected light, and abrasion also yields *glacial striations* consisting of rather straight scratches rarely more than a few millimeters deep (▣ Fig. 6.5). Rocks pulverized by abrasion yield an aggregate of clay- and silt-sized particles having the consistency of flour, hence the name *rock flour.* Rock flour is so common in streams discharging from glaciers that the water has a milky appearance.

Continental glaciers can derive sediment from mountains projecting through them, and windblown dust settles on their surfaces. Otherwise, most of their sediment is derived from the surface they move over and is transported in the lower part of the ice sheet. In contrast, valley glaciers carry sediment in all parts of the ice, but it is concentrated at the base and along the margins. Some of the marginal sediment is derived by abrasion and plucking, but much of it is supplied by mass wasting of slopes above the glacier.

Erosional Landforms of Valley Glaciers

Many mountain ranges are scenic to begin with, but when modified by valley glaciers, they take on a unique aspect of jagged, angular peaks and ridges in the midst of broad valleys (▣ Fig. 6.6). Mountain valleys eroded by running water have valley walls that descend steeply to a narrow valley bottom, giving them a V-shape in cross section (Fig. 6.6a). Valleys scoured by glaciers are deepened, widened, and straightened until they exhibit a U-shaped profile and are referred to as **U-shaped glacial troughs** (Figs. 6.6c and ▣ 6.7).

During the Ice Age, when glaciers were more extensive, sea level was about 130 m (425 ft) lower than at present, so glaciers flowing into the sea eroded their valleys below current sea level. When the glaciers melted at the end of the Ice

▣ FIGURE 6.5 Glacial polish and striations on rocks at Devil's Postpile National Monument, California.

Age, sea level rose, and the ocean filled the lower ends of the glacial troughs forming long, steep-walled embayments known as *fiords.*

Some of the world's highest and most spectacular waterfalls are found in recently glaciated areas. They plunge from **hanging valleys,** which are tributary valleys whose floors are at a higher level than that of the main valley (▣ Fig. 6.8). As Figure 6.6 shows, the large glacier in the main valley vigorously erodes, whereas the smaller glaciers in the tributary valleys are less capable of large-scale erosion. When the glaciers disappear, the small tributary valleys remain as hanging valleys.

Valley glaciers form and move out from steep-walled, bowl-shaped depressions called **cirques** at the upper end of their troughs (Fig. 6.6c). Cirques are typically steep-walled on three sides, but one side is open and leads into a glacial trough. The details of cirque origin are not fully understood, but they apparently form by glacial erosion of a preexisting depression on a mountainside.

The fact that cirques expand laterally and by headward erosion accounts for the origin of two other distinctive erosional features, arêtes and horns. **Arêtes**—narrow, serrated ridges—can form in two ways. If cirques form on opposite sides of a ridge, headward erosion may reduce the ridge until only a thin partition of rock remains (Fig. 6.6c). The same effect is produced when erosion in two parallel glacial troughs reduces the intervening ridge to a thin spine of rock.

The most majestic of all mountain peaks are steep-walled, pyramid-shaped **horns.** In order for a horn to form, a mountain peak must have at least three cirques on its flanks, all of which erode headward (Fig. 6.6c). Excellent examples of horns include Mount Assiniboine in the Canadian Rockies, the Grand Teton in Wyoming, and the most famous of all, the Matterhorn in Switzerland (▣ Fig. 6.9).

EARTHFACT

Glacial Surges

The flow rates of glaciers are complicated by *glacial surges,* which are bulges of ice moving through a glacier several times faster than normal flow velocity. During a surge, a glacier's terminus may advance rapidly; in 1993, for example, the terminus of the Bering Glacier in Alaska advanced more than 1,500 m (4,920 ft) in just three weeks. Some surges are caused by unusually heavy snowfall in the zone of accumulation. Others develop when large amounts of snow and ice are dislodged from mountain peaks and fall onto the upper parts of a glacier. In either case, rapid changes in the glacial budget take place.

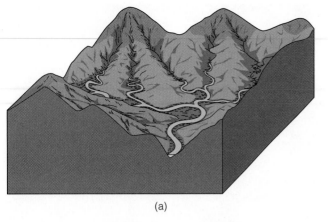

(a)

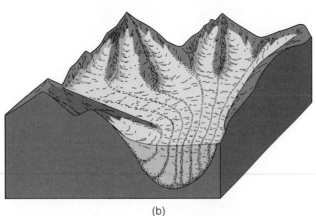

(b)

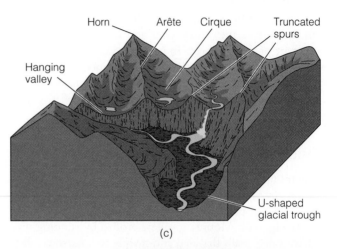

(c)

■ FIGURE 6.6 Erosional landforms produced by valley glaciation. (*a*) A mountain area before glaciation. (*b*) The same area during the maximum extent of valley glaciers. (*c*) After glaciation. (*d*) Glacial landscape at Glacier National Park, Montana.

(d)

Erosional Landforms of Continental Glaciers

Areas eroded by continental glaciers tend to be smooth and rounded because the glaciers bevel and abrade high areas that projected into the ice. Rather than yielding the sharp, angular landforms typical of valley glaciation, continental glaciers produce a landscape of rather flat topography interrupted by rounded hills.

Another feature of areas eroded by continental glaciers is deranged drainage (see Fig. 5.13e). Other distinctive features include numerous lakes and swamps, low relief, extensive exposures of striated and polished bedrock, and little or no soil. Areas possessing these erosional features, generally referred to as *ice-scoured plains,* are common in a large part of Canada.

■ FIGURE 6.7 *Above:* U-shaped glacial trough in northwestern Montana.

■ FIGURE 6.8 *Right:* Yosemite Falls plunge from a hanging valley in Yosemite National Park, California. The falls plunge 435 m (1,427 ft) vertically, cascade down a steep slope for another 205 m (672 ft), and then fall vertically 97 m (318 ft), for a total descent of 737 m (2,417 ft).

■ FIGURE 6.9 *Above:* The Matterhorn in Switzerland is a well-known horn.

 GLACIAL DEPOSITS

All sediment deposited as a result of glacial activity is called **glacial drift.** Two distinct types of glacial drift are recognized, *till* and *outwash*. Till, which is deposited directly by glacial ice, is a chaotic jumble of all sizes of particles with no sorting or layering. Till may even include boulders that have been transported far from their source areas; these are re-

ferred to as *glacial erratics* (▣ Fig. 6.10). In contrast, outwash, which is deposited by meltwater discharged from glaciers, does exhibit layering and sorting.

Landforms Composed of Till

The terminus of a glacier may become stabilized in one position for some time, perhaps a few years or even decades. Stabilization of the ice front does not mean that the glacier has ceased flowing, only that it has a balanced budget. Flow within the glacier continues, and sediment transported within or upon the ice is deposited as an **end moraine,** a pile of rubble at the glacier's terminus (▣ Fig. 6.11). End moraines of valley glaciers are commonly crescent-shaped ridges of till spanning the valley occupied by the glacier. Those of continental glaciers similarly parallel the ice front but are much more extensive.

Following a period of stabilization, a glacier may advance or retreat, depending on changes in its budget. If it advances, the ice front overrides and modifies its former moraine. Should a glacier's budget be negative, however, the ice front retreats, and as it does so, till liberated from the melting ice is deposited and forms a layer of *ground moraine* (Fig. 6.11b). Ground moraine has an irregular topography, whereas end moraine consists of long ridgelike accumulations of sediment. After a glacier has retreated for some time, its terminus may once again stabilize, and it will deposit another end moraine known as a **recessional moraine** (Fig. 6.11b). The outermost end moraine, marking the greatest extent of the glacier, goes by the special name *terminal moraine*.

▣ FIGURE 6.11 (*a*) The origin of an end moraine. (*b*) The outermost end moraine marking the greatest extent of a glacier is a terminal moraine, whereas a recessional moraine forms after a glacier's terminus has retreated, stabilized once again, and deposited another end moraine.

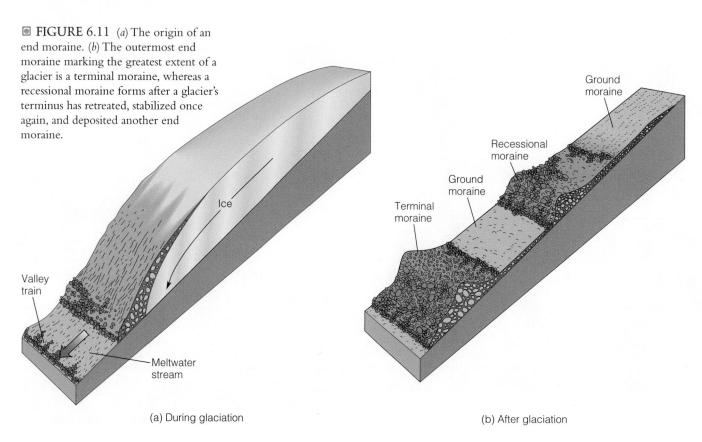

(a) During glaciation

(b) After glaciation

■ **FIGURE 6.12** *Left:* Lateral and medial moraines on a glacier in Alaska. Where two glaciers join, lateral moraines merge to form a medial moraine.

Valley glaciers transport considerable sediment along their margins, much of it abraded and plucked from the valley walls, but a significant amount falls or slides onto the glacier's surface by mass wasting processes. This sediment is deposited as long ridges of till called **lateral moraines** along the margin of the glacier (■ Fig. 6.12). Where two lateral moraines merge, as when a tributary glacier flows into a larger glacier, a *medial moraine* forms (Fig. 6.12). In fact, a large valley glacier often has several dark stripes of sediment on its surface, each of which is a medial moraine.

In many areas where glaciers have deposited till, the till has been reshaped into elongated hills called **drumlins.** Some drumlins measure as much as 50 m (164 ft) high and 1 km (0.6 mi) long, but most are much smaller. From the side, a drumlin looks like an inverted spoon with the steep end on the side from which the glacial ice advanced, and the gently sloping end pointing in the direction of ice movement (■ Fig. 6.13b).

One hypothesis for the origin of drumlins holds that they formed in the zone of plastic flow as glacial ice modified pre-existing till into streamlined hills. According to another hypothesis, drumlins formed when huge floods of glacial meltwater modified deposits of till. Drumlins rarely occur as single, isolated hills; instead hundreds or thousands of drumlins are found together in drumlin fields.

Landforms Composed of Outwash

Glaciers discharge sediment-laden meltwater most of the time, except perhaps during the coldest months. This meltwater forms a series of braided streams that radiate out from

■ **FIGURE 6.13** *Below:* Two stages in the origin of kettles, kames, drumlins, eskers, and outwash. (*a*) During glaciation. (*b*) After glaciation.

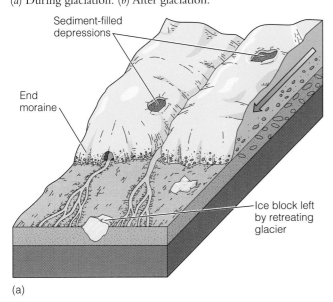

(a)

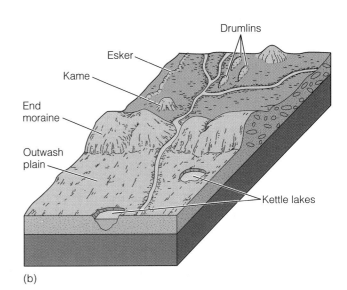

(b)

the front of continental glaciers over a wide region. So much sediment is supplied to the streams that much of it is deposited as sand and gravel bars (Fig. 6.13b). Valley glaciers also have braided streams extending from them, but their outwash is confined to the lower parts of glacial troughs.

Outwash deposits commonly contain numerous circular to oval depressions, many with small lakes. These depressions are **kettles,** which form when a retreating ice sheet or valley glacier leaves a block of ice that is subsequently partly or wholly buried (Fig. 6.13). When the ice block eventually melts, it leaves a depression that may extend below the water table and be the site of a small lake.

Kames, conical hills as much as 50 m (164 ft) high, form when a stream deposits sediment in a depression on a glacier's surface (Fig. 6.13). As the ice melts, the deposit is lowered to the ground surface. Long sinuous ridges of sand and gravel are **eskers,** some of which are as much as 100 m (328 ft) high and more than 100 km (62 mi) long. Observations of present-day glaciers indicate that eskers form by sediment deposition in tunnels beneath stagnant ice (Fig. 6.13).

 PLEISTOCENE GLACIATION

Many naturalists of the last century invoked the biblical flood to account for the large boulders throughout Europe that are found far from their sources, while others believed that the boulders were rafted to their present positions by icebergs floating in floodwaters. It was not until 1837 that the Swiss naturalist Louis Agassiz argued convincingly that the displaced boulders, many deposits, polished and striated bedrock, and many of the valleys of Europe resulted from huge glaciers moving over the land.

We know today that the Pleistocene Epoch, or what is commonly called the Ice Age, began about 1.6 million years ago and consisted of several intervals of glacial expansion separated by warmer interglacial periods when the glaciers disappeared. At least four major episodes of Pleistocene glaciation have been recognized in North America (■ Fig. 6.14), and six or seven major glacial advances and retreats are recognized in Europe. Based on the best available evidence, it appears that the Pleistocene ended about 10,000 years ago. However, scientists do not know if the present interglacial period will persist indefinitely, or whether we will enter another glacial interval.

At their greatest extent, Pleistocene glaciers covered about three times as much of Earth's surface as glaciers do now (■ Fig. 6.15). Large areas of North America were covered by glaciers more than 3 km (1.8 mi) thick as were Greenland, Scandinavia, Great Britain, Ireland, and a large area in northern Russia. Mountainous areas also experienced an expansion of valley glaciers and the development of ice caps.

Pluvial and Proglacial Lakes

During the Pleistocene, many of the basins in the western United States contained large lakes that formed as a result of greater precipitation and overall lower temperatures (especially during the summer), which lowered the evaporation rate. The largest of these *pluvial lakes,* as they are called, was Lake Bonneville, which attained a maximum size of 50,000 km^2 (19,245 mi^2) and a depth of at least 335 m (1,100 ft). The vast salt deposits of the Bonneville Salt Flats west of Salt Lake City, Utah, formed as parts of this ancient lake dried up: Great Salt Lake is simply the remnant of this once vast lake.

Another large pluvial lake existed in Death Valley, California, which is now the hottest, driest place in North America. When the lake evaporated, dissolved salts were precipitated on the valley floor; some of these deposits, especially borax, are important mineral resources.

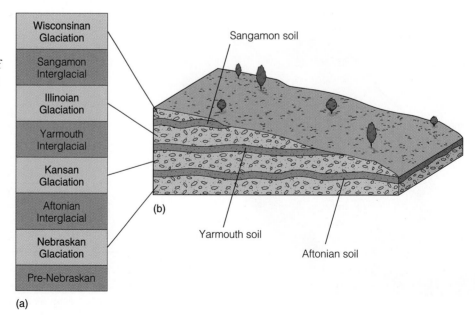

■ FIGURE 6.14 (*a*) Standard terminology for Pleistocene glacial and interglacial stages in North America. (*b*) Idealized succession of deposits and soils developed during the glacial and interglacial stages.

| Wisconsinan Glaciation |
| Sangamon Interglacial |
| Illinoian Glaciation |
| Yarmouth Interglacial |
| Kansan Glaciation |
| Aftonian Interglacial |
| Nebraskan Glaciation |
| Pre-Nebraskan |

(a)

Sangamon soil

Yarmouth soil

Aftonian soil

(b)

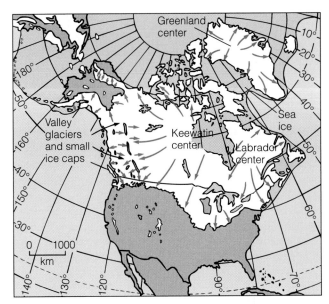

FIGURE 6.15 Centers of ice accumulation and maximum extent of Pleistocene glaciers in North America.

In contrast to pluvial lakes, which form far from glaciers, *proglacial lakes* are formed by meltwater accumulating along the margins of glaciers (see ▢ Perspective 6.1). Lake Agassiz, named in honor of the Swiss naturalist Louis Agassiz, was a large proglacial lake covering about 250,000 km² (96,225 mi²), mostly in Canada but extending into North Dakota. It persisted until the glacial ice along its northern margin melted, at which time the lake drained northward into Hudson Bay.

Changes in Sea Level

More than 70 million km³ (17 million mi³) of snow and ice covered the continents during the maximum extent of glaciers during the Pleistocene. This storage of ocean waters in glaciers lowered sea level about 130 m (425 ft) and exposed extensive areas of the continental shelves, which were quickly covered with vegetation. Indeed, a land bridge existed across the Bering Strait from Alaska to Siberia. Native Americans crossed the Bering land bridge, and various animals migrated between the continents; the American bison, for example, migrated to North America from Asia. The British Isles were connected to Europe during the glacial intervals because the shallow floor of the North Sea was above sea level. When the glaciers disappeared, these areas were again flooded, drowning the plants and forcing the animals to migrate farther inland.

Lowering of sea level during the Pleistocene also affected the base level of most major streams. As rivers adjusted to the new lower base level, they eroded their valleys deeper and extended their valleys across the emergent continental shelves (see Chapter 5). When sea level rose at the end of the Pleistocene, the lower ends of the stream valleys along the east coast of North America were flooded and are now important harbors.

A tremendous quantity of water is still stored on land in glaciers (see Table 5.1). If these glaciers should melt completely, sea level would rise about 70 m (230 ft), flooding all of the coastal areas where many of the world's large population centers are located.

 ## CAUSES OF GLACIATION

For more than a century, scientists have been attempting to develop a comprehensive theory explaining all aspects of ice ages, but they have not been completely successful. One reason for the lack of success is that the climatic changes responsible for glaciation, the cyclic occurrence of glacial-interglacial episodes, and short-term events such as the Little Ice Age operate on vastly different time scales.

Only a few periods of glaciation are recognized in the geologic record, each separated from the others by long intervals of mild climate. The long-term climatic changes responsible for these widely separated glacial episodes were probably caused by slow geographic changes related to plate tectonic activity. Moving plates can carry continents to high latitudes where glaciers can exist, provided they receive enough precipitation as snow. Plate collisions, uplift of vast areas far above sea level, and changing atmospheric and oceanic circulation patterns caused by the changing shapes and positions of plates also contribute to long-term climatic change.

Intermediate-term climatic events, such as the glacial-interglacial episodes of the Pleistocene, occur on time scales of tens to hundreds of thousands of years. A particularly interesting hypothesis for intermediate-term climatic events was put forth by the Yugoslavian astronomer Milutin Milankovitch during the 1920s. He proposed that minor irregularities in Earth's rotation and orbit are sufficient to alter the amount of solar radiation received at any given latitude and hence can affect climatic changes. Now called the **Milankovitch theory,** it was initially ignored, but has received renewed interest during the last 20 years.

Milankovitch attributed the onset of the Ice Age to variations in three parameters of Earth's orbit (▣ Fig. 6.16 on page 122): orbital eccentricity, which is the degree to which the orbit departs from a circle; the angle between Earth's axis and a line perpendicular to the plane of the ecliptic; and the precession of the equinoxes, which causes the position of the equinoxes and solstices to shift slowly around Earth's elliptical orbit. Continuous changes in these three parameters cause the amount of solar radiation received at any latitude to vary slightly over time, but the total heat received by the planet remains little changed. Milankovitch proposed, and now many scientists agree, that the interaction of these three parameters provided the triggering mechanism for the glacial-interglacial episodes of the Pleistocene.

Climatic events having durations of several centuries, such as the Little Ice Age, are too short to be accounted for by plate tectonics or Milankovitch cycles. Several hypotheses have been proposed, including variations in solar energy

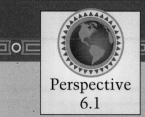

Perspective 6.1

Glacial Lake Missoula and the Channeled Scablands

The term *scabland* is used in the Pacific Northwest to describe areas where surface deposits have been scoured, exposing the underlying rock. A scabland area exists in a large part of eastern Washington where numerous deep and generally dry channels are present. Some of these channels are cut more than 70 m (230 ft) deep into basalt lava flows, and their floors are covered by gigantic "ripple marks"

as much as 10 m (33 ft) high and 70 to 100 m (230 to 328 ft) apart.

In 1923, J Harlan Bretz proposed that these channeled scablands formed during a single gigantic flood of glacial meltwater that lasted only a few days. His proposal was not well received by other scientists, most of whom preferred a more traditional explanation involving stream erosion over a long

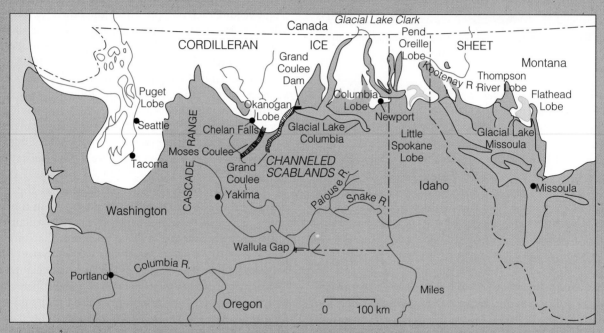

☐ **FIGURE 1** Location of glacial Lake Missoula and the channeled scablands of eastern Washington.

and volcanism. Variations in solar energy could result from changes within the Sun itself or from anything that would reduce the amount of energy received from the Sun. The latter could result from the solar system passing through clouds of interstellar dust and gas or from substances in Earth's atmosphere reflecting solar radiation back into space. Records kept for the past 75 years indicate that during this time the amount of solar radiation has varied only slightly. So although variations in solar energy may influence short-term climatic events, such a correlation has not been demonstrated.

During large volcanic eruptions, tremendous amounts of ash and gases, particularly sulfur gases, are spewed into the atmosphere where they reflect solar radiation and reduce at-

mospheric temperatures. Several large-scale volcanic events, such as the 1815 eruption of Tambora and the 1991 eruption of Pinatubo, are known to have had climatic effects. However, no relationship between periods of volcanic activity and periods of glaciation has yet been established.

 ## DESERTS

Any region receiving less than 25 cm (10 in) of precipitation per year is a **desert,** so in addition to the hot, arid areas one typically thinks of as desert, the polar regions are also technically deserts. Here, however, we are interested mostly in the hot, arid deserts, which typically have poorly developed period of time. The problem with Bretz's hypothesis was that he could not identify an adequate source for the floodwaters. He knew that glaciers had advanced far south into Washington, but he could not explain how so much ice melted so rapidly. The answer to his dilemma came from western Montana where an enormous ice-dammed lake (Lake Missoula) formed when an advancing glacier plugged a river valley in Idaho (☐ Fig. 1). At its highest level, Lake Missoula contained about 2,090 km³ (502 mi³) of water, about the same volume as present-day Lake Erie and Lake Ontario combined.

When the ice dam impounding Lake Missoula failed, the water rushed out at tremendous velocity and drained south and southwest across Idaho and into Washington. The maximum rate of flow is estimated to have been nearly 11 million m³/sec, about 55 times greater than the average discharge of the Amazon River. When these raging floodwaters reached eastern Washington, they stripped away the soil and most of the surface sediment, carving out huge valleys in solid bedrock. The currents were so powerful and turbulent that they plucked out and moved pieces of basalt measuring 10 m (33 ft) across. Within the channels, sand and gravel were shaped into huge ridges or so-called giant ripple marks (☐ Fig. 2).

Bretz originally thought that one massive flood formed the scablands, but we now know that Lake Missoula formed, drained catastrophically, and refilled at least four times and perhaps seven times. The largest lake formed 18,000 to 20,000 years ago, and its draining produced the last great flood. It is estimated that about one month passed from the time the last great flood burst forth from Lake Missoula until the streams in eastern Washington returned to their normal flow.

☐ FIGURE 2 These gravel ridges are the so-called giant ripple marks that formed when glacial Lake Missoula, drained across this area near Camas Hot Springs, Montana.

soils and are mostly or completely devoid of vegetation. Also, as opposed to polar deserts, hot, arid deserts have evaporation rates that far exceed the amount of annual precipitation (see ☐ Perspective 6.2 on page 124).

In North America, parts of the southwestern United States and northern Mexico are hot, dry deserts, while in South America deserts are largely restricted to coastal Chile and Peru (☐ Fig. 6.17). The Sahara in northern Africa and the Arabian Desert in the Middle East, along with most of Pakistan and western India, form the largest essentially unbroken desert environment in the Northern Hemisphere. More than 40% of Australia is desert, and most of the rest is semiarid. It is not surprising that Australia is commonly called the "desert continent."

The deserts discussed in the preceding paragraphs are found in belts between about 20° and 30° north and south of the equator. Air heated in the equatorial region rises, releases most of its moisture as rain, then moves northward and southward toward the poles. By the time it reaches 20° to 30° north or south latitude, the air has become cooler and denser and begins to descend. Compression warms the descending air masses and produces warm, dry areas, providing the perfect conditions for the formation of low-latitude deserts in both the Northern and Southern Hemispheres (Fig. 6.17).

Many low-latitude deserts have average summer temperatures ranging from 32° to 38°C (90° to 100°F) for several months. The highest temperature ever recorded was 58°C

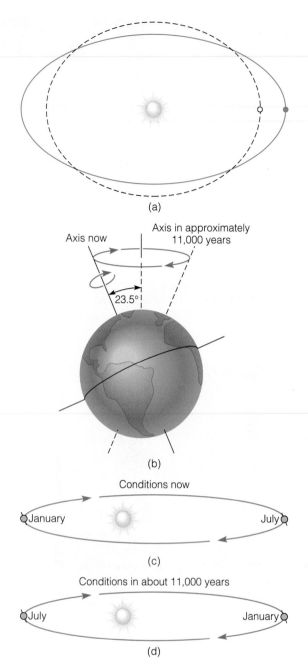

Death Valley

Death Valley in southeastern California and part of western Nevada is the hottest, driest, and lowest place in North America; at its lowest point, it is 86 m (282 ft) below sea level. Death Valley receives less than 5 cm (2 in) of rain per year and has normal daytime temperatures above 42°C (107°F)—a record high temperature reached 57°C (134°F) in the shade! Yet during the Pleistocene, when the climate of the region was more humid, numerous pluvial lakes spread across the valley. Lake Manly, the largest of these lakes, dried up about 10,000 years ago, when the climate became arid.

■ FIGURE 6.16 The Milankovitch theory. (*a*) Earth's orbit varies from nearly a circle (dashed line) to an ellipse (solid line) and back again in about 100,000 years. (*b*) The angle between Earth's axis and a line perpendicular to the plane of the ecliptic varies from 1.5° to its current 23.5° during a 41,000-year cycle. (*c*) At present, Earth is closest to the Sun in January when the Northern Hemisphere experiences winter. (*d*) In about 11,000 years, as a result of precession, Earth will be closer to the Sun in July, during summer in the Northern Hemisphere.

(136°F) in Libya in 1922. The amount of rainfall each year is unpredictable. It is not uncommon for a desert area to receive more than an entire year's average in one cloudburst and then receive little or no rain for several years.

Middle- and high-latitude deserts are found mostly within continental interiors in the Northern Hemisphere (Fig. 6.17). Many of these areas are dry because they are far from sources of moist marine air or because the presence of mountains produces a **rainshadow desert** (■ Fig. 6.18). When moist air moving inland meets a mountain range, it is forced upward, and as it rises, it cools, forms clouds, and releases precipitation on the windward side of the mountains. The air descending on the lee side of the mountains is much warmer and drier, producing a rainshadow desert.

PROCESSES IN DESERTS

Despite the great contrast between deserts and more humid areas, the same processes are at work, only operating under different climatic conditions. And contrary to popular belief, deserts are not simply sand-covered wastelands, but rather consist of vast areas of rock exposures or gravel-covered desert floor. Sand-covered areas constitute less than 25% of all deserts.

Although wind is more important in deserts than in humid regions, even in deserts it is subordinate to running water as a geologic agent. Most of a desert's annual precipitation comes in brief, heavy, localized cloudbursts, during which the infiltration capacity of surface materials is quickly exceeded and runoff begins. Dry stream channels quickly fill with raging torrents of muddy water and carve out steep-sided gullies. During these brief but high-energy floods, a considerable amount of sediment is eroded, transported, and deposited downstream.

Most desert streams flow only intermittently, and many of them never reach the sea because they lose water by evaporation and infiltration. Furthermore, they are not replenished by groundwater because the water table is generally far below the level of the streams. Accordingly, most desert streams flow into low areas where they deposit their sediment load, but they have insufficient water to exit from these areas.

Both mechanical and chemical weathering take place in deserts, but the importance of the latter is considerably reduced because of the aridity and scarcity of organic acids produced by the sparse vegetation. Desert soils, if developed,

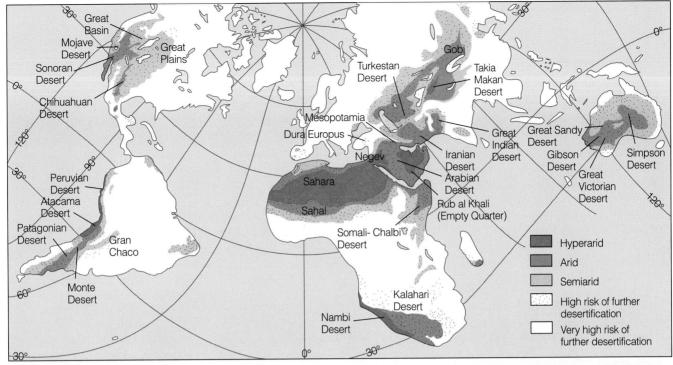

FIGURE 6.17 The distribution of Earth's arid (desert) and semiarid regions.

Map legend:
- Hyperarid
- Arid
- Semiarid
- High risk of further desertification
- Very high risk of further desertification

are mostly thin and patchy because soil-forming processes are restricted by the aridity. Furthermore, the sparseness of vegetation enhances wind and water erosion of the little soil that does form.

DESERTS AND WIND

Wind typically flows at a greater velocity than running water but has a much lower density, so its sediment-carrying capacity is much less than that of water. Clay- and silt-sized particles are transported as suspended load, creating clouds of dust or even dust storms. Once these fine particles are lifted into the atmosphere, they may be carried thousands of kilometers from their source. Indeed, dust that originated in the Sahara of Africa has been collected on the Caribbean island of Barbados.

Sand and gravel are too large to be carried by wind, so they are moved as *bed load* either by surface creep, which involves rolling and sliding, or by a process of intermittent bouncing. Wind erodes by abrasion and deflation. *Abrasion* occurs from the impact of sand grains on an object and is analogous to sandblasting. Because sand is rarely carried more than 1 m (3.3 ft) above the surface, wind abrasion merely modifies existing features by etching, pitting, smoothing, and polishing (Fig. 6.19).

FIGURE 6.18 Many deserts in the middle and high latitudes are rainshadow deserts, so named because they form on the lee side of mountain ranges. When moist marine air moving inland meets a mountain range, it rises, cools, and forms clouds that produce rain or snow that falls on the windward side of the mountains. The air descending on the lee side of the mountains is much warmer and drier, producing a rainshadow desert.

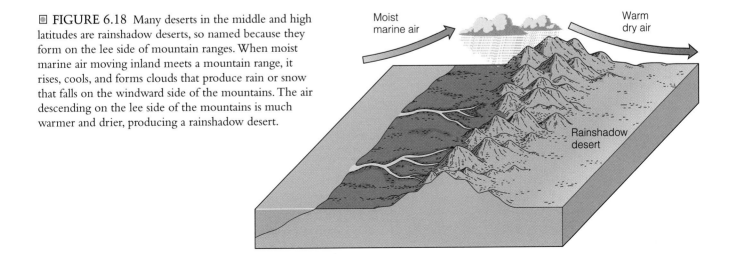

Focus on the Environment: Desertification

During the last few decades, deserts have been expanding across formerly productive land at a rate of about 70,000 km² (27,000 mi²) per year. Because of this phenomenon, known as *desertification*, hundreds of thousands of people have died of starvation or have been forced to migrate as "environmental refugees" from their homelands to camps where most are severely malnourished.

Most desertification takes place along the fringes of existing deserts (☐ Fig. 1). These fringes have a delicately balanced ecosystem that serves as a buffer between the desert on the one side and a semiarid environment on the other. Their potential to adjust to increasing environmental pressures from natural causes and human activity is limited.

Natural processes such as climatic change result in gradual expansion or contraction of deserts, but recent desertification has been greatly accelerated by human activities. In many areas, the natural vegetation has been cleared as crop cultivation has expanded into desert fringe areas to support growing populations. Because these areas are especially prone to droughts, crop failures are common, leaving the land bare and susceptible to increased wind and water erosion.

In the past, grass-covered fringe areas used for grazing achieved a natural balance between vegetation and livestock as nomadic herders grazed their animals and then moved on to another area. More recently, however, livestock numbers have been increasing, and they now far exceed the land's capacity to support them. As a result, the vegetation protecting the soil has diminished, causing the soil to dry and crumble and leading to accelerated soil erosion.

Drilling wells also contributes to desertification because human and livestock activity around a well destroys the vegetation, forming a bare area that merges with the expanding desert. Using well water for irrigation also contributes to desertification because the evaporating water deposits a small amount of salt in the soil that is not flushed out as it would be in areas with more rain. Eventually, the salt concentration becomes high enough to inhibit or prevent plant growth.

Collecting firewood contributes to desertification, particularly in countries where wood remains an important fuel. In the Sahel of Africa, a belt 300 to 1,100 km (185 to 680 mi) wide south of the Sahara, the expanding population has removed all trees and shrubs in the areas surrounding many towns. Journeys of several days on foot to collect firewood are common there.

With the emergence of new nations and increased foreign aid to the Sahel during the 1950s and 1960s, the movements of nomads and their herds were restricted, and large areas of grazing land were converted to cash crops such as peanuts and cotton. Expanding human and animal populations and more intensive agriculture put increasing demands on the land until the worst drought of the century brought untold misery to the people of the Sahel. Nearly 250,000 people and 3.5 million cattle died of starvation between 1968 and 1973. The crops failed and livestock stripped the land of what little vegetation remained. As a result, the adjacent Sahara expanded southward as much as 150 km (93 mi).

Deflation is the removal of loose surface sediment by wind. Characteristic features of deflation in many dry areas include *deflation hollows,* which are circular to oval depressions scoured out by wind erosion (☐ Fig. 6.20d). As deflation removes sand-sized and smaller particles, the remaining surface material is mostly various-sized gravel particles, which form a mosaic of close-fitting rocks known as *desert pavement* (Fig. 6.20a, b, and c). Once a desert pavement forms, it protects the underlying material from further deflation.

☐ FIGURE 6.19 A wind-abraded stone known as a ventifact along the Michigan shore of Lake Michigan.

□ **FIGURE 1** A sharp line marks the boundary between pasture and an encroaching dune in Niger, Africa. As the dune continues to advance, more land will be lost to desertification.

 ## WIND DEPOSITS

Mounds or ridges of wind-deposited sand known as **dunes** form when wind flows over and around an obstruction, resulting in two wind shadows (zones of quiet air) immediately in front of and behind the obstruction (□ Fig. 6.21a). As sand grains accumulate in these wind shadows, a sand dune grows and forms an ever-larger wind barrier. Most dunes have an asymmetrical profile, with a gentle windward slope and a steeper downwind or lee slope (Fig. 6.21b). A dune slowly moves in the prevailing wind direction as sand eroded from its windward side periodically slides down its lee slope (Fig. 6.21b). Even though dunes are common in deserts, they can also form wherever abundant sand is avail-

able, such as along the upper parts of many beaches in humid regions.

Barchan dunes are crescent-shaped with the tips of the crescent pointing downwind (□ Fig. 6.22a on page 128). They form on flat, dry surfaces with little vegetation, a limited sand supply, and a nearly constant wind direction. The largest barchan dunes are about 30 m (100 ft) high. Barchans are the most mobile dunes, moving at rates exceeding 10 m (33 ft) per year.

Longitudinal dunes (also known as *seif dunes*) are long, parallel ridges of sand aligned generally parallel to the direction of the prevailing winds (Fig. 6.22b). They form where sand is somewhat limited and winds converge from slightly different directions to produce a prevailing wind. Longitudinal

■ FIGURE 6.20 Deflation and the origin of desert pavement.
(a) Fine-grained material is removed by wind, (b) leaving a concentration of larger particles that form desert pavement.
(c) Desert pavement in the Mojave Desert, California. Several ventifacts can be seen in the lower left of the photo. (d) A deflation hollow in Death Valley, California.

Deflation

Wind

(a)

Desert pavement
(deflation ends)

Wind

(b)

(c)

(d)

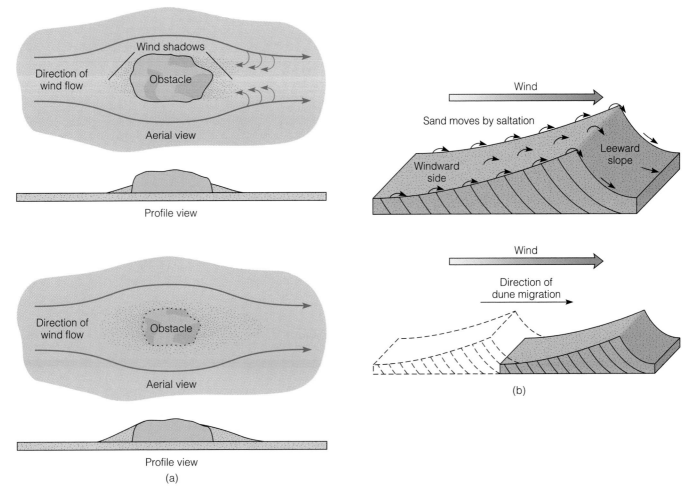

 FIGURE 6.21 (*a*) A dune forms as sand accumulates in wind shadows on the sides and downwind of an obstacle. (*b*) Profile view of a dune showing how it migrates downwind.

dunes range from 3 m to more than 100 m (10 to 328 ft) high, and some extend for more than 100 km (62 mi).

Transverse dunes form long ridges perpendicular to the prevailing wind direction in areas with abundant sand and little or no vegetation (Fig. 6.22c). When viewed from above, they have a wavelike appearance, so areas covered by these dunes are sometimes called sand seas. The crests of transverse dunes may be 200 m (656 ft) high, and some of them are up to 3 km (2 mi) wide. Along the margins of some fields of transverse dunes, where less sand is present, some of the dunes develop a barchanlike shape and are referred to as *barchanoid dunes.*

Parabolic dunes are most common in coastal areas with abundant sand, strong onshore winds, and a partial cover of vegetation (Fig. 6.22d). They have a crescent shape similar to barchan dunes, but their tips point upwind. Parabolic dunes tend to form on the downwind margin of deflation hollows.

In addition to sand, wind transports and deposits silt and clay, which settle to form deposits known as **loess.** Because such small particles can be carried great distances by wind, loess is generally deposited far from its desert source area. Other sources of loess include Pleistocene outwash deposits

and floodplains of rivers in semiarid regions. Loess-derived soils are some of the most fertile. It is, therefore, not surprising that major grain-producing regions, such as the Northern European Plain and the Great Plains of the United States and southern Canada, correspond to the distribution of extensive loess deposits.

DESERT LANDFORMS

After an infrequent and particularly intense desert rainstorm, excess water not absorbed into the ground accumulates in low areas as **playa lakes.** These lakes are temporary, lasting for a few hours to several months. Most of them are shallow, contain very salty water, and have shorelines that fluctuate widely as water flows in or leaves by evaporation and seepage into the ground. When a playa lake evaporates, the dry lake bed is a *playa* or *salt pan,* characterized by mud layers and various salt deposits (☐ Fig. 6.23).

Other common desert features include fan-shaped deposits known as **alluvial fans,** which were described in Chapter 5 (see Fig. 5.11). Several alluvial fans along a moun-

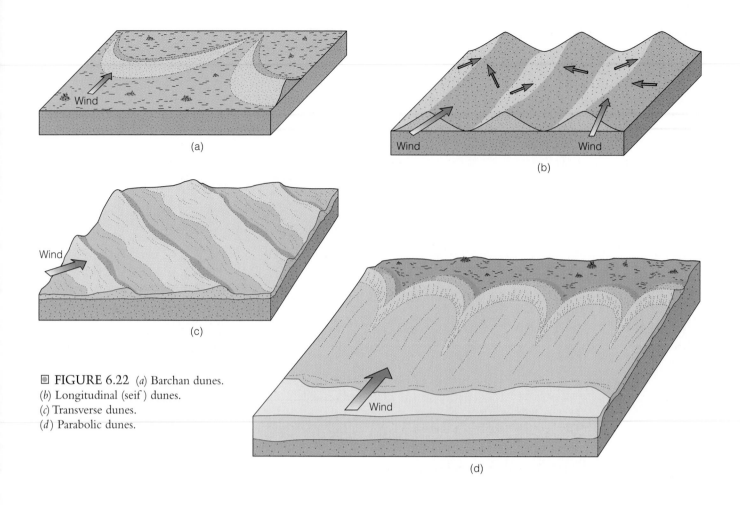

FIGURE 6.22 (*a*) Barchan dunes.
(*b*) Longitudinal (seif) dunes.
(*c*) Transverse dunes.
(*d*) Parabolic dunes.

FIGURE 6.23 A small playa in western Utah.

tain front may overlap, forming a broad apron known as a *bajada* (□ Fig. 6.24). Some large alluvial fans and bajadas are important sources of groundwater.

Most mountains in deserts rise abruptly from surfaces of low relief, known as **pediments,** that slope gently away from the mountain base (□ Fig. 6.25). Although the origin of pediments remains controversial, it appears that the combined activities of erosion by streams, sheet flooding, and various weathering processes along a mountain front are responsible for them. Rising conspicuously above many pediments are isolated steep-sided **inselbergs** (Fig. 6.25), which

are simply remnants of eroded mountains that were more resistant to weathering and erosion than adjacent rocks.

Other easily recognized erosional remnants common in arid and semiarid regions are mesas and buttes (□ Fig. 6.26). A *mesa* is a broad, flat-topped erosional remnant bounded on all sides by steep slopes, whereas *buttes* are isolated pillars. A mesa forms when a resistant rock layer is breached by erosion, allowing rapid erosion of the less resistant underlying rocks. Further weathering and erosion of a mesa reduce it until it is a butte (Fig. 6.26).

□ FIGURE 6.24 Coalescing alluvial fans forming a bajada at the base of the Black Mountains, Death Valley, California.

□ FIGURE 6.25 The gently sloping surface along the front of these mountains is a pediment.

■ **FIGURE 6.26** Mesa and butte near Moab, Utah.

CHAPTER SUMMARY

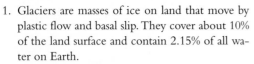

1. Glaciers are masses of ice on land that move by plastic flow and basal slip. They cover about 10% of the land surface and contain 2.15% of all water on Earth.

2. Valley glaciers are confined to mountain valleys and move from higher to lower elevations, whereas continental glaciers cover vast areas and flow outward in all directions from a zone of accumulation.

3. Glaciers form anywhere that winter snowfall exceeds summer melt and thus accumulates year after year. When an accumulating body of snow and ice is about 40 m (130 ft) thick, pressure causes it to flow.

4. The behavior of a glacier depends on its budget, which is the relationship between accumulation and wastage. A glacier's terminus advances, remains stationary, or retreats depending on whether it has a positive, balanced, or negative budget.

5. Erosion by valley glaciers yields several sharp, angular landforms including cirques, arêtes, and horns. Valley glaciers also erode U-shaped glacial troughs, fiords, and hanging valleys.

6. Continental glaciers abrade and bevel high areas, producing a smooth, rounded ice-scoured plain.

7. All sediment deposited as a result of glacial activity is referred to as glacial drift. Some drift, called till, is deposited by glacial ice, whereas outwash is deposited by meltwater streams discharging from glaciers. Landforms of till include moraines and drumlins, whereas kames and eskers are composed of outwash.

8. Several intervals of widespread glaciation separated by interglacial periods occurred in North America during the Pleistocene when glaciers covered about 30% of Earth's land surface.

9. Glacial intervals separated by tens or hundreds of millions of years are probably caused by changing positions of plates. The Milankovitch theory is widely accepted as an explanation for glacial-interglacial intervals.

10. Deserts receive less than 25 cm (10 in) of rain per year, have poorly developed soils, and have sparse vegetation. Most deserts are between 20° and 30° north and south latitude.

11. The dryness of deserts results from a belt of dry air descending at about 30° north or south latitude, remoteness from a source of precipitation, or a rainshadow effect.

12. Running water accounts for most erosion, transport, and deposition in deserts, but wind has a greater effect than it does in humid regions.

13. Wind erosion by deflation lowers a desert surface by selectively removing fine material, leaving a desert pavement.

14. Dunes are mounds or ridges of wind-deposited sand that migrate in the direction of the wind. The amount of sand available, the prevailing wind direction, wind velocity, and vegetation determine which of the four major types of dunes will form.

15. Loess, wind-deposited silt and clay, is derived from deserts, glacial outwash, and floodplains in semiarid regions. It covers about 10% of the land surface and weathers to form fertile soils.

16. Important desert landforms include playas, which result from evaporation of playa lakes, and alluvial fans, which may overlap with one another to form a bajada. Pediments, inselbergs, mesas, and buttes are erosional landforms typical of desert regions.

IMPORTANT TERMS

alluvial fan	dune	inselberg	plastic flow
arête	end moraine	kame	playa lake
basal slip	esker	kettle	rainshadow desert
cirque	glacial budget	lateral moraine	recessional moraine
continental glacier	glacial drift	loess	U-shaped glacial trough
deflation	glacier	Milankovitch theory	valley glacier
desert	hanging valley	pediment	
drumlin	horn		

REVIEW QUESTIONS

1. The onset of glacial-interglacial cycles is explained by the:
a. ____ plate tectonic theory; b. ____ magnetic field hypothesis; c. ____ Milankovitch theory; d. ____ rainshadow desert theory; e. ____ valley glaciation hypothesis.

2. If a glacier has a negative budget:
a. ____ its terminus will retreat; b. ____ its accumulation rate is greater than its wastage rate; c. ____ all flow ceases; d. ____ the glacier's volume increases; e. ____ crevasses will no longer form.

3. Which of the following is not an erosional landform?
a. ____ horn; b. ____ arête; c. ____ lateral moraine; d. ____ cirque; e. ____ U-shaped glacial trough.

4. The most recent ice age occurred during the:
a. ____ Archean Eon; b. ____ Pleistocene Epoch; c. ____ Mesozoic Era; d. ____ Cambrian Period; e. ____ Jurassic Period.

5. Pressure on ice at depth in a glacier causes it to move by:
a. ____ slump; b. ____ fracture; c. ____ basal slip; d. ____ mechanical weathering; e. ____ plastic flow.

6. Glacial drift is a general term for:
a. ____ the erosional landforms of continental glaciers; b. ____ all the deposits of glaciers; c. ____ icebergs floating at sea; d. ____ the movement of glaciers by basal slip; e. ____ the annual wastage rate of a glacier.

7. A knifelike ridge separating glaciers in adjacent U-shaped glacial troughs is a(n):
a. ____ fiord; b. ____ horn; c. ____ arête; d. ____ cirque; e. ____ recessional moraine.

8. Between what latitudes in both hemispheres are most hot, arid deserts found?
a. ____ 10° and 20°; b. ____ 20° and 30°; c. ____ 30° and 40°; d. ____ 40° and 60°; e. ____ 60° and 80°.

9. One way that wind transports bed load is:
a. ____ suspension; b. ____ infiltration; c. ____ surface creep; d. ____ precipitation; e. ____ abrasion.

10. Which particle size constitutes most of a wind's suspended load?
a. ____ sand; b. ____ silt; c. ____ clay; d. ____ answers (a) and (b); e. ____ answers (b) and (c).

11. Which of the following is a feature produced by wind erosion?
a. ____ playa; b. ____ loess; c. ____ dune; d. ____ ventifact; e. ____ esker.

12. A cresent-shaped dune with its tips pointing downwind is a _____ dune.
    a. ____ parabolic; b. ____ transverse; c. ____ mesa; d. ____ barchan; e. ____ longitudinal.

13. A dry lake bed in a desert is a(n):
    a. ____ bajada; b. ____ playa; c. ____ inselberg; d. ____ pediment; e. ____ mesa.

14. How does glacial ice form, and why is ice considered to be a mineral?

15. Explain how glaciers erode by abrasion and plucking.

16. Explain in terms of a glacial budget how a once active glacier becomes a stagnant glacier.

17. Discuss the process whereby terminal, recessional, and lateral moraines form.

18. Describe the processes responsible for the origin of a cirque, U-shaped glacial trough, and hanging valley.

19. How does a medial moraine form, and how can one determine the number of tributaries a valley glacier has by its medial moraines?

20. Explain how wind transports sediment.

21. Describe the two ways that wind erodes. How effective an erosional agent is wind?

22. How do sand dunes form and migrate?

23. What is the dominant type of weathering in deserts, and why is it so effective?

24. Explain the differences between a butte and a mesa, and describe how each forms.

POINTS TO PONDER

1. Much of northern Canada and northern Asia have climates in which continental glaciers could exist, yet none are present. Why?

2. What are the consequences of prolonged drought? What human activities exacerbate the effects of drought?

3. In North America, valley glaciers are most common in Alaska and western Canada, followed by Washington, Oregon, and California. Furthermore, no valley glaciers exist in North America east of the Rocky Mountains. How can you explain this distribution of glaciers?

4. Compare the effects of wind erosion and deposition in humid and arid regions.

ADDITIONAL READINGS

Agnew, C., and A. Warren. 1990. Sand trap. *The Sciences* March/April: 14–19.

Broecker, W. S., and G. H. Denton. 1990. What drives glacial cycles? *Scientific American* 262, no. 1: 49–56.

Brookfield, M. E., and T. S. Ahlbrandt. 1983. *Eolian sediments and processes.* New York: Elsevier.

Cover, C. 1984. The Earth's orbit and the ice ages. *Scientific American* 250, no. 2: 58–66.

Drewry, D. J. 1986. *Glacial geologic processes.* London: Edward Arnold.

Ellis, W. S. 1987. Africa's Sahel: The stricken land. *National Geographic* 172, no. 2: 140–79.

Greeley, R., and J. Iversen. 1985. *Wind as a geologic process.* Cambridge, Mass.: Cambridge University Press.

Grove, J. M. 1988. *The Little Ice Age.* London: Methuen.

Kurten, B. 1988. *Before the Indians.* New York: Columbia University Press.

Sharp, R. P. 1988. *Living ice: Understanding glaciers and glaciation.* New York: Cambridge University Press.

Thomas, D. S. G., ed. 1989. *Arid zone geomorphology.* New York: Halsted Press.

Wells, S. G., and D. R. Haragan. 1983. *Origin and evolution of deserts.* Albuquerque: University of New Mexico Press.

Williams, R. S., Jr. 1983. *Glaciers: Clues to future climate?* U.S. Geological Survey.

WORLD WIDE WEB SITES

For current updates and exercises, log on to
http://www.wadsworth.com/geo.

GLACIER RESEARCH GROUP
http://www.grg.sr.unh.edu/

This site is maintained by the Climate Change Research Center,
University of New Hampshire, Institute for the Study of Earth,
Oceans, and Space. It contains an overview of what it is and what
it does, as well as a listing of its current projects and links to other
glacially related sites.

U.S. GEOLOGICAL SURVEY ICE AND CLIMATE PROJECT
http://orcapaktcm.wr.usgs.gov/

This site contains links to the various Ice and Climate Projects being undertaken by the USGS.

THE NATIONAL PARK SERVICE DEATH VALLEY NATIONAL PARK
http://www.nps.gov/deva/

This site contains general information about Death Valley National
Park.

DEATH VALLEY NATIONAL PARK
http://www.desertusa.com/dv/du_dvpdesc.htm/

This site, operated by DesertUSA: THE ULTIMATE DESERT
RESOURCE, is a compendium of information about Death
Valley National Park.

DesertUSA
http://www.desertusa.com/flow.html

This site is devoted to information about deserts. It contains a
tremendous amount of information about the fauna, flora, geology,
physical environment, and other items of interest about deserts.

CHAPTER

7

Plate Tectonics

The space shuttle Discovery backdropped against an oblique view of the Kamchatka Peninsula.

Prologue

On August 2, 1990, Iraq invaded the oil-rich nation of Kuwait. In response, a multinational military force, including nearly half a million American troops, deployed to Saudi Arabia to protect that nation and its oil fields from attack. In February 1991, during Operation Desert Storm, allied forces drove the Iraqis from Kuwait, but not before they had released huge oil spills into the Persian Gulf and set fire to 700 Kuwaiti oil wells (☐ Fig. 7.1). These were the worst acts of environmental terrorism ever committed.

Politics and oil are inextricably intertwined in the Persian Gulf. The Iran-Iraq War of 1980–1989 was fought in part for control of the oil-rich Gulf coast of Iran. The Strait of Hormuz at the mouth of the Gulf, less than 50 km (31 mi) wide in places, is a vulnerable choke point through which half the world's oil passes.

Why is there so much oil in the Persian Gulf region? The answer lies in the slow movement of Earth's lithospheric plates (☐ Fig. 7.2). The Arabian plate is moving northeast away from Africa, opening the Red Sea and

Gulf of Aden and pushing under Iran. Before the Arabian plate converged on Iran, it probably had a continental shelf with the usual source rocks for petroleum. When the collision began, the continental shelf rocks were heated as they were pushed deeper into the Earth under Iran. The heating broke down organic molecules and hastened the transition of these molecules to oil. The Arabian block was tilted northeast, lowering the leading edge of the plate below sea level to form the Persian Gulf (Fig. 7.2).

The tilting also allowed the newly formed oil to migrate up the slope into the interior of the Arabian plate. Finally, the collision folded the rocks, creating traps to catch and retain the oil (see Perspective 10.2). Oil floats upward on water in the pore spaces in rocks and collects at the tops of up-arched folds known as *anticlines* if it is trapped by impervious rock layers above. The nations of Bahrain and Qatar are at the tops of gentle anticlines that protrude above the surface of the Persian Gulf, and the oil in Saudi Arabia and Kuwait is in similar folds.

The good news is that plate tectonics can explain why the Gulf region is so rich in oil. The bad news is

☐ FIGURE 7.1 The Kuwaiti night skies were illuminated by 700 blazing oil wells set afire by Iraqi troops during the 1991 Gulf War. The fires continued for 9 months.

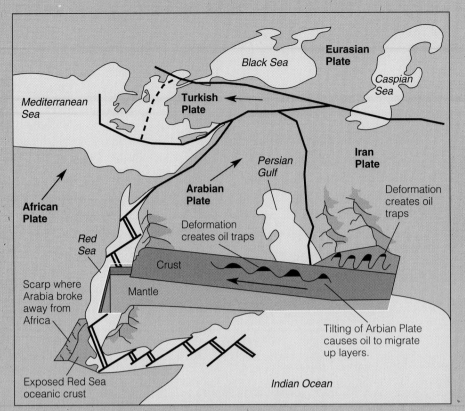

□ **FIGURE 7.2** Oil geologists divide oil-producing regions into two categories: the Middle East and everything else. In this oblique view, an imaginary trench has been cut across Arabia and the Persian Gulf. The collision between Arabia and Iran has tilted the Arabian Plate and crumpled rocks on the edges of both plates. The tilting of Arabia allows oil to migrate upslope to accumulate in traps created by folding. Along almost the entire length of the plate boundary, conditions are ideal for maximum accumulation of oil. Elsewhere, the convergence of Arabia and the Eurasian Plate is squeezing Turkey westward, much like a watermelon seed pinched between your fingers. Also, the opposing coastlines of the Red Sea fit almost perfectly, except at the southern end. There, a portion of Red Sea ocean floor is exposed, one of the few places oceanic crust is exposed on dry land. The true edge of the African crust is marked by a steep scarp, which matches the southwestern corner of the Arabian Plate almost perfectly.

that it seems to be unique. The Arabian plate is colliding broadside with Iran, and the entire collision zone, from northern Iraq to the Strait of Hormuz, is oil producing. Also, the collision between Arabia and Iran appears to be at just the right stage of development: not so early that the petroleum is still dispersed in its source rocks nor so late that the oil has been destroyed by metamorphism or lost due to erosion. Other plate collisions are in progress, but none match the ideal conditions of the Persian Gulf.

 ## INTRODUCTION

Since about 1960, the Earth sciences have experienced sweeping changes because of the confirmation of plate tectonic theory. For the first time, scientists have the conceptual tools to treat Earth as a system, to understand global processes and explain relationships among seemingly unrelated geologic phenomena, and to make rigorously testable predictions. Plate tectonics is one of the most important mechanisms operating in Earth and affects every other system (○ Table 7.1). For instance, interactions among moving plates determine the location of continents, ocean basins, and mountain systems, which in turn affects the atmospheric and oceanic circulation patterns that ultimately determine global climates. Plate movements have also profoundly influenced the geographic distribution, evolution, and extinction of plants and animals.

Although plate tectonic processes are slow, they nevertheless have profound effects on our lives. Geologists now realize that most earthquakes and volcanic eruptions take place at or near plate boundaries and are not merely random occurrences. Furthermore, the formation and distribution of many geologic resources, such as metal ores, are related to plate tectonic processes, and scientists are now incorporating plate tectonic theory into their prospecting efforts. Plate

TABLE 7.1 Plate Tectonics and Earth Systems

Solid Earth	Plate tectonics is driven by convection in the mantle and in turn drives mountain-building processes and associated igneous and metamorphic activity.
Atmosphere	Arrangement of continents affects solar heating and cooling and thus winds and weather systems. Rapid plate spreading and hot-spot activity may release volcanic carbon dioxide and affect global climate.
Hydrosphere	Continental arrangement affects ocean currents. Rate of spreading affects volume of mid-ocean ridges and hence sea level. Placement of continents may contribute to onset of ice ages.
Biosphere	Movement of continents creates corridors or barriers to migration and thus creates ecological niches. Habitats may be transported into more or less favorable climates.
Extraterrestrial	Arrangement of continents affects free circulation of ocean tides and influences tidal slowing of Earth's rotation.

tectonic theory has led to a greater understanding of how Earth has evolved and continues to do so.

PLATE TECTONIC THEORY

Plate tectonic theory is based on a model of Earth in which rigid **lithosphere,** consisting of oceanic and continental crust as well as about 100 km of the underlying upper mantle, is made up of a number of pieces known as plates. Lithospheric plates move over the hotter and weaker semiplastic **asthenosphere** below. As plates move, they separate, mostly at mid-oceanic ridges, while colliding in other areas. When plates collide, one plate generally dives beneath the other, although in some cases the plates simply slide past one another.

Because plate tectonic theory relies on many kinds of evidence, it is useful to approach the evidence as it was discovered. Not only is the story interesting, but a historical approach makes it easier to see how the different pieces of evidence fit together. Next we will take a brief overall look at the present concept of plate tectonics and examine in detail the processes that operate as plates move. Finally, we will apply plate tectonics to understanding the geologic evolution of a few selected areas.

Continental Fit

During the nineteenth century and possibly earlier, a number of people noticed the apparent fits of now-distant coastlines and suggested that these widely separated lands had once been together (■ Fig. 7.3). However, Alfred

Wegener (1880–1930), a meteorologist interested in determining the distribution of ancient climates, was the first to develop a coherent scientific theory in which all these continents had at one time been united into a single supercontinent.

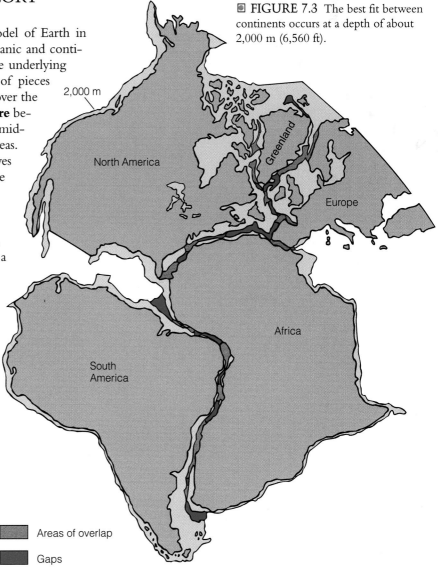

■ FIGURE 7.3 The best fit between continents occurs at a depth of about 2,000 m (6,560 ft).

2,000 m

North America

Greenland

Europe

Africa

South America

Areas of overlap

Gaps

Wegener, like those before him, was impressed by the close resemblance between the coastlines of continents on opposites sides of the Atlantic, particularly between South America and Africa. He cited this as partial evidence for his theory, but as his critics pointed out, coastlines are continually modified by erosion and deposition. Therefore, even if the continents had separated as Wegener proposed, it would not be likely that their coastlines would still fit together exactly.

Rather than simply using the shoreline for fitting continents together, it is more realistic to fit them along their true geologic margins, that is, where continental crust changes to oceanic crust, which is below sea level. In 1965, Sir Edward Bullard, an English geophysicist, and two associates showed that the best fit occurs along the continental margin at a depth of about 2,000 m (6,560 ft). When reconstructed in this way, the continents fit with very few gaps and overlaps (Fig. 7.3).

Glacial Evidence

Wegener's interest was drawn to glacial deposits in the Southern Hemisphere dating from the Permian Period about 220 to 250 million years ago. Evidence for these glaciers includes layers of till (sediments deposited by glaciers) and striations (scratch marks) in bedrock beneath the till. Fossils and sedimentary rocks of the same age from the Northern Hemisphere, however, give no indication of glaciation. In fact, it appears that the Northern Hemisphere had a tropical climate during the time that the Southern Hemisphere was glaciated.

All of the Southern Hemisphere continents except Antarctica are currently near the equator in subtropical or tropical climates. Furthermore, glacial striations in bedrock indicate that the glaciers moved from the areas of present-day oceans onto land. This would be highly unlikely because large glaciers on land flow outward from their central areas of accumulation toward the ocean.

If the continents did not move during the past, one would have to explain how glaciers moved from the oceans onto land and how large-scale glaciers existed near the equator. But if the continents are reassembled as a single landmass with South Africa located at the South Pole, one large area of glaciation emerges, and the direction of movement of this glacier makes sense (▣ Fig. 7.4). Furthermore, this geographic arrangement places the northern continents nearer the tropics, which is consistent with the fossil and sedimentary rock evidence.

Fossil Evidence

Before the Permian ice age, the southern continents had been home to a remarkably uniform assemblage of land plants and animals. The **Glossopteris flora,** for instance, is characterized by a seed fern of that name as well as by many other distinctive land plants. The present climates of the southern continents range from polar to tropical and are much too diverse to support the types of plants comprising

▣ FIGURE 7.4 (*a*) If the continents are brought together so that South Africa is located at the South Pole, then the glacial movements indicated by the striations make sense. In this situation, the glacier, located in a polar climate, moved radially outward from a thick central area toward its periphery. (*b*) Permian-aged glacial striations in bedrock exposed at Hallet's Cove, Australia, indicate the direction of glacial movement more than 200 million years ago.

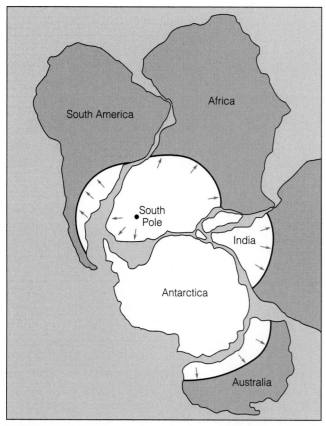

Glaciated area

Arrows indicate the direction of glacial movement based on striations preserved in bedrock.

(a)

(b)

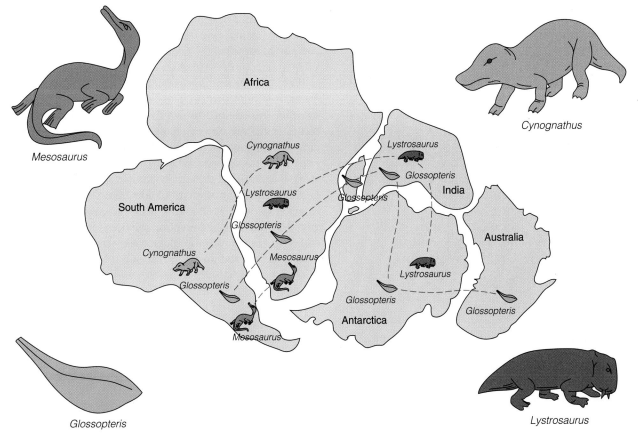

■ **FIGURE 7.5** Some of the animals and plants whose fossils are found today on the widely separated continents of South America, Africa, India, Australia, and Antarctica. These continents were joined together during the Late Paleozoic to form Gondwana, the southern landmass of Pangaea. *Glossopteris* and similar plants are found in Pennsylvanian- and Permian-aged deposits on all five continents. *Mesosaurus* is a freshwater reptile whose fossils are found in Permian-aged rocks in Brazil and South Africa. *Cynognathus* and *Lystrosaurus* are land reptiles that lived during the Early Triassic Period. Fossils of *Cynognathus* are found in South America and Africa, while fossils of *Lystrosaurus* have been recovered from Africa, India, and Antarctica.

the *Glossopteris* flora. Furthermore, the seeds of these plants are too heavy to be carried great distances by wind, and it seems unlikely that they floated across an ocean and remained viable. Wegener reasoned that these continents must have been joined so that the plants of these widely separated localities were all in the same latitudinal climatic belt and occupied a continuous geographic area (■ Fig. 7.5).

Remains of animals also provide strong evidence for a connection between the southern continents. Fossils of *Mesosaurus,* a freshwater reptile, are found in Permian-aged rocks in small areas of Brazil and South Africa, areas that match up when the continents are reassembled as in Figure 7.5. It seems unlikely that a small freshwater reptile could have swum across an ocean if the continents were in their present-day positions. *Lystrosaurus* and *Cynognathus* are both Triassic-aged land-dwelling reptiles whose remains also indicate a former connection between Southern Hemisphere continents (Fig. 7.5).

Similarities of Rocks and Mountain Ranges

If the continents were joined together at one time, then the rocks and mountain ranges of the same age in adjoining lo-

cations on opposite continents should closely match. Indeed, some mountain ranges seemingly end at the coastline of one continent only to apparently continue on another continent across an ocean (■ Fig. 7.6). The Appalachian Mountains of North America trend northeastward through the eastern United States and Canada and terminate abruptly at the Newfoundland coastline. Mountains of the same age and with the same structure are also found in eastern Greenland, Ireland, Great Britain, and Norway. East-west trending mountain belts in Argentina match closely with the Cape Mountains of South Africa. Even though these mountains are separated by the Atlantic Ocean, they form a continuous range when the continents are positioned next to each other (Fig. 7.6).

◉ ALFRED WEGENER AND CONTINENTAL DRIFT THEORY

Evidence from fossils, past glaciers, and close similarities between rocks and mountains on now widely separated continents convinced Wegener that all of the present continents were at one time amalgamated into a supercontinent he

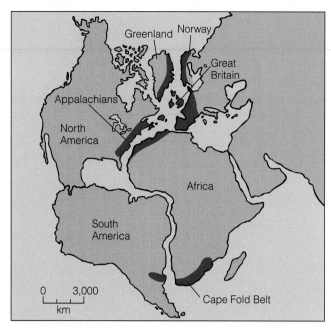

■ FIGURE 7.6 When continents are brought together, their mountain ranges form a single continuous range of the same age and style of deformation throughout. Such evidence indicates the continents were at one time joined together and were subsequently separated.

called **Pangaea,** from the Greek meaning "all land." In 1912 Wegener published his theory of **continental drift,** according to which the supercontinent Pangaea had fragmented during the Triassic Period, and its various parts, constituting today's continents, began drifting apart. The idea was highly controversial, of course, and most scientists rejected it. Wegener had no satisfactory explanation for how continents moved, and in a few cases, his arguments were simply wrong. Also, much of the most critical evidence for the movement of continents was found using technology that was not even dreamed of in 1912.

The responses to Wegener's theory show an interesting pattern. The most favorable response came from Southern Hemisphere geologists, who were most familiar with Wegener's key evidence. The harshest reaction came from Americans, who were more concerned with finding resources for American industry than with theoretical problems. Also, Americans tended to be preoccupied with the vast task of mapping the geology of North America, and few were familiar with the geology of the Southern Hemisphere.

The *International Geophysical Year* (1957–1958) was a key event in winning acceptance for continental drift. Oceanographic expeditions found that the Mid-Atlantic Ridge (Chapter 12) was part of a globe-girdling submarine mountain range. Along the crest of the mid-oceanic ridge was a rift valley, strong evidence that the lithosphere was pulling apart. Furthermore, it is now known that the age of the oceanic crust increases with distance from the mid-oceanic ridges, exactly as it should if plates are separating as suggested. In addition, sediment on the ocean floors turned out to be far thinner than geologists expected it to be if the oceans had existed for billions of years. However, sediment thickness on the sea floor increases with distance from mid-oceanic ridges because it has had more time to accumulate than it has at or near ridges (■ Fig. 7.7).

PALEOMAGNETISM AND POLAR WANDERING

Some of the technological innovations developed during World War II (1941–1945) made the confirmation of plate tectonics possible. Most significant perhaps was the development of highly sensitive devices for detecting magnetic fields. Although such devices, called *magnetometers,* are useful for mapping rocks, they were originally developed for locating submarines.

The study of Earth's ancient magnetic field is known as **paleomagnetism.** When a magma cools, its iron-bearing minerals align themselves with the magnetic field, thus forming a record of the intensity and directional properties of the field at the time the magma solidified. Research conducted during the 1950s showed that the location of the magnetic pole has varied through time. Paleomagnetic data from lava flows in Europe indicated that the north magnetic pole had wandered from the Pacific Ocean northward through eastern and then northern Asia to its present location near the geographic North Pole (■ Fig. 7.8). These data could be interpreted in three ways: the continents re-

■ FIGURE 7.7 The total thickness of deep-sea sediments increases away from oceanic ridges. This is because oceanic crust becomes older away from oceanic ridges, and there has been more time for sediment to accumulate.

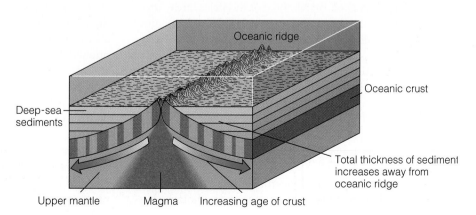

mained fixed and the north magnetic pole moved, the north magnetic pole stood still and the continents moved, or both the continents and the pole moved.

This problem was resolved when it was found that paleomagnetic readings from North America pointed to different magnetic pole locations than did readings of the same age from Europe (Fig. 7.8). Furthermore, analyses of paleomagnetic data from all continents indicated that each continent had a different magnetic pole, which seems highly unlikely. Accordingly, the best explanation for this apparent **polar wandering** is that the continents have moved. When the continents are fitted together so that the paleomagnetic data point to only one magnetic pole, we find, just as Wegener did, that the rock sequences, mountain ranges, and glacial deposits match.

Scientists also found evidence showing that Earth's magnetic field reversed from time to time, meaning that the north and south magnetic poles switched positions. (It is important to understand that only the *magnetic field* of the Earth flips; the Earth's rotation axis remains steady.) Times of normal polarity, as at present, alternated with times of reversed polarity of the magnetic field. The meaning of these data became clear when survey ships made magnetic surveys of a number of the mid-oceanic ridges. The surveys showed a remarkable pattern of stripes parallel to the crests of the ridges, which recorded normal and reversed polarity events. But the key discovery was that the stripes are not only parallel, but symmetrical—every stripe on one side of a ridge is matched by a corresponding stripe on the opposite side (Fig. 7.9). Even more remarkable, the stripe patterns were

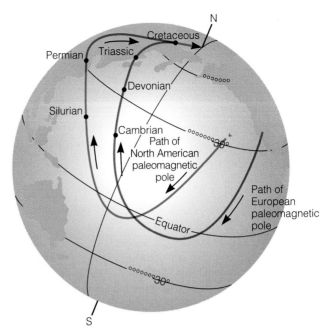

 FIGURE 7.8 The apparent paths of polar wandering for North America and Europe. The apparent location of the north magnetic pole is shown for different periods on each continent's polar wandering path.

similar for ridges around the world; thus, whatever created the pattern was global in its effects. The concept of **sea-floor spreading** accounts for all these observations. Magma rises from Earth's interior, intrudes the mid-oceanic ridges,

 FIGURE 7.9 The sequence of magnetic anomalies preserved within the oceanic crust on both sides of an oceanic ridge is identical to the sequence of magnetic reversals already known from continental lava flows. Magnetic anomalies are formed when magma intrudes into oceanic ridges; when the magma cools, it records the Earth's magnetic polarity at the time. Subsequent sea-floor spreading splits the previously formed crust in half, so that it moves laterally away from the oceanic ridge. Repeated intrusions record a symmetrical series of magnetic anomalies that reflect periods of normal and reversed polarity. The magnetic anomalies are recorded by a magnetometer, which measures the strength of the magnetic field.

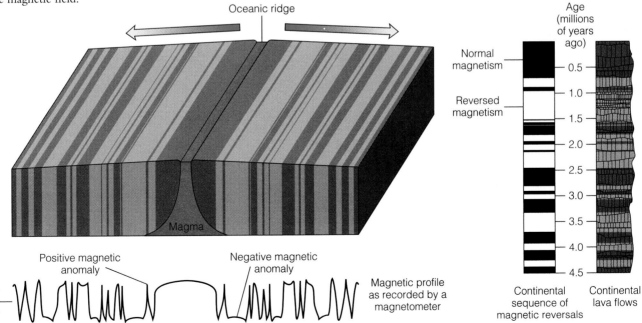

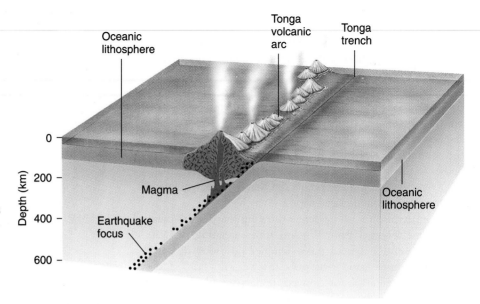

Oceanic
lithosphere

Tonga
volcanic
arc

Tonga
trench

Depth (km)
0 —
200 —
400 —
600 —

Magma

Earthquake
focus

Oceanic
lithosphere

 FIGURE 7.10 Focal depth increases in a well-defined zone that dips approximately 45° beneath the Tonga volcanic arc in the South Pacific. Dipping seismic zones are common features of island arcs and deep-ocean trenches.

cools, and is magnetized by Earth's magnetic field. As the lithosphere spreads apart and new material is added at the ridge, a "tape recording" of normal and reversed polarity events forms (Fig. 7.9).

If new lithosphere is being created at mid-oceanic ridges, where does it go? The answer to this question was already at hand. Two seismologists, Hugo Benioff and Kiroo Wadati, working independently, had plotted the distribution of earthquakes near deep-ocean trenches and showed that the earthquakes all lay along a thin plane sloping from the trench into Earth (Fig. 7.10). We now recognize that these zones mark the areas where oceanic lithosphere descends back into the mantle to be melted and recycled. We call this phenomenon **subduction,** and the place where this occurs is a **subduction zone.** Thus, new oceanic lithosphere is continually created at mid-oceanic ridges, or what are also known as **spreading ridges,** and it is consumed at subduction zones. The fact that oceanic crust is eventually subducted accounts for the fact that none is more than 180 million years old. In contrast, continental lithosphere is too thick and too light to be subducted, so rocks as old as 3.96 billion years are known on continents.

PLATES AND PLATE BOUNDARIES

The *World-Wide Standardized Seismic Network,* developed in the early 1960s, enabled scientists to refine Wegener's ideas into their present form, which we now call *plate tectonic theory.* Maps of earthquake locations made with the new system had unprecedented accuracy (see Fig. 8.4); they showed that earthquakes in the ocean basins were confined to a narrow zone along the crests of mid-oceanic ridges and virtually nowhere else. Earth's surface consists of great areas, or **plates,** that are geologically quiet, but are separated by nar-

row zones of seismic and volcanic activity where the plates separate, collide, or slide past one another (Fig. 7.11).

Several large plates and a number of smaller ones are now recognized (Fig. 7.11). A plate does not simply correspond to a continent or an ocean. Indeed, a plate might consist of continental crust and upper mantle, oceanic crust and upper mantle, or both types of crust and upper mantle. The North American plate includes most of North America plus most of the western North Atlantic, the western two-thirds of the Arctic Ocean, and part of eastern Siberia. All of this is a single unit of lithosphere moving as one plate. The Pacific plate, on the other hand, is made up of only oceanic crust and upper mantle, and it too is a single moving unit of lithosphere.

Rates of plate motions range from about 2 cm/yr at the Mid-Atlantic Ridge to 10 cm/yr at the East Pacific Rise. The Atlantic will widen by about a person's height over an average lifetime. Recently, it has become possible to measure the movements of plates directly using laser ranging to satellites or intercontinental linkages between radio telescopes trained on distant objects in space. Plate velocities measured with these methods agree well with geologic estimates.

Plates move relative to one another such that their boundaries can be characterized as *divergent, convergent,* and *transform* (○ Table 7.2). Interaction of plates at their boundaries accounts for most of Earth's earthquake and volcanic activity, as well as the locations of deep-ocean trenches and the deformation and volcanism associated with mountain building.

Divergent Plate Boundaries: Where New Oceans Form

Studies of the sea floor from submersibles carrying scientists, notably the FAMOUS Project (French-American Oceanographic Undersea Survey), have given us detailed views of what happens at the crests of mid-oceanic ridges. At the very

TABLE 7.2   Types of Plate Boundaries

TYPE	EXAMPLE	LANDFORMS	VOLCANISM
Divergent			
Oceanic	Mid-Atlantic Ridge	Mid-oceanic ridge with axial rift valley	Basalt
Continental	East African Rift Valley	Rift valley	Basalt and rhyolite, no andesite
Convergent			
Oceanic-oceanic	Aleutian Islands	Volcanic island arc, offshore oceanic trench	Basalt
Continental-oceanic	Andes	Offshore oceanic trench, volcanic mountain chain, mountain belt	Andesite
Continental-continental	Himalayas	Mountain belt	Minor
Transform	San Andreas fault	Fault valley	Minor

crest of a ridge is a narrow rift valley or graben. Within the graben, fissures split the ocean floor, and bulbous masses of lava known as pillow lava occur everywhere. In the center of the graben is a line of volcanic vents. These areas mark the sites of **divergent plate boundaries** where plates move apart (■ Fig. 7.12). Remarkably, the processes forming lithospheric plates thousands of kilometers across seem to be almost entirely concentrated in a narrow zone a few hundred meters wide.

The newly formed lithosphere at a mid-oceanic ridge is hot and less dense than oceanic lithosphere elsewhere; therefore the ridge rises high above the adjacent sea floor. As the lithosphere moves away from a spreading ridge, it cools, becomes denser, and sinks. Accordingly, oceanic depths are

■ FIGURE 7.11 The earth is divided into a dozen or so large lithospheric plates and many smaller ones. Note how India is colliding with Asia, breaking the crust of eastern Asia into pieces and pushing them eastward. The Australia-India Plate is breaking in two as a result of stresses caused by the collision. North of Australia, small island chains are colliding with New Guinea. The plate boundary is becoming jammed and earthquakes in the Pacific suggest that the crust of the Pacific is beginning to fracture, perhaps creating a new plate boundary to the north.

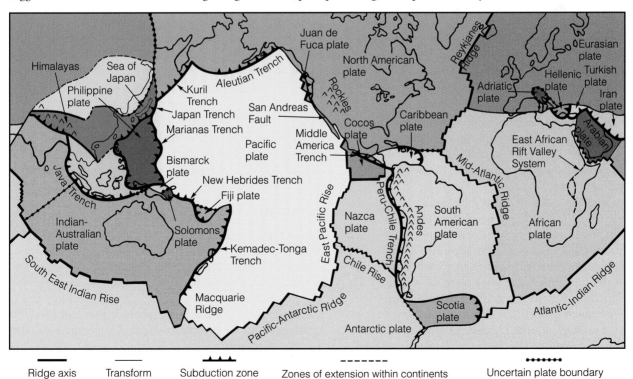

Ridge axis   Transform   Subduction zone   Zones of extension within continents   Uncertain plate boundary

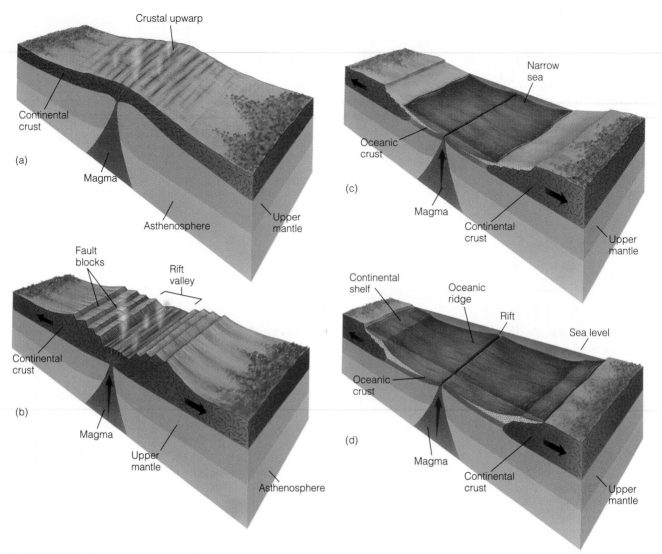

◉ FIGURE 7.12 History of a divergent plate boundary. (*a*) Rising magma beneath a continent pushes the crust up, producing numerous cracks and fractures. (*b*) As the crust is stretched and thinned, rift valleys develop, and lava flows onto the valley floors. (*c*) Continued spreading further separates the continent until a narrow seaway develops. (*d*) As spreading continues, an oceanic ridge system forms, and an ocean basin develops and grows.

greater with increasing distance from ridge crests, and, as noted earlier, the age of the lithosphere also increases with increasing distance from a ridge. After about 80 million years, it has cooled completely and sunk to about 5.5 km (3.4 mi) below the sea.

Plate divergence leaves its marks on the continents as well (◉ Fig. 7.13). When divergence begins beneath a continent, faults develop along the zone of separation, known as a rift, and large amounts of gabbro are intruded into the crust. If the magma comes from the mantle, basalt is erupted, whereas melting of continental crust results in eruptions of rhyolite. The rift system in East Africa is in this earliest stage of plate separation. Finally, the segments of continental crust on opposite sides of the rift separate, new oceanic lithosphere forms between the two halves of the continent, and a new ocean basin in the form of a long narrow sea develops. The Red Sea is a good example of this stage of plate sep-

aration (Fig. 7.13). With continued divergence, a broad ocean basin forms, such as the Atlantic, and the rifted continental margins that are now within plates rather than at their margins show little seismic or volcanic activity.

Convergent Plate Boundaries

A **convergent plate boundary** occurs where two plates are moving toward one another. A subduction zone forms where either two oceanic plates converge or an oceanic and a continental plate converge (◉ Figs. 7.14, ◉ 7.15 on page 146). The descending plate at subduction zones is oceanic lithosphere because continental lithosphere is too thick and light to descend into the mantle. As the descending plate dips into the mantle, it creates a deep trench on the ocean floor. The descending plate is gradually heated as it descends into the mantle, and it begins to melt by the time it reaches a depth

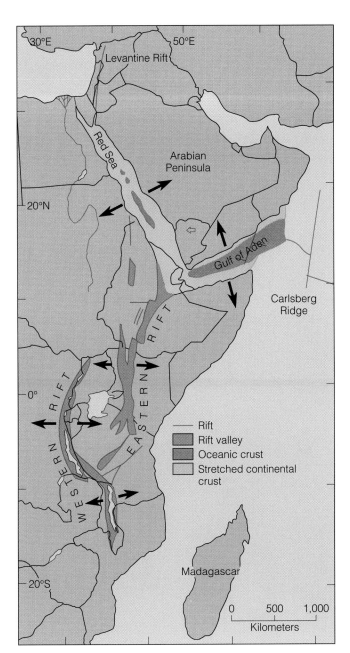

■ FIGURE 7.13 The East African rift valley is being formed by the separation of eastern Africa from the rest of the continent along a divergent plate boundary. The Red Sea and Gulf of Aden represent more advanced stages of rifting in which two continental blocks are separated by a narrow sea.

of about 100 km (60 mi). Molten rock from the descending plate and mantle invade the overriding plate, possibly melting some of the crust. The magma may break through the overriding plate, forming a chain of volcanoes, or solidify within the crust as intrusive igneous bodies, especially batholiths (see Chapter 9).

The overriding plate of a subduction zone also experiences changes. When the overriding plate is also oceanic crust, a chain of volcanic islands, or **island arc,** usually forms 100 km (62 mi) or so landward of the trench (Fig. 7.14). The Aleutian Islands are a typical island arc resulting from subduction of an oceanic plate at the Aleutian Trench.

The island arcs of the Pacific are often separated from a nearby continent by small seas like the Sea of Japan. These small seas or **marginal basins** form in a different way from most oceanic crust—not at a ridge, but over a broad area. Magma from the descending plate rises and creates new oceanic crust that eventually separates the island arc from the continent.

Hot Spots

The island of Hawaii lies at the southeast end of a chain of islands and submerged volcanoes (*seamounts*) that extend all the way to the Aleutian Trench. The islands show a steady age progression from southeast to northwest along the chain; active volcanism occurs on the island of Hawaii itself, whereas the northernmost seamount in the chain is about 70 million years old.

This chain is the product of a **hot spot,** a long-lived source of magma in the mantle, and marks the track of the Pacific plate as it moved over the hot spot. Hot spots are thought to be due to rising streams, or **plumes,** of hot material that originate deep in the mantle. As a plate moves over a hot spot, magma breaks through the crust occasionally to form a volcano. We can thus use these hot spots and resulting volcanoes to track plate motions.

Hot spots can also occur beneath continents, the best-known example being Yellowstone National Park. Hot spots may also be responsible for great basalt eruptions, called *flood basalts* (Chapter 9), like those of the Columbia Plateau in the northwestern United States, or the Deccan Plateau in western India.

In other cases, the overriding plate may be a continent. The convergent plate boundary where an oceanic plate is

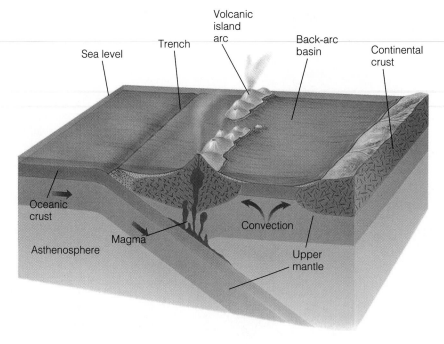

■ **FIGURE 7.14** Oceanic–oceanic plate boundary. An oceanic trench forms where one oceanic plate is subducted beneath another. On the overriding oceanic plate, a volcanic island arc forms from the rising magma generated from the subducting plate.

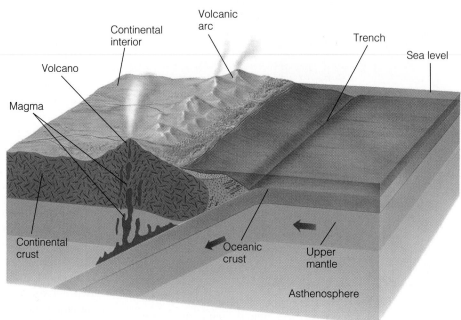

■ **FIGURE 7.15** Oceanic–continental plate boundary. When an oceanic plate is subducted beneath a continental plate, a volcanic mountain range is formed on the continental plate as a result of rising magma.

subducted beneath South America is the clearest example (Fig. 7.15). Magma from the descending plate penetrates the crust to form a chain of volcanoes, whereas the crust of the continent is deformed and metamorphosed to create a mountain range. In some cases, plate convergence may cause two continents to collide, such as the collision between India and Asia that formed the Himalayas (■ Fig. 7.16). Because each of these plates is continental lithosphere, neither is subducted. Rather they pile up, thickening Earth's crust.

Transform Faults

Lithosphere is neither created nor destroyed at **transform faults,** which are boundaries where plates slide past one an-

other (■ Fig. 7.17). These faults are generally free of volcanic activity or metamorphism. Often the plate boundary consists of many parallel fractures, with slivers of crust between the fractures moving and being deformed in complex ways. By far the most common transform faults are the fracture zones on the ocean floor between segments of mid-oceanic ridge, but they can occur on the continents as well. Perhaps the best-known continental transform fault is the San Andreas fault in California (see Chapter 8), where the Pacific plate is carrying a piece of California along the boundary of the North American plate. Much of the earthquake activity in California is caused by movements of the plates on opposite sides of the San Andreas fault and related faults.

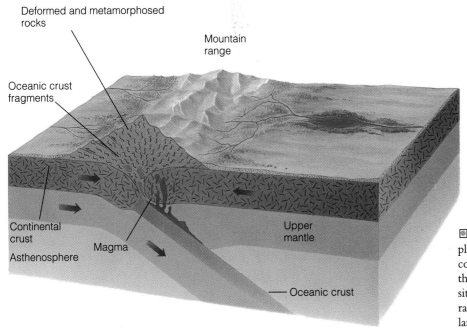

Deformed and metamorphosed rocks

Mountain range

Oceanic crust fragments

Continental crust

Asthenosphere

Magma

Upper mantle

Oceanic crust

■ FIGURE 7.16 Continental-continental plate boundary. When two continental plates converge, neither is subducted because of their great thickness and low and equal densities. As the two plates collide, a mountain range is formed in the interior of a new and larger continent.

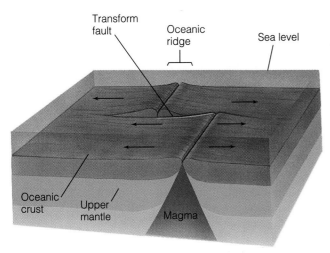

Transform fault

Oceanic ridge

Sea level

Oceanic crust

Upper mantle

Magma

■ FIGURE 7.17 Horizontal movement between plates occurs along a transform fault. The majority of transform faults connect two oceanic ridge segments. Note that relative motion between the plates occurs only between the two ridges.

 ## THE DRIVING MECHANISM OF PLATE TECTONICS

A major obstacle to the acceptance of Wegener's continental drift theory was the lack of a driving mechanism to explain continental movement. When it was shown that continents and ocean floors move together as parts of large plates and that new lithosphere is formed at spreading ridges by rising magma, most scientists accepted some type of **thermal convection** as the basic process responsible for plate motion. According to this concept, hot material within Earth rises toward the surface, but as it does, it cools and spreads laterally beneath the lithosphere and then flows to greater depths as a result of changing density. Divergent plate boundaries are located above rising convection currents, whereas convergent plate boundaries are located over descending convection currents (■ Fig. 7.18).

Some geologists think that in addition to thermal convection, plate movement also occurs because of a mechanism involving "slab-pull" or "ridge-push," both of which are gravity driven but still depend on temperature differences within the Earth (■ Fig. 7.19). In "slab-pull," the subducting cold slab of lithosphere, being denser than the surrounding warmer asthenosphere, pulls the rest of the plate along with it as it descends. In support of the "slab-pull" concept, convergent boundaries around large parts of their perimeter tend to move faster than other plates. Operating in conjunction with "slab-pull" is "ridge-push," which involves rising magma at spreading ridges; the magma pushes upward, making the ridges higher than the surrounding sea floor. It is thought that gravity pushes the oceanic lithosphere toward trenches (Fig. 7.19).

Currently, scientists are reasonably certain that some type of thermal convection is involved in plate movements, but the extent to which other mechanisms such as "slab-pull" and "ridge-push" are involved remains unresolved. Consequently, a comprehensive theory of plate movement has not been developed as yet, and much still remains to be learned about Earth's interior (see ■ Perspective 7.1 for more about the way geologists work).

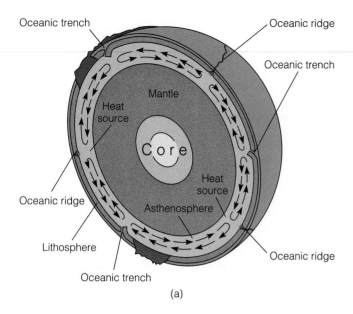

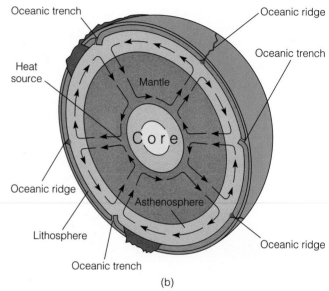

FIGURE 7.18 Two models involving thermal convection cells have been proposed to explain plate movement. (*a*) In one model, thermal convection cells are restricted to the asthenosphere. (*b*) In the other model, thermal convection cells involve the entire mantle.

FIGURE 7.19 Plate movement is also thought to occur because of gravity-driven "slab-pull" or "ridge-push" mechanisms. In "slab-pull," the edge of the subducting plate descends into Earth's interior, and the rest of the plate is pulled downward. In "ridge-push," rising magma pushes the oceanic ridges higher than the rest of the oceanic crust. Gravity thus pushes the oceanic lithosphere away from the ridges and toward the trenches.

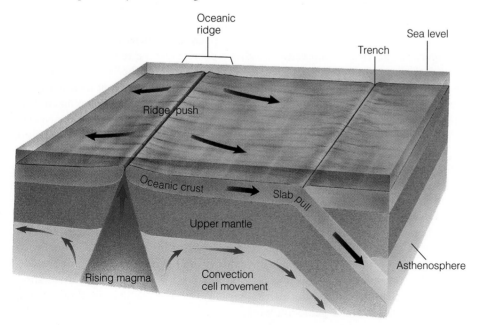

DETERMINING PAST PLATE POSITIONS

The magnetic stripes on the ocean floor offer the most precise way of determining the past positions of continents. As new oceanic crust is created at spreading ridges and magnetized by the Earth's magnetic field, successive bands of crust can be recognized by their magnetic stripes. When the stripes formed, they were together at the ridge crest. To determine the positions of the continents when a certain set of stripes formed, all we need to do is bring the stripes back together, moving the continents with them (■ Fig. 7.20). Unfortunately, when oceanic crust is subducted, the record it carries disappears, much like erasing a tape recording. Thus, we have a very detailed knowledge of the motions of the plates since the breakup of Pangaea, but a much poorer understanding before that time.

It is often possible to find the original positions of continents by fitting them together like pieces of a jigsaw puzzle (Fig. 7.3). The fit of the continents on opposite sides of the Atlantic, although not perfect, is very good. Most of the small gaps and overlaps are unimportant because we know

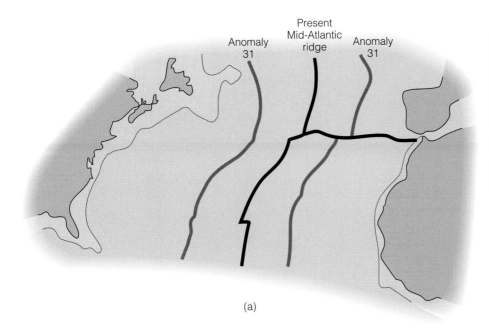

(a)

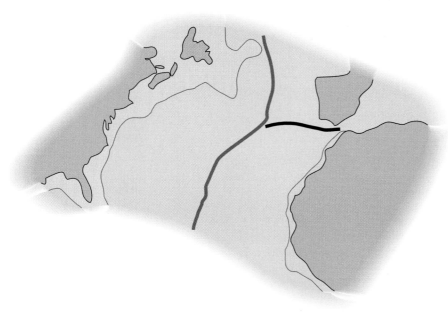

(b)

■ FIGURE 7.20 Reconstructing plate positions using magnetic anomalies. (a) The present North Atlantic, showing the present ridge and magnetic anomaly 31, which formed 67 million years ago. (b) The Atlantic as the last dinosaurs knew it. Anomaly 31 marks the plate boundary 67 million years ago. By moving the anomalies back together, along with the plates they are on, we reconstruct the former positions of the continents.

How Geologists See the World

There is a famous poem about blind men who attempt to describe an elephant. Each feels a part of the elephant and decides the entire elephant is like the part he touches, with amusing results. Few geologists are blind, literally or figuratively, but they must describe things that are far larger than elephants, and how they describe the world often depends on what they already know. We have already mentioned how geologists from different parts of the world responded to Wegener's theory of continental drift. Geologists familiar with Wegener's evidence were more receptive to the theory than geologists who were not. Plate tectonics also allows us to look back and understand why geologists responded to other problems the way they did.

For example, in the early years of this century, it was perfectly obvious to American geologists that earthquakes and faulting were related. It was hard to doubt such a connection with the San Andreas fault cutting California in two. To Japanese geologists, however, the connection was not at all obvious. Japan has more great earthquakes than California, yet few of those earthquakes are associated with faults that rupture the ground surface. Now we understand that California and Japan are at two different types of plate boundary and that the faults that produce earthquakes in Japan are far below the surface. We would not expect earthquakes in Japan to rupture the ground the way the San Andreas fault does.

Russia has often been isolated from the rest of the scientific world by language, politics, and geography, and Russian geologists have always had their own way of approaching problems. Russian geologists often spoke of "deep faults" that cut entirely through the crust, a concept that puzzled many scientists elsewhere. It is now clear that much of Asia formed through the collision of many small blocks of crust and that many of the "deep faults" identified by Russian geologists are actually boundaries between these crustal blocks. Other "deep faults" are great faults where blocks of Asian crust slide past one another in response to the collision of India with Asia.

Another novel Russian concept was the *aulacogen,* an inactive, sediment-filled rift valley that branches off from the edge of a continent. Russian geologists identified many aulacogens connected with ancient mountain belts in Russia, but geologists outside Russia took a long time to understand and apply the concept fully. It appears that when a continental plate breaks up, the first breaks form a radiating fracture system.

Some of the fractures join into a continent-spanning line of separation, and the others remain as aulacogens (□ Fig. 1).

The moral of these stories is that diversity and independent evaluation of the evidence are good because often someone with a fresh point of view can see important relationships that other scientists missed. The first result of contact between two different points of view in science is often confusion—and sometimes conflict—but in the end science is enriched.

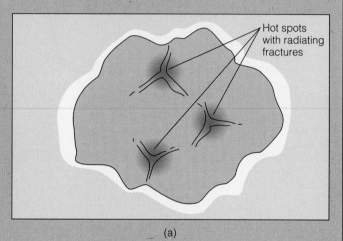

(a)

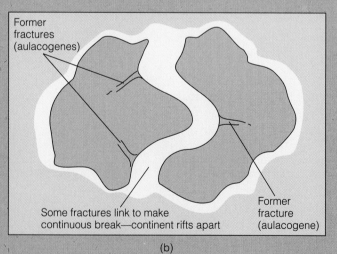

(b)

□ FIGURE 1 Aulacogenes. (*a*) Hot spots beneath a continent create a series of radiating fracture patterns. (*b*) Some fractures link to make a complete break across the continent. Remaining fractures are preserved as dead-end rift valleys, or aulacogenes.

the edges of continents are only approximate. A few of the misfits can be easily explained. For example, the eastern tip of Brazil overlaps Africa, but the overlap on the African side is the Niger River delta, which formed after the continents parted. Also, Iceland is missing, but it consists entirely of young volcanic rocks that formed after Pangaea began fragmenting.

Other problems are not so neatly explained. For instance, Central America was probably a series of small blocks either on the Pacific coast or in what is now the Gulf of Mexico.

Unfortunately, clear magnetic stripes do not exist in the Gulf to aid in determining the motions of these plates. The plate movements of the Gulf of Mexico and Central America are an active area of research with important economic implications because of the Gulf's oil resources.

Paleomagnetism is very helpful in deciphering plate positions because Earth's magnetic field is much like that of any magnet, with a north and a south pole. The magnetic field has both a horizontal direction (which is why a compass needle points north) and a vertical direction called dip, which varies from 0° at the equator to 90° at the poles (▣ Fig. 7.21). From the dip recorded by magnetic minerals in rocks, it is possible to determine their distance from the magnetic pole when they were magnetized. From this data we can determine the approximate latitude of the rock when it formed.

Other paleolatitude indicators such as coal, fossil coral, and evaporite deposits in the geologic record indicate low latitude, and glacial deposits indicate high latitude. There is, unfortunately, no reliable way of determining ancient longitudes. Paleomagnetism and climatic indicators can tell how far apart two continents were in a north-south direction, but not in an east-west direction.

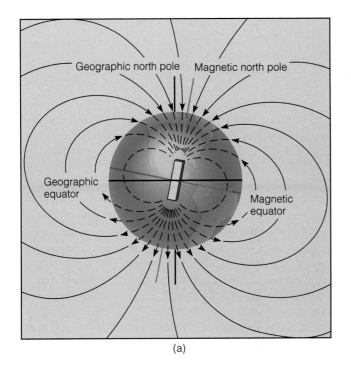

(a)

▣ FIGURE 7.21 (a) The magnetic field of the Earth has lines of force just like those of a bar magnet. (b) The strength of the magnetic field changes uniformly from the magnetic equator to the magnetic poles. This change in strength causes a dip needle to parallel the Earth's surface only at the magnetic equator, whereas its inclination with respect to the surface increases to 90° at the magnetic poles.

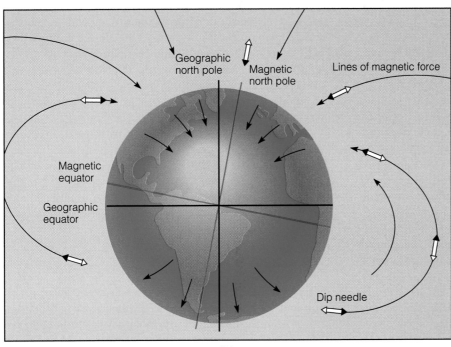

(b)

The Fragile Continents

Why did supercontinents, which occupied only 30 percent of Earth's area, break up into so many pieces? Continental crust is thick, but weak. At high temperatures, the rocks of continental crust are only about a tenth as strong as the rocks of the mantle. A slab of continental crust 30 km (20 mi) thick is much weaker than a comparable slab of oceanic crust and the mantle rock beneath. Pangaea and its predecessors broke up because they were weaker than the oceanic crust that made up the rest of Earth.

■ **FIGURE 7.22** World paleogeography at the end of the Paleozoic Era.

PLATE TECTONICS AND EARTH HISTORY

The emergence of plate tectonic theory is recognized as a major milestone in the Earth sciences. It provides a framework for interpreting Earth processes on a global scale, rather than viewing geologic events as separate and unrelated. It has led to the realization that the continents and ocean basins are part of a lithosphere-atmosphere-hydrosphere system that evolved together with Earth's interior.

The Assembly and Breakup of Pangaea

Many scientists now agree that plate tectonics has operated for at least 2 billion years in very much its present form. In fact, it appears that the continents have assembled into supercontinents and then broken apart several times during that time. This pattern of supercontinent formation and fragmentation is called the *Wilson Cycle,* after Canadian geologist J. T. Wilson, who first described it. About a billion years ago, according to one interpretation, North America was at the center of a supercontinent, with Australia and Antarctica to the west and the rest of Gondwanaland to the east. This supercontinent, which has been called *Rodinia,* broke apart and the pieces dispersed. The break along the west side of North America opened to become the ancestor of the Pacific, and the break on the east opened to become an early version of the Atlantic called the *proto-Atlantic.* By about 500 million years ago, several continents existed, all of them scattered along the equator with oceans at both poles.

About 450 million years ago, North America and Europe collided, and the Southern Hemisphere continents, already existing as a single landmass known as Gondwanaland, drifted over the South Pole, experiencing ice ages during the Ordovician and Permian periods, and then drifted north again. About 250 million years ago, North America and Europe collided with Gondwanaland, creating a belt of mountains and eliminating the proto-Atlantic. At nearly the same time, Siberia collided with the eastern part of Europe.

These collisions resulted in the assembly of the supercontinent Pangaea. A large wedge-shaped sea, the *Tethys,* was between what is now Europe-Asia and Gondwanaland (■ Fig. 7.22). The rest of Earth was a vast ocean, about which we know almost nothing because all its crust has been subducted.

During the Late Triassic Period, about 180 million years ago, an episode of fragmentation of Pangaea began that continues even now (■ Fig. 7.23). Separation began along divergent plate margins that were originally within the supercontinent but later developed into mid-oceanic ridges. For example, the Mid-Atlantic Ridge is the line along which North and South America separated from Europe and Africa. As divergence continued between these formerly connected continents, a long narrow sea formed and expanded to become the present Atlantic Ocean basin. A large part of Gondwanaland consisting of Madagascar, India, Australia, and Antarctica separated from Africa and began moving south. India eventually separated from Gondwanaland and moved progressively north. In doing so, India, along with the northward-moving African plate, caused the closure of the Tethys sea. Sometime between 40 and 50 million years ago, India collided with Asia resulting in the formation of the Himalayas. Indeed, India is still moving north at about 5 cm (2 in) per year, pushing the crust of China eastward and changing the outline of Asia.

Until about 20 million years ago, narrow remnants of the Tethys seaway existed at both ends of the Mediterranean. These seaways closed as Africa and Eurasia converged. When Africa and Spain collided about 6 million years ago, the Mediterranean was cut off from the Atlantic, resulting in one of the most remarkable episodes in Earth history (see □ Perspective 7.2).

What of the Future?

Plate movements are not predictable in detail, but if plates continue to move for the next 50 million years as they are now, Australia will collide with eastern Asia, crushing the islands of Indonesia into a complex mountainous belt similar to that in southern Europe. The Mediterranean will disap-

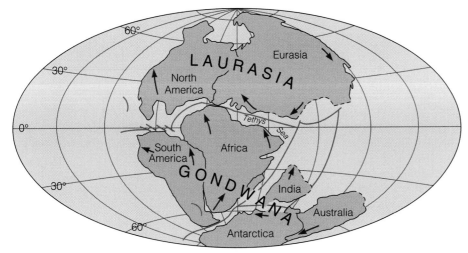

(a) Triassic Period (245–208 M.Y.A.)

■ **FIGURE 7.23** (*a*) Paleogeography of the world during the Late Triassic Period. Blue arrows show the direction of movement for the continents. Paleogeography of the world during (*b*) the Jurassic Period and (*c*) the Cretaceous Period.

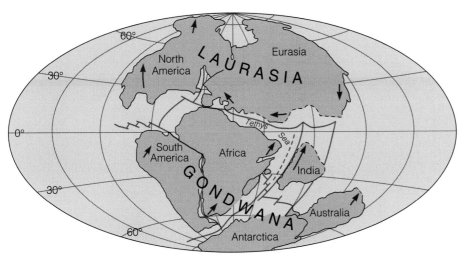

(b) Jurassic Period (208–144 M.Y.A.)

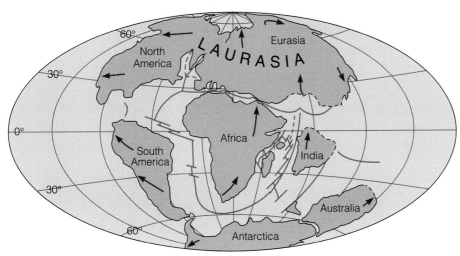

(c) Cretaceous Period (144–66 M.Y.A.)

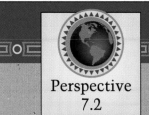
History of the Mediterranean

Today the Mediterranean Sea is in an arid region where the rate of evaporation exceeds the rate at which water is added to the sea by precipitation and runoff from the land. If it were not for the connection between the Mediterranean and the Atlantic Ocean at the Strait of Gibraltar (☐ Fig. 1), the Mediterranean would eventually dry up and become a vast desert lying far below sea level. According to one interpretation of Mediterranean history, this is precisely what happened about 6 million years ago during the Late Miocene.

Evaporite deposits 2 km (1.2 mi) thick have been discovered beneath the floor of the Mediterranean, and studies indicate that they were deposited in shallow-water environments rather than a deep-ocean basin. Observations such as this have led to the hypothesis that the Mediterranean periodically lost its connection with the Atlantic during the Late Miocene. With its supply of water cut off, the Mediterranean evaporated to near dryness in as little as 1,000 years and became a vast desert basin lying 3,000 m (9,840 ft) below sea level (☐ Fig. 2).

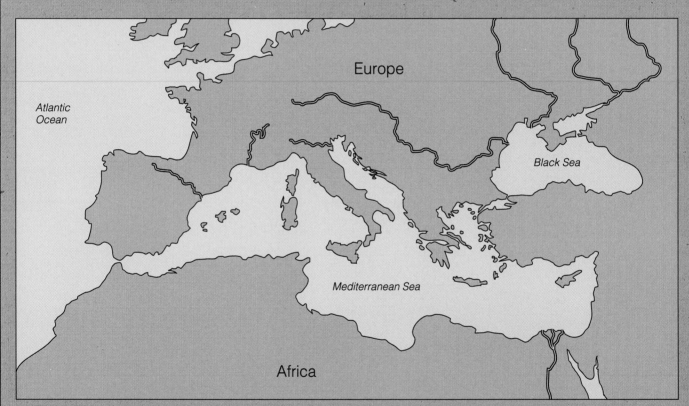

☐ FIGURE 1 The present Mediterranean is linked to the Atlantic only by the narrow Strait of Gibraltar. River discharge from the few rivers feeding the Mediterranean is insufficient to maintain its level.

Apparently, periods of isolation and evaporation of the Mediterranean alternated with periods when an oceanic connection was reestablished at the Strait of Gibraltar. When an oceanic connection existed, gravel and silt were deposited around the basin margin, and deep-sea sediments accumulated near the basin center. But when the oceanic connection was lost, deposition of carbonates followed by evaporites, especially rock gypsum and rock salt, took place in a sea that became progressively shallower and saltier.

Additional evidence for this view of Mediterranean history comes from buried stream valleys incised into bedrock far below sea level in Africa and southern Europe. Drilling more than 900 m (2,950 ft) into the Rhone delta in southern Europe has revealed a buried stream valley, and several stream-eroded submarine canyons have been traced to depths of more than 2,000 m (6,560 ft) below the present level of the Mediterranean. These must have been eroded when the level of the Mediterranean was much lower than at present. In Egypt at the Aswan High Dam, 465 km (288 mi) upstream on the Nile, Russian geologists discovered a buried stream valley 200 m (656 ft) below present sea level. Apparently, it too was eroded when sea level was much lower.

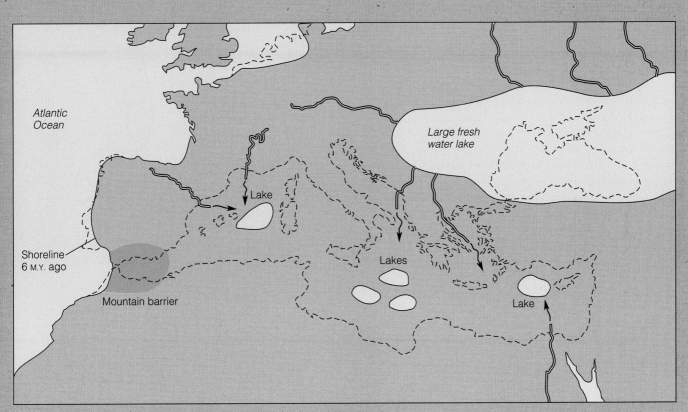

□ **FIGURE 2** About 6 million years ago, Spain and Africa collided to create a barrier that blocked off the Mediterranean, which soon dried out. Rivers fed small playa lakes on the dry sea floor. A large freshwater lake sprawled across much of southeastern Europe and covered the present Black, Caspian, and Aral Seas. This lake overflowed into the Mediterranean Basin and also fed small playa lakes.

pear as Africa and Eurasia collide. The Hawaiian Islands will erode and become coral atolls and then submerged seamounts, and new islands will appear to replace them. The Atlantic will widen and the Pacific will narrow, and coastal California will move north, eventually colliding with Alaska. Eventually, the continents may reassemble into a new Pangaea, completing yet another Wilson Cycle.

PLATE TECTONICS AND GEOLOGIC RESOURCES

A strong correlation exists between plate tectonics and several resource-forming processes (◉ Fig. 7.24). Subduction is responsible for many geologic processes because of the associated igneous, hydrothermal, and metamorphic processes. Fluids emanating from intrusions are responsible for deposits of gold, silver, copper, molybdenum, and many other metals. Geothermal heat is a potential resource in most subduction zones.

Stable continental interiors, or cratons, are sites for other resources, especially sedimentary resources such as coal, salt, and gypsum. Exposed Precambrian rocks within continents are rich storehouses of resources, many of which formed during ancient plate collisions. Spreading ridges and transform faults are relatively minor resource areas. Iceland taps geothermal heat from the Mid-Atlantic Ridge, and there has been exploratory geothermal drilling at the Salton Sea in California, which is located where the East Pacific Rise meets the San Andreas fault system. Copper has been mined for thousands of years from rocks on Cyprus that appear to have formed near hot water vents at divergent plate boundaries. Some of these vents are known as black smokers because dissolved minerals precipitate quickly in the cold sea water, creating dense clouds of particles. Such vents on the mid-oceanic ridges indicate that oceanic crust may be rich in metallic resources. Petroleum occurs in a wide variety of geologic settings, some related to plate tectonics (Perspective 10.2).

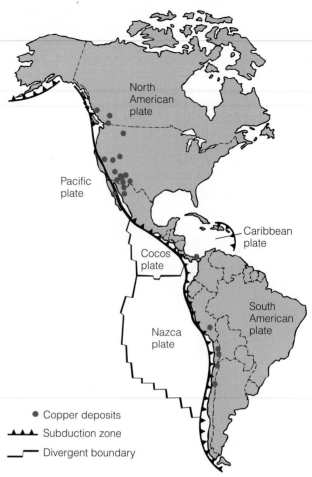

● Copper deposits
▲▲▲ Subduction zone
‾‾ Divergent boundary

◉ FIGURE 7.24 Important copper deposits are located along the west coasts of North and South America.

CHAPTER SUMMARY

1. Plate tectonics is a unifying theory in the Earth sciences because it allows scientists to explain otherwise seemingly unrelated geologic phenomena. Plate tectonics is also important because moving plates affect other Earth systems such as the atmosphere, hydrosphere, and biosphere.

2. Alfred Wegener developed his theory of continental drift, the predecessor of plate tectonics, based on the similarities of continental outlines and the distribution of fossils and past glaciers.

3. According to Wegener, a supercontinent known as Pangaea, consisting of most of Earth's present continents, existed at the end of the Paleozoic Era. Fragmentation of this supercontinent into moving plates accounts for the present distribution of land and sea.

4. Convincing evidence for moving plates was discovered when it was learned that the mid-oceanic ridges possess a central rift, and that the age of the oceanic crust increases with distance from ridge crests.

5. Studies of Earth's ancient magnetic field revealed an apparent paradox known as polar wandering. This paradox was resolved by moving the continents into different positions, thus making the paleomagnetic data consistent with a single magnetic north pole.

6. Surveys of the ocean floors revealed parallel and symmetric belts of magnetic normal and reversed polarity recorded in rocks of the oceanic crust. Sea-floor spreading accounts for these observations because new crust is generated at mid-oceanic ridges where it acquires its magnetism. These newly formed rocks then separate, and part of the record is carried on each of two diverging plates.

7. Oceanic lithosphere is formed at mid-oceanic ridges and consumed at subduction zones where one plate plunges beneath another.

8. Plate tectonic theory now holds that the outer part of Earth consists of plates of lithosphere that move over the partially molten asthenosphere, which is part of the upper mantle.

9. All lithosphere consists of that part of the upper mantle overlying the asthenosphere, but its uppermost part may be either continental crust or oceanic crust.

10. Three types of plate boundaries are recognized: divergent boundaries, where plates move away from each other; convergent boundaries, where two plates collide; and transform faults, where two plates slide past one another.

11. Although a comprehensive theory of plate movement has yet to be developed, scientists think that some type of thermal convection is involved. The degree to which "slab-pull" and "ridge-push" are involved in plate motions is unresolved.

12. The past positions of plates can be determined using paleomagnetic data, the fit of continental outlines, fossils, and the distribution of some sedimentary deposits such as coal and evaporites. In all cases, however, it is possible only to determine paleolatitude.

13. Plate tectonic theory is not only useful in explaining geologic phenomena such as earthquake activity and volcanism, but it is essential if one is to understand Earth history. Scientists now think that plate tectonics in its present form has operated for at least 2 billion years.

14. A close relationship exists between the formation of some geologic resources and plate boundaries. Some metal deposits are found at convergent plate boundaries, and hydrothermal activity at divergent boundaries is responsible for other metal deposits.

IMPORTANT TERMS

asthenosphere
continental drift
convergent plate boundary
divergent plate boundary
Glossopteris flora

hot spot
island arc
lithosphere
marginal basins
paleomagnetism

Pangaea
plate tectonic theory
plumes
polar wandering
sea-floor spreading

spreading ridge
subduction
subduction zone
thermal convection
transform fault

REVIEW QUESTIONS

1. The southern part of Pangaea, consisting of South America, Africa, India, Australia, and Antarctica, is known as:
   a. ___ Atlantis; b. ___ Gondwanaland; c. ___ Pacifica;
   d. ___ Laurentia; e. ___ Eurasia.

2. A segment of the outer part of the Earth consisting of continental crust, oceanic crust, and upper mantle is a(n):
   a. ___ continent; b. ___ ocean basin; c. ___ lithospheric plate; d. ___ subduction zone; e. ___ transform fault.

3. Divergent plate boundaries are where:
   a. ___ new continental lithosphere forms; b. ___ two plates collide; c. ___ two plates slide past one another; d. ___ new oceanic lithosphere forms; e. ___ oceanic lithosphere is subducted beneath continental lithosphere.

4. Subduction occurs along a _____ plate boundary.
   a. ___ transform; b. ___ lithospheric; c. ___ dynamic;
   d. ___ continental-continental; e. ___ convergent.

5. The San Andreas fault in California is a good example of a _____ plate boundary.
   a. ___ divergent; b. ___ transform; c. ___ extensional;
   d. ___ convergent; e. ___ hot spot.

6. Thermal convection within the Earth is probably at least partly responsible for:
   a. ___ the shape of the continents; b. ___ the amount of water in the oceans; c. ___ plate movements; d. ___ magnetic reversals; e. ___ the composition of Earth's core.

7. At what type of plate boundary would you expect earthquake activity but little or no volcanism?
   a. ___ divergent; b. ___ transform; c. ___ convergent;
   d. ___ answers (a) and (b); e. ___ all of these.

8. The *Glossopteris* flora is a(n):
   a. ___ type of freshwater reptile; b. ___ process during which new oceanic crust is formed by volcanism; c. ___ continent that existed at the beginning of the Mesozoic Era; d. ___ ancient plant assemblage _____ _____ Hemisphere continents; e. ___ sp_____ ____ grated from Europe to North Ame____

9. Which one of the following is cor___
   a. ___ plates generally move at sev___ Charles Darwin was the first to p__ continental drift; c. ___ continen___

than oceanic lithosphere; d. ___ considerable volcanism occurs at transform plate boundaries; e. ___ the oldest rocks on Earth are found in the ocean basins.

10. What kinds of evidence persuaded Wegener and other scientists that continents had moved?
11. What are some of the techniques used to determine the former positions of continents?
12. Explain how improvements in technology played a part in providing evidence for plate tectonics.
13. What is the *Glossopteris* flora and how can its present-day distribution be explained?

14. Why are the oldest rocks on Earth on the continents whereas rocks of the oceanic crust are less than 200 million years old?
15. What is thermal convection, and what role does it play in plate tectonics?
16. How can the magnetic patterns preserved in rocks of the oceanic crust be used to show that the sea floor has been spreading?
17. Summarize the geologic processes occurring at the three different types of plate boundaries.

POINTS TO PONDER

1. What features would an astronaut look for on another planet to see if plate tectonics is currently active or if it was active during the past?
2. If movement along the San Andreas fault, which separates the Pacific plate from the North American plate, averages 5.5 cm/yr, how long will it take before Los Angeles is opposite San Francisco?

3. In the Hawaiian Islands, the island of Kauai has a maximum age of 5.6 million years. Kauai is about 550 km from the hot spot where the island originally formed. How fast in centimeters per year has the Pacific plate been moving over this hot spot?

ADDITIONAL READINGS

Allegre, C. 1988. *The behavior of the Earth.* Cambridge, Mass.: Harvard University Press.

Bonatti, E. 1987. The rifting of continents. *Scientific American* 256, no. 3: 96–103.

———. 1994. The Earth's mantle below the oceans. *Scientific American* 270, no. 3: 44–51.

Condie, K. 1989. *Plate tectonics and crustal evolution,* 3d ed. New York: Pergamon Press.

Kearey, P., and F. J. Vine. 1990. *Global tectonics.* Palo Alto, Calif.: Blackwell Scientific Publishers.

Klein, G. D. ed. 1994. Pangaea: Paleoclimate, tectonics, and sedimentation during accretion, zenith, and breakup of a supercontinent. *Geological Society of America Special Paper* 288. Boulder, Colo.: Geological Society of America.

Murphy, J. B., and R. D. Nance. 1992. Mountain belts and the supercontinent cycle. *Scientific American* 266, no. 4: 84–91.

White, R. S., and D. P. McKenzie. 1989. Volcanism at rifts. *Scientific American* 252, no. 1: 62–71.

WORLD WIDE WEB SITES

For current updates and exercises, log on to http://www.wadsworth.com/geo.

TUTORIAL
http://www.seismo.unr.edu/ftp/pub/louie/class/100/plate-tectonics.html

This site offers an introduction to seismology and plate tectonics from the Seismological Laboratory at the University of Nevada–Reno.

TECTONIC PLATE MOTION
http://cddis.gsfc.nasa.gov/926/slrtecto.html

Maintained by NASA, this site describes how space-age techniques like satellite laser-ranging are used to track present-day plate motions.

CHAPTER

8

Earthquakes and Earth's Interior

...d during the Northridge, California, earthquake of 1994. The pillars, held together by iron reinforcing bars, ...otice the tension fractures in the concrete on the outside of the bends.

Prologue

In the early morning hours of January 17, 1994, southern California was rocked by a devastating earthquake measuring 6.7 on the Richter Magnitude Scale. The earthquake was centered on the city of Northridge just north of Los Angeles. When the initial shock was over, about 61 people had been killed and thousands were injured. Nine freeways had been severely damaged, thousands of homes and buildings were either destroyed or damaged, and 250 ruptured gas lines ignited numerous fires (☐ Fig. 8.1). Thousands of aftershocks followed the main earthquake; many of them added to the damage, which has been estimated at $15 to $30 billion.

Geologists know that southern California is riddled with active faults capable of producing strong earthquakes, but it was a previously unknown fault 15 km (9.3 mi) beneath Northridge that caused the tragic January 1994 earthquake. Many people are aware that movements on the San Andreas fault and its subsidiary faults have been responsible for many earthquakes in southern California. But only recently have geologists begun to realize that a network of interconnected faults that do not break the surface may be causing much of the earthquake activity in the Los Angeles area. Earthquakes on these hidden faults, just as on the obvious ones, are related to movements between the North American and Pacific plates.

How did the Los Angeles area fare during and after this most recent earthquake? Older, unreinforced masonry buildings and more modern wood-frame apartments built over ground-floor garages generally sustained the most damage. Structures built to the stricter building standards in force during the last five years typically escaped unscathed or with only minor damage.

The state transportation department (Caltrans) instituted a program of reinforcing bridges and freeway overpasses soon after the 1971 Sylmar earthquake and began a second phase of reinforcing structures after the 1989 Loma Prieta earthquake. Most of the reinforced structures suffered little or no damage during the Northridge earthquake, but several awaiting reinforcing collapsed, including part of the Santa Monica Freeway, the busiest highway in the world.

Regulations designed to protect utility lines have not yet been implemented, in part because of cost. As a consequence, numerous power and gas lines were ruptured, and three water aqueducts were severed, cutting off water to at least 40,000 people and power to an estimated 3.1 million residents.

Emergency measures had been well planned and rescue operations went well. Shelters were established, people fed and clothed, and disaster relief offices opened in the area shortly after the earthquake. Based on their experiences in other earthquakes, rescue agencies had invested in better rescue equipment including high-pressure air bags that can lift up to 72 tons, fiber-optics search cameras, and specially trained dogs that can sniff out buried victims. This experience and the up-to-date equipment helped rescue workers locate and extricate victims from the earthquake wreckage.

The Northridge earthquake was tragic, but it was not the "Big One" that Californians have been waiting for. And even though rescue and relief agencies operated efficiently, the earthquake reminds us that much still remains to be done in terms of earthquake prediction and preparedness.

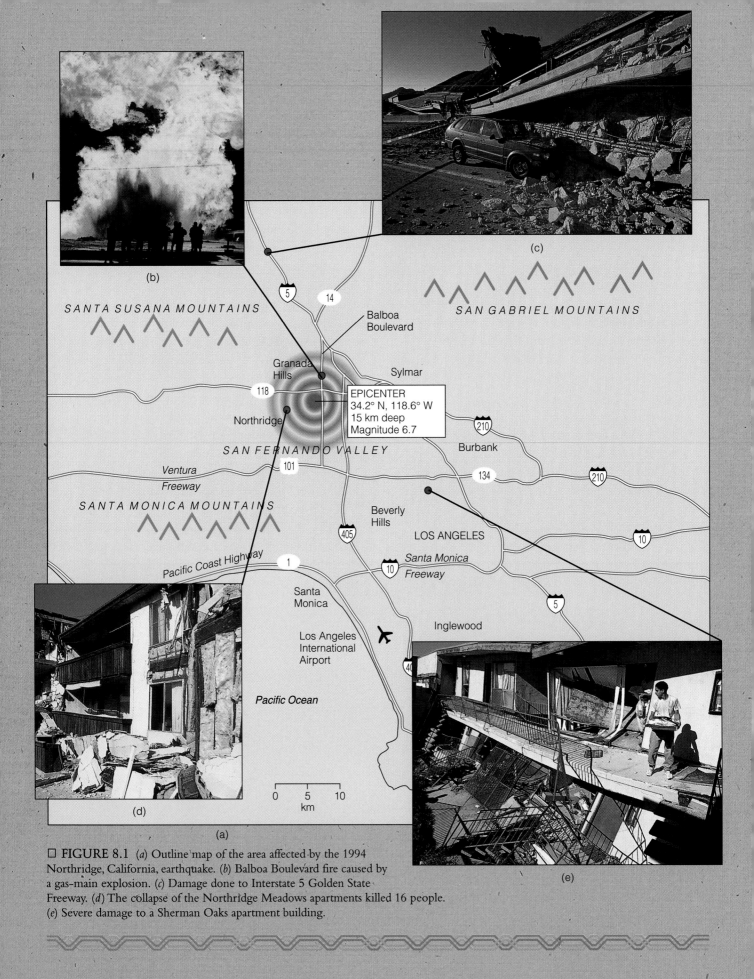

(b)

(c)

SANTA SUSANA MOUNTAINS

SAN GABRIEL MOUNTAINS

Balboa
Boulevard

Granada
Hills

Sylmar

EPICENTER
34.2° N, 118.6° W
15 km deep
Magnitude 6.7

118

Northridge

SAN FERNANDO VALLEY

Burbank

Ventura

101

134

Freeway

SANTA MONICA MOUNTAINS

Beverly
Hills

LOS ANGELES

405

Santa Monica

Pacific Coast Highway

1

10

Freeway

Santa
Monica

Inglewood

Los Angeles
International
Airport

Pacific Ocean

0   5   10
km

(d)

(e)

(a)

□ FIGURE 8.1 (a) Outline map of the area affected by the 1994 Northridge, California, earthquake. (b) Balboa Boulevard fire caused by a gas-main explosion. (c) Damage done to Interstate 5 Golden State Freeway. (d) The collapse of the Northridge Meadows apartments killed 16 people. (e) Severe damage to a Sherman Oaks apartment building.

INTRODUCTION

As one of nature's most frightening and destructive phenomena, earthquakes have always aroused a sense of fear. Even when an earthquake begins, there is no way to tell how strong the shaking will be or how long it will last. It is estimated that more than 13 million people have died as a result of earthquakes during the past 4,000 years, and approximately 1 million of these deaths occurred during the last century (○ Table 8.1). As cities located in earthquake-prone regions such as Tokyo and Los Angeles grow larger, more and more people will be in danger from devastating earthquakes.

An **earthquake** is defined as the vibration of Earth caused by the sudden release of energy beneath Earth's surface, usually as a result of displacement of rocks along fractures known as faults. Following an earthquake, adjustments along a fault commonly generate a series of earthquakes referred to as **aftershocks.** Most of these are smaller than the main shock, but they can cause considerable damage to al-

ready weakened structures. Indeed, much of the destruction from the 1755 earthquake in Lisbon, Portugal, was caused by aftershocks. After a small earthquake, aftershock activity usually ceases within a few days, but it may persist for months following a large earthquake.

Early humans and cultures explained earthquakes in a more imaginative and colorful way, often attributing them to the movements of some organism on which Earth rested. In Japan, the organism was a giant catfish; in Mongolia, a giant frog; in China, an ox; in India, a giant mole; in parts of South America, a whale; and to the Algonquin Indians of North America, an immense tortoise.

The Greek philosopher Aristotle offered what he considered to be a natural explanation for earthquakes. He thought that atmospheric winds were drawn into Earth's interior where they caused fires and swept around the various subterranean cavities trying to escape. This movement of underground air caused earthquakes and occasional volcanic eruptions. Today, geologists know that the majority of earthquakes result from faulting associated with plate movements.

TABLE 8.1 Significant Earthquakes of the World

YEAR	LOCATION	MAGNITUDE (Estimated before 1935)	DEATHS (Estimated)
893	India		180,000
1201	Syria		1,000,000
1556	China (Shanxi Province)	8.0	1,000,000
1730	Japan (Hokkaidō Prefecture)		137,000
1737	India (Calcutta)		300,000
1755	Portugal (Lisbon)	8.6	70,000
1811–12	USA (New Madrid, Missouri)	7.5	20
1857	USA (Fort Tejon, California)	8.3	1
1868	Chile and Peru	8.5	25,000
1872	USA (Owens Valley, California)	8.5	27
1886	USA (Charleston, South Carolina)	7.0	60
1905	India (Punjab-Kashmir region)	8.6	19,000
1906	USA (San Francisco, California)	8.3	700
1908	Italy (Messina)	7.5	83,000
1920	China (Gansu)	8.6	100,000
1923	Japan (Tokyo)	8.3	143,000
1950	India (Assam) and Tibet	8.6	1,530
1964	USA (Alaska)	8.6	131
1970	Peru (Chimbote)	7.8	25,000
1971	USA (San Fernando, California)	6.6	65
1976	Guatemala	7.5	23,000
1976	China (Tangshan)	8.0	242,000
1985	Mexico (Mexico City)	8.1	9,500
1988	Armenia	7.0	25,000
1989	USA (Loma Prieta, California)	7.1	63
1990	Iran	7.3	40,000
1993	India	6.4	30,000+
1994	USA (Northridge, California)	6.7	61
1995	Japan (Kobe)	7.2	5,000+
1995	Russia	7.6	2,000+

The study of earthquakes not only helps us to understand their causes and distribution and to mitigate some of their hazards, but it also provides us with a sensitive tool for investigating Earth's interior.

SEISMOLOGY AND SEISMOGRAPHS

Seismology, the study of earthquakes, began emerging as a true science around 1880 with the development of seismographs that effectively recorded earthquake waves. A *seismograph* is an instrument that detects, records, and measures the various vibrations produced by an earthquake (Fig. 8.2).

Measuring ground motion during an earthquake is a problem because any instrument on Earth moves as the ground moves. Notice in Figure 8.2a that the seismograph employs a weight suspended on a cable, which is essentially a pendulum. Contrary to intuition, ground motion does not make the pendulum swing. Instead the pendulum tends to remain stationary while the instrument moves. A seismograph of this type responds to ground motions at right angles to the pendulum, so a complete seismic station must have at least two such instruments, one sensitive to north-south motions and one to east-west motions. In addition, vertical movements are detected by a seismograph with a spring-supported weight (Fig. 8.2b). The record made by a seismograph is a *seismogram*.

When an earthquake occurs, energy in the form of *seismic waves* radiates outward in all directions from the point of release (Fig. 8.3). Most earthquakes result when rocks in Earth's crust rupture along a fault because of the buildup of excessive pressure, which is usually caused by plate movement. Once a rupture begins, it moves along the fault at a velocity of several km/sec for as long as conditions for failure exist. The length of the fault along which rupture occurs can range from a few meters to several hundred kilometers. Kinks or bends in the fault often limit the length of the fault rupture. The longer the rupture, the more time it takes for all of the stored energy in the rocks to be released, and therefore the longer the ground will shake.

The location within the crust where rupture initiates, and thus where the energy is released, is referred to as the **focus;** the point on Earth's surface vertically above the focus is the **epicenter,** which is the location that is usually given in news reports on earthquakes (Fig. 8.3).

Seismologists recognize three categories of earthquakes based on the depth of their foci. *Shallow-focus* earthquakes have a focal depth of less than 70 km (43 mi). Earthquakes with foci between 70 and 300 km (43–186 mi) are referred to as *intermediate focus,* and those with foci greater than 300 km (>186 mi) are *deep focus.* Earthquakes are not evenly distributed among these three categories. Approximately 90% of all earthquake foci occur at a depth of less than 100 km (62 mi), whereas only about 3% of all earthquakes are deep. Shallow-focus earthquakes are, with few exceptions, the most destructive.

The Frequency and Distribution of Earthquakes

Most earthquakes (almost 95%) take place in seismic belts that correspond to plate boundaries where stresses develop as plates converge, diverge, and slide past each other. Earthquake activity distant from plate margins is minimal, but can be devastating when it occurs. The relationship between plate margins and the distribution of earthquakes is readily apparent when the locations of earthquake epicenters are superimposed on a map showing the boundaries of Earth's plates (Fig. 8.4).

 FIGURE 8.2 (*a*) A horizontal-motion seismograph. Because of its inertia, the heavy mass that contains the marker will remain stationary while the rest of the structure moves along with the ground during an earthquake. As long as the length of the arm is not parallel to the direction of ground movement, the marker will record the earthquake waves on the rotating drum. (*b*) A vertical-motion seismograph. This seismograph operates on the same principle as a horizontal-motion instrument and records vertical ground movement.

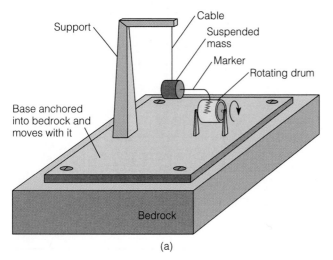

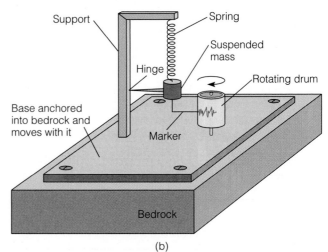

(a)

(b)

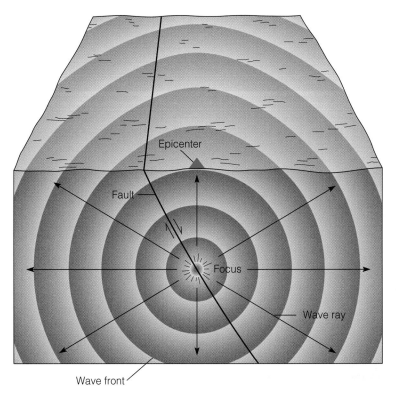

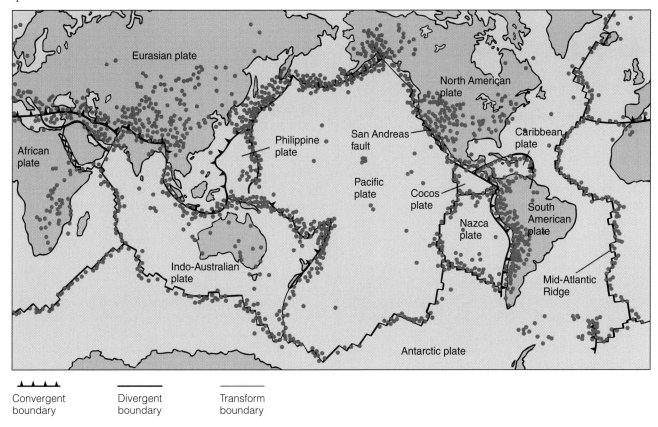

FIGURE 8.3 The focus of an earthquake is the location where rupture begins and energy is released. The place on Earth's surface vertically above the focus is the epicenter. Seismic wave fronts move outward in all directions from their source, the focus of an earthquake. Wave rays are lines drawn perpendicular to wave fronts.

FIGURE 8.4 The relationship between earthquake epicenters and plate boundaries. Approximately 80% of earthquakes occur within the circum-Pacific belt, 15% within the Mediterranean–Asiatic belt, and the remaining 5% within the interiors of plates or along oceanic spreading ridge systems. Each dot represents a single earthquake epicenter.

FIGURE 8.5 Some of the damage in Kobe, Japan, caused by the January 1995 earthquake. More than 5,000 people died, 25,000 were injured, and 300,000 had to be evacuated. About 50,000 buildings were destroyed. Property damage was estimated at more than $30 billion.

FIGURE 8.6 Damage in Oakland, California, resulting from the October 1989 Loma Prieta earthquake. The columns supporting part of double-decked Interstate 880 failed, causing the upper deck to collapse onto the lower one.

The majority of all earthquakes (approximately 80%) occur in the *circum-Pacific belt,* a zone of seismic activity nearly encircling the Pacific Ocean basin. Most of these earthquakes result from convergence along plate margins, as in the case of the 1995 earthquake at Kobe, Japan (Fig. 8.5). Likewise, earthquakes in extreme northern California, Oregon, and Washington result from plate convergence. Along most of the U.S. Pacific coast, however, plates slide past one another rather than converge. The October 17, 1989 Loma Prieta earthquake in the San Francisco area (Fig. 8.6) and the January 17, 1994 Northridge earthquake happened along this plate boundary.

The second major seismic belt, the *Mediterranean-Asiatic belt,* accounts for approximately 15% of all earthquakes. This belt extends westerly from Indonesia through the Himalayas, across Iran and Turkey, and westerly through the Mediterranean region of Europe. The devastating earthquake that struck Armenia in 1988 killing 25,000 people and the 1990 earthquake in Iran that killed 40,000 are recent examples of the destructive earthquakes that strike this region (Table 8.1).

The remaining 5% of earthquakes occur mostly in the interiors of plates and along oceanic spreading ridge systems, that is, at divergent plate boundaries. Most of these earthquakes are not very strong although several major intraplate earthquakes are worthy of mention. For instance, the 1811 and 1812 earthquakes near New Madrid, Missouri, killed approximately 20 people and nearly destroyed the town of New Madrid. So strong were these earthquakes that they were felt from the Rocky Mountains to the Atlantic Ocean and from the Canadian border to the Gulf of Mexico. Another major intraplate earthquake struck Charleston, South Carolina, on August 31, 1886, killing 60 people and causing $23 million in property damage (Fig. 8.7).

The cause of intraplate earthquakes is not well understood, but geologists think they arise from localized stresses caused by the compression that most plates experience along their margins. The release of these stresses and hence the resulting intraplate earthquakes are related to local factors. Interestingly, many intraplate earthquakes are associated with very ancient and presumed inactive faults that are reactivated at various intervals.

More than 150,000 earthquakes strong enough to be felt by someone are recorded every year by the worldwide network of seismograph stations. In addition, seismologists estimate that about 900,000 earthquakes are recorded annually by seismographs, but are too small to be individually cataloged. These small earthquakes result from the energy released as continual adjustments take place between and within Earth's various plates.

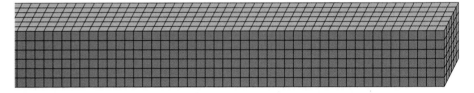

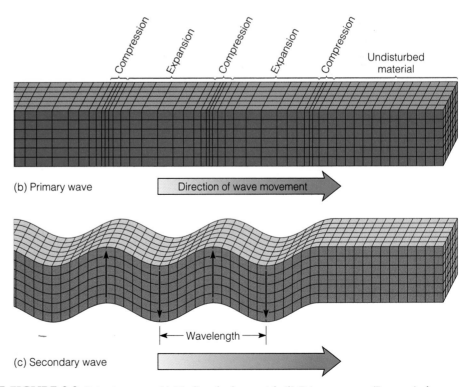

☐ FIGURE 8.7 Damage done to Charleston, South Carolina, by the earthquake of August 31, 1886. This earthquake is the largest reported in the eastern United States.

Seismic Waves

The shaking and destruction resulting from earthquakes are caused by two different types of seismic waves: *body waves,* which travel through the Earth and are somewhat like sound waves; and *surface waves,* which travel only along the ground surface and are analogous to ocean waves.

An earthquake generates two types of body waves: P-waves and S-waves (☐ Fig. 8.8). **P-waves,** or *primary waves,* are the fastest seismic waves and can travel through solids, liquids, and gases. P-waves are compressional, or push-pull, waves and are similar to sound waves in that they move material forward and backward along a line in the same direction that the waves themselves are moving (Fig. 8.8b). Thus, the material P-waves travel through is alternately expanded and compressed as the wave moves through it and returns to its original volume after the wave passes by.

(a) Undisturbed material

Compression Expansion Compression Expansion Compression Undisturbed material

(b) Primary wave   Direction of wave movement

Wavelength

(c) Secondary wave

☐ FIGURE 8.8 Seismic waves. (*a*) Undisturbed material. (*b*) Primary waves (P-waves) alternately compress and expand material in the same direction as the wave movement. (*c*) Secondary waves (S-waves) move material perpendicular to the direction of wave movement.

S-waves, or *secondary waves,* are somewhat slower than P-waves and can only travel through solids. S-waves are *shear waves* because they move the material perpendicular to the direction of travel, thereby producing shear stresses in the material they move through (Fig. 8.8c). S-waves deform materials out of shape as they pass. Because liquids (as well as gases) are not rigid, they have no shear strength and S-waves cannot be transmitted through them.

The velocities of P- and S-waves are determined by the density and elasticity of the materials through which they travel. For example, seismic waves travel more slowly through rocks of greater density, but more rapidly through rocks with greater elasticity. *Elasticity* is a property of solids, such as rocks, and means that once they have been deformed by an applied force, they return to their original shape when the force is no longer present. Because P-wave velocity is greater than S-wave velocity in all materials, P-waves always arrive at seismic stations first.

Surface waves travel along the surface of the ground, or just below it along planes between rock layers, and are slower than body waves. Unlike the sharp jolting and shaking that body waves cause, surface waves generally produce a rolling or swaying motion, much like the experience of being on a boat. In fact, surface waves may cause lakes and water levels in wells to oscillate far beyond the area where the earthquake is felt, and in very large earthquakes, they can make the entire Earth oscillate for days.

Locating an Earthquake

The various seismic waves travel at different speeds and thus arrive at a seismograph at different times. As ◙ Figure 8.9 illustrates, the first waves to arrive are the P-waves, which travel at nearly twice the velocity of the S-waves that follow. Both the P- and S-waves travel directly from the focus to the seismograph through Earth's interior. The last waves to arrive are the surface waves, which are the slowest and also travel the longest route along Earth's surface.

Seismologists have determined the average travel times of P- and S-waves for any specific distance. These P- and S-

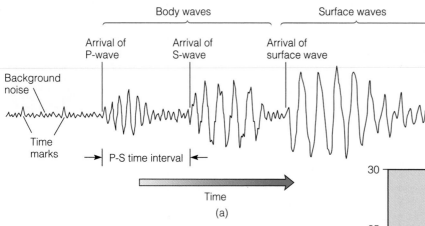

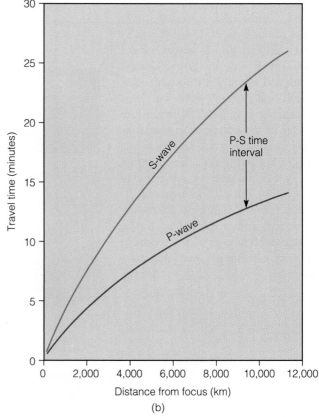

◙ **FIGURE 8.9** (*a*) A schematic seismogram showing the arrival order and pattern produced by P-, S-, and surface waves. When an earthquake occurs, body and surface waves radiate outward from the focus at the same time. Because P-waves are the fastest, they arrive at a seismograph first, followed by S-waves, and then by surface waves, which are the slowest waves. (*b*) A time-distance graph showing the average travel times for P- and S-waves. The farther away a seismograph station is from the focus of an earthquake, the longer the interval between the arrivals of the P- and S-waves, and hence the greater the distance between the curves on the time-distance graph as indicated by the P-S time interval.

wave travel times are published as *time-distance graphs* and illustrate that the difference between the arrival times of the P- and S-waves is a function of the distance of the seismograph from the focus (Fig. 8.9b).

As Figure 8.10 demonstrates, the epicenter of any earthquake can be determined by using a time-distance graph and knowing the arrival times of the P- and S-waves at any three seismograph locations. Subtracting the arrival time of the first P-wave from the arrival time of the first S-wave gives the time interval between the arrivals of the two waves for each seismograph location. Each time interval is then plotted on the time-distance graph, and a line is drawn straight down to the distance axis of the graph, indicating how far away each station is from the focus of the earthquake. Then a circle whose radius equals the distance shown on the time-distance graph from each of the three seismograph locations is drawn on a map (Fig. 8.10). The intersection of the three circles is the location of the earthquake's epicenter. A minimum of three locations is needed because two locations will provide two possible epicenters and one location will provide an infinite number of possible epicenters.

Measuring Earthquake Intensity and Magnitude

Scientists measure the strength of an earthquake in two different ways. The first, *intensity*, is a qualitative assessment of the kinds of damage done by an earthquake. The second, *magnitude*, is a quantitative measurement of the amount of energy released by an earthquake. Each method provides important data about earthquakes and their effects.

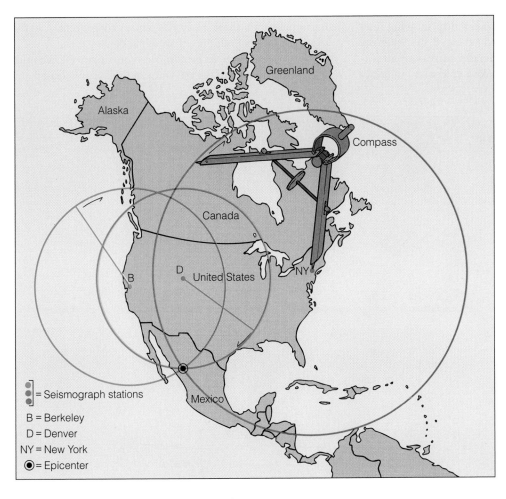

FIGURE 8.10 Three seismograph stations are needed to locate the epicenter of an earthquake. The P-S time interval is plotted on a time-distance graph for each seismograph station to determine the distance that station is from the epicenter. A circle with that radius is drawn from each station, and the intersection of the three circles is the epicenter of the earthquake.

= Seismograph stations

B = Berkeley
D = Denver
NY = New York
= Epicenter

TABLE 8.2   The Mercalli Intensity Scale

INTENSITY	DESCRIPTION	APPROXIMATE MAGNITUDE EQUIVALENT*
I	Felt only by a very few favorably situated people.	2
II	Felt by a few favorably situated people, such as those on upper floors of buildings.	2.5
III	Widely felt by people in quite surroundings.	3
IV	Felt by most people indoors; objects rock.	3.5
V	Almost everyone notices shaking. People awakened from sleep.	4.5
VI	Felt by all. Trees sway, heavy objects overturned.	5
VII	Walls crack, general alarm.	6
VIII	Chimneys fall, many walls crack. Automobile steering affected.	6.5
IX	Widespread ground failure, landslides. Most masonry buildings damaged.	7
X	Widespread ground failure, landslides. Most masonry buildings damaged.	7.5
XI	Most buildings destroyed and services disrupted. Bridges destroyed.	8
XII	Waves seen on ground, total destruction. Ground acceleration exceeds gravity so that objects are thrown into air. Rarely recorded.	8.5

*The magnitude equivalent is the approximate magnitude of a shallow earthquake expected to produce a given intensity near its epicenter. Actual effects vary with distance to the focus, local geology, construction materials, and other factors.

Intensity is a subjective measure of the kind of damage done by an earthquake as well as people's reaction to it. The most common intensity scale used in the United States is the **Modified Mercalli Intensity Scale,** which has values ranging from I to XII (○ Table 8.2). If enough intensity reports are available after an earthquake, a map showing lines of equal intensity can be constructed (▣ Fig. 8.11). Intensity varies from place to place and generally decreases with increasing distance from an earthquake's epicenter. Intensity also depends on the geology of the area. Shaking is more intense in unconsolidated sediment than in hard bedrock (▣ Fig. 8.12), and seismic waves travel more effectively through flat rock layers than through complex, folded rocks. Intensity also depends on an observer's location and activity. Earthquakes are more easily felt while sitting quietly on the top floor of a tall building than while driving a car.

If earthquakes are to be compared quantitatively, we must use a scale that measures the amount of energy released and

▣ FIGURE 8.11 Modified Mercalli Intensity map for the 1971 San Fernando Valley, California, earthquake showing the region divided into intensity zones based on the kind of damage done. This earthquake had a magnitude of 6.6.

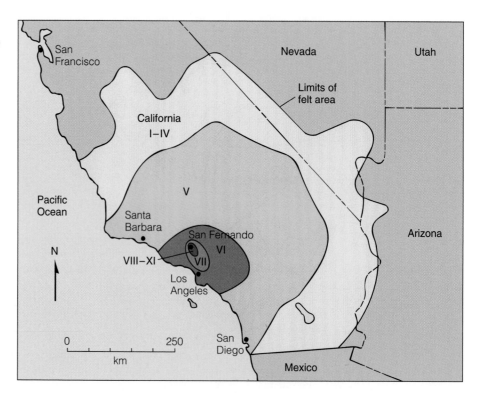

170   CHAPTER 8   Earthquakes and Earth's Interior

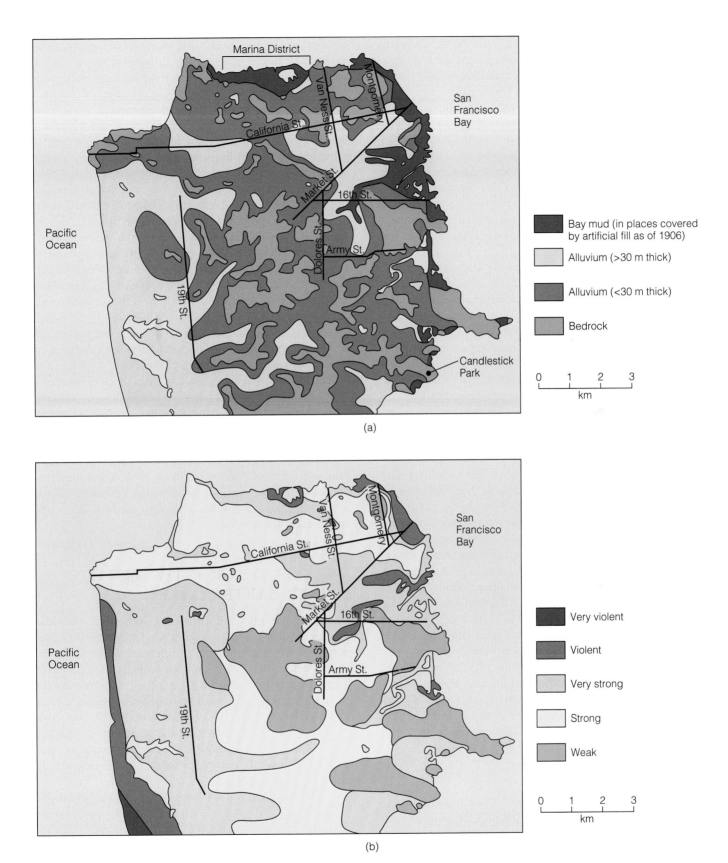

■ **FIGURE 8.12** The geology of San Francisco (*a*) corresponds closely to the intensity of shaking during the 1906 San Francisco earthquake (*b*). The thicker and less consolidated the sedimentary cover, the stronger the shaking. Note particularly along 16th Street that intensity varied from violent in a filled-in valley to weak on a bedrock hill only a couple of hundred meters away. During the 1989 Loma Prieta earthquake, centered over 100 km away, damage was severe in the Marina district, built on artificial fill. In contrast, Candlestick Park, where the third game of the World Series was about to begin, is built on bedrock and suffered almost no damage.

is independent of intensity. Such a scale was developed in 1935 by Charles F. Richter, a seismologist at the California Institute of Technology. The **Richter Magnitude Scale** measures earthquake **magnitude,** which is the total amount of energy released by an earthquake at its source. It is an open-ended scale. The largest magnitude recorded has been 8.6, and though values greater than 9 are theoretically possible, they are highly improbable because rocks are not able to store the energy necessary to generate earthquakes of this magnitude. The scale is designed so that the smallest earthquakes likely to be felt will have positive magnitudes, but very small earthquakes can have zero or even negative magnitude. A magnitude zero earthquake is about equal in energy to a stick of dynamite.

Each whole-number increase in the Richter Magnitude Scale represents a 10-fold increase in the ground motion during an earthquake. Accordingly, an earthquake with a magnitude of 5 produces 10 times as much ground motion as a 4, and so on. However, this 10-fold increase in ground motion corresponds to roughly a 30-fold increase in the amount of energy released. Thus, an earthquake with a magnitude of 5 releases 30 times the energy of a 4, a 6 releases 900 times the energy, and a 7 releases 27,000 times the energy of a 4.

The Richter Magnitude Scale underestimates the energy of very large earthquakes because it measures the highest peak recorded on a seismogram, which records only an instant during an earthquake. During great earthquakes, however, the energy might be released over several minutes and along hundreds of kilometers of a fault. A modification of the Richter Magnitude Scale, known as the *seismic-moment magnitude scale,* is now used for these earthquakes. On this scale large earthquakes can exceed magnitude 9.

There is a direct relationship between the area of slip on a fault surface and the resulting earthquake. Thus, there is also a close relationship between the plate tectonic setting of an earthquake and its magnitude. Earthquakes in oceanic crust, where the crust is thin, tend to be no larger than magnitude 7. Earthquakes in continental crust, where most people live and where the crust is much thicker, can be magnitude 8 or greater because the thicker crust allows a larger fault area. The greatest earthquakes occur along subduction zones in continental crust, where the crust is thick and the fault dips very gently. In such settings, the area of fault that can slip is very great, and magnitudes can exceed 9 on the seismic-moment scale (8.5 on the Richter Scale).

HOW AND WHY EARTHQUAKES OCCUR

The cause of most earthquakes is movement of Earth's tectonic plates (see Chapter 7). Most of the world's shallow earthquakes, and virtually all its deeper earthquakes (below 100 km, or 62 mi), occur where two tectonic plates are in contact. A minority, though a significant minority, of quakes occur far from plate boundaries. The New Madrid, Missouri earthquakes of 1811–1812 and the Charleston, South

Carolina earthquake of 1886 are examples of *intraplate earthquakes.* The causes of intraplate earthquakes are still not as well understood as the causes of earthquakes along plate boundaries, but a useful analogy might be that of moving a house. Regardless of how careful the movers are, it is all but impossible to move something as large as a house without its internal parts shifting slightly. Similarly, objects as large as crustal plates are not likely to move without some internal stresses that occasionally are relieved by earthquakes.

The mechanism responsible for earthquakes was first detailed by H. H. Read after the 1906 San Francisco earthquake. Read proposed the **elastic rebound theory** (■ Fig. 8.13), which holds that stresses in the crust near a fault deform the rocks out of shape. Initially, friction prevents the fault from slipping. Eventually, however, the stress becomes too great to withstand, and the fault slips, or ruptures, resulting in an earthquake. During the slippage, the rocks along the fault spring back to an undeformed condition. In 1906, slippage along the San Andreas fault in California reached 6 m (20 ft). Usually, the first slippage does not fully relieve the pent-up stresses, and additional slippages take place, causing aftershocks, some of which can rival the main shock in severity.

Rocks are not perfectly rigid, and faults slip a bit at a time rather than all at once. A familiar analogy might be the problem of adjusting a mattress on a bed. When we pull on the mattress, only part of it slides because the mattress can deform out of shape. We have to pull and push the mattress from a number of different spots to move it in its entirety. Similarly, faults generally do not slip all at once; instead only part of the fault slips, and the crust near the fault deforms to adjust to the movement.

THE DESTRUCTIVE EFFECTS OF EARTHQUAKES

The destructive effects of earthquakes include ground shaking, fire, seismic sea waves, and landslides, as well as disruption of vital services, panic, and psychological shock (see ▢ Perspective 8.1). The amount of property damage, loss of life, and injury depends on the time of day an earthquake occurs, its magnitude, the distance from the epicenter, the geology of the area, the type of construction of various structures, population density, and the duration of shaking. Generally speaking, earthquakes occurring during working and school hours in densely populated urban areas are the most destructive.

Ground shaking usually causes more damage and results in more loss of life and injuries than any other earthquake hazard. Structures built on solid bedrock generally suffer less damage than those built on poorly consolidated material such as water-saturated sediments or artificial fill. Structures on poorly consolidated or water-saturated material are subjected to ground shaking of longer duration and greater amplitude than those on bedrock. In addition, fill and water-saturated sediments tend to liquefy, or behave as a fluid, a process known as *liquefaction*. When shaken, the individual grains lose cohesion and the ground flows. Dramatic examples of damage resulting from liquefaction include Niigata, Japan, where large apartment buildings were tipped to their sides after the water-saturated soil of a hillside collapsed in 1964, and Mexico City, which is built on soft lake bed sediments (Fig. 8.14).

During the Loma Prieta earthquake that caused the third game of the 1989 World Series to be postponed, those districts in the San Francisco–Oakland Bay area built on artificial fill or reclaimed bay mud suffered the most damage. In the Marina district of San Francisco, numerous buildings were destroyed, and a fire, fed by broken gas lines, lit up the night sky. The failure of the columns supporting a portion of the two-tiered Interstate 880 freeway in Oakland sent the upper tier crashing down onto the lower one, killing 42 unfortunate motorists (see Fig. 8.6). The shaking lasted less than 15 seconds but resulted in 63 deaths, 3,800 injuries, and $6 billion in property damage and left at least 12,000 people homeless.

In addition to the magnitude of an earthquake and the regional geology, the material used and the type of construction also affect the amount of damage done. Adobe and mud-walled structures are the weakest of all and almost always collapse during an earthquake. Unreinforced brick

EARTHFACT

The Greatest Deep Earthquake

A magnitude 8.2 earthquake with a focal depth of 640 km (397 mi) was recorded on June 10, 1994, beneath Bolivia, making it the greatest deep earthquake ever recorded. Because of its great depth, the earthquake caused no major damage and no fatalities. It was felt as far north as Toronto, Canada. Scientists are hopeful that information from this earthquake may help resolve the controversy over what causes deep earthquakes.

structures and poorly built concrete structures are also particularly susceptible to collapse. The 6.4 magnitude earthquake that struck India in 1993 killed about 30,000 people whereas the 6.7 magnitude Northridge earthquake resulted in only 55 deaths. Both earthquakes were in densely populated regions, but in India the brick and stone buildings could not withstand ground shaking; most collapsed entombing their occupants (Fig. 8.15).

FIGURE 8.13 (a) According to the elastic rebound theory, when rocks are deformed, they store energy and bend. When the inherent strength of the rocks is exceeded, they rupture, releasing the energy in the form of earthquake waves that radiate outward in all directions. Upon rupture, the rocks rebound to their former undeformed shape. (b) During the 1906 San Francisco earthquake, this fence in Marin County was displaced 2.5 m (8.2 ft).

(b)

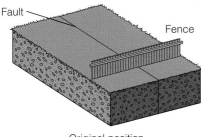

Original position

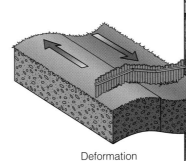

Deformation

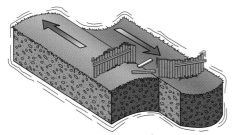

Rupture and release of energy

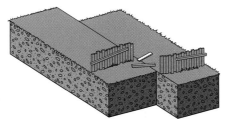

Rocks rebound to original undeformed shape

(a)

Safe Buildings for Dangerous Places

It has been said that "earthquakes don't kill people, buildings do." Building collapse is the major cause of death and injury in urban areas, but earthquake-safe buildings can be built (☐ Fig. 1). In fact, earthquake-safe buildings come in two types: low-tech and high-tech. Low-tech buildings are wood-frame dwellings, whether of Western or Japanese construction. These are perhaps safest of all, because they flex during an earthquake and are small enough to ride out ground shaking as a small boat rides out waves. At the high-tech end of the technological spectrum, steel-frame skyscrapers can be designed to withstand even the worst earthquakes. Well-designed skyscrapers in Mexico City rode out the 1985 disaster undamaged.

Using common sense, it is possible to design earthquake-resistant buildings, without great cost, in even the worst settings. A classic example was the Imperial Hotel in Tokyo, designed by Frank Lloyd Wright. Wright knew the site of the hotel was underlain by clay that would shake strongly in an earthquake, so he designed the hotel in sections that could move independently. He even insisted on a pond in the courtyard for fire fighting. The hotel survived the great earthquake of 1923 and it served as a center for earthquake relief efforts.

Reinforced concrete is less earthquake-resistant than steel-frame construction. A number of freeways collapsed in California during earthquakes in 1989 and 1994 despite having reinforced concrete pillars. Unreinforced concrete is easily damaged. Buildings with adjoining wings are particularly prone to damage. In effect, the earthquake takes two buildings and slams them together repeatedly, with results that are easy to imagine. Many old buildings also have ornamental facings and ledges that could fall during an earthquake. Even with well-designed buildings, falling glass can be a lethal hazard.

At the bottom of the safety list is *adobe*, or dried clay. This material is widely used in less-developed nations because it is cheap, but it is also heavy and has almost no strength to resist side-to-side, or *shear*, motions. Roofs of such buildings are often tile, supported by inadequate rafters. Concrete in less-developed nations is often of poor or uneven quality and not much safer than adobe. Adobe and poor concrete construction are one reason why the death tolls from earthquakes in less-developed nations are so high. For example, the magnitude 7.1 Loma Prieta earthquake in California in 1989 killed 63 people. An earthquake of similar magnitude in Armenia in 1988 killed 25,000. Wood, the safest material for home construction, is at a premium in many countries that are subject to earthquakes. Many of the forests in these countries were cut long ago, and what wood remains is often harvested for fuel. Even adobe buildings, however, can be made much safer by constructing buttresses to help support the walls.

Fire is a danger in urban areas. Furnaces and stoves may be overturned, gas mains broken, and electric lines shorted out.

■ FIGURE 8.14 This 15-story reinforced concrete building collapsed due to the ground shaking that occurred during the 1985 Mexico City earthquake. The soft lake bed sediments on which Mexico City is built amplified the seismic waves as they passed through.

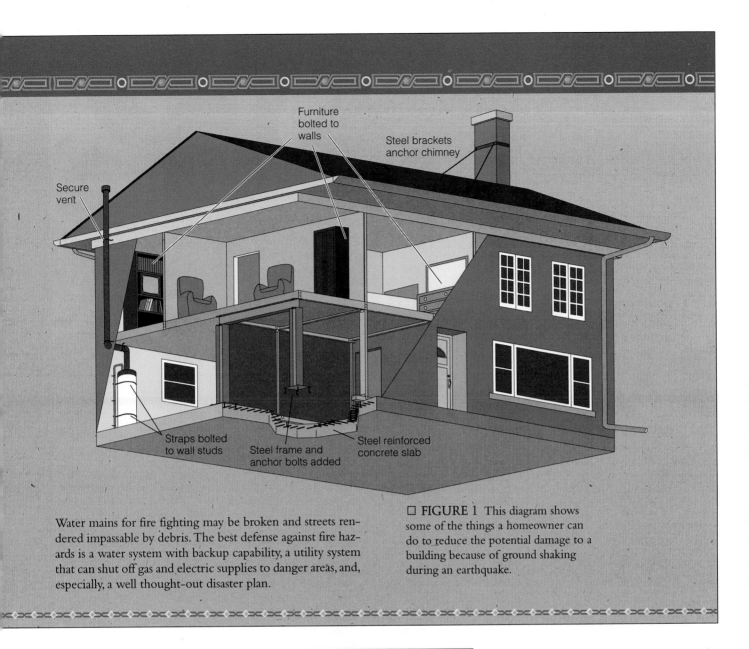

Furniture
bolted to
walls

Steel brackets
anchor chimney

Secure
vent

Straps bolted
to wall studs

Steel frame and
anchor bolts added

Steel reinforced
concrete slab

Water mains for fire fighting may be broken and streets rendered impassable by debris. The best defense against fire hazards is a water system with backup capability, a utility system that can shut off gas and electric supplies to danger areas, and, especially, a well thought-out disaster plan.

☐ **FIGURE 1** This diagram shows some of the things a homeowner can do to reduce the potential damage to a building because of ground shaking during an earthquake.

▣ **FIGURE 8.15** In 1993, India experienced its worst earthquake in more than 50 years. Thousands of brick and stone houses collapsed, killing at least 30,000 people.

 FIGURE 8.16 San Francisco Marina district fire caused by broken gas lines during the 1989 Loma Prieta earthquake.

In many earthquakes, particularly in urban areas, fire is a major hazard (◙ Fig. 8.16). Almost 90% of the damage done in the 1906 San Francisco earthquake was caused by fire. The shaking severed many of the electrical and gas lines, which touched off flames and started numerous fires all over the city. Because water mains were ruptured by the earthquake, there was no effective way to fight the fires. Hence, they raged out of control for three days, destroying much of the city.

Seismic sea waves or **tsunami** are destructive sea waves that are usually produced by earthquakes but can also be caused by submarine landslides or volcanic eruptions (◙ Fig. 8.17). Tsunami are popularly called tidal waves, although they are not related to tides. Instead, most tsunami result from the sudden movement of the sea floor, which sets up waves within the water that travel outward, much like the ripples that form when a stone is thrown into a pond.

Tsunami travel at several hundred km/hr and are commonly not felt in the open ocean because their wave height is usually less than 1 m (3.3 ft) and the distance between wave crests is typically several hundred kilometers. When tsunami approach shorelines, though, the waves slow down and water piles up to heights of up to 65 m (213 ft).

Following a 1946 tsunami that killed 159 people and caused $25 million in property damage in Hawaii, the U.S. Coast and Geodetic Survey established a Tsunami Early Warning System in Honolulu, Hawaii, in an attempt to min-

imize tsunami devastation. This system combines seismographs and instruments that can detect earthquake-generated waves. Whenever a strong earthquake occurs anywhere within the Pacific basin, its location is determined, and instruments are checked to see if a tsunami has been generated. If it has, a warning is sent out to evacuate people from low-lying areas that may be affected.

Earthquake-triggered landslides are particularly dangerous in mountainous regions and have been responsible for tremendous amounts of damage and many deaths. For example, the 1970 Peru earthquake caused an avalanche that completely destroyed the town of Yungay, resulting in 25,000 deaths.

EARTHQUAKE PREDICTION

Can earthquakes be predicted (see ◙ Perspective 8.2)? A successful prediction must include a time frame for the occurrence of the earthquake, its location, and its strength. In spite of the tremendous amount of information about the cause of earthquakes, successful predictions are still quite rare. Nevertheless, if reliable predictions can be made, they can greatly reduce the number of deaths and injuries.

From an analysis of historic records and the distribution of known faults, *seismic risk maps* can be constructed that indicate the likelihood and potential severity of future earth-

■ **FIGURE 8.17** As a tsunami crashes into the street behind them, residents of Hilo, Hawaii, run for their lives. This tsunami was generated by an earthquake in the Aleutian Islands and resulted in considerable property damage to Hilo and the deaths of 154 people.

quakes based on the intensity of past earthquakes (■ Fig. 8.18). Although such maps cannot predict when the next major earthquake will occur, they are useful in helping people plan for future earthquakes.

One long-range prediction technique used in seismically active areas involves plotting the location of major earth-

quakes and their aftershocks to detect areas that have had major earthquakes in the past but are currently inactive. Such regions are locked and not releasing energy, making these *seismic gaps* prime locations for future earthquakes. Several seismic gaps along the San Andreas fault have the potential for future major earthquakes. The 1989 Loma Prieta

■ **FIGURE 8.18** A 1969 seismic risk map for the United States based on intensity data collected by the U.S. Coast and Geodetic Survey.

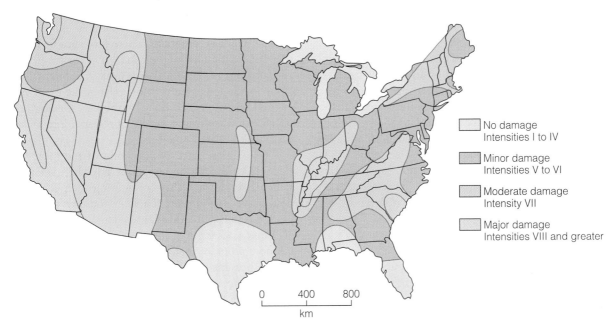

No damage
Intensities I to IV

Minor damage
Intensities V to VI

Moderate damage
Intensity VII

Major damage
Intensities VIII and greater

0   400   800
km

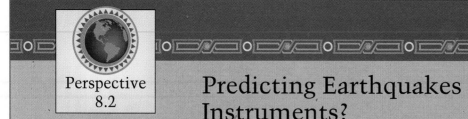

Predicting Earthquakes Without Instruments?

The biggest news story on December 3, 1990, in the New Madrid, Missouri area was that there was no story. A great earthquake predicted for that date by Dr. Iben Browning did not happen. Browning, a zoologist and climatologist, had developed a theory that tidal forces (Chapter 19) acting on Earth could trigger earthquakes. When the Earth, Moon, and Sun are aligned in a straight line, they exert greater than normal gravitational forces (although the forces are still relatively weak) that some think could trigger fault movement. Browning was hardly the first unsuccessful predictor. Brian Brady had predicted a great earthquake in Peru in 1981, David Stewart predicted a magnitude 6 earthquake for North Carolina in 1975, and Henry Minturn predicted a damaging earthquake for southern California in 1975. All three predictions were highly publicized; none was correct.

Responding to predictions like these is a no-win situation for scientists. If they do not comment, they leave the field open for the predictor and the news media. Often the predictor claims a long string of "successful predictions," which turn out, on closer examination, not to exist. For example, Browning was widely but incorrectly reported to have predicted the 1989 Loma Prieta earthquake with great accuracy.

All he said, however, was, "There will probably be several earthquakes around the world. Richter 6-plus magnitude. . . ."[*] As it turns out, Browning's predictions of earthquakes corresponding to tidal forces rate no better than random guessing. If scientists do respond to the prediction, they generate additional publicity for it and may create the impression they are trying to suppress unorthodox ideas.

Animals are widely supposed to be capable of sensing impending earthquakes; one California predictor tallies the number of advertisements for lost pets on the grounds that pets will be more prone to run away before an earthquake. Scientists who have studied this method find that lost-pet ads fluctuate randomly, but when a peak in ads coincides with an earthquake, it is counted as a "prediction." It is possible that some animals can detect ground vibrations or low-frequency sounds that humans cannot, but so far instruments have not identified what, if anything, the animals might be sensing.

---

[*]Quoted in Spence, W. *et al.* 1993. Responses to Iben Browning's prediction of a 1990 New Madrid, Missouri, earthquake. *U.S. Geological Survey Circular 1083.*

---

earthquake and the 1985 Mexico City earthquake both occurred in previously identified seismic gaps.

Changes in elevation and tilting of the land surface have frequently preceded earthquakes and may be warnings of impending quakes. Extremely slight changes in the angle of the ground surface can be measured by tiltmeters. Tiltmeters have been placed on both sides of the San Andreas fault to measure tilting of the ground surface that is thought to result from increasing pressure in the rocks. Data from measurements in central California indicate significant tilting immediately preceding small earthquakes. Furthermore, extensive tiltmeter work performed in Japan prior to the 1964 Niigata earthquake clearly showed a relationship between increased tilting and the main shock. While more research is needed, such changes appear to be useful in making short-term earthquake predictions.

Other earthquake precursors include fluctuations in the water level of wells, changes in Earth's magnetic field, and the electrical resistance of the ground. These fluctuations are thought to result from changes in the amount of pore space in rocks due to increasing pressure. A change in animal behavior prior to earthquakes also is frequently mentioned. It

may be that animals are sensing small and subtle changes in Earth prior to a quake that humans simply do not sense.

Currently, only four nations—the United States, Japan, Russia, and China—have government-sponsored earthquake prediction programs. These programs include laboratory and field studies of the behavior of rocks before, during, and after major earthquakes as well as monitoring activity along major active faults. Most earthquake prediction work in the United States is done by the U.S. Geological Survey (USGS) and involves a variety of research into all aspects of earthquake-related phenomena.

The Chinese have perhaps one of the most ambitious earthquake prediction programs anywhere in the world, which is understandable considering their long history of destructive earthquakes. Their earthquake prediction program was initiated soon after two large earthquakes occurred at Xingtai (300 km southwest of Beijing) in 1966. The Chinese program includes extensive study and monitoring of all possible earthquake precursors. In addition, the Chinese also emphasize changes in phenomena that can be observed by seeing and hearing without the use of sophisticated instruments. They have had remarkable success in pre-

 FIGURE 8.19 Many of the approximately 242,000 people who died in the 1976 earthquake in Tangshan, China, were killed by collapsing structures. Many of the buildings were constructed of unreinforced brick, which has no flexibility, and quickly fell down during the earthquake.

dicting earthquakes, particularly in the short term, such as the 1975 Haicheng earthquake. They failed, however, to predict the devastating 1976 Tangshan earthquake that killed about 242,000 people (Fig. 8.19).

Great strides are being made toward dependable, accurate earthquake predictions, and studies are underway to assess public reactions to long-, medium-, and short-term earthquake warnings. Unless short-term warnings are actually followed by an earthquake, most people will probably ignore the warnings as they frequently do now for hurricanes, tornadoes, and tsunami.

EARTHQUAKE CONTROL

If earthquake prediction is still in the future, can anything be done to control earthquakes? Because of the tremendous forces involved, humans are certainly not going to be able to prevent earthquakes. However, there may be ways to dissipate the destructive energy of large earthquakes by releasing it in small amounts that will not cause extensive damage.

During the early to mid-1960s, Denver, Colorado, experienced numerous small earthquakes. This was surprising because Denver had not been prone to earthquakes in the past. In 1962, David M. Evans, a geologist, suggested that the earthquakes in Denver were directly related to the injection of contaminated waste water into a disposal well 3,674 m (12,050 ft) deep at the Rocky Mountain Arsenal, northeast of Denver. The U.S. Army initially denied that there was any connection, but a USGS study concluded that the pumping of waste fluids into fractured rocks beneath the disposal well decreased the friction on opposite sides of fractures and, in effect, lubricated them so that movement occurred, causing the earthquakes that Denver experienced.

Experiments conducted in 1969 at an abandoned oil field near Rangely, Colorado, confirmed the arsenal hypothesis. Water was pumped in and out of abandoned oil wells, the pore-water pressure in these wells was measured, and seismographs were installed in the area to measure any seismic activity. Monitoring showed that small earthquakes were occurring in the area when fluid was injected and that earthquake activity declined when fluids were pumped out. What the geologists were doing was starting and stopping earthquakes at will, and the relationship between pore-water pressures and earthquakes was established.

Based upon these results, some scientists have proposed that fluids be pumped into the locked segments of active faults to cause small- to moderate-sized earthquakes. They think that this would relieve the pressure on the fault and prevent a major earthquake from occurring. While this plan is intriguing, it also has many potential problems. For instance, there is no guarantee that only a small earthquake might result. Instead a major earthquake might occur, causing tremendous property damage and loss of life. Who would be responsible? Certainly, a great deal more research is needed before such an experiment is performed, even in an area of low population density.

It appears that until such time as earthquakes can be accurately predicted or controlled, the best means of defense is careful planning and preparation.

EARTH'S INTERIOR

During most of historic time, Earth's interior was perceived as an underground world of vast caverns, heat, and sulfur gases, populated by demons. By the 1860s, scientists knew that Earth's average density is 5.5 g/cm^3 and that pressure and temperature increase with depth. And even though Earth's interior is hidden from direct observation, scientists now have a reasonably good idea of its internal structure and composition.

Earth consists of concentric layers that differ in composition and density and are separated from adjacent layers by rather distinct boundaries (see Fig. 1.12). Recall that the outermost layer, or the **crust,** is Earth's very thin skin. Below the crust and extending about halfway to the center is the **mantle,** which comprises more than 80% of Earth's volume. The central part of Earth consists of a **core,** which is divided into a solid inner core and a liquid outer part.

As we noted earlier, the velocities of P- and S-waves are determined by the density and elasticity of the materials through which they travel. Both the density and elasticity of rocks increase with depth, but elasticity increases faster than density, resulting in a general increase in the velocity of seismic waves. P-waves travel faster than S-waves through all materials, but unlike P-waves, S-waves cannot be transmitted through a liquid because liquids have no shear strength (rigidity)—they simply flow in response to a shear stress.

As a seismic wave travels from one material into another of different density and elasticity, its velocity and direction of travel change. That is, the wave is bent, a phenomenon known as **refraction** (■ Fig. 8.20). Because seismic waves pass through materials of differing density and elasticity, they are continually refracted so that their paths are curved; the only exception is that wave rays are not refracted if their direction of travel is perpendicular to a boundary (Fig. 8.20). In that case they travel in a straight line.

In addition to refraction, seismic rays are also **reflected,** much as light is reflected from a mirror. Some of the energy

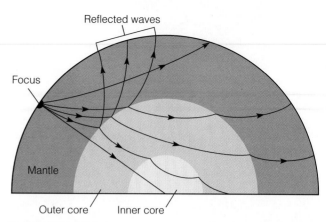

■ **FIGURE 8.20** Refraction and reflection of P-waves. When seismic waves pass through a boundary separating Earth materials of different density or elasticity, they are refracted, and some of their energy is reflected back to the surface.

of seismic rays that encounter a boundary separating materials of different density or elasticity within Earth is *reflected* back to Earth's surface (Fig. 8.20). If we know the wave velocity and the time required for it to travel from its source to the boundary and back to the surface, we can calculate the depth of the reflecting boundary. Such information is useful in determining not only the depths of the various layers within Earth, but also the depths of sedimentary rocks that may contain petroleum.

The Core

In 1906, R. D. Oldham of the Geological Survey of India postulated the existence of a core that transmits seismic waves at a slower rate than shallower Earth materials. We now know that P-wave velocity decreases markedly at a depth of 2,900 km (1,798 mi), indicating a major discontinuity now recognized as the core-mantle boundary (■ Fig. 8.21).

The sudden decrease in P-wave velocity at the core-mantle boundary causes P-waves entering the core to be refracted in such a way that very little P-wave energy reaches Earth's surface in the area between 103° and 143° from an earthquake focus (■ Fig. 8.22a). This area in which little P-wave energy is recorded by seismometers is a **P-wave shadow zone.**

In 1926, the British physicist Harold Jeffreys realized that S-waves were not simply slowed by the core, but were completely blocked by it. So in addition to a P-wave shadow zone, a much larger and more complete **S-wave shadow zone** exists (Fig. 8.22b). At locations greater than 103° from an earthquake focus, no S-waves are recorded, indicating that S-waves cannot be transmitted through the core. S-waves will not pass through a liquid, so it seems that the outer core must be liquid or behave as a liquid.

We can estimate the core's density and composition by using seismic evidence and laboratory experiments. Further-

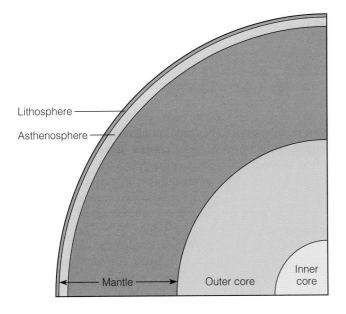

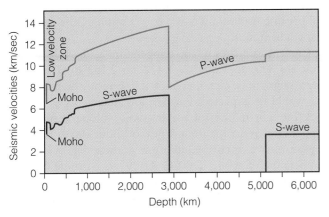

■ FIGURE 8.21 *Above:* Profiles showing seismic wave velocities versus depth. Several discontinuities are shown across which seismic wave velocities change rapidly.

more, meteorites, which are thought to represent remnants of the material from which the solar system formed, can be used to make estimates of density and composition. For example, meteorites composed of iron and nickel alloys may represent the differentiated interiors of large asteroids and approximate the density and composition of Earth's core. The density of the outer core varies from 9.9 to 13.0 g/cm^3. At Earth's center, the pressure is equivalent to about 3.5 million times normal atmospheric pressure.

The core cannot be composed of the minerals common at Earth's surface because even under the tremendous pressures at great depth, they would still not be dense enough to yield an average density of 5.5 g/cm^3. Both the outer and inner core are thought to be composed largely of iron, but pure iron is too dense to be the sole constituent of the outer core. Thus, it must be "diluted" with elements of lesser density. Laboratory experiments and comparisons with iron meteorites indicate that about 12% of the outer core may consist of sulfur and perhaps some silicon and small amounts of nickel and potassium.

In contrast, pure iron is not dense enough to account for the estimated density of the inner core. Most scientists think that perhaps 10 to 20% of the inner core also consists of nickel. These metals form an iron-nickel alloy that under the pressure at that depth is thought to be sufficiently dense to account for the density of the inner core. When the core formed during early Earth history, it was probably entirely molten and has since cooled to the point that its interior has crystallized.

The Mantle

Another significant discovery about Earth's interior was made in 1909 when the Yugoslavian seismologist Andrija Mohorovičić detected a discontinuity at a depth of about 30 km (19 mi). While studying arrival times of seismic waves

■ FIGURE 8.22 *Below:* (*a*) P-waves are refracted so that no direct P-wave energy reaches Earth's surface in the P-wave shadow zone. (*b*) The presence of an S-wave shadow zone indicates that S-waves are being blocked within Earth.

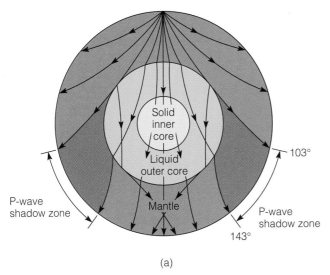

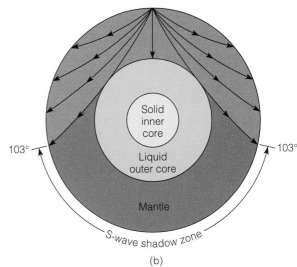

(a)

(b)

from Balkan earthquakes, Mohorovičić noticed that two distinct sets of P- and S-waves were recorded at seismic stations a few hundred kilometers from an earthquake's epicenter. He reasoned that one set of waves traveled directly from the epicenter to the seismic station, whereas the other waves had penetrated a deeper layer where they were refracted.

From his observations Mohorovičić concluded that a sharp boundary separating rocks with different properties exists at a depth of about 30 km. He postulated that P-waves below this boundary travel at 8 km/sec, whereas those above the boundary travel at 6.75 km/sec. When an earthquake occurs, some waves travel directly from the focus to the seismic station, while others travel through the deeper layer and some of their energy is refracted back to the surface. Waves traveling through the deeper layer travel farther to a seismic station, but they do so more rapidly than those in the shallower layer. The boundary identified by Mohorovičić separates the crust from the mantle and is now called the **Mohorovičić discontinuity,** or simply the **Moho.** It is present everywhere except beneath spreading ridges, but its depth varies: beneath the continents it averages 35 km (22 mi), but ranges from 20 to 90 km (12 to 56 mi); beneath the sea floor it is 5 to 10 km (3 to 6 mi) deep (Fig. 8.21).

Although seismic wave velocity in the mantle generally increases with depth, several discontinuities also exist. Between depths of 100 and 250 km (62 and 155 mi), both P- and S-wave velocities decrease markedly. This layer between 100 and 250 km deep is the **low-velocity zone;** it corresponds closely to the *asthenosphere,* a layer in which the rocks are close to their melting point and thus are less elastic; this decrease in elasticity accounts for the observed decrease in seismic wave velocity. The asthenosphere is an important zone because it may be where some magmas are generated. Furthermore, it lacks strength and flows plastically and is thought to be the layer over which the outer, rigid *lithosphere* moves.

The mantle's density varies from 3.3 to 5.7 g/cm³ and can be inferred rather accurately from seismic waves. The igneous rock *peridotite,* which contains mostly ferromagnesian minerals, is considered the most likely component of the mantle. Laboratory experiments indicate that it possesses physical properties that would account for the mantle's density and observed rates of seismic wave transmissions. Peridotite also forms the lower parts of igneous rock sequences thought to be fragments of the oceanic crust and upper mantle emplaced on land. In addition, peridotite occurs as inclusions in volcanic rock bodies such as *kimberlite pipes* that are known to have come from depths of 100 to 300 km (62 to 186 mi). These inclusions appear to be pieces of the mantle.

 # EARTH'S INTERNAL HEAT

During the nineteenth century, scientists realized that Earth's temperature in deep mines increases with depth. More recently, the same trend has been observed in deep drill holes, but even in these we can measure temperatures directly down to a depth of only a few kilometers. The temperature increase with depth, or **geothermal gradient,** near the surface is about 25°C/km, although it varies from area to area. In areas of active or recently active volcanism, the geothermal gradient is greater than in adjacent nonvolcanic areas, and temperature rises faster beneath spreading ridges than elsewhere beneath the sea floor.

Unfortunately, the geothermal gradient is not useful for estimating temperatures deep in Earth. If we were simply to extrapolate from the surface downward, the temperature at 100 km (62 mi) would be so high that in spite of the great pressure, all known rocks would melt. Yet except for pockets of magma, it appears that the mantle is solid rather than liquid because it transmits S-waves. Accordingly, the geothermal gradient must decrease markedly.

Current estimates of the temperature at the base of the crust are 800° to 1,200°C (1,472° to 2,192°F). The latter figure seems to be an upper limit: if it were any higher, melting would be expected. Furthermore, fragments of mantle rock brought to the surface by some volcanic eruptions, thought to have come from depths of about 100 to 300 km, appear to have reached equilibrium at these depths and at a temperature of about 1,200°C. At the core-mantle boundary, the temperature is probably between 3,500° and 5,000°C (6,332° and 9,032°F); the wide spread of values indicates the uncertainties of such estimates. If these figures are reasonably accurate, the geothermal gradient in the mantle is only about 1°C/km.

Considering that the core is so remote and so many uncertainties exist regarding its composition, only very general estimates of its temperature can be made. The maximum temperature at the center of the core is thought to be about 6,500°C (11,732°F), very close to the estimated temperature for the surface of the Sun!

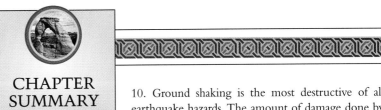
1. Earthquakes are vibrations of Earth caused by the sudden release of energy, usually along a fault.

2. The elastic rebound theory states that pressure builds in rocks on opposite sides of a fault until the inherent strength of the rocks is exceeded and rupture occurs. When the rocks rupture, stored energy is released as they snap back to their original position.

3. Seismology is the study of earthquakes. Earthquakes are recorded on seismographs, and the record of an earthquake is a seismogram.

4. The focus of an earthquake is the point where energy is released. Vertically above the focus on Earth's surface is the epicenter.

5. Most earthquakes occur within seismic belts. Approximately 80% of all earthquakes occur in the circum-Pacific belt, 15% within the Mediterranean-Asiatic belt, and the remaining 5% mostly in the interior of the plates or along oceanic spreading ridge systems.

6. The two types of body waves are P-waves and S-waves. Both travel through Earth, although S-waves do not travel through liquids. P-waves are the fastest waves and are compressional, while S-waves are shear. Surface waves travel along or just below Earth's surface.

7. The distance to the epicenter of an earthquake can be determined by the use of a time-distance graph of the P- and S-waves. Three seismographs are needed to locate the epicenter of an earthquake.

8. Intensity is a measure of the kind of damage done by an earthquake and is expressed by values from I to XII in the Modified Mercalli Intensity Scale.

9. Magnitude measures the amount of energy released by an earthquake and is expressed in the Richter Magnitude Scale. Each increase in the magnitude number represents about a 30-fold increase in energy released.

10. Ground shaking is the most destructive of all earthquake hazards. The amount of damage done by an earthquake depends upon the geology of the area, the type of building construction, the magnitude of the earthquake, and the duration of shaking. Tsunami are seismic sea waves that are usually produced by earthquakes. They can do tremendous damage to coastlines.

11. Earth is concentrically layered into an iron-rich core with a solid inner core and a liquid outer part, a rocky mantle, and an oceanic crust and continental crust.

12. Much of the information about Earth's interior has been derived from studies of P- and S-waves that travel through Earth. Laboratory experiments, comparisons with meteorites, and studies of inclusions in volcanic rocks provide additional information.

13. Density and elasticity of Earth materials determine the velocity of seismic waves. Seismic waves are refracted when their direction of travel changes. Wave reflection occurs at boundaries across which the properties of rocks change.

14. The behavior of P- and S-waves within Earth and the presence of P- and S-wave shadow zones allow scientists to estimate the density and composition of the interior and to estimate the size and depth of the core and mantle.

15. Earth's inner core is thought to be composed of iron and nickel, whereas the outer core is probably composed mostly of iron with 10 to 20% sulfur and other substances in lesser quantities. Peridotite is the most likely component of the mantle.

16. The geothermal gradient of 25°C/km cannot continue to great depths, otherwise most of Earth would be molten. The geothermal gradient for the mantle and core is probably about 1°C/km. The temperature at Earth's center is estimated to be 6,500°C.

IMPORTANT TERMS

aftershock
core
crust
earthquake
elastic rebound theory
epicenter
focus

geothermal gradient
intensity
low-velocity zone
magnitude
mantle
Modified Mercalli Intensity
   Scale

Mohorovičić discontinuity
   (Moho)
P-wave
P-wave shadow zone
reflection
refraction

Richter Magnitude Scale
seismology
surface wave
S-wave
S-wave shadow zone
tsunami

1. According to the elastic rebound theory:
   a. _____ earthquakes originate deep within Earth; b. _____ earthquakes originate in the asthenosphere where rocks are plastic; c. _____ earthquakes occur where the strength of the rock is exceeded; d. _____ rocks are elastic and do not rebound to their former position; e. _____ none of these.

2. The majority of all earthquakes occur:
   a. _____ in the Mediterranean-Asiatic belt; b. _____ in the interior of plates; c. _____ in the circum-Atlantic belt; d. _____ in the circum-Pacific belt; e. _____ along spreading ridges.

3. An epicenter is:
   a. _____ the location where rupture begins; b. _____ the point on Earth's surface vertically above the focus; c. _____ the same as the hypocenter; d. _____ the location where energy is released; e. _____ none of these.

4. A qualitative assessment of the kinds of damage done by an earthquake is expressed by:
   a. _____ seismicity; b. _____ dilatancy; c. _____ magnitude; d. _____ intensity; e. _____ none of these.

5. How much more energy is released by a magnitude 5 earthquake than by one of magnitude 2?
   a. _____ 2.5 times; b. _____ 3 times; c. _____ 30 times; d. _____ 1,000 times; e. _____ 27,000 times.

6. A tsunami is a:
   a. _____ measure of the energy released by an earthquake; b. _____ seismic sea wave; c. _____ precursor to an earthquake; d. _____ locked portion of a fault; e. _____ seismic gap.

7. The average density of Earth is _____ g/cm^3.
   a. _____ 12.0; b. _____ 5.5; c. _____ 6.75; d. _____ 1.0; e. _____ 2.5.

8. When seismic waves travel through materials having different properties, their direction of travel changes. This phenomenon is wave:
   a. _____ elasticity; b. _____ energy dissipation; c. _____ refraction; d. _____ deflection; e. _____ reflection.

9. A major seismic discontinuity at a depth of 2,900 km is the:
   a. _____ core-mantle boundary; b. _____ oceanic crust–continental crust boundary; c. _____ Moho; d. _____ inner core–outer core boundary; e. _____ lithosphere-asthenosphere boundary.

10. How does the elastic rebound theory explain how energy is released during an earthquake?

11. What is the relationship between plate boundaries and earthquakes?

12. Explain the difference between intensity and magnitude and between the Modified Mercalli Intensity Scale and the Richter Magnitude Scale.

13. Explain how tsunami are produced and why they are so destructive.

14. Explain how seismic waves are refracted and reflected.

15. What is the significance of the S-wave shadow zone?

16. On a map of the United States, locate an earthquake that is 1,600 km (1,000 mi) from each of these cities: Chicago, Houston, Seattle.

17. Why do seismic stations need more than one type of seismograph?

18. What are some of the factors that influence earthquake intensity?

19. How can scientists determine the composition and density of Earth's interior?

1. If Earth were completely solid and had the same composition and density throughout, how would P- and S-waves behave as they traveled through Earth?

2. Some geologists think that by pumping fluids into locked segments of active faults, small- to moderate-sized earthquakes can be generated. These earthquakes would relieve the built-up pressure along the fault and thus prevent a major earthquake. Discuss whether this idea is really feasible. Are there any guarantees that the earthquake will be limited to only the locked segments along the fault? Can geologists be sure that the resulting earthquake will not be a major one? Considering the social, political, and economic consequences, do you think such an experiment will ever be tested on an active fault?

3. Using the arrival times of P- and S-waves shown in the chart and the graph in Figure 8.9, calculate how far away from each seismograph station the earthquake occurred. How would you determine the epicenter of this earthquake?

	ARRIVAL TIME OF P-WAVE	ARRIVAL TIME OF S-WAVE
Station A:	2:59:03 P.M.	3:04:03 P.M.
Station B:	2:51:16 P.M.	3:01:16 P.M.
Station C:	2:48:25 P.M.	2:55:55 P.M.
Distance from the earthquake:		
Station A: _____		
Station B: _____		
Station C: _____		

4. One of the goals of earthquake prediction is to prevent loss of life and injury by evacuating people from an area before an earthquake occurs. Discuss the social, economic, and political

problems associated with such evacuations. For example, who would pay for lost earnings to businesses and individuals because they would not be able to operate or work during the time of evacuation? Who is responsible if looting occurs in the evacuated area? What will happen to public confidence if the predicted earthquake fails to occur?

5. Suppose that the energy along a fault is sufficient to yield a magnitude 7 earthquake. How many earthquakes of magnitude 4 would be necessary to release all of this energy?

ADDITIONAL READINGS

Bolt, B. A. 1996. *Earthquakes.* New York: W. H. Freeman.

Canby, T. Y. 1990. California earthquake—prelude to the big one? *National Geographic* 177, no. 5: 76–105.

Davidson, K. 1994. Predicting earthquakes: Can it be done? *Earth* 3, no. 3: 56–63.

Dawson, J. 1993. CAT scanning the Earth. *Earth* 2, no. 3: 36–41.

Dvorak, J., and T. Peek. 1993. Swept away: The deadly power of tsunamis. *Earth* 2, no. 4: 52–59.

Fischman, J. 1992. Falling into the gap: A new theory shakes up earthquake predictions. *Discover* October 1992: 56–63.

Fowler, C. M. R. 1990. *The solid Earth.* New York: Cambridge Univ. Press.

Frohlich, C. 1989. Deep earthquakes. *Scientific American* 260, no. 1: 48–55.

Hanks, T. C. 1985. *National earthquake hazard reduction program: Scientific status.* U.S. Geological Survey Bulletin 1659.

Jeanloz, R. 1990. The nature of the Earth's core. *Annual Review of Earth and Planetary Sciences* 18: 357–86.

Jeanloz, R., and T. Lay. 1993. The core-mantle boundary. *Scientific American* 268, no. 5: 48–55.

Johnston, A. C., and L. R. Kanter. 1990. Earthquakes in stable continental crust. *Scientific American* 262, no. 3: 68–75.

Lomnitz, C. 1994. *Fundamentals of earthquake prediction.* New York: John Wiley & Sons.

Ritchie, D. 1994. *Encyclopedia of earthquakes and volcanoes.* New York: Facts On File.

Wysession, M. 1995. The inner workings of the Earth. *American Scientist* 83, no. 2: 134–46.

WORLD WIDE WEB SITES

For current updates and exercises, log on to http://www.wadsworth.com/geo.

SEISMICITY MAPS
http://quake.wr.usgs.gov/QUAKES/WEEKREPS/weekly.html

Weekly seismic maps and reports from the U.S. Geological Survey are provided for California, the United States, and the world.

SAN FRANCISCO QUAKES
http://www.slip.net/~dfowler/1906/museum.html

This Web page for the Museum of the City of San Francisco gives information on the 1906 earthquake and the 1989 Loma Prieta earthquake.

CHAPTER
9

Volcanoes and Igneous Activity

The 1989 eruption of Mount Redoubt, Alaska, seen at sunset.

Prologue

One of history's best-known volcanic eruptions was the eruption of Mount Vesuvius that destroyed Pompeii, Herculaneum, and Stabiae in A.D. 79 (□ Fig. 9.1).

The eruption was described in detail by the Roman writer Pliny the Younger, whose uncle, Pliny the Elder, died while investigating the disaster. The three cities were buried by ash, which later hardened into rock, and were forgotten until 1709 when a well was dug into the ruins. A frenzy of mining for ancient artifacts ensued, and today the excavated cities are great tourist attractions.

Pompeii's most famous relics are molds of human bodies, often with startling detail. After the victims were buried, the ash hardened around them before the bodies decayed. Of the estimated 20,000 residents of Pompeii, only about 2,000 were preserved in the ruins, leading most geologists and historians to assume that the rest had time to escape. In recent years, new evidence has changed this picture dramatically. In 1982, the ancient waterfront of Herculaneum was excavated. The excavators found hundreds of skeletons in postures that show clear evidence of sudden and violent death. The eruption of Vesuvius was far more violent and deadly than anyone had previously guessed.

The ash layers over the ruins indicate that the first stages of the eruption were powered by gas pressures great enough to throw the ash high into the air. By the time the ash settled to the ground, it had cooled to a safe temperature. Later in the eruption, the gas pressure began to falter. As the gas pressure dropped, the column of hot ash above the volcano collapsed and fell. By the time the ash reached the ground, it was moving at high speed and rolled down the volcano as a *nuée ardente,* or *pyroclastic flow,* a lethal cloud of hot ash and superheated gases. This sequence of events happened several times. Possibly, many refugees from Pompeii were also caught in flight by hot ash flows and lie buried beyond the area that has been excavated. Stabiae, the third town destroyed, was out of reach of the ash flows but was buried by windblown ash.

Vesuvius is one of a chain of volcanoes along the southern coast of Italy, where the Mediterranean crust is

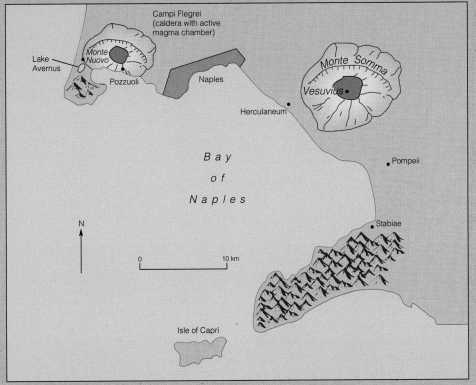

□ FIGURE 9.1 Mount Vesuvius and its vicinity.

being consumed in a subduction zone. The city of Naples lies easily within reach of a violent eruption of Vesuvius. Perhaps no major city in the world is in more imminent risk from a volcanic eruption. Not far to the northwest of Naples lies a volcanic area called the Campi Flegrei, which last erupted in 1538. The Campi Flegrei is most interesting for the pulsations of its magma chamber. Within the Campi Flegrei is the small coastal town of Pozzuoli, which has risen and sunk a number of times through the centuries as the magma beneath it has ebbed and flowed. In Roman times Pozzuoli stood high, but by the Middle Ages it had sunk several meters, allowing former temples to be flooded. By 1800 it had risen again. Holes bored by marine animals in the formerly sunken temples were among the first clear evidence for active movements of Earth's crust to be found by early geologists.

More recently, Pozzuoli has been rising, but of more concern to the townspeople were the nearly incessant earthquakes that literally shook the town apart in 1983 and 1984, forcing much of the populace to leave. Although the earthquakes have quieted, many of the buildings in Pozzuoli are now unsafe.

 # INTRODUCTION

Volcanism is the process by which molten rock from Earth's interior reaches the surface. Molten rock beneath the surface is called *magma,* whereas molten rock erupted onto the surface is termed *lava*. **Volcanoes,** the vents through which lava erupts, are spectacular and important geological features, but they can also present tremendous hazards to human life. Because of volcanoes' danger, mystery, and obvious connections to Earth's interior, many cultures since earliest times have associated them with death, divine wrath, and the underworld. But volcanoes are just one part of the process by which molten rock from Earth's interior is added to the crust. For every lava flow that erupts from a volcano, far larger quantities of molten rock harden beneath the surface to form *igneous rocks* (see Chapter 3) that may be exposed long afterward by erosion. These masses of subsurface igneous rocks are called **intrusions.** Table 9.1 lists a few of the ways igneous activity and volcanism affect Earth and its systems. In this chapter we will explore intrusions, volcanoes, and the processes that form and sometimes destroy volcanoes.

 # MOLTEN ROCK IN EARTH

Two fundamental questions about volcanoes are, Why do rocks melt in the first place? and Why does molten rock rise? As Chapter 8 explained, the breakdown of radioactive elements deep in Earth produces considerable heat. However, the high pressure in Earth's interior inhibits melting. The melting point temperature of most materials increases as pressure increases because the atoms in a material must be able to move apart freely if the material is to melt. Even though temperatures deep in Earth are far hotter than any lavas that erupt at the surface, most of the interior is solid. Here and there, however, small amounts of rock deep in Earth melt. As warm, light material slowly rises and cooler, denser material sinks, Earth's mantle flows slowly—a process called *convection* (see Chapter 7). If hot material rises quickly to a shallow level where the melting point is low, the material may melt. The melting points of igneous rocks are also lowered by water. Thus, water deep in the crust and upper

TABLE 9.1 Volcanoes, Igneous Activity, and Earth Systems	
Solid Earth	Plate tectonics drives most igneous activity. Igneous activity concentrates silica, potassium, and other elements in the crust. Igneous activity along mid-ocean ridges creates the crust of the ocean floors.
Atmosphere	Volcanic carbon dioxide may affect global climate. Fine particulate matter may intercept sunlight and cool Earth.
Hydrosphere	Volcanic eruptions release water from Earth's interior. Chemicals in volcanic ash can pollute water bodies. Heating of subterranean water creates hot springs and geysers.
Cryosphere	Volcanic carbon dioxide may warm global climate and inhibit glaciation. Fine particulate matter may intercept sunlight and cool Earth, contributing to glaciation. High volcanoes often support glaciers.
Biosphere	Volcanic eruptions destroy life, but they also contribute soil nutrients as igneous rocks are weathered. Effects on atmosphere and hydrosphere also have an impact on life.
Extraterrestrial	Particulate matter from large eruptions affects Earth's *albedo,* the fraction of sunlight reflected back into space.

mantle is important in lowering the melting point of rocks and generating magma.

Magma rises to the surface because it is less dense than solid rock. The weight of overlying rocks on the magma chamber helps force magma upward. Gases like water vapor in the magma are also under high pressure and provide the pressure to power explosive volcanic eruptions.

PLATE TECTONICS AND VOLCANOES

Volcanoes and igneous activity occur in three principal plate tectonic settings: spreading centers or rifts, hot spots, and subduction zones (Fig. 9.2). The type of magma and volcanic eruption style depend on the plate tectonic setting and also on whether the crust is oceanic or continental. Earth's mantle consists mostly of iron and magnesium silicate min-erals. When mantle material melts, it usually forms magma of basalt composition. Volcanoes fed by such magma erupt mostly basalt, and intrusions of such magma solidify to form gabbro. Whether at spreading centers, hot spots, or subduction zones, volcanoes on oceanic crust, which is mostly basalt, erupt almost entirely basalt.

Igneous activity on continental crust is far more complex. Magma on the continents may be derived directly from the mantle, from melting of continental crust, or most likely from both sources. Thus, a much greater variety of igneous rock types are found on the continents than in the ocean basins.

Spreading centers and hot spots on continents evolve in either of two ways. If spreading is fast, the crust breaks apart. New basaltic magma comes from the mantle. If spreading is slow, rising magma has a long time to melt the lower crust and the magma is more silica rich. Basalt volcanism is straight from the mantle; rhyolite results from melted crust.

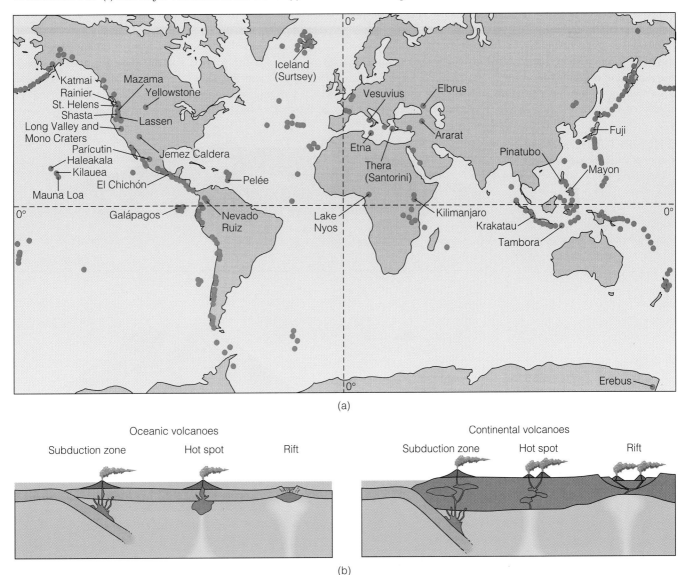

FIGURE 9.2 (a) The major volcanoes of the world. (b) Plate tectonic settings of volcanoes.

(a)

(b)

The earliest igneous activity is usually basaltic volcanism and intrusion of gabbro. The crust of the continent may eventually be broken completely apart so that new oceanic crust forms between the two pieces and a new ocean begins to form. This process happened to all the present continents as Pangaea split apart, and it is just beginning now in the Red Sea. Sometimes, as during the separation of Africa and South America, **flood basalts,** which are basalt lava flows of great volume, may be erupted. If the spreading occurs slowly, however, and especially if the continent does not split apart, the lower part of the continental crust may also melt. In such cases, large amounts of rhyolite volcanism may occur as well as basaltic volcanism.

Hot spots on continents display a similar pattern. Where the hot spot produces magma at high rates, and the magma can reach the surface quickly, volcanism tends to be basaltic. Great flood basalts in Oregon and Washington, the Deccan region of western India, and northeastern Siberia probably mark the former locations of such hot spots. In other cases, hot spots may melt the continental crust and generate huge quantities of rhyolite and granite. Yellowstone National Park is an example of a hot spot that has caused rhyolite volcanism. About 1,400 million years ago, a large area extending from Kansas and Oklahoma to Illinois was affected by a crustal heating episode that resulted in widespread granitic intrusions and rhyolite volcanism.

Subduction zones on the continents display the widest variety of igneous activity. When a subduction zone first forms, the descending plate begins to melt, and some of the resulting magma penetrates the continental crust and reaches the surface. As one might expect, this magma derived from the melting of basaltic oceanic crust is basaltic. As time goes on, rising magma heats the continental crust and melts some of it. Thus, the magma in a continental-oceanic subduction zone becomes progressively more silica rich. Volcanoes at continental-oceanic subduction zones erupt predominantly andesite and eventually rhyolite, while vast quantities of granitic magma invade the crust to form **batholiths,** the largest intrusions.

IGNEOUS ROCKS AND VOLCANOES

It first appears that natural phenomena could hardly be farther apart than the linking of silica tetrahedra on the atomic scale and the 1980 eruption of Mount St. Helens or the construction of the Hawaiian Island chain. Yet volcanic phenomena are directly related to the type of magma that feeds the volcano, and magma's physical properties depend, to a great degree, on its composition. The silica content of magma determines not only the type of igneous rock that forms but also the sort of volcano that the magma builds (▣ Fig. 9.3) and the style of volcanic eruption that occurs. Many volcanic phenomena can be related to *Bowen's series* (▣ Fig. 9.4), which was introduced in Chapter 3. The more

▣ FIGURE 9.3 Different types of volcanoes and their relative sizes.

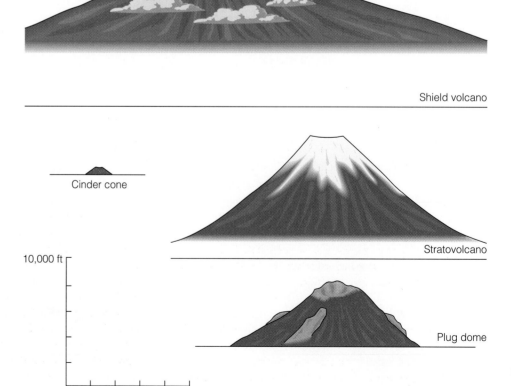

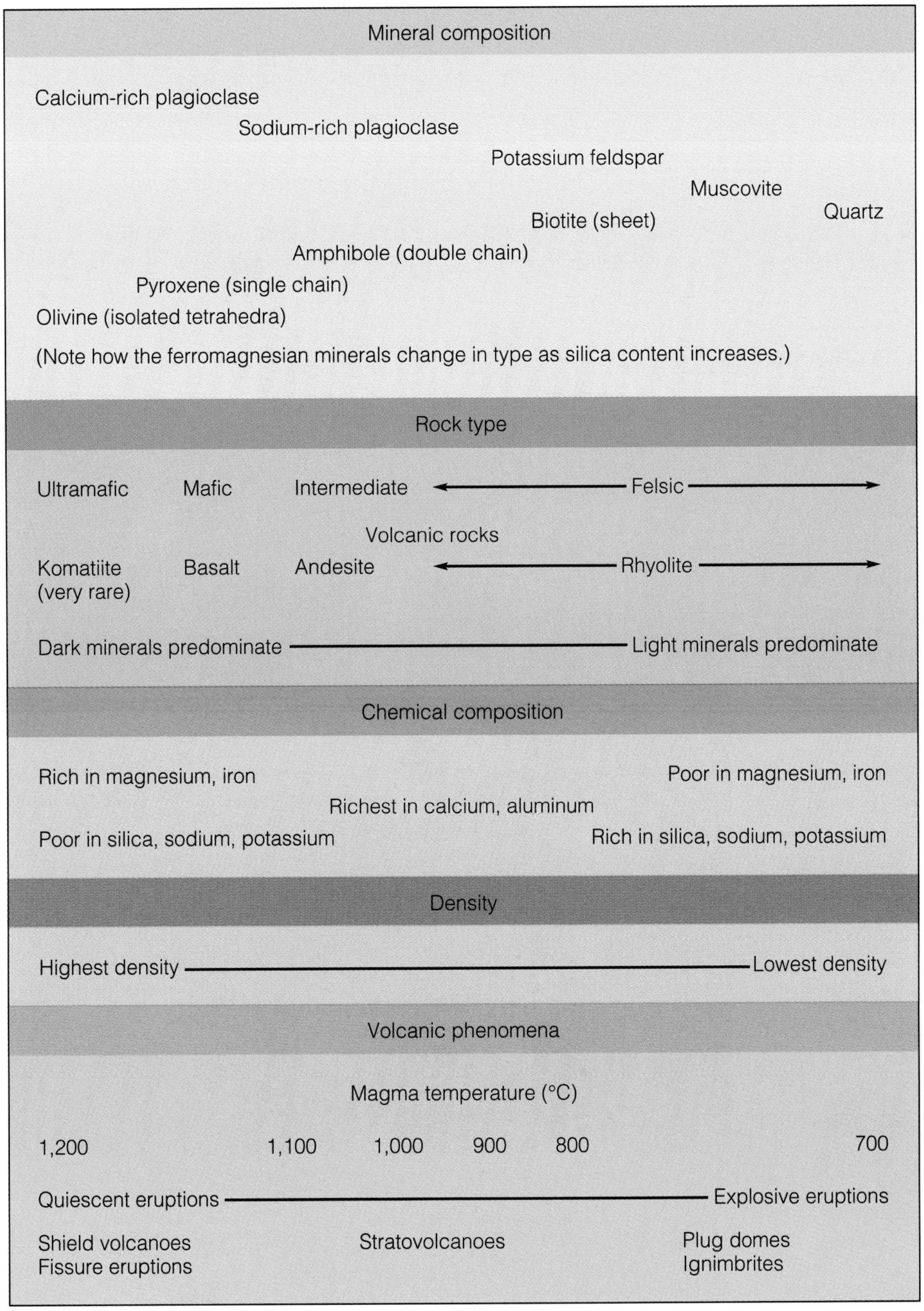

Mineral composition						
Calcium-rich plagioclase						
	Sodium-rich plagioclase					
		Potassium feldspar				
				Muscovite		
			Biotite (sheet)		Quartz	
	Amphibole (double chain)					
	Pyroxene (single chain)					
Olivine (isolated tetrahedra)						

(Note how the ferromagnesian minerals change in type as silica content increases.)

Rock type

Ultramafic   Mafic   Intermediate ◄——————— Felsic ——————————►

Volcanic rocks

Komatiite   Basalt   Andesite ◄——————— Rhyolite ——————————►
(very rare)

Dark minerals predominate ——————————— Light minerals predominate

Chemical composition

Rich in magnesium, iron   Poor in magnesium, iron

Richest in calcium, aluminum

Poor in silica, sodium, potassium   Rich in silica, sodium, potassium

Density

Highest density ————————————————— Lowest density

Volcanic phenomena

Magma temperature (°C)

1,200   1,100   1,000   900   800   700

Quiescent eruptions ——————————————— Explosive eruptions

Shield volcanoes   Stratovolcanoes   Plug domes
Fissure eruptions   Ignimbrites

FIGURE 9.4 Bowen's series and its relation to volcanic phenomena.

silica rich a magma is, the higher its resistance to flow, or **viscosity.** (A good illustration of viscosity in everyday life is the way water pours easily compared to a milk shake.) Although the linking of silica tetrahedra on the molecular scale and the destruction of Pompeii seem enormously removed from one another, they are actually directly connected.

As igneous rocks crystallize, silica tetrahedra in the magma link to form chains and networks. The higher the silica content, the more numerous are the chains. As the magma flows, the chains tangle and interfere with each other, and the magma becomes stiffer and flows less easily. Thus, the higher a magma's silica content, the higher its viscosity. Silica-poor magmas are fluid and allow gases to bubble out and escape easily, but silica-rich magmas are viscous. They trap gases and often allow pressures to build up to catastrophic levels. Volcanoes that erupt silica-poor lavas like Kilauea in Hawaii are sometimes so nonviolent that tourists can safely approach within a few hundred meters of an eruption. In contrast, when Mount St. Helens, with a very silica-rich magma, erupted, being 25 km away (15 mi) was not safe (see □ Perspective 9.1). Silica-rich magmas are associated with the most violent eruptions; unfortunately, these are the magmas that occur on the continents where people live.

Silica-poor mafic rocks like *basalt* have low viscosity. Because basalt is highly fluid and allows gases to escape readily, basaltic volcanoes tend to have mild eruptions and to

The Eruption of Mount St. Helens

Mount St. Helens in southwestern Washington is the youngest of the Cascade Range volcanoes, being about 30,000 years old. Before 1980 it was a medium-sized Cascade peak (2,950 m or 9,677 ft) with the most nearly perfect cone of any volcano in the Pacific Northwest (☐ Fig. 1). The favorite vantage point for photographing the volcano was Spirit Lake, a small lake at the north foot of the mountain created about 3,000 years ago when mudflows from the volcano dammed the headwaters of the Toutle River. In 1975, three scientists of the U.S. Geological Survey studied Mount St. Helens and predicted that it would probably erupt before the end of the twentieth century. As it turned out, they had only five years to wait.

When swarms of earthquakes began under Mount St. Helens, geologists converged on the mountain in time for its first eruption on March 28, 1980. After a few weeks of mild steam explosions, activity ceased. The north side of the mountain began to swell ominously, pushing outward by 100 m (300 ft). In mid-May 1980, earthquakes resumed, and observers began watching the volcano closely. On Sunday, May 18, at 8:32 A.M. a strong earthquake shook the volcano. The bulging north slope broke up and rushed down the mountain as a great landslide (☐ Fig. 2).

☐ **FIGURE 1** Mount St. Helens before its eruption in 1980.

☐ **FIGURE 2** On May 18, 1980, the north side of Mount St. Helens collapsed in a huge landslide, releasing a tremendous lateral blast.

☐ FIGURE 3 Mount St. Helens after its eruption.

As the landslide rushed down the mountain, it opened the vent of the volcano, and a tremendous lateral blast of hot ash and gas rushed out. The blast, moving even faster than the landslide, broke off trees 60 cm (2 ft) in diameter. Victims caught in the blast were burned to death or died when their lungs clogged with ash.

The eruption melted much of the glacial ice and fresh snow on the peak, and this water mixed with the volcanic ash and landslide debris to make huge quantities of mud. Mudflows poured down the Toutle River and other rivers, destroying bridges as they went. The mud poured into the Columbia River, blocking the shipping channel to Portland.

After the initial lateral blast, Mount St. Helens vented about 1 km³ (0.25 mi³) of ash, which fell in measurable amounts as far away as Wyoming. Spokane and Yakima received several centimeters. Fortunately, the ashfall was not heavy enough to collapse many structures, nor did it last long enough to disrupt essential services or seriously damage crops. Had the ashfall lasted more than a few days, the effects could have been far worse.

After the eruption, Mount St. Helens was radically changed. Where the summit had been, there was now a huge amphitheater, whose floor lay 640 m (2,100 ft) below the old summit elevation (☐ Fig. 3). The former contents of the amphitheater were now spread across the floor of Spirit Lake and down the valleys leading away from the volcano. Pasty rhyolite lava oozed slowly out of the vent to build a dome, punctuated by occasional steam explosions. Since 1980, the dome has grown to 850 m (2,800 ft) in diameter and 250 m (810 ft) high. It may very well continue to grow until it fills the amphitheater and restores the conical shape of the mountain.

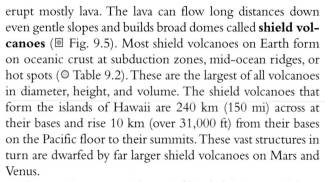

FIGURE 9.5 Haleakala on the Hawaiian island of Maui shows the characteristic broad profile of a shield volcano. The vertical relief in this picture is more than 3,050 m (10,000 ft).

erupt mostly lava. The lava can flow long distances down even gentle slopes and builds broad domes called **shield volcanoes** (Fig. 9.5). Most shield volcanoes on Earth form on oceanic crust at subduction zones, mid-ocean ridges, or hot spots (Table 9.2). These are the largest of all volcanoes in diameter, height, and volume. The shield volcanoes that form the islands of Hawaii are 240 km (150 mi) across at their bases and rise 10 km (over 31,000 ft) from their bases on the Pacific floor to their summits. These vast structures in turn are dwarfed by far larger shield volcanoes on Mars and Venus.

On occasion, great volumes of basalt have erupted from fissures in the crust without building any sort of volcanic cone. Individual flows in such eruptions may contain more than 100 km^3 (24 mi^3) of lava and extend for hundreds of kilometers. Such vast outpourings are called flood basalts (Fig. 9.6). As noted earlier, flood basalts are usually related

FIGURE 9.6 The Columbia Plateau of Oregon and Washington was covered by thick flood basalts about 10 million years ago.

to hot spots or the rifting of continents and are found in Oregon and Washington, India, and Siberia; they also make up several large plateaus on the ocean floor (see Chapter 18). Some individual lava flows in flood basalt regions contain hundreds of cubic kilometers of lava that all erupted within a week or so.

The commonest intermediate volcanic rock is *andesite*. Andesite volcanoes often have violent eruptions and produce large quantities of pyroclastic debris, which piles up near the vent. The lava is viscous, solidifies rapidly, and also accumulates near the vent. The resulting mountain, a **stratovolcano** or composite volcano (Fig. 9.7), slopes steeply

FIGURE 9.7 Mount Shasta, in northern California, is an excellent example of a stratovolcano with a parasitic cone. The vertical relief of this figure and Figure 9.5 are comparable.

TABLE 9.2   Plate Tectonics and Volcanoes

PLATE TECTONIC SETTING	PRINCIPAL VOLCANIC ROCKS	PRINCIPAL TYPES OF VOLCANISM	EXAMPLE
Spreading Center			
Oceanic crust (mid-ocean ridge)	Basalt	Shield volcanoes Flood basalts	Iceland
Continental crust (rapid rifting)	Basalt	Flood basalts	Mesozoic flood basalts in Africa and South America
Continental crust (slow rifting)	Rhyolite	Plug domes Stratovolcanoes	East African Rift Valley
Hot Spot			
Oceanic crust	Basalt	Shield volcanoes Flood basalts	Hawaii Undersea volcanic plateaus
Continental crust (rapid production of magma)	Basalt	Flood basalts	Columbia Plateau in Oregon-Washington
Continental crust (slow production of magma)	Rhyolite	Plug domes Ignimbrites	Yellowstone; central U.S. 1,400 million years ago
Subduction Zone			
Ocean-ocean	Basalt	Shield volcanoes Stratovolcanoes	Western Aleutians, Marianas
Continent-ocean	Andesite Rhyolite	Andesite and rhyolite Stratovolcanoes	Cascades, Andes, Indonesia, Japa

near the summit and more gently near the base, with the characteristic profile most people associate with volcanoes. Stratovolcanoes are composed of alternating layers of lava, mudflow deposits, and pyroclastic debris. They can be more than 30 km (20 mi) across at the base and 4,500 m (14,000 ft) high. Mount Fuji (Japan), Vesuvius (Italy), and Mount Rainier (Washington State) are all stratovolcanoes. Most andesite stratovolcanoes occur at subduction zones.

Rhyolite, the commonest silica-rich volcanic rock, is such a viscous lava that it flows little or not at all when erupted, and it tends to be accompanied by violent eruptions. Sometimes, rhyolite emerges from a new vent and oozes out to form a *plug dome* (☐ Fig. 9.8) up to a few hundred meters high and a few kilometers across. Other rhyolite volcanoes build up cones that externally look much like stratovolcanoes, but are built of overlapping domes with much pyroclastic debris. Mount St. Helens and Mount Lassen in California, the two most recently active volcanoes in the continental United States, are of this type. Mount St. Helens is now building a dome within its crater that may eventually fill the crater and restore the conical appearance of the mountain. Rhyolite volcanoes form at subduction zones and also during continental rifting and some hot spot activity.

☐ FIGURE 9.8 The Mono Craters in California are good examples of small plug domes.

VOLCANIC ERUPTIONS AND VOLCANIC HAZARDS

Volcanoes produce a wide variety of materials. Foremost in most people's minds is **lava** (▣ Fig. 9.9). Lava can range in consistency from a free-flowing liquid to a pasty mass that barely spreads beyond the eruption vent. Although lava and volcanoes are synonymous to many people, lava is actually only a small part of the output of volcanoes and in itself constitutes a comparatively minor safety hazard. Lava flows tend to move slowly, and there is usually enough advance warning to permit escape.

Volcanoes erupt huge quantities of broken rock (▣ Fig. 9.10), collectively called **pyroclastic** (from the Greek words for "fire" and "broken"). Large fragments several centimeters across or larger are called *bombs*. Volcanic rock dust is called ash. **Volcanic ash** has nothing to do with burning; it is nothing more than pulverized rock. Small volcanic eruptions frequently build up small cones of pyroclastic debris around the vent called **cinder cones** (▣ Fig. 9.11). Cinder cone eruptions may or may not be accompanied by lava flows.

Pyroclastic materials can be a serious hazard. Ash can pile up on roofs and cause buildings to collapse, be a respiratory hazard, render water sources undrinkable, bury crops, and damage aircraft engines. Before the advent of rapid communication and transportation for relief efforts, starvation due to environmental disruption was probably the greatest danger from volcanoes. Close to the volcano, larger blocks of broken rock can be a danger.

The 1980 eruption of Mount St. Helens made many nonscientists aware, for the first time, of how important *mud* is in volcanic eruptions. Volcanic ash mixes with water from several sources to make enormous quantities of mud. The

▣ FIGURE 9.9 A lava flow.

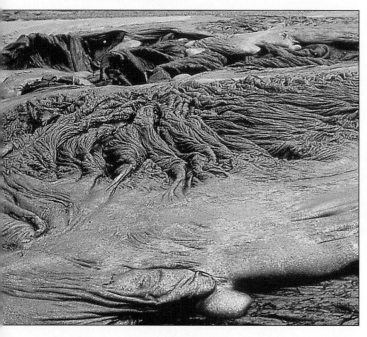

▣ FIGURE 9.10 Pyroclastic rocks. (*a*) Volcanic bombs. (*b*) Tuff. These pinnacles in Crater Lake National Park, Oregon, are carved from tuff that formed from an ash flow. Note the large lumps of pumice protruding from the tuff. (*c*) Volcanic ash. Here, ash is being cleaned off an airplane in Alaska.

(a)

(b)

(c)

FIGURE 9.11 This cinder cone eroded by wave action in Hawaii shows the dipping layers that make up a typical cinder cone.

Triggering Volcanic Eruptions

People frequently wonder whether drilling into a volcano or a magma chamber might trigger a volcanic eruption. It has already happened. In 1977, a drill hole was bored in the Krafla geothermal field in Iceland to tap high-pressure steam for electric power generation. The steam at Krafla is produced when underground water is heated by magma near the surface. Suddenly, to the astonishment of the drillers, lava began spurting out of the hole. Evidently, the drill hole intersected a fissure that connected with the magma chamber. The lava spurted for a few minutes; then the drill hole returned to venting steam. About a cubic meter of lava erupted, making this the smallest recorded "volcanic eruption" in history.

The conduits in volcanoes that feed lava to the surface are tens, even hundreds, of meters in diameter. A drill hole a few centimeters in diameter is just not big enough to affect a volcano significantly; even basalt magma is much stiffer than cold motor oil or honey and cannot flow rapidly through a small opening. Drilling into a volcano does not trigger large eruptions, nor can it be used to control or prevent eruptions.

FIGURE 9.12 This volcanic mudflow formed when the Colombian volcano Nevado Ruiz erupted in 1985, melting its summit icecap. The flow swept over the town of Armero, killing 20,000 in minutes.

volcano itself erupts huge quantities of water vapor that often condenses as rain. If the volcano has a snow cover or glaciers, the frozen water can melt suddenly. This happened at Mount St. Helens and also when Nevado Ruiz in Colombia erupted in 1985, creating mudflows that killed 20,000 people (Fig. 9.12).

Focus on the Environment: Volcanoes and Global Climate

Volcanic eruptions can have global and regional effects caused by widespread ash deposition, aerosols, and gases. Mount St. Helens deposited 10 cm (4 in) of ash up to 300 km (186 mi) away, but far greater ash eruptions have occurred in the geologic past. One of the largest ashfalls known occurred over eastern North America during the Ordovician Period about 450 million years ago. Several hundred cubic kilometers of ash blanketed the shallow seafloor in a layer tapering from a few meters thick in Alabama to a few centimeters thick in Minnesota. There are also thick volcanic ash layers in northern Europe, which at that time was close to southeastern North America. Although the ashfall certainly exterminated marine life over a very large area, it did not result in any major extinctions. Evidently, life quickly recolonized the devastated area.

Volcanic particles high in the atmosphere can reflect sunlight back to space, cooling Earth. Although volcanic ash receives the most attention in news reports, the most significant global effects of volcanic eruptions are actually produced by aerosols, tiny droplets of liquid, especially sulfuric acid. The best-documented global cooling by a volcanic eruption occurred after the eruption of the Indonesian volcano Tambora in 1815. The year 1816 was called "the year without a summer"; both North America and Europe experienced unusually cool temperatures and widespread crop failures. Summer frosts and snowfalls occurred in the mountains of New England. Observers often describe the weakening of sunlight by thick volcanic particulates as persistent haze or "dry fog." Accounts of unusual cold and persistent haze have been recorded a number of times in history and may indicate other major volcanic eruptions.

The eruption of Tambora vented about 150 km^3 (36 mi^3) of ash, but this material is only a fraction of the many thousands of cubic kilometers of ash vented by caldera collapses in the geologic past. Such huge eruptions could cause serious global cooling. For example, a great eruption in Indonesia about 75,000 years ago coincided roughly with the beginning of the last major Pleistocene glacial advance (see Chapter 11), and ejecta from this eruption may have played a part in the global cooling that led to the ice advance.

Flood basalts may also have very significant global effects. The sulfur content of basaltic lava is about 10 times that of rhyolite, and some flood basalt flows contain hundreds of cubic kilometers of basalt, all erupted within a few days or weeks. Thus, flood basalt eruptions could have great global climatic effects. Major flood basalt eruptions in India about 65 million years ago occurred near the time of the mass extinction at the end of the Cretaceous period (see Chapter 1), and another episode of flood basalt eruption in Siberia about 248 million years ago coincided with a mass extinction at the end of the Permian Period. The Permian extinction was the greatest mass extinction in Earth's history.

Volcanic gases may change the chemistry of the atmosphere. Large eruptions release large amounts of carbon dioxide, sulfur dioxide, hydrogen chloride, and hydrogen fluoride, all of which could have significant global effects. Large flood basalt eruptions about 120 million years ago may have released enough carbon dioxide into the atmosphere to contribute to the warm climates that prevailed globally at that time. Besides the direct addition of volcanic gases to the atmosphere, volcanic hydrogen chloride may affect Earth's ozone layer (see Chapter 12). The eruption of Mount Pinatubo in the Philippines in 1991 was followed by a drop in the ozone content of the atmosphere, although human-made fluorocarbons and related chemicals are a considerably more potent force in causing ozone depletion.

Mudflows are a much more serious hazard than lava flows. Unlike lava flows, mud can move rapidly and it does not harden on the move. Mudflows frequently travel at more than 80 km (50 mi) an hour. They often appear without warning and allow no time for escape. In addition to doing direct damage, mudflows can cause floods. A mudflow that enters a reservoir or lake can displace the water suddenly. A mudflow can also block a valley, damming a lake behind the flow. If the mudflow dam is unstable, the lake can drain catastrophically and without warning. Mudflows can fill river valleys, leaving no channel for normal runoff. After the eruption of Mount St. Helens in 1980, the Army Corps of Engineers devoted a great deal of effort to stabilizing lakes dammed by mudflows and reopening filled stream channels.

Volcanoes also emit great quantities of gases. In fact, gas pressure is what powers volcanic eruptions. Most of the gas erupted by a volcano is ordinary steam, and most of the remainder is carbon dioxide. It is likely that much of Earth's atmosphere and oceans first reached the surface as volcanic gases. But mixed with these relatively innocuous gases are sulfur dioxide (SO_2), hydrogen sulfide (H_2S), hydrogen chloride (HCl), and hydrogen fluoride (HF). These gases can be harmful to life near the volcano. Extremely large eruptions can emit enough gases and fine dust to affect global climate (see □ Perspective 9.2).

■ FIGURE 9.13 About 1,700 people died in Cameroon when volcanic carbon dioxide dissolved in the water of Lake Nyos was catastrophically released.

Volcanic gases can be exceedingly dangerous because of their high temperature, but they are rarely a toxic hazard because they usually are vented in the open air. However, gases that collect in enclosed spaces can be a hazard. The worst volcanic gas disaster known occurred in the African nation of Cameroon on August 21, 1986, when some 1,700 people in the vicinity of Lake Nyos were mysteriously and suddenly killed (■ Fig. 9.13).

Lake Nyos occupies the crater of a former volcano. Investigators immediately suspected volcanic activity as the cause of the disaster, but there was no evidence of an eruption. Instead, it appears that volcanic gases, principally carbon dioxide, had been seeping into Lake Nyos for a long time. Lake Nyos is small, only 1.6 km (1 mi) across, but quite deep (208 m or 680 ft). The pressure deep in the lake was great enough to allow large amounts of gas to dissolve in the water. Eventually, the water contained more gas than it could hold. As gas bubbled to the surface, the pressure began to drop and more gas bubbles began to form—exactly the way a soda bottle bubbles when the cap is removed. Within minutes the gas in the lake escaped and flowed along the ground. Carbon dioxide is odorless and colorless, and most victims simply lost consciousness before they had any idea of danger. The deep waters of the lake are so highly charged with carbon dioxide that samples brought to the surface literally fizz like soda water.

The most dramatic hazard of volcanoes is the sudden, explosive release of hot gases. Clouds of volcanic gas can often be at temperatures over 500°C (900°F) and can move at over 160 km (100 mi) an hour. Victims caught by the blast are burned, asphyxiated by gases or ash-laden air, struck by flying or falling debris, or thrown into obstacles themselves. The 60 people who died in the Mount St. Helens eruption of 1980 all died from these causes.

EARTHFACT

High Technology and Volcanoes

High technology and volcanic ash do not mix. When Mount Vesuvius erupted in 1944, during World War II, the ash created serious problems for Allied aircraft. Fine abrasive ash wears down machinery rapidly and can be deadly for engines whose parts require accuracies of thousandths of an inch. For this reason, heavy ashfalls from the 1991 eruption of Mount Pinatubo in the Philippines forced the abandonment of Clark Air Force Base. Ash is no friendlier to computers: a single speck can lodge between a disk drive head and the disk, possibly destroying data or causing the computer to crash. When Mount St. Helens erupted, computers all over the Pacific Northwest experienced problems, and businesses had to institute special protective measures. Aircraft flying from the United States to Asia have suffered engine damage on a number of occasions from flying through clouds of volcanic ash in the Aleutian Islands or Russia's Kamchatka Peninsula.

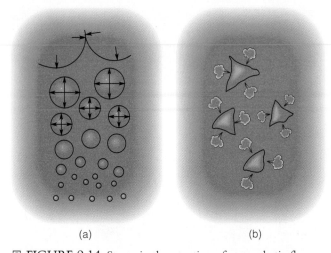

(a)   (b)

■ FIGURE 9.14 Stages in the eruption of a pyroclastic flow: (*a*) Silica-rich magma rises in the neck of a volcano. As the magma nears the surface, the gases begin to expand. (*b*) Expanding gases break the magma into fragments, which remain suspended by the force of their escaping gases. (*c*) The cloud of hot gas and rock shards rolls down the flank of the volcano as a fast-moving, ground-hugging ash flow.

Even deadlier and more spectacular are eruptions called *nuées ardentes* (French for "burning cloud") or **pyroclastic flows.** In a pyroclastic flow, gas–charged magma rises in a volcano. As the magma rises, the pressure drops and the gases start to expand. The expansion forces lava out of the volcano, causing a further drop in pressure and still more expansion. The result is very much like taking the top off a warm, well-shaken bottle of soda. As the lava comes out, it breaks into fragments that are supported by the force of the escaping gas. The mass of hot gas and rock fragments forms a dense cloud with almost no friction that can flow at high speeds for long distances (■ Fig. 9.14).

In 1902, the volcano Mount Pelée on the island of Martinique in the West Indies unleashed a deadly pyroclastic flow. In the weeks before the fatal eruption, the volcano had vented a number of nuées ardentes that flowed harm-

■ FIGURE 9.15 St. Pierre, on the island of Martinique, after the devastating pyroclastic flow of 1902. All of the estimated 30,000 inhabitants were killed almost instantly.

(c)

lessly to the sea. Then, on the morning of May 8, 1902, the side of the volcano gave way, and a pyroclastic flow rolled down the mountain. Most of the cloud flowed down an un-inhabited valley and out to sea, but part of the cloud over-flowed the valley and spread along the coast and over the capital of St. Pierre, killing all its 30,000 inhabitants instantly. Only a few observers on ships and a couple of survivors on land lived to give an account of the disaster (■ Fig. 9.15).

Some pyroclastic flow eruptions, like that of Mount Pelée in 1902, are largely gas with comparatively little ash. Others, like that of Mount Katmai in Alaska in 1912, produce enormous quantities of ash. Ash flows from Mount Katmai accumulated to depths of 100 m (300 ft) in just a few hours. Pyroclastic flows on a scale unmatched in historic times have happened many times in the geologic past. About 700,000 years ago, an enormous eruption of ash in California just east of the Sierra Nevada vented about 600 km^3 (150 mi^3) of ash. The ash layer piled up over 120 m (400 ft) deep near the vent and has been identified as far east as Kansas. Pyroclastic flow deposits of great thickness are called *ignimbrites* (from the Latin words meaning "fire" and "storm"). Few terms in science are more aptly and vividly descriptive.

LIFE HISTORY OF VOLCANOES

Volcanoes have limited lifetimes. Sooner or later the magma source stops producing magma, or the vent of the volcano becomes too solidly plugged to permit eruptions. Even as a volcano is being built, erosion modifies it, and when a volcano dies, erosion rapidly wears it away. During its lifetime, a volcano can be suddenly and catastrophically changed by eruptions.

A few volcanoes have actually been observed from birth. In 1943, a Mexican farmer was astonished to see steam coming from a crack in the ground in his cornfield. The area had been rattled for several weeks by hundreds of small earthquakes, one indication of magma movement. Over the next decade, the new volcano, Parícutin, grew to a height of over 300 m (1,000 ft) and covered the surrounding countryside with lava. Almost every moment of its life, the volcano was observed, studied, and photographed, and there were no casualties. In 1963, a submarine volcano, Surtsey, began erupting off the southern coast of Iceland and remained active until 1967. Every other volcano on Earth also began, at some time, by erupting where no volcano existed before.

In addition to growing through repeated eruptions, volcanoes can change abruptly during their lifetimes. The vents of many older volcanoes become so plugged with hardened lava that it becomes easier for lava to emerge from the side of the volcano in a flank eruption. The vent itself may be permanently relocated and a new cone built on the flanks of the old one. Such a cone is called a *parasitic cone.* Mount Shasta in California has a spectacular parasitic cone called Shastina (▣ Figure 9.16).

The magma beneath a volcano supports part of the volcano's weight. If the magma pressure drops, either because the magma drains back deeper into the crust, or because the magma is erupted as lava, the volcano may subside to form a basin called a **caldera** (Spanish for "cauldron"). A caldera is different from a volcanic *crater,* which is the small depression in the summit of a volcano created by explosive eruptions. Craters form by explosion and are rarely more than a few hundred meters across; calderas form by collapse and can be kilometers across (▣ Fig. 9.17). Often the subsidence of a caldera is relatively peaceful. On Mauna Loa and Kilauea in Hawaii, basaltic magma swells the summit area of the volcano before an eruption, often by several meters. When the magma is erupted, the pressure that had been supporting the summit area lessens, and a portion of the summit area subsides to form a summit caldera.

In other cases, the caldera collapse is sudden and violent. If the volcano erupts huge quantities of ash, it may empty its magma chamber and collapse. If the collapsing cone is in the sea, the results can be cataclysmic. In 1883, the volcano Krakatau, between Java and Sumatra, collapsed in this way after several days of violent eruption. The sea poured into the former magma chamber, came into contact with the hot rock, and vaporized in a tremendous series of steam explosions. Enormous sea waves washed over nearby coasts, killing

▣ FIGURE 9.16 Mount Shasta in northern California has a well-developed parasitic cone called Shastina.

about 30,000 people and washing ships a kilometer or more inland. The blasts were heard over 5,000 km (3,000 mi) away, almost to the coast of Africa. An even more devastating undersea caldera collapse destroyed the volcano of Thera or Santorini in the Aegean Sea in about 1500 B.C. The ashfall and sea wave from the eruption may have destroyed the Minoan civilization on the nearby island of Crete. Memories of the cataclysm may have spawned the myth of the Lost Continent of Atlantis.

Calderas often fill with water to form lakes, commonly but inaccurately called crater lakes. For example, Crater Lake in Oregon (▣ Fig. 9.18) was once the site of a volcano called Mount Mazama, estimated to have been about 3,700 m (12,000 ft) high. About 4650 B.C., Mount Mazama erupted about 60 km^3 (15 mi^3) of ash and caved in. The cavity filled with water to form a lake 604 m (1,982 ft) deep, the deepest lake in the United States, but only 10 km (6 mi) across. Renewed volcanic activity has built several volcanic cones in the caldera, but so deep is the lake that only one, Wizard Island, rises above the surface. Lake Nyos, described earlier, is also a crater lake.

■ **FIGURE 9.17** How calderas form: (*a*) Collapse calderas form when a volcano erupts much of its near-surface magma and collapses. (*b*) The Jemez Caldera in New Mexico stands out plainly on this satellite image. (*c*) Summit calderas form when magma drains away from the summit area of a volcano. (*d*) The summit caldera of Mauna Loa in Hawaii.

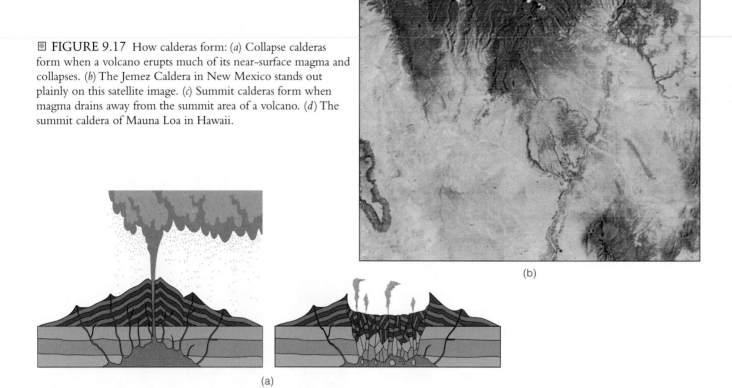

(b)

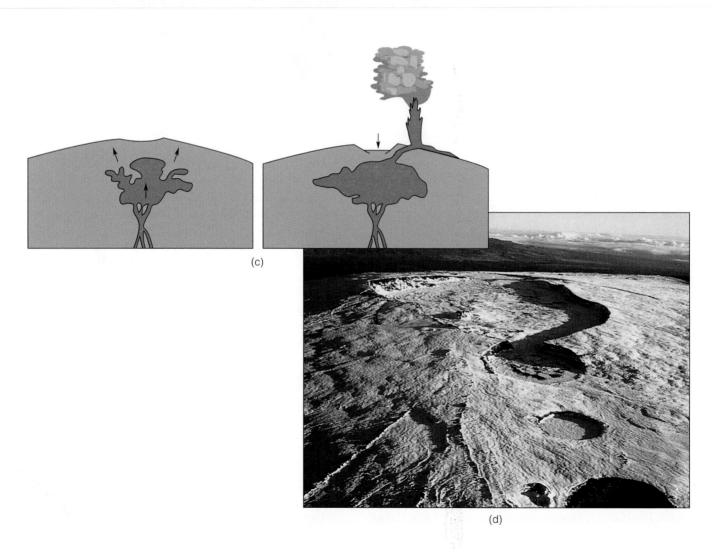

(a)

(c)

(d)

■ FIGURE 9.18 Crater Lake in Oregon is the best-known collapse caldera in the United States.

Sometimes magma may erupt directly from its magma chamber without building a volcanic cone. One such collapsed magma chamber lies beneath Yellowstone National Park. A gigantic caldera 80 km (50 mi) across formed when the roof of the Yellowstone magma chamber subsided. This caldera erupted three huge ash flows over the last 2 million years, each about 600,000 years apart. Somewhat ominously, the last one occurred about 600,000 years ago.

At first glance, nothing could look more solid than a mountain like Mount Rainier or Mauna Loa. Yet dramatic new evidence has come to light that volcanic mountains are much more fragile than anyone realized. The apparently solid mass of a stratovolcano is actually a very weakly cemented pile of lava flows alternating with ash layers and mudflow deposits. In addition, acidic gases and solutions circulating through the rocks weaken the volcano from within. The eruption of Mount St. Helens made geologists aware of the possibility that volcanoes could suddenly collapse. Armed with this new understanding, geologists began examining other volcanoes for signs of ancient collapse. Examples turned up rapidly. Mount Shasta, in northern California, collapsed about 300,000 years ago in an enormous landslide much greater in volume than the landslide that took place at Mount St. Helens (■ Fig. 9.19). North of the mountain, a broad expanse of hummocky hills, once believed to be ancient lava flows covered by glacial deposits, turned out to be deposits from this landslide.

Shield volcanoes may collapse as well. Interestingly, some of the best evidence is on Mars, where vast shield volcanoes with steep cliffs at their bases are surrounded by broad expanses of flowlike deposits that look much like landslides (■ Fig. 9.20a). The shield volcanoes of Hawaii have also undergone collapses (Fig. 9.20b). A number of steep submarine cliffs on the submerged flanks of the Hawaiian chain are large landslide scars, and sonar images of the surrounding seafloor clearly show the landslide debris (Fig. 9.20c). More than 50 landslides have been mapped.

Volcanoes erupt for varying lengths of time. Cinder cones remain active for only a few decades at most and rarely resume activity once they stop erupting. Shield volcanoes and large stratovolcanoes may remain active for a million years or

■ FIGURE 9.19 The hummocky terrain in the foreground resulted from the collapse of part of Mount Shasta about 300,000 years ago.

Shasta

Black Butte
(small plug dome)

■ **FIGURE 9.20** Great volcanic landslides. (*a*) Olympus Mons on Mars, showing its summit caldera, basal scarp, and possible landslide debris. (*b*) The steep north sides of Molokai and Oahu Islands, Hawaii, are due in part to collapse of shield volcanoes. This sonar image shows huge landslide deposits extending away from both islands. White indicates areas of rough topography. Tuscaloosa Seamount is a single large landslide block more than 20 km across. Some of the submarine landslides may have moved slowly, others moved quickly and generated large sea waves. (*c*) The steep cliffs on the north coast of the Hawaiian island of Molokai are an ancient landslide scar where the north half of a shield volcano collapsed. The small peninsula, Kalaupapa, was built by small eruptions after the landslide. Kalaupapa was the site of the infamous leper colony of the nineteenth century; the steep cliffs effectively prevented escape (though trails up the cliffs do exist).

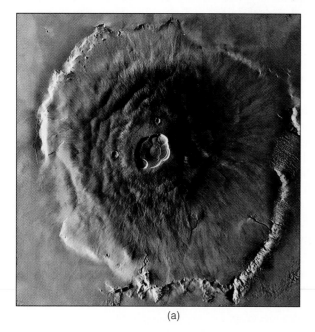

(a)

(b)

(c)

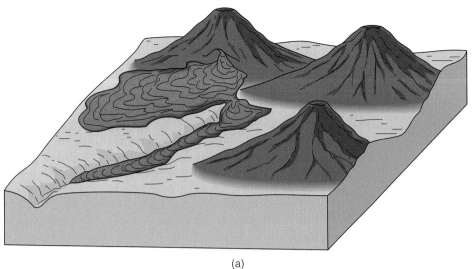

(a)

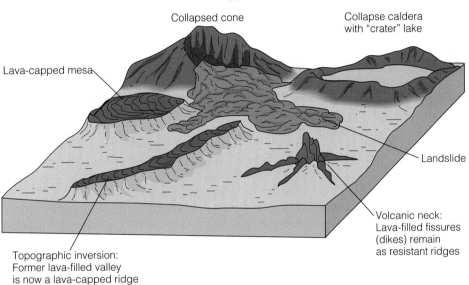

Collapsed cone

Collapse caldera
with "crater" lake

Lava-capped mesa

Landslide

Volcanic neck:
Lava-filled fissures
(dikes) remain
as resistant ridges

Topographic inversion:
Former lava-filled valley
is now a lava-capped ridge

(b)

FIGURE 9.21 Evolution of a volcanic landscape. (*a*) An active volcanic landscape. (*b*) The same landscape after perhaps a million years, showing some of the landforms that develop from volcanoes.

more. A volcano that has not erupted in decades or longer, but has the potential for future eruptions, is termed *dormant*. A volcano that has ceased activity entirely is said to be *extinct*. There is no sure way to determine if a volcano is extinct. Volcanoes have resumed activity after centuries of dormancy. Some volcanoes have remained quiet for so long that local inhabitants had no idea that the mountain was a volcano.

Eventually, volcanoes do become extinct, and erosion tears them down as it does any other mountain (Fig. 9.21). Often the flanks of the volcano are softer than the solidified lava in the vent of the volcano, and erosion leaves the solid lava standing as a steep *volcanic neck* (Fig. 9.22). After a few million years, all topographic evidence of a volcano is gone, but evidence for the former existence of a volcano remains in the form of volcanic rocks. Eventually, erosion may actually expose the solidified former magma chamber of the volcano. Magma that hardens within Earth's crust forms an intrusion, which may later be exposed on the surface by erosion.

INTRUSIONS

Geologists divide intrusions into several classes. Intrusions are *concordant* if they are parallel to structures in the adjacent rocks; they are *discordant* if they cut across structures. The rock invaded by an intrusion is called the *country rock*.

Tabular intrusions are sheetlike in form. They form when magma invades cracks and fissures in the host rock (Fig. 9.23). **Dikes** are discordant; they cut across layers in the host rock. **Sills** are concordant, or parallel to layers in the host rock (Fig. 9.24). *Laccoliths* are shallow sill-like intrusions that force the overlying rocks upward to create a blisterlike uplift. *Lopoliths* are enormous sill-like intrusions, usually of gabbro or related rocks, that may be kilometers thick and cover hundreds of square kilometers. Two examples of lopoliths are the Duluth Gabbro that covers much of the Minnesota shore of Lake Superior and the Bushveld Complex in South Africa, source of much of the world's platinum.

■ FIGURE 9.22 Ship Rock in northwestern New Mexico is one of the best-known volcanic necks in the world.

Irregular intrusions come in all sizes (■ Fig. 9.25). The general term for a nontabular intrusion is a **pluton.** Small discordant intrusions are called *stocks.* Very large discordant intrusions, usually of granite, are called *batholiths.* The Sierra Nevada Batholith of California, 160 km (100 mi) wide and 640 km (400 mi) long, is one of the most famous batholiths. Most batholiths are actually made up of dozens or even hundreds of smaller intrusions. On a geologic map, a batholith is

enormous, but geophysical methods (see Perspective 10.1 in Chapter 10) show that most batholiths are lens-like masses only a few kilometers thick. Often remnants of the roof of the batholith are preserved within and on top of the batholith.

The Sierra Nevada Batholith contains many thousands of cubic kilometers of granite. How could such an enormous mass move into Earth's crust? One might guess that magma could melt its way upward through the crust, but it takes a great deal of energy to melt rock, energy that can only be supplied by cooling the magma. Magma is rarely much above its melting point and has little heat to spare, and it is losing heat to the surrounding cooler rocks all of the time. For every gram of crustal rock melted, at least a gram of magma must solidify. Thus, very few intrusions melt their way through the crust.

Many intrusions simply force their way into the crust. Tabular intrusions usually form when magma under pressure is squeezed into cracks, or between layers in the rocks, and simply forces the rocks aside. Some larger intrusions also force the country rocks aside, crumpling and contorting the rocks adjacent to the intrusion. The force that drives the magma upward is usually the pressure in the magma chamber far below (■ Fig. 9.26). Many intrusions move upward because magma is less dense than most solid rocks. Other intrusions "nibble" their way through the crust by a process called *stoping.* Stoping is probably the most important mechanism for emplacing large intrusions. Pieces of the roof of the intrusion break off and sink slowly through the magma. Often the intrusion hardens while pieces of country rock are still sinking, and chunks of rock are trapped in the intrusion as *xenoliths* (Greek for "foreign stone"). Once intrusions have invaded the crust, they cool and solidify. A small sill a meter or so thick solidifies in a few days; a sill 100 m (328 ft) thick might take a few decades, but a large batholith can take thousands of years to solidify.

■ FIGURE 9.23 Principal types of concordant and tabular intrusions.

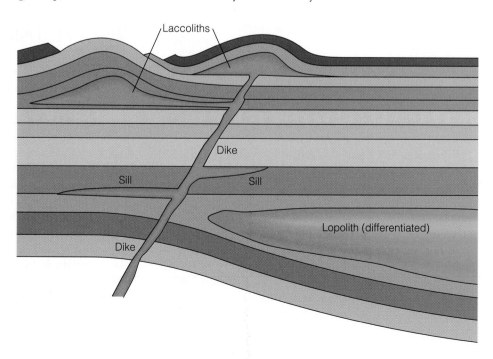

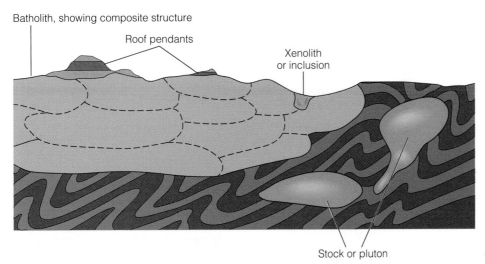

■ FIGURE 9.24 The Palisades Sill forms prominent cliffs along the Hudson River opposite New York City.

Batholith, showing composite structure

Roof pendants

Xenolith or inclusion

Stock or pluton

■ FIGURE 9.25 Principal types of discordant and irregular intrusions.

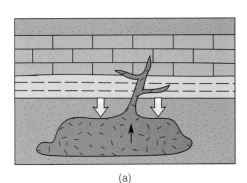

(a)

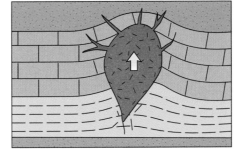

(b)

(c)

■ FIGURE 9.26 Several ways intrusions can invade the crust. (*a*) Tabular intrusions are often squeezed into cracks by pressure originating in the magma chamber. This pressure is largely due to the weight of the rocks above the magma chamber. (*b*) A diapiric intrusion rises through the crust because it is lighter than the adjacent rocks. (*c*) Stoping: the intrusion breaks off roof rocks, which sink through the magma. Stoping is probably the most common intrusion mechanism.

1. Volcanoes erupt not only lava flows, but vast quantities of gases, pyroclastic debris or broken rock fragments, pulverized rock or ash, and mudflows.
2. Silica-poor lava like basalt tends to build low, broad shield volcanoes, whereas silica-rich lava like rhyolite builds steeper stratovolcanoes or plug domes. Silica-rich lavas tend to be associated with the most violent eruptions.
3. Major hazards of volcanoes include lava flows, mudflows, floods, ashfalls, and flying ejecta.
4. Pyroclastic flows are fast-moving clouds of hot gas and rock fragments that can be extremely dangerous.
5. In addition to direct effects of eruptions, volcanoes can affect the global environment by changing the carbon dioxide content of the atmosphere or by ejecting into the atmosphere fine particles that reflect sunlight and cool Earth.
6. Volcanoes can evolve in a variety of ways. Sometimes a new vent forms on the side of a volcano to produce a parasitic cone. Often the summit of a volcano collapses to form a caldera.
7. Recently, geologists have discovered that many volcanoes collapse catastrophically.
8. Eventually, all volcanoes cease activity and become extinct.
9. Intrusions are masses of igneous rock that harden beneath the surface, later to be exposed by erosion.
10. Tabular intrusions are sheetlike and include dikes, sills, laccoliths, and lopoliths.
11. Irregular intrusions include stocks, plutons, and batholiths. Batholiths are usually granitic and are among the largest intrusions.

batholith
caldera
cinder cone
dike

flood basalts
intrusion
lava
pluton

pyroclastic
pyroclastic flow
shield volcano
sill

stratovolcano
viscosity
volcanic ash
volcano

1. The pressure to power volcanic eruptions comes from:
   a. ____ oxygen; b. ____ sulfur; c. ____ chlorine; d. ____ steam.
2. Silica-rich lavas are associated with the most violent eruptions because:
   a. ____ silica is highly chemically reactive; b. ____ silica explodes on contact with air; c. ____ silica-rich lava is viscous and traps gases; d. ____ volcanoes with silica-rich lava are the largest volcanoes.
3. Which of the following is the least significant hazard of volcanoes?
   a. ____ mudflows; b. ____ lava flows; c. ____ pyroclastic flows.
4. Shield volcanoes are the largest volcanoes by what measure?
   a. ____ height; b. ____ diameter; c. ____ volume; d. ____ all of these.
5. Which of the following plate tectonic settings would likely have shield volcanoes?
   a. ____ continental subduction zone; b. ____ continental hot spot; c. ____ continental spreading center; d. ____ oceanic hot spot.
6. Regardless of plate tectonic setting, magma that forms directly from melting of mantle rocks tends to be:
   a. ____ basalt; b. ____ andesite; c. ____ rhyolite; d. ____ granite.
7. The viscosity of magma determines:
   a. ____ the type of volcano that forms; b. ____ the relative amounts of lava and pyroclastic material; c. ____ the explosiveness of the eruption; d. ____ all of these.
8. Which of the following statements is true of mudflows and lava flows?
   a. ____ lava flows are more dangerous because they move faster; b. ____ lava flows are more dangerous because they are so hot; c. ____ mudflows are more dangerous because they contain explosive gases; d. ____ mudflows are more dangerous because they move faster and do not harden as they flow.
9. The two most abundant gases in volcanic emissions are:
   a. ____ hydrogen sulfide and hydrogen chloride; b. ____ hydrogen sulfide and water vapor; c. ____ sulfur dioxide and carbon dioxide; d. ____ water vapor and carbon dioxide.
10. An outcrop contains sedimentary layers standing on end. A tabular gabbro intrusion is intruded between the layers. This intrusion is:
    a. ____ a dike because it is vertical; b. ____ a sill because it is vertical; c. ____ a sill because it is parallel to the layers; d. ____ a dike now, though it was originally a sill.
11. Which of the following is a discordant intrusion?
    a. ____ laccolith; b. ____ lopolith; c. ____ sill; d. ____ stock.

12. How does the composition of lava relate to the type of volcano that erupts it?
13. What types of materials are produced by volcanoes?
14. What role does mud play in volcanic eruptions? How does it form?
15. What happens when a pyroclastic flow erupts? Why are nuées ardentes dangerous?
16. Why is Crater Lake technically misnamed?

17. Describe the two major ways calderas form and give an example of each.
18. What are some of the lines of evidence that have convinced geologists that volcanoes collapse more often than had been suspected?
19. What is the difference between dormant and extinct volcanoes? Can these terms be precisely defined? Why or why not?
20. What is a batholith? What rock type usually makes up a batholith?

POINTS TO PONDER

1. In an area where erosion long ago leveled any ancient topography, what are some things you would look for to decide whether there had ever been volcanic activity in that region?
2. The 1997 film *Volcano* depicted an unexpected volcanic eruption in Los Angeles. How might you detect a magma chamber that had not yet produced a volcano? Hint: reread the facts in Chapter 8 about S-waves.
3. After a volcanic eruption, how might you go about estimating the total amount of material erupted by the volcano?

ADDITIONAL READINGS

Core R., and O. L. Mazzatenta. 1984. The dead do tell tales at Vesuvius. *National Geographic* 165, no. 5: 557–613.

———. 1984. A prayer for Pozzuoli. *National Geographic* 165, no. 5: 614–25.

Decker R., and B. Decker. 1981. The eruptions of Mount St. Helens. *Scientific American* 244, no. 3: 68–91.

Decker, R. W., T. L. Wright, and P. H Stauffer. 1986. *Volcanism in Hawaii.* U.S. Geological Survey Professional Paper 1350.

Francis, P. 1983. Giant volcanic calderas. *Scientific American* 248, no. 6: 60–70.

Francis, P., and S. Self. 1983. The eruption of Krakatau. *Scientific American* 249, no. 5: 172–87.

———. 1987. Collapsing volcanoes. *Scientific American* 256, no. 6: 90–99.

Harris, S. L. 1980. *Fire and ice: The Cascade volcanoes.* Seattle: Pacific Search Press.

Hekenian, R. 1984. Undersea volcanoes. *Scientific American* 251, no. 1: 46–55.

Judge, J. 1982. A buried Roman town gives up its dead. *National Geographic* 162, no. 6: 678.

Kittleman, L. R. 1979. Tephra [pyroclastic debris]. *Scientific American* 241, no. 6: 160–77.

Kling, G. W., M. A. Clark, H. R. Compton, and others. 1987. The 1986 Lake Nyos gas disaster in Cameroon, West Africa. *Science* 236: 169–75.

Lipman, P. W., and D. R. Mullineaux. 1981. *The 1980 eruption of Mount St. Helens.* U.S. Geological Survey Professional Paper 1250.

Macdonald, G. A., A. A. Abbott, and F. L. Peterson. 1983. *Volcanoes in the sea.* Honolulu: University of Hawaii Press.

Peck, D. L., T. L. Wright, and R. W. Decker. 1979. The lava lakes of Kilauea. *Scientific American* 241, no. 4: 114–29.

Simkin, T., L. Siebert, L. McClelland, and others. 1981. *Volcanoes of the world.* Stroudsburg, Pa.: Smithsonian Institution and Hutchinson Ross.

Smith, R. B., and R. L. Christiansen. 1980. Yellowstone Park as a window on Earth's interior. *Scientific American* 242, no. 2: 104–17.

WORLD WIDE WEB SITES

For current updates and exercises, log on to http://www.wadsworth.com/geo.

VOLCANOES OF THE WORLD
http=//volcano.und.nodak.edu/volc.of.world.html

WHAT'S ERUPTING NOW!
http=//volcano.und.nodak.edu/vwdocs/current_volcs/current.html

The two sites listed above are operated from one of the least volcanic places on Earth—the University of North Dakota.

USGS HAWAII VOLCANO WATCH REPORTS
http://www.soest.hawaii.edu/hvol

ALASKA VOLCANO OBSERVATORY
http:/www.avo.alaska.edu/volcs/shishaldin/shishaldin.html
Information on the volcanoes of Hawaii two sites listed above.

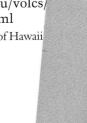

CHAPTER

10

Mountains and Mountain Building

Owens Valley and the east face of the Sierra Nevada, California. The Sierra Nevada has been uplifted along a huge fracture so that it now stands more than 3,000 m above the valleys to the east.

Prologue

After completing the monumental task of extending a survey network the length of India between 1802 and 1843, the British rulers of India turned their attention to the Himalayas in then-forbidden Nepal and Tibet. Indian scouts, at great risk to their lives, traveled into these regions disguised as religious pilgrims and brought back priceless geographical information, but the altitudes of the mountain peaks could only be measured from afar.

Until the British began measuring the Himalayas, the Andes of South America were considered the highest mountains in the world. The Himalayas easily surpassed the Andes. Although the Himalayas could be seen from a number of places in India, it was impossible to tell which summits seen from one spot corresponded with summits seen from elsewhere. The approach the British adopted was to measure the directions and apparent altitudes of every summit as seen from six different locations and plot *all* the lines of sight on a single map. Where the six lines intersected was the true location of a given peak. One day Radhanath Sikdar, the Indian survey assistant, calculated the elevation of an apparently unremarkable peak. It did not seem to be very high, but that was only because it was farther away than the others. When Sikdar finished the calculation, he ran to the survey director and announced, "Sir, I have found the highest mountain in the world." The peak was named Mount Everest (☐ Fig. 10.1), after Sir George Everest, who had directed the survey, often under great hardship, from 1830 to 1843. The peak is also called Sagarmatha in Nepali and Qomolungma in Tibetan. (There is a certain irony in the name *Mount Everest*. Unlike many colonial authorities, Everest had insisted on preserving local place-names on maps, yet the highest of all Himalayan peaks is best known by a foreign name. For many years, the area around Mount Everest was off-limits to foreigners, and the local name of the mountain was unknown.) Later surveys established the height of Mount Everest as 8,848 m (29,028 ft).

For many years, the exciting possibility lingered that a peak even higher than Everest might be hidden somewhere in an unexplored region, but as blanks on the world map were filled in, that hope dwindled. During World War II (1941–1945), pilots flying from India to China repeatedly claimed that a peak named Amne Maqin, in central China, was higher than Everest. Surveys eventually found the height of Amne Maqin to be only 7,162 m (23,490 ft), not even close.

In 1986, a new challenge to Mount Everest came from satellite range-finding. The peak K2 (☐ Fig. 10.2), also called Mount Godwin-Austen, on the India-Pakistan border was claimed to be higher than Everest. The peak's altitude above its base was well known, but measurements using satellite positioning devices indicated that the *base* of the peak might be

☐ **FIGURE 10.1** Mount Everest, the highest peak on Earth.

about 300 m (1,000 ft) higher than previously believed. More detailed studies showed that the initial measurements were due to a faulty instrument. K2, at 8,613 m (28,250 ft) is still the second-highest peak in the world. Further studies of K2 are likely to be some time in coming because of a quite different altitude record. India and Pakistan contest the border near K2. Both sides have established military posts in the area and fought skirmishes above 6,100 m (20,000 ft), the highest known ground military actions in history. Both sides have lost far more casualties to cold and altitude sickness than to fighting.

Mount Everest is the highest point on Earth above sea level, but not the most distant point from Earth's center. Because of Earth's equatorial bulge, the summit of Chimborazo in Ecuador at 6,269 m (20,561 ft) is about 2 km (1.2 mi) farther from Earth's center than is the summit of Mount Everest.

☐ FIGURE 10.2 K2, or Mount Godwin-Austen, the second-highest peak on Earth.

 INTRODUCTION

Mountains form in a variety of ways (☐ Fig. 10.3). Many like Mauna Loa and Mount Rainier are built up by volcanic activity. Other mountains form when erosion strips away soft rocks and leaves resistant remnants behind as hills, or *monadnocks*. Indeed, all mountains on Earth are shaped to a great extent by erosion. Many mountains owe their existence to movements of the crust. Mountains raised by faulting are very common and include the Sierra Nevada and the mountain ranges of the Great Basin. But the greatest mountain ranges like the Alps or the Rocky Mountains exist because of warping and flexing of the crust caused by plate tectonics (see Chapter 7). Mountain building profoundly affects many Earth systems (○ Table 10.1). To understand the great mountain ranges of the world, we must first examine the forces that shape Earth's crust and how rocks respond to these forces. To do so, we must also deal with important physical concepts like force, pressure, stress, and strain.

 EARTH FORCES AND THE DEFORMATION OF ROCKS

In popular speech, we often use the words *stress, pressure,* and *strain* interchangeably, as in "I've been under a lot of stress (or strain or pressure) lately." In geology, stress and strain are important and quite different concepts that are essential to understanding the movements of Earth's crust.

In an example of an emotionally stressful situation that also illustrates the physical meaning of stress, a person is walking across a frozen lake when the ice suddenly begins to crack. What should the person do? Most safety experts say to lie down flat on the ice, but why? The *force* the person exerts on the ice due to Earth's gravitational pull on the person is the same whether the person is standing, sitting, or lying. But the amount of force per unit area, or **stress,** is different. When the person is standing, force is concentrated in the small area beneath the person's feet, and the stress on the ice is large. If the person lies flat, the force is spread over a much larger area, and the stress is much smaller, maybe small enough to prevent an icy swim.

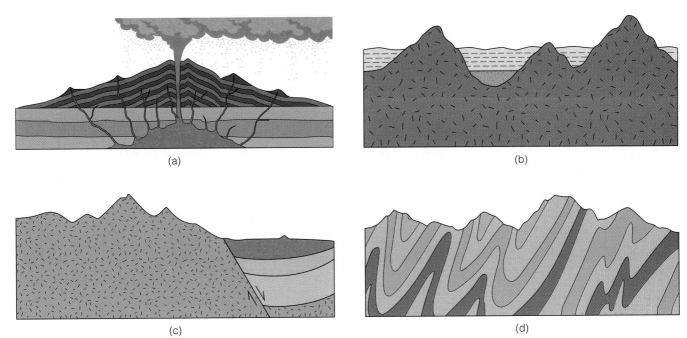

■ FIGURE 10.3 Important types of mountains: (*a*) volcanic; (*b*) erosional remnant (monadnock); (*c*) fault-block; (*d*) folded or orogenic.

TABLE 10.1 Mountain Building and Earth Systems

Solid Earth	Plate tectonics drives mountain building and associated igneous and metamorphic activity. Erosion of mountains provides sediment for sedimentary rocks.
Atmosphere	Mountains divert winds and weather systems. Exposure and weathering of carbonate rocks may affect atmospheric carbon dioxide and warming or cooling of the Earth.
Hydrosphere	Mountains intercept moisture, resulting in rainy upwind sides and dry downwind sides (rainshadows). Mountain building diverts drainage systems and creates drainage divides. Materials dissolved during weathering and erosion may affect chemistry of oceans. Crustal movements create some large lake basins.
Cryosphere	High mountains may support glaciers. Uplift of mountains may affect global air circulation and contribute to onset of ice ages. Exposure and weathering of carbonate rocks may affect atmospheric carbon dioxide and global climate.
Biosphere	Effects on atmosphere and hydrosphere have impacts on life. Mountains create altitude habitat zones and barriers to migration. Weathering of rocks releases nutrients.
Extraterrestrial	Little effect.

Stress behaves in a variety of ways. Stress that is uniform in all directions is called *hydrostatic* (■ Fig. 10.4a). Stress that is stronger in some directions than others is *directed* (Fig. 10.4b). Stresses that act to compress objects are called *compression* (Fig. 10.4c), whereas those that stretch objects are called *tension* (Fig. 10.4d). The common term *pressure* usually refers to compressional stresses. A somewhat more complex kind of stress called *shear* causes rocks on the opposite sides of faults to deform out of shape and eventually slip past each other (Fig. 10.4e). Singer Billy Joel, who was in Kobe, Japan, during the earthquake of January 17, 1995 (coincidentally, one year to the day after the 1994 Northridge earthquake), gave a perfect description of shear: "I saw the wall turn into a parallelogram."

EARTHFACT

Not-so-Solid Rocks

It's very hard to picture something as solid as a rock deforming like putty, but compared to the pressures in Earth, rocks are weak, in fact, downright squishy. Some geologists study the structures of mountain ranges by deforming scale models. A mountain range 100 km (62 mi) across might be reduced 100,000 times to a model 1 m (3.3 ft) across. The strength of the materials in the model must be reduced by a factor of 100,000 as well, to the strength of soft waxes and petroleum jellies.

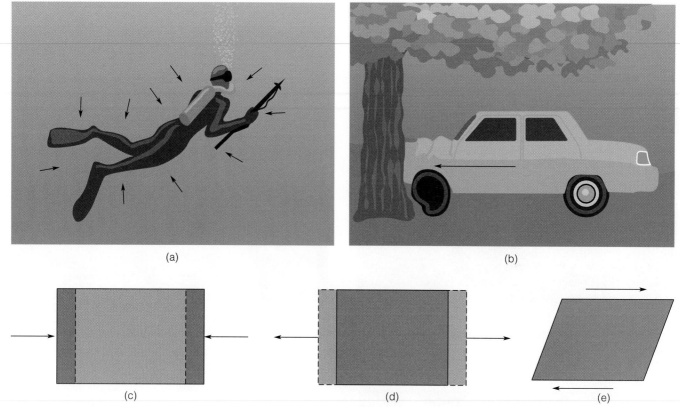

(a)

(b)

(c)

(d)

(e)

■ FIGURE 10.4 Types of stress: (*a*) hydrostatic stress; (*b*) directed stress; (*c*) compression; (*d*) tension; (*e*) shear.

Strain is the amount a material deforms under stress, compared to its original dimensions (■ Fig. 10.5). In *elastic strain,* stress deforms a material temporarily out of shape, but when the stress is removed, the material snaps back to its original dimensions (Fig. 10.5a). If the stress is great enough, however, other kinds of strain occur in which the material deforms permanently out of shape. Many materials simply break. Such behavior, called **brittle,** is a common behavior for rocks near Earth's surface where pressures and temperatures are low (Fig. 10.5b). Other materials may deform permanently out of shape without breaking. Such materials are said to be **ductile** (Fig. 10.5c). The same material can behave in a brittle or ductile manner depending on how much it is deformed, how quickly, and how much stress is applied. A common example of a material that is both brittle and ductile is glacial ice (see Chapter 11). Near the surface, ice in a glacier is brittle, but about 100 m (328 ft) or so below the surface, glacial ice is ductile. Many metals deform in a ductile manner up to a point, then break in a brittle manner if deformed too much.

Generally, rocks are brittle near Earth's surface, but become ductile at high pressures and temperatures. A rock may break in brittle fashion if subjected to a sudden stress like the rupture of a fault, but deform in a ductile manner if subjected to slow, steady stresses. Rocks vary greatly in their tendency for ductile behavior. Rock salt and gypsum are extremely ductile, and even the weight of overlying rocks is enough to make them deform. Salt, in particular, is able to

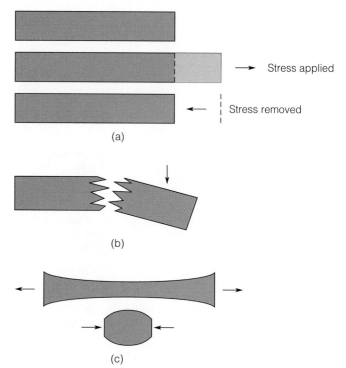

Stress applied

Stress removed

(a)

(b)

(c)

■ FIGURE 10.5 Types of strain: (*a*) elastic strain; (*b*) brittle strain; (*c*) ductile strain.

squeeze up through weak spots in the overlying rocks to form *salt domes* (Fig. 10.6). Shale, limestone, and marble are also very ductile rocks. Quartzite, granite, and basalt are among the least ductile rocks.

BRITTLE DEFORMATION: FAULTS AND JOINTS

Fractures in Earth are classified according to the sorts of movement that occur on them (Fig. 10.7). Most rocks have fractures, called **joints,** that form because of local or regional stresses in the rocks without movement of the rocks on either side. Often two sets of joints form at the same time and intersect to form X-shaped patterns. Studies of rocks in the laboratory show that the acute angle of the X marks the direction of greatest compression while the obtuse angle marks the direction of least compression. This relationship makes joints useful indicators of stresses in the crust.

Faults are fractures in rocks where movement occurs. When the relative motion of the rocks on opposite sides of the fault is mostly vertical, the fault is called a *dip-slip fault.* **Normal faults** are associated with extension of the crust, **thrust faults** with compression. Not surprisingly, spreading plate boundaries tend to be associated with normal faults, and convergent plate boundaries with thrust faults. In mountain belts, rocks often break into great thin sheets called *nappes,* bounded above and below by thrust faults, that have slid long distances across the underlying rocks. In some cases, a thrust fault may abut another fault and not reach the surface. A fault of this sort is called a *blind thrust* (Fig. 10.8). The 1994 Northridge earthquake occurred on a blind thrust fault. Blind thrusts are now recognized as serious potential earthquake hazards in the Los Angeles area.

Transcurrent faults are faults whose motion is primarily horizontal. If you look across the fault as it moves, and the opposite side of the fault moves to your right, the fault is called *right-lateral*. The San Andreas is a right-lateral fault. If the opposite side of the fault moves to your left, the fault is called *left-lateral*. The Garlock fault, a great fault north of Los Angeles, is left-lateral. This classification system does not depend on which side of the fault the observer is on.

Faults create many distinctive landforms (Fig. 10.9 on page 218). Active faults can offset the land surface, creating steep *fault scarps*. As the opposite sides of a fault grind against one another, they create zones of shattered and broken rock that are especially susceptible to weathering and erosion, so major faults are often marked by linear valleys. Ancient faults with rocks of differing hardness on either side are sometimes revealed when erosion strips away the softer rocks.

Often, blocks of crust bounded by faults can move up or down. An uplifted block bounded by faults is called a *horst*. Most of the mountain ranges of Nevada are horsts. A block of tilted crust bounded on one side by a fault, like the Sierra Nevada, is called a *tilted fault-block*. A block that has dropped downward along faults is called a *graben*. Most grabens are bounded by normal faults. When the crust pulls apart, a graben block drops to fill the gap just like the keystone in an arch. *Rift valleys* are systems of grabens that extend for long distances, such as the East African Rift, the Rio Grande Rift that extends most of the length of New Mexico, and the rift valleys along the crests of the mid-ocean ridges (see Chapter 18). Rift valleys and grabens are among the few continental landforms that can extend below sea level. The Dead Sea, the lowest dry land on Earth (390 m or 1,280 ft below sea level), is a graben, as is Death Valley (86 m or 282 ft below sea level), the lowest point in North America. The world's deepest lakes, Baikal in Siberia (1,650 m or 5,500 ft) and Tanganyika in Africa (1,200 m or 4,000 ft), lie in grabens.

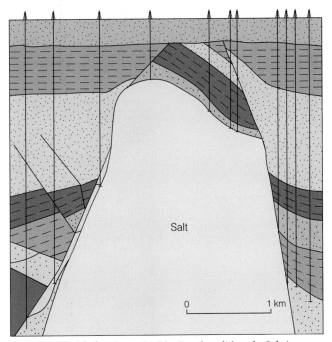

Salt

0   1 km

 FIGURE 10.6 Salt is a highly ductile solid rock. Salt is so ductile that it flows upward through weak spots in the overlying rocks to form salt domes, as in the White Castle Dome of Louisiana.

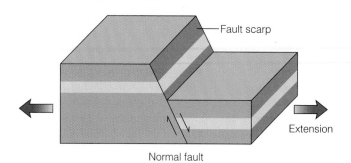

Fault scarp

Extension

Normal fault

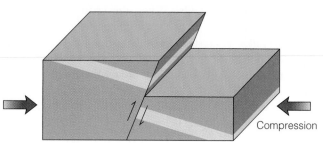

Compression

Reverse fault

Shear

Right-lateral strike-slip fault

Lower Paleozoic sediments in Big Bend Country, Texas

Death Valley area, California

Imperial Valley, California, lettuce field

Great transcurrent faults like the San Andreas create their own distinctive landforms. Frequently, erosion forms a deep, straight valley along the fault trace. Streams that cross the fault are offset by fault motion to create *offset streams.* Transcurrent faults in the oceans are marked by great *fracture zones* (see Chapter 18), fault scarps that offset the ocean floor for hundreds or even thousands of kilometers.

DUCTILE DEFORMATION: FOLDS

Many structures in rocks are due to compression. On scales ranging from centimeters to kilometers across, rock layers buckle and *fold* (Fig. 10.10). **Anticlines,** or folds that arch upward, have older rocks in their centers and younger rocks on their flanks. More or less equidimensional anticlines are called *domes.* **Synclines,** or folds that bend downward, have younger rocks in their centers and older rocks on their flanks. Roughly equidimensional synclines are called *basins.* The plane along the middle of the fold is called the **axial plane.** If you push on a bedspread so that it wrinkles, you will find that the axial planes of the folds you created are perpendicular to the direction of compression, and the same is true of rocks.

 FIGURE 10.7 (b) Joints.

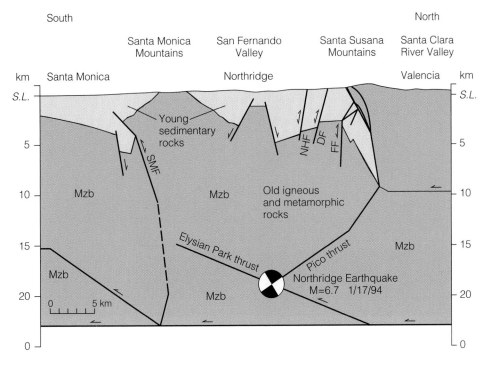

 FIGURE 10.8 The Northridge earthquake of 1994 occurred on the Pico thrust, a blind thrust. Such faults are dangerous because their existence is often unsuspected until they generate an earthquake.

(a)

| Simpson Park Range | Pine Valley | Sulphur Springs Range | Diamond Valley | Diamond Range | Newark Valley | Ruby Mountains |

Roberts Mountain thrust

(b)

(c)

■ **FIGURE 10.9** Some important fault landforms: (*a*) an offset stream along the San Andreas fault in southern California; (*b*) cross section of horsts and grabens in Nevada; (*c*) a horst—the Humboldt Range, Nevada.

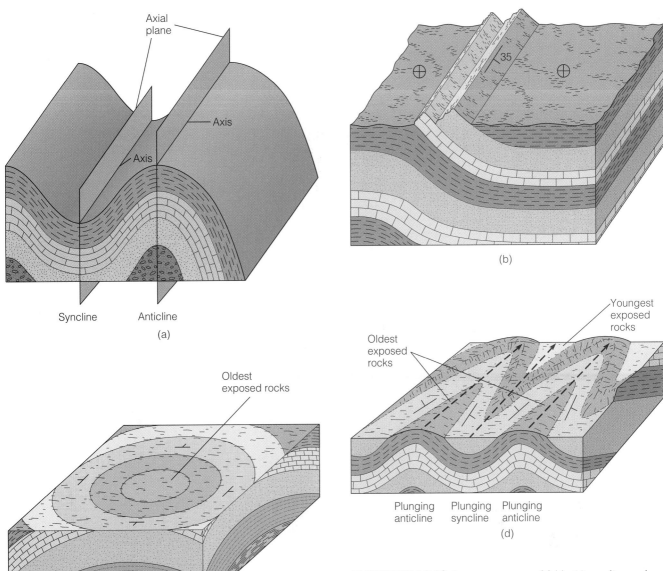

Axial plane

Axis

Axis

Syncline   Anticline

(a)

Oldest exposed rocks

Dome

(b)

35

Oldest exposed rocks

Youngest exposed rocks

Plunging anticline   Plunging syncline   Plunging anticline

(d)

■ **FIGURE 10.10** Important types of folds: (*a*) syncline and anticline showing axial planes; (*b*) monocline; (*c*) dome and basin; (*d*) when folds are not horizontal, they are said to plunge; plunging folds create zigzag outcrop patterns.

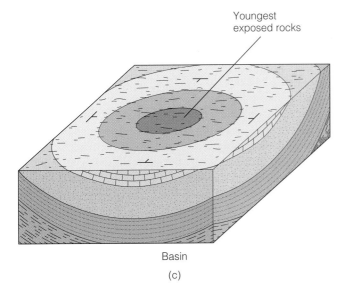

Youngest exposed rocks

Basin

(c)

On scales ranging from atomic to a few cm across, deformed rocks often develop a platy or sheetlike texture called **foliation** that can form in many different ways (■ Fig. 10.11). Objects in the rock can simply be flattened by compression. Mineral grains can rotate into parallel alignments as the rock is flattened, much the way toothpicks strewn on a floor would be aligned by a push broom as they are swept up. New mineral grains that form during metamorphism often grow more easily outward at right angles to the compression rather than against it. All of these mechanisms result in foliation forming perpendicular to the direction of compression. Merely seeing foliation in a rock allows

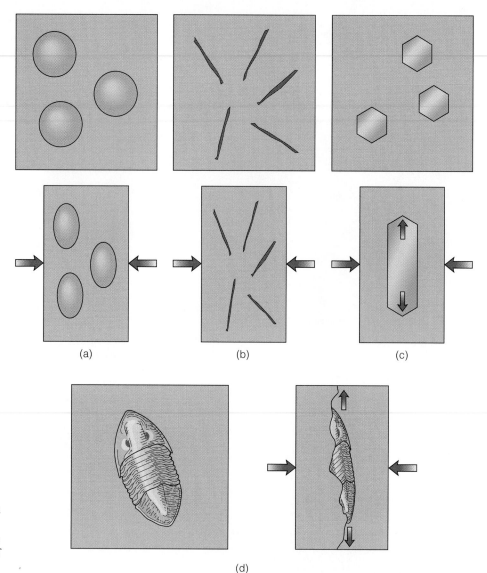

 FIGURE 10.11 Foliation forms in a variety of ways: (*a*) flattening of grains; (*b*) rotation of grains; (*c*) growth of new minerals; (*d*) removal of part of the rock by solution.

a geologist to know immediately that the rock has been deformed and also to know the direction of the greatest compression.

Thus, on a small scale foliation in rocks forms perpendicular to the direction of compression; on a large scale, the axial planes of folds also form perpendicular to the direction of compression. Thus, *foliation is usually parallel to the axial planes of folds* (◾ Fig. 10.12). This consistent relationship puts a very powerful tool in the hands of the geologist. A geologist who observes foliation and bedding in an outcrop can mentally picture the overall shape of the fold (◾ Fig. 10.13).

◾ LARGE-SCALE GEOLOGIC STRUCTURES AND GEOLOGIC MAPS

A geologist must study small outcrops of rock and use this information to figure out the size, shape, and relationships of masses of rock that may be many miles across. The most common technique geologists use to make sense out of large areas is to summarize the information on a **geologic map.** Locations and boundaries of rock units are plotted on the map along with other useful information. For example, when rocks are tilted or folded, the geologist must also note their orientation. The orientation of a tilted layer is defined by its **strike** and **dip** (◾ Fig. 10.14 on page 222). The strike is the direction of a horizontal line in the layer. The dip of the layer is simply the angle the layer makes with the horizontal. Mapping rocks on the surface enables geologists to visualize the structure of the rocks near the surface, but to map the structure of deeply buried rocks, other techniques are needed (see ◻ Perspective 10.1 on page 224).

Where rocks have been folded, erosion rapidly cuts into the fold to expose rocks of differing ages. Different kinds of structures in the rocks result in characteristic patterns on geologic maps (◾ Fig. 10.15). One of the most important practical applications of understanding rock structures is in the search for petroleum (see ◻ Perspective 10.2 on page 228).

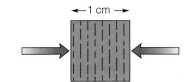

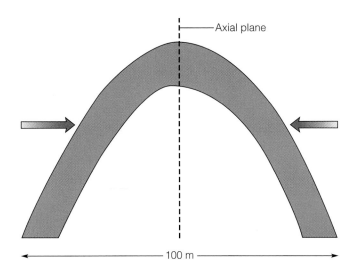
Axial plane

100 m

In many cases, folding can actually tilt rock layers through more than 90 degrees so that older rocks lie on younger ones (Fig. 10.16), the reverse of the normal principle of superposition (see Chapter 11). This situation is usually easy to recognize because most sedimentary rocks contain sedimentary structures like ripple marks or cross-bedding (see Chapter 3). These structures are all governed in some way by gravity and have a distinctive "right-side-up" appearance. It is easy to tell whether these structures have been overturned.

LARGE-SCALE MOVEMENTS OF THE CRUST

Isostasy

Earth is not perfectly rigid. Because its interior is ductile, Earth's rotation causes the planet to bulge at the equator by about 1/298 of its diameter, or about 40 km (28 m) out of a total diameter of 12,800 km (7,900 m). The ductile interior also causes the crust to sink or rise in response to changing loads on the crust. This process is called **isostasy.**

Foliation

■ FIGURE 10.12 *Above:* Foliation is a small-scale deformation structure in rocks, whereas folding is a larger-scale structure. Nevertheless, they both form perpendicular to the direction of compression. Foliation therefore is usually parallel to the axial planes of folds.

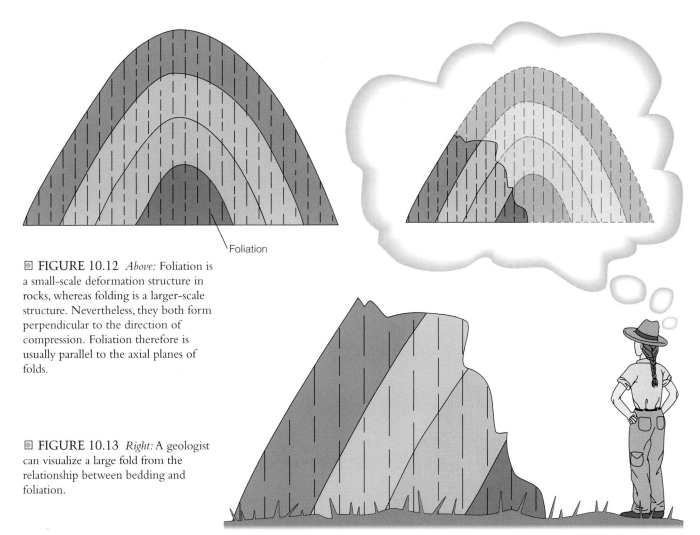

■ FIGURE 10.13 *Right:* A geologist can visualize a large fold from the relationship between bedding and foliation.

FIGURE 10.14 Strike and dip. (*a*) A Brunton compass, used by geologists to record the orientation of rocks in the field. (*b*) A geologist using a Brunton compass to measure the dip of a layer. (*c*) These conglomerate layers in Michigan slope into Lake Superior. The water line shows the strike of the beds, and the angle the beds make with the water surface indicates the dip.

(a)

(b)

(c)

FIGURE 10.15 Folds can be recognized by the distinctive patterns they make on geologic maps.

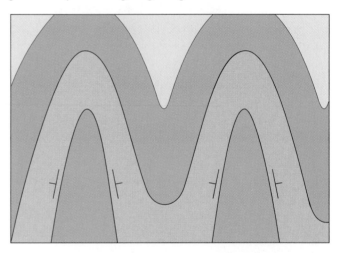

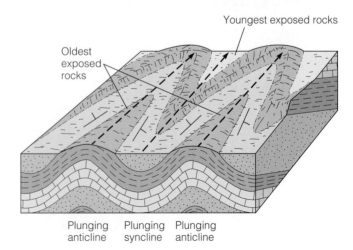

Oldest exposed rocks

Youngest exposed rocks

Plunging anticline   Plunging syncline   Plunging anticline

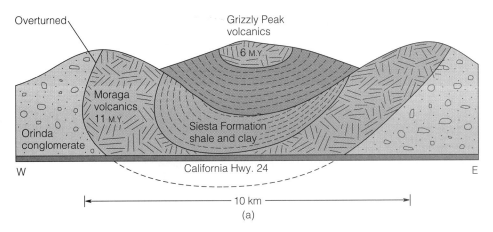

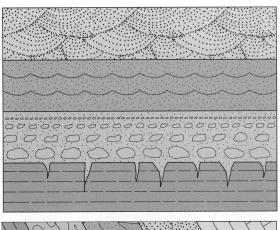

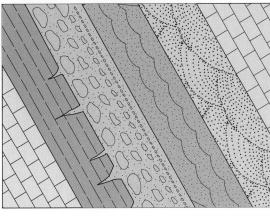

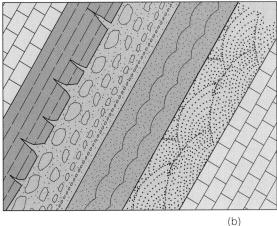

■ FIGURE 10.16 (*a*) In the Berkeley Hills east of San Francisco, California, rock layers have been overturned within the last 6 million years. (*b*) Sedimentary structures in rocks have a distinctive "right-way-up" appearance and make it easy to tell whether rock layers have been overturned.

Overturned

Grizzly Peak volcanics

6 M.Y.

Moraga volcanics
11 M.Y.

Orinda conglomerate

Siesta Formation shale and clay

California Hwy. 24

W   E

⟵ 10 km ⟶

(a)

Undisturbed layers

Cross-beds: Often concave upward and cut off at top

Ripple marks: Sharp crests and smooth troughs

Graded bedding: Coarsest particles on bottom

Mud cracks: Open toward top of layer

Tilted 60° but not overturned

Tilted 120° (overturned)

(b)

Large-Scale Movements of the Crust   223

Geophysics: Seeing Underground

The science of applying the methods of physics to the problems of geology is called *geophysics.* Seismology (see Chapter 8) is a special branch of geophysics. Geophysical techniques enable geologists to study otherwise inaccessible parts of Earth, for example, to locate petroleum traps (see Perspective 10.2).

Gravity measurements can be used to probe the subsurface. A gravity meter, or *gravimeter,* is basically a sensitive spring balance. Instead of using gravity to determine weight, a gravimeter uses a known weight to find the force of gravity to within one part in a million, or .0001%. Maps of gravity data show departures from the average strength of gravity, or *anomalies* (□ Fig. 1). Dense rocks like basalt cause *positive anomalies,* or gravity measurements slightly above average. Low-density rocks, such as granite, salt, and porous sandstones, produce unusually low gravity readings, or *negative anomalies.*

□ FIGURE 1 A gravity anomaly map of the area around the Chicxulub impact crater (see Chapter 1). The crater has a central positive anomaly surrounded by a zone of low gravity readings. The central positive anomaly probably represents uplifted rock in the center of the crater; the low gravity readings around it are probably due to shattered rock and sediments filling the crater. Contour intervals represent 1/500,000 of the average strength of gravity.

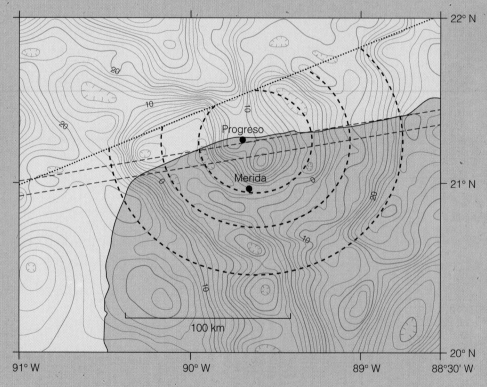

When a large load is placed on the crust, the crust sinks. For example, around the Hawaiian Islands there is a prominent trough on the Pacific Ocean floor, where the tremendous mass of the Hawaiian shield volcanoes has depressed the crust (▣ Fig. 10.17a). One human activity that puts large loads on the crust is the filling of a reservoir. Studies around large new dams often show that the crust has tilted and small earthquake activity has increased due to loading of the crust by the water in the reservoir. This matter is of interest to dam builders because there is always the chance that the loading may trigger an earthquake strong enough to damage the dam.

In the last few thousand years, the continents have been relieved of a tremendous load as the Pleistocene ice sheets melted (see Chapter 6). The ice sheets over North America and Europe were probably as thick as the Greenland and Antarctic ice sheets, enough to depress the crust by about a kilometer (Fig. 10.17b). As the ice melted and removed its load from the crust, the crust rose, and mantle rocks flowed in to fill the space formerly occupied by the crust. Because mantle material flows slowly, the ice melted much faster than the crust could rise. Although the ice sheets are gone, the crust is still depressed and will continue to rise until isostatic

Studies of Earth's magnetism, using *magnetometers,* can also be used to probe the subsurface (☐ Fig. 2). Magnetometers were first developed during World War II to detect submarines, but they can also detect the magnetic fields of rocks mostly due to their magnetite content. Magnetic studies of the ocean floor provided some of the most important evidence for plate tectonics (see Chapter 7).

Gravity and magnetic maps complement one another because the two techniques measure different rock properties. Having the two maps together is much better than having either separately. In addition, geologists also usually try to compare their interpretations with actual rock samples collected in the field or brought to the surface during well drilling.

Electrical sounding works by detecting electric currents that are either generated artificially or are of natural origin. This method is widely used for studies of aquifers, because water with dissolved ions is a good electrical conductor. It is also useful in detecting ore bodies, which frequently contain electrically conducting minerals. Finally, it is used for detecting buried artificial structures like pipelines and old buried storage tanks.

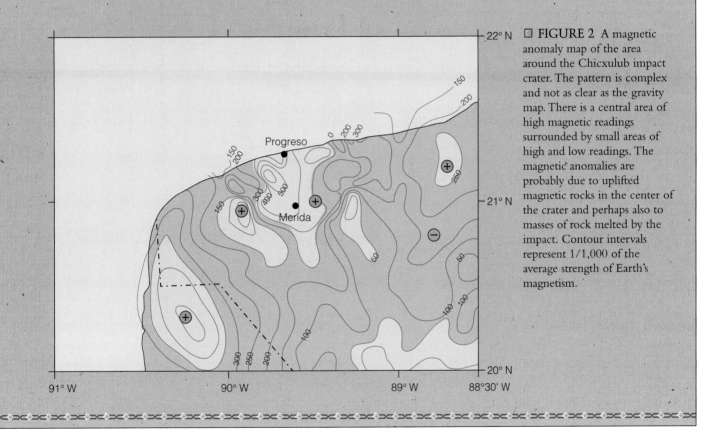

☐ FIGURE 2 A magnetic anomaly map of the area around the Chicxulub impact crater. The pattern is complex and not as clear as the gravity map. There is a central area of high magnetic readings surrounded by small areas of high and low readings. The magnetic anomalies are probably due to uplifted magnetic rocks in the center of the crater and perhaps also to masses of rock melted by the impact. Contour intervals represent 1/1,000 of the average strength of Earth's magnetism.

balance is reestablished. There are ancient beaches around Hudson Bay that are now 300 m (1,000 ft) above sea level, indicating how much the crust has risen since the sea first entered the bay about 8,000 years ago.

Isostasy is largely responsible for the uplift of both mid-ocean ridges and mountain ranges. The lithosphere of the mid-ocean ridges, being hotter than most lithosphere, is also lighter and floats higher in the mantle. During mountain building, plate tectonic processes thicken the crust in a variety of ways, causing the crust to rise.

Orogeny

In addition to isostasy, other forces also cause crustal movements. In otherwise stable parts of the continents, gentle uplift or subsidence occurs to create domes and basins (☐ Fig. 10.18). In Arizona, one such crustal uplift raised a small, gentle anticline across the course of the Colorado River, which cut downward through the rising crust to form the Grand Canyon. These movements generally do not deform or metamorphose the crust greatly and usually are not accompanied by volcanic or igneous activity. Their causes are not clearly

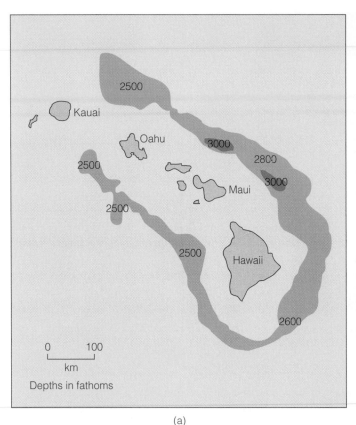

(a)

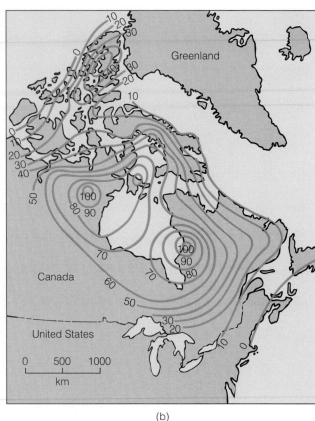

(b)

 FIGURE 10.17 (*a*) The tremendous mass of the Hawaiian shield volcanoes has depressed the surrounding crust, creating a shallow depression around the base of the volcanoes. (*b*) Isostatic rebound in Canada, following the retreat of the Pleistocene ice sheets.

understood, although they often seem to be related to episodes of rapid plate movement and mountain formation.

The process that created the great mountain belts of the world is called **orogeny** (from Greek *oros,* or "mountain," and *gen-,* or "formation"). Orogeny creates elongate belts of intensely deformed and metamorphosed rocks and is usually accompanied by vigorous igneous activity. Geologists refer to mountain ranges formed this way as *orogenic belts.*

Even while the crust is being uplifted, erosion begins carving the upland to produce mountain topography. In areas of active uplift, like Japan and southern California, surveys show rates of uplift of 1 cm per year or more (1 in in 2.5 years), enough to raise a 3 km (10,000 ft) mountain range in only 300,000 years, and much faster than the rate of erosion. Once active uplift ceases, erosion rapidly levels the mountains, usually within a few tens of millions of years. The evidence of former uplift persists, however, in the form of unconformities or interruptions in the sedimentary rock record (see Chapter 11) and as belts of deformed and metamorphosed rocks. Ancient orogenies can be recognized by the imprints they leave in the rocks long after the mountains that formed during the orogeny are gone. In many places, uplift and erosion have brought to the surface ancient rocks that bear witness to orogenies over two billion years ago.

SUBDUCTION AND OROGENY

Ocean-Ocean Subduction Zones

Convergent plate boundaries are of three types: ocean–ocean, continent–ocean, and continent–continent (Fig. 10.19 on page 230). Ocean–ocean subduction zones like the Aleutian Islands or the Marianas are fairly simple in structure compared to the other types (Fig. 10.20). They consist of a trench and a basaltic volcanic island arc, or **igneous arc,** where magma rising from the descending plate reaches the surface. Sediment erodes off the volcanic islands and partially fills the trench. Continued convergence of the plates deforms and metamorphoses the sediment in the trench, creating an **accretionary prism.** Fragments of oceanic crust, called *ophiolites,* may also break off and be added to the prism. As the volume of sediment increases, the sediment is scraped off against the overriding plate, creating a wedge of deformed sediment much like the wedge of snow that forms ahead of an advancing snowplow. Sedimentary rocks adjacent to the descending plate are thrust beneath the overlying rocks. Some rocks may reach depths of more than 20 km (12 mi) yet still be relatively cool, experiencing *blueschist metamorphism* (see Chapter 3). In the zone where the two opposing plates slide past each other, crustal rocks and sediment from

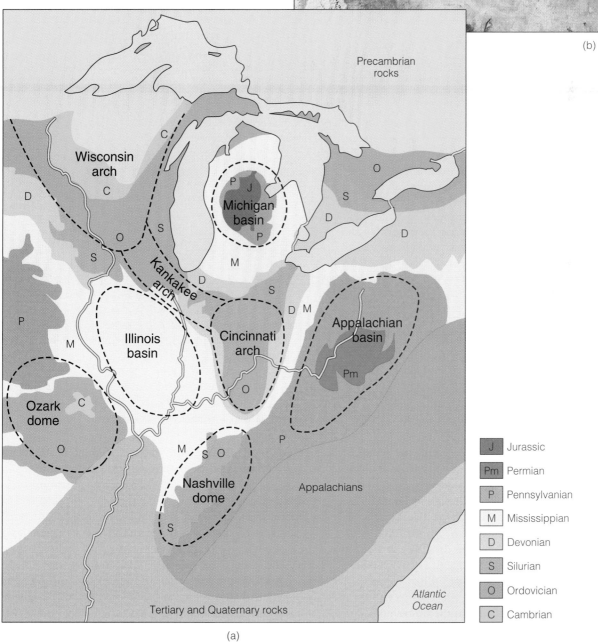

■ FIGURE 10.18 (*a*) A geologic map of the central United States, showing domes and basins due to gentle crustal movements. (*b*) In this satellite view, the Grand Canyon cuts across the middle of a gentle anticline. The higher, cooler elevation of the anticline is evident from its forest cover (red in this image) and the snow cover north of the canyon.

(b)

Precambrian rocks

Wisconsin arch

C

C

D

O

S

Michigan basin

P

J

P

Kankakee arch

S

D

M

S

Illinois basin

Cincinnati arch

D M

Appalachian basin

Pm

P

M

Ozark dome

C

O

M

S

O

Nashville dome

P

Appalachians

S

Atlantic Ocean

Tertiary and Quaternary rocks

(a)

J	Jurassic
Pm	Permian
P	Pennsylvanian
M	Mississippian
D	Devonian
S	Silurian
O	Ordovician
C	Cambrian

Focus on the Environment: Crustal Movements and Fossil Fuels

Although the parent materials for fossil fuels are deposited in sedimentary rocks, heat within Earth and deformation of the crust are key processes in making fossil fuels accessible and usable.

The parent material of coal is woody material. Fossils of woody plants are abundant in coal-bearing formations, and often the coal itself preserves obvious traces of its woody past. The change from peat to so-called soft or *bituminous* coal takes place at the low temperatures found in most thick accumulations of sedimentary rocks, but anthracite coal is a true metamorphic rock. Anthracite is coal at just the right stage of metamorphism: less metamorphism and the coal is still bituminous, more and the coal becomes unburnable graphite. In the United States, anthracite is mined in a very narrow zone in Pennsylvania where just the right metamorphic conditions occurred (☐ Fig. 1). Anthracite is a hot- and clean-burning coal, because metamorphism has removed much of the water and other impurities found in lower-grade coals, but its rarity creates serious economic disadvantages. Furthermore, anthracite coal beds are often folded and thus cannot be mined with conventional coal-mining machinery designed for work in horizontal beds. Thus, despite its advantages anthracite is too expensive for most coal users.

The precursor molecules of petroleum are probably fatty acids in marine microorganisms. Heat within Earth breaks these molecules down into petroleum and makes them more easily mobile within the rocks. Like groundwater (see Chapter 5), underground oil is trapped in tiny pores in the rock, like water in a sponge. Despite the term "oil pool," oil floats upward on subterranean water until it either reaches the surface or is trapped beneath some impervious structure, or *trap*. Usually, natural gas, the lightest component, is at the top of the trap, then a layer of oil, then finally water (☐ Fig. 2). Sometimes oil reaches the surface as a natural petroleum

seep; such seeps were used on a small scale for thousands of years for medicine, lubricants, waterproofing, and fuel.

The commercial search for petroleum is the underground search for traps, often with the aid of geophysics (see Perspective 10.1). The search for petroleum is by far the most economically important application of knowledge of Earth structures.

Some petroleum traps are sedimentary structures. For example, a buried river channel might be preserved as a thin

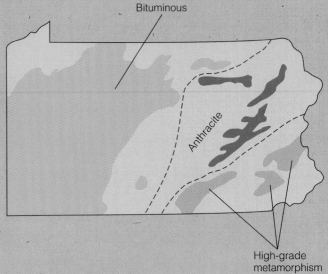

☐ **FIGURE 1** The anthracite mining district of Pennsylvania is a narrow belt where temperatures were just right for the formation of anthracite. To the east, metamorphism converted coal to graphite; to the west, the coal is still bituminous. Anthracite is a rare coal because it forms under such a narrow range of conditions.

---

the accretionary prism are churned together to create a chaotic mixture of rocks called *melange*. A very long-lived ocean-ocean subduction zone may eventually create a substantial landmass consisting largely of basaltic volcanic rocks and sedimentary rocks derived from them. The Isthmus of Panama and possibly the Caribbean islands of Cuba, Hispaniola, and Puerto Rico formed in this way.

Continent-Ocean Subduction Zones

Continent-ocean plate margins, such as the western coast of South America, are the principal places where orogeny oc-

curs (☐ Fig. 10.21 on page 231). As the mountain belt forms, sediment erodes off the continent into the oceanic trench and is deformed by plate convergence to form an accretionary prism, with the subduction zone being marked by ophiolites, blueschist metamorphism, and melange. A couple of hundred kilometers inland from the trench, magma from the descending plate invades the crust and reaches the surface. This zone is the igneous arc, the zone of most intense igneous activity, metamorphism, and deformation. The first magma, largely derived from the descending plate, is basaltic in composition, but as time goes on, the magma reacts with the rocks of the continental crust to become progressively band of porous sand encased in otherwise impervious shale or limestone. Ancient reefs, which are rich in cavities and often overlain by impervious layers, are also common petroleum traps. The greatest petroleum traps, however, are created by crustal movements. When commercial drilling for petroleum began in the late nineteenth century, well drillers quickly noted a connection between petroleum and anticlines, which are one of the most common petroleum traps. Salt domes also are important petroleum traps; petroleum migrates up the tilted layers near the dome and is trapped against the impervious salt. Petroleum can also be trapped beneath faults and unconformities. See the Prologue to Chapter 7 for a discussion of oil in the Middle East.

☐ FIGURE 2 Important types of petroleum traps: (*a*) anticline; (*b*) buried reef; (*c*) buried channel sand; (*d*) pinch-out; (*e*) unconformity; (*f*) thrust fault; (*g*) salt dome.

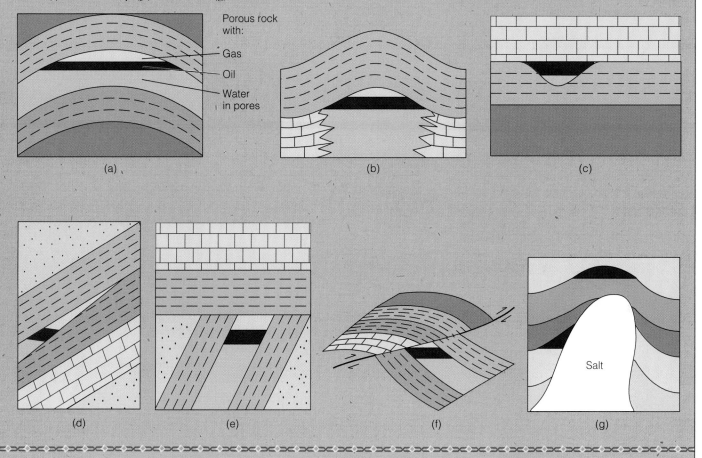

more silica rich. Andesite and rhyolite volcanism occur, and granitic batholiths invade the crust. Heating of the crust metamorphoses its rocks to amphibolite grade in the hottest and deepest areas and to greenschist grade in cooler and shallower areas. Heating of the crust also makes the rocks softer and easier to deform.

Mountain ranges are high mostly because orogenic processes thicken the crust and isostasy causes the thick crust to rise in the mantle. A variety of processes thicken the crust in orogenic belts (☐ Fig. 10.22). The convergence of plates in itself compresses and thickens the crust, as does the scraping of sediment in the accretionary prism. Heating of the crust makes it less dense and also more ductile and easier to deform. Intrusion of magma into the crust increases its volume, and a great deal of magma simply accumulates at the base of the crust, a process called *underplating*. Finally, collision of two continents may thrust one slab of continental crust over another, doubling its thickness. All these processes result in thicker continental crust, which rises because of isostasy.

Farther into the continent, deformation and metamorphism become less intense in the **foreland.** Here, the rocks consist mostly of a thin veneer of sedimentary rocks lying on the ancient igneous and metamorphic rocks of the continent.

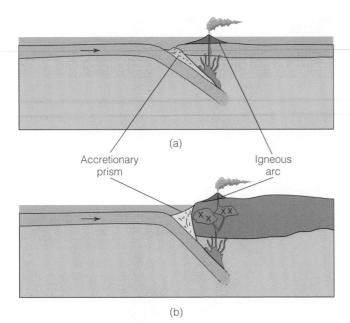

Accretionary prism — Igneous arc

(a)

(b)

Continent 1 — Continent 2

(c)

◉ FIGURE 10.19 (*a*) Ocean-ocean subduction zone.
(*b*) Continent-ocean subduction zone. (*c*) Continent-continent
subduction zone.

◉ FIGURE 10.20 A typical ocean-ocean subduction zone.

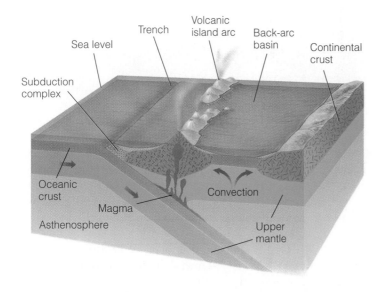

Sea level — Trench — Volcanic island arc — Back-arc basin — Continental crust — Subduction complex — Oceanic crust — Magma — Asthenosphere — Convection — Upper mantle

As orogeny proceeds, this thin veneer can be detached from the underlying crust and folded much the way a carpet rumples when a heavy piece of furniture is pushed across it. In other cases, thin sheets of rock called *nappes* may slide along nearly horizontal thrust faults. Sediment eroded off the rising mountain range is also deposited on the foreland as thick deposits of terrestrial sandstone and conglomerate. Beyond the foreland is the **craton,** the stable interior of the continent that is little affected by orogeny. The craton may be covered by a thin veneer of sedimentary rocks or may have ancient exposed igneous and metamorphic rocks at the surface. These rocks are themselves the products of ancient orogenies. Large expanses of ancient Precambrian igneous and metamorphic rocks are exposed on all the continents and are termed *shields.*

Continent-ocean orogenic belts are not symmetrical, and it is easy to determine which side of the orogenic belt faced the ocean, even long after the orogeny has ceased. The lack of symmetry of these orogenic belts is important not only in deciphering the history of ancient orogenies, but in interpreting orogenies that result from the collision of landmasses.

Collisional Orogenies

Plate convergence can eventually bring two landmasses into contact. The landmasses can range from very small to continents. On a small scale, the ocean floor is not smooth. When a volcanic seamount enters a subduction zone, it may be broken off and added to the accretionary prism. Larger blocks of crust may also be added to orogenic belts. For example, an island arc may collide with a continent; the "tail" of New Guinea is one place where such a collision has recently taken place. The complex chains of islands north and east of Australia are island arcs in the process of accreting to the Australian plate (◉ Fig. 10.23a). In other cases, a microcontinent may collide with an orogenic

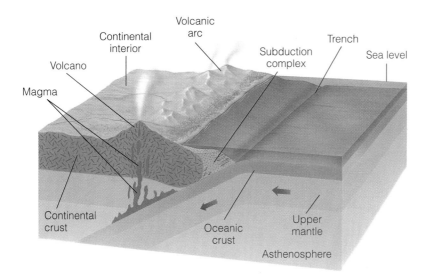

FIGURE 10.21 A typical continent–ocean subduction zone.

belt. For example, the "toe" of Italy is a former fragment of the Alps that became wedged between Sicily (which is actually part of the African plate) and Italy (Fig. 10.23b). Small fragments that become added to orogenic belts are called **exotic terranes** (not "terrains"). When an existing subduction zone becomes too clogged with exotic terranes, the descending plate may break in a new location and start the subduction process over again. In fact, some geologists suspect that a band of seismic activity across the southwestern corner of the Pacific plate marks the future location of a new subduction zone.

Small collisions of this sort have played a fundamental role in shaping the continents. Almost all of Alaska is a mosaic of exotic terranes; in fact, all of western North America is a mosaic of over 200 exotic terranes added to the continent in the last 600 million years (Fig. 10.24). Many other parts of the world, including Greece, Turkey, Iran, and most of China, seem to have formed through this process, called *continental accretion*. Precambrian continental accretion took place as well. For example, between about two billion and one billion years ago, most of North America apparently was assembled from many large and small pieces (Fig. 10.25).

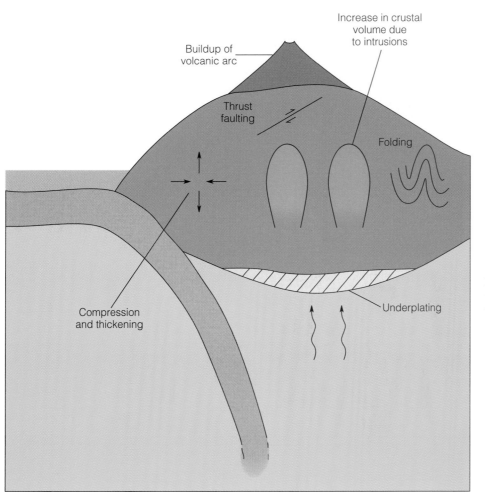

FIGURE 10.22 A variety of processes thicken the crust in orogenic belts. Isostatic uplift causes the thickened crust to rise.

FIGURE 10.23 Recent terrane accretions. (*a*) Island arc accretion in progress: the southwest Pacific northeast of Australia. The complex chains of islands between the Australian and Pacific plates are exotic terranes in the process of accretion. The band of earthquakes across the corner of the Pacific plate may represent the beginnings of a new subduction zone.
(*b*) A microcontinent collision: Calabria, the "toe" of Italy, is a former fragment of the Alps wedged between Italy and Sicily (which is actually a corner of the African plate). Most of the large islands of the western Mediterranean are also microcontinents that were formerly attached to Europe.

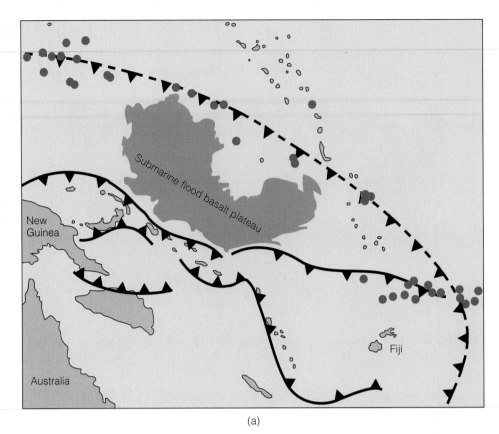

(a)

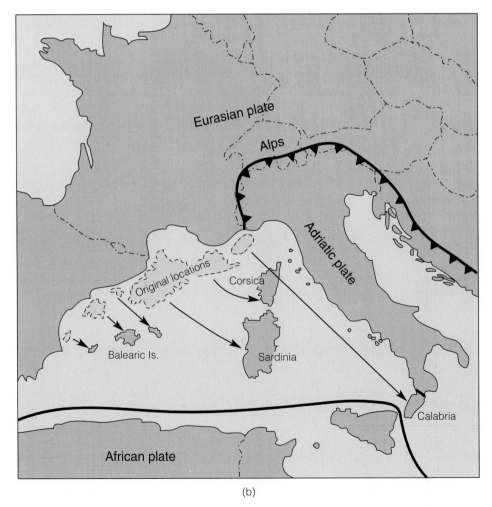

(b)

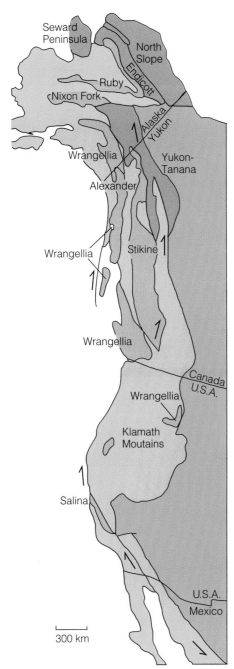

FIGURE 10.24 Some of the larger exotic terranes of western North America.

FIGURE 10.25 North America itself accreted from smaller terranes during the Precambrian.

Legend for Figure 10.25:
- 1.0–1.3 Billion
- 1.6–1.8 Billion
- 1.8–2.0 Billion
- >2.5 Billion

Subduction can also cause two large continents to collide with each other, for example, in the Persian Gulf (see the Prologue to Chapter 7), between Italy and Europe to form the Alps, or between India and Eurasia to raise the Himalayas (Fig. 10.26). The boundary between the continents is identified by a *suture,* marked by such typical subduction zone features as melanges, ophiolites, or blueschist metamorphism. In many cases, one continent was adjacent to a subduction zone while the other had a passive continental margin. After the collision, the rocks on one side of the suture reflect the igneous and metamorphic history of a typical orogenic belt, while the other side may reflect an un-

eventful precollision history. In other cases, both continents may have faced subduction zones; in this case, both sides of the suture reflect orogeny before the collision.

Whatever the history of the opposing continental margins, during the collision they undergo intense deformation and metamorphism. The continent that is being carried into the subduction zone cannot be subducted, because continental crust is too light to sink into the mantle. One continent can be thrust under the other for long distances, however, as India has been thrust beneath the Tibetan Plateau.

It takes only a few minutes to read about mountain formation, but in reality, the process takes millions of years. The uplift may proceed at a centimeter per year, and erosion may occur nearly as fast, so a person would see little change in a lifetime. Millions of people in Indonesia, Japan, and Peru live out their lives without ever realizing that they are living in an active orogenic belt.

In Chapters 7 through 10, we have examined Earth's interior systems and the processes and forces that change the planet from within. Rocks become changed to new forms, folded and recrystallized, and raised into mountains. In the next chapter, we will begin examining another of Earth's systems: its atmosphere.

The Discovery of Isostasy

In the early nineteenth century, the British launched the survey of India that eventually measured the height of Mount Everest (see the Prologue). As the surveyors worked north from southern India, the locations of places determined by two different survey methods began to disagree with each other. By the time the surveyors reached northern India, the two methods differed by 150 m (500 ft).

It seemed obvious that, since the problems became worse the farther north the surveyors went, the cause of the difficulty lay in northern India. Could the mass of the Himalayas be affecting the survey? The astronomical instruments used for checking latitude were leveled using spirit levels and plumb lines. If the Himalayas were exerting a sideways gravitational pull, those instruments would not be truly level (■ Fig. 1). Surprisingly, calculations showed that the errors should actually have been much greater than observed. Evidently, there must be light rocks deep in Earth to offset the extra mass of the mountains.

One scientist, G. B. Airy, reasoned that because Earth is not perfectly rigid, the continents must float in the

■ FIGURE 1 (a) The survey of India ran into problems because the gravitational pull of the Himalayas caused astronomical instruments to be slightly out of level. The latitudes determined with the instruments were incorrect. (b) When the gravitational effect of the Himalayas was actually calculated, it turned out that the survey error should have been much greater than it actually was. The attraction of the Himalayas was partially offset by low-density rocks deep in Earth where high-density rocks would normally have been.

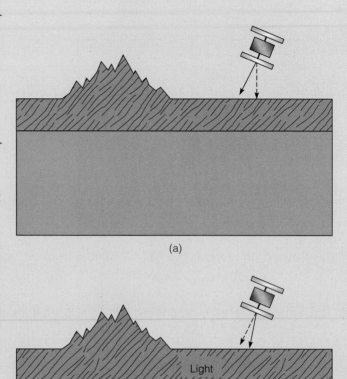

(a)

(b)

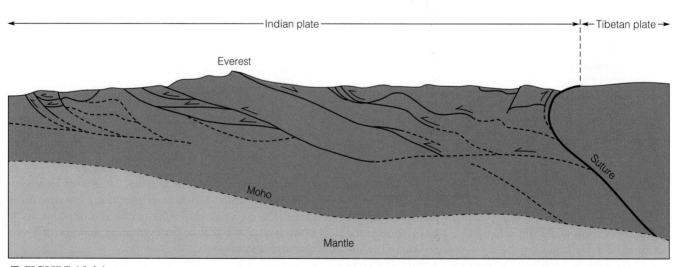

■ FIGURE 10.26 A continent-continent collision: the Himalayas. The cross section passes through Mount Everest.

ductile material of the mantle. Airy reasoned that thicker blocks of crust would float higher, but also extend deeper into the mantle, than crust of average thickness (Fig. 2a). Another scientist, J. H. Pratt, argued that the crust differed in density from place to place. In Pratt's view, the Himalayas were high because they were made of lighter rock than the rest of India (Fig. 2b).

Airy and Pratt were both right: both thickness and density govern the elevation of the crust. Seismic soundings show that the continental crust is thicker under mountain ranges, so Airy was correct in the case of the Himalayas. But there are also places where the crust differs in elevation because it differs in density, as Pratt believed. The mid-ocean ridges, for example, owe their elevation to the fact that the crust beneath them is hotter and less dense than old, cool oceanic crust (see Chapter 18). The greatest elevation contrast on Earth, the contrast between the ocean basins and the continents, is due both to density and to thickness. The continents are high both because continental crust is light and because it is thick. The ocean basins are low because oceanic crust is thin and dense (Fig. 2c).

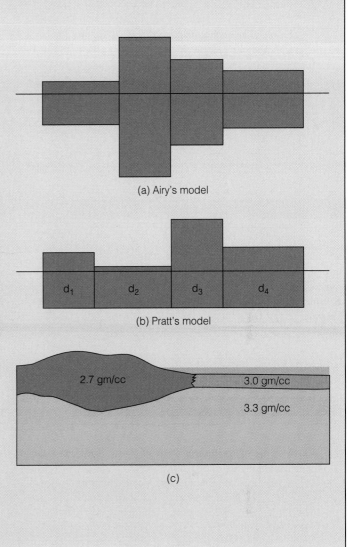

(a) Airy's model

(b) Pratt's model

(c)

FIGURE 2 Isostasy. (*a*) Airy's concept: Elevation is due to differences in the crust's thickness. (*b*) Pratt's concept: Elevation is due to differences in the crust's density. (*c*) In reality, Earth's surface relief is due to differences in both the thickness and the density of the crust. The continents stand high because continental crust is both thicker and lighter than oceanic crust.

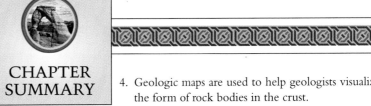

CHAPTER SUMMARY

1. Stress is force per unit area. Strain is the amount of deformation of a rock.

2. Elastic strain is temporary deformation that disappears when the stress is removed. Rocks that break above a certain stress are termed brittle; if they deform without breaking, they are ductile. Rocks can exhibit both brittle and ductile behavior, depending on temperature, pressure, and how rapidly stress is applied.

3. Deformation produces folds and foliation in rocks. Foliation is usually perpendicular to the direction of greatest compression and usually parallel to the axial plane of folds. Using this fact, geologists can use foliation as a valuable clue in understanding and mapping large folds.

4. Geologic maps are used to help geologists visualize the form of rock bodies in the crust.

5. Orientations of rocks are defined in terms of strike, the direction of a horizontal line in the rock layer, and dip, the angle the layer makes with the horizontal.

6. Geologic structures such as anticlines, domes, basins, and synclines can be recognized by the distinctive outcrop patterns they produce on geologic maps.

7. Isostasy is the process in which the crust rises or sinks in the mantle by buoyancy. Thick crust tends to float upward and produce highlands. Heavy loads like seamounts or ice sheets can cause the crust to sink downward.

8. Orogeny is the process of mountain formation. It produces elongated belts of intense deformation, metamorphism, and igneous activity.
9. Orogenic belts form at subduction zones. A typical orogenic belt consists of an accretionary prism, island arc, and foreland.
10. A typical orogenic belt on the edge of a continent consists of an accretionary prism, igneous arc, and foreland.
11. Mountains rise because orogenic processes thicken the crust, which then rises because of isostasy.
12. The stable continental interior is called the craton. Ancient exposed areas of igneous and metamorphic rocks are termed shields.
13. Plate convergence can eventually cause landmasses to collide. Small landmasses such as island arcs or microcontinents are called terranes. Terrane accretion has built up large areas of the present continents.
14. Continents can collide to form orogenic belts. This process is happening now between India and Asia and between Arabia and Iran.
15. Petroleum accumulates beneath traps, some of which are sedimentary structures, but most of which are created by crustal movements.

IMPORTANT TERMS

accretionary prism	ductile	igneous arc	stress
anticline	exotic terrane	isostasy	strike
axial plane	fault	joint	syncline
brittle	foliation	normal fault	thrust fault
craton	foreland	orogeny	transcurrent fault
dip	geologic map	strain	

REVIEW QUESTIONS

1. Which of the following is *not* true of elastic, brittle, and ductile behavior?
   a. ____ the bend of a diving board is an example of elastic behavior; b. ____ the shattering of glass is an example of brittle behavior; c. ____ the molding of modeling clay is an example of elastic behavior; d. ____ the bending of metal is an example of ductile behavior; e. ____ the formation of joints in rock is an example of brittle behavior.
2. Why is foliation usually parallel to the axial planes of folds?
   a. ____ foliation makes rocks weaker in some directions than others; b. ____ both structures form as a result of brittle behavior in the rocks; c. ____ both structures form perpendicular to the direction of greatest compression; d. ____ both structures form parallel to the direction of greatest compression.
3. Why are continents high and ocean basins low?
   a. ____ continental crust is thick, and oceanic crust is thin; b. ____ continental crust is light, and oceanic crust is dense; c. ____ the difference is due entirely to thickness variations; d. ____ the difference is due entirely to density variations; e. ____ answers (a) and (b) are both correct.
4. How can an anticline be distinguished from a syncline on a geologic map?
   a. ____ the older rocks are in the center of the fold; b. ____ the younger rocks are in the center of the fold; c. ____ the rocks form a U pattern on the map; d. ____ there is no way to distinguish them.
5. If folding can overturn rock layers, how can a geologist determine whether the rocks are right side up?

   a. ____ geologists have to use radiometric dating; b. ____ there is no way to tell; c. ____ geologists can only tell if there are fossils in the rocks; d. ____ sedimentary structures have a distinct "right way up."
6. Which part of an orogenic belt is most likely to display blueschist metamorphism?
   a. ____ accretionary prism; b. ____ igneous arc; c. ____ foreland.
7. Which of the following rocks is most ductile?
   a. ____ granite; b. ____ basalt; c. ____ sandstone; d. ____ rock salt.
8. Why is the expression "oil pool" inaccurate?
   a. ____ oil forms in sedimentary rocks and occurs in layers, not pools; b. ____ oil moves upward and accumulates beneath traps; it does not sink into depressions; c. ____ like groundwater, oil does not collect in open cavities but occurs in the pore spaces in rocks; d. ____ answers (b) and (c) are both correct.
9. What is the difference between hydrostatic and directed pressure? Which more nearly describes stresses deep in Earth?
10. Describe the ways foliation forms.
11. Why do geologists construct geologic maps? What kinds of information do they put on these maps?
12. Describe some of the ways a geologist can use structures seen in an outcrop to understand large structures that may be miles across.
13. Name the three types of convergent plate boundaries.
14. Describe some of the processes that cause uplift in mountain ranges.

15. Put a small object under this book so that it is tilted. Measure its dip with a protractor. Find or mark a horizontal line on the cover to identify the direction of strike.

16. Where have terrane and continental collisions occurred in the recent geologic past?

17. What are some of the methods geologists have developed to gain information about deeply buried and inaccessible rocks?

18. If erosion wears away a mountain range entirely, how can geologists still learn about ancient orogenies?

POINTS TO PONDER

1. You can drive a nail a centimeter or more into a board with a single blow. Striking the board just as hard with the hammer alone makes only a dent. What accounts for the difference?

2. If the Antarctic ice cap were to melt entirely, what would happen to the crust of Antarctica? Where would the water go?

ADDITIONAL READINGS

Appenzeller, T. 1989. Just a veneer: The upper crust of continents can slip and deform on its own. *Scientific American* 261 (November): 26.

Berthon S., and A. Robinson. 1991. *The shape of the world.* Chicago: Rand McNally.

Davis, G. H. 1984. *Structural geology of rocks and regions.* New York: Wiley.

Dennis, J. G. 1987. *Structural geology: An introduction.* Dubuque, Iowa: William C. Brown.

Horgan, J. 1988. Ice house: Did a growth spurt of mountain ranges initiate the ice ages? *Scientific American* 259 (November): 22.

Jordan, T. H., and J. B. Minster. 1988. Measuring crustal deformation in the American West. *Scientific American* 259 (August): 48–55.

Molnar, P. 1986. The structure of mountain ranges. *Scientific American* 255 (July): 70–79.

Ruddiman, W. F., and J. E. Kutzback. 1991. Plateau uplift and climatic change (Tibetan and Colorado Plateaus). *Scientific American* 264 (March): 66–72.

Suppe, J. 1985. *Principles of structural geology.* Englewood Cliffs, N.J.: Prentice-Hall.

Talbot, C. J., and M. P. A. Jackson. 1987. Salt tectonics. *Scientific American* 257 (August): 70–79.

WORLD WIDE WEB SITES

For current updates and exercises, log on to http://www.wadsworth.com/geo.

THE AMERICAN GEOPHYSICAL UNION
http//www.agu.org/

LAMONT DOHERTY EARTH OBSERVATORY
http://www.ldeo.columbia.edu/

These two sites have material of general as well as professional interest. Most of the Web sites for crustal deformation are oriented toward professional scientists rather than the general public.

CHAPTER
11

Geologic Time and the History of Life

...ckerboard Mesa in Zion National Park, Utah, belong to the Jurassic-aged Navajo Sandstone which represents an ancient wind-blown ...es intersect large cross–beds, giving this cliff its checkerboard appearance.

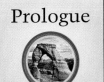

Prologue

What is time? We organize our lives around it with the help of clocks, calendars, and appointment books, yet most of us feel we have too little time—we are always running "behind" or "out of time." According to biologists and psychologists, children less than two years old and animals have no conscious concept of past or future and exist in a "timeless" present. In some belief systems, time is circular: it has no beginning or end, and like a circle, all things are destined to return to where they once were.

For most people, though, time is linear and moves like a flowing stream. It can be measured and subdivided; we can place events in a chronology in which there is a history of past events and expectations for the future. Timekeeping in one fashion or another has been practiced for thousands of years, but precise timekeeping was not possible until the fifteenth century when the Dutch scientist Christian Huygens constructed the first pendulum clock.

Today the quartz watch is the most popular timepiece. Powered by a battery, a quartz crystal vibrates about 100,000 times per second, and an integrated circuit counts these vibrations and converts them into a digital or dial reading on your watch face. An inexpensive quartz watch is remarkably accurate, and precision-manufactured quartz clocks are accurate to within one second per 10 years.

Precise timekeeping has become increasingly important in our technological world. Ships and aircraft plot their locations by satellite, relying on an extremely accurate time signal. Deep-space probes require precisely timed radio commands, and physicists exploring the motion within an atom's nucleus deal in trillionths of a second. To achieve such accuracy, scientists use atomic clocks. First developed in the 1940s, these clocks rely on the oscillations of an atom's electrons, whose rhythm is so regular that the clocks are accurate to within a few thousandths of a second per day.

While physicists deal with incredibly short intervals of time, astronomers and Earth scientists are concerned with time measured in millions and billions of years. When astronomers look at a distant galaxy, they are seeing what it looked like billions of years ago. Earth scientists deal with the concept of **geologic time,** the interval of time since Earth formed 4.6 billion years ago. When they investigate rocks in the walls of the Grand Canyon, they are deciphering about 2 billion years of Earth history.

Most people have difficulty comprehending geologic time because they tend to view time from the perspective of their own existence. From this vantage point, hundreds or thousands of years seem incomparably long. But when considering geologic time, an entirely different frame of reference is needed. Thousands of years is but an instant in geologic time, and a "sudden" event might occur over millions of years.

INTRODUCTION

An appreciation of the immensity of geologic time is necessary for an understanding of both the physical and the biologic history of Earth (Fig. 11.1). To gain some insight into the duration of geologic time, consider this. If 1 second represents 1 year, it would take more than 11½ days to count out 1 million years and nearly 146 years to count out the 4.6 billion years of Earth history!

Two different frames of reference are used when speaking of geologic time. **Relative dating** involves placing events in their proper chronological sequence, that is, in the order of their occurrence. Relative dating tells us that one event preceded another, but does not tell us how many years ago a particular event took place or how long it lasted. **Absolute dating** gives specific dates for events expressed in years before the present. Various dating techniques based on radioactive decay are used to determine absolute ages.

EARLY CONCEPTS OF GEOLOGIC TIME

The concept of geologic time and its measurement have changed over the course of human history. Many Christian scholars and clerics tried to establish the date of creation by

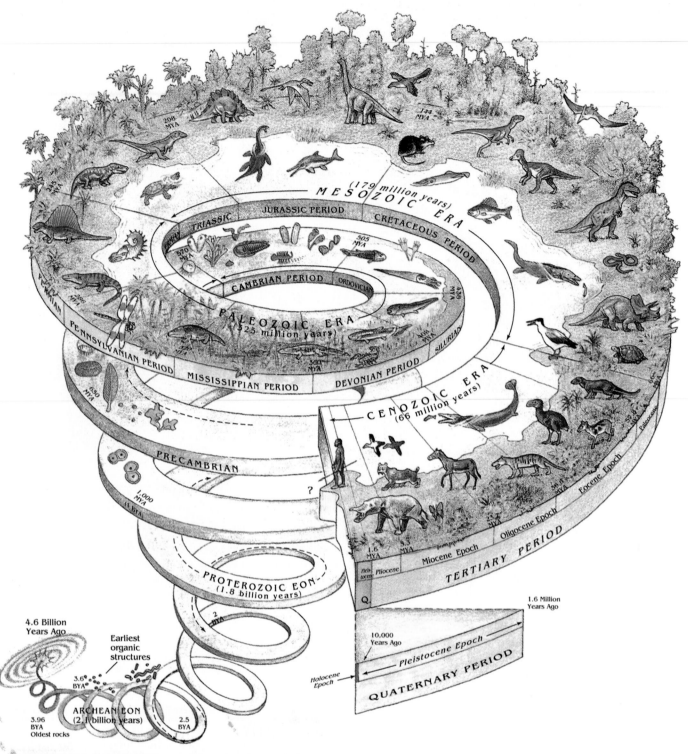

■ FIGURE 11.1 Geologic time is depicted in this spiral history of Earth from its time of formation 4.6 billion years ago to the present. (BYA = billion years ago; MYA = million years ago).

analyzing historical records and the genealogies found in Scripture. James Ussher (1581–1665), archbishop of Armagh, in Ireland, is credited as being the first to use this method. In 1650 he announced that Earth had been created on October 22, 4004 B.C.

The idea of a very young Earth, perhaps no more than 6,000 years old, as proposed by Archbishop Ussher, required

that all of Earth history be explained within this context. Consequently, scholars developed the concept of *catastrophism,* which explained Earth history as resulting from a series of sudden widespread catastrophic events. Eventually, catastrophism was abandoned as an untenable hypothesis because it clearly was not supported by evidence from the geologic record.

 FIGURE 11.2 According to the principle of uniformitarianism, present-day processes have operated throughout geologic time. These ripple marks preserved in ancient rocks give every indication that they formed just as wave-formed ripple marks do today.

JAMES HUTTON AND THE RECOGNITION OF GEOLOGIC TIME

Observing present-day processes such as sediment transport, wave action, and erosion by running water, the Scottish geologist James Hutton (1726–1797) concluded that given enough time these processes could account for the geologic features of Earth. His idea that present-day processes have operated throughout geologic time was the basis for the **principle of uniformitarianism.** Despite Hutton's work, catastrophism continued to be widely accepted until 1830, when Charles Lyell published his book *Principles of Geology,* which effectively discredited catastrophism and firmly established uniformitarianism as a fundamental principle in geology. With the acceptance of uniformitarianism came the idea that Earth is old, certainly much older than 6,000 years.

Uniformitarianism is a powerful principle that allows us to decipher Earth history by interpreting the geologic record in terms of known processes (Fig. 11.2). We should keep in mind that uniformitarianism does not exclude sudden or large-scale events such as volcanic eruptions, earthquakes, landslides, flooding, and meteorite impacts. As a matter of fact, many of the features preserved in the geologic record probably resulted from short-term events like these, a view that is certainly in keeping with the principle of uniformitarianism.

RELATIVE DATING METHODS

Before the development of dating techniques based on radioactive decay, scientists had no reliable means of absolute dating and depended solely on relative dating methods. These methods allow us to decipher geologic history by placing events in sequential order, but they cannot tell us how long ago these events occurred.

Fundamental Principles of Relative Dating

The Danish anatomist Nicolas Steno (1638–1686) noted that during floods, streams spread across their floodplains and deposit layers of sediment that bury organisms dwelling there. Later floods produce younger layers of sediments that are deposited or superposed over previous deposits. From these observations, Steno arrived at the **principle of superposition.** According to this principle, in an undisturbed succession of sedimentary rock layers, the oldest layer is at the bottom and the youngest layer is at the top (Fig. 11.3a). Accordingly, the relative ages of all layers in the se-

(a)

(b)

 FIGURE 11.3 (*a*) The oldest rocks are at the bottom of these rock layers, and the youngest rocks are at the top, illustrating the principle of superposition. Furthermore, the layers are horizontal as they were when originally deposited. (*b*) These rocks in Utah are inclined from horizontal (dipping). We can infer that they were deposited horizontally or nearly so and were tilted to their present position after they were lithified.

(a)

(b)

◼ FIGURE 11.4 (a) Cross-cutting relationships are shown by the small faults cutting through these layers of volcanic ash in Oregon. (b) This light-colored granite in northern Wisconsin is younger than the dark-colored basalt. The relative ages of these two rock bodies can be determined because the granite contains inclusions of basalt.

quence can be determined, as well as the relative ages of any fossils they contain.

Steno also observed that when sediment settles from water, it forms essentially horizontal layers, illustrating the **principle of original horizontality** (Fig. 11.3b). Therefore, sedimentary rock layers steeply inclined from horizontal must have been tilted after deposition and lithification.

James Hutton recognized that a fracture or an igneous intrusion such as a dike must be younger than the rocks it cuts or intrudes (◼ Fig. 11.4a). Likewise, a valley eroded by a stream is younger than the rocks that have been cut or disrupted by erosion. This **principle of cross-cutting relationships,** as it is called, is very important in relative dating and interpreting Earth history.

Another way to determine relative ages is by using the **principle of inclusions,** which holds that inclusions, or fragments of one rock contained within another, are older than the rock layer containing the inclusions (Fig. 11.4b). A granite batholith, for example, containing pieces of

an adjacent sandstone must have been intruded into the sandstone and is thus younger than the sandstone. If, on the other hand, the sandstone contains granite rock fragments, the granite must have been the source rock for the fragments and is therefore older than the sandstone.

Unconformities

Our discussion so far has been concerned with sequences of rocks in which no depositional breaks of any consequence occur. Such rock sequences are said to be *conformable,* meaning that they preserve a more or less continuous record of the events that occurred during their time of origin. Actually, deposition in most environments is episodic rather than continuous, so most deposits record only a fraction of the total time represented. In short, the geologic record contains numerous gaps, but many of these can be ignored because they represent insignificant intervals of time when considered in the context of geologic time.

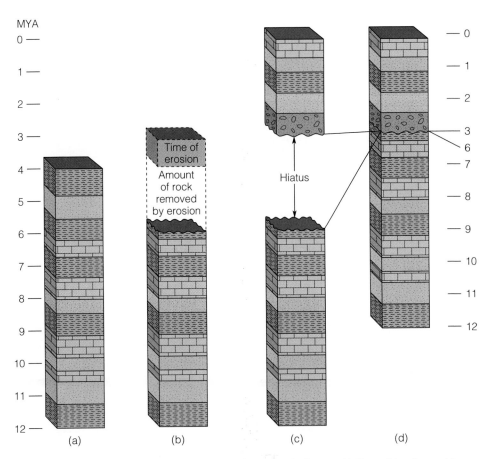

■ FIGURE 11.5 Development of an unconformity and a hiatus. (*a*) Deposition began 12 million years ago (MYA) and continued until 4 MYA. (*b*) A 1-million-year episode of erosion took place, and during that time rocks representing 2 million years of geologic time were eroded. (*c*) A hiatus of 3 million years exists between the older rocks and the rocks that formed during a renewed episode of deposition that began 3 MYA. (*d*) The actual geologic record. The unconformity represents a major break in our record of geologic time.

Surfaces separating rock layers in the geologic record that encompass significant amounts of time are **unconformities,** and any interval of time not represented by rocks in a particular area is a *hiatus* (■ Fig. 11.5). Thus, an unconformity is a surface of nondeposition or erosion or both that separates younger rocks from older rocks. As such, it is a break in our record of geologic time.

Three types of unconformities are recognized. A **disconformity** is an erosion surface between rock layers that are parallel with one another (■ Fig. 11.6a and b). An erosion surface on tilted or folded rocks over which younger rock layers have been deposited is an **angular unconformity** (Fig. 11.6c and d). The third type of unconformity is a **nonconformity,** an erosion surface cut into metamorphic or igneous rocks that is covered by sedimentary rocks (Fig. 11.6e and f).

Once the nature of unconformities and the various relative dating principles are understood, the geologic history of any area can be deciphered (□ Perspective 11.1).

 CORRELATION

The various principles of relative dating allow us to decipher the geologic history of a local area, but if we are to interpret the history of an entire region, we must have some method of determining the relative ages of rocks in widely separated areas. Demonstrating that rocks in different areas are the same age is known as **correlation.**

That fossils are the remains of once living organisms has been known for centuries, yet their utility in relative dating and correlation was not fully appreciated until the early 1800s. William Smith (1769–1839), an English civil engineer involved in surveying and building canals in southern England, independently recognized the principle of superposition by reasoning that fossils at the bottom of a series of sedimentary rock layers are older than those at the top.

Smith's observations served as the basis for the **principle of fossil succession** (also called the principle of faunal succession), according to which groups or assemblages of fossils succeed one another through time in a regular and

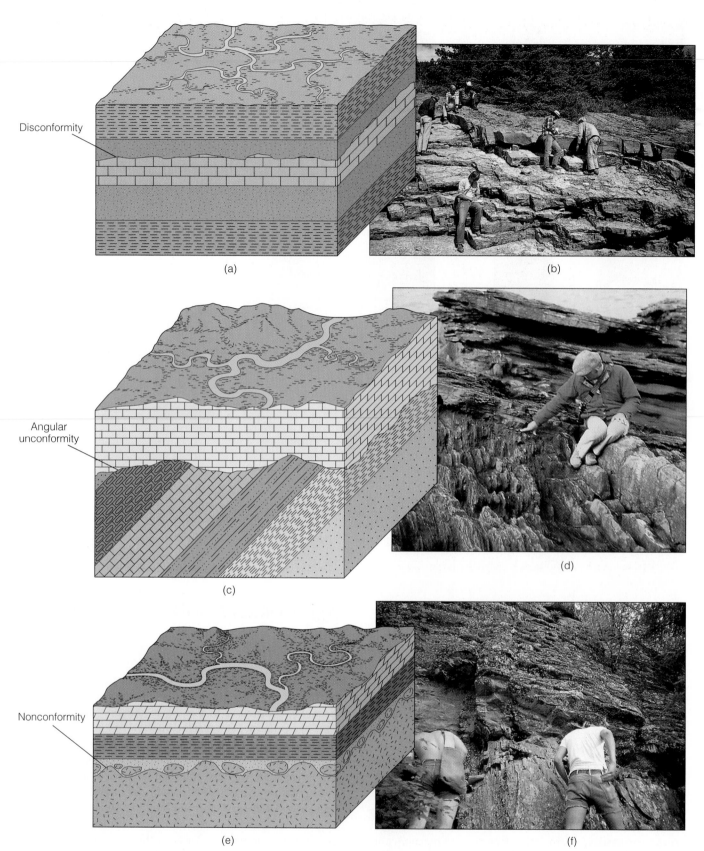

■ FIGURE 11.6 (*a*) A disconformity. (*b*) Disconformity between Mississippian and Jurassic rocks in Montana. The geologist at the upper left is sitting on Jurassic rocks and his foot is resting on Mississippian rocks. (*c*) An angular unconformity. (*d*) Angular unconformity at Siccar Point, Scotland. (*e*) A nonconformity. (*f*) Nonconformity between Precambrian metamorphic rocks and overlying Cambrian-aged sandstone in South Dakota.

Perspective 11.1

Applying the Principles of Relative Dating

We can decipher the geologic history of the area represented in □ Figure 1 by applying the various principles of relative dating. According to the principles of superposition and original horizontality, beds A, B, C, D, E, F, and G were deposited horizontally; then they were either tilted, faulted (H), and eroded, or they were faulted (H), tilted, and then eroded. We cannot determine whether the faulting was before or after the tilting, but we do know that the fault (H) occurred after beds A–G were deposited, because it cuts through them (principle of cross-cutting relationships).

Beds J, K, and L were then deposited horizontally over the erosion surface, producing an angular unconformity (I). Following deposition of these three beds, the entire sequence, beds A through L, was intruded by a dike (M), which, according to the principle of cross-cutting relationships, must be younger than all the rocks it intrudes.

The entire area was then uplifted and eroded; next beds P and Q were deposited, producing a disconformity (N) between beds L and P and a nonconformity (O) between igneous intrusion M and bed P. We know that the relationship between igneous intrusion M and bed P is a nonconformity because of the presence of inclusions of M in P (principle of inclusions).

According to the principle of cross-cutting relationships, dike R must be younger than bed Q because it intrudes into it. It could have intruded anytime *after* bed Q was deposited, but we cannot determine whether R was formed right after Q was deposited or after S and/or T formed. For purposes of this history, we will assume it intruded after the deposition of bed Q.

Following the intrusion of dike R into bed Q, lava S flowed over bed Q, followed by deposition of bed T. Although the lava flow (S) is not a sedimentary unit, the principle of superposition still applies because it flowed on Earth's surface and was then buried beneath younger materials.

In this example, we established the sequence of events by using the principles of relative dating. Remember, however, that we have no way of knowing how many years ago these events occurred unless we obtain absolute ages for the igneous rocks.

□ FIGURE 1 Block diagram of a hypothetical area to which the various relative dating principles can be applied to decipher geologic history.

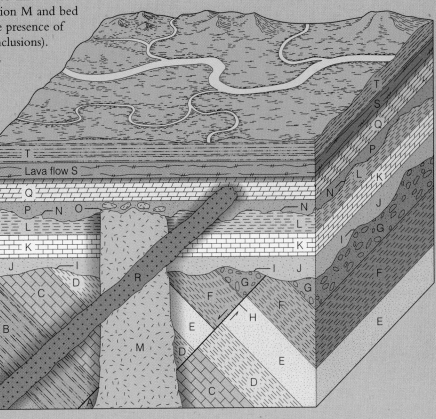

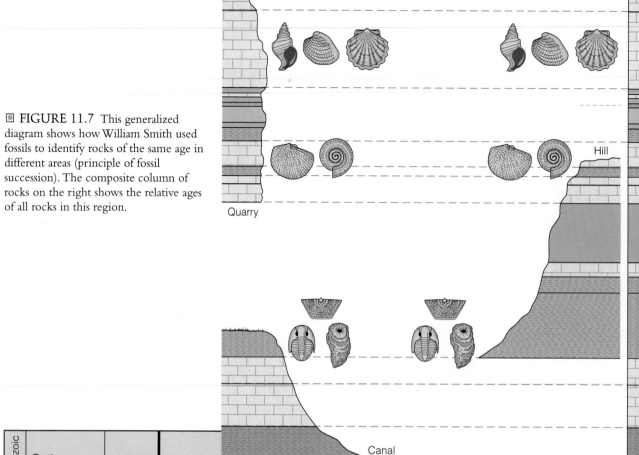

FIGURE 11.7 This generalized diagram shows how William Smith used fossils to identify rocks of the same age in different areas (principle of fossil succession). The composite column of rocks on the right shows the relative ages of all rocks in this region.

		Lingula	Inoceramus	Isotelus
Cenozoic	Tertiary			
Mesozoic	Cretaceous			
Mesozoic	Jurassic	*Lingula*	*Inoceramus*	
Mesozoic	Triassic			
Paleozoic	Permian			
Paleozoic	Pennsylvanian			
Paleozoic	Mississippian			
Paleozoic	Devonian			
Paleozoic	Silurian			
Paleozoic	Ordovician			*Isotelus*
Paleozoic	Cambrian			

predictable order (Fig. 11.7). This principle is important because while superposition can be used to determine relative ages in a single locality, it cannot be used to determine the relative ages of rocks in distant areas. Nor can rock type be used in correlation, because the same type of rock, sandstone, for instance, has formed repeatedly through geologic time.

Assemblages of fossils, however, are unique; that is, they occur only once. Accordingly, if rocks containing similar fossils are matched up, we can assume that they are of the same relative age (Fig. 11.7). In short, Smith discovered a method that allowed the relative ages of sedimentary rocks in widely separated areas to be determined regardless of their composition.

Fossils that are easily identified, are geographically widespread, and existed for a rather short geologic time are particularly useful in correlation. Such fossils are called **guide fossils** (or index fossils) (Fig. 11.8). Two of the fossils

FIGURE 11.8 The time of existence, or geologic ranges, of three marine invertebrate animals are shown by the dark, vertical lines. The brachiopod *Lingula* is of little use in correlation because of its long geologic range. The trilobite *Isotelus* and the clam *Inoceramus* are good guide fossils because they are geographically widespread, are easily identified, and have short geologic ranges.

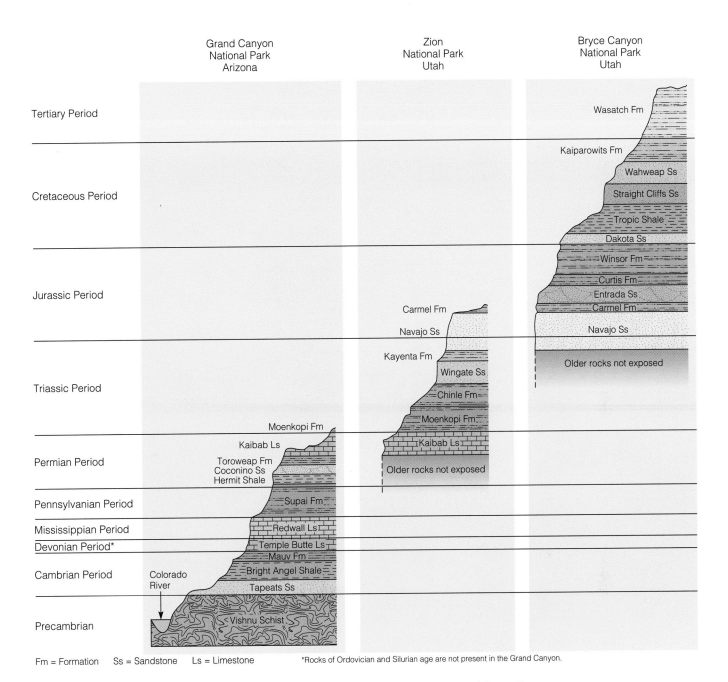

Grand Canyon
National Park
Arizona

Zion
National Park
Utah

Bryce Canyon
National Park
Utah

Tertiary Period

Wasatch Fm

Kaiparowits Fm

Cretaceous Period

Wahweap Ss

Straight Cliffs Ss

Tropic Shale

Dakota Ss

Winsor Fm

Curtis Fm

Jurassic Period

Carmel Fm

Navajo Ss

Entrada Ss

Carmel Fm

Navajo Ss

Older rocks not exposed

Kayenta Fm

Wingate Ss

Triassic Period

Chinle Fm

Moenkopi Fm

Moenkopi Fm

Kaibab Ls

Kaibab Ls

Permian Period

Toroweap Fm
Coconino Ss
Hermit Shale

Older rocks not exposed

Pennsylvanian Period

Supai Fm

Mississippian Period

Redwall Ls

Devonian Period*

Temple Butte Ls

Mauv Fm

Cambrian Period

Colorado
River

Bright Angel Shale

Tapeats Ss

Precambrian

Vishnu Schist

Fm = Formation   Ss = Sandstone   Ls = Limestone   *Rocks of Ordovician and Silurian age are not present in the Grand Canyon.

 FIGURE 11.9 Correlation of rocks within the Colorado Plateau of the western United States. By correlating rocks from various locations, the history of the entire region can be deciphered.

shown in Figure 11.8 meet these criteria, but one, the brachiopod *Lingula,* has a geologic range from Ordovician to Recent, making it of little use in correlation. A good example of correlation is provided by the rocks of the Colorado Plateau of the western United States (■ Fig. 11.9). About 2 billion years of geologic history are recorded in this region, but because of unconformities, the entire record is not preserved at any single location. By correlating rocks from one area to another, the history of the entire region can be deciphered.

THE GEOLOGIC TIME SCALE

The **geologic time scale** is a chart arranged with the designation for the earliest part of geologic time at the bottom, followed upward by designations for progressively younger intervals of time (■ Fig. 11.10). It was developed primarily during the nineteenth century in Great Britain and western Europe by using relative dating methods such as superposition and fossil succession in the study of rock exposures. Using these principles, scientists were able to

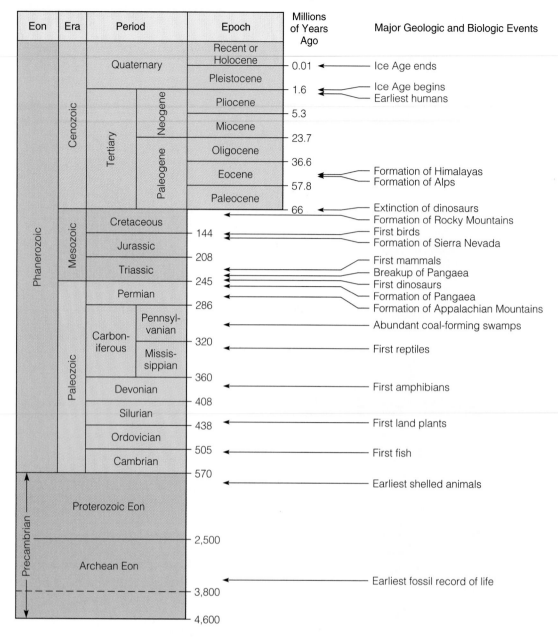

Eon	Era	Period		Epoch	Millions of Years Ago	Major Geologic and Biologic Events
Phanerozoic	Cenozoic	Quaternary		Recent or Holocene	0.01	Ice Age ends
				Pleistocene	1.6	Ice Age begins / Earliest humans
		Tertiary	Neogene	Pliocene	5.3	
				Miocene	23.7	
			Paleogene	Oligocene	36.6	
				Eocene	57.8	Formation of Himalayas / Formation of Alps
				Paleocene	66	
	Mesozoic	Cretaceous			144	Extinction of dinosaurs / Formation of Rocky Mountains / First birds / Formation of Sierra Nevada
		Jurassic			208	
		Triassic			245	First mammals / Breakup of Pangaea / First dinosaurs / Formation of Pangaea / Formation of Appalachian Mountains
	Paleozoic	Permian			286	
		Carboniferous	Pennsylvanian		320	Abundant coal-forming swamps
			Mississippian		360	First reptiles
		Devonian			408	First amphibians
		Silurian			438	First land plants
		Ordovician			505	First fish
		Cambrian			570	Earliest shelled animals
Precambrian		Proterozoic Eon			2,500	
		Archean Eon			3,800	Earliest fossil record of life
					4,600	

■ FIGURE 11.10 The geologic time scale. Some important geologic and biologic events are indicated along the right margin.

determine relative ages and correlate rocks of similar age from area to area and thus piece together a relative geologic time scale (Fig. 11.10).

The earliest part of geologic time is designated by the widely used but informal term *Precambrian,* which refers to all time before the Cambrian Period (Fig. 11.10). Included in the Precambrian are two eons, the *Archean* and the *Proterozoic;* the interval of time before the Archean has no formal designation. Following the Precambrian is the *Phanerozoic* Eon, which consists of three eras, the *Paleozoic,* the *Mesozoic,* and the *Cenozoic;* each era is further divided into *periods,* and the periods are divided into *epochs* (Fig. 11.10). Thus, the hierarchy of time designations is eon, era, period, and epoch.

Most of the geologic time scale has existed in its present form since the 1840s, yet only since the beginning of the twentieth century have scientists been able to assign absolute ages to the time scale. Before absolute age dates were available, scientists knew, for example, that Devonian rocks and fossils were younger than those of the Silurian, but they had no way of telling when the Devonian began or how long it lasted.

 ABSOLUTE DATING METHODS

During the late 1700s and 1800s, several attempts were made to determine Earth's age on the basis of scientific evidence

Fossils and the History of Earth's Rotation

It appears that Earth's rotation rate has been slowing about 2.5 seconds per 100,000 years, mostly from the frictional effects of tides. If true, then both the number of hours in a day and the number of days in the year have been decreasing as well. Supporting evidence for this hypothesis comes from fossil organisms that while alive added daily and yearly growth-rings. Present-day living corals add about one growth-ring a day, or 365 growth-rings per year, whereas Pennsylvanian corals added about 390 growth-rings per year and those from the Devonian added about 400. Studies of fossil corals and other organisms have generally confirmed this hypothesis.

rather than the methods employed by Archbishop Ussher. Some scholars tried to determine the deposition rates for various sediments and then calculate how long it took to deposit rock layers. They also reasoned that if the total thickness of all sedimentary rock in Earth's crust were known, the time required for these rocks to form could be calculated.

This attempt at absolute dating yielded ages ranging from less than 1 million years to more than 1 billion years for the same rocks! Obviously, such dating was not very reliable. The reason was that the underlying assumptions were in error; sedimentation rates are not constant, and no one could tell how much the original deposits had compacted or how much may have been lost to erosion. In short, although such attempts were ingenious, they all failed.

Not until the discovery of radioactivity did scientists have a tool that would allow absolute dating of geologic events. Some isotopes of the naturally occurring chemical elements are radioactive, meaning that they spontaneously change or decay to other, more stable isotopes, releasing energy in the process. Accordingly, **radioactivity** is the process whereby an unstable atomic nucleus is spontaneously transformed into an atomic nucleus of a different element.

Radioactive Decay and Half-Lives

When discussing radioactive decay rates, it is convenient to refer to the **half-life** of a radioactive element, which is the time it takes for one-half of the atoms of the original unstable **parent element** to decay to atoms of a new, stable **daughter element.** One-half of the atoms of a parent element such as uranium 238, for example, decay to a stable daughter product known as lead 206 in 4.5 billion years (○ Table 11.1).

The half-life of a radioactive element is constant regardless of external conditions and can be precisely measured in the laboratory. During each half-life, the number of radioactive atoms is reduced by one-half, so after one half-life, only half of the original parent atoms are still present, after two half-lives, one-quarter of the original parent atoms are present, and so on (▣ Fig. 11.11).

By measuring the parent-daughter ratio and knowing the half-life of the parent element, scientists can calculate the age of a sample containing a radioactive element. To obtain accurate radiometric dates, however, scientists must be sure they are dealing with a *closed system,* meaning that the ratio between parent and daughter atoms results only from radioactive decay. If parent or daughter atoms have been added or removed from the system, a calculated age will be inaccurate. When the age is inaccurate, it is most commonly too young rather than too old.

It is sometimes possible to cross-check a date by measuring the parent-daughter ratios of two different radioactive elements in the same sample. If samples containing uranium 235 and uranium 238, which have different decay rates, have

TABLE 11.1   Five of the Principal Radioactive Isotopes Used in Radiometric Dating

ISOTOPES		HALF-LIFE OF PARENT (years)	EFFECTIVE DATING RANGE (years)	MINERALS AND ROCKS THAT CAN BE DATED
Parent	Daughter			
Uranium 238	Lead 206	4.5 billion	10 million to 4.6 billion	Zircon Uraninite
Uranium 235	Lead 207	704 million		
Thorium 232	Lead 208	14 billion		
Rubidium 87	Strontium 87	48.8 billion	10 million to 4.6 billion	Muscovite Biotite Potassium feldspar Whole metamorphic or igneous rock
Potassium 40	Argon 40	1.3 billion	100,000 to 4.6 billion	Glauconite Hornblende Muscovite Whole volcanic rock Biotite

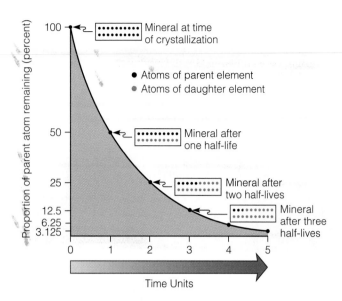

■ **FIGURE 11.11** Geometric radioactive decay curve, in which each time unit represents one half-life, and each half-life is the time it takes for one-half of the parent element atoms to decay to daughter element atoms.

remained closed systems, the ages obtained from each parent-daughter ratio should be in close agreement and therefore yield a reliable age. If the ages do not agree closely, other samples must be taken and the ratios measured to see which, if either, date is correct.

Carbon 14 Dating

Carbon, one of the basic elements in all organisms, has three isotopes; two of these, carbon 12 and 13, are stable, whereas carbon 14 is radioactive and has a half-life of 5,730 years plus or minus 30 years. The short half-life of carbon 14 makes this dating technique practical only for specimens less than about 70,000 years old. Though it is the most widely known dating method, it is the least used for dating events in Earth history because of its limited application. It is, however, very useful in archaeology and has greatly aided in unraveling the events of the latter part of the Ice Age (Pleistocene Epoch).

Carbon 14 is constantly formed in the upper atmosphere where nitrogen is bombarded by high-energy particles, mostly neutrons (■ Fig. 11.12). When a neutron strikes a nitrogen nucleus (atomic number 7, atomic mass number 14), it may be absorbed into the nucleus and a proton emitted. When this occurs, the atomic number of the nitrogen atom decreases by one, the atomic mass number remains unchanged, and an atom of carbon 14 (atomic number 6, atomic mass number 14) is formed. The newly formed carbon 14, along with carbon 12 and 13, is absorbed in a nearly constant ratio by all living organisms (Fig. 11.12). When an organism dies, however, carbon 14 decays back to nitrogen, and the ratio of carbon 14 to carbon 12 decreases.

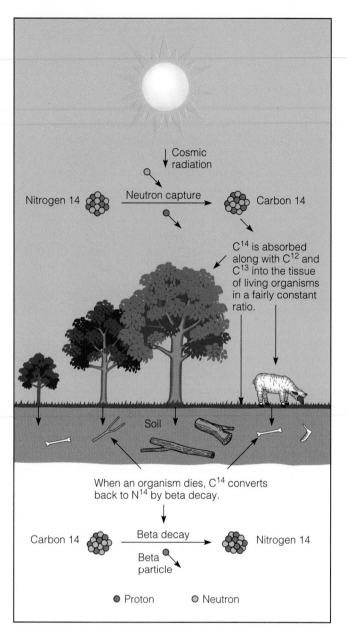

■ **FIGURE 11.12** The carbon cycle showing the formation, dispersal, and decay of carbon 14.

Unlike all other methods of radioactive decay dating, this dating technique does not consider the daughter product. Carbon 14 dating is based solely on the ratio of carbon 14 to carbon 12 and is generally used to date once living material such as wood and bone.

Quantifying the Geologic Time Scale

Applying absolute ages to the relative geologic time scale is not as straightforward as it might seem. A number of minerals in sedimentary rocks are radioactive and can yield absolute ages, but the ages tell only the age of the source rock that supplied the mineral to the deposit. Most absolute ages of sedimentary rocks must be determined indirectly by dating associated igneous rocks.

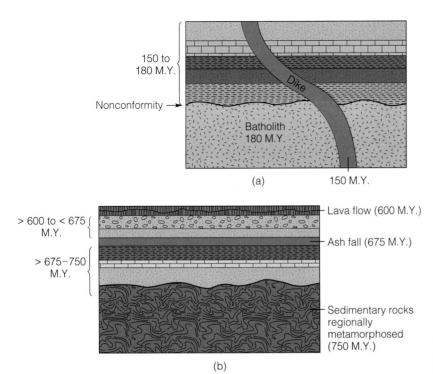

Nonconformity →

Dike

Batholith 180 M.Y.

(a)

150 M.Y.

> 600 to < 675 M.Y.

> 675–750 M.Y.

Lava flow (600 M.Y.)

Ash fall (675 M.Y.)

Sedimentary rocks regionally metamorphosed (750 M.Y.)

(b)

FIGURE 11.13 Absolute ages of sedimentary rocks can be determined by dating associated igneous and metamorphic rocks.

According to the principle of cross-cutting relationships, an igneous intrusion is younger than the rock it intrudes (Fig. 11.4a). Therefore, an accurately dated igneous intrusion provides a minimum age for the sedimentary rock it intrudes and a maximum age for any sedimentary rocks unconformably overlying it (Fig. 11.13a). A date for regionally metamorphosed rocks gives a maximum age for any overlying sedimentary rocks (Fig. 11.13b). The best way to get good dates for sedimentary rocks is by using interlayered volcanic rocks. A volcanic ash fall or lava flow is an excellent time-equivalent surface, providing a minimum age for the sedimentary rocks below and a maximum age for the rocks above (Fig. 11.13b).

Thousands of absolute ages are now known for sedimentary rocks of known relative ages, and these absolute ages have been added to the relative time scale. In this way, scientists have been able to determine the absolute ages for the various time designations such as Proterozoic and Jurassic (Fig. 11.10).

 ## THE OLDEST ROCKS AND EARTH'S AGE

Ernest Rutherford, an English physicist, proposed that radioactive decay could be used to date rocks, and by 1907 Earth's age had been estimated at somewhere between 400 million and 2 billion years. More sophisticated instruments and a more thorough understanding of radioactive decay have since allowed scientists to date geological events more accurately.

Currently, 3.96-billion-year-old gneisses from the Northwest Territories of Canada are the oldest known rocks on Earth. And because these rocks are metamorphic, they must

have formed from older rocks. Even older rocks must have existed in Australia where sedimentary rocks contain mineral fragments 4.2 billion years old, indicating that source rocks at least that old were present. The age of Earth is accepted as 4.6 billion years, or 640 million years older than the oldest known rocks. For this earliest interval in Earth history—nearly 14% of all geologic time—no geologic record exists.

The fact that Earth materials are continuously changed by weathering, metamorphism, and melting makes it unlikely that rocks dating from Earth's earliest history will ever be found. Accordingly, the age of Earth must be based on other considerations. Most evidence for Earth's great antiquity comes from the study of Moon rocks and meteorites; meteorites are probably fragments of asteroids and represent material that originated when the solar system formed. Absolute ages for both the oldest Moon rocks and meteorites cluster at about 4.6 billion years, indicating a similar age for Earth.

 ## FOSSILS AND THE HISTORY OF LIFE

Fossils, the remains or traces of ancient organisms preserved in rocks, are mostly the hard skeletal parts of organisms such as shells, bones, and teeth, but under exceptional conditions even the soft-part anatomy may be preserved. Several frozen woolly mammoths, for example, have been discovered in Alaska and Siberia with hair, flesh, and internal organs preserved (Fig. 11.14).

For any potential fossil to be preserved ravages of destructive processes such as ru scavengers, exposure to the atmosphere,

FIGURE 11.14 Frozen baby mammoth found in Siberia in 1977. The baby was six or seven months old, 1.15 m (3.8 ft) long, and 1.0 m (3.3 ft) tall. When alive, it had a hairy coat, but most of the hair has fallen out except around the feet. The carcass is about 40,000 years old.

Obviously, the soft parts of organisms are devoured or decomposed most rapidly, but even bones and shells will be destroyed unless buried and protected in mud, sand, or volcanic ash. Even if buried, they may be dissolved by groundwater or destroyed by alteration of the host rock during metamorphism. Nevertheless, fossils are quite common. Shells of marine animals are particularly common and easily collected in many areas, and even the bones and teeth of dinosaurs are much more common than most people realize. The remains of organisms themselves are known as *body fossils* to distinguish them from *trace fossils* such as tracks, trails, and burrows of ancient organisms (Fig. 11.15a and b).

Some fossils retain their original composition and structure and are thus preserved as unaltered remains, but many have been altered in some way. Dissolved mineral matter may precipitate in the pores of bones, teeth, and shells or fill the spaces within cells of wood. Wood may be preserved by the replacement of the woody tissues by silica; it then is referred to as *petrified,* a term that means "to become stone" (Fig. 11.15c). The calcium carbonate ($CaCO_3$) shells of some marine animals have been replaced by silicon dioxide (SiO_2) or iron sulfide (FeS_2) (Fig. 11.15d). Insects and leaves, stems, and roots of plants are commonly preserved as a thin carbon film that shows the details of the original organism (Fig. 11.15e). Shells in sediment may be dissolved leaving a cavity called a *mold* that is shaped like the shell. If a mold is filled in, a *cast* is formed (Fig. 11.15f).

If it were not for fossils, we would have no knowledge of extinct animals such as trilobites and dinosaurs, so fossils constitute our only record of ancient life. The quality of the fossil record is variable, however. For some groups of organisms, it is very good, but for others, it is poor or, in some cases, nonexistent. The fossil record is also biased, being best for organisms with preservable skeletons. Nevertheless, it provides us with an overview of the history of life on Earth.

Precambrian Life

Prior to the mid-1950s, we had very little knowledge of fossils older than Paleozoic. Scientists had long assumed that the fossils so abundant in Cambrian sedimentary rocks must have had a long earlier history, but no record of these organisms was known. Some enigmatic Precambrian fossils had been reported, but they were mostly dismissed as inorganic features of some sort. In fact, all of Precambrian time was once referred to as the *Azoic,* meaning without life.

In the early 1900s, Charles Walcott described layered, moundlike structures from the Precambrian of Ontario, Canada—structures that are now called *stromatolites.* Walcott proposed that they represented reefs constructed by algae, but paleontologists did not demonstrate that stromatolites are the products of organic activity until 1954. Studies of present-day stromatolites in such areas as Shark Bay, Australia, show that they originate by entrapment of sediment grains on sticky mats of photosynthesizing bacteria, or what are commonly called blue-green algae. Rocks 3.3 to 3.5 billion years old in Australia contain the oldest known stromatolites.

The earliest life-forms were all varieties of single-celled bacteria that lacked a cell nucleus and apparently reproduced asexually, much like bacteria today. More complex cells with a distinct nucleus and capable of sexual reproduction did not appear in the fossil record until sometime between 1.8 and 1.4 billion years ago. These were still single-celled organisms, but once they appeared, organic diversity increased markedly. Multicelled algae are known from rocks at least 1 billion years old, and during the Late Precambrian, about 700 million years ago, the first multicelled animals evolved (Fig. 11.16).

Even though multicelled animals were present by the Late Precambrian, they have a poor fossil record because they lacked durable skeletons. Near the end of the Precambrian, however, animals possessing skeletons existed, as indicated by minute scraps of shell-like material and spicules, presumably from sponges. Nevertheless, animals with durable skeletons of chitin (a complex organic substance), silica (SiO_2), and calcium carbonate ($CaCO_3$) were not abundant until the beginning of the Paleozoic Era.

Life of the Paleozoic

Long ago, scientists observed that the remains of animals with skeletons appeared rather suddenly in the fossil record at the beginning of the Paleozoic. The evolution of numerous invertebrates (animals lacking a segmented vertebral column) with durable skeletons is sometimes referred to as an explosive development of new types of animals. In fact, their appearance was rapid only in the context of geologic time, occurring over tens of millions of years during the

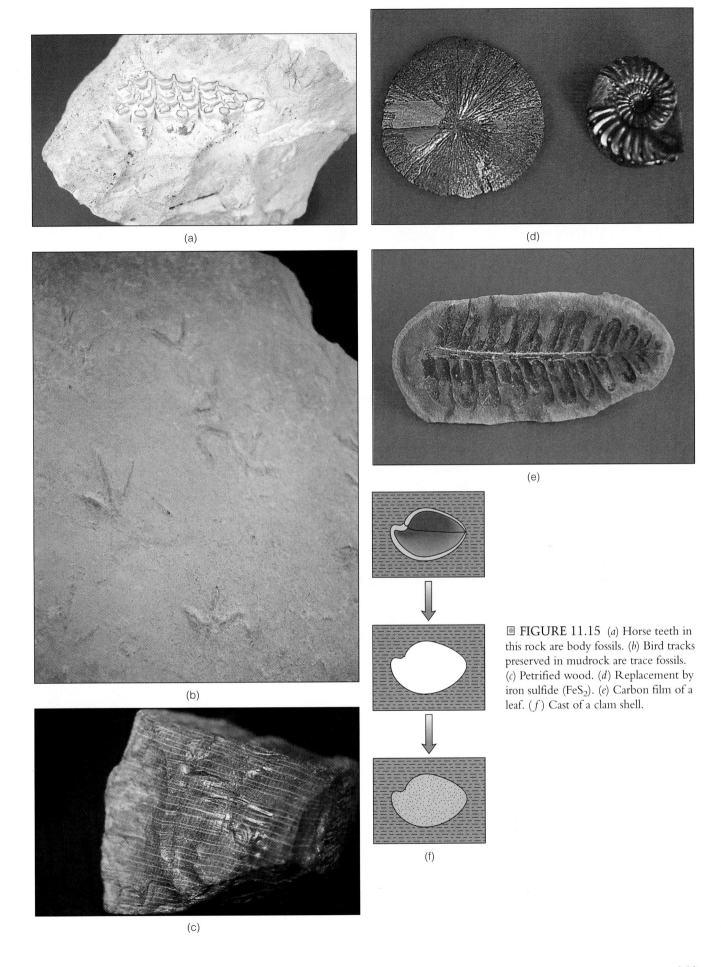

■ FIGURE 11.15 (a) Horse teeth in this rock are body fossils. (b) Bird tracks preserved in mudrock are trace fossils. (c) Petrified wood. (d) Replacement by iron sulfide (FeS$_2$). (e) Carbon film of a leaf. (f) Cast of a clam shell.

(a)

(b)

(c)

(d)

(e)

(f)

◉ FIGURE 11.16 Impressions of two multicelled animals from the Late Precambrian of Australia.

Cambrian and on into the Ordovician. Had humans been present to observe this event, it would have seemed incredibly gradual.

Trilobites and *brachiopods* were by far the most conspicuous animals of the Cambrian marine invertebrate community, making up more than half of the total (◉ Fig. 11.17). But it is important to remember that the fossil record is much better for organisms with durable skeletons, so we know little about the soft-bodied animals of that time. Following a decline of trilobites at the end of the Cambrian Period, brachiopods became the most commonly preserved fossils in Paleozoic marine faunas, but many other organisms were present as well, including corals, clams, snails, sea lilies, and nautiloids and ammonites, which are extinct relatives of the squid and octopus (Fig. 11.17).

In addition to the numerous invertebrate groups, vertebrates (animals with a segmented vertebral column) also evolved and diversified during the Paleozoic. Remains of jawless fishes with poorly developed fins and external bony armor are found in Late Cambrian–aged rocks in Wyoming. All are now extinct except for their distant relatives the lamprey and slime hag.

During the Silurian, two more groups of fishes evolved. One of these, known as placoderms, included one of the largest marine predators to ever exist. The second group, the acanthodians, included the probable ancestors of the present-day bony fishes and the cartilaginous fishes. Among the bony fishes, one group known as lobe-fins was particularly important because they included the ancestor of the amphibians, the first land-dwelling vertebrate animals.

◉ FIGURE 11.17 A Middle Ordovician sea-floor fauna. Trilobites, nautiloids, brachiopods, sea lilies, and colonial corals are shown.

(a)

(b)

■ FIGURE 11.18 (a) Amphibians are thought to have evolved from a lobe-finned fish similar to this one known as *Eusthenopteron.* (b) A Late Devonian landscape from eastern Greenland showing one of the oldest known amphibians, *Ichthyostega,* which grew to about 1 m (3.3 ft) long.

Although amphibians were living on land by the Late Devonian (■ Fig. 11.18), they were not the first organisms to do so. Indeed, plants made the transition to land during the Ordovician, and several invertebrates including insects, millipedes, spiders, and snails invaded the land before amphibians. Early land-dwelling animals encountered several critical problems including drying out, reproduction, the effects of gravity, and the extraction of oxygen from the atmosphere by lungs rather than from water by gills. These problems were partly overcome by some of the lobe-finned fishes; they already had a backbone and limbs that could be used for support and walking on land and lungs to extract oxygen from the atmosphere.

Amphibians were limited in colonizing the land because they had to return to water to lay their eggs, and they never completely overcame the problem of drying out. In contrast, reptiles developed a tough skin or scales to preserve their internal moisture and an egg that had to be laid on land, thus freeing them from water. The oldest known reptiles, which

appeared during the Mississippian Period, were small, agile animals known as *stem reptiles* because they were the ancestors of all other reptiles (■ Fig. 11.19). One of the descendant groups of the stem reptiles was the pelycosaurs, primitive mammal-like reptiles, which by the Late Paleozoic were the first land-dwelling vertebrates to become diverse and widespread.

Plants encountered many of the same problems faced by animals as they made the transition to land: particularly drying out, the effects of gravity, and reproduction. Plants adapted by evolving a variety of structural features that allowed them to invade the land during the Ordovician and later periods. Most experts agree that the ancestors of land plants were green algae that first evolved in a marine

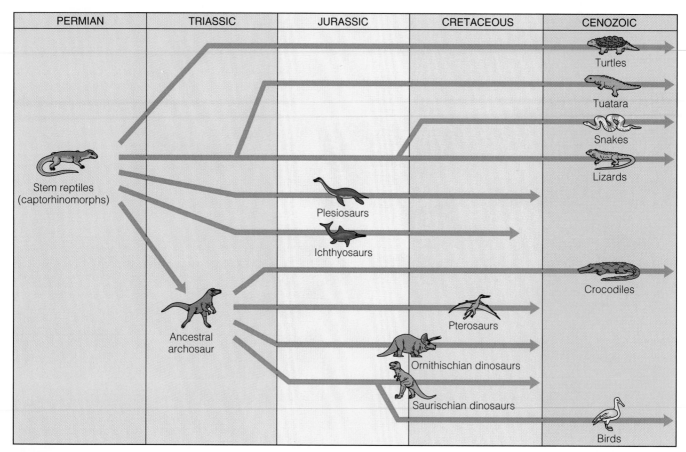

FIGURE 11.19 Evolutionary relationships among fossil and living reptiles and birds.

environment, then moved into freshwater, and then finally onto land.

The commonest and most widespread land plants are characterized as vascular, meaning they have a tissue system of specialized cells for the movement of water and nutrients. Nonvascular plants, such as mosses and fungi, lack these specialized cells and are typically small and usually live in low, moist areas. The fungi were probably the first to make the transition to land, but their fossil record is poor.

The earliest known, well-preserved vascular land plants were small plants with Y-shaped stems from the Silurian. They are known as seedless vascular plants because they did not produce a seed. Although these plants lived on land, they were restricted to moist areas, as are many of their living descendants, such as ferns.

In addition to the evolution of diverse seedless vascular plants by the Late Devonian, another significant floral event took place. The evolution of the seed at this time liberated vascular plants from their dependence on moist conditions and allowed them to spread over all parts of the land. The first to do so were the flowerless seed plants, which include the living cycads, conifers, and ginkgoes.

Global changes at the end of the Paleozoic Era resulted in the greatest extinction in Earth history. Particularly hard hit by this extinction were marine invertebrates; two groups of primitive corals, two groups of bryozoans (moss animals),

many brachiopods, all remaining trilobites, and several other groups either became extinct or were greatly diminished. On land, many amphibian and reptile families also died out.

Two hypotheses for the marine invertebrate extinctions are (1) a reduction of living area related to widespread withdrawal of shallow seas from the continents and suturing of the continents to form Pangaea, and (2) decreased ocean salinity due to widespread deposition of evaporites on land.

Life of the Mesozoic

Following the wave of extinctions at the end of the Paleozoic, marine invertebrates repopulated the seas during the Mesozoic. Clams, oysters, and snails became increasingly diverse and abundant, and ammonites and nautiloids were among the most important Mesozoic invertebrate groups (◙ Fig. 11.20). On the other hand, brachiopods never recovered from their near extinction and have remained a minor invertebrate group ever since. Where marine waters were warm, clear, and shallow, corals again proliferated, but these corals were of a new and more familiar type. Single-celled animals known as foraminifera were also very important. Floating or planktonic forms in particular were extremely common, but many of them became extinct at the end of the Mesozoic, and only a few types survived into the Cenozoic.

FIGURE 11.20 Mesozoic ammonite. Ammonites were abundant during the Mesozoic but became extinct at the end of that era.

The Mesozoic Era is commonly referred to as the "Age of Reptiles," alluding to the fact that reptiles were the most diverse and abundant land-dwelling vertebrates (Fig. 11.19). Many people find this the most interesting chapter in the history of life because among the reptiles were dinosaurs, flying reptiles, and marine reptiles.

A common but erroneous perception is that dinosaurs were poorly adapted animals that had trouble surviving. True, they became extinct, but to consider this a failure is to ignore the more than 150 million years when they were the dominant land vertebrates. During their existence they diversified into numerous types and adapted to a wide variety of environments. Nor were dinosaurs the lethargic beasts often portrayed in various media. Recent evidence indicates that some may have been very active and possibly warm-blooded. It also appears that some species cared for their young long after hatching, a behavioral characteristic most often associated with birds and mammals. Although some dinosaurs were quite large, many were smaller than present-day elephants and one was no larger than a chicken.

Among the dinosaurs, two groups are generally recognized, based on the shape of the pelvis. One of these, the saurischians, included the largest dinosaurs such as *Brachiosaurus* (■ Fig. 11.21) and the carnivores *Tyrannosaurus* and

FIGURE 11.21 Late Jurassic scene showing *Brachiosaurus,* the largest known dinosaur, and the plated dinosaur *Stegosaurus.*

Dinosaur Graveyard

In 1981 a remarkable concentration of duck-billed dinosaur bones was discovered in northwestern Montana. Paleontologist John R. Horner and his crews dug pits at various places in the area and estimated that the remains of at least 10,000 dinosaurs were preserved there! Apparently, a vast herd of dinosaurs was overcome by ash and gases from a nearby volcano and subsequently buried in ash. Individual dinosaurs ranged from juveniles 3.6 m (12 ft) long to adults 7.6 m (25 ft) in length.

Velociraptor of *Jurassic Park* fame. The other group, the ornithischians, was more diverse. It included the stegosaurs (Fig. 11.21) and duck-billed dinosaurs and a variety of others, all of whom were herbivores.

Although dinosaurs attract the most attention, several other Mesozoic reptiles were common. Flying reptiles and marine reptiles were contemporaries of dinosaurs as were crocodiles, turtles, lizards, and snakes (Fig. 11.19). In addition, it appears that during the Jurassic Period, small carnivorous dinosaurs gave rise to birds, which, although warm-blooded and feather-covered, retain a number of reptilian characteristics, such as the type of egg.

During the Late Triassic, mammals evolved from one group of mammal-like reptiles known as therapsids. The transition from mammal-like reptile to mammal is well documented by fossils and is so gradational that classification of some fossils as either reptile or mammal is difficult. Even though mammals evolved at the same time as dinosaurs, their diversity remained low during the rest of the Mesozoic, and all were small animals.

Triassic and Jurassic land-plant communities, like those of the Late Paleozoic, were composed of seedless vascular plants and various flowerless seed plants. During the Cretaceous, these two groups were largely replaced by the flowering plants, which now account for about 95% of all vascular plant species. Since they first evolved, flowering plants have adapted to nearly every terrestrial habitat from mountains to deserts to shallow coastal waters.

The mass extinction at the end of the Mesozoic was second in magnitude only to the Paleozoic extinctions. Casualties of this extinction included dinosaurs, flying reptiles, marine reptiles, and several types of marine invertebrates. Among the latter were the ammonites (Fig. 11.20) that had been so abundant throughout the Mesozoic, a type of reef-building clam, and some species of floating foraminifera.

Some scientists think a large meteorite or comet hit Earth at the end of the Mesozoic Era, setting in motion a chain of events that led to widespread extinctions (see the Prologue in Chapter 1). According to this idea, enough material was ejected into the atmosphere to block out sunlight for several months, resulting in the cessation of photosynthesis and the collapse of the food chain. Certainly, an impact of this size would have had worldwide effects, and recently a possible impact crater has been discovered centered on the north end of the Yucatán Peninsula of Mexico.

Many scientists concede that the evidence for an impact of some sort is compelling, but question whether it caused extinctions. Some scientists are of the opinion that a variety of organisms, including dinosaurs, were already on the decline and headed for extinction before the end of the Mesozoic. In that case, a meteorite or comet impact may have simply hastened the process.

Life of the Cenozoic

The world's flora and fauna continued to evolve during the Cenozoic as more familiar types of plants and animals appeared. Flowering plants continued to diversify, but some flowerless seed plants, especially conifers, remained abundant, and seedless vascular plants still occupied many habitats. The Cenozoic marine environment was populated by plants and animals that survived the Mesozoic extinction. Floating foraminifera comprised a major component of the marine invertebrate community, but corals, clams, and snails also proliferated.

For more than 100 million years, mammals had coexisted with dinosaurs, yet the fossil record indicates that even during the Cretaceous Period only a limited number of varieties of mammals existed. The Mesozoic mass extinction eliminated the dinosaurs and their relatives, thereby creating numerous adaptive opportunities that were quickly exploited by mammals. In fact, the Cenozoic is called the "Age of Mammals."

Among the mammals only monotremes lay eggs, whereas the marsupials and placentals give birth to live young. Marsupials are born in a very immature, almost embryonic condition, and then undergo further development in the mother's pouch. In the placentals, a placenta nourishes the embryo, permitting the young to develop much more fully before birth. The fossil record of the monotremes is so poor that their relationships to other mammals are not clear. The marsupials and placentals, on the other hand, diverged from a common ancestor during the Cretaceous and since then have had separate evolutionary histories.

The phenomenal success of placental mammals is related in part to their reproductive method. A measure of this success is that more than 90% of all mammals, fossil and living, are placental. The only long-term success for marsupials has been in the Australian region where most of the present-day species live. Marsupials were also common in South America during much of the Cenozoic before a land connection existed between that continent and North America. Once a land connection formed a few million years ago, however, many placentals migrated south, and all species of indigenous marsupials except opossums became extinct.

If we could go back and visit the Early Cenozoic, we would probably not recognize many of the mammals. Most

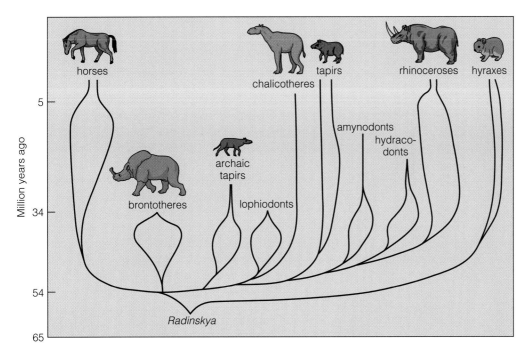

FIGURE 11.22 Evolution of horses and their relatives. They all evolved from a common ancestor and thus share a number of characteristics yet differ markedly in appearance. The titanotheres and chalicotheres are extinct.

of them had not yet become clearly differentiated from their ancestors, and the differences between herbivores and carnivores were slight. Some would be at least vaguely familiar, but the ancestors of horses, camels, whales, and elephants would bear little resemblance to their living descendants. By the Middle Cenozoic, though, all of the major groups of living placental mammals had evolved, and many of these would be easily recognized.

Numerous groups of mammals evolved during the Cenozoic, and some such as horses and their relatives have excellent fossil records. Horses and their living relatives, the rhinoceroses and tapirs, evolved from small Early Cenozoic ancestors (■ Fig. 11.22). Horses and rhinoceroses were common in North America, but died out here during the Pleistocene and now survive only in the Old World (see □ Perspective 11.2).

Primates, the order of mammals to which humans belong, were undoubtedly present by the Early Cenozoic. The hominids (family Hominidae), the primate family that includes present-day humans and their extinct ancestors, have a fossil record extending back only about 4 million years. Several features distinguish them from other primates: hominids walk on two legs rather than four, they show a trend toward a large, complex brain, and they manufacture and use sophisticated tools.

The fossil record of humans can be traced back to the genus *Australopithecus,* which lived in Africa more than 4 million years ago (■ Fig. 11.23). Most individuals were small and had a brain size of 380 to 600 cubic centimeters (cc) as compared to the average of 1,350 cc for present-day

humans. Their bones and preserved trackways indicate that they walked fully upright as humans do today. *Australopithecus* was succeeded by *Homo habilis* and *Homo erectus.* The latter had a much larger brain, 800 to 1,300 cc, but still possessed a thick-walled skull and prominent brow ridges; otherwise they closely resembled *Homo sapiens.* Sometime between 300,000 and 200,000 years ago, *Homo sapiens* evolved from *Homo erectus,* but the transition seems to have been gradual.

The most famous of all fossil humans are the *Neanderthals* who inhabited Europe and the Near East from about 150,000 to 32,000 years ago. Neanderthals differed from present-day humans mostly in the skull, which was long and low with heavy brow ridges, a projecting mouth, and a weak receding chin; their bodies were also somewhat more massive and more heavily muscled than ours.

About 35,000 years ago, humans closely resembling present-day Europeans moved into the region inhabited by the Neanderthals and replaced them. These *Cro-Magnons,* as they are called, lived from about 35,000 to 10,000 years ago, and during this time, they developed art and technology to a much greater level than their ancestors had.

One of the most remarkable aspects of the Cenozoic history of mammals is that so many very large species existed during the Pleistocene Epoch. Mastodons and mammoths, giant bison, huge ground sloths, giant camels, and beavers 2 m (6.5 ft) tall lived in North America. Giant kangaroos, wombats the size of rhinoceroses, leopard-sized marsupial lions, and large platypuses inhabited Australia. In Europe and parts of Asia lived cave bears, elephants, and the giant deer

Fossil Rhinoceroses in Nebraska

Ten million years ago, in what is now north-eastern Nebraska, a vast grassland was inhabited by short-legged, aquatic rhinoceroses, camels, three-toed horses, saber-toothed deer, land turtles, and many other animals. This habitat was a temperate savanna with life as varied and abundant as it is now on the savannas of East Africa. But many of these animals perished when a vast cloud of volcanic ash rolled in, probably from the southwest.

Michael Voorhies of the University of Nebraska and his crews have recovered the remains of hundreds of victims of this catastrophe, especially rhinoceroses (☐ Fig. 1). The magnitude of this event is difficult to imagine because the source of the ash cloud may have been in New Mexico, more than 1,000 km (620 mi) away. We know from historic eruptions that ash can be carried this far. During the 1815 eruption of Tambora in Indonesia, for instance, an ash layer 22 cm (8.6 in) thick accumulated 400 km (248 mi) from the volcano, and some ash was carried more than 1,500 km (930 mi).

The Nebraska ash fall was probably 10 to 20 cm (4–8 in) thick, although it was redistributed by the wind and collected to 1 m (3.3 ft) deep in a depression where the fossils were recovered. This depression was the site of a water hole where horses, camels, deer, turtles, and birds perished when the ash fell. Their skeletons show signs of partial decomposition, scavenging, and trampling, an indication that they laid at the surface before they were completely buried.

Very soon after the initial ash fall, the depression was visited by herds of rhinoceroses and a few horses and camels. These, too, were suffocated by ash, but in this case the ash was probably blown into the depression by the wind. In any case, these animals were quickly buried, as indicated by the large number of complete skeletons.

One of the most remarkable things about these fossils is the preserved detail including such delicate structures as tiny middle ear bones. Additionally, the association of rhinoceroses of all ages from babies to mature adults gives us a better understanding of herd structure; a herd was probably made up of a single male and several females with their calves.

Only rarely are paleontologists fortunate enough to find and recover so many well-preserved vertebrate animals. An ancient catastrophe turns out to be our good fortune because it provides a unique glimpse of what life was like 10 million years ago.

☐ **FIGURE 1** Paleontologists excavating rhinoceros (foreground) and horse (background) skeletons from volcanic ash near Orchard, Nebraska.

■ FIGURE 11.23 Recreation of a Pliocene landscape showing members of *Australopithecus afarensis* gathering and eating various fruits and seeds.

commonly called the Irish elk (■ Fig. 11.24). In addition to mammals, giant birds up to 3.5 m (11.5 ft) tall and weighing 585 kg (1,285 lb) existed in New Zealand, Madagascar, and Australia, and giant vultures with a wingspan of 3.6 m (12 ft) are known from California.

Many smaller mammals were also present, many of which still exist. The major evolutionary trend in mammals, however, was toward large body size, which may have been an adaptation to the cooler temperatures of the Pleistocene. Large animals have proportionately less surface area compared to their volume and therefore retain heat more effectively than do smaller animals.

Near the end of the Pleistocene, almost all of the large terrestrial mammals of North America, South America, and Australia became extinct. Extinctions also occurred on the other continents, but they had considerably less impact. This extinction was modest by comparison to earlier ones, but it was unusual in that it affected, with few exceptions, only large land-dwelling mammals. The debate over the cause of this extinction continues with some arguing that the large mammals could not adapt to the rapid climatic changes at the end of the Ice Age and others maintaining that these mammals were killed off by human hunters, a hypothesis known as *prehistoric overkill*.

■ FIGURE 11.24 Restoration of the giant deer *Megaloceros giganteus,* commonly called the Irish elk. It lived in Europe and Asia during the Pleistocene. Large males had an antler spread of about 3.35 m (11 ft).

CHAPTER SUMMARY

1. Relative dating involves determining the order of occurrence of geologic events, whereas absolute dating results in assigning specific ages for events in years before the present.

2. Until well into the 1800s, the prevailing idea was that Earth is young, and catastrophism was invoked to explain Earth history.

3. James Hutton proposed that present-day processes operating over vast amounts of time could account for Earth's geologic features. His observations were essential in establishing the principle of uniformitarianism, which displaced catastrophism as a fundamental concept in geology.

4. The principles of superposition, original horizontality, cross-cutting relationships, inclusions, and fossil succession are essential in relative dating.

5. Numerous unconformities are present in the geologic record; all of them represent times of erosion, nondeposition, or both. Three types of unconformities are recognized: disconformity, angular unconformity, and nonconformity.

6. Correlation is the process of demonstrating that rocks in different areas are of the same age; it is commonly done by using fossils.

7. Radioactive decay dating techniques allow scientists to determine absolute ages for events in Earth history. All radioactive decay dating, except carbon 14, depends on knowing the ratio of parent atoms to daughter atoms and the half-life of the parent material.

8. In carbon 14 dating, the ratio of carbon 14 to carbon 12 is used to determine the age of a specimen. Generally, only organic materials are dated by this method, and it is effective back to only about 70,000 years ago.

9. The geologic time scale is a chart showing the designations for intervals of geologic time. The hierarchy of terms for geologic time are eon, era, period, and epoch.

10. The oldest known rocks on Earth are 3.96 billion years old, but an age of 4.6 billion years is indicated by absolute ages of Moon rocks and meteorites.

11. Fossils, the remains or traces of organisms preserved in rocks, provide our only record of prehistoric life.

12. During much of the Precambrian, all organisms were single-celled varieties of bacteria. Multicelled algae are known from rocks at least 1 billion years old, and multicelled animals first appeared about 700 million years ago.

13. Marine invertebrate animals with durable skeletons appeared in abundance during the Cambrian and Ordovician periods, diversified throughout the rest of the Paleozoic, and then suffered mass extinctions at the end of the era. Survivors of this extinction repopulated the seas during the Mesozoic, but

some of these died out during the Cretaceous extinction. The present-day marine invertebrate fauna evolved during the Cenozoic.

14. Among the vertebrates, various fishes, amphibians, and reptiles evolved during the Paleozoic. The Mesozoic was the time of dominance of dinosaurs, flying reptiles, and marine reptiles, all of which became extinct at the end of the Mesozoic. Mammals and birds also evolved during the Mesozoic and became dominant elements of the vertebrate fauna during the Cenozoic.

15. Nonvascular and vascular plants populated the land by the Silurian. Among the latter, seedless plants appeared first, followed later during the Paleozoic by flowerless seed plants.

During the Cretaceous, these two groups were largely replaced by flowering plants.

16. The fossil record of *Homo sapiens* can be traced back about 4 million years to *Australopithecus*. *Australopithecus*, which was small-brained but walked upright was ancestral to *Homo habilis*, who was succeeded by *Homo erectus* and finally by *Homo sapiens*.

17. Many Pleistocene mammals and some birds were quite large. Most of these large species of mammals in North and South America and Australia became extinct at the end of the Pleistocene.

IMPORTANT TERMS

absolute dating
angular unconformity
correlation
daughter element
disconformity
fossil
geologic time

geologic time scale
guide fossil
half-life
nonconformity
parent element
principle of cross-cutting
  relationships

principle of fossil succession
principle of inclusions
principle of original
  horizontality
principle of superposition

principle of
  uniformitarianism
radioactivity
relative dating
unconformity

REVIEW QUESTIONS

1. In which type of unconformity are sedimentary rock layers parallel to one another?
   a. ____ nonconformity; b. ____ angular unconformity; c. ____ hiatus; d. ____ disconformity; e. ____ uniformitarianism.

2. Placing geologic events in a sequential order as determined by their position in the geologic record is:
   a. ____ absolute dating; b. ____ superposition; c. ____ relative dating; d. ____ correlation; e. ____ radioactive decay.

3. If a rock is heated during metamorphism and daughter atoms migrate out of a mineral that is subsequently dated, an inaccurate date will be obtained. This date will be _____ the actual date:
   a. ____ younger than; b. ____ 1 million years older than; c. ____ the same as; d. ____ equal to; e. ____ older than.

4. Which one of the following principles can be used to show that rocks in different areas are the same relative age?
   a. ____ cross-cutting relationships; b. ____ superposition; c. ____ fossil succession; d. ____ original horizontality; e. ____ uniformitarianism.

5. Which geologic principle holds that the oldest layer is on the bottom in a succession of sedimentary rocks and the youngest is on top?
   a. ____ lateral continuity; b. ____ Phanerozoic Eon; c. ____ carbon 14 dating; d. ____ superposition; e. ____ fossil content.

6. The era younger than the Mesozoic is the:
   a. ____ Archean; b. ____ Paleozoic; c. ____ Silurian; d. ____ Jurassic; e. ____ Cenozoic.

7. Which one of the following is a trace fossil?
   a. ____ dinosaur tooth; b. ____ frozen mammoth; c. ____ bird bone; d. ____ worm burrow; e. ____ clam shell.

8. The oldest known rocks on Earth are _____ years old.
   a. ____ 4.6 billion; b. ____ 1.65 million; c. ____ 3.96 billion; d. ____ 600,000; e. ____ 570 million.

9. Durable skeletons appear in abundance in the fossil record at the:
   a. ____ end of the Mesozoic Era; b. ____ middle of the Cenozoic Era; c. ____ beginning of the Cretaceous Period; d. ____ end of the Archean Eon; e. ____ beginning of the Paleozoic Era.

10. The first vertebrate animals to live on land were:
    a. ____ insects; b. ____ dinosaurs; c. ____ reptiles; d. ____ snakes; e. ____ amphibians.

11. The most diverse group of land plants is known as:
    a. ____ flowering plants; b. ____ mosses; c. ____ lichens; d. ____ nonvascular plants; e. ____ seedless vascular plants.

12. The work of _____ is the basis for the principle of uniformitarianism.
    a. ____ Archbishop Ussher; b. ____ James Hutton; c. ____ Charles Darwin; d. ____ Lord Kelvin; e. ____ Nicolas Steno.

13. Which one of the following statements is correct?
a. _____ catastrophism replaced superposition as a fundamental principle in geology; b. _____ an igneous intrusion is younger than the rocks it intrudes; c. _____ uniformitarianism does not allow for sudden, large-scale events such as earthquakes; d. _____ similar rock types in different areas must be the same age; e. _____ the first land-dwelling organisms were amphibians.

14. The largest dinosaurs and the meat-eaters belong to the group known as:
a. _____ stromatolites; b. _____ saurischians; c. _____ correlation; d. _____ amphibians; e. _____ ammonites.

15. Describe the contributions to the development of Earth science made by James Hutton, Charles Lyell, and Nicolas Steno.

16. Explain how the relative ages of a granite batholith and an overlying sandstone can be determined.

17. What kinds of problems were encountered by animals and plants as they moved from water to land, and how successful were the early members of these groups in solving these problems?

18. Why is the principle of uniformitarianism so important to Earth scientists?

19. How was Earth's age determined? Also, explain why no rocks dating from the earliest Earth are likely to be found.

20. Explain how fossils can be used to demonstrate that rocks in different areas are the same relative age.

21. What is radioactive decay, and how can it be used to determine absolute ages?

22. What is the significance of absolute ages determined for igneous, metamorphic, and sedimentary rocks?

23. How does carbon 14 dating differ from other absolute dating methods, and what are its limitations?

24. Give a brief overview of the evolution of vertebrate animals during the Paleozoic Era.

25. How does relative dating differ from absolute dating?

POINTS TO PONDER

1. If a radioactive element has a half-life of 4 million years, the amount of parent material remaining after 12 million years will be what fraction of the original amount?

2. How many half-lives are required to yield a mineral with 625 parent atoms and 19,375 daughter atoms?

3. How can geologic events such as explosive volcanic eruptions, earthquakes, and landslides be encompassed by the principle of uniformitarianism?

4. How is it possible for scientists to determine that dinosaurs became extinct 66 million years ago since sedimentary rocks containing dinosaur fossils cannot be reliably dated?

5. In a sequence of sedimentary rocks, you observe limestone at the bottom followed upward by sandstone, basalt, sandstone, and conglomerate. What evidence would you look for to determine whether the basalt is a buried lava flow or a sill?

ADDITIONAL READINGS

Albritton, C. C., Jr. 1984. Geologic time. *Journal of Geological Education* 32, no. 1: 29–37.

Bakker, R. T. 1993. Bakker's field guide to Jurassic dinosaurs. *Earth* 2, no. 5: 33–43.

Berry, W. B. N. 1987. *Growth of a prehistoric time scale,* 2d ed. Palo Alto, Calif.: Blackwell Scientific Publications.

Boslough, J. 1990. The enigma of time. *National Geographic* 177, no. 3: 109–32.

Colbert, E. H., and M. Morales. 1991. *Evolution of the vertebrates,* 4th ed. New York: John Wiley & Sons.

Cowan, R. 1995. *History of life,* 2d ed. Boston: Blackwell Scientific Publications.

Donovan, S. K., ed. 1989. *Mass extinctions.* New York: Columbia University Press.

Geyh, M. H., and H. Schleicher. 1990. *Absolute age determinations.* New York: Springer-Verlag.

Gore, R. 1993. Dinosaurs. *National Geographic* 183, no. 1: 2–53.

Gould, S. J. 1987. *Time's arrow, time's cycle.* Cambridge, Mass.: Harvard University Press.

Gould, S. J., ed. 1993. *The book of life.* New York: W. W. Norton.

Gray, J., and W. Shear. 1992. Early life on land. *American Scientist* 80, no. 5: 444–56.

Lucas, S. G. 1996. *Dinosaurs: The textbook.* Dubuque, Iowa: W. C. Brown.

Ramsey, N. F. 1988. Precise measurement of time. *American Scientist* 76, no. 1: 42–49.

Schopf, J. W., ed. 1992. *Major events in the history of life.* Boston: James & Bartlett.

Thomas, B., and R. Spicer. 1987. *Evolution and palaeobiology of land plants.* Portland, Ore.: Dioscorides Press.

Wicander, R., and J. S. Monroe. 1993. *Historical geology: Evolution of the Earth and life through time,* 2d ed. St. Paul, Minn.: West Publishing Company.

For current updates and exercises, log on to http://www.wadsworth.com/geo.

UNIVERSITY OF CALIFORNIA MUSEUM OF PALEONTOLOGY
http://ucmpl.berkeley.edu:80/welcome.html

Visit this site for an excellent description of any aspect of geologic time, paleontology, and evolution.

U.S. GEOLOGICAL SURVEY: RADON IN EARTH, AIR, AND WATER
http://sedwww.cr.usgs.gov:8080/radon/radonhome.html

This site is maintained by the U.S. Geological Survey and contains information about the geology of radon, radon potential in the United States, and USGS radon publications as well as links to other sites with information about radon.

RADIOCARBON WEB-INFO
http://www2.waikato.ac.nz/c14/webinfo/index.html

This site is maintained by the radiocarbon labs of Waikato and Oxford Universities. The site contains information about the basis of carbon 14 dating, applications, measurement methods, and other carbon 14 web sites.

CHAPTER

12

Seawater and the Sea Floor

View of an atoll, an oval or circular reef enclosing a lagoon, in the Pacific Ocean. Reefs are composed of skeletons of corals and various other marine animals.

Prologue

When the submersible *Alvin* descended to the Galapagos Rift in 1979, scientists observed mounds of metal-rich sediments. Near these mounds the researchers saw what they called black smokers, chimneylike vents discharging plumes of hot, black water (□ Fig. 12.1). Since 1979 similar vents have been observed at or near spreading ridges in several other areas. These vents form near spreading ridges as seawater seeps through cracks and fissures into the oceanic crust, where it is heated by hot rocks and then rises and is discharged onto the sea floor.

These submarine hot water vents or hydrothermal vents are interesting for several reasons. Near the vents live communities of organisms, including bacteria, crabs, mussels, starfish, and tubeworms, many of which had never been seen before (Fig. 12.1). In most biological communities, photosynthesizing organisms form the base of the food chain and provide nutrients for the herbivores and carnivores. In vent communities no sunlight is available for photosynthesis, and the base of the food chain consists of bacteria that practice chemosynthesis; they oxidize sulfur compounds from the hot vent waters, thus providing their own nutrients and nutrients for other members of the food chain.

Another interesting aspect of submarine hydrothermal vents is their economic potential. When seawater circulates down through the basaltic oceanic crust, it is heated to as much as 400°C. The hot water reacts with the basalt and is transformed into a metal-bearing solution that rises and discharges onto the sea floor, where it cools and forms a chimney consisting of iron, copper, and zinc sulfides and other minerals (Fig. 12.1).

Apparently the chimneys grow rapidly. A 10 m high chimney accidentally knocked over in 1991 by the submersible *Alvin* grew to 6 m in just three months. They are ephemeral, however; one observed in 1979 was inactive six months later. When their activity ceases, the vents eventually collapse and are incorporated into a moundlike mineral deposit.

Scientists aboard *Alvin* in April 1991 saw the results of a submarine eruption on the East Pacific Rise that they missed by less than two weeks. Fresh lava and ash covered the area as well as the remains of tubeworms killed during the eruption. In a nearby area, a new fissure opened in the sea floor, and by December 1993, a new hydrothermal vent community had become established with tubeworms fully 1.2 m long.

The economic potential of hydrothermal vent deposits is tremendous. Deposits in the Atlantis II Deep of the Red Sea contain an estimated 100 million tons of metals, including iron, copper, zinc, silver, and gold. These deposits are fully as large as the major deposits mined on land, which formed by hydrothermal vent activity on what was then the sea floor. None of the Red Sea deposits are currently being mined, but the technology to exploit them exists. In fact, the Saudi Arabian and Sudanese governments have determined that it is feasible to recover these deposits and are making plans to do so.

□ FIGURE 12.1 The submersible *Alvin* sheds light on hydrothermal vents at the Galapagos Rift, a branch of the East Pacific Rise. Seawater seeping through the oceanic crust is heated and then rises and builds chimneys on the sea floor. Communities of organisms, including tubeworms, giant clams, crabs, and several types of fish, live near the vents.

INTRODUCTION

During most of historic time, people knew little of the oceans and believed that the sea floor was flat and featureless. In 450 B.C. Herodotus compiled a map that shows the ancient Greeks' perception of the oceans (▣ Fig. 12.2). According to their limited view of world geography, the Mediterranean Sea was surrounded by the three known landmasses, Libya (northern Africa), Europe, and Asia, all of which were surrounded by vast oceans. Despite their misconceptions about geography, the ancient Greeks did determine Earth's size rather accurately, but Western Europeans were not aware of the vastness of the oceans until the fifteenth and sixteenth centuries when various explorers sought new trade routes to the Indies.

When Christopher Columbus set sail on August 3, 1492, in an attempt to find a route to the Indies, he greatly underestimated the width of the Atlantic Ocean. Contrary to popular belief, Columbus was not attempting to demonstrate that Earth is spherical—Earth's spherical shape was well accepted by then. The controversy was over Earth's circumference and what was the shortest route to China. On this controversial point, Columbus's critics were correct. In any event, during this and other voyages, Europeans sailed to the Americas, the Pacific Ocean, Australia, New Zealand, the Hawaiian Islands, and many other areas previously unknown to them.

These voyages added considerably to our knowledge of the oceans, but truly scientific investigations of the oceans did not begin until the late 1700s. Great Britain was then the dominant maritime power, and to maintain that dominance, the British sought to increase their knowledge of the oceans. The earliest British scientific voyages were led by Captain James Cook in 1768, 1772, and 1777. In 1872, the converted British warship H.M.S. *Challenger* began a four-year voyage, during which seawater was sampled and analyzed, oceanic depths were determined at nearly 500 locations, rocks and sediments were recovered from the sea floor, and more than 4,000 new marine species were classified. Even Charles Darwin, although best known for his ideas on organic evolution, contributed to our knowledge of the oceans by proposing a theory explaining the evolution of coral reefs.

Continuing exploration of the oceans revealed that the sea floor is not flat and featureless as formerly believed. Indeed, scientists discovered that the sea floor possesses varied topography including oceanic trenches, submarine ridges, broad plateaus, hills, and vast plains.

▣ **FIGURE 12.2** This map, compiled by Herodotus, shows the Greeks' perception of the world in 450 B.C. The three known continents are surrounded by vast oceans.

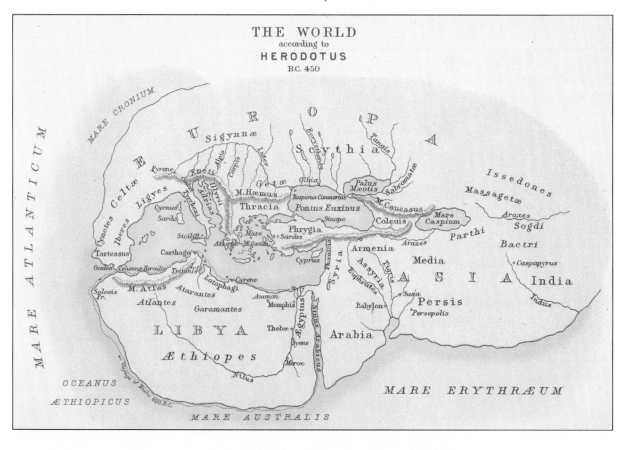

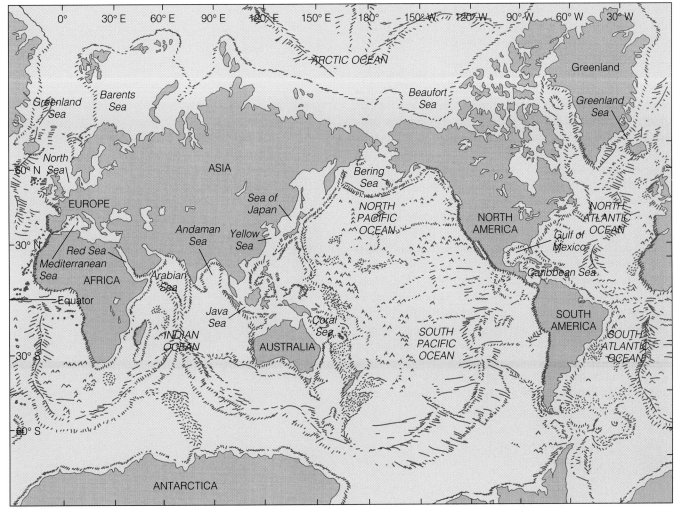

■ FIGURE 12.3 The geographic limits of the oceans and various seas.

 OCEANS AND SEAS

About 71% of Earth's surface is covered by an interconnected body of saltwater we refer to as *oceans* and *seas*. Four very large bodies of water are sufficiently distinct from one another to be recognized as separate oceans: the Pacific, the Atlantic, the Indian, and the Arctic (■ Fig. 12.3). The Pacific is by far the largest in area and volume and has the greatest depth (○ Table 12.1); its area is nearly equal to the areas of the other three oceans combined, and its volume accounts for almost 53% of all water on Earth. In 1513, Vasco Núñez de Balboa named the Pacific Ocean after the word meaning tranquil or peaceful. The Romans used the name Atlantic Ocean, probably from the Atlas Mountains in northwestern Africa, to designate the large, unexplored ocean that

TABLE 12.1   Numerical Data for the Oceans

OCEAN*	SURFACE AREA (million km²)	WATER VOLUME (million km³)	AVERAGE DEPTH (km)	MAXIMUM DEPTH (km)
Pacific	180	700	4.0	11.0
Atlantic	93	335	3.6	9.2
Indian	77	285	3.7	7.5
Arctic	15	17	1.1	5.2

*Excludes adjacent seas.
SOURCE: Pinet, P. R. 1992. *Oceanography*. St. Paul, Minn.: West Publishing Company.

lay to the west of the Mediterranean Sea. In contrast to the Pacific, which is roughly circular and is 13,000 km (8,065 mi) wide on average, the Atlantic is long and narrow, averaging 6,600 km (4,095 mi) in width.

The Indian Ocean is confined mostly to the Southern Hemisphere (Fig. 12.3). Its southern boundary is Antarctica, and it is bounded on the west by the Atlantic and on the east by the Pacific. The Arctic Ocean is the smallest and shallowest of the oceans (Table 12.1). Some oceanographers consider the Arctic Ocean to be a northern extension of the Atlantic, but it is separated from the Atlantic by a large but discontinuous submarine ridge trending east-west through Iceland. Furthermore, the Arctic is distinct from the other oceans in that it is nearly landlocked and much of it is covered by sea ice.

Sea is a term designating a smaller body of water such as one occupying an indentation into a continent (the Caribbean Sea and Sea of Japan, for example) (Fig. 12.3). These seas are simply marginal parts of oceans. Some seas, such as the Mediterranean, which is a marginal part of the Atlantic, are nearly encircled by land (Fig. 12.3); only a narrow passage, the Strait of Gibraltar, connects the Mediterranean Sea and the Atlantic Ocean. *Sea* is also used for some bodies of water completely enclosed by land, such as the Caspian Sea and Dead Sea; these are actually lakes and will not be considered in this chapter.

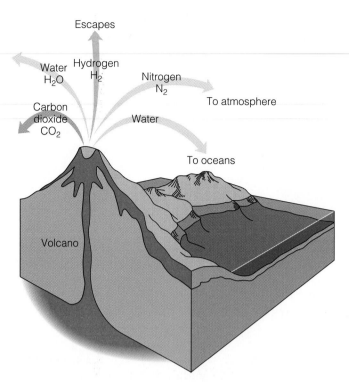

 FIGURE 12.4 Gases derived from within Earth by outgassing probably formed the early atmosphere and surface waters.

THE ORIGIN OF SEAWATER

During its earliest history, Earth was probably hot, airless, and lacked surface water. Volcanic activity was more common then than it is at present, however, because Earth possessed more heat. As we noted in previous chapters, volcanoes emit a variety of gases, the most abundant of which is water vapor. The atmosphere and surface waters were probably derived from the mantle and emitted at the surface as volcanic gases in a process known as **outgassing** (Fig. 12.4). As Earth cooled, water vapor began condensing and fell as rain, which accumulated to form the surface waters. The mantle is thought to contain about 0.5% water by weight, so if the surface waters were in fact derived by outgassing, only about one-fifteenth of this mantle water would have to have been released. Another hypothesis for the origin of the surface water—that much of it was derived from comets—is not yet widely accepted.

Geologic evidence indicates that oceans were present at least 3.5 billion years ago, although their volumes and extent are unknown. We can envision an early hot Earth with considerable volcanic activity and rapid accumulation of surface waters by outgassing, a process that continues to the present. Is the volume of oceanic waters still increasing? Probably it is, but if so, the rate has slowed considerably because the amount of heat available to generate magma has decreased. As a result, the quantity of water now being added to the oceans is trivial compared to their volumes.

COMPOSITION OF SEAWATER

Basically, seawater consists of water containing various ions in solution. More than 70 chemical elements have been identified in seawater, but the most common are the chloride ion (Cl) and the sodium ion (Na). More than 85.6% of all dissolved substances in seawater are chloride and sodium ions, and these two ions account for seawater's most distinctive characteristic—its saltiness. Six ions make up more than 99% of all dissolved solids in seawater (Table 12.2).

Seawater's saltiness, or its **salinity,** is a measure of the total quantity of dissolved solids. On the average, 1,000 g, or 1 kg, of seawater contains 35 g of dissolved solids. Accordingly, 96.5% of a kilogram of seawater consists of water, whereas the remaining 3.5% represents the weight of dissolved solids. Another way of stating this, and the one preferred by oceanographers, is in parts per thousand shown by the symbol ‰. Using this notation, seawater has an average salinity of 35‰.

The salinity of seawater in the open oceans varies from 32‰ to 37‰. The highest salinity values are in two belts; one between 20° and 30° north latitude, and the other between 15° and 20° south latitude. These higher-than-average values are caused by higher evaporation rates and hence the concentration of dissolved substances. Areas near the equator have lower salinity values because of greater precipitation and lower evaporation rates. In the polar regions, salinity fluctuates during the year as sea ice forms and melts.

TABLE 12.2	Major Ions in Seawater
ION	**PERCENTAGE**
Chloride (Cl⁻)	55.04
Sodium (Na⁺)	30.61
Sulfate (SO₄²⁻)	7.68
Magnesium (Mg²⁺)	3.69
Calcium (Ca²⁺)	1.16
Potassium (K⁺)	1.10
Total	99.28

SOURCE: Pinet, P. R. 1992. *Oceanography.*

In some marginal seas, especially in hot, dry areas nearly isolated from the open ocean, evaporation rates are high and salinity values may be in excess of 40‰.

In addition to the major constituents in seawater (Table 12.2), a number of other substances are also present. Seawater contains various nutrients such as nitrogen and phosphorus that are used by organisms. It also contains gases, especially nitrogen (N₂), oxygen (O₂), and carbon dioxide (CO₂), and a number of other elements in minute quantities. Zinc, for example, has a concentration of 10 parts per billion (ppb), and gold's concentration is 0.004 ppb, or 4 parts per trillion.

The continents are the source of many of the chemical substances dissolved in seawater, although some are derived from within Earth by outgassing. During chemical weathering, various minerals yield ions in solution, which are transported as the dissolved load of rivers. Rivers from the continents add about 4 billion tons of dissolved solids to the oceans each year.

In one early attempt to determine absolute ages, it was assumed that the oceans were originally fresh and that their salinity has gradually increased to the present value. By estimating the volume of ocean water and its salinity, and determining the rate of additions by streams, one could presumably calculate the age of the oceans. Although this attempt at absolute dating was ingenious, the underlying assumptions were wrong, so the dates were meaningless. For example, there was no way to calculate how much salt had been recycled or the amount of salt stored in continental salt deposits or in sea-floor clay deposits. We know that the oceans were salty during their very early history and that their salinity has changed little over at least the last 1.5 billion years. It appears then that seawater is in a state of equilibrium, meaning that additions are offset by losses.

For the salinity of seawater to remain rather constant, a continuous recycling of ions must be occurring. In other words, various ions in solution must be extracted from seawater at about the same rate they are supplied by runoff and hydrothermal vent activity; otherwise seawater would indeed be getting saltier with time. To maintain the balance, ions are removed from seawater when evaporites such as

rock salt (NaCl) and rock gypsum (CaSO₄·2H₂O) are precipitated, when salt spray is blown onshore, when magnesium is used in dolomite and clay minerals, and when organisms use calcium and silica to construct their shells.

 # LIGHT PENETRATION AND THE COLOR OF SEAWATER

Some of the sunlight striking the ocean surface is reflected back into the atmosphere, but the remainder enters the water and is absorbed by water molecules. When visible light, one form of energy radiated by the Sun, enters seawater, most is absorbed in the upper 1 m (3.3 ft) and elevates the surface-water temperature. However, seawater absorbs the longer wavelengths of visible light, the reds and yellows, more readily than the shorter wavelengths, the greens and blues (◫ Fig. 12.5a). The blue color of seawater in the open ocean is related to this selective absorption of visible light.

Even in the clearest seawater, less than 1% of the light reaches a depth of 100 m (328 ft), but even at 200 m (656 ft), some blue and green light can be perceived. In shallow nearshore water, suspended solids such as particulate matter and tiny floating plants (phytoplankton) make the water turbid, so light cannot penetrate deeper than about 20 m (65 ft) (Fig. 12.5b). Under these conditions, the yellow and green wavelengths of visible light penetrate the deepest, giving these waters their characteristic greenish color (Fig. 12.5c).

Based on the decreasing intensity of light with depth, two vertical layers are recognized in seawater. The upper layer, known as the **photic zone,** is generally 100 m (328 ft) or less thick and receives enough light for plants to photosynthesize (Fig. 12.5a). Below is the **aphotic zone** where too little light is available for plants to survive. With few exceptions, life in the oceans depends directly or indirectly on sunlight and the organic productivity in the photic zone. Even most animals that live at great oceanic depths depend on organic substances produced within the photic zone that

EARTHFACT

Desalinization of Seawater

Seawater must be desalinized to make it suitable for human consumption or irrigation. Unfortunately, desalinization requires considerable energy, thus making the process quite costly. The reason for this is the nature of water molecules, which are asymmetric (see Fig. 4.5) and possess a slight negative electrical charge at one end and a slight positive charge at the other end. This asymmetry is what makes water such a good solvent, but it also accounts for the fact that water molecules bond tightly to other compounds and ions. The energy needed to disrupt these bonds is what makes desalinization so expensive.

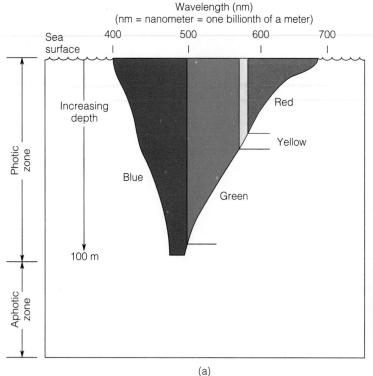

Wavelength (nm)
(nm = nanometer = one billionth of a meter)

400   500   600   700

Sea surface

Increasing depth

Red

Yellow

Blue

Green

100 m

Photic zone

Aphotic zone

(a)

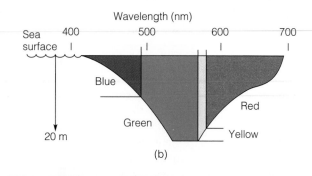

Wavelength (nm)

400   500   600   700

Sea surface

Blue

Green

Red

Yellow

20 m

(b)

(c)

  FIGURE 12.5 (*a*) The longer blue and green wavelengths of visible light penetrate much deeper into seawater than the shorter red and yellow wavelengths, thus giving seawater in the open ocean a blue color. (*b*) Suspended sediment and organic matter in nearshore water limit the depth of light penetration. Yellow and green wavelengths penetrate deepest, so these waters are generally greenish. (*c*) Nearshore greenish seawater along the coast of California.

settle to the sea floor, the most notable exception being the organisms living near hydrothermal vents (see the Prologue).

INVESTIGATING THE SEA FLOOR

The Deep Sea Drilling Project, an international program sponsored by several oceanographic institutions and funded by the National Science Foundation, began in 1969. Its first research vessel, the *Glomar Challenger,* was capable of drilling in water more than 6,000 m (19,685 ft) deep and recovering long cores of sea-floor sediment and oceanic crust. During the next 15 years, the *Glomar Challenger* drilled more than 1,000 holes in the sea floor.

The Deep Sea Drilling Project came to an end in 1983 when the *Glomar Challenger* was retired. However, an international project, the Ocean Drilling Program, continued where the Deep Sea Drilling Project left off, and a larger, more advanced research vessel, the JOIDES *Resolution,* made its first voyage in 1985. (JOIDES is an acronym for Joint Oceanographic Institutions for Deep Earth Sampling.)

In addition to surface vessels, submersibles, both remotely controlled and carrying scientists, have been added

to the research arsenal of oceanographers. In 1985, the *Argo,* towed by a surface vessel and equipped with sonar and television systems, provided the first views of the British ocean liner R.M.S. *Titanic* since it sank in 1912. The U.S. Geological Survey is using a towed device with a sonar system to produce sea-floor images resembling aerial photographs. Researchers aboard the submersible *Alvin* have observed submarine hydrothermal vents (see the Prologue) and have explored parts of the oceanic ridge system. During a dive in 1991, scientists observed what appeared to be evidence of very recent submarine volcanism.

The first measurements of oceanic depths were made by lowering a weighted line to the sea floor and measuring the length of the line. Now an instrument called an *echo sounder* is used. Sound waves from a ship are reflected from the sea floor and detected by instruments on the ship, yielding a continuous profile of the sea floor. Depth is determined by knowing the velocity of sound waves in water and the time it takes for a wave to reach the sea floor and return to the ship.

Seismic profiling is similar to echo sounding but even more informative. Strong waves are generated at an energy source, the waves penetrate the layers beneath the sea floor, and some of the energy is reflected from various horizons back to the surface (◫ Fig. 12.6). Seismic profiling has been particularly useful in mapping the structure of the oceanic crust beneath sea-floor sediments.

Although scientific investigations of the oceans have been yielding important information for more than two hundred years, much of our current knowledge has been acquired since World War II. This statement is particularly true with respect to the sea floor, because only in recent decades has instrumentation been available to study this largely hidden domain. The data collected are not only important in their own right but also have provided much of the evidence supporting plate tectonic theory.

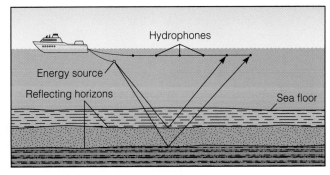

◫ **FIGURE 12.6** Seismic profiling and the detection of buried layers beneath the sea floor. Some of the energy generated at the energy source is reflected from various horizons back to the surface where it is detected by hydrophones.

STRUCTURE AND COMPOSITION OF THE OCEANIC CRUST

Most of the oceanic crust is consumed at subduction zones, but a tiny amount of it is not subducted. Small slivers of oceanic crust along with deep-sea sediments and pieces of the upper mantle, called **ophiolites,** are emplaced in mountain ranges on land, usually by moving along large fractures known as thrust faults.

An ophiolite consists of a layer of deep-sea sedimentary rocks underlain successively by a layer of pillow lavas and lava flows and a complex of vertical basaltic dikes. Further downward is gabbro and then rock representing the upper mantle. Thus, a complete ophiolite sequence consists of deep-sea sedimentary rock, oceanic crust, and upper mantle rocks (◫ Fig. 12.7).

Sampling and direct observations reveal that the upper part of the oceanic crust is indeed composed of a layer of pillow lavas and lava flows underlain by a sheeted dike complex. Confirmation of what lay below the dike complex was not obtained until 1989, when scientists in a submersible descended to the walls of a sea-floor fracture (transform fault) in the North Atlantic. There rocks of the upper mantle and lower oceanic crust were close to the sea floor, and so, for the first time, scientists observed these parts of the sequence.

SEA-FLOOR TOPOGRAPHY

The sea floor is not monotonous and flat as was once believed. Indeed, it is as varied as the land's surface, but for centuries people were unaware of this fact. In 1959, Maurice Ewing, Bruce Heezen, and Marie Tharp published a spectacular three-dimensional map of the North Atlantic showing vast plains and conical seamounts, as well as the Mid-Atlantic Ridge with its central rift valley. As more of the world's ocean floors were explored, this original map was

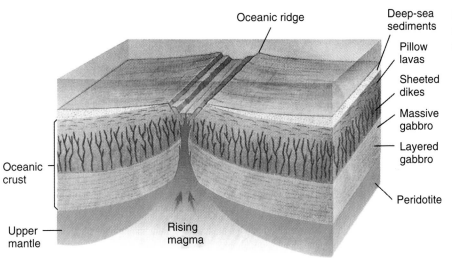

◫ **FIGURE 12.7** New oceanic crust consisting of the layers shown here forms as magma rises beneath oceanic ridges.

FIGURE 12.8 This map of the sea floor resulted from the work of Maurice Ewing, Bruce Heezen, and Marie Tharp.

expanded to reveal numerous other ocean-floor features (■ Fig. 12.8).

 CONTINENTAL MARGINS

Continental margins, zones separating a continent above sea level from the deep-sea floor, consist of a gently sloping *continental shelf,* a more steeply inclined *continental slope,* and, in some cases, a deeper, gently sloping *continental rise* (■ Fig. 12.9). At its outer limit, the continental margin merges with the deep-sea floor or descends into an oceanic trench.

Most people perceive continents as the land areas outlined by sea level. However, the true geologic margin of a continent—that is, where continental crust changes to oceanic crust—is below sea level, generally somewhere beneath the continental slope. Accordingly, marginal parts of continents are submerged.

The Continental Shelf

Between the shoreline and continental slope of all continents lies the **continental shelf,** an area where the sea floor slopes much less than 1° seaward; the seafloor averages about 2 m/km (10 ft/mi), or 0.1°. The outer edge of the continental shelf is generally taken to correspond to the point at an average depth of 135 m (425 ft) where the inclination of the sea floor increases rather abruptly to several degrees (Fig. 12.9). Continental shelves range in width from a few tens of meters to more than 1,000 km (620 mi). The shelf along the east coast of North America is as much as several hundred kilometers across in some places, whereas along the west coast it is only a few kilometers wide.

FIGURE 12.9 A generalized profile of the sea floor showing features of the continental margins. The vertical dimensions of the features in this profile are greatly exaggerated because the vertical and horizontal scales differ.

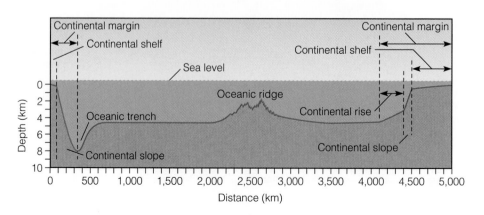

Deep, steep-sided **submarine canyons** are more characteristic of the continental slope, but some of them extend well up onto the continental shelf. Some of these canyons lie offshore from the mouths of large rivers. At times during the Pleistocene (1.6 million to 10,000 years ago), sea level was about 130 m (425 ft) lower than at present, so much of the continental shelves was above sea level. Rivers flowed across these exposed shelves and eroded deep canyons that were subsequently flooded when sea level rose at the end of the Pleistocene. However, most submarine canyons extend to depths far greater than can be explained by river erosion during periods of lower sea level. Furthermore, many submarine canyons are not associated with rivers on land. They are discussed more fully in a later section.

As a result of lower sea level during the Pleistocene, much of the sediment on continental shelves accumulated in river channels and on floodplains. In fact, in areas such as northern Europe and parts of North America, glaciers extended onto the exposed shelves and deposited gravel, sand, and mud. Since the Pleistocene, sea level has risen, submerging the shelf sediments, which are now being reworked by marine processes. That these sediments were, in fact, deposited on land is indicated by evidence of human settlements and fossils of mammoths and mastodons (extinct relatives of elephants) and other land-dwelling animals.

The Continental Slope and Rise

The seaward margin of the continental shelf is marked by the shelf-slope break where the relatively steep **continental slope** begins (Fig. 12.9). The inclination of continental slopes averages about 4°, but ranges from 1° to 25°. In many places, especially around the margins of the Atlantic, the continental slope merges with the more gently sloping **continental rise.** In other places, such as around the Pacific Ocean, slopes commonly descend directly into an oceanic trench, and a continental rise is absent (Fig. 12.9).

Turbidity Currents, Submarine Fans, and Submarine Canyons

Turbidity currents are sediment-water mixtures denser than normal seawater that flow downslope to the deep-sea floor, generally through submarine canyons (▣ Fig. 12.10a and b). An individual turbidity current flows onto the relatively flat sea floor where it slows and begins depositing sediment; the coarsest particles are deposited first, followed by progressively smaller particles, thus forming *graded bedding* (Fig. 12.10c). These deposits accumulate as a series of overlapping **submarine fans,** which constitute a large part of the continental rise (Fig. 12.10a).

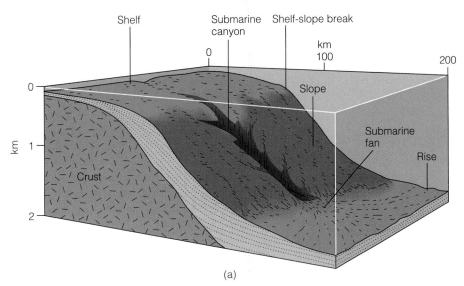

▣ FIGURE 12.10 (*a*) Submarine fans formed by deposition of sediments carried down submarine canyons by turbidity currents. Much of the continental rise is composed of overlapping submarine fans. (*b*) Turbidity currents flow downslope along the sea floor (or lake bottom) because of their density. (*c*) Graded bedding formed by deposition from a turbidity current.

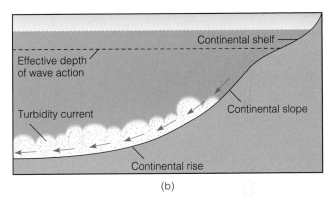

(b)

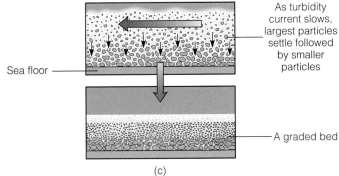

(c)

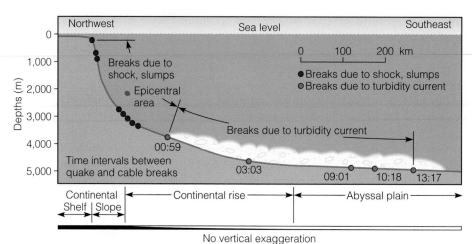

 FIGURE 12.11 Submarine cable breaks caused by an earthquake-generated turbidity current south of Newfoundland. This profile of the sea floor shows the locations of the cables and the times when they were severed. The vertical dimension is highly exaggerated. The profile labeled "no vertical exaggeration" shows what the sea floor actually looks like in this area.

No one has ever observed a turbidity current in progress, so for many years there was considerable debate about their existence. Perhaps the most compelling evidence for the existence of turbidity currents is the pattern of trans-Atlantic cable breaks that took place south of Newfoundland on November 18, 1929 (Fig. 12.11). Initially, it was assumed that an earthquake on that date had ruptured several trans-Atlantic telephone and telegraph cables. While the breaks on the continental shelf near the epicenter occurred when the earthquake struck, cables farther seaward were broken later and in succession. The last cable to break was 720 km (445 mi) from the source of the earthquake, and it did not snap until 13 hours after the first break (Fig. 12.11).

In 1949, scientists realized the earthquake had generated a turbidity current that moved downslope, breaking the cables in succession. The precise time that each cable broke was known, so it was a simple matter to calculate the velocity of the turbidity current. It apparently moved at about 80 km/hr (50 mi/hr) on the continental slope, but slowed to about 27 km/hr (17 mi/hr) when it reached the continental rise.

As mentioned previously, submarine canyons are found on the continental shelves but are best developed on continental slopes (Fig. 12.10). Some submarine canyons can be traced across the shelf to rivers on land, but many have no such association. It is known that turbidity currents move through submarine canyons, and these currents are now thought to be the primary agent responsible for the erosion of the canyons.

TYPES OF CONTINENTAL MARGINS

Two types of continental margins are generally recognized: *active* and *passive*. An **active continental margin** develops at the leading edge of a continental plate where oceanic lithosphere is subducted (Fig. 12.12a). The west coast of South America is a good example. Here the continental margin is characterized by earthquakes, a geologically young mountain range, and volcanism. Additionally, the continental shelf is narrow, and the continental slope descends directly into the Peru-Chile Trench.

The configuration and geologic activity of the eastern continental margins of North and South America differ considerably from those of the western margins. The eastern continental margins developed as a result of rifting of the supercontinent Pangaea. The continental crust was stretched, thinned, and fractured as rifting proceeded, and as plate separation took place, the newly formed continental margins became the sites of deposition of land-derived sediments. These **passive continental margins** are within rather than at the edge of a plate (Fig. 12.12b). They possess broad continental shelves and a continental slope and rise; vast, flat *abyssal plains* are commonly present adjacent to the rises. Passive continental margins also lack the intense seismic and volcanic activity characteristic of active continental margins.

Active continental margins lack a continental rise because the slope descends directly into an oceanic trench (Fig. 12.12a). Just as on passive continental margins, sediment is transported down the slope by turbidity currents, but it simply fills the trench rather than forming a rise. The proximity of the trench to the continent also explains why the continental shelf is so narrow. In contrast, the continental shelf of a passive continental margin is much wider because land-derived sedimentary deposits build outward into the ocean (Fig. 12.12b). The type of continental margin varies through time depending on the stage of development of the ocean basin (see Perspective 12.1 on page 278).

THE DEEP-OCEAN BASIN

On average, the oceans are more than 3,800 m (12,465 ft) deep, and as a result, most of the sea floor is completely dark, no plant life exists, the temperature is generally just above 0°C (32°F), and pressure varies from 200 to more than 1,000 atmospheres depending on depth. Although scientists in submersibles have descended to the greatest oceanic depths, much of the deep-ocean basin has been studied only by echo sounding, seismic profiling, drilling, and remote submersibles.

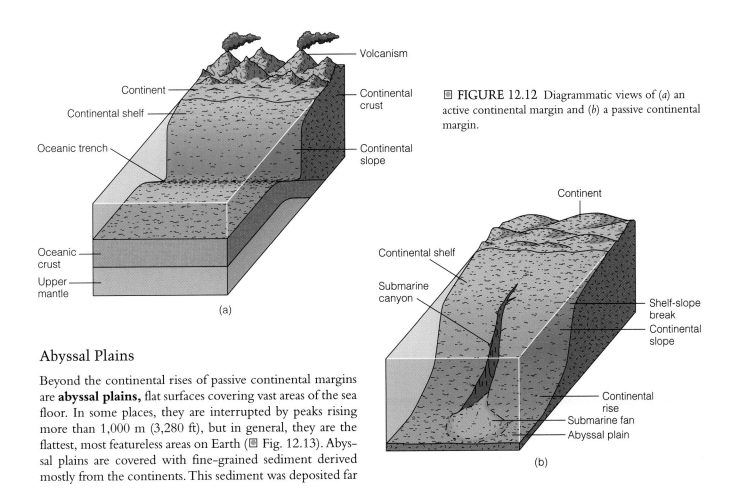

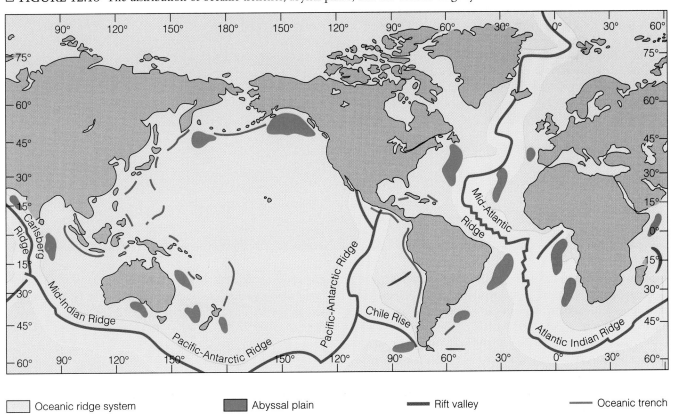

FIGURE 12.12 Diagrammatic views of (*a*) an active continental margin and (*b*) a passive continental margin.

Abyssal Plains

Beyond the continental rises of passive continental margins are **abyssal plains,** flat surfaces covering vast areas of the sea floor. In some places, they are interrupted by peaks rising more than 1,000 m (3,280 ft), but in general, they are the flattest, most featureless areas on Earth (▣ Fig. 12.13). Abyssal plains are covered with fine-grained sediment derived mostly from the continents. This sediment was deposited far

FIGURE 12.13 The distribution of oceanic trenches, abyssal plains, and the oceanic ridge system.

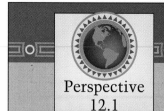

The Opening and Closing of Oceans

According to Canadian geologist J. Tuzo Wilson, oceans evolve through a distinct series of stages beginning with the rifting of a continent and culminating with the closure of a once expansive ocean. The generalized life cycle of ocean basins is divided into six stages (□ Fig. 1). This cycle of opening and closing ocean basins is now known as the *Wilson cycle*.

The onset of a Wilson cycle is marked by rifting of a continent and the origin of a complex system of linear rift valleys. As rifting proceeds, Earth's crust is stretched and thinned, and large blocks are displaced along normal faults paralleling the rift margins (Fig. 1). Rift valleys or basins formed in this manner are referred to as *embryonic,* since they represent the first stage in the origin of an ocean basin. Perhaps the best example of this initial stage in the Wilson cycle is the East African rift system.

Continued stretching and thinning of the continental crust mark the second stage of ocean-basin evolution. During this stage, oceanic crust is formed from basaltic magma, and the rift valley subsides below sea level, forming a long, narrow sea (Fig. 1). The Red Sea and Gulf of Aden are in this young or juvenile stage. Stage 3 or the development of a mature ocean basin results from the continuation of the processes just described. Divergence of plates and the generation of oceanic crust at a spreading ridge continue until eventually an expansive ocean basin develops (Fig. 1). As the ocean expands, sediments eroded from the continents on either side of the ocean basin are deposited in the sea, producing broad passive continental margins. The Atlantic Ocean is in this stage of development.

Following stage 3, the basin enters a declining stage as the oceanic crust becomes cooler and denser and begins to subduct at the margins of the ocean, thereby transforming passive continental margins into active ones (Fig. 1). Oceanic lithosphere is consumed by subduction, and eventually even the spreading ridge itself is subducted. Parts of the East Pacific Rise are now being subducted beneath Central and South America as the Pacific Ocean basin becomes smaller.

Continuing closure of an ocean basin leads to a *terminal* stage in which island arcs and continents that were once widely separated begin to collide (Fig. 1). The Mediterranean Sea is in this stage, which, if continued, results in *suturing* as two colliding plates become welded together. An example of this stage is provided by the Himalayas of Asia, which began forming 40 or 50 million years ago when India collided with Asia. In fact, this collision is continuing, and India is currently being thrust beneath Asia at a rate of about 5 cm (2 in.) per year.

□ FIGURE 1 *Right:* The Wilson cycle portrays ocean-basin development in six stages.

from land by the settling of tiny particles in seawater. The flat topography of the abyssal plains results from the deposition of sediment in sufficient quantities to bury the otherwise rugged sea floor.

Abyssal plains are invariably found seaward of continental rises, which are composed mostly of overlapping submarine fans that owe their origin to deposition by turbidity currents (Fig. 12.10a). Along active continental margins, sediments derived from the shelf and slope are trapped in an oceanic trench, and abyssal plains fail to develop. Accordingly, abyssal plains are common in the Atlantic Ocean but rare in the Pacific Ocean basin (Fig. 12.13).

Oceanic Trenches

Although **oceanic trenches** constitute a small percentage of the sea floor, they are very important, for it is here that lithospheric plates are consumed by subduction (see Chapter 7).

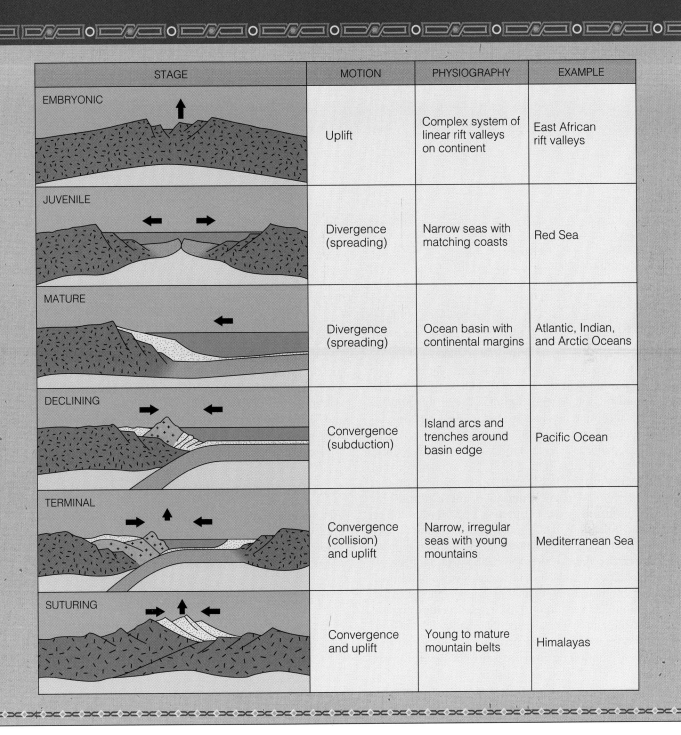

STAGE	MOTION	PHYSIOGRAPHY	EXAMPLE
EMBRYONIC	Uplift	Complex system of linear rift valleys on continent	East African rift valleys
JUVENILE	Divergence (spreading)	Narrow seas with matching coasts	Red Sea
MATURE	Divergence (spreading)	Ocean basin with continental margins	Atlantic, Indian, and Arctic Oceans
DECLINING	Convergence (subduction)	Island arcs and trenches around basin edge	Pacific Ocean
TERMINAL	Convergence (collision) and uplift	Narrow, irregular seas with young mountains	Mediterranean Sea
SUTURING	Convergence and uplift	Young to mature mountain belts	Himalayas

Oceanic trenches are long, narrow features restricted to active continental margins; thus, they are common around the margins of the Pacific Ocean basin (Fig. 12.13). The Peru-Chile Trench west of South America is 5,900 km (3,660 mi) long, but only 100 km (62 mi) wide. Oceanic trenches are also the sites of the greatest oceanic depths; the Peru-Chile Trench is more than 8,000 m (26,245 ft) deep, and a depth of more than 11,000 m (36,000 ft) has been recorded in the Challenger Deep of the Marianas Trench.

Oceanic trenches show anomalously low heat flow compared to the rest of the oceanic crust, so it appears that the crust here is cooler and slightly denser than elsewhere. Seismic activity also occurs at or near trenches. In fact, a pattern emerges when the focal depths of earthquakes near convergent plate boundaries and their adjacent oceanic trenches are plotted. Notice in ▣ Figure 12.14 that a characteristic narrow, dipping zone of earthquake foci is well defined.

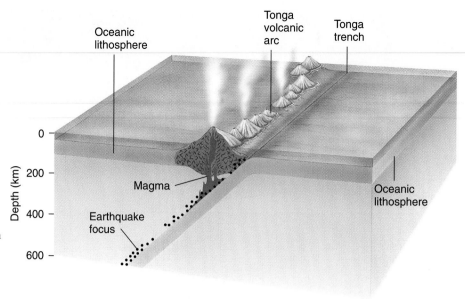

FIGURE 12.14 Focal depth increases in a well-defined zone that dips approximately 45° beneath the Tonga volcanic arc in the South Pacific. These dipping earthquake zones occur where one plate is subducted beneath another.

Oceanic Ridges

A feature called the Telegraph Plateau was discovered in the Atlantic Ocean basin during the late nineteenth century when the first submarine cable was laid between North America and Europe. Scientists proposed that this plateau was actually a continuous ridge extending the length of the Atlantic Ocean basin. Subsequent investigations revealed that this proposal was correct, and we now call this feature the Mid-Atlantic Ridge (Fig. 12.13). Some people claim that the Mid-Atlantic Ridge and other features of the Atlantic sea floor represent remnants of the mythical lost continent of Atlantis (see ▣ Perspective 12.2 on page 282).

The Mid-Atlantic Ridge is more than 2,000 km (1,240 mi) wide, and because it is hot and more buoyant than oceanic crust elsewhere, it rises about 2.5 km (1.5 mi) above the sea floor adjacent to it. It is, in fact, part of a much larger system of submarine mountainous topography at least 65,000 km (40,000 mi) long. This **oceanic ridge** system runs from the Arctic Ocean through the middle of the Atlantic, curves around South Africa, and passes into the Indian Ocean, continuing from there into the Pacific Ocean basin (Fig. 12.13). Its length surpasses that of the largest mountain range on land. Mountain ranges on land are typically composed of granitic and metamorphic rocks and sedimentary rocks that have been folded and fractured by compressional forces. Oceanic ridges, on the other hand, are composed of volcanic rocks (mostly basalt) and have features produced by tensional forces.

Along the crests of some ridges is a rift that appears to have opened in response to tensional forces (▣ Fig. 12.15), although in portions of the East Pacific Rise, this feature is absent or very small. These rifts are commonly 1 to 2 km (0.6 to 1.2 mi) deep and several kilometers wide. Rifts open as sea-floor spreading occurs; they are characterized by shallow-focus earthquakes, basaltic volcanism, and high heat flow.

As a part of Project FAMOUS (French-American Mid-Ocean Undersea Study), submersible craft descended into the rift of the Mid-Atlantic Ridge, and more recent dives have investigated other rifts (Fig. 12.15). Although no active volcanism was observed, the researchers did see pillow lavas and sheet lava flows, some of which appeared to have formed very recently. In addition, hydrothermal vents such as black smokers have been observed (see the Prologue).

Fractures in the Sea Floor

Oceanic ridges are not continuous features winding without interruption around the globe. They abruptly terminate where they are offset along major fractures oriented more or less at right angles to ridge axes (▣ Fig. 12.16). These large-

---

EARTHFACT

Aseismic Ridges

Long ridges and broad plateaulike features lacking seismic activity and rising 2 or 3 km (1.2 to 1.8 mi) above the surrounding sea floor are *aseismic ridges*. A few of these ridges are referred to as microcontinents, which are thought to be small fragments separated from continents during rifting. The Jan Mayan Ridge in the North Atlantic is a good example. Most aseismic ridges form as a linear succession of hot spot volcanoes, each of which was carried laterally with the plate upon which it originated. The result is a sequence of submarine peaks extending from an oceanic ridge.

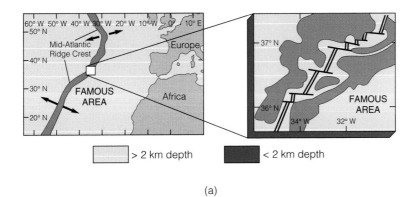

> ☐ **FIGURE 12.15** The FAMOUS
> expedition. (*a*) During the early 1970s,
> detailed studies were conducted within
> a small area of the Mid-Atlantic Ridge
> crest, the so-called FAMOUS area.
> (*b*) Observations made from a
> submersible revealed that recent
> volcanic outpourings were
> concentrated into moundlike deposits
> near the center of the axial rift valley.
> The high, steep shoulders of the axial
> rift are bounded by fault scarps, which
> splinter the crustal layer into large
> blocks.

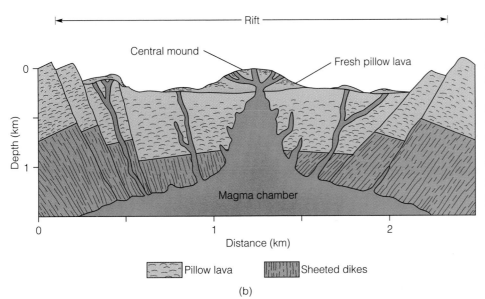

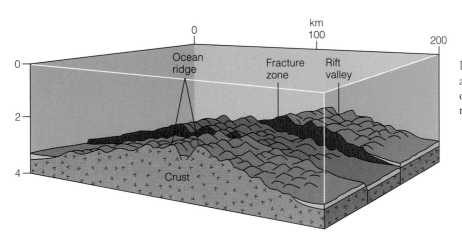

> ☐ **FIGURE 12.16** Diagrammatic view of
> a fracture offsetting a ridge. Earthquakes
> occur in the segments between the offset
> ridge crests.

scale fractures run for hundreds of kilometers, although they are difficult to trace if buried beneath sea-floor sediments. Many scientists are convinced that some geologic features on the continents can best be accounted for by the extension of these fractures into continents.

Where fractures offset oceanic ridges, they are characterized by shallow seismic activity only in the areas between the displaced ridge segments (Fig. 12.16). Furthermore, because

ridges are higher than the sea floor adjacent to them, the offset segments yield vertical relief on the sea floor. For example, nearly vertical escarpments 3 or 4 km (1.8 to 2.5 mi) high develop, as illustrated in Figure 12.16. Recall from earlier discussions that these fractures are referred to as transform faults in plate tectonics terminology, and on land they are recognized as strike-slip faults such as the San Andreas fault in California.

Perspective 12.2

Lost Continents

Most people have heard of the mythical lost continent of Atlantis, but few are aware of the source of the Atlantis legend or the evidence cited for the existence of this continent. Only two known sources of the Atlantis legend exist, both written about 350 B.C. by the Greek philosopher Plato. In two of his philosophical dialogues, the *Timaeus* and the *Critias,* Plato tells of Atlantis, a large island continent in the Atlantic Ocean west of the Pillars of Hercules, which we now call the Strait of Gibraltar (☐ Fig. 1). Plato also wrote that after its conquest by Athens, Atlantis disappeared:

> . . . there were violent earthquakes and floods and one terrible day and night came when . . . Atlantis . . . disappeared beneath the sea. And for this reason even now the sea there has become unnavigable and unsearchable, blocked as it is by the mud shallows which the island produced as it sank.★

If the destruction of Atlantis was a real event, rather than one conjured up by Plato to make a philosophical point, he nevertheless lived long after it was supposed to have occurred. According to Plato, Solon, an Athenian who lived about two hundred years before Plato, heard the Atlantis story from Egyptian priests who claimed the event had happened 9,000 years before their time. Solon reportedly told the story to his grandson, Critias, who in turn told it to Plato.

Present-day proponents of the Atlantis legend generally cite two types of evidence to support their claim that Atlantis did once exist. First, they point to supposed cultural resemblances on opposite sides of the Atlantic Ocean, such as the

similar shapes of pyramids in Egypt and Central and South America. They contend that these similarities are due to cultural diffusion from a highly developed civilization of Atlantis. According to archaeologists, however, few similarities actually exist, and those that do can be explained as the independent development of similar features by different cultures.

Second, supporters of the legend assert that remnants of the sunken continent can be found. No "mud shallows" exist in the Atlantic as Plato claimed, but the Azores, Bermuda, the Bahamas, and the Mid-Atlantic Ridge are alleged to be

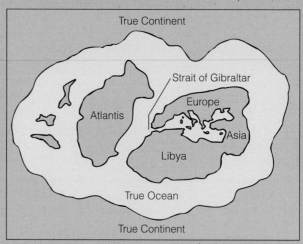

☐ FIGURE 1 According to Plato, Atlantis was a large continent west of the Pillars of Hercules, which we now call the Strait of Gibraltar.

★From Plato, *Timaeus,* quoted in E. W. Ramage, ed., *Atlantis: Fact or Fiction?* (Bloomington, Ind.: Indiana University Press, 1978), p. 13.

Seamounts and Guyots

As noted previously, the sea floor is not a flat, featureless plain, except for the abyssal plains, and even these are underlain by rugged topography. In fact, a large number of volcanic hills, seamounts, and guyots rise above the sea floor. These features are present in all ocean basins, but are particularly abundant in the Pacific. All are of volcanic origin, differing from one another mostly in size.

Seamounts rise more than 1 km (0.6 mi) above the sea floor; if they are flat topped, they are called **guyots** rather than seamounts (☐ Fig. 12.17 on page 284). Guyots are volcanoes that originally extended above sea level. But as the plate upon

which they were situated continued to move, they were carried away from a spreading ridge, and the oceanic crust cooled and descended to greater oceanic depths. Thus, what was once an island was eroded by waves as it slowly sank beneath the sea, giving it the typical flat-topped appearance of a guyot.

 ## DEEP-SEA SEDIMENTS

Deep-sea sediments consist mostly of fine-grained deposits because few mechanisms exist that can transport coarse-grained sediment (sand and gravel) far from land. Coarse

remnants of Atlantis. If a continent had actually sunk in the Atlantic, however, it could be easily detected by a gravity survey. Recall that continental crust has a granitic composition and a lower density than oceanic crust. If a continent were actually present beneath the Atlantic Ocean, it would be indicated by a huge negative gravity anomaly, but no such anomaly has been detected. Furthermore, the crust beneath the Atlantic has been drilled in many places, and all samples recovered indicate that its composition is the same as that of oceanic crust elsewhere.

In short, there is no geological evidence for Atlantis. Nevertheless, some archaeologists think that the legend may be based on a real event. About 1390 B.C., a huge volcanic eruption destroyed the island of Thera in the Mediterranean Sea, which was an important center of Minoan civilization. The eruption was one of the most violent during historic time, and much of the island disappeared when it subsided to form a caldera. Most of the island's inhabitants escaped (☐ Fig. 2), but the eruption probably contributed to the demise of the Minoan culture on Crete. At least 10 cm (4 in) of ash fell on parts of Crete, and the coastal areas of the island were probably devastated by tsunami. It is possible that Plato used an account of the destruction of Thera, but fictionalized it for his own purposes, thereby giving rise to the Atlantis legend.

☐ FIGURE 2 ·An artist's rendition of the volcanic eruption on Thera in about 1390 B.C. that destroyed most of the island. Most of the island's inhabitants escaped the devastation.

sediment in icebergs is transported into the ocean basins, however, and a broad band of *glacial-marine sediment* has been deposited adjacent to Antarctica and Greenland (☐ Fig. 12.18).

Most of the fine-grained sediment in the deep sea is windblown dust and volcanic ash from the continents and oceanic islands and shells of microscopic organisms that live in the photic zone. Most of the deeper parts of the ocean basins are covered by brown or reddish clay derived from the continents and oceanic islands. The remaining sea-floor sediment is composed mostly of shells of microscopic marine animals and plants.

Other sources of sediment include cosmic dust and deposits resulting from chemical reactions in seawater. The manganese nodules that are fairly common in all ocean basins are a good example of the latter (☐ Fig. 12.19). These nodules are composed mostly of manganese and iron oxides, but also contain copper, nickel, and cobalt. The contribution of cosmic dust to deep-sea sediment is negligible. Even though some researchers estimate that as much as 40,000 metric tons of cosmic dust may fall to Earth each year, this is a trivial quantity compared to the volume of sediments derived from other sources.

■ FIGURE 12.17 Submarine volcanoes may build up above sea level to form seamounts. As the plate upon which these volcanoes rest moves away from a spreading ridge, the volcanoes sink beneath sea level and become guyots.

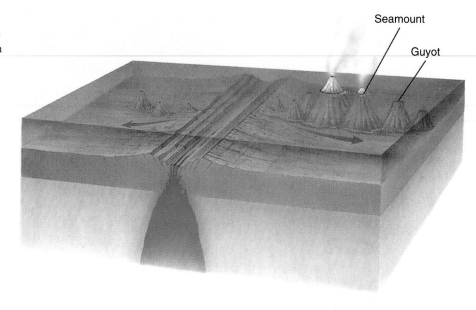

Seamount

Guyot

■ FIGURE 12.18 The formation of glacial-marine sediments by ice rafting.

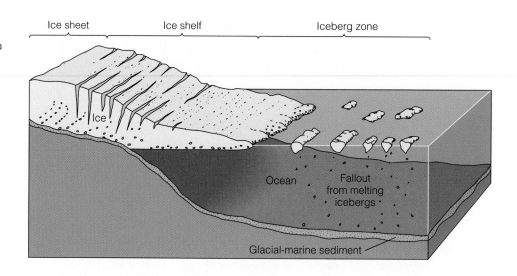

Ice sheet   Ice shelf   Iceberg zone

Ice

Ocean   Fallout from melting icebergs

Glacial-marine sediment

■ FIGURE 12.19 Manganese nodules on the sea floor south of Australia.

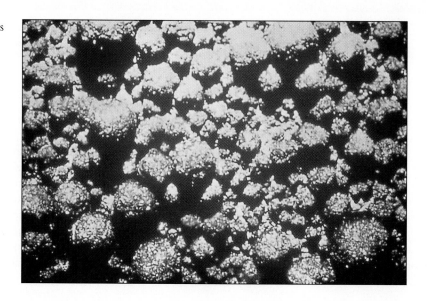

REEFS

Moundlike, wave-resistant structures composed of the skeletons of organisms are known as **reefs** (▣ Fig. 12.20). They are commonly called coral reefs, but in addition to corals, reefs consist of a solid framework of skeletons of clams and such encrusting organisms as algae and sponges. Reefs grow to a depth of about 45 or 50 m (145 to 165 ft) and are restricted to shallow tropical seas where the water is clear, and the water temperature does not fall below about 20°C (68°F).

Three types of reefs are recognized: fringing, barrier, and atoll (▣ Fig. 12.21). *Fringing reefs* lie close to the shore of an island or continent. They have a rough, tablelike surface, are as much as 1 km (0.6 mi) wide and, on their seaward side, slope steeply down to the sea floor. *Barrier reefs* are similar to fringing reefs, except that they are separated from the mainland by a lagoon. Probably the best-known barrier reef in the world is the 2,000 km (1,240 mi) long Great Barrier Reef of Australia.

The last type of reef is an *atoll,* which is a circular to oval reef surrounding a lagoon (Fig. 12.21). These reefs form around volcanic islands that subside below sea level as the plate upon which they rest moves progressively farther from an oceanic ridge (Fig. 12.17). As subsidence occurs, organisms construct the reef upward so that the living part of the reef remains in shallow water. Eventually, however, the island subsides below sea level, leaving a circular lagoon surrounded by a more or less continuous reef (Fig. 12.21). Such reefs are particularly common in the western Pacific Ocean basin, where many began as fringing reefs but, as subsidence occurred, evolved first into barrier reefs and finally into atolls. This particular scenario for the evolution of reefs from fringing to barrier to atoll was proposed more than 150 years ago by Charles Darwin while he was serving as a naturalist on the British research ship H.M.S. *Beagle.* Drilling into

▣ FIGURE 12.20 Reefs such as this one in the Red Sea are wave-resistant structures composed of the skeletons of organisms.

▣ FIGURE 12.21 Three-stage development of an atoll. In the first stage, a fringing reef forms, but as the island sinks, a barrier reef becomes separated from the island by a lagoon. As the island disappears beneath the sea, the reef continues to grow upward, thus forming an atoll. An oceanic island carried into deeper water by plate movement can account for this sequence.

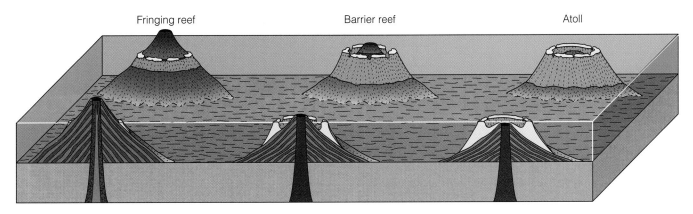

Fringing reef   Barrier reef   Atoll

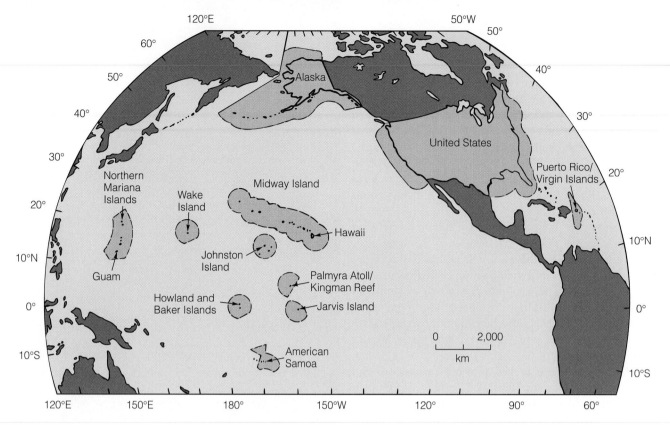

◉ FIGURE 12.22 The Exclusive Economic Zone (EEZ) includes a vast area adjacent to the United States and its possessions.

atolls has revealed that they do indeed rest upon a basement of volcanic rocks, thus confirming Darwin's hypothesis.

 RESOURCES FROM THE SEA

Seawater contains many elements in solution, some of which are extracted for various industrial and domestic uses. In many places sodium chloride (table salt) is produced by the evaporation of seawater, and a large proportion of the world's magnesium is produced from seawater. Numerous other elements and compounds can be extracted from seawater, but for many, such as gold, the cost is prohibitive.

In addition to substances in seawater, deposits on the sea floor are becoming increasingly important. Many of these potential resources lie well beyond the margins of the continents, so their ownership is a political and legal problem that has not yet been resolved. The United States by a presidential proclamation issued on March 10, 1983, claims sovereign rights over an area designated as the **Exclusive Economic Zone (EEZ),** which extends seaward 200 nautical miles (371 km) from the coast (◉ Fig. 12.22). In short, the United States claims all resources in an area about 1.7 times larger than its land area. A number of other nations also claim sovereign rights to resources within 200 nautical miles of their coasts.

Numerous resources are found within the EEZ, some of which have been exploited for many years. Sand and gravel for construction are mined from the continental shelf in several areas, and about 17% of U.S. oil and natural gas production comes from wells on the continental shelf. Several sedimentary basins within the EEZ are known to contain hydrocarbons whereas others are areas of potential hydrocarbon production. Ancient shelf deposits in the Persian Gulf region contain the world's largest reserves of oil (see the Prologue to Chapter 7).

Other resources of interest include mineral deposits formed by submarine hydrothermal activity at spreading ridges (see the Prologue). Such deposits containing iron, copper, zinc, and other metals have been identified within the EEZ at the Gorda Ridge off the coast of California and Oregon, and similar deposits at the Juan de Fuca Ridge are within the Canadian EEZ.

Other potential resources include the manganese nodules discussed previously (Fig. 12.19) and metal oxide crusts found on seamounts. Manganese nodules contain manganese, cobalt, nickel, and copper; the United States is heavily dependent on imports of the first three of these elements. Within the EEZ, manganese nodules are found near Johnston Island in the Pacific Ocean and on the Blake Plateau off the east coast of South Carolina and Georgia. Seamounts within the EEZ in the Pacific are known to have metal oxide crusts several centimeters thick from which cobalt and manganese could be mined.

1. Oceans and their marginal seas cover about 71% of Earth's surface. All surface water probably originated by outgassing from the mantle.

2. On the average, 1,000 g of seawater contains 35 g of ions in solution, most of which are chloride (Cl) and sodium (Na) ions. The salinity of seawater is usually expressed as 35‰, meaning 35 parts per thousand.

3. Ocean salinity remains rather constant because the rate of input of ions by runoff and outgassing is equaled by losses from such processes as deposition of evaporites and the use of ions by organisms.

4. Continental margins separate the continents above sea level from the deep ocean basin. They consist of a continental shelf, continental slope, and in some cases a continental rise.

5. Continental shelves slope gently seaward and range in width from a few tens of meters to more than 1,000 km. The continental slope begins at an average depth of 135 m where the inclination of the sea floor increases rather abruptly from less than 1° to several degrees.

6. Submarine canyons are characteristic of the continental slope, but some extend well up onto the shelf and lie offshore from large rivers. River erosion during the Pleistocene may account for some submarine canyons, but most were probably eroded by turbidity currents.

7. Turbidity currents commonly move through submarine canyons and deposit overlapping submarine fans that constitute a large part of the continental rise.

8. Active continental margins are characterized by a narrow shelf and a slope that descends directly into an oceanic trench with no rise present. They are also areas of volcanism and earthquakes.

9. Passive continental margins lack volcanism and exhibit little earthquake activity. They have a wide shelf and a slope descending to a rise, which, in some places, merges with abyssal plains.

10. Long, narrow features known as oceanic trenches are the sites where oceanic lithosphere is subducted. They are also the sites of the greatest oceanic depths.

11. Oceanic ridges are made up of volcanic rocks, and many ridges possess a central rift caused by tensional forces. Basaltic volcanism and shallow-focus earthquakes occur at ridges. Large fractures interrupt and offset ridges in many areas.

12. Other important features of the sea floor include seamounts rising more than 1 km high and guyots, which are submerged, flat-topped seamounts.

13. Deep-sea sediments are composed mostly of fine-grained particles derived from continents and oceanic islands and the microscopic shells of organisms.

14. Reefs, which occur in shallow, warm seas, are wave-resistant structures composed of animal skeletons, particularly corals. Fringing and barrier reefs and atolls are the three types of reefs.

15. The United States claims rights to all resources within 200 nautical miles (371 km) of its shores. Numerous resources including various metals are found within this Exclusive Economic Zone. Many other nations have made similar claims to resources in their coastal waters.

IMPORTANT
TERMS

abyssal plain	continental slope	outgassing	seamount
active continental margin	Exclusive Economic Zone	passive continental margin	submarine canyon
aphotic zone	guyot	photic zone	submarine fan
continental margin	oceanic ridge	reef	turbidity current
continental rise	oceanic trench	salinity	
continental shelf	ophiolite		

1. Much of the continental rise is composed of:
   a. _____ seamounts; b. _____ submarine fans; c. _____ fringing reefs; d. _____ pillow lavas; e. _____ ophiolite.
2. The greatest oceanic depths are at:
   a. _____ aseismic ridges; b. _____ guyots; c. _____ the shelf-slope break; d. _____ oceanic trenches; e. _____ passive continental margins.
3. Abyssal plains are most common:
   a. _____ adjacent to the Mid-Atlantic Ridge; b. _____ near the East Pacific Rise; c. _____ along the west coast of South America; d. _____ around the margins of the Atlantic; e. _____ in the rift valleys of oceanic ridges.
4. A circular reef enclosing a lagoon is a(n):
   a. _____ barrier reef; b. _____ seamount; c. _____ aseismic ridge; d. _____ guyot; e. _____ atoll.
5. Submarine canyons are most characteristic of:
   a. _____ continental shelves; b. _____ abyssal plains; c. _____ continental slopes; d. _____ rift valleys; e. _____ fractures in the sea floor.
6. Earth's surface waters probably originated through the process of:
   a. _____ dewatering; b. _____ subduction; c. _____ outgassing; d. _____ crustal fracturing; e. _____ submarine erosion.
7. Continental shelves:
   a. _____ are composed of submarine lava flows; b. _____ lie between continental slopes and rises; c. _____ descend to an average depth of 1,500 m; d. _____ slope gently from the shoreline to the shelf-slope break; e. _____ are widest along active continental margins.
8. The flattest, most featureless areas on Earth are:
   a. _____ oceanic ridges; b. _____ abyssal plains; c. _____ continental slopes; d. _____ continental margins; e. _____ seamounts.
9. Sediment-water mixtures having a greater density than seawater flow along the sea floor as:
   a. _____ submarine eruptions; b. _____ turbidity currents; c. _____ aphotic synthesis; d. _____ ophiolites; e. _____ seismic profiles.
10. Which of the following statements is correct?
    a. _____ most of the continental margins around the Atlantic are passive; b. _____ oceanic ridges are composed of deformed sedimentary rocks; c. _____ the deposits of turbidity currents consist of clay; d. _____ most intermediate- and deep-focus earthquakes occur at oceanic ridges; e. _____ the salinity of seawater is expressed in parts per billion.
11. Which pair of ions are the most common in seawater?
    a. _____ potassium and magnesium; b. _____ chloride and sodium; c. _____ zinc and calcium; d. _____ calcium and sulfate; e. _____ fluoride and chloride.
12. The most useful method of determining the structure of the oceanic crust beneath continental shelf sediments is:
    a. _____ echo sounding; b. _____ observations from submersibles; c. _____ dredging; d. _____ seismic profiling; e. _____ underwater photography.
13. Which of the following is not characteristic of an active continental margin?
    a. _____ volcanism; b. _____ earthquakes; c. _____ oceanic trench; d. _____ continental rise; e. _____ mountains.
14. Graded bedding is typical in:
    a. _____ coral reefs; b. _____ manganese nodules; c. _____ turbidity current deposits; d. _____ deep-sea clay; e. _____ glacial-marine deposits.
15. How do mineral deposits form adjacent to hydrothermal vents on the sea floor?
16. Describe the continental rise, and explain why a rise is found at some continental margins and not at others.
17. Where do abyssal plains most commonly develop? What are they composed of?
18. What is the significance of oceanic trenches, and where are they found?
19. How do oceanic ridges differ from mountain ranges on land?
20. Describe the sequence of events leading to the origin of an atoll.
21. What is the Exclusive Economic Zone? What kinds of resources are found within it?

1. The concentration of dissolved solids in the Dead Sea is about 315,000 parts per million. What is its salinity? How much saltier is it than normal seawater?
2. Which ocean basin, the Pacific or Atlantic, is becoming larger? Explain.
3. How does a passive continental margin form, and why are such margins so much wider than active continental margins?
4. How is it possible for dissolved substances to be added to the oceans continuously and yet for the salinity of ocean water to remain constant?

Anderson, R. N. 1986. *Marine geology.* New York: John Wiley & Sons.

Bishop, J. M. 1984. *Applied oceanography.* New York: John Wiley & Sons.

Davis, R. A. 1987. *Oceanography: An introduction to the marine environment.* Dubuque, Iowa: W. C. Brown.

Edmond, J. M., and K. Von Damm. 1983. Hot springs on the ocean floor. *Scientific American* 248, no. 4: 78–93.

Gass, I. G. 1982. Ophiolites. *Scientific American* 247, no. 2: 122–31.

Kennett, J. P. 1982. *Marine geology.* Englewood Cliffs, N.J.: Prentice-Hall.

Lutz, R. A., and R. M. Haymon. 1994. Rebirth of a deep-sea vent. *National Geographic,* 186, no. 5: 114–26.

MacDonald, J. B., and P. J. Fox. 1990. The mid-ocean ridge. *Scientific American* 262, no. 6: 72–79.

Mark, K. 1976. Coral reefs, seamounts, and guyots. *Sea Frontiers* 22, no. 3: 143–49.

Nicolas, A. 1995. *The mid-oceanic ridges: Mountains below sea level.* New York: Springer-Verlag.

Pinet, P. 1992. *Oceanography: An introduction to the planet oceanus.* St. Paul, Minn.: West Publishing Company.

Rona, P. A. 1986. Mineral deposits from sea-floor hot springs. *Scientific American* 254, no. 1: 84–93.

Ross, D. A. 1988. *Introduction to oceanography.* Englewood Cliffs, N.J.: Prentice-Hall.

Thurman, H. V. 1988. *Introductory oceanography,* 5th ed. Columbus, Ohio: Merrill Publishing Company.

Tolmazin, D. 1985. *Elements of dynamic oceanography.* Boston, Mass.: Allen & Unwin.

Yulsman, T. 1996. The seafloor laid bare. *Earth* 5, no. 3: 42–51.

WORLD WIDE WEB SITES

For current updates and exercises, log on to http://www.wadsworth.com/geo.

WORLD DATA CENTER-A MARINE GEOLOGY & GEOPHYSICS
http://www.ngdc.noaa.gov/mgg/aboutmgg/wdcarngg.html

This site mainly provides databases and database searches on the world's oceans. It also provides links to other related sites concerned with marine geology.

LAMONT-DOHERTY EARTH OBSERVATORY OF COLUMBIA UNIVERSITY
http://www.ldeo.columbia.edu/

This site contains extensive amounts of information and images of their many and varied ongoing research projects in marine geology and geophysics.

CHAPTER

13

Ocean Circulation, Shorelines, and Shoreline Processes

Waves pounding the shoreline of Maui in the Hawaiian Islands.

Prologue

Wind-generated waves, especially those formed during storms, are responsible for most geologic work on shorelines, but tsunami and landslide surges can have disastrous effects. As explained in Chapter 8, tsunami are generated by fault displacement of the sea floor, submarine slides and slumps, and explosive volcanic eruptions. One of the most destructive tsunami ever recorded was caused by the explosive eruption of Krakatoa in 1883.

In the open ocean, tsunami may pass unnoticed, but when they enter shallow coastal waters, their wave height increases to as much as 65 m (213 ft)! The first indication of an approaching tsunami is a rapid withdrawal of the sea from coastal regions, followed a few minutes later by destructive waves. In some cases, tsunami come in as rapidly rising tides, and their backwash, which undermines structures and carries loose objects out to sea, causes most of the damage and fatalities.

The largest of all waves occur in restricted bodies of water, such as bays or lakes, when water is suddenly displaced by large landslides or rockfalls. Nearly 3,000 people drowned in Italy in 1963 when a slide into the Vaiont Reservoir caused a huge wave (see Perspective 4.2). The largest of these so-called *landslide surges* took place on July 9, 1958, in Lituya Bay, Alaska, when an estimated 30.5 million m^3 of rock plunged into the bay from a height of more than 900 m (2,950 ft). The sudden displacement of water caused a 536 m (1,758 ft) high surge on the opposite side of the bay.

Several coastal areas in the United States have been devastated by storm surges during hurricanes. In 1900, Galveston, Texas, a town of 38,000 on a long narrow barrier island a short distance from the mainland, was nearly destroyed. Storm waves surged inland and eventually covered the entire island, killing between 6,000 and 8,000 people. To protect the city from future flooding, a colossal two-part project was begun in 1902. First, a huge seawall was constructed to protect the city from waves (☐ Fig. 13.1). The second part of the project entailed raising parts of the city to the level of the top of the seawall (☐ Fig. 13.2).

To gain some perspective on the magnitude of the problem of coastal flooding during storms, consider this: of the nearly $2 billion paid out by the federal government's National Flood Insurance program since 1974, most has gone to owners of beachfront homes.

☐ FIGURE 13.1 Construction of this seawall to protect Galveston, Texas, from storm waves began in 1902.

☐ FIGURE 13.2 Some of the nearly 3,000 buildings in Galveston, Texas, that were raised and supported on stilts until sand fill was pumped beneath them.

INTRODUCTION

The ocean's surface is rarely calm; its waters are nearly always in motion as a result of surface currents, waves, and tides. Wind generates surface currents resulting in large-scale circulation as enormous quantities of water are transported great distances. Surface currents, such as the Gulf Stream, also transfer heat from the equatorial regions toward the poles and thus have a modifying effect on climate.

Deep-ocean waters are also constantly in motion. Unlike the wind-driven surface currents, deep-ocean currents are caused mostly by density differences between adjacent water masses. Most deep-ocean circulation carries water from the polar regions toward the equator. Surface currents and deep-ocean circulation both result in horizontal movement of water, but circulation also occurs vertically where water is transferred from the surface to depth, or from depth to the surface. Water transfer from depth to the surface, or *upwelling,* has important biological and economic consequences.

Wind blowing over the ocean's surface generates waves that transfer energy in the direction of wave advance. When waves approach a shoreline, they commonly become higher and steeper and eventually curl over as *breakers,* thereby expending their energy on the shoreline. Waves also produce nearshore currents that flow parallel to shorelines; the combined actions of waves and these nearshore currents have profound effects on shorelines. Accordingly, shorelines are in a continuous state of change as their materials are eroded, transported, and deposited.

The twice daily rise and fall of the ocean's surface, or the tides, are controlled by the gravitational attraction of the Sun and especially the Moon. Most coastal areas experience two high and two low tides daily, but the number of daily tidal fluctuations and the tidal range are complicated by the combined effects of the Sun and Moon and by shoreline configuration. Compared to waves and wave-generated nearshore currents, tides have little effect in modifying shorelines.

WIND-DRIVEN SURFACE CURRENTS

When wind blows over water, some of its energy is transferred to the water, causing both surface currents and oscillations of the water surface known as waves. The mechanism whereby energy is transferred from wind to water is related to the frictional drag resulting from one fluid (air) moving over another fluid (water). The friction or stress created in this manner generates a surface current with a velocity of only about 3 or 4% of the wind speed. The influence of wind diminishes rapidly with depth, and no direct effect of wind is felt at depths exceeding about 100 m (328 ft).

A map showing global surface-water current patterns is simplified because it represents surface flow averaged over a long time (▣ Fig. 13.3). At any one place at any particular time, the actual flow may deviate markedly from the average because of temporary effects such as local storms. In any case, a global pattern of surface currents is generated by wind and modified by the Coriolis effect and the distribution of land and sea.

If Earth did not rotate, ocean surface currents would move toward the equator in a straight line, but because it does rotate, they are deflected to the right of their direction of motion (clockwise) in the Northern Hemisphere and to the left of their direction of motion (counterclockwise) in the Southern Hemisphere (see Fig. 17.6). This deflection is known as the Coriolis effect. The combination of wind stress and the Coriolis effect produces large-scale water circulation systems known as **gyres** between the 60° parallels in the North and South Atlantic and Pacific Oceans and in the Indian Ocean (Fig. 13.3).

Unlike wind, which can blow unimpeded across both land and water, ocean currents are restricted to the ocean basins and are strongly controlled by the distribution of land and sea. When they approach land, currents may be split in complex ways and be deflected and flow in directions unrelated to the prevailing wind direction. So, in addition to wind and the Coriolis effect, another major control on ocean surface currents is the configuration of the ocean basins.

Examination of Figure 13.3 reveals that the gyres in the Northern and Southern Hemispheres are somewhat similar except that they rotate in opposite directions. As will be discussed in more detail in Chapter 17, air circulation resulting from differential heating of the ocean surface and Earth's rotation results in latitudinal wind belts. The gyres shown in both hemispheres in Figure 13.3 are driven by the westerlies of the mid-latitudes and the trade winds of the subtropics (see Fig. 17.14). The currents within these gyres vary considerably, however. The *Gulf Stream* in the northwestern Atlantic and the Kuroshio Current in the northwestern Pacific are the strongest with velocities of 3 to 4 km/hr (1.8 to 2.5 mi/hr) and are typically narrow, usually measuring no more than 50 to 75 km (31 to 46 mi) wide. By contrast, the eastern arms of these gyres, the Canary and California Currents, respectively, are hundreds of kilometers wide and rarely flow at more than 1 km/hr (0.6 mi/hr).

The **Gulf Stream,** one of the best-studied currents, was recognized as early as 1513 by Ponce de León, and by 1519 it was well known to those sailing between Spain and America. During the 1600s and 1700s, maps and charts of the Gulf Stream were published in varying quality, including one in 1777 by Timothy Folger and Benjamin Franklin.

The Gulf Stream along the east coast of North America is part of a much larger surface current system, the North Atlantic gyre (Fig. 13.3). In the South Atlantic, the South Equatorial Current is split into two parts by the eastward projection of Brazil. The part flowing north merges with the North Equatorial Current, which then splits into two water masses that rejoin as they exit the Gulf of Mexico between Florida and Cuba and become the Florida Current (Fig. 13.3). As this current flows northeasterly from about Cape

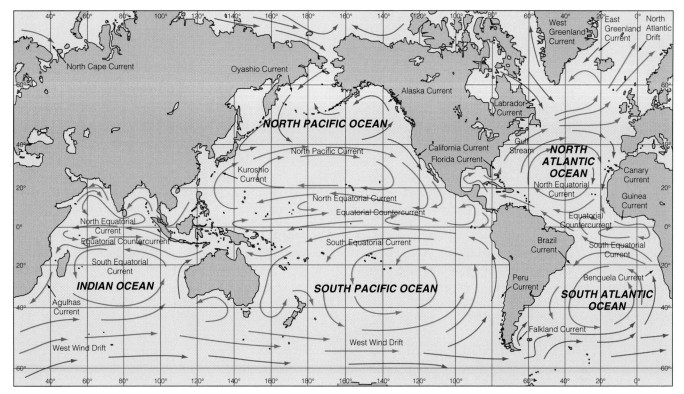

■ FIGURE 13.3 Global surface ocean currents. Warm currents are shown in red; cold currents in blue. The centers for the major current circulation patterns, or gyres, roughly correspond to the oceanic centers of the subtropical highs.

Hatteras, North Carolina, it becomes the Gulf Stream, which in this area flows as fast as 9 km/hr (5.5 mi/hr). Near Chesapeake Bay it carries a volume of water exceeding 90 million m³/sec, although this figure falls to about 40 million m³/sec by the time it reaches southern Newfoundland.

The Gulf Stream becomes the North Atlantic Current at about 40° north latitude and 45° west longitude and flows easterly across the North Atlantic. Mixing of the cold waters of the Labrador Current with the warm water of the North Atlantic Current results in persistent fog at their juncture. Mixing also results in the splitting of the North Atlantic Current into two branches. One branch continues northeastward past Iceland and Norway, while a southerly flowing branch called the Canary Current flows along the west coasts of Spain and North Africa and merges with the North Atlantic Equatorial Drift, thus completing the circulation of the North Atlantic gyre (Fig. 13.3). Within the central part of the North Atlantic gyre lies a vast area known as the Sargasso Sea (see □ Perspective 13.1).

In the Southern Hemisphere, the gyres in the South Pacific and South Atlantic are similar to those of the north, except for their opposite deflection in a counterclockwise direction by the Coriolis effect. These gyres are also narrowest and flow most rapidly along their western margins and are broad and sluggish along their northerly flowing arms. The gyre in the Indian Ocean varies more than the others as a result of seasonal reversal of monsoon winds.

In addition to the gyres in the South Pacific and South Atlantic Oceans, the prevailing westerlies also generate the

Antarctic Circumpolar Current. Rather than rotating in a gyre, this current encircles Antarctica because unlike other currents it is not deflected by land.

Wind-driven surface currents are also important as a major world temperature control (see Chapter 17). Seawater absorbs huge quantities of heat and transports it to high latitudes in warm currents. Cold currents, on the other hand, originate at high latitudes and flow toward the equator. In general, the temperature-modifying effects of warm currents are most notable on the east coasts of continents, whereas cold currents tend to flow toward the equator along west coasts (Fig. 13.3).

 UPWELLING AND DOWNWELLING

In the preceding section, we discussed surface currents, which involve horizontal circulation. Vertical circulation takes place when **upwelling** slowly transfers cold water from depth to the surface and when **downwelling** transfers warm surface water to depth. Upwelling is the more important of these processes, so we will consider it in the most detail.

Upwelling not only transfers water from depth to the surface, but also carries nutrients, particularly nitrates and phosphate, into the photic zone. Here, high plankton concentrations are sustained that in turn support other organisms. In fact, other than the continental shelves and areas adjacent to hydrothermal vents on the sea floor, areas of upwelling are the only parts of the oceans where biologic productivity is

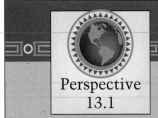

The Sargasso Sea

The Sargasso Sea is a lens of slowly clockwise-rotating warm water about 500 m (1,640 ft) thick bounded by currents of the North Atlantic gyre (☐ Fig. 1). Its name is derived from the brown seaweed *Sargassum* that floats over much of its surface. It measures roughly 3,200 km by 1,600 km (1,985 by 990 mi) and experiences less wind and rain than adjacent parts of the Atlantic.

Christopher Columbus is the first known traveler to pass through the Sargasso Sea, and since that time it has taken on an air of mystery that persists to the present. Although Columbus reported the floating seaweed accurately, stories told by superstitious sailors soon gave the area a reputation for mysterious events. It was believed that ships were becalmed in the sea not because of the lack of wind, but because thick, impenetrable mats of floating seaweed stalled and grew over the ships, holding them fast. The crews then reportedly starved or died of thirst and only their skeletons remained, or they were consumed by sea monsters.

Adding to the mystery were tales of sea monsters that could drag a ship to the depths, and the lack of air in the area so that crews reportedly suffocated. As recently as 1970, one author mentioned in a footnote that "The Sargasso Sea is still imperfectly known, since propeller-driven ships do not venture into it, because of the accumulation of seaweed it contains."[*] Another contemporary report claims that the lost continent of Atlantis lies beneath the mysterious waters of the Sargasso Sea.

Despite such claims, there is little mysterious about the Sargasso Sea. The legend that the sea is a graveyard of ships arose because a number of derelict ships have, in fact, been found floating there—but not because they were trapped in

seaweed. Because the Sargasso Sea lies within the North Atlantic gyre, debris that floats into the sea, including derelict ships, usually remains there until it sinks. Despite claims to the contrary, propeller-driven ships and even sailing ships have no difficulty passing through the sea. The floating seaweed is so dispersed that it impedes ship traffic not at all. The seaweed itself is derived from shallow waters in the Caribbean Sea, where it is ripped loose from the bottom and set adrift. The two species of *Sargassum* that constitute most of the seaweed are buoyed by gas-filled floats, so the seaweed remains in the photic zone and provides a habitat for numerous creatures that have adapted to a drifting lifestyle.

Even though the myths about the Sargasso Sea are unfounded, it is still interesting for several reasons. One is the floating seaweed with its peculiar assemblage of organisms, which might give the impression that such biological diversity indicates richly productive waters. In fact, the Sargasso Sea is the marine equivalent of a barren wasteland. Sinking water at the center of the sea inhibits mixing with deep nutrient-rich waters, so its warm, blue waters are nearly devoid of the microscopic plants essential to organic productivity. If the small crabs, snails, worms, and other animals left the seaweed, they would find themselves in a vast sea with few nutrients.

The Sargasso Sea is also the only known breeding ground of North American and European eels, both of which live in fresh water but reproduce in the ocean. These eels migrate downstream to the ocean and then travel by an unknown route to the Sargasso Sea. There they reproduce at depths of 400 to 700 m (1,130 to 2,995 ft); the adults then die, and the larvae float in the surface currents of the Sargasso Sea for one to three years. When the larvae finally reach coastal waters, they metamorphose into elvers, which resemble the adults, and then migrate upstream to lakes and ponds where they stay until they repeat the reproductive cycle.

[*]S. Hutin, *Alien Races and Fantastic Civilizations* (New York: Berkeley Publishing Co., 1970), p. 148.

---

very high. It is no accident that some of the most productive fisheries are in regions of upwelling. Less than 1% of the ocean surfaces are areas of upwelling, yet they support more than 50% (by weight) of all fishes.

Oceanographers generally recognize three types of upwelling: polar, coastal, and equatorial (☐ Fig. 13.4). Near Antarctica, floating ice shelves 100 to 200 m (328 to 656 ft) thick cover broad areas such as the Weddell and Ross Seas. As water freezes to the base of these ice sheets the remaining near-surface water becomes saltier and denser. Much of this

denser water sinks or downwells and is replaced by upward-circulating Circumpolar Deep Water (Fig. 13.4a). These upwelling, nutrient-rich, cold waters support a thriving biota including abundant phytoplankton and zooplankton, which in turn support large populations of fishes, squid, penguins, seals, and whales.

Most coastal upwelling takes place along the west coasts of Africa, North America, and South America, although one notable exception is present in the Indian Ocean. Coastal upwelling occurs when water moves offshore and is replaced

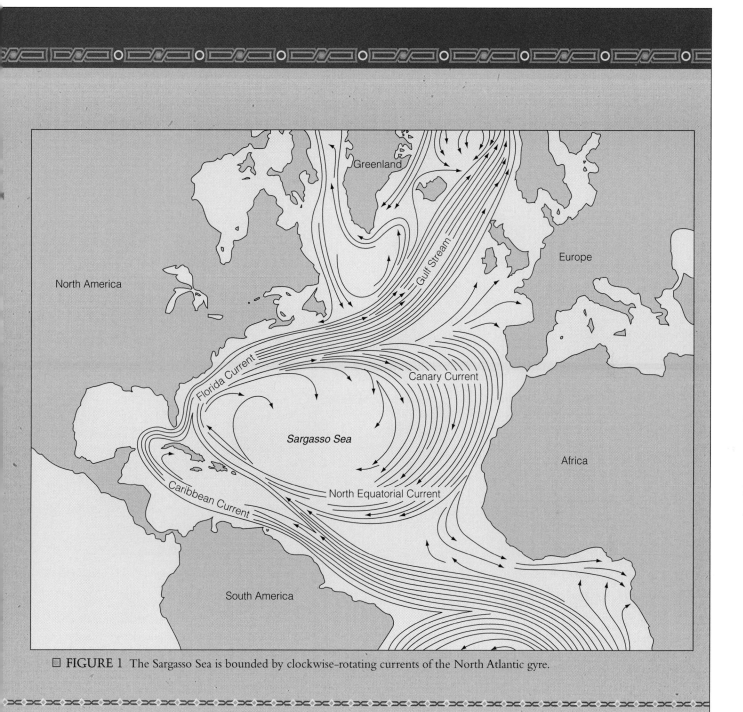

□ FIGURE 1 The Sargasso Sea is bounded by clockwise-rotating currents of the North Atlantic gyre.

by water rising from depth (Fig. 13.4b). For example, the southerly winds blowing along the coast of Peru coupled with the Coriolis effect transport water in a seaward direction, and cold, nutrient-rich water from depth rises to replace it. This area, too, is a major fishery. Every three to seven years changes in the surface water in this area are associated with the onset of El Niño, a weather phenomenon with far-reaching consequences (see Chapter 17).

The circulation in the equatorial regions is very complex, but equatorial upwelling is caused mostly by the westward-flowing North and South Equatorial currents (Fig. 13.4c). The North Equatorial Current is deflected to the right in the Northern Hemisphere, whereas the South Equatorial Current is deflected to the left in the Southern Hemisphere. As a result, the warm equatorial waters diverge; as horizontal flow occurs to both the north and the south, cold water rises to replace it (Fig. 13.4c). By contrast, in the central parts of gyres, such as the Sargasso Sea, currents converge, a bulge develops on the water surface, and downwelling takes place.

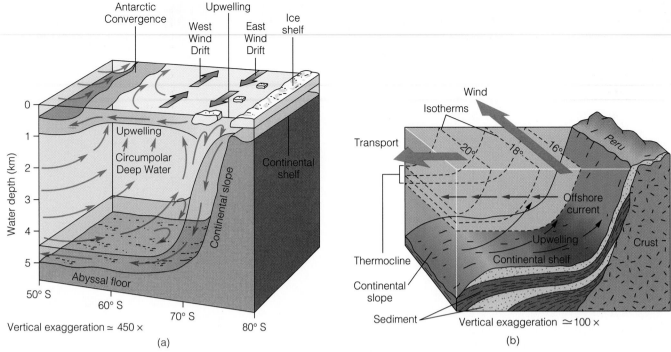

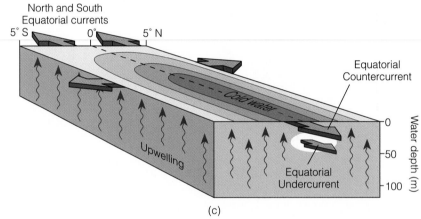

FIGURE 13.4 Three types of upwelling. (*a*) As ice freezes to the base of floating ice sheets near Antarctica, the water becomes saltier and denser. It then sinks or downwells and is replaced by upwelling cold water. (*b*) Coastal upwelling. As warm surface water is blown offshore by the wind, it is replaced by upwelling cold bottom waters. (*c*) Upwelling where currents diverge.

 DEEP-OCEAN CIRCULATION

Wind-generated gyres are restricted to the surface layers of water, but even deep-ocean water, though not directly affected by wind, is constantly in motion. Density differences caused by temperature and salinity changes generate these subsurface currents, a phenomenon known as **thermohaline circulation.** Circulation resulting from density differences operates on a simple principle: a water mass of greater density (colder or more saline) will displace and flow beneath a water mass of lesser density. Deep-ocean circulation affects about 90% of all ocean water, yet because studying it is expensive and time-consuming, oceanographers know less about it than about other ocean movements.

Either the gain or the loss of water at the surface can result in density changes. For instance, where large rivers enter the ocean, salinity is reduced as freshwater and seawater mix, and salinity is also reduced in equatorial areas of high rainfall. When water evaporates, the remaining water becomes more saline, and its density increases. Likewise, water

chilled in the polar regions becomes denser and displaces warmer water. Once water becomes denser and begins to sink, its temperature and salinity remain remarkably constant, except at the margins of adjacent water masses where slight mixing occurs.

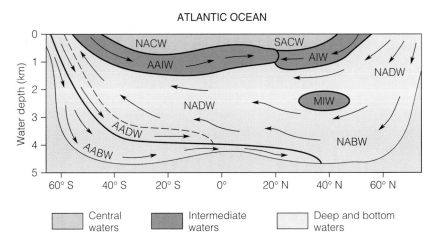

ATLANTIC OCEAN

▣ **FIGURE 13.5** North-south profile showing water-mass structure in the Atlantic Ocean. Deep ocean water transport results mainly from the formation of cold, dense water in the polar regions, which sinks, flows toward the equator, and displaces less dense water. NACW and SACW = North and South Atlantic central water; AAIW and AIW = Antarctic and Atlantic intermediate water; MIW = Mediterranean intermediate water; NADW and AADW = North Atlantic and Antarctic deep water; and NABW and AABW = North Atlantic and Antarctic bottom water.

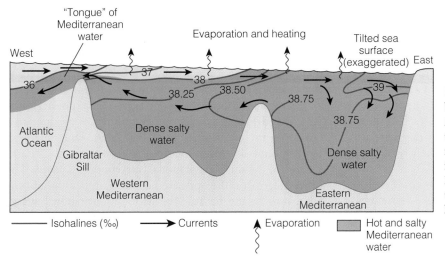

▣ **FIGURE 13.6** Pattern of water circulation between the Atlantic Ocean and the Mediterranean Sea. Because of high evaporation rates, water in the Mediterranean becomes more saline and denser. As a result, it sinks and flows into the Atlantic over the Gibraltar Sill. Less dense water from the Atlantic flows into the Mediterranean at the surface. Isohalines are lines of equal salinity.

Ocean water masses identified by their uniform conditions of temperature and salinity include *central waters, intermediate waters, deep waters,* and *bottom waters* (▣ Fig. 13.5). Central waters are the least dense and warmest; they extend down to the underside of the *thermocline,* which is a layer of water in which the temperature changes rapidly in a vertical direction. Ideally, these waters are underlain successively by intermediate waters, which extend to about 2 km (1.2 mi) deep, deep waters, and finally bottom waters, which are very cold and the densest of all oceanic waters (Fig. 13.5).

In the Atlantic Ocean, the central waters are underlain by Antarctic Intermediate Water and Arctic Intermediate Water, both of which are chilled in their respective polar regions, become denser, and move beneath the warmer central waters (Fig. 13.5). These intermediate waters in turn are underlain by North Atlantic Deep Water and Antarctic Deep Water, and finally by Antarctic and North Atlantic Bottom Waters (Fig. 13.5).

Both surface and subsurface circulation occur between the nearly landlocked Mediterranean Sea and the North Atlantic. Seawater in the Mediterranean evaporates in the intense Sun and is replenished by surface-water flow from the North Atlantic through the Strait of Gibraltar. Nevertheless, the water in the eastern end of the sea becomes more saline and thus denser. This denser water sinks and collects on the sea floor until it flows over the submarine ridge at the Strait of Gibraltar and moves as a subsurface current into the Atlantic as a tongue of warm salty water (▣ Fig. 13.6).

The fourfold division of water masses in a vertical dimension recognized in the Atlantic is absent in the Pacific Ocean. Here, as in the Atlantic, the central waters are underlain by intermediate waters, but below 2,000 m (6,560 ft), temperature and salinity are rather constant, indicating the absence of well-defined water masses (▣ Fig. 13.7). However, water that was mixed in the South Atlantic Ocean is injected into the Pacific by the powerful Antarctic Circumpolar Current. In any case, this water moves northward, some of it reaching about 20° N latitude. The rest of the water in the Pacific is designated as Pacific Subarctic Water (Fig. 13.7).

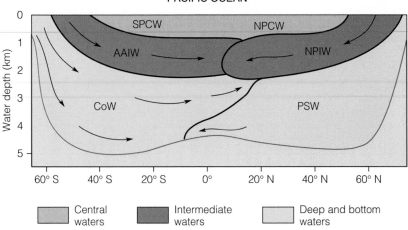

FIGURE 13.7 North-south profile showing water-mass structure in the Pacific Ocean. As in the Atlantic, cold, dense water sinks and flows along the sea floor toward the equator. SPCW and NPCW = South and North Pacific central water; AAIW and NPIW = Antarctic and North Pacific intermediate water; CoW = common water; and PSW = Pacific Subarctic water.

SHORELINES AND SHORELINE PROCESSES

Shorelines are the areas between low tide and the highest level on land affected by storm waves. We are concerned mostly with ocean shorelines where tides, waves, and nearshore currents continually modify existing shoreline features. However, waves and nearshore currents are also effective processes in large lakes, the shorelines of which exhibit many of the same features as seashores. The most notable differences are that waves and nearshore currents are more energetic on seashores, and even the largest lakes lack appreciable tides.

The continents possess more than 400,000 km (248,160 mi) of shorelines. They vary from rocky, steep shorelines, such as those in Maine and much of the western United States and Canada, to broad sandy beaches as in eastern North America from New Jersey southward. Whatever their type, on all shorelines a continual interplay exists between the energy levels of shoreline processes and the shoreline materials. In areas where energy levels are particularly high, erosion predominates and the shoreline may retreat landward. Where sediment supply from the land is great, deposition dominates and the shoreline may build seaward. On shorelines with broad sandy beaches, beach sand is continually shifted from one area to another by waves and nearshore currents.

TIDES

On sea coasts the surface of the ocean rises and falls once or twice daily in response to the gravitational attraction of the Moon and Sun. These regular fluctuations in the sea's surface are **tides.** Most shoreline areas experience two high tides and two low tides daily as sea level rises and falls anywhere from a few centimeters to more than 15 m (49 ft) (▣ Fig. 13.8). During rising or *flood tide,* more and more of the shoreline area is flooded until high tide is reached. During *ebb tide,* currents flow seaward exposing some of the shoreline area.

Both the Moon and the Sun have sufficient gravitational attraction to exert tide-generating forces strong enough to deform the solid body of Earth, but they have a much greater influence on the oceans. The Sun is 27 million times more massive than the Moon, but it is 390 times as far from Earth, and its tide-generating force is only 46% as strong as that of the Moon. Accordingly, the tides are dominated by the Moon, but the Sun plays an important role as well.

If we consider only the Moon acting on a spherical, water-covered Earth, the tide-generating forces produce two bulges on the ocean surface (▣ Fig. 13.9). One bulge extends toward the Moon because it is on the side of Earth where the Moon's gravitational attraction is greatest. The other bulge is on the opposite side of Earth because of centrifugal force and because the Moon's gravitational attraction is least there. These two bulges point toward and away from the Moon (Fig. 13.9), so as Earth rotates and the Moon's position changes, an observer at a particular shoreline location experiences the rhythmic rise and fall of tides twice daily. The heights of two successive tides may vary depending on the Moon's inclination with respect to the equator.

The Moon revolves around Earth in about 28 days, so its position with respect to any latitude changes slightly each day. That is, as the Moon moves in its orbit and Earth rotates on its axis, it takes the Moon 50 minutes longer each day to return to the same position it was in the previous day. Thus, an observer would experience a high tide at 1:00 P.M. one day, for example, and at 1:50 P.M. on the following day.

Tides are also complicated by the combined effects of the Moon and Sun. Even though the Sun's tide-generating force is weaker than the Moon's, when the Moon and Sun are aligned every two weeks, their forces are added together and generate *spring tides,* which are about 20% higher than average tides (Fig. 13.9b). When the Moon and Sun are at right angles to one another, also at two-week intervals, the Sun's tide-generating force cancels some of the Moon's, and shorelines experience *neap tides,* which are about 20% lower than average (Fig. 13.9c).

Tidal ranges are also affected by shoreline configuration. Broad, gently sloping continental shelves as in the Gulf of

FIGURE 13.8 (*a*) Low tide and (*b*) high tide.

(a)

(b)

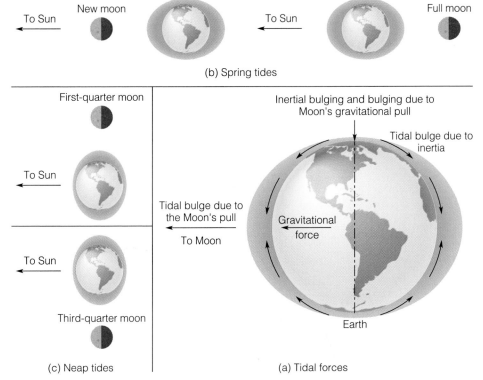

New moon

To Sun

Full moon

(b) Spring tides

To Sun

First-quarter moon

To Sun

To Sun

Third-quarter moon

(c) Neap tides

Inertial bulging and bulging due to
Moon's gravitational pull

Tidal bulge due to
inertia

Tidal bulge due to
the Moon's pull

To Moon

Gravitational
force

Earth

(a) Tidal forces

FIGURE 13.9 (*a*) Tides are
caused by the gravitational pull of the
Moon and, to a lesser degree, the Sun.
The Earth-Moon-Sun alignments at
the times of (*b*) spring and (*c*) neap
tides.

Mexico have low tidal ranges, whereas steep, irregular shorelines experience a much greater rise and fall of tides. Tidal ranges are greatest in some narrow, funnel-shaped bays and inlets, as in the Bay of Fundy, Nova Scotia, where a tidal range of 16.5 m (54 ft) has been recorded. Tides have an important impact on shorelines because the area of wave attack constantly shifts onshore and offshore as the tides rise and fall. Tidal currents themselves, however, have little modifying effect on shorelines, except in some narrow passages where tidal current velocity is great enough to erode and transport sediment.

 # WAVES

Oscillations of a water surface, called **waves,** occur on all bodies of water, but are most significant in large lakes and the oceans. Many of the erosional and depositional features of the world's shorelines form and are modified by the energy of incoming waves. ▣ Figure 13.10 shows a typical series of waves in deep water and the terminology applied to them. The highest part of a wave is its *crest,* whereas the low point between crests is the *trough.* *Wave length* is the distance between successive wave crests (or troughs), and *wave height* is the vertical distance from trough to crest. The speed at which a wave advances is actually a measure of the velocity of the wave form rather than a measure of the speed of the molecules of water. In fact, the water in deep-water waves moves forward and back as a wave passes but has little net forward movement.

As waves move across a water surface, the water "particles" rotate in circular orbits and transfer energy in the direction of wave advance (Fig. 13.10). The diameters of the orbits followed by water particles in waves diminish rapidly with depth, and at a depth of about one-half wave length, called **wave base,** they are essentially zero (Fig. 13.10). At depths exceeding wave base, the water and sea floor, or lake floor, are unaffected by surface waves. The significance of wave base will be explored more fully in later sections.

Wave Generation

When wind blows over water, some of its energy is transferred to the water, causing the water surface to oscillate. In an area of wave generation, as beneath a storm center at sea, sharp-crested, irregular waves called *seas* develop. Seas are an aggregate of waves of various heights and lengths, and one wave cannot be clearly distinguished from another. As seas move out from the area of wave generation, they are sorted into broad *swells* that have long rounded crests and are all about the same size (▣ Fig. 13.11).

As one would expect, the harder and longer the wind blows, the larger are the waves generated. Wind velocity and duration, however, are not the only factors controlling the size of waves. High-velocity wind blowing over a small pond will never generate large waves regardless of how long it blows. In fact, waves are present on ponds and most lakes only while the wind is blowing; once the wind stops, the water surface quickly smooths out. In contrast, the ocean's surface is almost constantly in motion, and waves with

▣ **FIGURE 13.10** Waves and the terminology applied to them. The water in waves moves in circular orbits that decrease in size with depth. At wave base, which is at a depth of one-half wave length, water is not disturbed by surface waves. As deep-water waves move toward shore, the orbital motion of water within them is disrupted when they enter water shallower than wave base. Wave length decreases while wave height increases, causing the waves to oversteepen and collapse or plunge forward as breakers.

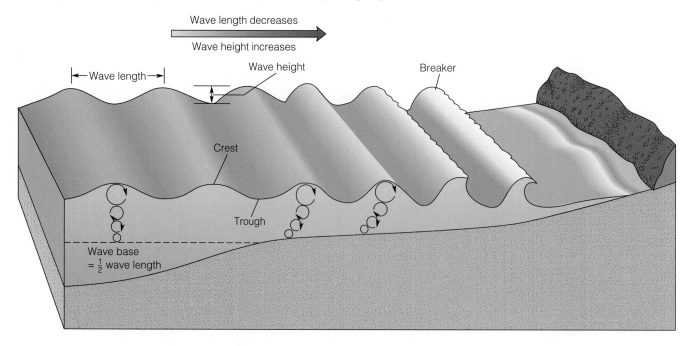

(a)

FIGURE 13.11 (a) Small swells in the Atlantic Ocean near Massachusetts. (b) Plunging breaker on the north shore of Oahu, Hawaii.

(b)

heights of 20 to 30 m (65 to 98 ft) have been recorded during storms.

The reason for the disparity between wave sizes on ponds and lakes and on oceans is the **fetch,** which is the distance the wind blows over a continuous water surface. Fetch is limited by the available water surface, so on ponds and lakes it corresponds to their length or width, depending on wind direction. A wind blowing the length of Lake Superior, about 560 km (345 mi), can generate large waves, but even larger ones develop in the oceans where the fetch is much greater.

Shallow-Water Waves and Breakers

Waves moving out from the area of generation form swells and lose only a small amount of energy as they travel across the ocean. In deep-water swells, the water surface oscillates and water particles move in orbital paths, with little net displacement of water in the direction of wave advance (Fig. 13.10). As these waves enter progressively shallower water, however, the water is displaced in the direction of wave advance.

When deep-water waves enter shallow water, they are transformed from broad, undulating swells into sharp-crested waves. This transformation begins at a water depth of wave base; that is, it begins where wave base intersects the sea floor. At this point, the waves "feel" the sea floor, and the orbital motions of water particles within the waves are disrupted (Fig. 13.10). As the waves move further shoreward, the speed of wave advance and wave length decrease, and wave height increases. In effect, as waves enter shallower water, they become oversteepened; the wave crest advances faster than the wave form, until eventually the crest collapses or plunges forward as a **breaker** (Fig. 13.11b). Breakers are

commonly several times higher than deep-water waves, and when they collapse or plunge forward, their kinetic energy is expended on the shoreline.

NEARSHORE CURRENTS

It is convenient to identify the *nearshore zone* as the area extending from the shoreline to just beyond the place where waves break. It includes a *breaker zone* and a *surf zone,* which is where breaking waves rush forward onto the shore followed by seaward movement of the water as backwash (Fig. 13.10). The width of the nearshore zone varies depending on the wave length of approaching waves, because long waves break at greater depth, and thus farther offshore, than do short waves.

Deep-water waves have long, continuous crests, but rarely are their crests parallel with the shoreline (Fig. 13.12). Accordingly, one part of a wave enters shallow water before other parts of the same wave. As a wave begins breaking, its velocity diminishes, but the part of the wave still in deep water races ahead, partly catching up, until it too encounters wave base. The net effect of this oblique

 FIGURE 13.12 Wave refraction. These waves approaching a shoreline at an angle are refracted and more nearly parallel the shore as they enter progressively shallower water. Refracted waves generate a longshore current that flows in the direction of wave approach, from right to left in this example.

approach is that waves bend, a phenomenon called **wave refraction,** so that they more nearly parallel the shoreline (Fig. 13.12).

Even though waves are refracted, they still usually strike the shoreline at some angle, causing the water between the breaker zone and the beach to flow parallel to the shoreline. These **longshore currents,** as they are called, are long and narrow and flow in the same general direction as the approaching waves (Fig. 13.12). Longshore currents are particularly important agents of transport and deposition in the nearshore zone.

SHORELINE DEPOSITION

Depositional features of shorelines include *beaches, spits, baymouth bars, tombolos,* and *barrier islands.* The nature of these deposits is determined largely by wave energy and longshore currents.

Beaches

Beaches are the most familiar of all coastal landforms, attracting millions of visitors each year and providing the economic base for many communities. By definition, a **beach** is a deposit of unconsolidated sediment, commonly sand, extending landward from low tide to a change in topography such as a line of sand dunes, a sea cliff, or where permanent vegetation begins. Depending on shoreline configuration and wave intensity, beaches may be discontinuous, existing only as *pocket beaches* in protected areas such as embayments, or they may be continuous for long distances (Fig. 13.13).

Some of the sediment on beaches is derived by weathering and wave erosion of the shoreline, but most is transported

 FIGURE 13.13 (*a*) A pocket beach in California. (*b*) The Grand Strand of South Carolina, shown here at Myrtle Beach, is 100 km (62 mi) of nearly continuous beach.

(a)

(b)

to the coast by streams and redistributed along the shoreline by **longshore drift,** which is the phenomenon of sand transport along a shoreline by longshore currents (▣ Fig. 13.14). As previously noted, waves usually strike beaches at some angle, causing sand grains to move up the beach at a similar angle; as the sand grains are carried seaward in the backwash, however, they move perpendicular to the long axis of the beach, so individual sand grains move in a zigzag pattern in the direction of longshore currents. This longshore drift is not restricted to the beach; it involves sand from the outer margin of the breaker zone to the upper limit of wave swash (Fig. 13.14a).

In an attempt to widen a beach or prevent erosion, shoreline residents often build *groins,* structures that project seaward at right angles to the shoreline (Fig. 13.14b). Groins interrupt the flow of longshore currents, causing sand deposition on their upcurrent sides and widening the beach at that location. However, erosion inevitably occurs on the downcurrent side of a groin.

A beach's profile (▣ Fig. 13.15) can be thought of as a profile of equilibrium; that is, all parts of the beach are adjusted to the prevailing conditions of wave intensity and longshore currents. Tides and longshore currents affect beaches to some degree, but by far the most important agent modifying their equilibrium profile is storm waves. In many areas beach profiles change with the seasons, so we can recognize *summer beaches* and *winter beaches,* each of which is adjusted to the conditions prevailing at these times (Fig. 13.15). Summer beaches are generally wide, covered with

sand, and characterized by a gently sloping, smooth offshore profile. Winter beaches, on the other hand, tend to be narrower, coarser grained, and steeper, and their offshore profiles reveal sand bars paralleling the shoreline (Fig. 13.15).

Seasonal changes in beach profiles are related to changing wave intensity. During winter, sand eroded from the beach by storm waves is transported offshore where it is stored in sand bars (Fig. 13.15a). The same sand eroded from the beach during winter returns the next summer when it is driven onshore by more gentle swells. The volume of sand in

▣ FIGURE 13.14 (*a*) Longshore drift involves sediment transport by longshore currents along the shoreline between the breaker zone and the upper limit of wave action. (*b*) These groins at Cape May, New Jersey, interrupt the flow of longshore currents so sand is trapped on their upcurrent side. On the downcurrent side of the groins, sand is eroded.

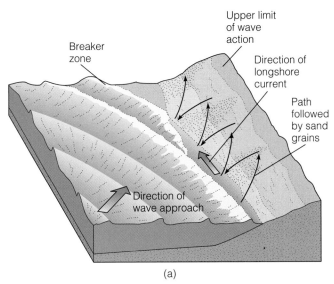

(a)

(b)

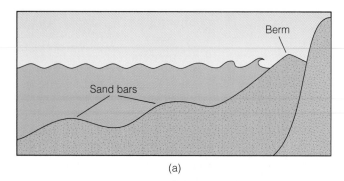

(a)

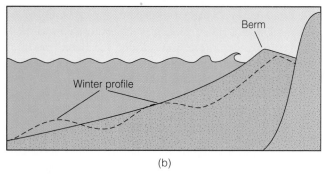

(b)

■ FIGURE 13.15 Seasonal changes in beach profiles. (a) A winter beach with offshore sand bars. (b) A wider summer beach with its more gently sloping profile.

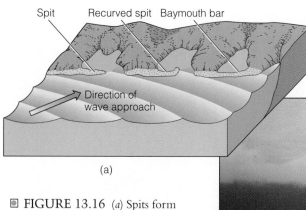

(a)

■ FIGURE 13.16 (a) Spits form where longshore currents deposit sand in deeper water, as at the entrance to a bay. A baymouth bar is simply a spit that has grown until it extends across the mouth of a bay. (b) A spit at the mouth of the Klamath River in California.

the system remains more or less constant; it simply moves offshore or onshore depending on wave energy.

Spits, Baymouth Bars, and Tombolos

Other common depositional landforms on shorelines are *spits* and *baymouth bars*, both of which are variations of the same feature. A **spit** is simply a continuation of a beach forming a point, or "free end," projecting into a body of water, commonly a bay. A **baymouth bar** is a spit that has grown until it completely closes off a bay from the open sea (■ Fig. 13.16).

Both spits and baymouth bars form and grow as a result of longshore drift. Where currents are weak, as in the deeper water at the opening to a bay, longshore current velocity diminishes, and sediment is deposited, forming a sand bar. The free ends of many spits are curved by wave refraction or waves approaching from a different direction. Such spits are called *hooks* or *recurved spits* (Fig. 13.16). A rarer type of spit, called a **tombolo,** extends out into the sea and connects an island to the mainland. Wave refraction around an island causes converging currents that turn seaward and deposit a sand bar connecting the shore with the island (■ Fig. 13.17).

Although spits, baymouth bars, and tombolos are most commonly found on seacoasts, many examples of the same features are found in large lakes. Whether along seacoasts or lakeshores, these sand deposits present a continuing problem where bays must be kept open for pleasure boating or commercial shipping. The entrances to such bays must either be protected or regularly dredged. The most common way to protect bay entrances is to build *jetties,* which are structures extending seaward (or lakeward) that protect the bay from deposition by longshore currents (■ Fig. 13.18).

(b)

(a)

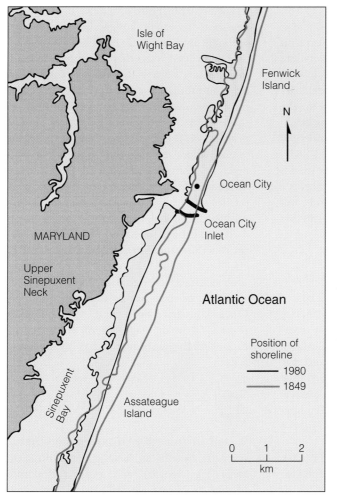

(b)

■ FIGURE 13.17 (*a*) Origin of a tombolo. Wave refraction around an island causes longshore currents to converge and deposit a sand bar that joins the island with the mainland. (*b*) A tombolo in Oregon.

■ FIGURE 13.18 The seaward-projecting heavy, black lines represent jetties constructed in the 1930s to protect the inlet at Ocean City, Maryland. The jetties have protected the inlet, but they also disrupt the net southerly longshore drift. As a result, Assateague Island has been starved of sediment and has migrated about 500 m (1,640 ft) landward and is now offset from Fenwick Island to the north. Both islands are barrier islands.

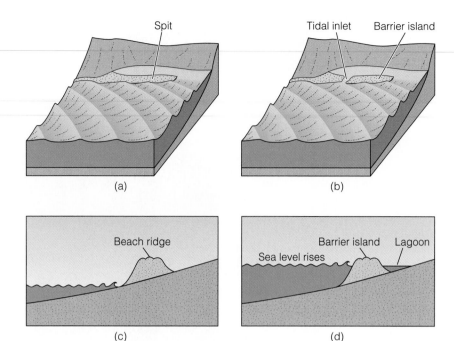

Spit

Tidal inlet   Barrier island

(a)

(b)

Beach ridge

(c)

Sea level rises   Barrier island   Lagoon

(d)

■ FIGURE 13.19 Two models for the origin of barrier islands. (*a*) Longshore currents extend a spit along a coast. (*b*) During a storm, the spit is breached, forming a tidal inlet and a barrier island. (*c*) A beach ridge forms on land, and when sea level rises (*d*) it is partly submerged, forming a barrier island.

Barrier Islands

Barrier islands are long, narrow islands composed of sand and separated from the mainland by a lagoon (Fig. 13.18). The origin of barrier islands has been long debated and is still not completely resolved. It is known that they form on gently sloping continental shelves with abundant sand where both tidal fluctuations and wave energy levels are low, as along the Gulf Coast of Texas. According to one model, barrier islands formed as spits that became detached from the land, while another model proposes they formed as beach ridges on coasts that were subsequently partly submerged when sea level rose (■ Fig. 13.19).

Because sea level is currently rising, most barrier islands are migrating in a landward direction. Such migration is a natural consequence of the evolution of these islands, but it poses a problem for island residents and communities (see □ Perspective 13.2 on page 308).

The Nearshore Sediment Budget

We can think of the gains and losses of sediment in a coastal zone in terms of a **nearshore sediment budget** (■ Fig. 13.20). If a nearshore system has a balanced budget, sediment is supplied as fast as it is removed, and the volume of sediment remains more or less constant, although sand may shift offshore and onshore with the changing seasons (Fig. 13.15). A positive budget means gains exceed losses, whereas a negative budget results when losses exceed gains. If a negative budget prevails long enough, a nearshore system is depleted and beaches may eventually disappear.

Erosion of sea cliffs provides some sediment to beaches, but in most areas probably no more than 5 to 10% of the total sediment supply is derived from this source. Most of the sediment on typical beaches is transported to the shoreline by streams and then redistributed along the shoreline by longshore drift. Thus, longshore drift also plays a role in the nearshore sediment budget because it continually moves sediment into and away from beach systems (Fig. 13.20).

The primary ways that a nearshore system loses sediment include longshore drift, offshore transport, wind, and deposition in submarine canyons (Fig. 13.20). Offshore transport mostly involves seaward transport of fine-grained sediment that eventually settles in deeper water. Wind is an important process because it removes sand from beaches and blows it inland where it commonly piles up as sand dunes. If the heads of submarine canyons are near shore, huge quantities of sand are funneled into them and deposited in deeper water. In most areas, though, submarine canyons are too far offshore to interrupt the flow of sand in the nearshore zone.

If a nearshore system is in equilibrium, its incoming supply of sediment exactly offsets its losses, and this delicate balance is maintained unless the system is somehow disrupted. One common change that affects this balance is the construction of dams across the streams supplying sediment. Once dams have been built, all sediment from the upper reaches of the drainage systems is trapped in reservoirs and cannot reach the shoreline, thereby depleting the nearshore system.

SHORELINE EROSION

Where erosion rather than deposition predominates along seacoasts, beaches are lacking or are found only in protected areas (Fig. 13.13a), and a sea cliff commonly develops (■ Fig. 13.21 on page 310). Sea cliffs are frequently pounded

Inputs	+	Outputs	=	Balance
Longshore transport into beach		Longshore transport out of beach		Accretion
River supply		Offshore transport		Erosion
Cliff erosion		Wind transportation into dunes		Steady state
Onshore transport				

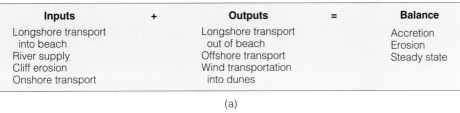

(a)

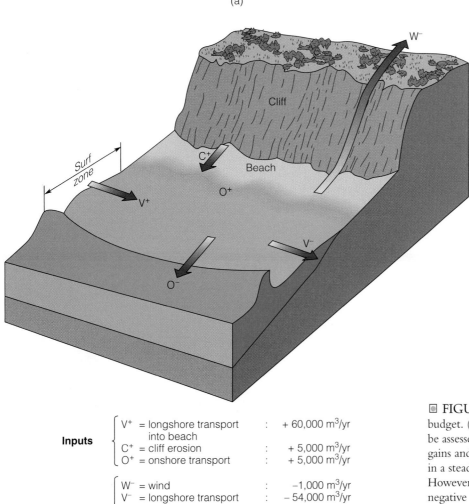

Inputs
- V^+ = longshore transport into beach : + 60,000 m^3/yr
- C^+ = cliff erosion : + 5,000 m^3/yr
- O^+ = onshore transport : + 5,000 m^3/yr

Outputs
- W^- = wind : –1,000 m^3/yr
- V^- = longshore transport out of beach : – 54,000 m^3/yr
- O^- = offshore transport (includes transport to submarine canyons) : – 20,000 m^3/yr

Balance : – 5,000 m^3/yr (net erosion)

(b)

■ FIGURE 13.20 The nearshore sediment budget. (*a*) The long-term sediment budget can be assessed by considering gains versus losses. If gains and losses are equal, a nearshore system is in a steady state or state of equilibrium. However, if losses exceed gains, the budget is negative and beaches are eroded. Accretion takes place when the nearshore system has a positive budget, meaning that gains exceed losses. (*b*) A hypothetical example of a negative nearshore sediment budget. In this example, net erosion occurs at the rate of 5,000 m^3/yr.

by waves, especially during storms, and the cliff is eroded as a result of *corrosion, hydraulic action,* and *abrasion.* Corrosion is an erosional process involving the wearing away of rock by chemical processes, especially the solvent action of seawater. The force of the water itself, called hydraulic action, is a particularly effective erosional process. Waves exert tremendous pressure on shorelines by direct impact, but are most effective on sea cliffs composed of sediment or rocks that are highly fractured. Abrasion is an erosion process involving the grinding action on rocks by sand and gravel carried by waves.

Wave-Cut Platforms and Associated Landforms

The rate at which a sea cliff erodes and retreats landward depends on wave intensity and the resistance of the coastal rocks or sediments. Most sea cliff retreat takes place during storms and, as one would expect, occurs most rapidly in sea cliffs composed of sediment. A sea cliff consisting of glacial drift on Cape Cod, Massachusetts, retreats as much as 30 m (98 ft) per century, and some parts of the White Cliffs of Dover in Great Britain are retreating at more than 100 m

Focus on the Environment: Rising Sea Level and Coastal Management

Shorelines in many parts of the world are eroding as sea level rises. According to one study, 54% of U.S. shorelines have yearly erosion rates ranging from millimeters to more than 10 m (33 ft) in a few areas (☐ Fig. 1). As sea level rises, buildings that were once some distance from the ocean are now being undermined and destroyed.

During the last century, sea level rose about 12 cm (4.7 in) worldwide, and all indications are that it will continue to rise. The absolute rate of sea-level rise in a particular shoreline region depends on two factors. The first is the volume of water in the ocean basins, which is increasing as a result of melting of glacial ice and thermal expansion of near-surface seawater. Many scientists think that sea level will continue to rise because of global warming resulting from concentrations of greenhouse gases in the atmosphere.

The second factor controlling sea level is the rate of uplift or subsidence of a coastal area. In some areas, uplift is occurring so fast that sea level is actually falling with respect to the land. In other areas, however, sea level is rising while the coastal region is simultaneously subsiding, resulting in a net change in sea level of as much as 30 cm (12 in) per century. Perhaps such a "slow" rate of sea-level rise seems insignificant; after all, it amounts to only a few millimeters per year. But in gently sloping coastal areas, as along much of the eastern and Gulf coasts of the United States, even a slight rise in sea level would eventually have widespread effects.

Many of the nearly 300 barrier islands along the east and Gulf coasts of the United States are migrating landward as a result of rising sea level. During storms, their seaward sides are eroded by large waves that carry sand over the islands and into their lagoons, resulting in a gradual landward shift of the entire island complex (☐ Fig. 2). During the last 120 years, Hatteras Island, North Carolina, has migrated nearly 500 m (1,640 ft) landward so that Cape Hatteras lighthouse, which was 460 m (1,510 ft) from the shoreline when it was built in 1870, now stands on a promontory in the Atlantic Ocean.

Rising sea level also directly threatens many beaches that communities depend on for revenue. The beach at Miami

☐ FIGURE 1 Shoreline erosion in the United States. No data are available for the uncolored parts of the shoreline.

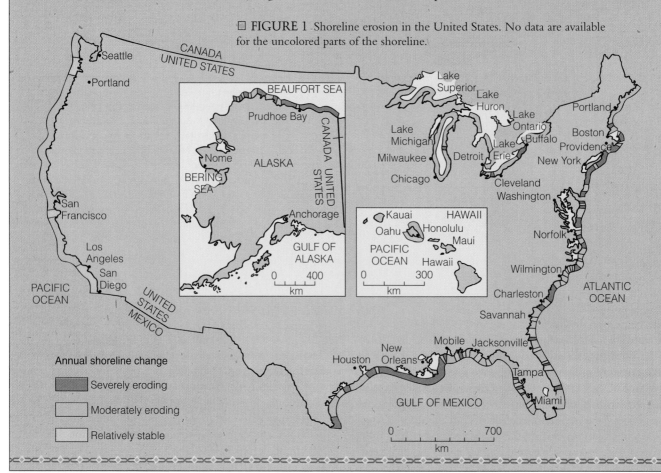

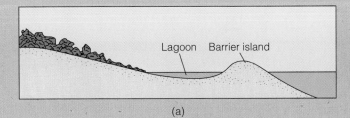

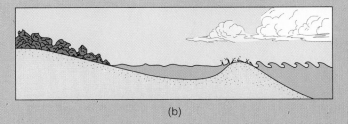

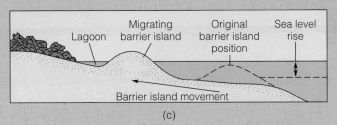

□ **FIGURE 2** Rising sea level and the landward migration of barrier islands. (*a*) Barrier island before landward migration in response to rising sea level. (*b*) Landward movement occurs when storm waves wash sand from the seaward side of the islands and deposit it in the lagoon. (*c*) Over time, the entire complex migrates landward.

Beach, Florida, was disappearing at an alarming rate until the Army Corps of Engineers began replacing the eroded beach sand (□ Fig. 3). The problem is even more serious in other countries. A rise in sea level of only 2 m (6.5 ft) would inundate large areas of the east and Gulf coasts, but would cover 20% of the entire country of Bangladesh. Other problems associated with sea-level rise include increased coastal flooding during storms and saltwater incursions that may threaten groundwater supplies.

Since nothing can be done to prevent sea level from rising, engineers and scientists must examine what can be done to minimize the effects of shoreline erosion. At present, only a few options exist. One is to put strict controls on coastal development. North Carolina permits large structures to be sited no closer to the shoreline than 60 times the annual erosion rate. Although a growing awareness of shoreline processes has resulted in similar legislation elsewhere, some states have virtually no restrictions on coastal development.

Regulating coastal development is commendable, but it has no impact on existing structures and coastal communities. Moving individual dwellings and small communities inland is possible, though expensive, but it is impractical for large population centers. Communities such as Atlantic City, New Jersey, Miami Beach, Florida, and Galveston, Texas, have adopted one of two strategies to combat coastal erosion. One is to build protective barriers such as seawalls. Seawalls, such as the one at Galveston, Texas (see the Prologue), can be effective, but they are tremendously expensive to construct and maintain. Furthermore, seawalls retard erosion only in the area directly behind them; Galveston Island west of the seawall has been eroded back about 45 m (148 ft) in the several decades since it was completed.

Another option, adopted by both Atlantic City and Miami Beach, is to pump sand onto beaches to replace that lost to erosion (Fig. 3). This, too, is expensive as the sand must be replenished periodically because erosion is a continuing process. Ocean City, Maryland, has used both options; more than $50 million was spent in just five years to replenish the beach and build a seawall.

□ **FIGURE 3** The beach at Miami Beach, Florida (*a*) after and (*b*) before the U.S. Army Corps of Engineers' beach nourishment project.

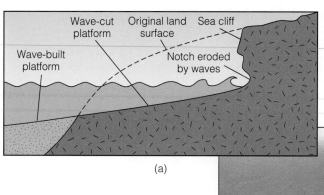

(a)

(b)

■ FIGURE 13.21 (*a*) Wave erosion of a sea cliff produces a gently sloping surface called a wave-cut platform. Deposition at the seaward margin of the wave-cut platform forms a wave-built platform. (*b*) A sea cliff and wave-cut platform.

(328 ft) per century. By comparison, sea cliffs of dense igneous or metamorphic rocks retreat much more slowly.

Sea cliffs erode mostly as a result of hydraulic action and abrasion at their bases. As a sea cliff is undercut by erosion, the upper part is left unsupported and susceptible to mass wasting processes. Thus, sea cliffs retreat little by little, and as they do, they leave a beveled surface known as a **wave-cut platform** that slopes gently seaward (Fig. 13.21). Broad wave-cut platforms exist in many areas, but invariably the water over them is shallow because the abrasive planing action of waves is only effective to about 10 m (33 ft) deep.

Sea cliffs do not erode uniformly because some of their materials are more resistant to erosion than others. *Headlands* are seaward-projecting parts of the shoreline resulting from this differential erosion. Once formed, headlands erode on both sides as waves are refracted around them. *Sea caves* may form on opposite sides of a headland, and if these join, they form a *sea arch* (■ Fig. 13.22). Continued erosion generally causes the span of an arch to collapse, yielding isolated *sea stacks* on wave-cut platforms (Fig. 13.22). In the long run, shoreline processes tend to straighten an initially irregular shoreline. They do so because wave refraction causes more wave energy to be expended on headlands and less on embayments, so headlands are eroded, and some of the sediment yielded by erosion is deposited in the embayments. The net effect of these processes is to straighten the shoreline.

 TYPES OF COASTS

Coasts can be classified in several ways, none of which is completely satisfactory. *Depositional coasts,* such as the U.S. Gulf Coast, are characterized by an abundance of detrital sediment, long sandy beaches, well-developed deltas, and barrier islands. *Erosional coasts* are generally steep and irregular, typically lack beaches except in protected areas (Fig. 13.13a), and are further characterized by erosional features such as sea cliffs, wave-cut platforms, and sea stacks (Figs. 13.21 and 13.22). Much of the shoreline along the west coast of North America falls into this category.

The following section examines coasts in terms of their changing relationships to sea level. Some coasts, such as those in southern California, which are described as emergent (uplifted), may be erosional as well. In other words, coasts commonly possess features allowing them to be categorized in several ways.

Submergent and Emergent Coasts

If sea level rises with respect to the land or the land subsides, coastal regions are flooded and said to be **submergent** or *drowned* (■ Fig. 13.23). Much of the east coast of North America from Maine southward through South Carolina was flooded by the rise in sea level following the Pleistocene, so it is now extremely irregular. Recall that during the

FIGURE 13.22 (*a*) Erosion of a headland and the origin of sea caves, sea arches, and sea stacks. (*b*) This sea stack in Australia has an arch developed in it. (*c*) Sea stacks on Australia's south coast.

Pleistocene expansion of glaciers, sea level was as much as 130 m (425 ft) lower than at present, and streams eroded their valleys more deeply as they adjusted to a lower base level. When sea level rose, the lower ends of these valleys were flooded and formed *estuaries* such as Delaware and Chesapeake Bays (Fig. 13.23), which are the seaward ends of river valleys where seawater and freshwater mix.

Emergent coasts are found where the land has risen with respect to sea level (Fig. 13.24). Emergence can take place when water is withdrawn from the oceans as occurred during the Pleistocene expansion of glaciers. At present coasts are emerging as a result of isostasy or tectonism. In northeastern Canada and the Scandinavian countries, the coasts are irregular because isostatic rebound is elevating formerly glaciated terrain from beneath the sea.

Coasts rising in response to tectonism tend to be rather straight because the sea-floor topography being exposed as uplift proceeds is smooth. The west coasts of North and South America are emergent and are rising as a result of plate tectonics. Distinctive features of these coasts are **marine terraces,** which are old wave-cut platforms now elevated above sea level (Fig. 13.25). Uplift in such areas appears to be episodic rather than continuous, as indicated by the multiple levels of terraces in some areas. In southern California, for instance, several terrace levels are present, each of which appears to represent a period of stability followed by uplift. The highest of these terraces is now about 425 m (1,395 ft) above sea level.

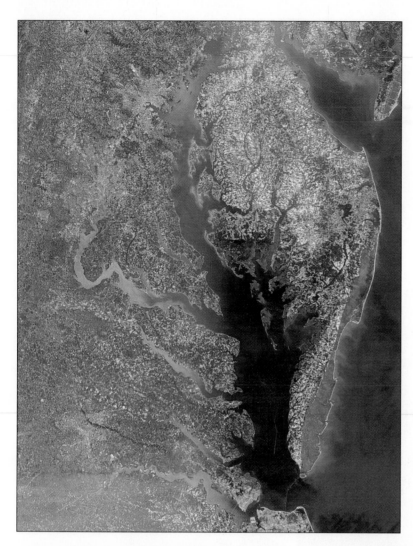

◻ FIGURE 13.23 Much of the east coast of the United States is submergent. Chesapeake Bay, the enormous indentation of the shoreline, is a large estuary. It formed when this area was flooded as sea level rose following the Pleistocene.

◻ FIGURE 13.24 An emergent coast in California. Such coasts are characterized by sea cliffs, and they tend to be straighter than submergent coasts.

■ FIGURE 13.25 This gently sloping surface along the Pacific coast of California is a marine terrace. Notice the old sea stacks rising above this terrace.

CHAPTER SUMMARY

1. Surface currents or gyres in the oceans are driven by wind but are modified by the Coriolis effect and the distribution of continents.

2. The pattern of gyres in the Northern and Southern Hemispheres is somewhat similar, except that in the Northern Hemisphere they rotate clockwise whereas in the Southern Hemisphere they rotate counterclockwise.

3. Gyres are important in modifying climate because warm currents transport heat from the equator toward the poles, and cold currents flow from the poles toward the equator.

4. Upwelling currents carry nutrient-rich, cold water from depth to the ocean's surface, whereas downwelling transfers warm surface water to depth. Three types of upwelling are recognized: polar, coastal, and equatorial.

5. Deep-ocean or thermohaline circulation is driven by density differences in water masses caused by temperature and salinity changes.

6. Shorelines are continually modified by the energy of waves and longshore currents and, to a lesser degree, by tidal currents.

7. Tides result from the gravitational attraction of the Moon and Sun. Most shoreline areas experience two high tides and two low tides daily.

8. Waves are oscillations on water surfaces that transmit energy in the direction of wave movement. Waves affect the water and sea floor only to wave base, which is a depth equal to one-half wave length.

9. Breakers form when waves enter shallow water where they become oversteepened and collapse or plunge forward, thus expending their kinetic energy.

10. Waves approaching a shoreline at an angle generate longshore currents that are capable of considerable erosion, transport, and deposition.

11. Beaches, the most common coastal landforms, are continually modified by nearshore processes, and their profiles generally exhibit seasonal changes.

12. Spits, baymouth bars, and tombolos all form and grow as a result of longshore current transport and deposition. Barrier islands are nearshore sediment deposits of uncertain origin that parallel the shoreline but are separated from it by a lagoon.

13. The nearshore sediment budget in a nearshore system is generally balanced unless the system is somehow disrupted as when dams are built across streams supplying sand to the system.

14. Coasts dominated by deposition have long sandy beaches, deltas, and barrier islands, whereas erosional coasts have sea cliffs, wave-cut platforms, and sea stacks.

15. Submergent and emergent coasts are defined on the basis of their relationship to changes in sea level. Submergent coasts are characterized by estuaries and are extremely irregular. Emergent coasts possess sea cliffs, sea stacks, and marine terraces.

barrier island
baymouth bar
beach
breaker
downwelling
emergent coast

fetch
Gulf Stream
gyre
longshore current
longshore drift
marine terrace

nearshore sediment budget
shoreline
spit
submergent coast
thermohaline circulation
tide

tombolo
upwelling
wave
wave base
wave-cut platform
wave refraction

REVIEW QUESTIONS

1. The limb of the North Atlantic gyre flowing northward along the east coast of North America is the:
   a. _____ Canary Current; b. _____ Kuroshio Current; c. _____ Gulf Stream; d. _____ Western Boundary Undercurrent; e. _____ Greenland Current.

2. Areas of upwelling in the oceans account for:
   a. _____ deflection of surface currents to the right in the Northern Hemisphere; b. _____ high biologic productivity; c. _____ circulation in the Sargasso Sea; d. _____ westerly flowing surface currents in the Antarctic region; e. _____ deep-ocean circulation from the poles to the equator.

3. Deep-ocean circulation is known as _____ circulation.
   a. _____ salinity; b. _____ compositional; c. _____ circumpolar; d. _____ vertical; e. _____ thermohaline.

4. Which of the following is not a depositional landform?
   a. _____ spit; b. _____ tombolo; c. _____ baymouth bar; d. _____ beach; e. _____ sea stack.

5. Wave base is:
   a. _____ the distance offshore that waves break; b. _____ the width of a longshore current; c. _____ the depth at which the orbital motion in surface waves dies out; d. _____ the distance the wind blows over a water surface; e. _____ the height of storm waves.

6. Waves approaching a shoreline at an angle generate:
   a. _____ flood tides; b. _____ longshore currents; c. _____ tidal currents; d. _____ marine terraces; e. _____ emergent coasts.

7. Erosion of a sea cliff produces a gently sloping surface known as a:
   a. _____ wave-cut platform; b. _____ depositional coast; c. _____ barrier island; d. _____ sea cave; e. _____ baymouth bar.

8. Islands composed of sand and separated from the mainland by a lagoon are:
   a. _____ barrier islands; b. _____ atolls; c. _____ tombolos; d. _____ pocket beaches; e. _____ spits.

9. The force of waves impacting on shorelines is:
   a. _____ corrosion; b. _____ wave oscillation; c. _____ hydraulic action; d. _____ terracing; e. _____ translation.

10. The distance the wind blows over a water surface is the:
    a. _____ spit; b. _____ wave length; c. _____ trough; d. _____ fetch; e. _____ wave height.

11. The bending of waves so that they more nearly parallel the shoreline is:
    a. _____ wave compaction; b. _____ wave oscillation; c. _____ wave deflection; d. _____ wave refraction; e. _____ wave translation.

12. A sand deposit extending into the mouth of a bay is a:
    a. _____ headland; b. _____ beach; c. _____ spit; d. _____ wave-cut platform; e. _____ sea stack.

13. The gyres in both the Northern and Southern Hemispheres flow fastest and are widest along their _____ arms.
    a. _____ northern; b. _____ southern; c. _____ western; d. _____ eastern; e. _____ central.

14. Most beaches receive most of their sediment from:
    a. _____ wave erosion of sea cliffs; b. _____ erosion of offshore reefs; c. _____ streams; d. _____ wind transport; e. _____ coastal emergence.

15. Explain the three ways that upwelling occurs.

16. What generates and modifies surface currents in the oceans?

17. How do deep- and shallow-water waves differ?

18. What is wave base, and how does it affect waves as they enter shallow water?

19. Explain how a longshore current is generated.

20. Sketch a north-south shoreline along which several groins have been constructed. Assume that waves approach from the northwest.

21. Sketch the profiles of a summer beach and a winter beach, and explain why they differ.

22. How does a tombolo form?

23. Why does an observer at a shoreline experience two high and two low tides each day?

24. What are the characteristics of submergent and emergent coasts?

25. Describe the circulation of seawater between the Atlantic Ocean and the Mediterranean Sea.

26. How do wave-cut platforms and wave-built platforms develop?

POINTS TO PONDER

1. An initially straight shoreline is composed of granite flanked on both sides by glacial drift. Diagram and explain this shoreline's probable response to erosion.
2. How might burning fossil fuels have some long-term effect on shoreline erosion?
3. Why are long, broad sandy beaches more common in eastern North America than in western North America?
4. A hypothetical nearshore area has a balanced sand budget, but a dam is constructed on the stream supplying most of the land-derived sand and a wall is built to protect sea cliffs from erosion. What would likely happen to the beach in this area?

ADDITIONAL READINGS

Abrahamson, D. E., ed. 1989. *The challenge of global warming.* Washington, D.C.: Island Press.

Bird, E. C. F. 1984. *Coasts: An introduction to coastal geomorphology.* New York: Blackwell.

Bird, E. C. F., and M. L. Schwartz. 1985. *The world's coastline.* New York: Von Nostrand Reinhold.

Flanagan, R. 1993. Beaches on the brink. *Earth* 2, no. 6: 24–33.

Fox, W. T. 1983. *At the sea's edge.* Englewood Cliffs, N.J.: Prentice-Hall.

Garrett, C., and L. R. M. Maas. 1993. Tides and their effects. *Oceanus* 36, no. 1: 27–37.

Hecht, J. 1988. America in peril from the sea. *New Scientist* 118: 54–59.

Komar, P. D. 1976. *Beach processes and sedimentation.* Englewood Cliffs, N.J.: Prentice-Hall.

_____. 1983. *CRC handbook of coastal processes and erosion.* Boca Raton, Fla.: CRC Press.

Pethick, J. 1984. *An introduction to coastal geomorphology.* London: Edward Arnold.

Schneider, S. H. 1990. *Global warming: Are we entering the greenhouse century?* San Francisco: Sierra Club Books.

Snead, R. 1982. *Coastal landforms and surface features.* Stroudsburg, Pa.: Hutchinson Ross Publishing Co.

Sunamara, T. 1992. *Geomorphology of rocky coasts.* New York: John Wiley & Sons.

Viles, H., and T. Spencer. 1995. *Coastal problems.* London: Edward Arnold.

Walden, D. 1990. Raising Galveston. *American Heritage of Invention and Technology* 5: 8–18.

Williams, S. J., K. Dodd, and K. K. Gohn. 1990. Coasts in crisis. *U.S. Geological Survey Circular 1075.*

WORLD WIDE WEB SITES

For current updates and exercises, log on to http://www.wadsworth.com/geo.

COASTAL PROGRAM DIVISION COASTAL ZONE MANAGEMENT PROGRAM

http://www.nos.noaa.gov/ocrm/cpd/

This government site is dedicated to the coastal zone management program of the National Ocean Service.

U.S. GEOLOGICAL SURVEY THE NATIONAL MARINE AND COASTAL GEOLOGY PROGRAM

http://www.thomson.com/marine.usgs.gov/marine-plan/index.html

As stated at the top of this web site, The National Marine and Coastal Geology Program is "a plan for geologic research on environmental, hazards, and resources issues affecting the Nation's coastal realms and marine federal lands."

CHAPTER

14

The Atmosphere: Basic Properties

The atmosphere is a relatively thin envelope of gases, tiny particles, and clouds that encircles Earth.

Prologue

A freeze in Brazil drives up the price of coffee in the grocery store; a sudden summer thunderstorm washes out a baseball game; heavy snow and icy roads close schools (☐ Fig. 14.1); thick fog restricts visibility and cancels flights; a heat wave triggers a run on air conditioners; and an early season snowfall allows a ski resort to open weeks ahead of schedule. These events underscore the sensitivity of all sectors of society to weather and its variability. Weather affects transportation, agriculture, recreation, commerce and industry, and the demand for power for home heating and cooling.

At times, the weather becomes extreme, and the impact can be costly in terms of lives lost, personal injuries, and property damage. Tornadoes smash houses and turn two-by-fours into lethal projectiles; hurricanes produce damaging winds and torrential rains that cause devastating floods (☐ Fig. 14.2); lightning claims the life of a golfer trying to get in one last hole; and fog shrouding an interstate highway contributes to a multivehicle pileup.

In the effort to anticipate weather extremes, atmospheric scientists rely on advanced technologies. They monitor weather systems using satellites, Doppler radar systems, and automated weather stations. Weather data received instantaneously from around the globe are processed by supercomputers that are programmed with numerical models that simulate the behavior of the Earth-atmosphere system. (The *Earth-atmosphere system* is the Earth's surface and atmosphere considered together.)

Society has benefited greatly from the technological advances of modern atmospheric science. For example, the probability of a surprise snowstorm creeping up the eastern seaboard and burying the northeastern megalopolis is much less today than it was before satellites came into widespread use. Heavy snows still paralyze large cities, but at least residents receive some advance warning. Atmospheric scientists can also provide advance warning of an approaching hurricane for timely evacuation of low-lying coastal areas. Nevertheless, despite these advances, society is still vulnerable to the vagaries of weather as evidenced by recent weather disasters (○ Table 14.1).

☐ FIGURE 14.1 Snow creates hazardous driving and walking conditions.

☐ FIGURE 14.2 A portion of South Carolina's coastal plain is flooded by the storm surge produced by Hurricane Hugo in September 1989.

TABLE 14.1   Recent Weather Extremes in the United States

EVENT	DATE	LOCATION	FATALITIES[a]	DAMAGE ($ billions)
Heat and humidity	July 1995	Midwest	1000	★
Hurricane Opal	Oct. 1995	Florida	21	2.1
Wildfires	Fall 1993	Southern California	4	>1.0
Flooding	Summer 1993	Midwest	48	12.0
Drought and heat	Summer 1993	Southeast	★	1.0
Blizzard	March 1993	East	270	>3.0
Hurricane Iniki	Sept. 1992	Kauai, Hawaii	6	1.8
Hurricane Andrew	Aug. 1992	Florida, Louisiana	29	25.0
Hurricane Bob	Aug. 1991	North Carolina to New England	18	1.5
Hurricane Hugo	Sept. 1989	North and South Carolina	21	7.0

[a]Includes both direct and indirect deaths.
★Undetermined.
SOURCE: National Climatic Data Center.

INTRODUCTION

Variability is a fundamental characteristic of weather; weather varies from place to place and through time. It is usually warmer in the southern United States than in southern Canada, wetter in the Pacific Northwest than in Arizona, and snowier on a mountaintop than in a mountain valley. If the weather has been unusually warm and dry, it may turn cold and wet. Because of its basic geographical and temporal variability, **weather** is defined as the state of the atmosphere, in terms of such variables as temperature, humidity, and cloud cover, at a given place and time. *Meteorology* is the study of the atmosphere and the characteristics and life cycles of weather systems.

Closely related to weather is climate. **Climate** refers to the weather conditions at some locality averaged over a specified time period plus extremes in weather. Climate governs the supply of fresh water, the demand for fuel for home heating and cooling, and the types of crops that can be cultivated. The probability of weather extremes (such as drought or flooding rains) is also an important aspect of climate. *Climatology* is the study of climate, including its controls and variability.

Weather is the state of the **atmosphere,** a thin envelope of gases and tiny suspended particles that surrounds the planet. Although thin, the atmosphere is essential for life on Earth and the external geological forces that shape the land surface. The atmosphere protects organisms from exposure to hazardous levels of solar radiation, contains the gases necessary for respiration and photosynthesis, and supplies the water needed by all forms of life. As discussed in Chapters 4, 5, and 6, the atmosphere plays an important role in physical and chemical weathering, soil formation, and erosion by wind, running water, and glacial ice.

This chapter surveys the basic properties of the atmosphere: the evolution and composition of the atmosphere, air temperature and its variation with altitude, heat transport within the Earth-atmosphere system, air pressure and its variations, and some characteristics of the upper atmosphere.

ORIGINS OF THE ATMOSPHERE

The Sun and the solar system developed out of an immense cloud of dust and gases within the Milky Way Galaxy. About 4.6 billion years ago, the solid Earth and its atmosphere began to evolve. Initially, Earth was an aggregate of dust and cosmic debris. Between 4.6 and 4.4 billion years ago, the planet grew via accretion as it swept up cosmic dust in its path and was bombarded by planetesimals.

The early evolution of the atmosphere was influenced largely by the gases emitted as by-products of surface igneous activity, a process known as **outgassing** (Fig. 14.3). As much as 85% of the total outgassing that has occurred took place within the first million years of Earth's existence, but slow outgassing has continued ever since. Outgassing produced an early atmosphere that was rich in carbon dioxide (CO_2) and about 10 to 20 times denser than the modern atmosphere. In addition to carbon dioxide, the early atmosphere contained some nitrogen (N_2) and water vapor (H_2O) and trace amounts of methane (CH_4), ammonia (NH_3), and sulfur dioxide (SO_2). Free oxygen (O_2) was absent although oxygen was combined with other elements in compounds such as water vapor and carbon dioxide.

Astronomers believe that during the Archean Eon (4.5 to 2.5 billion years ago), the Sun's energy output was about 75% of what it is today. Although this would imply a cooler planet, the cooling effect of the weaker solar rays was likely countered by the warming effect of atmospheric carbon dioxide. By slowing the escape of Earth's heat to space, carbon dioxide elevates the temperature at Earth's surface.

◉ **FIGURE 14.3** Gaseous emissions from volcanic activity contributed to the evolution of Earth's atmosphere. This is the spectacular eruption of Mount Pinatubo in the Philippines in July 1991.

All other factors being equal, increasing amounts of carbon dioxide in the atmosphere would cause the climate to become progressively warmer. All other factors were not equal, however, and with time the carbon dioxide concentration in the atmosphere declined. According to a theory proposed in the early 1980s by Earth scientists J. C. G. Walker, J. F. Kasting, and P. B. Hayes, interactions within the Earth-atmosphere system put a brake on global warming by reducing the amount of carbon dioxide in the atmosphere. Early on, rising temperatures altered the hydrologic cycle by increasing the rate of evaporation of water. More water vapor in the atmosphere meant more clouds and more rainfall. Some atmospheric carbon dioxide dissolved in the rainwater producing weak carbonic acid that interacted with crustal rock. Through this chemical weathering (see Chapter 4), carbon combined with silicate minerals and was stored in sediments and sedimentary rocks. Inevitably then, as more and more carbon became sequestered in bedrock, the concentration of atmospheric carbon dioxide declined and the global temperature fell.

From time to time in the geologic past, the concentration of atmospheric carbon dioxide has fluctuated, with important implications for global climate. Geological evidence points to a burst of volcanic activity on the floor of the Pacific Ocean about 120 to 100 million years ago. Some of the carbon dioxide released by that activity eventually escaped to the atmosphere triggering an episode of global warming. The result was tropical conditions even at high lat-

itudes. During the Pleistocene, atmospheric carbon dioxide levels declined during episodes of glacial expansion and rose during episodes of glacial recession. And in modern time, burning of fossil fuels and deforestation are increasing atmospheric carbon dioxide levels, with potential impacts for global climate (Chapter 15).

Living organisms also played an important role in the evolution of Earth's atmosphere, primarily through photosynthesis. **Photosynthesis** is the process whereby plants use sunlight, water from the soil, and carbon dioxide from the air to manufacture their food. A by-product of photosynthesis is oxygen, which is released to the atmosphere. Hence, plants are a sink for carbon dioxide and a source of oxygen.

Photosynthesis began about 3.5 billion years ago with the appearance of the first living organisms, blue-green marine algae. Some of the post-Archean decline in carbon dioxide was due to photosynthesis, although most can be attributed to the geochemical processes described earlier. Photosynthesis was much more important in increasing the oxygen content of the atmosphere. For the first 1 to 2 billion years, most of the oxygen released by marine algae combined with minerals in the sea with little reaching the atmosphere. About 2 billion years ago, oxidation of available minerals was nearly complete, and the flow of oxygen into the atmosphere accelerated, so that by 1.5 billion years ago oxygen became the second most abundant atmospheric gas after nitrogen. Geological evidence suggests another abrupt rise in oxygen levels about 1.1 billion to 700 million years ago. This more recent pulse in oxygen concentration apparently triggered the evolution of large animals.

The availability of free oxygen in the atmosphere made it possible for life to evolve on land. Previously, potentially lethal intensities of solar ultraviolet radiation had flooded Earth's surface, confining the evolution of life to the sea where the water provided some protection from the radiation. Abundant oxygen in the atmosphere now made possible chemical reactions, powered by ultraviolet radiation, that convert oxygen to ozone (O_3) and ozone to oxygen. In this way, the atmosphere shields Earth's surface from high-intensity ultraviolet radiation. This so-called *ozone shield* is discussed further in the next chapter.

Especially since the Industrial Revolution, human activities have contributed to the evolution of the atmosphere. As the by-products of industrial and agricultural processes have been emitted into the atmosphere, they have altered its composition (see ◻ Perspective 14.1).

 ## COMPOSITION OF THE ATMOSPHERE

Outgassing and photosynthesis ultimately produced the modern atmosphere, a mixture of mostly nitrogen and oxygen plus many trace gases and suspended particles. Because the lower atmosphere undergoes continuous mixing, the chief atmospheric gases occur in nearly the same relative proportions everywhere up to an altitude of about 80 km (50 mi). This lower portion of the atmosphere is known as the *homosphere.*

Excluding water vapor (which has a variable concentration), the homosphere is 78.08% nitrogen (N_2) and 20.95% oxygen (O_2) by volume. The next most abundant gases are argon (0.93%) and carbon dioxide (0.035%). As ◯ Table 14.2 shows, air also includes small amounts of helium, methane, ozone, and several other gases. The percentage by volume of some of these trace gases varies with time and location within the homosphere.

Suspended particles within the atmosphere are collectively called **aerosols.** Aerosols are both solid and liquid and occur in wide ranges of size, shape, and chemical composition. Some are too small to be visible, but when others such as water droplets and ice crystals are clustered together, they can be seen as clouds. Aerosols tend to concentrate in the lower atmosphere near their sources on Earth's surface. They derive from soil and sediment eroded by the wind, tiny crystals of sea salt formed from breaking sea waves, volcanic emissions, and by-products of industrial processes.

The significance of an atmospheric gas is not always related to its relative concentration. Although water vapor and carbon dioxide occur in the atmosphere in very low concentrations, both gases are essential for life. Even in the most

TABLE 14.2   The Relative Proportions of Gases Composing Dry Air in the Lower Atmosphere (below 80 km)

GAS	PERCENTAGE BY VOLUME	PARTS PER MILLION
Nitrogen	78.08	780,840.0
Oxygen	20.95	209,460.0
Argon	0.93	9,340.0
Carbon dioxide	0.035	350.0
Neon	0.0018	18.0
Helium	0.00052	5.2
Methane	0.00014	1.4
Krypton	0.00010	1.0
Nitrous oxide	0.00005	0.5
Hydrogen	0.00005	0.5
Ozone	0.000007	0.07
Xenon	0.000009	0.09

Focus on the Environment:
Human Activity and Air Quality

Over time, but especially since the Industrial Revolution, humans have been agents of change in the evolution of Earth's atmosphere. Their impact has come primarily through the use of the atmosphere as a depository for some of the waste by-products of industry, transportation, and agriculture. Ironically, those waste gases and aerosols are potentially harmful for human health and well-being. Fortunately, the efforts of the United States and other industrialized nations to curb atmospheric pollution over the past 25 years have met with some success.

An *air pollutant* is a gas or aerosol that occurs in the atmosphere at a concentration that threatens the health and well-being of living organisms (especially people) or that disrupts the orderly workings of the environment. Interestingly, most air pollutants are natural components of the atmosphere that become pollutants when their concentrations rise to harmful levels. An example is sulfur dioxide (SO_2), a normal trace component of the atmosphere, that is also released as a by-product of fossil fuel (coal and oil) burning and paper making. Unless emissions are controlled, the local concentration of sulfur dioxide may increase to levels that adversely affect the human respiratory system. Other air pollutants do not occur naturally in the atmosphere and may cause problems even in low concentrations. Chlorofluorocarbons (CFCs), which have been implicated in the thinning of the stratospheric ozone shield (see Chapter 15), are an example.

The internal combustion engine that burns petroleum and powers most motor vehicles is the single most significant human-related source of air pollutants. The U.S. Environmental Protection Agency (EPA) reports that the nation's transportation sector emits almost 60 million metric tons (63 million tons) of the major air pollutants annually (principally, carbon monoxide, nitrogen oxides, and volatile organic compounds). This represents about 45% of total U.S. emissions of the chief air pollutants. Besides human-related sources, air pollutants are emitted by natural sources including brush and forest fires, volcanic eruptions, wind erosion of soil and sediment, decay of dead organisms, and pollen dispersal.

Often a distinction is made between primary and secondary air pollutants. *Primary air pollutants* are contaminants immediately upon their entry into the atmosphere through a chimney or exhaust pipe. Carbon monoxide (CO) and nitrogen oxides in automobile exhaust are examples. *Secondary air pollutants* are produced when gases and/or aerosols undergo chemical reactions in the atmosphere. An example is *photochemical smog,* a visibility-restricting mixture of pollutants (▢ Fig. 1). It is the product of complex chemical reactions that take place when solar radiation interacts with motor vehicle emissions and volatile organic compounds of industrial and biological origin.

Although strict laws have cut emissions of the major air pollutants and improved air quality over much of the United States and Canada, problems remain. Many of the largest cities still have serious problems with photochemical smog, and some have failed to meet federally mandated air quality standards for carbon monoxide. Furthermore, the perplexing problem of so-called hazardous air pollutants persists. These are contaminants that pose a health risk even in very low concentrations. Removing them from smokestack emissions is usually extremely difficult, however, because of high costs and technological limitations.

Scientists are concerned about the effect humans are having on the atmosphere for two reasons. Not only does air pollution pose a potential threat to human health, but certain air pollutants that contribute to the so-called *greenhouse effect* also have global climatic implications (see Chapter 15).

▢ FIGURE 1 Photochemical smog restricts visibility over the Los Angeles basin.

humid regions of the globe, water vapor is never more than about 4% of the air. Without water vapor, however, there would be no hydrologic cycle (see Chapter 5). On average, carbon dioxide accounts for only 0.035% of the homosphere by volume, and yet without it, there would be no photosynthesis and no plants and animals.

 HEAT AND TEMPERATURE

Air temperature is one of the most important variables used to describe the state of the atmosphere (weather). Experience tells us that air temperature varies with time and location: summers are warmer than winters; days are usually warmer than nights; lowlands and lower latitudes are normally warmer than highlands and higher latitudes. Experience also indicates that air temperature strongly influences personal comfort and that extremely low or high temperatures can be hazardous (see □ Perspective 14.2 on page 324).

Heat Versus Temperature

Temperature and heat are closely related concepts. When a kettle of water is heated on a stove, the temperature of the water rises. When ice cubes are dropped into a beverage, the temperature of the beverage falls. The precise distinction between heat and temperature requires a look at what is happening at the atomic or molecular scale. The atoms or molecules that compose all substances are always in rapid, random motion. Hence, atoms and molecules possess *kinetic energy,* the energy of motion. **Heat** is the *total* kinetic energy of the atoms or molecules composing a substance. The individual atoms or molecules of a substance do not all move at the same rate, however; there is always a range of kinetic energy among atoms or molecules. **Temperature** is directly proportional to the *average* kinetic energy of individual atoms or molecules.

An illustration will help to clarify the distinction between heat and temperature. A cup of water at 65°C (149°F) is much hotter than a bathtub of water at 35°C (95°F); that is, the average kinetic energy of water molecules is greater at 65°C than at 35°C. The bathtub contains much more water than the cup, however, so the total kinetic-molecular energy (heat) in the bath water is much greater than in the cup of water. Considerably more heat must be removed from the bath water than from the cup of water to cool both to the same temperature. In other words, the cup of water cools to room temperature more rapidly than does the bath water.

Although temperature is a convenient way to describe the relative hotness or coldness of something, heat energy can be quantified directly. For some meteorological purposes, heat is still measured in calories where 1 *calorie (cal)* is defined as the quantity of heat that must be added to raise the temperature of 1 gram of water 1 Celsius degree. The conventional unit of heat energy is the *joule (J);* 1 cal equals 4.1868 J.

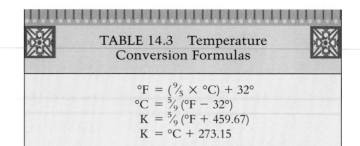

TABLE 14.3   Temperature Conversion Formulas

$$°F = (\tfrac{9}{5} \times °C) + 32°$$
$$°C = \tfrac{5}{9}(°F - 32°)$$
$$K = \tfrac{5}{9}(°F + 459.67)$$
$$K = °C + 273.15$$

Temperature Scales and Measurement

For most scientific purposes, temperature is graduated on the Celsius (°C) scale, which features a convenient 100-degree interval between the freezing and boiling points of pure water. The United States is one of only a few nations that still use the Fahrenheit (°F) temperature scale for weather observations. If a thermometer marked in both scales is placed in a mixture of ice and water, the Celsius scale reads 0°C, and the Fahrenheit scale reads 32°F. In boiling water at sea level, the readings are 100°C and 212°F.

As heat is removed from an object, the average kinetic-molecular energy decreases; that is, the temperature of the object drops. Theoretically, with sufficient cooling a point is reached where the average kinetic-molecular energy is zero; that is, all molecular motion ceases. That point is known as **absolute zero** and corresponds to a temperature of −273.15°C (−459.67°F). Actually, though, some atomic-level motion occurs at absolute zero.

A third temperature scale, the Kelvin scale, begins at absolute zero. Whereas temperature is designated as *degrees Celsius (°C)* and *degrees Fahrenheit (°F)* on the Celsius and Fahrenheit scales, respectively, temperature is simply *kelvins (K)* on the Kelvin scale. Because nothing can be colder than absolute zero, the Kelvin scale has no negative temperatures. An increment of 1 kelvin equals an increment of 1 Celsius degree. ▣ Figure 14.4 compares the three scales, and ○ Table 14.3 presents conversion formulas.

▣ FIGURE 14.4 A comparison of the Kelvin, Celsius, and Fahrenheit temperature scales.

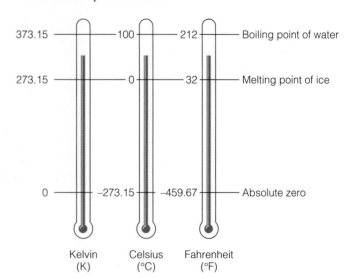

A *thermometer* monitors air temperature. A familiar type of thermometer is a liquid-in-glass tube attached to a graduated scale (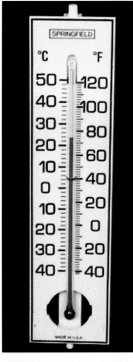 Fig. 14.5). The liquid is either mercury (which freezes at −39°C or −38°F) or alcohol (which freezes at −117°C or −179°F). Heating expands both the glass tube and the liquid but the liquid more than the glass. As air warms, the liquid expands and rises in the tube; as air cools, the liquid contracts and drops in the tube.

Another type of thermometer utilizes a diode (a semiconductor) as a sensor. In these electronic thermometers, air temperature is calibrated against changes in electrical resistance. Some electronic thermometers are designed to automatically record the highest and lowest temperatures over a specified time interval, say, 24 hours. As part of the current modernization of the National Weather Service (NWS), electronic thermometers are replacing liquid-in-glass thermometers at official weather stations.

Thermometers should be ventilated and shielded from precipitation, direct sunlight, and the night sky. Enclosing thermometers (and other weather instruments) in a shelter, such as the one shown in  Figure 14.6, is standard practice for official air temperature measurements. An instrument shelter should be located in an open grassy area well away from buildings or other obstacles. If a shelter is not available, a thermometer should be mounted outside a window on the shady north side of a building.

It is useful to monitor variations of air temperature with altitude. To this end, meteorologists launch a hydrogen- or helium-filled balloon carrying an instrument package known as a **radiosonde** ( Fig. 14.7). This device is equipped with

◨ FIGURE 14.5 A liquid-in-glass thermometer.

◨ FIGURE 14.6 A standard louvered wooden shelter protects weather instruments from exposure to direct sunlight, precipitation, and the night sky.

◨ FIGURE 14.7 A National Weather Service meteorologist holds a radiosonde (*a*) that will be carried aloft by a hydrogen-filled balloon (*b*). A radiosonde is an instrument package that transmits to a ground station continuous profiles of air temperature, pressure, and humidity up to an altitude of about 30 km (19 mi).

(a)

(b)

Temperature Extremes and Human Well-Being

Human well-being depends partially on the exchange of heat between the body and its surroundings. Excessive heat loss or heat gain not only makes a person uncomfortable but may be hazardous to health and, in extreme cases, may cause death.

Heat is generated in the body by metabolic processes. Heat is exchanged between the body and its environment in response to temperature gradients between the skin and the surrounding (ambient) air. If the skin is warmer than the air, heat flows from the body to the environment via conduction, convection, radiation, and vaporization of perspiration. The first three processes are described elsewhere in this chapter. As for the last process, water molecules are more energetic in the vapor phase than in the liquid phase, so heat is required for perspiration to evaporate. That heat is supplied by the skin, a process known as *evaporative cooling*. On the other hand, if the air is warmer than the skin, heat is supplied to the body through conduction, convection, and radiation.

For human health and well-being, the key consideration is not so much the temperature of the skin but rather the temperature of vital internal organs such as the heart and lungs. Normally, the body regulates heat loss and heat gain so that the temperature of these organs is within a few degrees of 37°C (98.6°F), the body's *core temperature*.

When ambient air temperatures are between 20 and 25°C (68 and 77°F), most fully clothed people feel comfortable at rest. Within this air temperature range, the body readily maintains a constant core temperature. As the air temperature departs from this range, the body uses various *thermoregulatory processes,* including changes in the circulatory system, enhanced perspiration, and shivering, to keep its core temperature in check.

Outside on a hot, sunny day, the core temperature may begin to rise, and in response, the body perspires more profusely, thus increasing the rate of evaporative cooling. Also, circulation changes deliver more heat to the skin surface where it is dissipated to the environment via conduction, convection, and radiation. This gives the skin a flushed appearance. Outside on a bitter cold, windy day, the core temperature may begin to fall, and in response, the body begins to shiver—a muscular activity that produces heat. Also, blood circulation is restricted so that less heat is delivered to the skin surface and lost to the environment. This gives the skin a pale appearance.

During periods of extreme heat or cold, thermoregulatory processes may fail to keep the core temperature in check. For example, a person whose auto breaks down in the desert sets off without water in search of help. Or an inadequately clothed person ventures outside in a blizzard and becomes disoriented and lost. In the first case, the person risks *hyperthermia,* a life-threatening condition that occurs when the core temperature rises unchecked and tops 39°C (102°F). In the second case, the person risks *hypothermia,* a life-threatening condition that develops when the core temperature falls unchecked and drops below 35°C (95°F).

Initial symptoms of hyperthermia (popularly referred to as heatstroke or sunstroke) include muscle cramps or spasms; the victim slips into unconsciousness and, if untreated, may die within hours. Treatment entails lowering the core temperature from outside the body (e.g., moving the victim to cooler surroundings). Symptoms of hypothermia include violent shivering followed by muscle rigidity, impaired mental ability, and lethargy. The victim becomes unconscious, and if the core temperature drops below about 30°C (86°F), death is likely. Treatment entails preventing further heat loss (e.g., finding shelter) and adding heat to the body.

Hyperthermia is more likely when high air temperatures are combined with high humidity. The higher the relative humidity (see Chapter 16), the less readily will perspiration evaporate from the skin. Hence, evaporative cooling is greatly impaired on hot, humid days. Because of the potential health hazard, the U.S. National Weather Service issues the *heat index* during the summer months. The heat index expresses the combination of temperature and relative humidity as an *apparent temperature,* essentially what the air feels like (○ Table 1).

At high humidities, the apparent temperature is higher than the actual air temperature, and symptoms of heat stress are likely especially among the elderly and the ill (○ Table 2). At an air temperature of 38°C (100°F) and a relative humidity of 50%, the apparent temperature is 49°C (120°F). Thus, a person exposed to such conditions experiences the same heat stress that accompanies an actual air temperature of 49°C.

In mid-July 1995, an unusually humid heat wave produced exceptionally high apparent temperatures over the Midwest. Between July 10 and 16, maximum apparent temperatures ranged up to 120°F (▱ Fig. 1). Chicago was particularly hard-hit with apparent temperatures peaking at 119°F on July 13 and 116°F on July 14. According to the city's Department of Health, the excessive heat and humidity were the chief reason the month's mortality count was 726 above normal. The county medical examiner certified heat exhaustion as the cause of death of 525 people.

Hypothermia is more likely when low air temperature is accompanied by strong winds. Wind transports heat away from the body at a faster rate than air temperature alone

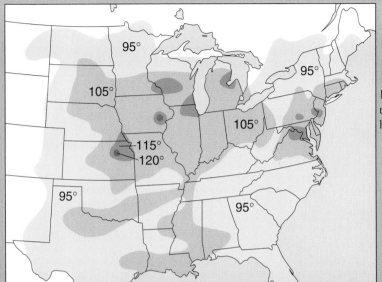

□ FIGURE 1 Maximum apparent temperatures (°F) during the oppressive heat wave of July 10–16, 1995.

would indicate. Because of the potential health hazard at northern latitudes (and high elevations), the U.S. National Weather Service reports the *windchill equivalent temperature,* or simply the *windchill,* during winter months. ○ Table 3 gives the windchill for ranges of air temperature and wind speed.

If the air temperature is 3°C (37°F) and there is no wind, the windchill is the same as the actual air temperature, that is, 3°C. At the same air temperature with the wind blowing at 9 m per second, the windchill is −10°C (14°F). Skin temperature does not actually drop to −10°C; skin temperature can drop no lower than the temperature of the ambient air. However, any exposed body part would lose heat at the same rate it would at an air temperature of −10°C with no wind.

Although low windchills heighten the possibility of hypothermia, often the most immediate danger is *frostbite,* the freezing of body tissue. Recall that one of the body's natural responses to low air temperature is to reduce blood circulation to the skin. This cools the skin, and any further heat loss caused by low air temperatures and strong winds may lead to frostbite. Frostbite is most likely for parts of the body having a relatively high surface area to volume ratio, such as the nose, fingers, and ears.

TABLE 1 Heat Index

RELATIVE HUMIDITY (%)	AIR TEMPERATURE (°F)										
	70	75	80	85	90	95	100	105	110	115	120
	Apparent temperature (°F)										
0	64	69	73	78	83	87	91	95	99	103	107
10	65	70	75	80	85	90	95	100	105	111	116
20	66	72	77	82	87	93	99	105	112	120	130
30	67	73	78	84	90	96	104	113	123	135	148
40	68	74	79	86	93	101	110	123	137	151	
50	69	75	81	88	96	107	120	135	150		
60	70	76	82	90	100	114	132	149			
70	70	77	85	93	106	124	144				
80	71	78	86	97	113	136					
90	71	79	88	102	122						
100	72	80	91	108							

SOURCE: National Weather Service, NOAA.

TABLE 2 Hazards Posed by Heat Stress by Range of Apparent Temperature

CATEGORY	APPARENT TEMPERATURE[a]	HEAT SYNDROME
I	54°C or higher (130°F or higher)	Heatstroke or sunstroke *imminent*
II	41 to 54°C (105 to 130°F)	Sunstroke, heat cramps, or heat exhaustion *likely* Heatstroke *possible* with prolonged exposure and physical activity
III	32 to 41°C (90 to 105°F)	Sunstroke, heat cramps, or heat exhaustion *possible* with prolonged exposure and physical activity
IV	27 to 32°C (80 to 90°F)	Fatigue *possible* with prolonged exposure and physical activity

[a]Apparent temperature combines the effects of heat and humidity on human comfort.
SOURCE: National Weather Service, NOAA.

TABLE 3 Windchill Equivalent Temperatures (°C and °F)

WIND SPEED (m/sec)	AIR TEMPERATURE (°C)															
	6	3	0	−3	−6	−9	−12	−15	−18	−21	−24	−27	−30	−33	−36	−39
3	3	−1	−4	−7	−11	−14	−18	−21	−24	−28	−31	−34	−38	−41	−45	−48
6	−2	−6	−10	−14	−18	−22	−26	−30	−34	−38	−42	−46	−50	−54	−58	−62
9	−6	−10	−14	−18	−23	−27	−31	−35	−40	−44	−48	−53	−57	−61	−65	−70
12	−8	−12	−17	−21	−26	−30	−35	−39	−44	−48	−53	−57	−62	−66	−71	−75
15	−9	−14	−18	−23	−27	−32	−37	−41	−46	−51	−55	−60	−65	−69	−74	−79
18	−10	−14	−19	−24	−29	−33	−38	−43	−48	−52	−57	−62	−67	−71	−76	−81
21	−10	−15	−20	−25	−29	−34	−39	−44	−49	−53	−58	−63	−68	−73	−77	−82
24	−10	−15	−20	−25	−30	−35	−39	−44	−49	−54	−59	−63	−68	−73	−78	−83

WIND SPEED (mi/hr)	AIR TEMPERATURE (°F)																		
	45	40	35	30	25	20	15	10	5	0	−5	−10	−15	−20	−25	−30	−35	−40	−45
5	43	37	32	27	22	16	11	6	1	−5	−10	−15	−20	−26	−31	−36	−41	−47	−52
10	34	28	22	16	10	4	−3	−9	−15	−21	−27	−33	−40	−46	−52	−58	−64	−70	−76
15	29	22	16	9	2	−5	−11	−18	−25	−32	−38	−45	−52	−58	−65	−72	−79	−85	−92
20	25	18	11	4	−3	−10	−17	−25	−32	−39	−46	−53	−60	−67	−74	−82	−89	−96	−103
25	23	15	8	0	−7	−15	−22	−29	−37	−44	−52	−59	−66	−74	−81	−89	−96	−104	−111
30	21	13	5	−2	−10	−18	−25	−33	−41	−48	−56	−63	−71	−79	−86	−94	−102	−109	−117
35	19	11	3	−4	−12	−20	−28	−35	−43	−51	−59	−67	−74	−82	−90	−98	−106	−113	−121
40	18	10	2	−6	−14	−22	−29	−37	−45	−53	−61	−69	−77	−85	−93	−101	−108	−116	−124
45	17	9	1	−7	−15	−23	−31	−39	−47	−55	−62	−70	−78	−86	−94	−102	−110	−118	−126

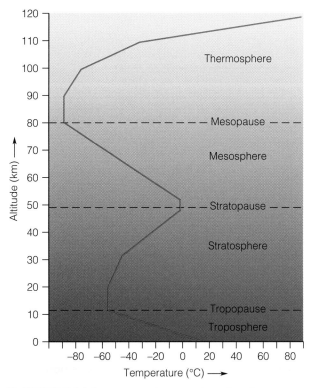

FIGURE 14.8 Based on the vertical profile of average air temperature, the atmosphere is divided into the troposphere, stratosphere, mesosphere, and thermosphere.

a radio transmitter that sends continuous measurements of air temperature, pressure, and relative humidity with altitude back to a ground station. The balloon bursts at an altitude of about 30 km (19 mi), and the instrument package descends to Earth's surface under a parachute.

Temperature Profile of the Atmosphere

The average vertical profile of air temperature is the basis for subdividing the atmosphere into four concentric layers: troposphere, stratosphere, mesosphere, and thermosphere (Fig. 14.8).

The lowest layer, the **troposphere,** is between Earth's surface and an average altitude that ranges from about 6 km (4 mi) at the poles to about 16 km (10 mi) at the equator. The troposphere is where most weather takes place. On average, air temperature drops 6.5 Celsius degrees for every 1,000 meters of ascent within the troposphere (3.5 Fahrenheit degrees per 1,000 feet). Hence, highlands are usually colder than lowlands (Fig. 14.9). The *tropopause* marks the boundary between the troposphere and the next higher layer, the stratosphere.

The **stratosphere** extends from the tropopause up to about 50 km (31 mi). On average, the temperature is constant in the lower portion of the stratosphere. Above about 20 km (12 mi), the temperature rises with increasing altitude up to the top of the stratosphere, the *stratopause.* At the

FIGURE 14.9 Within the troposphere, air temperature usually falls with increasing altitude, a major reason why mountaintops are colder than lowlands.

stratopause, the average temperature is about the same as it is at sea level. The stratosphere is above the weather but is nevertheless very important for life on Earth because it contains the protective ozone shield.

The *stratopause* is the transition zone between the stratosphere and the next higher layer, the *mesosphere.* Within this layer, the air temperature again falls with increasing altitude. The mesosphere extends up to the *mesopause,* about 85 km (53 mi) above Earth's surface. The average temperature near the mesopause is the lowest within the atmosphere, about −95°C (−139°F). Above this is the *thermosphere,* where the temperature climbs rapidly with increasing altitude.

TRANSPORT OF HEAT

Air temperatures usually vary from place to place within the Earth-atmosphere system. A change in temperature with distance is called a *temperature gradient.* For example, a temperature gradient always prevails between the relatively hot equator and the relatively cold polar regions and between the warmer surface of Earth and the colder tropopause.

In response to a temperature gradient, heat is transported from areas of higher temperatures toward areas of lower temperatures. Heat transport obeys the **second law of thermodynamics,** which states that all systems tend toward disorder. The presence of any gradient within a system indicates order, so the gradients are eliminated as the system shifts toward disorder. Hence, the *second law* predicts that where a temperature gradient exists, heat flows from warm to cold so as to eliminate the gradient. Furthermore, the steeper the temperature gradient, the more rapid is the rate of heat flow. In response to temperature gradients within the Earth-atmosphere system, heat is transported via conduction, convection, and radiation.

Conduction

Heat conduction occurs within an object or between objects that are in direct physical contact. With **conduction,** heat is transferred via collisions between neighboring atoms or molecules. This explains why a metal spoon heats up when placed in a cup of hot coffee. The more energetic molecules of the warmer coffee collide with the less energetic atoms of the cooler spoon, thereby transferring some kinetic-molecular energy to the atoms of the spoon. These atoms then transfer some of their heat energy, via collisions, to neighboring atoms, so that heat is conducted up the handle of the spoon and the handle becomes hot to the touch.

Some substances are better conductors of heat than others. In general, solids are better conductors than liquids, and liquids are better conductors than gases. ○ Table 14.4 lists heat conductivities of some common substances. Note that air is a relatively poor conductor of heat.

People take advantage of air's poor heat conductivity when they install storm windows or insulation in their homes to conserve energy. Still air has a lower heat conductivity than air in motion. Hence, to take maximum advantage of air's insulating property, air is confined between the panes of glass of a storm window or between the fibers that make up a layer of household insulation.

When and where Earth's surface is warmer than the overlying air, heat is conducted from the surface to the thin layer of air in immediate contact with the surface. The process is very slow, however, because air's heat conductivity is relatively low. Convection is much more important than conduction in transporting heat vertically within the troposphere.

Convection

Convection is the transport of heat within a fluid through motions of the fluid itself. Unlike conduction, which takes place in solids, liquids, and gases, convection generally is confined to liquids and gases. The convection currents in Earth's mantle are an important exception to this rule (see Chapters 7 and 8).

TABLE 14.4   Heat Conductivities of Some Familiar Substances

SUBSTANCE	HEAT CONDUCTIVITY[a]
Copper	0.92
Aluminum	0.50
Iron	0.16
Ice (at 0°C)	0.0054
Limestone	0.0048
Concrete	0.0022
Water (at 10°C)	0.0014
Dry sand	0.0013
Air (at 20°C)	0.000061
Air (at 0°C)	0.000058

[a]Heat conductivity is defined as the quantity of heat (calories) that would flow through a unit area of a substance (square centimeter) in one second in response to a temperature gradient of one Celsius degree per centimeter. Hence, heat conductivity is in units of cal per cm^2 per sec per C° per cm.

Convection currents develop within the atmosphere in response to contrasts in air density. Cold air is denser than warm air, so when cold air and warm air come in contact, the cold air pushes under and lifts the warm air. In other words, cold air sinks and warm air rises. A similar effect takes place when cold tap water flows into a tub of hot water: the cold water sinks to the bottom of the tub forcing the warm water upward. Hence, air heated by conduction from the warmer ground rises and is replaced by cooler air subsiding from above. That air, in turn, is heated by the ground and the process is repeated. The net result is a convective circulation as shown in ▣ Figure 14.10. Space heaters take advantage of heat circulation by convection currents; this is why heaters are placed along the baseboard (rather than the ceiling) of a room.

▣ FIGURE 14.10 Convective circulation of air transports heat from Earth's surface and the lower atmosphere to higher altitudes. Warm air is less dense than cold air, so the cold air sinks and forces the warm air to rise.

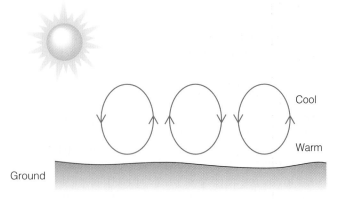

Radiation

Radiation is both a form of energy and a means whereby energy is transferred from one place or object to another. Although radiation takes many forms (see Chapter 15), they all consist of electromagnetic waves traveling at the speed of light (Fig. 14.11). Unlike conduction and convection, radiation does not require a material medium; that is, radiation can pass through a vacuum. Although interplanetary space is not a true vacuum, it is so highly rarefied that conduction and convection play no role in transporting heat from the Sun to the Earth-atmosphere system. Rather, radiation is the principal means whereby planet Earth is heated by the Sun. Radiation is also the means whereby heat escapes from Earth to space.

All objects both absorb and emit radiation. When radiation is absorbed, electromagnetic energy is converted to heat. When radiation is emitted, heat is converted to electromagnetic energy and lost to the environment. If an object absorbs more radiation than it emits (*radiational heating*), its temperature rises, and if an object emits more radiation than it absorbs (*radiational cooling*), its temperature falls. At equilibrium, when rates of absorption and emission are the same, the object's temperature remains constant.

The Earth-atmosphere system is in radiative equilibrium with its surroundings, but this does not mean that the various components of the system (for example, oceans, land, glaciers) have constant temperatures. Heat may be redistributed among the components of the Earth-atmosphere system so that air temperature at a particular location may undergo significant short- and long-term variations. Hence, global radiative equilibrium does not preclude changes in climate.

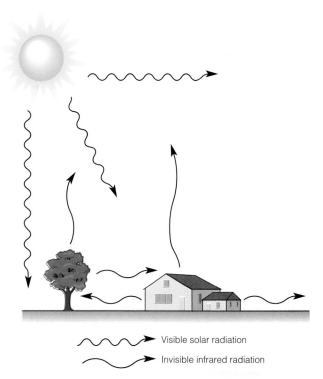

Visible solar radiation

Invisible infrared radiation

 FIGURE 14.11 Radiation is both a form of energy and a means of energy transfer; it consists of electromagnetic waves that propagate at the speed of light.

 HEAT TRANSFER AND SPECIFIC HEAT

When heat is transported from one place to another within the Earth-atmosphere system, whether by conduction, convection, or radiation, temperatures change. But the magnitude of the temperature response to an input (or output) of a specified quantity of heat varies from one substance to another depending on the specific heat. **Specific heat** is the amount of heat needed to change the temperature of 1 gram of a substance by 1 Celsius degree. Table 14.5 lists the specific heats of some familiar materials. Note that the specific heat of water is 1 calorie per gram per Celsius degree.

When the same quantity of heat is supplied to two substances having different specific heats, the substance with the lower specific heat will experience the greater temperature rise. This helps explain why the beach sand feels warmer than the water when you walk barefoot along a beach in summer. The specific heat of moist quartz beach sand is about one-quarter the specific heat of water. Hence, 1 calorie of heat raises the temperature of 1 gram of water 1 Celsius degree, compared to about 4 Celsius degrees for the same quantity of beach sand.

Differences in specific heat also help explain why the temperature of a land surface varies much more over time than does the temperature of the surface of a body of water. The materials that compose a land surface (e.g., soil, vegetation, asphalt) have much lower specific heats than water does. Hence, in response to the same input or output of heat, a land surface warms up more during the day (and in

TABLE 14.5 Specific Heats of Some Familiar Substances	
SUBSTANCE	**SPECIFIC HEAT** (cal per gm per C°)
Water	1.000
Wet mud	0.600
Ice (at 0°C)	0.478
Wood	0.420
Moist sand	0.240
Aluminum	0.214
Brick	0.200
Granite	0.192
Dry sand	0.188
Dry air	0.171
Copper	0.093
Silver	0.056
Gold	0.031

summer) and cools down more at night (and during the winter). Resistance to a change in temperature is referred to as *thermal inertia*.

Differences in the mechanisms and rate of heat transport also contribute to the greater thermal inertia of a water surface versus a land surface. Land is opaque to solar radiation so solar heating is confined to the land's surface. Because soil and air are poor conductors of heat, the heat is conducted away from the surface very slowly. In contrast, sunlight readily penetrates water to some depth and is absorbed (converted to heat) through great volumes of water. Also, circulation of sea or lake water readily transports heat throughout the entire volume of the water body. For all these reasons, during the day, the temperature of a land surface rises more than does the temperature of the surface of an adjacent body of water. At night, land surfaces typically radiate heat to space more readily than do water surfaces. Consequently, at night, the land surface cools down more than the water surface does.

To a large extent, air temperature is governed by the temperature of the surface over which the air resides or travels. Thus, air over the ocean or large lakes exhibits smaller temperature changes with time than does air over land. The climatic implications of this are significant. Places that are immediately downwind of an ocean have a *maritime climate* with relatively little difference between average winter and summer temperatures. Places that are well inland have a *continental climate* with much greater differences between average winter and summer temperatures. Compare, for example, the average monthly temperatures for San Francisco and St. Louis (Fig. 14.12). Summers are warmer and winters are colder in continental St. Louis than in maritime San Francisco even though the two cities are at about the same latitude.

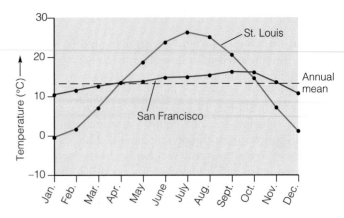

FIGURE 14.12 Average monthly temperatures for San Francisco and St. Louis. Note the greater contrast between winter and summer at continental St. Louis versus maritime San Francisco.

in contact with air. In a tiny fraction of a second, billions upon billions of gas molecules bombard every square centimeter of Earth's surface. The cumulative force produced by those impacts is the total air pressure.

For our purposes, it is convenient to think of **air pressure** as the weight per unit area of a column of air. The column of air stretches from where we are measuring the air pressure (Earth's surface, for example) to the top of the atmosphere. Hence, air pressure always decreases with increasing altitude.

Air Pressure Measurement and Units

The standard weather instrument for measuring air pressure is the *barometer,* which is available in two basic types: mercurial and aneroid. The *mercurial barometer* consists of a glass tube about a meter (39 in) in length, open at one end and sealed at the other, and nearly filled with mercury. The tube is oriented vertically with its open end extending into a small reservoir of mercury (Fig. 14.13). The mercury column settles down the glass tube until the weight of the column just balances the weight of the atmosphere (air pressure) acting on the surface of the mercury reservoir.

At sea level, the average air pressure is about 1.0 kg per square cm (14.7 lb per in^2), which supports the mercury column to a height of 760 mm (29.92 in). This is the origin of the common practice of expressing air pressure in units of length (so many inches or millimeters of mercury). The height of the mercury column varies directly with air pressure: rising air pressure forces the column to rise, and falling air pressure allows the column to drop.

AIR PRESSURE

Television and radio weathercasters routinely report the latest air pressure reading along with local air temperature, relative humidity, wind speed and direction, and sky conditions. The significance of air pressure becomes apparent when its variations are tracked over a period of many days. Significant changes in weather accompany relatively small fluctuations in air pressure. Rising air pressure usually means that the weather stays fair or improves, and falling air pressure usually signals increasing cloudiness and stormy weather. What is air pressure and why is it linked to changes in the weather?

Pressure is the same as a *stress,* that is, a force acting over some surface area. In this case the force is exerted by the gas molecules that compose air. Always in rapid and random motion, gas molecules impact the surfaces of all objects that come

Partially evacuated

Mercury

Mercury column

Weight of mercury column

Atmospheric pressure

FIGURE 14.13
A mercurial barometer.

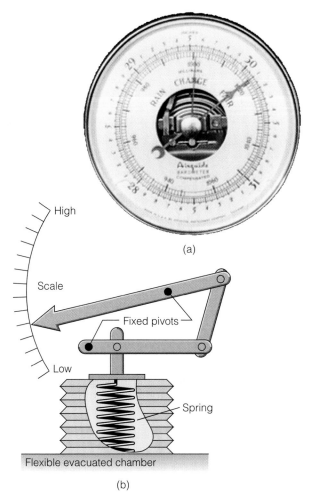

(a)

(b)

■ **FIGURE 14.14** (*a*) An external view of an aneroid barometer. (*b*) An internal view.

An *aneroid barometer* is more portable but less accurate than a mercurial barometer. The heart of this instrument is a flexible metal chamber from which much of the air has been removed (■ Fig. 14.14). A spring prevents the chamber from collapsing. Rising air pressure compresses the chamber, and falling air pressure allows it to expand. Gears and levers translate the flexing of the chamber to a pointer on a dial that is graduated in millimeters (or inches) of mercury or other units of air pressure.

Aneroid barometers for use in the home often have markings on the dial such as *fair, changeable,* and *stormy* corresponding to certain ranges of air pressure. Never take these markings literally because a specific type of weather does not always accompany a particular barometer reading. A more useful approach for barometer-based weather forecasting is to determine the *air pressure tendency,* or the variation in air pressure over some interval of time. For this purpose, aneroid barometers are equipped with a second pointer that functions as a reference marker. First turn the knob on the barometer face so that the pointer corresponds to the current air pressure reading. Then, at some later time (say, in two or three hours), compare the new pressure reading to the reading set earlier. The difference in the two readings is the air pressure tendency.

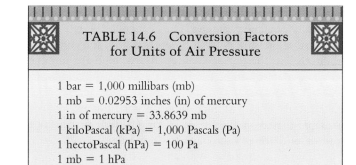

TABLE 14.6   Conversion Factors for Units of Air Pressure

1 bar = 1,000 millibars (mb)
1 mb = 0.02953 inches (in) of mercury
1 in of mercury = 33.8639 mb
1 kiloPascal (kPa) = 1,000 Pascals (Pa)
1 hectoPascal (hPa) = 100 Pa
1 mb = 1 hPa
1 in of mercury = 33.8639 hPa
1 in of mercury = 25.4 millimeters (mm) of mercury

Physically, it is more precise to express air pressure in units of pressure rather than units of length (millimeters or inches of mercury). The metric unit of pressure is the *Pascal (Pa)*. Thus, the average air pressure at sea level is 101,325 Pa, 1013.25 hectoPascals (hPa), or 101.325 kiloPascals (kPa). Traditionally, U.S. atmospheric scientists use the *millibar (mb)* as the official unit of air pressure where 1 mb equals 1 hPa or 100 Pa. The average sea-level air pressure is 1013.25 mb. Conversion factors for the various units of air pressure appear in ◎ Table 14.6.

Changes with Altitude

As ▣ Figure 14.15 shows, air pressure drops with increasing altitude. Air pressure is the weight of a column of air, and with ascent into the atmosphere, there is progressively less air above. But the rate of fall is not uniform: air pressure drops

■ **FIGURE 14.15** The fall in air pressure with increasing altitude.

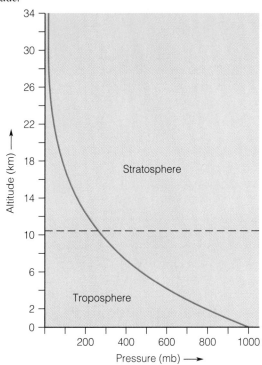

rapidly with altitude in the lower atmosphere and then more gradually at higher altitudes. The change in the rate of decline of air pressure with altitude stems from the compressibility of air; that is, air's density and volume are variable. Because gravity compresses the atmosphere, the number of gas molecules per unit volume is maximum at Earth's surface and declines with increasing altitude. *Thinning* of air is so rapid that at an altitude of 16 km (10 mi), air density is only about 10% of its average sea-level value.

Rapid thinning of air with altitude has important implications for human health. Especially above an altitude of 2,400 m (8,000 ft), the number of oxygen molecules per unit volume is so low that many people develop distressing symptoms of altitude sickness (shortness of breath, headache, and nausea). After a few days, the body adjusts to some extent, a process known as *acclimatization*. The heart beats more rapidly, breathing is deeper, and the number of red blood cells—transporters of oxygen in the bloodstream—increases. Acclimatization has limits, however. Even genetically endowed people have not been able to acclimatize permanently to elevations above about 5,200 m (17,000 ft). For this reason, airliners must be pressurized.

Although air pressure and air density drop rapidly with altitude, Earth's atmosphere has no clearly defined top. About 99% of the atmosphere's mass is below an altitude of 32 km (20 mi), but traces occur hundreds of kilometers beyond. Thus, no single altitude marks the boundary between the atmosphere and interplanetary space. Instead, Earth's atmosphere gradually merges with the highly rarefied interplanetary gases, hydrogen and helium.

Spatial and Temporal Changes

At any point in time, barometer readings at Earth's surface differ from one locality to another. Spatial variations in air pressure are not entirely due to differences in elevation of weather stations. In fact, meteorologists are much more interested in factors other than elevation that influence barometer readings. Hence, they routinely eliminate the influence of station elevation on air pressure by adjusting local air pressure readings upward to what the air pressure would be if the weather station were actually at sea level. This process is known as a *reduction to sea level*.

After barometer readings are reduced to sea level, air pressure at a specific point in time always varies from one place to another. For example, on the surface weather map in ◉ Figure 14.16, *isobars,* lines of equal air pressure, join localities reporting the same air pressure. Air pressure (reduced

◉ **FIGURE 14.16** *Isobars* are lines on a weather map that join locations having the same air pressure (in millibars reduced to sea level). "H" indicates a region of relatively high air pressure, and "L" indicates a region of relatively low air pressure.

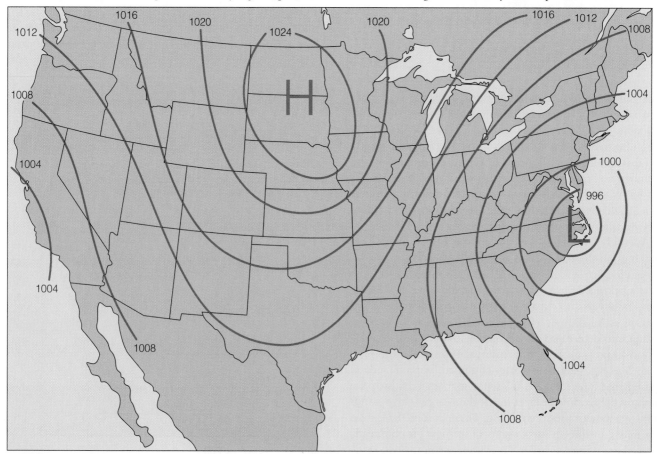

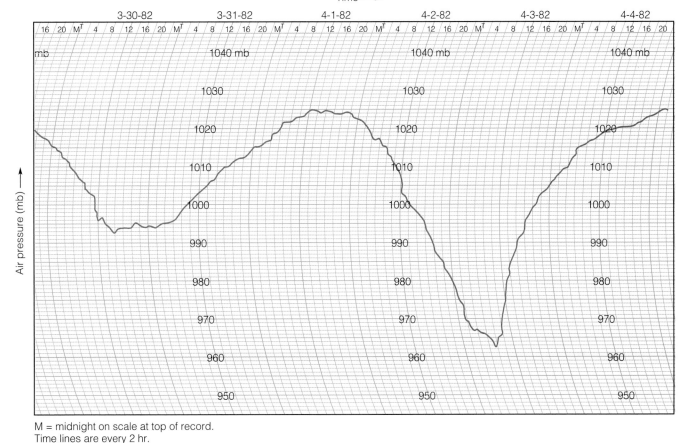

Time →

M = midnight on scale at top of record.
Time lines are every 2 hr.

■ FIGURE 14.17 An example of how air pressure (in millibars) varies with time at a particular location.

to sea level) at a particular location also varies with time as shown in ▣ Figure 14.17.

Spatial and temporal variations in air pressure are mostly due to the continuous procession of air masses that characterizes the weather of middle latitudes. Some air masses exert higher air pressure at Earth's surface than others do. An *air mass* is a huge volume of air, overlying thousands of square kilometers, that is relatively uniform horizontally in both temperature and water vapor concentration (humidity). The pressure exerted by an air mass depends upon air density, which, in turn, hinges mostly on air temperature and, to a lesser extent, on humidity. As air masses move from place to place, surface air pressures fall or rise, and the weather changes.

All other factors being equal, a cold air mass exerts higher surface air pressures than does a warm air mass. (*Surface air pressure* refers to the barometer reading at Earth's surface.) Recall that temperature is a measure of the average kinetic energy of individual molecules and that heating air causes its constituent molecules to move about more actively. The atmosphere is not confined by walls (except Earth's surface), so air is free to expand and contract. Thus, within the atmosphere, air density is variable. Heating air increases the space between neighboring gas molecules and thus reduces

air density. Conversely, cooling air reduces the space between neighboring gas molecules and thus increases air density. As the density of a column of air changes, so too does the pressure exerted by the column on Earth's surface. A column of warm air is less dense than a column of cold air of the same thickness and thus exerts less pressure.

Contrary to popular opinion, humid air is *less* dense than dry air at the same temperature. Water vapor reduces the density of air because the molecular weight of water is less than the average molecular weight of dry air. When water vaporizes into the atmosphere, water molecules take the place of other heavier gas molecules, mainly nitrogen (N_2) and oxygen (O_2). Adding water vapor thus makes the mixture of gases less dense.

A warm, dry air mass produces higher surface air pressures than does an equally warm but humid air mass. A cold, humid air mass is accompanied by higher surface air pressures than a mild and equally humid air mass. As one air mass replaces another, the air pressure changes. The air pressure rises, for example, as a surge of bitter cold air from Canada pushes out the milder air that was entrenched over an area. On the other hand, the arrival of a surge of warm, humid air from off the Gulf of Mexico leads to a fall in air pressure.

The Flight of a Baseball

In his fascinating book, *The Physics of Baseball,* Robert K. Adair describes the forces that influence the trajectory of a pitched or batted ball.★ Some of those forces arise from atmospheric conditions. As a ball leaves a bat or a pitcher's hand, it meets with air resistance whose magnitude varies directly with air density. Air density, in turn, is affected by air pressure, temperature, and humidity. As we have seen, air thins and air pressure drops with increasing altitude, while air density decreases with rising temperature and rising humidity.

On a calm day, a fly ball might travel 122 m (400 ft) in Boston's Fenway Park, which is near sea level. A baseball hit with the same force and at the same angle in Denver's mile-high baseball park would travel about 9 m (30 ft) farther. The difference occurs because the ball encounters slightly less air resistance at Denver due to lower air density and pressure. The average air pressure at Denver is about 83% of that at Boston. As a rule of thumb, a high fly ball travels about 1.8 m (6 ft) farther for every 25 mm (1.0 in) drop in barometer reading.

Air temperature has a greater effect on air density (and, hence, air resistance) than does humidity. Adair points out that our 122 m (400 ft) fly ball will travel about 6 m (20 ft) farther at an air temperature of 35°C (95°F) than at 7°C (45°F). By contrast, a baseball will travel less than a meter farther on a very humid day versus a day of very low humidity. To this point, we have assumed calm conditions. Depending on direction and speed, winds will either lengthen or shorten the trajectory of a baseball. For example, a 16 km (10 mi) per hour wind at a batter's back will add about 9 m (30 ft) to a 122 m (400 ft) home run.

Baseball is a game of inches so changes in weather sometimes make a difference in the outcome of a game. With different atmospheric conditions, a game-ending fly ball that is easily caught at the outfield fence might be a game-winning, out-of-the-park home run.

★Robert K. Adair, *The Physics of Baseball,* 2d ed. (New York: Harper Perennial, 1994).

 ## THE IONOSPHERE AND THE AURORA

The **ionosphere** is a portion of the upper atmosphere (mainly at altitudes of 80 to 400 km or 50 to 250 mi) that is named for its relatively high concentration of ions. An *ion* is an electrically charged atomic-scale particle. Highly energetic solar radiation that enters the upper atmosphere strips electrons from oxygen and nitrogen atoms and molecules, leaving them as positively charged ions.

The ionosphere has no apparent influence on day-to-day weather, but it reflects radio signals, thereby making possible long-distance radio transmissions. Radio signals are forms of electromagnetic radiation that travel in straight lines and bounce back and forth between Earth's surface and the ionosphere (Fig. 14.18). By repeated reflections, a radio signal may travel completely around the planet.

The ionosphere is also the site of the spectacular **aurora** that appears in the clear night sky as bands or overlapping curtains of greenish yellow light, occasionally fringed with pink or violet (Fig. 14.19). The *aurora borealis,* or *northern lights,* appears in the Northern Hemisphere, and the *aurora australis,* or *southern lights,* in the Southern Hemisphere.

FIGURE 14.18 Radio signals travel in straight lines in all directions away from a transmitter, but some signals are reflected back toward Earth's surface by the ionosphere. This ionospheric reflection is strongest at night and makes possible long-distance radio communication.

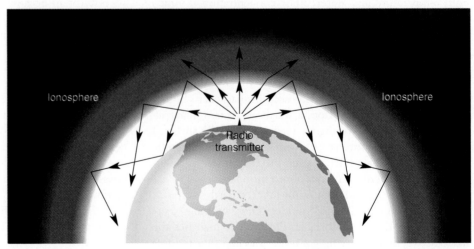

Ionosphere

Ionosphere

Radio transmitter

Earth

FIGURE 14.19 The aurora consists of spectacular curtains of colored light that appear in the clear night sky. The aurora is caused by an interaction between the solar wind and Earth's magnetic field.

An interaction between the solar wind and Earth's magnetic field is primarily responsible for the aurora. The *solar wind* is a stream of electrically charged subatomic particles (protons and electrons) that continually emanates from the Sun and travels off into space at speeds of 400 to 500 km (250 to 300 mi) per second. Like a rock in a stream, Earth's magnetic field deflects the solar wind and is thereby deformed into a teardrop-shaped cavity surrounding the planet known as the *magnetosphere* (Fig. 14.20). The solar wind's interaction with the magnetosphere generates beams of electrons that collide with atoms and molecules in the ionosphere. The collisions tear apart molecules, excite

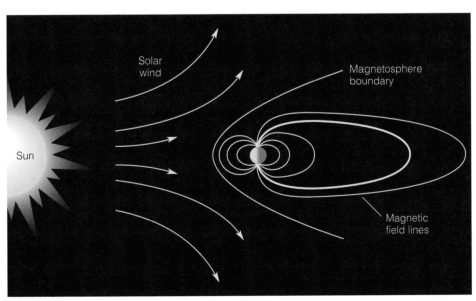

Solar wind

Magnetosphere boundary

Sun

Magnetic field lines

FIGURE 14.20 Like a rock in a stream, Earth's magnetic field deflects the solar wind and is thereby deformed into the magnetosphere, a teardrop-shaped cavity surrounding the planet.

■ FIGURE 14.21 A solar flare.

atoms, and increase ion and electron densities. As atoms shift down from their excited (energized) states and as ions combine with free electrons, they emit radiation, part of which is visible as the light of the aurora.

Earth's magnetic field channels some solar wind particles into two doughnut-shaped belts, called the *auroral ovals,* which are centered on the north and south geomagnetic poles. These belts of almost continuous aurora are situated within 20 and 30 degrees of latitude of the geomagnetic poles. The Northern Hemisphere auroral oval is centered on the northwest edge of Greenland at latitude 78.5° N and longitude 69° W. Consequently, the aurora is usually visible only at high latitudes.

Solar activity governs auroral activity. When the Sun is quiet, the auroral ovals shrink, and when the Sun is active, they expand equatorward, and the aurora is visible across the northern United States and sometimes even farther south. An active Sun is characterized by gigantic explosions called *solar flares* (■ Fig. 14.21). A solar flare is a brief event (lasting perhaps an hour) that produces a shock wave that surges at 500 to 1,000 km (300 to 600 mi) per second through the solar wind. Collision of the shock wave with the magnetosphere expands the auroral ovals equatorward.

1. Weather varies from one place to another and with time. Weather is the state of the atmosphere at a specific place and time. Climate encompasses average weather plus extremes in weather at a given location over some interval of time.

2. The modern atmosphere is the product of a lengthy evolutionary process that began about 4.6 billion years ago. Outgassing played a key role in this evolution by contributing water vapor, carbon dioxide, and nitrogen. Photosynthesis was the primary source of free oxygen (O_2).

3. Within the homosphere, the lowest 80 km (50 mi) of the atmosphere, the principal atmospheric gases, nitrogen (N_2) and oxygen (O_2), occur everywhere in the same relative proportion (about 4 to 1).

4. The significance of an atmospheric gas or aerosol is not always related to its relative concentration. Water vapor and carbon dioxide, for example, are *minor* in concentration but play important roles in the atmosphere.

5. Heat is the total kinetic energy of the atoms or molecules that compose a substance. Temperature, on the other hand, is directly proportional to the average kinetic energy of the individual atoms or molecules.

6. The Fahrenheit temperature scale continues to be used in the United States for weather reports. The Celsius temperature scale is more convenient in that a 100-degree interval separates the freezing and boiling points of pure water. The Kelvin scale is based on absolute zero and is a more direct measure of average kinetic-molecular activity.

7. The atmosphere is subdivided into four concentric layers (troposphere, stratosphere, mesosphere, and thermosphere) based on the vertical profile of average temperature. Almost all weather takes place in the troposphere, the lowest subdivision of the atmosphere.

8. In response to a temperature gradient, heat flows from areas of higher temperatures toward areas of lower temperatures. This is a consequence of the second law of thermodynamics. Heat flows via conduction, convection, and radiation.

9. Air (especially when still) is a relatively poor conductor of heat.

10. Convection is the transport of heat within a fluid via motions of the fluid itself. Convection is much more important than conduction in transporting heat within the troposphere.

11. All objects absorb and emit radiation. At radiative equilibrium, absorption balances emission, and the temperature of the object is constant. If absorption exceeds emission, the temperature rises, and if emission exceeds absorption, the temperature falls.

12. The temperature of a body of water is less variable than the temperature of a land surface for three reasons: the specific heat of water is higher than that of land, solar radiation readily penetrates water but land is opaque, and water circulates heat whereas land only conducts heat.

13. Because of the moderating influence of large bodies of water, the winter-to-summer temperature contrast is greater in continental climates than in maritime climates.

14. Air pressure and air density drop rapidly with increasing altitude in the lower atmosphere and then decline more gradually aloft.

15. Barometer readings are reduced to sea level to remove the influence of weather station elevation on air pressure.

16. Within the atmosphere, air density varies inversely with air temperature. Hence, all other factors being equal, cold air masses are denser and exert higher surface air pressures than do warm air masses.

17. Within the atmosphere, air density also varies inversely with water vapor concentration. Hence, at equivalent temperatures, dry air masses are denser and exert higher surface air pressures than do humid air masses.

18. As a general rule, high or rising air pressure signals fair weather, and low or falling air pressure means stormy weather.

19. The ionosphere is situated in the upper atmosphere and features a relatively high concentration of charged particles (ions and electrons).

20. An aurora (northern or southern lights) develops in the ionosphere when the solar wind interacts with Earth's magnetic field. The result is a spectacular display of curtains of color in the night sky at high latitudes.

IMPORTANT TERMS

absolute zero	conduction	radiation	stratosphere
aerosols	convection	radiosonde	temperature
air pressure	heat	second law of	troposphere
atmosphere	ionosphere	thermodynamics	weather
aurora	outgassing	specific heat	
climate	photosynthesis		

1. Distinguish between *weather* and *climate*. Explain why describing climate exclusively in terms of average weather may be misleading.
2. What role did outgassing play in the evolution of Earth's atmosphere?
3. An atmospheric gas that is a by-product of photosynthesis is:
   a. _____ carbon dioxide; b. _____ nitrogen; c. _____ argon; d. _____ oxygen; e. _____ ozone.
4. This gas, required for photosynthesis, was the most abundant atmospheric gas at one time in the geologic past:
   a. _____ oxygen; b. _____ nitrogen; c. _____ ozone; d. _____ carbon dioxide; e. _____ water vapor.
5. Most aerosols are by-products of activities that take place at Earth's surface. List several natural sources of aerosols and some sources related to human activity.
6. Explain how certain *minor* constituents of the atmosphere are essential for continuation of life on Earth.
7. What is the difference between heat and temperature?
8. Why is the Celsius temperature scale more convenient than the Fahrenheit temperature scale for most scientific purposes?
9. What is the significance of absolute zero?
10. Why is it a good idea to shelter a thermometer?
11. The thermal subdivision of the atmosphere in which most weather takes place is the:
    a. _____ troposphere; b. _____ stratosphere; c. _____ mesosphere; d. _____ thermosphere; e. _____ ionosphere.
12. Heat always flows from relatively warm areas to relatively cold areas. Explain how this illustrates the second law of thermodynamics.
13. Describe three processes whereby heat is transported in response to an air temperature gradient. Provide a common example of each process.
14. Air is a relatively poor conductor of heat. How is this property of air used in insulating a home?
15. Why is convection more important than conduction in the transport of heat within the troposphere?

16. Define specific heat.
17. What is the relationship between specific heat and thermal inertia?
18. Which one of the following cities has the most continental climate?
    a. _____ San Diego, California; b. _____ Los Angeles, California; c. _____ Boston, Massachusetts; d. _____ Chicago, Illinois; e. _____ Washington, D.C.
19. Define air pressure and describe how it varies with altitude.
20. If air pressure is not reduced to sea level, which one of the following cities would have the lowest average air pressure?
    a. _____ Boston, Massachusetts; b. _____ Dallas, Texas; c. _____ Chicago, Illinois; d. _____ Seattle, Washington; e. _____ Denver, Colorado.
21. Does Earth's atmosphere have a clearly defined top? Explain your response.
22. How does air temperature influence the pressure exerted by a column of air?
23. How does the concentration of water vapor influence the pressure exerted by a column of air?
24. Raising the concentration of the water vapor component of air decreases air density. Explain why.
25. As a general rule, how does the weather change as the air pressure at some location on Earth's surface rises or falls?
26. Which air mass exerts a greater surface air pressure, a warm, humid air mass or an equally warm, dry air mass? Explain your response.
27. List the advantages of an aneroid barometer over a mercurial barometer.
28. Explain why air pressure tendency can be useful in forecasting the weather.
29. Why is the aurora usually only visible at high latitudes?
30. How and why does auroral activity vary with activity on the Sun?

1. If air is such a poor conductor of heat, why do people go to the trouble of wrapping hot water pipes with insulating materials for energy conservation?
2. It is not necessary to construct a building so that it can withstand the weight exerted by the atmosphere. Explain why.
3. Suppose that an aneroid barometer is used as an altimeter, an instrument that measures altitude. How would changes in air temperature affect readings on the altimeter?

4. Within the homosphere, oxygen thins with increasing altitude, that is, the density of oxygen decreases with altitude. Yet, the concentration of oxygen (in parts per million) is uniform throughout the homosphere. Explain this apparent contradiction.

ADDITIONAL READINGS

Allegre, C. J., and S. H. Schneider. 1994. The evolution of the Earth. *Scientific American* 271, no. 4: 66–75.

Driscoll, D. M. 1987. Windchill: The 'brr' index. *Weatherwise* 40, no. 6: 321–26.

Hill, J. 1991. *Weather from above: America's meteorological satellites.* Washington, D.C.: Smithsonian Institution Press.

Houston, C. S. 1992. Mountain sickness. *Scientific American* 267, no. 4: 58–66.

Hughes, P., and D. LeComte. 1996. Tragedy in Chicago. *Weatherwise* 49, no. 1: 18–20.

Snow, J. T., M. E. Akridge, and S. B. Harley. 1992. Basic meteorological observations for schools: Atmospheric pressure. *Bulletin of the American Meteorological Society* 73: 781–94.

Snow, J. T., and S. B. Harley. 1987. Basic meteorological observations for schools: Temperature. *Bulletin of the American Meteorological Society* 68: 468–96.

WORLD WIDE WEB SITES

For current updates and exercises, log on to http://www.wadsworth.com/geo.

UNIVERSITY OF ILLINOIS: THE DAILY PLANET
http://www.atmos.uiuc.edu/

This site includes modules on a variety of weather topics such as air pressure and atmospheric optics and has a cloud catalog and climate summaries.

NATIONAL CENTER FOR ATMOSPHERIC RESEARCH (NCAR)
http://www.ucar.edu/metapage.html

NCAR research initiatives are summarized and links to real-time weather information are provided at this site.

USATODAY WEATHER PAGE
http://www.usatoday.com/weather/wlead.htm

A wealth of weather information and forecasts are featured. The site also includes answers to the most frequently asked questions about the weather, climate, and environmentally related issues.

CHAPTER 15

Solar and Terrestrial Radiation

OUTLINE

The sun is the ultimate source of energy that drives the circulation of the atmosphere and is responsible for weather.

Prologue

Two of today's most publicized environmental issues are the threats to the stratospheric ozone shield and the potential for global warming. Both issues are related to the impact that human activities are having on the flow of radiational energy into and out of the Earth-atmosphere system.

Within the stratosphere, one set of chemical reactions continuously generates ozone (O_3) from oxygen (O_2) while another set of reactions causes ozone to break down into oxygen. The net effect is a trace concentration of stratospheric ozone. Both sets of chemical reactions are powered by ultraviolet radiation, an energetic and invisible form of solar radiation. Its use in the formation and destruction of ozone prevents potentially dangerous levels of ultraviolet radiation from reaching Earth's surface. Indeed, without the protective *ozone shield,* life as we know it could not have evolved.

The principal concern regarding the ozone shield is the threat posed by a group of chemicals known as chlorofluorocarbons (CFCs). Specifically, CFCs break down in the stratosphere and liberate chlorine (Cl), a gas that reacts with and destroys ozone. Less stratospheric ozone means more intense ultraviolet radiation at Earth's surface and a greater risk of skin cancer.

As pointed out in Chapter 14, atmospheric carbon dioxide (CO_2) slows the escape of heat from Earth to space. Heat is transported as invisible infrared radiation. In this way, carbon dioxide (along with water vapor and certain other gases) contributes to the planet's so-called *greenhouse effect* and significantly elevates temperatures in the lower troposphere. Without this normal greenhouse effect, Earth's surface would be considerably colder than it is now.

The principal concern today is the upward trend in the concentration of atmospheric carbon dioxide due mainly to the burning of coal and oil for energy. Increased levels of carbon dioxide are likely to enhance the natural greenhouse effect and, if uncompensated, could cause global warming. Higher global temperatures are likely to disrupt agriculture, shrink glaciers, and raise sea level. In fact, some atmospheric scientists argue that a doubling of atmospheric carbon dioxide, which could occur by late in the next century, could trigger the greatest change in climate since the Pleistocene.

 ## INTRODUCTION

The Sun drives the atmosphere; that is, the Sun is the source of the energy that powers the circulation of the atmosphere that ultimately is responsible for weather and its variability. The Sun continually emits energy to space in the form of electromagnetic radiation. The tiny fraction of solar energy that is intercepted by Earth is converted into other forms of energy including heat and the kinetic energy of the winds.

This chapter describes how solar radiation heats the Earth-atmosphere system and how the system responds by emitting infrared radiation to space. A major focus is the role greenhouse gases play in making the surface of the planet warm enough to sustain life. On a global scale, imbalances in radiational heating and radiational cooling produce temperature gradients, and the atmosphere circulates in response. Thereby heat is transported from Earth's surface to the troposphere and from the tropics to the middle and high latitudes.

 ## ELECTROMAGNETIC RADIATION

All objects both emit and absorb *electromagnetic radiation,* so named because this form of energy has both electrical and magnetic properties. Radio waves, microwaves, infrared radiation, visible light, ultraviolet radiation, X rays, and gamma rays are all forms of electromagnetic radiation. Together, they comprise the **electromagnetic spectrum** (Fig. 15.1).

Electromagnetic radiation travels as waves that are distinguished by wavelength or frequency. *Wavelength* is the distance between successive wave crests (or, equivalently, wave troughs), as shown in Figure 15.2. *Wave frequency* is the number of wave crests (or troughs) that pass a point in a specified period of time, usually 1 second. Passage of one complete wave is a *cycle,* and a frequency of one cycle per second equals 1.0 hertz (Hz). Wave frequency varies inversely with wavelength, so the shorter the wavelength, the higher the frequency.

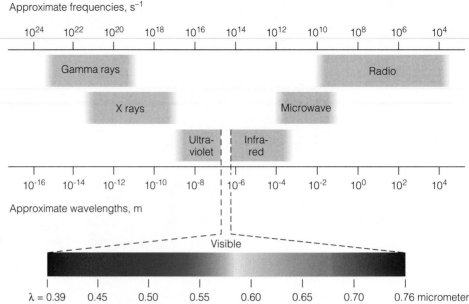

Approximate frequencies, s⁻¹

10^{24} 10^{22} 10^{20} 10^{18} 10^{16} 10^{14} 10^{12} 10^{10} 10^{8} 10^{6} 10^{4}

Gamma rays

Radio

X rays

Microwave

Ultra-violet

Infra-red

10^{-16} 10^{-14} 10^{-12} 10^{-10} 10^{-8} 10^{-6} 10^{-4} 10^{-2} 10^{0} 10^{2} 10^{4}

Approximate wavelengths, m

FIGURE 15.1 The electromagnetic spectrum. The various forms of electromagnetic radiation are distinguished on the basis of wavelength (λ), frequency, and energy level.

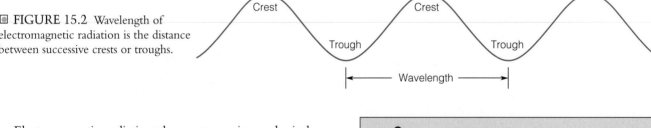

Visible

λ = 0.39   0.45   0.50   0.55   0.60   0.65   0.70   0.76 micrometer

FIGURE 15.2 Wavelength of electromagnetic radiation is the distance between successive crests or troughs.

Wavelength

Crest   Crest

Trough   Trough

Wavelength

Electromagnetic radiation does not require a physical medium. Radiation can travel through a vacuum as well as through gases, liquids, and solids. In a vacuum, all types of electromagnetic radiation travel at maximum speed, 300,000 km per second (186,000 mi per second). Electromagnetic radiation slows down when passing through materials, with its speed depending on the wavelength and the type of material. In passing from one substance to another, electromagnetic radiation may be reflected or refracted (bent) at the interface. Electromagnetic radiation may also be absorbed, that is, converted to heat.

Several physical laws describe the properties of electromagnetic radiation emitted by a perfect radiator, a so-called blackbody. At a specified temperature, a **blackbody** absorbs all incident radiation at every wavelength; a blackbody reflects no radiation. Although neither the Sun nor the Earth is a blackbody, they closely approximate perfect radiators, so blackbody radiation laws can be applied to them with useful results. Two important blackbody radiation laws are Wien's displacement law and the Stefan-Boltzmann law.

Wien's displacement law holds that the wavelength of the most intense radiation emitted by an object is inversely proportional to the object's absolute (kelvin) temperature. Hence, hot objects (such as the Sun) emit peak radiation at shorter wavelengths than do colder objects (such as the Earth-atmosphere system).

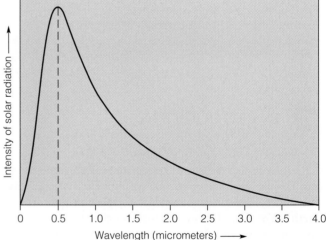

FIGURE 15.3 Blackbody radiation curve for the Sun. The most intense radiation is emitted at a wavelength of about 0.5 micrometer (in the visible portion of the electromagnetic spectrum).

The variation in radiation intensity with wavelength for a blackbody at the same radiating temperature as the Sun, about 6,000°C (11,000°F), is shown in Figure 15.3. Most of the radiation emitted by the Sun is at wavelengths between 0.25 and 2.5 micrometers, but it is most intense at a

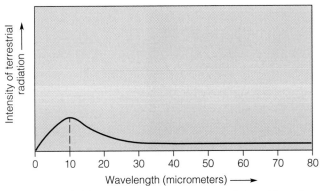

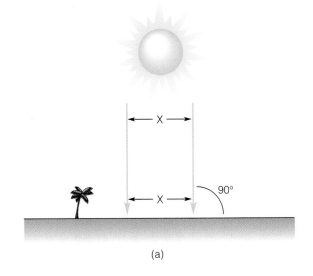

(a)

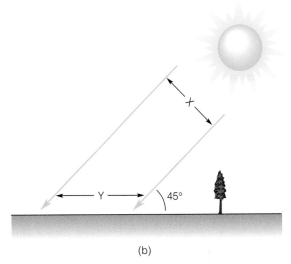

(b)

⊡ **FIGURE 15.4** Blackbody radiation curve for Earth's surface. The most intense radiation is emitted at a wavelength of about 10 micrometers (in the infrared portion of the electromagnetic spectrum).

wavelength of about 0.5 micrometer (green light). (A micrometer is one millionth of a meter.) The variation of radiation intensity with wavelength for a blackbody at the average radiating temperature of Earth's surface, about 15°C (59°F), is shown in ⊡ Figure 15.4. Earth's surface emits primarily infrared radiation (mostly at wavelengths between 4 and 24 micrometers) with maximum intensity at a wavelength of about 10 micrometers.

The blackbody curves in Figures 15.3 and 15.4 represent the total radiational energy emitted at all wavelengths per unit surface area. The vertical scale is not the same for the two figures because the Sun emits considerably more total radiational energy than does the Earth-atmosphere system. The **Stefan-Boltzmann law** describes this contrast in radiational energy emission: The total energy radiated by an object at all wavelengths is proportional to the fourth power of its absolute (kelvin) temperature (T^4). The Stefan-Boltzmann law predicts that the hotter Sun's energy output per square meter is about 160,000 greater than that of the cooler Earth-atmosphere system.

 INPUT OF SOLAR RADIATION

Earth intercepts only about one two-billionth of the enormous quantity of energy continually radiated by the Sun to space. Solar radiation that reaches Earth consists of mostly visible, infrared, and ultraviolet radiation. The input of solar radiation varies with latitude and time of the year and is the primary reason for differences in air temperature within the Earth-atmosphere system.

Solar Altitude

The intensity of the solar radiation striking Earth's surface varies significantly over the course of a year. The Sun is lower in the sky in winter than in summer, and its rays are less intense. Also, winter days are shorter than summer days. Significant changes in the intensity of solar radiation also oc-

⊡ **FIGURE 15.5** The intensity of the solar radiation received at Earth's surface depends on the solar altitude. (a) Maximum intensity occurs at a solar altitude of 90°; (b) the intensity decreases at lower solar altitudes. Before striking Earth's surface, both beams have the same intensity (the same cross section X). Upon striking the surface, the beam in panel (b) spreads out over a larger area and is less intense than the beam in panel (a). Cross section Y is greater than cross section X.

cur through the course of a day; the Sun's rays are more intense at noon than at sunrise or sunset.

The angle of the Sun above the horizon, called the **solar altitude,** influences the intensity of solar radiation received at Earth's surface. As ⊡ Figure 15.5 shows, where the noon Sun is directly overhead (solar altitude is 90°), the Sun's rays striking the surface are most concentrated (intense). With decreasing solar altitude, solar radiation spreads over a greater area of the surface and thus becomes less intense.

Solar altitude always varies with latitude. The Sun is so far from Earth—about 150 million km (93 million mi) on average—that solar radiation reaches the planet as parallel beams of uniform intensity. Earth is very nearly a sphere and

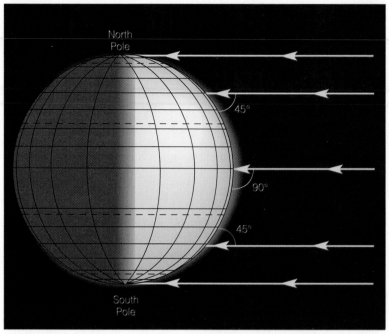

FIGURE 15.6 Solar altitude always varies with latitude and is maximum (90°) where the noon Sun is directly overhead.

presents a curved surface to the beams of incoming solar radiation. Hence, solar altitude always decreases north and south of the latitude belt where the noon Sun is directly overhead (☐ Fig. 15.6).

Another factor that affects the intensity of solar radiation received at Earth's surface is the length of the path sunlight travels through the atmosphere. The Sun's rays travel a longer path at low solar altitudes than at high solar altitudes. In addition, at low solar altitudes, more interaction occurs between incoming solar radiation and the component gases and aerosols of the atmosphere. For this reason also, the intensity of solar radiation diminishes as solar altitude decreases.

Whereas solar altitude influences the intensity of solar radiation striking Earth's surface, the length of day affects the total amount of radiational energy that is received. Variations in both solar altitude and day length accompany the planet's motions in space and seasonal changes.

Earth's Motions in Space

Once every 24 hours, Earth makes one complete rotation on its axis. Hence, at any hour, half the planet is dark (night), and the other half is illuminated by solar radiation (day).

Each year (actually 365.25 days), Earth completes one revolution about the Sun in a slightly elliptical orbit. The eccentricity of Earth's orbit is so slight that the Earth-to-Sun distance varies by only about 3.3% through the year. The planet is closest to the Sun (147 million km, or 91 million mi) on about January 3 and farthest from the Sun (152 million km, or 94 million mi) on about July 4, the dates of *per-*

ihelion and *aphelion,* respectively. In the Northern Hemisphere, Earth is closest to the Sun in winter and farthest from the Sun in summer. Hence, the eccentricity of Earth's orbit is not responsible for the seasons.

Seasons are caused by the 23°27′ tilt of Earth's equatorial plane to its orbital plane (☐ Fig. 15.7). This tilt causes Earth's orientation to the Sun to continually change as it orbits the Sun. The Northern Hemisphere *tilts away* from the Sun in winter and *tilts toward* the Sun in summer. When the Northern Hemisphere tilts away from the Sun, the Southern Hemisphere tilts toward the Sun. Thus, when the Northern Hemisphere is experiencing winter, the Southern Hemisphere is enjoying summer, and vice versa.

Changes in Earth's orientation to the Sun are accompanied by changes in solar altitude and day length. In the winter hemisphere, solar altitudes are lower, days are shorter, and the solar radiation striking Earth's surface is less intense. In the summer hemisphere, solar altitudes are higher, days are longer, and the solar radiation striking Earth's surface is more intense. Winters are colder than summers because the solar radiation is less intense and there are fewer hours of sunshine.

The latitude belt where the Sun's noon rays are most intense (solar altitude is 90°) shifts back and forth between 23°27′ south of the equator and 23°27′ north of the equator. On March 21 and September 23, the Sun's noon position is directly over the equator. Day and night are of equal length (12 hours) everywhere except at the poles (☐ Fig. 15.8a on page 346). These dates are the **equinoxes** (Latin for *equal nights*).

Following the equinoxes, the Sun continues its apparent shift toward its maximum poleward locations, its **solstice**

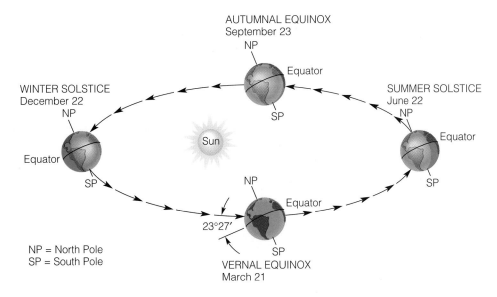

◙ **FIGURE 15.7** The seasons are due to the 23°27′ tilt of Earth's equatorial plane to the plane of its orbit about the Sun. The eccentricity of Earth's orbit is greatly exaggerated in this drawing.

latitudes. On June 22, the Sun's noon rays are directly overhead at 23°27′ N, the latitude belt called the *Tropic of Cancer.* As Figure 15.8b shows, on that date there are 24 hours of daylight north of the *Arctic Circle* (66°33′ N) and 24 hours of darkness south of the *Antarctic Circle* (66°33′ S). Elsewhere, days are longer than nights in the Northern Hemisphere, where the solstice is the first day of summer, and shorter than nights in the Southern Hemisphere, where it is the first day of winter.

On December 22, the noon Sun is directly overhead at 23°27′ S latitude, the *Tropic of Capricorn.* As Figure 15.8c shows, there are 24 hours of daylight south of the Antarctic Circle and 24 hours of darkness north of the Arctic Circle. Elsewhere, nights are longer than days in the Northern Hemisphere, where the solstice is the first day of winter, and days are longer than nights in the Southern Hemisphere, where it is the first day of summer.

In summary, the intensity of solar radiation is maximum along the latitude belt where the noon Sun is directly overhead. North and south of that latitude, the intensity of solar radiation diminishes as the solar altitude decreases. Furthermore, as ◙ Figure 15.9 shows, the seasonal (winter-to-summer) contrast in day length increases with latitude.

◈ INTERACTIONS OF SOLAR RADIATION

Solar radiation that is intercepted by Earth during its motions in space interacts with the atmosphere and the surface of the planet. As a result of these interactions, some solar radiation is returned to space and some is converted to heat. This section focuses on the nature of these interactions.

Solar Radiation and the Atmosphere

Solar radiation interacts with gases, aerosols, and clouds as it travels through the atmosphere. These interactions involve reflection, scattering, and absorption. Solar radiation that is not reflected or scattered back to space or absorbed by the atmosphere reaches Earth's surface.

Reflection occurs at the interface between two different media such as air and cloud when some of the radiation striking (incident on) that interface is sent back. The fraction of incident radiation reflected by a surface is the **albedo** of that surface:

$$albedo = \frac{reflected\ radiation}{incident\ radiation}$$

where albedo is usually expressed as a percentage. Surfaces having a relatively high albedo for sunlight appear light colored whereas those having a relatively low albedo appear dark colored.

Within the atmosphere, cloud tops are strong reflectors of solar radiation (◙ Fig. 15.10 on page 348). Cloud-top albedo varies from under 40% for thin clouds to 80% or more for thick clouds. The average albedo for all cloud types and thicknesses is about 55%, and at any time, clouds cover about 60% of the planet.

Scattering is the dispersal of radiation in all directions— up, down, and sideways. (Hence, reflection is a special case of scattering.) Both gas molecules and suspended aerosols scatter solar radiation but with some important differences. Scattering by nitrogen and oxygen molecules varies by wavelength. Blue-violet light is scattered much more than the other colors that compose visible light. This is the principal reason for the color of the daytime sky. On the other hand, water droplets and ice crystals in clouds scatter all

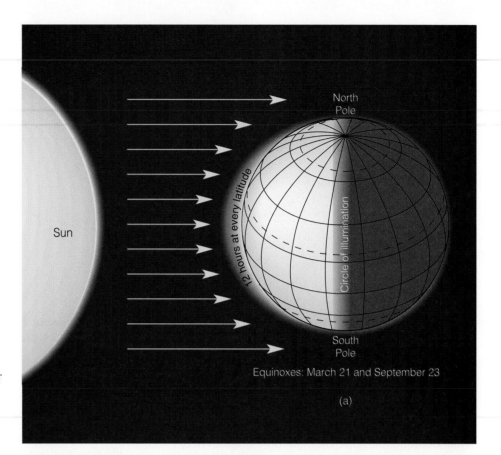

FIGURE 15.8 The distribution of solar radiation and day length at (*a*) the equinoxes, (*b*) the Northern Hemi–sphere summer solstice, and (*c*) the Northern Hemisphere winter solstice.

North Pole

Sun

12 hours at every latitude

Circle of illumination

South Pole

Equinoxes: March 21 and September 23

(a)

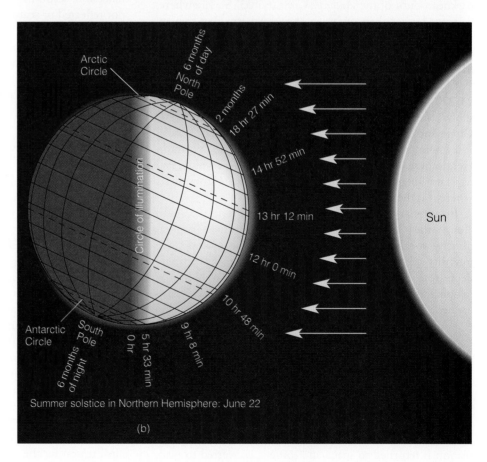

Arctic Circle

6 months of day

North Pole

2 months

18 hr 27 min

14 hr 52 min

13 hr 12 min

Sun

12 hr 0 min

Circle of illumination

10 hr 48 min

9 hr 8 min

Antarctic Circle

South Pole

5 hr 33 min 0 hr

6 months of night

Summer solstice in Northern Hemisphere: June 22

(b)

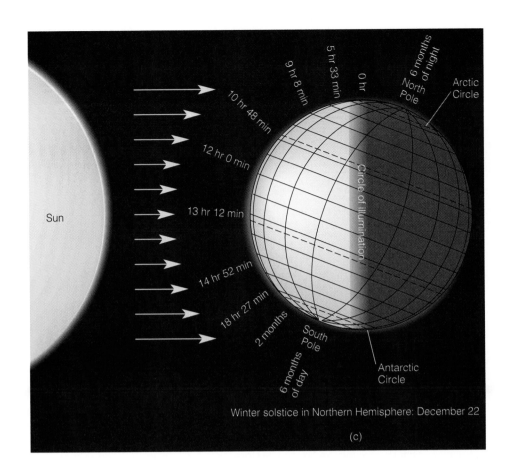

Winter solstice in Northern Hemisphere: December 22

(c)

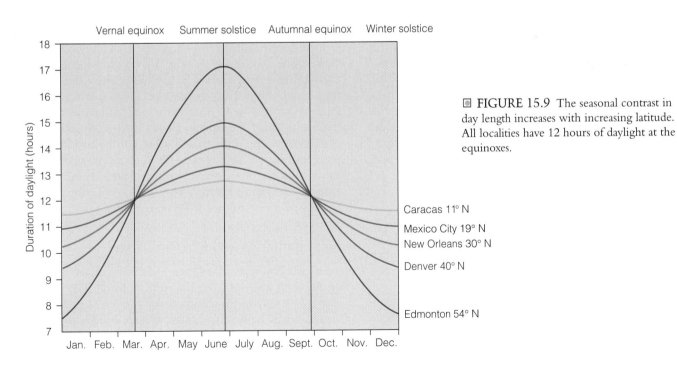

Vernal equinox   Summer solstice   Autumnal equinox   Winter solstice

Caracas 11° N
Mexico City 19° N
New Orleans 30° N
Denver 40° N
Edmonton 54° N

▣ FIGURE 15.9 The seasonal contrast in day length increases with increasing latitude. All localities have 12 hours of daylight at the equinoxes.

■ FIGURE 15.10 The top surfaces of clouds appear bright because they are strong reflectors of sunlight. Clouds appear white because water droplets and ice crystals scatter all wavelengths of visible solar radiation equally.

wavelengths of visible solar radiation equally, so clouds are white.

Reflection and scattering within the atmosphere alter the path of solar radiation. Through **absorption,** however, radiation is converted to heat. The principal absorbers of solar radiation are oxygen, ozone, water vapor, and aerosols. Gases absorb selectively by wavelength, absorbing strongly in some wavelengths and weakly or not at all in others. Clouds, on the other hand, are relatively poor absorbers of solar radiation, typically absorbing less than 10% of the radiation that strikes the cloud top.

Stratospheric Ozone Shield

Ozone (O_3) is a relatively unstable molecule consisting of three oxygen atoms. Ozone's impact on life can be either negative or positive depending on where it occurs in the atmosphere. Within the lower troposphere, ozone is an important component of photochemical smog and a serious air pollutant that is harmful to plants and animals. In the stratosphere, ozone shields organisms at Earth's surface from exposure to high intensities of solar ultraviolet (UV) radiation. Without this stratospheric **ozone shield,** terrestrial life could not exist on the planet.

Ultraviolet radiation is absorbed during both the formation and the destruction of ozone within the stratosphere. At altitudes between about 10 and 50 km (6 and 31 mi), UV breaks down oxygen molecules (O_2) into oxygen atoms (O), which subsequently combine with other oxygen molecules to form ozone molecules (O_3). At the same time, UV converts ozone to oxygen (■ Fig. 15.11). These competing chemical reactions produce a maximum ozone concentration of only about 10 ppm (parts per million) at altitudes of 20 to 25 km (12 to 16 mi). Some solar UV passes through the ozone shield and reaches Earth's surface where overexposure may cause sunburn and other serious human health problems, such as skin cancer. For this reason, recent indications that the stratospheric ozone shield is thinning have aroused concern and sparked an international effort to protect the ozone shield (see ■ Perspective 15.1).

Solar Radiation and Earth's Surface

Solar radiation that strikes Earth's surface is either reflected or absorbed depending on the surface albedo. The portion that is not reflected is absorbed and converted to heat. As noted earlier, light surfaces have higher albedos than dark surfaces. New snow's albedo ranges between 75 and 95%; that is, 75 to 95% of the solar radiation striking the surface of a fresh snow cover is reflected, and the rest (5 to 25%) is absorbed. At the other extreme, the albedo of a dark surface, such as a conifer forest, may be as low as 5%. ○ Table 15.1 lists the albedos of some common surfaces.

High-energy ultraviolet
radiation strikes an oxygen
molecule (O_2) . . .

Ozone (O_3) absorbs a range of
ultraviolet radiation . . .

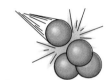

. . . and causes it to split into
two free oxygen atoms (O).

. . . splitting the molecule into one
free oxygen atom (O) and one
molecule of ordinary oxygen (O_2).

The free oxygen atoms collide
with molecules of oxygen . . .

The free oxygen atom then can
collide with an ozone
molecule . . .

. . . to form ozone
molecules (O_3).

. . . to form molecules of oxygen.

(a) Ozone production

(b) Ozone destruction

■ **FIGURE 15.11** (*a*) Chemical reactions involved in the natural production of ozone (O_3) within the stratosphere; (*b*) chemical reactions involved in ozone's destruction.

TABLE 15.1   Reflectivity (Albedo) of Some Common Surface Types for Visible Solar Radiation

SURFACE	ALBEDO (percentage reflected)
Grassland	16–20
Deciduous forest	15–20
Coniferous forest	5–15
Green crops	15–25
Tundra	15–20
Desert	25–30
Sand	18–28
Blacktopped road	5–10
Sea ice	30–40
Fresh snow	75–95
Old snow	40–60
Glacier ice	20–40
Water (high solar altitude)	3–10
Water (low solar altitude)	10–100
Cities	14–18

The albedo of some surfaces varies with the solar altitude. This variation is especially pronounced for the surface of bodies of water. With clear skies, the albedo of a lake or sea surface increases with decreasing solar altitude. The albedo sharply increases for solar altitudes of less than 30° and approaches 100% near sunrise and sunset. The average albedo of the ocean surface is only about 8%; hence, the ocean is a strong absorber of solar radiation. Low albedo is the reason the ocean appears dark in a visible satellite image such as ■ Figure 15.12.

■ **FIGURE 15.12** In this visible satellite image, the ocean surface appears dark because ocean water strongly absorbs solar radiation. A visible satellite image is like a black and white photograph in which the most reflective (highest albedo) surfaces such as clouds or snow cover appear bright white.

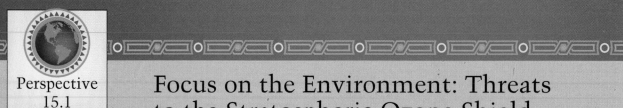

Focus on the Environment: Threats to the Stratospheric Ozone Shield

A group of chemicals, known as CFCs (for *chlorofluorocarbons*), may be eroding the stratospheric ozone shield. Until an international ban was imposed on their production and use in 1996, CFCs had been widely used for many decades as chilling agents in refrigerators and air conditioners and as blowing agents for manufacturing foam insulation. (Their use as propellants in aerosol sprays was banned in the 1970s.) Within the troposphere, CFCs are chemically nonreactive and are thus long-lived. But when atmospheric circulation transports CFCs into the stratosphere, intense UV radiation breaks them down, releasing chlorine (Cl). Chlorine acts as a catalyst in chemical reactions that convert ozone to oxygen. Through a chemical feedback loop, each chlorine atom can destroy tens of thousands of ozone molecules.

Less ozone in the stratosphere likely would mean more intense UV radiation at Earth's surface and a greater risk of skin cancer, eye damage including cataracts, and immune deficiencies. Various scientific studies suggest that a 2.5% thinning of the ozone shield could boost the incidence of human skin cancers by 10%.

The first signal that the stratospheric ozone shield was thinning came from Antarctica. For about six weeks during the Southern Hemisphere spring (mainly September and October), the ozone layer in the Antarctic stratosphere thins. By sometime in November (December at the latest), the ozone level recovers. The area of ozone depletion is about the size of the continental United States and is known as the *Antarctic ozone hole* (☐ Fig. 1). Record or near-record ozone depletion was measured in the Antarctic stratosphere every spring from 1989 through at least 1996.

Discovery of the Antarctic ozone hole prompted scientists to search for a reason. A field study conducted during the Antarctic spring of 1987 found ozone depletion of 50% and unusually high concentrations of chlorine monoxide (ClO), one of the products of chemical reactions known to destroy ozone. For many scientists, discovery of chlorine monoxide was proof-positive that, as had been suspected since the early 1970s, CFCs were eroding the stratospheric ozone shield.

Examination of trends in stratospheric ozone levels elsewhere around the globe revealed a downward trend in stratospheric ozone at midlatitude ground stations for all seasons since 1970. Shorter-term satellite measurements also revealed a negative trend in ozone concentration in both hemispheres except near the equator where no significant change was detected.

First reports of a significant increase in the UV radiation received at Earth's surface came from Toronto, Ontario. Scientists of Canada's Atmospheric Environment Service (AES) found that summer levels of UV increased about 7% annually between 1989 and 1993. In the same period, winter levels of UV climbed more than 5% each year.

The first international response to the threats to the stratospheric ozone shield was the *Montreal Protocol* of September 1987. Delegates from 23 nations agreed to reduce production and use of CFCs by 50% (from 1986 levels) by June 1998. Subsequently, the growing threat of ozone depletion prompted a change in this timetable. In November 1992, representatives of more than half the world's nations agreed to phase out CFCs and other ozone-depleting chemicals by the mid-1990s. The long residence time of CFCs, however, means that they will remain in the atmosphere for a century or longer before being cycled out by natural processes. Consequently, the chlorine concentration in the atmosphere is likely to peak around the year 2000 and then decline.

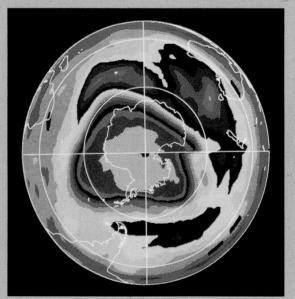

☐ FIGURE 1 The Antarctic ozone hole, an area where stratospheric ozone is depleted, appears violet in this satellite image.

INFRARED RESPONSE AND THE GREENHOUSE EFFECT

With continual absorption of solar radiation, the temperature of the Earth-atmosphere system would steadily rise. Actually, the average global air temperature varies little from one year to the next because solar heating of the planet is balanced by the escape of heat (through infrared radiation) to space. Although solar radiation is supplied to only about half the planet at any time, the entire Earth-atmosphere system emits infrared radiation ceaselessly, day and night.

Greenhouse Warming

Because solar radiation and terrestrial infrared (IR) radiation have different properties, they interact differently with the atmosphere. Although the atmosphere is nearly transparent to visible solar radiation, certain components of the atmosphere strongly absorb IR and radiate some back to Earth's surface where it is absorbed (▣ Fig. 15.13). Infrared radiation by the atmosphere slows the escape of Earth's heat to space and elevates the average temperature of the lower troposphere.

Like the atmosphere, window glass is more transparent to visible solar radiation than to IR radiation. Greenhouses that are used for growing plants take advantage of this property

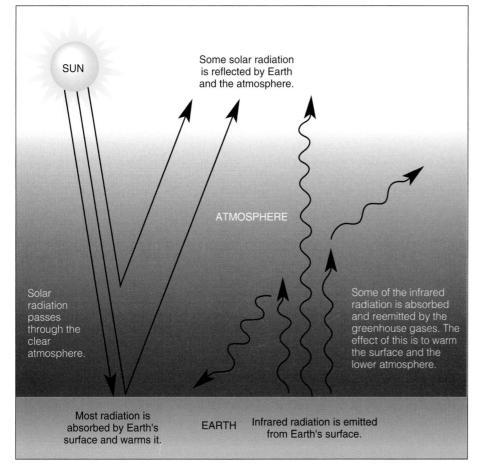

▣ FIGURE 15.13 The greenhouse effect involves the absorption and radiation of terrestrial IR radiation by water vapor, carbon dioxide, and several other gases.

■ FIGURE 15.14 The glass of a greenhouse is analogous to the IR-absorbing gases of the atmosphere in that glass is relatively transparent to sunlight but strongly absorbs IR radiation.

by being constructed mostly of glass panes (■ Fig. 15.14). Sunlight readily passes through the glass and is absorbed within the greenhouse. Some of the heat is radiated as IR, which is absorbed and radiated by the glass. Although sheltering from the wind is actually more important than radiation trapping in elevating the air temperature within the greenhouse, scientists and others often refer to the warming caused by IR-absorbing atmospheric gases as the **greenhouse effect,** and the contributing gases are known as *greenhouse gases.*

Because of the greenhouse effect, Earth's average surface temperature is much higher than it would be otherwise. Viewed from space, the planet radiates at about −18°C (0°F), whereas the average temperature at Earth's surface is about 15°C (59°F). Hence, the greenhouse effect raises Earth's surface temperature by

$$[15°C − (−18°C)] = 33 \text{ Celsius degrees}$$

or

$$(59°F − 0°F) = 59 \text{ Fahrenheit degrees}$$

The main greenhouse gas is water vapor; others are carbon dioxide, ozone, methane (CH_4), and nitrous oxide (N_2O). As Figure ■ 15.15 shows, the percentage of IR absorbed by the greenhouse gases varies with wavelength. Significantly, the percentage absorbed is very low near the wavelength of peak IR emission (10 micrometers). Most heat from Earth's surface eventually escapes to space through these so-called *atmospheric windows,* wavelength bands where there is little or no absorption of IR radiation.

Comparison of typical summer weather in the American Southwest and along the Gulf Coast illustrates the importance of water vapor in the greenhouse effect. Both regions receive about the same intensity of solar radiation and usually have afternoon temperatures above 30°C (86°F). At night, however, air temperatures can be very different. In the Southwest, where the air is less humid, IR more readily escapes to space, and surface air temperatures may drop below 15°C (59°F) by sunrise. Along the Gulf Coast, however, the air is usually more humid so that more IR is absorbed. Because a portion of this heat is radiated back to Earth's surface, nighttime temperatures may fall only into the 20s Celsius (the 70s Fahrenheit).

Clouds are composed of IR-absorbing water droplets and/or ice crystals and also contribute to the greenhouse effect. Hence, nights usually are warmer when the sky is covered with clouds than when it is clear.

Prospects for Global Warming

Today, considerable concern surrounds the potential global climatic consequences of rising concentrations of certain greenhouse gases, especially carbon dioxide (■ Fig. 15.16). Atmospheric carbon dioxide has been increasing since the Industrial Revolution primarily because of coal and oil combustion and, to a much lesser extent, because of deforestation that reduces removal of carbon dioxide by photosynthesis. More carbon dioxide in the atmosphere enhances the natural greenhouse effect and, unless compensated, may trigger significant global warming (see ■ Perspective 15.2).

Numerical models of the Earth-atmosphere system are used to estimate the amount of global warming that could accompany a continued climb in atmospheric carbon dioxide. Depending on the specific climate model, these experiments predict that Earth's average surface temperature will rise between 1 and 3.5 Celsius degrees (1.8 and 6.3 Fahrenheit degrees) with a doubling of carbon dioxide—possibly by the end of the next century. Such warming could match the globe's post-Pleistocene temperature rise, but at a rate up to 100 times faster.

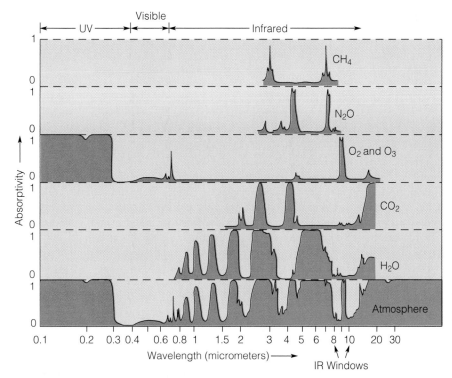

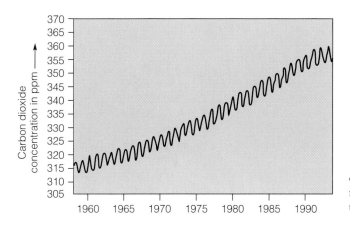

FIGURE 15.15 The percentage of solar and terrestrial IR radiation absorbed by gaseous components of the atmosphere varies with wavelength and the specific gas.

In addition to carbon dioxide, upward trends in other greenhouse gases may also contribute to global warming. Concentrations of methane (CH_4) and nitrous oxide (N_2O) and, until recently, CFCs (chlorofluorocarbons) have also been rising (○ Table 15.2). Although atmospheric concentrations of these gases are considerably less than carbon dioxide, they are more efficient absorbers of infrared radiation because they strongly absorb within *atmospheric windows.*

FIGURE 15.16 The upward trend in atmospheric carbon dioxide as recorded at Mauna Loa Observatory, Hawaii. Annual fluctuations in rates of photosynthesis and respiration account for the yearly cycle of carbon dioxide.

TABLE 15.2   Trends in Greenhouse Gases

GREENHOUSE GAS	PRESENT CONCENTRATION (ppb)[a]	PREINDUSTRIAL CONCENTRATION (ppb)	ANNUAL RATE OF INCREASE (%)
Carbon dioxide (CO_2)	353,000	280,000	0.5%
Methane (CH_4)	1,720	790	0.9
Nitrous oxide (N_2O)	310	288	0.3
Ozone (O_3)[b]	20–40	10	0.5–2.0
Chlorofluorocarbons (CFCs)	0.28–0.48	0	0.0[c]

[a]Parts per billion by volume at Earth's surface.
[b]Northern Hemisphere tropospheric ozone only.
[c]Worldwide production of CFCs ceased in 1996 by international agreement.

Focus on the Environment: Societal Impacts of Global Warming

As noted elsewhere in this chapter, a continued rise in the concentration of atmospheric carbon dioxide could cause global warming of 1.0 to 3.5 Celsius degrees (1.8 to 6.3 Fahrenheit degrees) by the year 2100. Such a change in climate would impact virtually every sector of society with many adverse and some beneficial effects.

In the fall of 1995, the Intergovernmental Panel on Climate Change (IPCC), sponsored by the United Nations, reported that if the magnitude of global warming at the high end of projections actually happened, North American climate zones would shift northward by perhaps 550 km (340 mi). The consequent disruption of ecosystems likely would cause the extinction of some vegetation species. In the tropics, higher temperatures would lead to expansion of the tropical regions where malaria-transmitting mosquitoes breed, thus increasing the incidence of that disease.

Global warming would have both pluses and minuses for agriculture. Higher temperatures at middle and high latitudes would lengthen the growing season, and more carbon dioxide in the air is known to spur plant growth. On the other hand, climate shifts in some areas would require farmers to change their cultivation practices and might lead to a drop in productivity. Higher summer temperatures and more frequent droughts in the North American grain belt, for example, would likely force farmers to switch from corn to wheat and increase their dependence on irrigation.

Amplification of warming in polar latitudes could melt enough glacial ice that sea level would rise and inundate low-lying coastal plains. Warming also causes seawater to expand, further adding to the rise in sea level. According to 1995 IPCC estimates, the combination of melting glaciers and expansion of seawater could raise sea level between 25 and 80 cm (10 and 30 in) by the year 2100. Storm waves and surges superimposed on this rise would likely greatly increase the incidence of shoreline flooding and erosion.

Projections of energy demand for space heating and cooling are mixed. Residents of middle and high latitudes would need less fuel for space heating in winter but would need more for summer air conditioning. Most Canadians are likely to have a net energy savings. One study estimated that milder winters in Ontario would cut energy demand for space heating by 45%, significantly offsetting an estimated 7% increase in energy demand for summer air conditioning.

Warmer winters would also affect transportation, lengthening the navigation season on lakes, rivers, and harbors where ice cover is a problem. Where higher temperatures are accompanied by drier conditions, however, river discharge will decrease and barge traffic may be curtailed.

Has the warming begun? Some atmospheric scientists point to recent trends in climate that are similar to the changes that global climate models predict will accompany enhanced greenhouse warming. These trends include the following:

1. Globally, the nine warmest years since the 1880s have occurred since 1979.
2. Between 1980 and 1994, the incidence of weather extremes (e.g., drought, heat, precipitation) increased in the United States.
3. Mountain glaciers in some tropical and subtropical latitudes are retreating at accelerated rates.

Although some recent climatic trends are consistent with predictions made by global climate models, there is no evidence of a cause-effect relationship between those trends and increases in greenhouse gas concentrations. Climate controls other than greenhouse gases such as volcanic activity or changes in temperature patterns of ocean surfaces may be responsible. The Earth-atmosphere system is complex and highly interactive, so other processes may compensate for an enhanced greenhouse effect. For example, the ocean has great thermal inertia, which may temper global warming. Furthermore, higher temperatures at polar latitudes may translate into higher humidities, more snowfall, and eventually more glacial ice.

HEAT IMBALANCES: ATMOSPHERE VERSUS EARTH'S SURFACE

The globally and annually averaged distribution of incoming solar radiation and outgoing terrestrial IR radiation is shown in ▣ Figure 15.17. Assume that 100 units of solar radiation enter the upper atmosphere. The Earth-atmosphere system reflects or scatters back to space about 31% of solar radiation intercepted by Earth; this is the Earth's **planetary albedo.** Of the solar radiation intercepted by the Earth-atmosphere system, only about 23% is absorbed by the atmosphere. The remaining 46% is absorbed by Earth's surface, chiefly because of the low average albedo of the ocean waters, which cover about 71% of the globe. Of the 115 units of IR emitted by Earth's surface, 106 units are absorbed by

Some scientists question the integrity of the global mean temperature record that shows a gradual warming trend from the 1880s into the 1990s (☐ Fig. 1). They point out that more sophisticated thermometers, the warming effect of urbanization, and the absence of temperature data from vast stretches of the ocean may bias the record. Criticism is also directed at climate models used to predict how rising levels of greenhouse gases might impact climate. Models differ, for example, in their ability to simulate the influence of water vapor, clouds, and ocean currents on climate. Furthermore, climate models generally do poorly in predicting regional responses to global climate change.

Although uncertainty surrounds the question of carbon dioxide–induced global warming, some experts argue that so much is at stake for society that international action should be taken immediately to cut carbon dioxide emissions. They recommend (1) less reliance on fossil fuels, (2) greater use of solar and wind power, (3) higher energy efficiencies in industries and motor vehicles (e.g., more miles per gallon), (4) ending deforestation, and (5) increased emphasis on reforestation. Even if greenhouse warming does not occur, such actions are advisable because they will help alleviate other serious environmental problems. For example, less reliance on coal will also cut emissions of sulfur dioxide, a serious air pollutant.

The first coordinated international response to the potential of global greenhouse warming took place at the United Nations' Earth Summit in Rio de Janeiro, Brazil, in June 1992. Delegates from 106 nations negotiated the world's first climate treaty, the *UN Framework Convention on Climate Change.* Their principal goal was to stabilize emissions of greenhouse gases. Provisions of the treaty called for developed (industrialized) nations to reduce carbon dioxide emissions to 1990 levels by the year 2000.

After the ratification of the treaty by 116 nations and the European Union, concern grew that these provisions might be inadequate to head off global warming. Hence, in the spring of 1995, delegates met again, this time in Berlin, Germany, and approved the so-called *Berlin Mandate,* an agreement to negotiate a new set of targets by 1997. Resistance to more stringent emissions goals is strong and is coming from oil-producing nations as well as many industrial interests. As of this writing, negotiations continue.

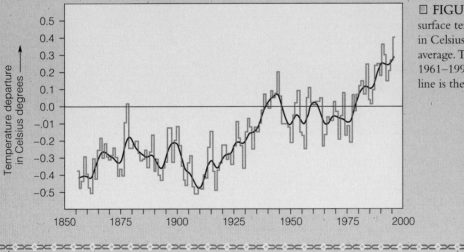

☐ FIGURE 1 Variation in global average surface temperature expressed as a departure in Celsius degrees from the 1961–1990 average. The horizontal line is the 1961–1990 mean temperature, and the red line is the trend.

clouds and greenhouse gases. Of the 106 units absorbed, 100 are radiated back to the surface (the greenhouse effect).

The global distribution of incoming solar radiation and outgoing terrestrial IR implies net warming of the surface and net cooling of the atmosphere (○ Table 15.3). Within the atmosphere, the rate of cooling due to the emission of IR to space is greater than the rate of warming due to the absorption of solar radiation. At Earth's surface, however, the rate of warming due to absorption of solar radiation exceeds the rate of cooling due to emission of IR.

The atmosphere is *not* cooling relative to Earth's surface. Heat is transferred from the surface to the atmosphere and compensates for imbalances in radiational heating and cooling. Heat transfer mechanisms involve a combination of sensible heating (conduction plus convection) and latent heating (phase changes of water). As Figure 15.17 shows, 31

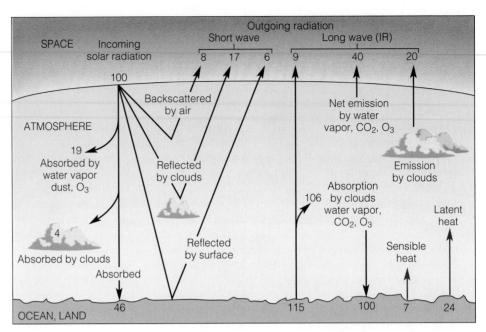

■ **FIGURE 15.17** The globally and annually averaged distribution of incoming solar radiation and outgoing terrestrial IR radiation. Assume that 100 units of solar radiation are intercepted by the Earth-atmosphere system.

TABLE 15.3   Global Radiation Balance

Solar radiation intercepted by Earth	100 units
Solar radiation budget	
Scattered and reflected to space (8 + 17 + 6)	31
Absorbed by the atmosphere (19 + 4)	23
Absorbed by Earth's surface	46
Total	100 units
Radiation budget at Earth's surface	
Infrared cooling (100 − 115)	−15
Solar heating	+46
Net heating	+31 units
Radiation budget of the atmosphere	
Infrared cooling (−40 −20 +6)	−54
Solar heating	+23
Net cooling	−31 units
Heat transfer: Earth's surface to atmosphere	
Sensible heating (conduction plus convection)	7
Latent heating (phase changes of water)	24
Net transfer	31 units

units of heat are transferred from Earth's surface to the atmosphere by these two processes: 7 units by sensible heating ($^7/_{31}$ or about 23%) and 24 units by latent heating ($^{24}/_{31}$ or about 77%).

In **sensible heating,** heat is transported by a combination of conduction and convection (see Chapter 14). During the day, heat is conducted from Earth's relatively warm surface to the cooler overlying air. Cool air is denser than warm air so the cooler air forces the warmer air to rise. Convective currents thus transport heat from the surface into the troposphere.

In **latent heating,** heat is transported through changes in the phase of water. Depending on the type of phase

change, water either absorbs heat from the environment or releases heat to it (■ Fig. 15.18). Heat energy is associated with phase changes because water's three phases represent different levels of molecular activity. Water molecules are relatively inactive in the solid phase (ice) and vibrate about fixed locations. This is why an ice cube retains its shape at subfreezing temperatures. In the liquid phase, water molecules move about with greater freedom, and water takes the shape of its container. In the vapor phase, water molecules exhibit maximum activity and readily diffuse throughout the entire volume of the container. A change in the phase of water is thus linked to a change in molecular activity, which is brought about by either an addition or a release of heat.

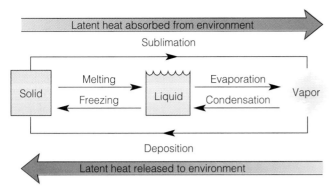

■ FIGURE 15.18 As water changes phase, latent heat is either absorbed from the environment or released to it.

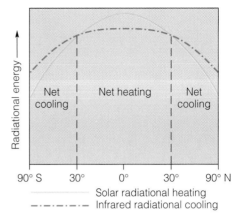

■ FIGURE 15.19 A schematic representation of the variation of incoming solar radiation and outgoing terrestrial IR radiation with latitude. Between 30° north and south, there is net radiational heating, and poleward of 30°, there is net radiational cooling.

Latent heat is involved exclusively with changing the phase of water and not with raising or lowering its temperature. Although the temperature of the environment changes, the temperature of the water undergoing a phase change remains constant until the change is complete.

As Earth's surface absorbs solar and IR radiation, some of the available heat is used to vaporize water from oceans, lakes, rivers, soil, and vegetation. This water vapor rises into the troposphere where some is converted to tiny liquid water droplets (*condensation*) or ice crystals (*deposition*) that are visible as clouds. Latent heat needed for vaporization of water is supplied at the surface, and that same heat is subsequently released to the atmosphere when clouds form. Through latent heating, then, heat is transferred from Earth's surface into the troposphere.

Sometimes heat is transported from the troposphere to Earth's surface, the opposite of the average global situation. This is the case, for example, when warm winds blow over cold, snow-covered ground. Heat transport from the lower troposphere to the surface usually occurs at night when radiational cooling causes the surface to become colder than the overlying air.

HEAT IMBALANCES: VARIATION BY LATITUDE

On a global scale, imbalances in rates of radiational heating and radiational cooling occur not only vertically, between Earth's surface and the atmosphere, but also horizontally, with latitude. The total solar radiation absorbed by the Earth-atmosphere system equals the total emission of IR radiation to space. Thus, the planet is in *radiative equilibrium* with its surroundings. Rates of radiational heating and radiational cooling vary with latitude, however, and the atmosphere circulates in response.

As noted earlier, parallel beams of incoming solar radiation strike lower latitudes more directly than higher latitudes. Hence, solar radiation is more intense per unit surface area at lower latitudes than at higher latitudes. The fall in average air temperature with increasing latitude also means

that the rate of IR emission declines with latitude. Over the course of a year, at middle and high latitudes, the rate of IR cooling exceeds the rate of solar heating. At low latitudes, the rate of solar heating exceeds the rate of IR cooling (■ Fig. 15.19).

In both hemispheres, the latitude circle at 30° roughly divides the region of *net* radiational cooling from the region of *net* radiational warming. Latitudes poleward of about 30° N and 30° S experience net radiational cooling over the course of a year, whereas tropical latitudes are sites of net warming. The tropics do not become progressively warmer relative to the middle and high latitudes, however, because heat is continually transported from the tropics into the middle and high latitudes. This **poleward heat transport** is brought about by (1) air mass exchange, (2) storms, and (3) ocean circulation.

An **air mass** is a huge volume of air covering thousands of square kilometers that is relatively uniform horizontally in temperature and humidity. The properties of an air mass are largely determined by the characteristics of the surface over which it resides or travels. Air masses that form at high latitudes over cold, snow-covered or ice-covered surfaces are relatively cold, and those that form at low latitudes are relatively warm. Air masses that develop over oceans are humid, and those that form over land are dry. Hence, the four basic types of air masses are cold and humid, cold and dry, warm and humid, and warm and dry.

Warm air masses exchange places with cold air masses. Warm air masses from lower latitudes flow poleward and are replaced by cold air masses that flow toward the equator from source regions at high latitudes. This exchange of air masses results in net transport of heat poleward.

The release of latent heat in storms is second in importance to air mass exchange in the poleward transport of heat. At low latitudes, water that evaporates from the warm ocean surface is drawn into the circulation of a developing storm. As the storm travels into higher latitudes, some of that wa-

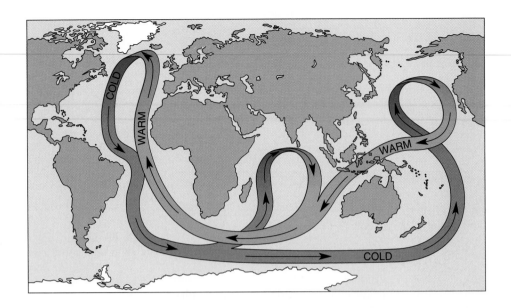

FIGURE 15.20 Huge conveyor belt–like currents that traverse the length of the ocean basins transport heat from low to middle and high latitudes.

ter vapor condenses into clouds, thereby releasing latent heat to the troposphere. Latent heat acquired at low latitudes is thereby transported into middle and high latitudes.

The ocean contributes to poleward heat transport primarily through huge conveyor belt–like currents that traverse the length of the ocean basins (Fig. 15.20). According to W. S. Broecker of Columbia University's Lamont-Doherty Earth Observatory, in the Atlantic conveyor system, warm surface waters flow northward from the tropics to near Greenland. Chilling of these waters in northern latitudes raises their density, which causes them to sink and form a bottom current that flows into the South Atlantic as far south as Antarctica. In that region, the bottom current

is warmer (and less dense) than the surface waters, so the current wells up to the surface, is chilled, and sinks again. Branches of that cold bottom current then spread northward into the Atlantic, Indian, and Pacific Oceans.

Surface water that is cooler than the overlying air is a *heat sink;* that is, heat is conducted from air to sea. And surface water that is warmer than the overlying air is a *heat source;* that is, heat is conducted from sea to air. The oceanic conveyor belts function as a heat sink in tropical latitudes and as a heat source at high latitudes, thereby contributing to poleward heat transport.

WEATHER: RESPONSE TO HEAT IMBALANCES

Imbalances in the rates of radiational heating and radiational cooling produce temperature gradients between (1) Earth's surface and the troposphere and (2) the tropics and the higher latitudes. In response, heat is transported within the Earth-atmosphere system via conduction, convection, cloud development, north-south exchange of air masses, and latent heat release in storms. In this way, the atmosphere circulates and brings about changes in weather. In summary, the Sun drives the atmosphere. Imbalances in solar heating spur atmospheric circulation (weather), which redistributes heat within the Earth-atmosphere system.

VARIATION IN AIR TEMPERATURE

Air temperature fluctuates from hour to hour, night to day, place to place, and with the seasons. The basis for atmospheric circulation described in this chapter helps explain the variability of air temperature. Together radiation and the movement of air masses regulate air temperature locally. Although these two influences on air temperature work together, it is useful initially to examine them separately.

EARTHFACT

Drought and Heat

Unusually high air temperatures often accompany *drought,* a lengthy period of extreme moisture deficit. All other factors being equal, in response to the same solar radiation, air over a dry surface warms up more than air over a moist surface. When the surface is dry, absorbed radiation is used primarily for sensible heating of the air, that is, for conduction and convection of heat from the surface into the overlying air. Hence, the air temperature is relatively high. On the other hand, when the surface is moist, much of the absorbed radiation is used to vaporize water, there is less sensible heating, and the air temperature is lower. As a drought progresses, soils dry out and lakes and other reservoirs shrink. Less moisture translates into higher air temperatures and higher rates of evaporation, and the combination of these stresses causes crops to wither and die.

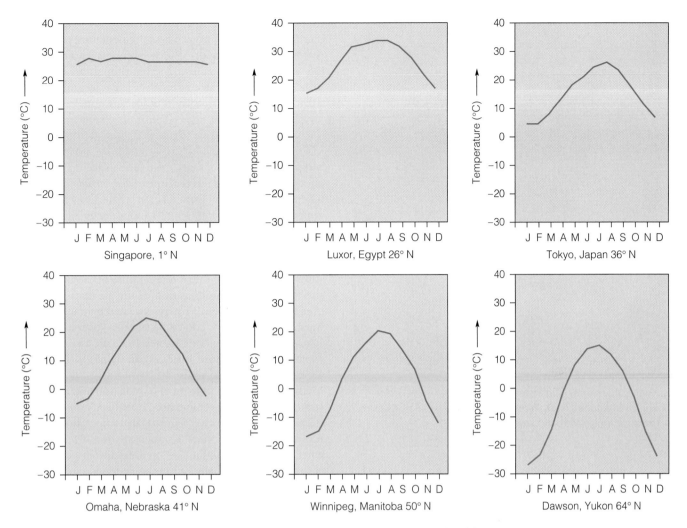

FIGURE 15.21 Average monthly air temperatures for various locations. The amplitude of the annual temperature curve generally increases with latitude because of increasing seasonal contrast in incoming solar radiation.

Radiational Controls

Variations in incoming solar radiation and outgoing IR radiation affect the local air temperature. These variations depend on (1) the time of day and the day of the year, which determine the solar altitude and the intensity and number of hours of incoming solar radiation; (2) surface characteristics that determine the amount of reflected and absorbed radiation (albedo) and the way available heat is divided between sensible and latent heating; and (3) sky conditions, because clouds affect the flow of both solar and terrestrial IR radiation. Air temperature is usually lower in January than in July (in the Northern Hemisphere), during the night than during the day, when the ground is snow covered rather than bare, when the ground is wet rather than dry, and under cloudy rather than clear afternoon skies.

The annual temperature cycle mirrors the systematic variations in solar radiation through the course of a year. Hence, the seasonal contrast in air temperature generally increases with increasing latitude. In the latitude belt between the Tropics of Cancer and Capricorn, solar altitude and day length vary little through the year, and the average monthly air temperature likewise shows little variation (Fig. 15.21). Near the equator, in fact, the temperature difference between night and day often is greater than the winter-to-summer temperature contrast.

Outside the tropics, solar radiation exhibits a pronounced maximum and minimum during the year and accounts for the distinct winter-to-summer temperature contrast observed in middle latitudes (Fig. 15.21). In areas downwind from the ocean, such as western Europe, the maritime influence tempers this seasonal temperature contrast. Poleward of the Arctic and Antarctic Circles, the seasonal difference in solar radiation is extreme, ranging from zero or very little in winter to a maximum in summer, and the annual temperature cycle reflects this contrast.

At middle and high latitudes, the warmest and coldest months of the year typically do not coincide with times of maximum and minimum solar radiation, respectively. On average, the warmest time of year occurs about a month af-

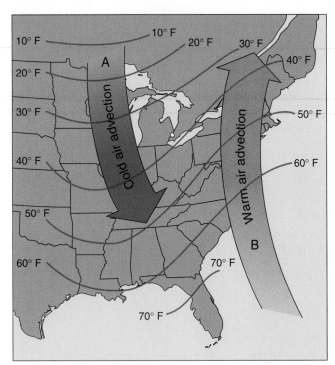

■ FIGURE 15.22 Cold air advection (arrow A) occurs when the wind blows from colder areas toward warmer areas, and warm air advection (arrow B) occurs when the wind blows from warmer areas toward colder areas.

ter the summer solstice, and the coldest time of year is about a month after the winter solstice. This lag is slightly longer in maritime climates.

A lag between radiation and the response of air temperature is also evident on a daily basis. Typically, the day's lowest air temperature is recorded near sunrise as the culmination of a night of radiational cooling. The air temperature does not begin to rise until the input of energy from the Sun exceeds the loss of energy by IR radiation. This occurs about one-half hour after sunrise. The air temperature continues to rise as long as the input of solar energy is greater than the loss of energy by IR radiation. Even though the intensity of solar radiation peaks around noon, the day's highest temperature usually occurs in early to midafternoon.

Air Mass Controls

Movement of an air mass from one place to another is known as **air mass advection.** *Cold air advection* occurs when the wind blows from a colder region to a warmer region (arrow A in ▣ Fig. 15.22), and *warm air advection* takes place when the wind blows from a warmer region to a colder region (arrow B in Fig. 15.22).

The influence of air mass advection on local air temperature depends on the initial temperature of the new air mass and the degree of modification of the air mass as it travels over Earth's surface. An air mass is heated by traveling over a warmer surface and is cooled by traveling over a colder surface. Also, as discussed in Chapter 16, as air ascends and descends, its temperature changes: air cools as it rises and warms as it descends.

To this point, the influences of radiation and air mass advecxtion on local air temperature have been considered separately. Actually, the two regulate air temperature together. Air mass advection may compensate for, or even overwhelm, radiational influences on local air temperature. The radiation balance usually causes the air temperature to climb from a minimum around sunrise to a maximum by early or midafternoon. This normal temperature behavior can change with an influx of cold air. Depending on how cold it is, the advecting air may cause air temperatures to climb more slowly than usual or remain steady. If cold air advection is extreme, air temperatures may drop throughout the day even though skies are bright and sunny.

CHAPTER SUMMARY

1. All objects emit electromagnetic radiation. The many forms of electromagnetic radiation make up the electromagnetic spectrum and can be distinguished on the basis of wavelength and frequency.
2. The Earth-atmosphere system and the Sun closely approximate perfect radiators, so blackbody radiation laws can be applied to them with useful results.
3. Wien's displacement law predicts that the wavelength of the most intense radiation varies inversely with the absolute (kelvin) temperature of the radiating object. Hence, solar radiation peaks in the visible, and Earth-atmosphere (terrestrial) radiation peaks in the infrared (IR).

4. According to the Stefan-Boltzmann law, the total energy radiated by an object at all wavelengths is directly proportional to the fourth power of its absolute temperature. Hence, the Sun emits considerably more radiational energy per unit area than does the Earth-atmosphere system.
5. Solar altitude, the angle of the Sun above the horizon, influences the intensity of the solar radiation that strikes Earth's surface. All other factors being equal, as the solar altitude increases, the intensity of solar radiation received at the surface also increases.

6. As a consequence of Earth's elliptical orbit, nearly spherical shape, and tilted rotational axis, solar radiation is distributed unevenly over Earth's surface and changes through the course of a year.

7. In the summer hemisphere, solar altitudes are higher, days are longer, there is more solar radiation, and air temperatures are higher. In the winter hemisphere, solar altitudes are lower, days are shorter, there is less solar radiation, and air temperatures are lower.

8. Solar radiation that is not reflected or scattered to space or absorbed by the atmosphere (converted to heat) reaches Earth's surface.

9. Absorption of ultraviolet (UV) radiation during the natural formation and destruction of ozone within the stratosphere shields us from potentially lethal levels of ultraviolet radiation.

10. Solar radiation that reaches Earth's surface is either reflected or absorbed depending on the surface albedo. Light-colored surfaces have a relatively high albedo for visible radiation, and dark-colored surfaces have a relatively low albedo.

11. The Earth-atmosphere system responds to solar heating by emitting infrared radiation to space. Water vapor, carbon dioxide, and other atmospheric gases absorb and reradiate some IR back toward Earth's surface, thereby significantly elevating the average temperature of the lower troposphere. Clouds also contribute to this so-called greenhouse effect.

12. Rising levels of atmospheric carbon dioxide and other IR-absorbing gases may enhance the natural greenhouse effect and cause global warming. Combustion of fossil fuels and, to a lesser extent, deforestation account for the upward trend in atmospheric carbon dioxide.

13. In the atmosphere, the rate of cooling due to infrared emission is greater than the rate of warming due to absorption of solar radiation. At Earth's surface, on the other hand, the rate of warming due to absorption of solar radiation is greater than the rate of cooling due to emission of infrared radiation. In re-sponse, heat is transported from the surface to the atmosphere via conduction, convection, and vaporization of water followed by cloud development.

14. Latent heat is either required or released when water changes phase. A phase change involves a change in molecular activity. Heat is released to the environment when water vapor condenses or when water freezes, and heat is absorbed from the environment when ice melts or water evaporates.

15. The ratio of heat used for conduction and convection (sensible heating) to that used for phase changes of water (latent heating) varies with the amount of moisture at Earth's surface.

16. The total energy (in the form of solar radiation) absorbed by the Earth-atmosphere system equals the total energy (in the form of infrared radiation) emitted to space by the Earth-atmosphere system.

17. Poleward of about 30° latitude, over the course of a year, the rate of cooling due to infrared emission to space is greater than the rate of warming due to absorption of solar radiation. In tropical latitudes, on the other hand, the rate of warming due to absorption of solar radiation is greater than the rate of cooling due to emission of infrared radiation. Poleward heat transport is the consequence.

18. Poleward heat transport is brought about by north/south exchange of air masses, release of latent heat in storms, and the flow of warm and cold ocean currents.

19. The Sun drives the circulation of the atmosphere by producing heat imbalances within the Earth-atmosphere system. Atmospheric circulation redistributes heat within the system.

20. Radiation and air mass advection regulate local air temperature. Radiation varies with time of day, day of the year, cloud cover, and properties of Earth's surface. Air mass advection occurs when winds blow from colder to warmer regions (cold air advection) or from warmer to colder regions (warm air advection).

IMPORTANT TERMS

absorption	electromagnetic spectrum	planetary albedo	solar altitude
air mass	equinoxes	poleward heat transport	solstice
air mass advection	greenhouse effect	reflection	Stefan–Boltzmann law
albedo	latent heating	scattering	Wien's displacement law
blackbody	ozone shield	sensible heating	

1. Which one of the following types of electromagnetic radiation has the longest wavelength (and lowest frequency)?
a. ___ gamma rays; b. ___ X-rays; c. ___ ultraviolet radiation; d. ___ violet light; e. ___ red light.

2. What is a blackbody?

3. In your own words define Wien's displacement law and the Stefan-Boltzmann law. Apply these laws to the Sun and the Earth-atmosphere system.

4. In the Northern Hemisphere, we are closer to the Sun during winter than during summer. Why then is winter colder than summer?

5. What is the significance of the Tropics of Cancer and Capricorn?

6. How and why does solar altitude affect the intensity of solar radiation received at Earth's surface?

7. Distinguish between scattering and reflection of radiation.

8. Which one of the following surfaces has the highest albedo for solar radiation?
a. ___ the ocean; b. ___ clouds; c. ___ asphalt pavement; d. ___ conifer forest; e. ___ fresh snow cover.

9. How does the albedo of clouds compare to the average albedo of the ocean surface?

10. In what way is the glass of a greenhouse analogous to the atmosphere?

11. Which one of the following is not a greenhouse gas?
a. ___ carbon dioxide; b. ___ nitrogen; c. ___ water vapor; d. ___ methane; e. ___ nitrous oxide.

12. How does cloudiness influence the rate at which air temperature falls at night?

13. An enhanced greenhouse effect appears likely to be the consequence of rising trends in certain greenhouse gases. Identify those gases.

14. Distinguish between radiational heating and radiational cooling.

15. On a global scale and over the course of a year, compare the rates of radiational heating and cooling of the atmosphere with the rates of radiational heating and cooling of Earth's surface.

16. How is excess heat at Earth's surface transported into the troposphere? Which of these processes is most important and why?

17. All of the following phase changes require an addition of latent heat with the exception of:
a. ___ freezing of water; b. ___ vaporization of water; c. ___ vaporization of snow; d. ___ melting of ice.

18. Explain how sensible heating and latent heating are both involved in the formation of cumulus (convective) clouds.

19. Explain why heat is released to the atmosphere when clouds form.

20. How might unusually low air temperatures be the consequence of a persistent snow cover?

21. On a globally averaged basis, which one of the following processes is most effective at cooling Earth's surface?
a. ___ conduction of heat; b. ___ convection of heat; c. ___ emission of infrared radiation; d. ___ vaporization of water; e. ___ condensation of water.

22. Under what conditions is the net flow of heat directed from the troposphere to Earth's surface?

23. Describe how radiational heating (due to absorption of solar radiation) and radiational cooling (due to emission of infrared radiation) vary with latitude.

24. What processes are involved in poleward heat transport?

25. Why is air temperature variable?

26. Although solar radiation reaches its maximum intensity around noon, air temperatures in the lower troposphere typically do not reach a maximum until several hours later. Explain why.

27. How does cloudiness influence air temperature?

28. Distinguish between cold air advection and warm air advection.

29. Is air mass advection possible with calm conditions? Explain your answer.

30. Under what conditions might the day's high temperature occur at 11 P.M.?

1. Suppose that the angle between Earth's equatorial plane and its orbital plane were to increase. How would this change affect the winter-to-summer temperature contrast in the Northern Hemisphere?

2. How does the albedo of the moon compare with the planetary albedo of Earth? Consider that the moon has no clouds and no oceans.

3. Some atmospheric scientists argue that the term "greenhouse effect" is misleading and should not be used in discussions of the planetary radiation balance. What is the basis for this argument?

4. In the absence of air mass advection, the early morning temperature does not begin to rise until the rate of absorption of solar radiation exceeds the rate of emission of infrared radiation. Explain why.

Brune, W. H. 1990. Ozone crisis: The case against chlorofluoro-carbons. *Weatherwise* 43, no. 3: 136–43.

Foukal, P. V. 1990. The variable Sun. *Scientific American* 262, no. 2: 34–41.

Graedel, T. E., and P. J. Crutzen. 1995. *Atmosphere, climate, and change.* New York: W. H. Freeman.

Lindzen, R. S. 1990. Some coolness concerning global warming. *Bulletin of the American Meteorological Society* 71: 288–99.

Stolarski, R., et al. 1992. Measured trends in stratospheric ozone. *Science* 256: 342–49.

WORLD WIDE WEB SITES

For current updates and exercises, log on to http://www.wadsworth.com/geo.

NATIONAL CLIMATIC DATA CENTER
http://www.ncdc.noaa.gov/

The data center is a source of climatic data and specialized products such as Great Lakes ice-coverage.

THE NATIONAL OCEANIC AND ATMOSPHERIC ADMINISTRATION NETWORK INFORMATION CENTER
http://nic.noaa.gov/ or http://www.noaa.gov/

This site is an entry point for many of the services provided by NOAA, including the National Weather Service and the Climate Prediction Center.

ENVIRONMENT CANADA HOME PAGE
http://www.ns.doe.ca/how.html

Weather and climate information for Canada, answers to the most frequently asked questions, and links to other Environment Canada Internet sites are available at this site.

CHAPTER

16

Humidity, Clouds, and Precipitation

On Earth, water occurs in all three phases, as invisible vapor, liquid, and ice and snow.

Prologue

Hot, dry *Santa Ana winds* blow off the high desert plateaus of Utah and Nevada and sweep southwestward around the Sierra Nevada onto the southern California coastal plain and Pacific Ocean. The winds further dry out vegetation already parched by the long, dry summer and increase the potential for wildfire. Compounding the fire threat in southern California is chaparral, a widespread shrub community whose oil-rich tissues virtually explode when ignited.

Summers are dry and winters are wet in southern California. Winter rains falling on nutrient-rich ashes of burned vegetation spur resprouting of shrubs that help stabilize hill slopes denuded by autumn wildfires. If heavy rains begin before vegetation is fully reestablished on the hillsides, soil, sediment, and debris are washed downslope into houses and streambeds and onto roads (□ Fig. 16.1).

On October 27, 1993, Santa Ana winds fanned 15 major wildfires from Ventura County south to San Diego County. Firefighters were unable to contain the flames until the winds finally died down three days later. After a respite of several days, Santa Ana winds again strengthened and whipped flames through previously unburnt areas of Malibu and Topanga. In the aftermath, government officials estimated total property damage in excess of $1 billion. Only two years earlier, on October 19 and 20, 1991, wildfires had raged across Oakland and Berkeley, killing 25 people and destroying nearly 3,000 dwellings. Fire weather in southern California illustrates the interplay among winds, precipitation regimes, local ecology, and erosion potential.

□ FIGURE 16.1 In southern California, mudflows often occur when heavy rains fall on hillsides denuded by autumn wildfires.

INTRODUCTION

Water occurs in the atmosphere in all three phases: water vapor (an invisible gas), ice crystals and water droplets in clouds, and solid and liquid precipitation. Most atmospheric water is concentrated in the lower portion of the troposphere. As part of the global hydrologic cycle, water cycles into the atmosphere as vapor from reservoirs of water at Earth's surface (oceans, rivers, lakes, soil, glaciers, vegetation). Water leaves the atmosphere and returns to the surface as rain, snow, and other forms of precipitation. The atmosphere is a relatively dynamic component of the hydrologic cycle—a single water molecule resides in the atmosphere an average of only 10 days before being cycled out.

The principal focus of this chapter is water in the atmosphere. This chapter covers the cycling of water between

Earth's surface and the atmosphere, ways of describing the water vapor concentration of air, the saturation of air through cooling, and the influence of atmospheric stability on cloud development. Clouds, precipitation types, and the formation of dew, frost, and fog are also surveyed.

HUMIDITY OF AIR

On a muggy summer afternoon, people often complain, "It's not the heat, it's the humidity." A combination of high humidity and high temperature can cause considerable personal discomfort by reducing evaporative cooling of the skin (see Perspective 14.2). Humidity varies with time and from one place to another. As the concentration of water vapor increases, clouds are more likely to form. This section addresses the usual ways of quantifying the water vapor concentration of air: vapor pressure and relative humidity.

Vapor Pressure

When water vaporizes, water molecules readily mix with the other atmospheric gases and contribute to the total air pressure. **Vapor pressure** is the pressure contributed by water vapor alone and varies directly with water vapor concentration.

Vapor pressure is very unlikely to exceed 40 mb anywhere. Even in the warm, humid air over tropical oceans, no more than about 4% of the lowest kilometer of the troposphere is water vapor. Hence, the maximum possible vapor pressure is about 4% of the mean air pressure (1013.25 mb) at sea level or 40 mb. In extremely cold localities such as interior Greenland, the vapor pressure in the lower troposphere is close to zero.

At a specified air temperature, there is an upper limit to the vapor pressure. At that maximum value, called the **saturation vapor pressure,** air is saturated with respect to water vapor. The saturation vapor pressure represents an equilibrium condition that varies with air temperature. At the interface between water and air (or between ice and air), water molecules continually shift between the liquid (or solid) and vapor phases. During *evaporation,* more water molecules enter the vapor phase than return to the liquid phase, and during *condensation,* more water molecules return to the liquid phase than enter the vapor phase. Eventually, a dynamic equilibrium may develop such that liquid water becomes vapor at the same rate that water vapor becomes liquid. At equilibrium, air is saturated, and the vapor pressure equals the saturation vapor pressure.

Elevating the temperature at the air/water interface upsets this equilibrium. The higher the temperature, the greater the average kinetic energy of individual molecules, and the more readily water molecules escape the water surface as vapor. Evaporation prevails, and a new equilibrium is established between water molecules becoming vapor and water molecules becoming liquid. Now, however, the system is at a higher temperature than before, and the water va-

TABLE 16.1 Saturation Vapor Pressure Versus Temperature

TEMPERATURE		SATURATION VAPOR PRESSURE (mb)	
°C	°F	Over Water	Over Ice
50	122	123.40	
45	113	95.86	
40	104	73.78	
35	95	56.24	
30	86	42.43	
25	77	31.67	
20	68	23.37	
15	59	17.04	
10	50	12.27	
5	41	8.72	
0	32	6.11	6.11
−5	23	4.21[a]	4.02[a]
−10	14	2.86	2.60
−15	5	1.91	1.65
−20	−4	1.25	1.03
−25	−13	0.80	0.63
−30	−22	0.51	0.38

[a]Note that for subfreezing temperatures, two different values are given: one over supercooled water and the other over ice. *Supercooled water* is liquid water at a temperature below 0°C.

por concentration is greater. Thus, the saturation vapor pressure is higher.

The saturation vapor pressure depends on the rate of vaporization of liquid water, which, in turn, is governed by temperature. As ◎ Table 16.1 shows, the saturation vapor pressure increases as the temperature rises. Hence, the water vapor concentration is higher in warm saturated air than in cold saturated air.

Relative Humidity

Television and radio weathercasters routinely report the water vapor concentration of air as relative humidity. **Relative humidity (RH)** compares the vapor pressure to the saturation vapor pressure and is expressed as a percentage:

$$RH = \frac{\text{vapor pressure}}{\text{saturation vapor pressure}} \times 100\%$$

When the vapor pressure equals the saturation vapor pressure, the relative humidity is 100%, and the air is saturated with respect to water vapor.

Suppose, for example, that the air temperature is 15°C (59°F) and the vapor pressure is 4 mb. From Table 16.1, at 15°C the saturation vapor pressure of air is about 17 mb; hence, the relative humidity is about 23.5%:

$$RH = \frac{4 \text{ mb}}{17 \text{ mb}} \times 100\% = 23.5\%$$

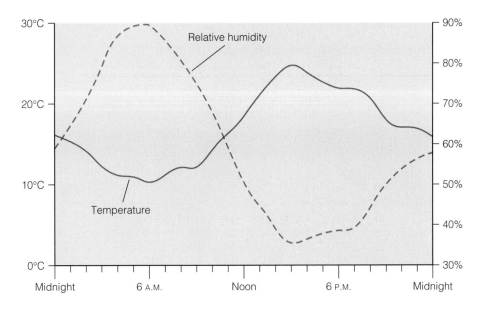

 FIGURE 16.2 Daily variation of air temperature and relative humidity. Relative humidity is highest when the air temperature is lowest and lowest when the air temperature is highest.

Relative humidity varies directly with vapor pressure and inversely with saturation vapor pressure. Because the saturation vapor pressure depends on air temperature, the relative humidity can change even though the actual concentration of water vapor does not.

As noted in Chapter 15, in the absence of cold or warm air advection, the air temperature usually rises from a minimum near sunrise to a maximum during early to midafternoon and falls thereafter. If the vapor pressure remains constant throughout the day, the relative humidity varies inversely with air temperature. Hence, the relative humidity is highest when air temperature is lowest and vice versa (◙ Fig. 16.2). Between sunrise and early afternoon, the relative humidity drops, not because the air is drying out, but because the rising air temperature increases the saturation vapor pressure.

Humidity Measurement

Atmospheric humidity is measured with a *psychrometer,* an instrument consisting of two mercury-in-glass thermometers mounted side by side on a plate that is attached to a handle (◙ Fig. 16.3). The bulb of one thermometer (*wet-bulb thermometer*) is covered by a muslin sock soaked in distilled water. The psychrometer is whirled about until the wet-bulb temperature reading steadies. Water evaporates from the muslin sock, and the latent heat required for evaporation lowers the wet-bulb temperature. The other thermometer (*dry-bulb thermometer*) measures air temperature. The drier the air, the greater the evaporation, and the lower the reading on the wet-bulb thermome-

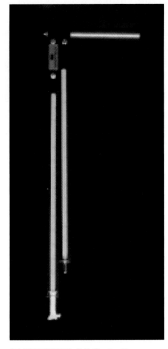

◙ FIGURE 16.3 A sling psychrometer is used to determine relative humidity.

ter compared to the dry-bulb reading. A psychrometric table calibrates the temperature difference between the two thermometers (*wet-bulb depression*) against percent relative humidity (○ Table 16.2).

Suppose, for example, that the air temperature is 10°C (50°F) and the wet-bulb depression is 3 Celsius degrees. From Table 16.2, the relative humidity is 65%. When the dry-bulb and wet-bulb readings are the same, no evaporation takes place because the air is saturated; that is, the relative humidity is 100%.

Other weather instruments, called *hygrometers,* measure relative humidity more directly and are more dependable than psychrometers at subfreezing temperatures. One type of hygrometer calibrates relative humidity against the lengthening and shrinkage of a sheaf of human hair. Another design calibrates humidity against changes in electrical resistance of certain chemicals as they adsorb water vapor from the air. This type of hygrometer is used in radiosondes (see Chapter 14).

◈ SATURATION PROCESSES

As the relative humidity approaches 100%, water vapor is more and more likely to change phase; that is, clouds are increasingly likely to form. Relative humidity increases when (1) air is cooled (causing the saturation vapor pressure to drop) or (2) water vapor is added to air (increasing the vapor pressure).

The most common means whereby the relative humidity approaches saturation and clouds form is by **expansional**

TABLE 16.2   Psychrometric Table: Relative Humidity (Percent)

DRY-BULB TEMPERATURE (°C)	WET-BULB DEPRESSION (Celsius degrees)														
	0.5	1.0	1.5	2.0	2.5	3.0	3.5	4.0	4.5	5.0	7.5	10.0	12.5	15.0	17.5
−10.0	85	69	54	39	24	10	—	—	—	—	—	—	—	—	—
−7.5	87	73	60	48	35	22	10	—	—	—	—	—	—	—	—
−5.0	88	77	66	54	43	32	21	11	1	—	—	—	—	—	—
−2.5	90	80	70	60	50	41	31	22	12	3	—	—	—	—	—
0.0	91	82	73	65	56	47	39	31	23	15	—	—	—	—	—
2.5	92	84	76	68	61	53	46	38	31	24	—	—	—	—	—
5.0	93	86	78	71	65	58	51	45	38	32	1	—	—	—	—
7.5	93	87	80	74	68	62	56	50	44	38	11	—	—	—	—
10.0	94	88	82	76	71	65	60	54	49	44	19	—	—	—	—
12.5	94	89	84	78	73	68	63	58	53	48	25	4	—	—	—
15.0	95	90	85	80	75	70	66	61	57	52	31	12	—	—	—
17.5	95	90	86	81	77	72	68	64	60	55	36	18	2	—	—
20.0	95	91	87	82	78	74	70	66	62	58	40	24	8	—	—
22.5	96	92	87	83	80	76	72	68	64	61	44	28	14	1	—
25.0	96	92	88	84	81	77	73	70	66	63	47	32	19	7	—
27.5	96	92	89	85	82	78	75	71	68	65	50	36	23	12	1
30.0	96	93	89	86	82	79	76	73	70	67	52	39	27	16	6
32.5	97	93	90	86	83	80	77	74	71	68	54	42	30	20	11
35.0	97	93	90	87	84	81	78	75	72	69	56	44	33	23	14
37.5	97	94	91	87	85	82	79	76	73	70	58	46	36	26	18
40.0	97	94	91	88	85	82	79	77	74	72	59	48	38	29	21

cooling of air. When gases expand, their temperature usually drops. For example, air released from an automobile tire is cool to the touch. Within the tire, the air pressure may be 34 lb per square inch whereas atmospheric pressure averages less than 15 lb per square inch. When the tire valve is opened, the escaping air expands and cools. The escaping air pushes aside the air that occupied the volume into which it expands. Hence, expanding air does work on its environment. Work requires energy, and the energy needed for expansion comes from the kinetic-molecular energy (heat) of the air so the air temperature drops.

In the atmosphere, expansional cooling occurs whenever air ascends. Air pressure falls with increasing altitude so air expands as it rises in the atmosphere. The same thing happens to a helium-filled balloon as it drifts skyward. Ascending air expands and cools, and if the air is unsaturated, its relative humidity increases. With sufficient ascent and expansional cooling, the relative humidity approaches 100% and clouds form.

On the other hand, when air is compressed, it warms, a process known as **compressional warming.** For example, the cylinder wall of a bicycle pump heats up as air is pumped (compressed) into a tire. The work of compressing air is converted to heat. Hence, as air descends within the atmosphere, it is compressed and warmed, and its relative humidity declines. Where air descends, clouds either vaporize or do not develop. It is no accident that the world's major deserts occur where the prevailing atmospheric circulation favors subsiding air.

Within the atmosphere, expansional cooling and compressional warming of unsaturated air are adiabatic processes. In an **adiabatic process,** no heat is exchanged between a mass and its surroundings. During ascent or descent within the atmosphere, clear air is neither heated nor cooled by radiation, conduction, phase changes of water, or mixing with its environment. Temperature changes of ascending or descending air are due only to expansion or compression.

Ascending unsaturated (clear) air cools at the rate of about 10 Celsius degrees per 1,000 m of ascent (5.5 Fahrenheit degrees per 1,000 ft). This is the *dry adiabatic lapse rate* (■ Fig. 16.4). Descending unsaturated air warms at the same rate.

If ascending air cools to saturation (100% relative humidity), water vapor condenses (or deposits), and latent heat is released to the environment. Latent heat *partially* offsets expansional cooling. Hence, ascending saturated (cloudy) air cools more slowly than ascending unsaturated (clear) air (■ Fig. 16.5). Ascending cloudy air cools at the *moist adiabatic lapse rate,* which ranges from about 4 Celsius degrees per 1,000 m (2.2 Fahrenheit degrees per 1,000 ft) for very warm saturated air to almost 9 Celsius degrees per 1,000 m (5 Fahrenheit degrees per 1,000 ft) for very cold saturated air. An average value of 6 Celsius degrees per 1,000 m (3.3 Fahrenheit degrees per 1,000 ft) is appropriate for most purposes.

The relative humidity of ascending saturated (cloudy) air remains steady at about 100%. As cloudy air expands and cools, its saturation vapor pressure drops. Water vapor converts to water droplets or ice crystals (cloud formation),

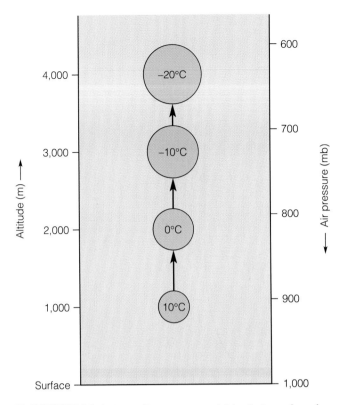

FIGURE 16.4 Ascending unsaturated (clear) air cools at the dry adiabatic lapse rate of 10 Celsius degrees for every 1,000 m of ascent.

causing the vapor pressure to fall at the same rate as the saturation vapor pressure.

ATMOSPHERIC STABILITY

Density differences within the atmosphere may either enhance or suppress vertical motion of air, depending on atmospheric stability. Think of ascending and descending currents of air as continuous streams of discrete mass units, called *air parcels*. An air parcel is subject to buoyant forces due to density differences between the parcel and the surrounding (*ambient*) air. The warmer the air parcel, the lower its density. Parcels that are warmer (lighter) than the ambient air tend to rise, and parcels that are cooler (denser) than the ambient air tend to sink. Air parcels continue to ascend or descend until they encounter ambient air having the same temperature (or density).

Atmospheric stability is determined by comparing the temperature change of an ascending or descending air parcel with the temperature profile of the ambient air in which the parcel ascends or descends. The rate of cooling of a rising air parcel depends on whether the parcel is clear (dry adiabatic lapse rate) or cloudy (moist adiabatic lapse rate); a descending air parcel warms at 10 Celsius degrees per 1,000 m. As described in Chapter 14, a radiosonde measures the vertical temperature profile (or sounding).

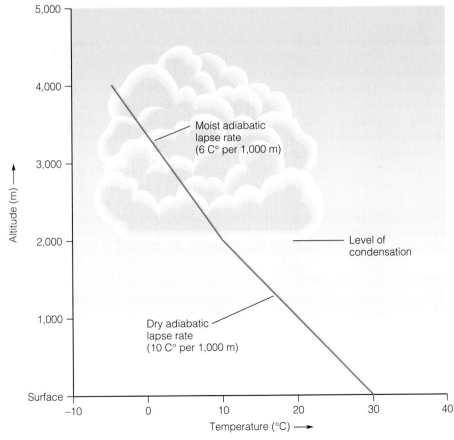

FIGURE 16.5 With sufficient ascent (and expansional cooling), air becomes saturated with water vapor. Any further ascent of the saturated (cloudy) air is accompanied by cooling at an average moist adiabatic lapse rate of 6 Celsius degrees per 1,000 m.

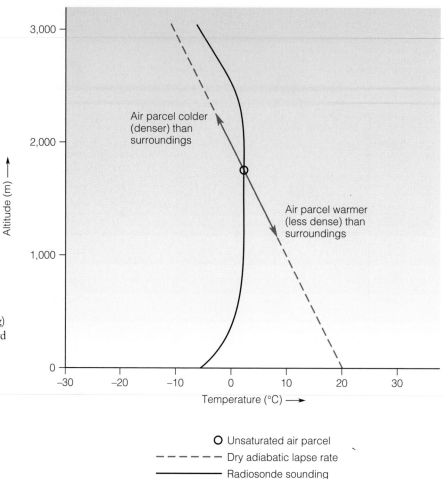

■ FIGURE 16.6 If the ambient (surrounding) air is stable, an air parcel that is displaced upward or downward returns to its original altitude. Stable air inhibits vertical motion of air.

A radiosonde sounding indicates whether the ambient air is stable, unstable, or neutral. An air parcel ascending within *stable air* becomes cooler (denser) than the ambient air, and a descending air parcel becomes warmer (less dense) than the ambient air (■ Fig. 16.6). Any upward or downward displacement of an air parcel in stable air encounters forces that return the parcel to its original altitude. An air parcel ascending within *unstable air* becomes warmer (less dense) than the ambient air and continues to ascend, and a descending air parcel becomes cooler (denser) than the ambient air and continues to descend (■ Fig. 16.7). An air parcel moving upward or downward in unstable air encounters forces that spur vertical motion. Air ascending or descending within *neutral air* has the same temperature (and density) as the ambient air, so neutral air neither inhibits nor spurs vertical motion of air parcels. Hence, clouds are most likely to develop in an unstable atmosphere.

Although lapse rates for both unsaturated (clear) and saturated (cloudy) air parcels are essentially fixed, soundings vary from day to day and even from one hour to the next. This means that atmospheric stability also varies. ■ Figure 16.8 is a plot of several sample soundings along with the dry adiabatic and (average) moist adiabatic lapse rates.

Comparison of a sounding with the dry and moist adiabatic lapse rates leads to the following conclusions:

1. If the temperature of the ambient air drops *more* rapidly with altitude than the dry adiabatic lapse rate, then the ambient air is unstable for both clear and cloudy air parcels.
2. Any sounding that plots between the dry adiabatic and moist adiabatic lapse rates indicates *conditional stability.* The ambient air is stable for clear air parcels and unstable for cloudy air parcels.
3. An air layer is stable for both cloudy and clear air parcels when the temperature of the ambient air drops more slowly with altitude than the moist adiabatic lapse rate, does not change with altitude (*isothermal*), or increases with altitude (*temperature inversion*).
4. A sounding that equals the dry adiabatic lapse rate is neutral for clear air parcels and unstable for cloudy air parcels.
5. A sounding that is the same as the moist adiabatic lapse rate is neutral for cloudy air parcels and stable for clear air parcels.

Atmospheric stability thus influences weather by affecting the vertical motion of air. Stable air suppresses vertical motion and cloud development, and unstable air enhances vertical motion, convection, expansional cooling, and cloud development. In addition, as discussed in ■ Perspective 16.1, atmospheric stability also influences air pollution potential.

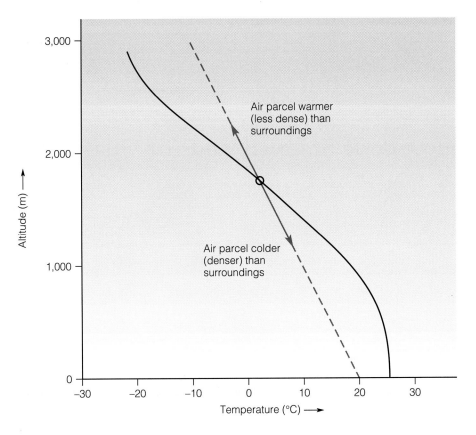

■ FIGURE 16.7 If the ambient (surrounding) air is unstable, an air parcel that is displaced upward will continue to move upward, and an air parcel that is displaced downward will continue to move downward. Unstable air enhances vertical motion of air.

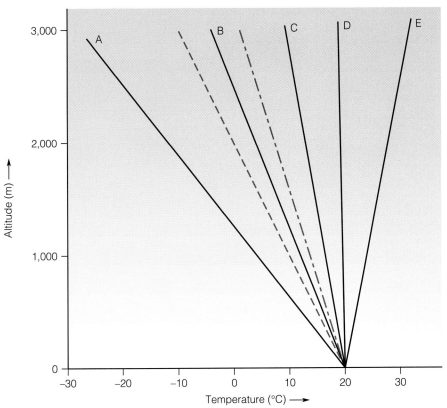

■ FIGURE 16.8 Sample radiosonde soundings and cases of atmospheric stability.

– – – – – Dry adiabatic lapse rate

–·–·–· Moist adiabatic lapse rate

Radiosonde sounding: A Absolute instability   D Absolute stability (isothermal)
                       B Conditional stability   E Absolute stability (inversion)
                       C Absolute stability (lapse)

Focus on the Environment:
Atmospheric Stability and Air Quality

Air pollutants are especially hazardous to human health when atmospheric conditions reduce their rate of dilution. Once pollutants enter the atmosphere through smoke stacks or exhaust pipes, their concentrations usually begin to decline as they mix with cleaner air. The more thorough the mixing, the more rapid the rate of dilution. When conditions in the atmosphere favor rapid dilution, the impact of air pollution is usually minor. On other occasions, called *air pollution episodes,* conditions in the atmosphere minimize dilution, and the impact can be serious. The rate of dilution depends upon wind speed and atmospheric stability.

Air mixes more thoroughly on a windy day than on a calm one. When it is windy, irregular movements of air mix polluted air and cleaner air and accelerate dilution. But when there is no wind, dilution occurs by the very slow process of *molecular diffusion,* that is, dispersal at the molecular scale. As a rule of thumb, doubling the wind speed cuts the concentration of air pollutants in half (☐ Fig. 1).

As discussed elsewhere in this chapter, atmospheric stability affects the up and down motion of air. Convection and mixing are enhanced when the ambient air is unstable and inhibited when the ambient air is stable. The stability of air thus influences the rate at which polluted air is diluted. A puff of smoke emitted into unstable air is diluted more than smoke emitted into stable air. Stable air inhibits the upward transport of air pollutants, and a stable air layer aloft may trap air pollutants below. Continual emission of contaminants into stable air results in the accumulation and concentration of pollutants.

Mixing depth is the distance between Earth's surface and the altitude reached by convection currents. When mixing depths are great (many kilometers, for example), air pollutants readily mix with abundant clear air, and dilution is enhanced. When mixing depths are shallow (less than 1,000 m or 3,300 ft, for example), air pollutants are restricted to a smaller volume of air, and concentrations may build. Mixing depths are greater in unstable air than in stable air. Because solar radiation governs convection, mixing depths usually are greater in the afternoon than in the morning, during the day than at night, and in summer than in winter.

The behavior of a plume of smoke belching from a smokestack can sometimes indicate the stability of the ambient air. A plume entering unstable air undulates, as in ☐ Figure 2a. Such plume behavior signals that polluted air is mixing with the ambient air, thus facilitating dilution. The result is improved air quality—except where the plume loops to the ground. A smoke plume that flattens and spreads slowly

downwind, as in Figure 2b, signals very stable conditions and minimal dilution.

An air pollution episode is most likely when a persistent temperature inversion develops. Within an air layer characterized by a *temperature inversion,* the ambient air temperature increases with altitude; that is, warmer air overlies cooler air. This extremely stable atmospheric layering inhibits mixing and dilution of pollutants. An inversion may occur aloft or at the surface.

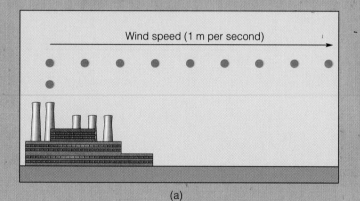

(a)

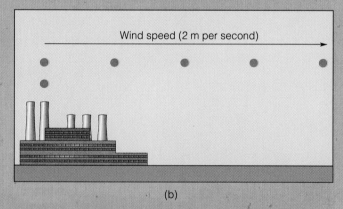

(b)

☐ **FIGURE 1** Dilution of a plume of smoke. In general, doubling the wind speed cuts the concentration of air pollutants in half. In both (*a*) and (*b*), pollutants are being emitted to the atmosphere at the same rate, that is, 1 unit per second.

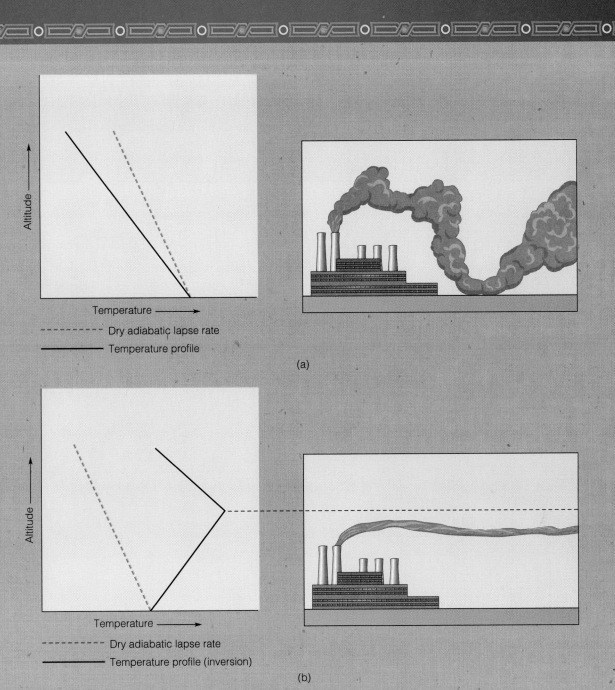

Altitude →

Temperature →

- - - - - Dry adiabatic lapse rate
——— Temperature profile

(a)

Altitude →

Temperature →

- - - - - Dry adiabatic lapse rate
——— Temperature profile (inversion)

(b)

☐ FIGURE 2 (*a*) A smoke plume emitted into unstable air loops and readily mixes with cleaner air (*b*) whereas a smoke plume emitted into stable air slowly flattens and spreads out with minimal dilution.

Extreme nighttime radiational cooling may produce a temperature inversion at Earth's surface. With clear night skies and light winds, radiational cooling chills the ground, which in turn chills the air in contact with it. Because the air near the surface is colder than the air aloft, a low-level temperature inversion develops (Fig. 2b). Such temperature inversions usually disappear within a few hours after sunrise because the Sun heats the ground, which heats the overlying air and eventually reestablishes the usual temperature profile in which air temperature drops with increasing altitude. In winter, however, when the Sun's rays are weak and the ground is covered by a highly reflective layer of snow, a radiational temperature inversion may persist for days or even weeks.

An elevated temperature inversion normally overlies the Los Angeles basin and contributes to the region's high air pollution potential (□ Fig. 3). Cool air that drifts inland from the Pacific Ocean is overlain by subsiding warm air aloft. The two contrasting air layers are separated by a temperature inversion. The inversion traps motor vehicle and industrial emissions, and bright sunshine powers chemical reactions that generate photochemical smog.

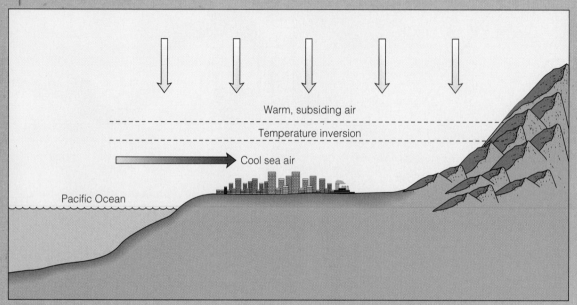

Warm, subsiding air

Temperature inversion

Cool sea air

Pacific Ocean

□ FIGURE 3 An elevated temperature inversion over Los Angeles contributes to that city's relatively high air pollution potential.

LIFTING PROCESSES

With sufficient expansional cooling, the relative humidity of ascending air approaches 100%, and clouds form. Air ascends (1) as the upward branch of a convective current, (2) along the surface of a front, (3) up the slopes of a hill or mountain, or (4) where surface winds converge.

Cumulus clouds form where convection currents ascend, and the sky is clear where convection currents descend (▣ Fig. 16.9). In general, the higher the altitude reached by convection currents, the greater the expansional cooling, and the more likely that clouds and precipitation will develop.

▣ FIGURE 16.9 Fair-weather cumulus clouds. Air is ascending where there are clouds and descending where the sky is blue.

When contrasting air masses meet along a front, clouds and precipitation often develop. A **front** is a narrow zone of transition between two air masses that differ in temperature and/or humidity. A warm air mass is less dense than a cold air mass. Hence, as warm air advances and cold air retreats, the warm air rides up and over the cold air (▣ Fig. 16.10).

This phenomenon is called *overrunning*. The leading edge of the warm air at Earth's surface is known as a *warm front*. Where cold air advances into warm air, the denser cold air slides under the less dense warm air and forces it upward (▣ Fig. 16.11). The leading edge of the cold air at Earth's surface is known as a *cold front*. Hence, along fronts, the

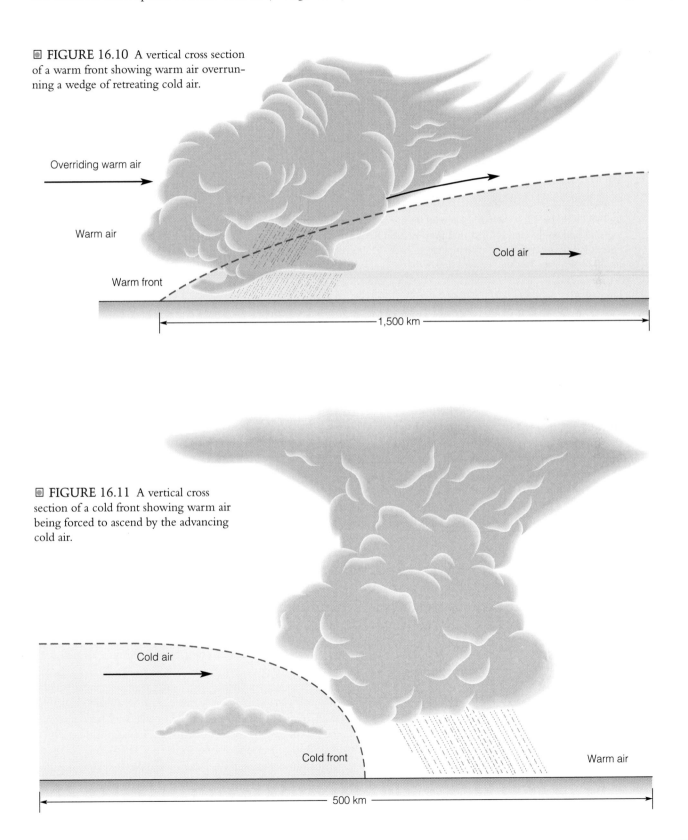

▣ FIGURE 16.10 A vertical cross section of a warm front showing warm air overrunning a wedge of retreating cold air.

▣ FIGURE 16.11 A vertical cross section of a cold front showing warm air being forced to ascend by the advancing cold air.

 FIGURE 16.12 As winds are forced up the windward slopes of a mountain range, the ascending air expands and cools, clouds form, and precipitation develops. On the mountain's leeward slopes, the air descends and is compressed, its temperature rises, and its relative humidity drops. Hence, clouds dissipate or fail to develop over the leeward slopes.

warmer air ascends, expands, and cools. Expansional cooling may lead to cloud development and perhaps rain or snow. Fronts and frontal weather are discussed further in Chapter 17.

As winds sweep across the landscape, hill and mountain slopes force air to ascend. This process is known as **orographic lifting.** Consequent expansional cooling may cause clouds and precipitation to develop. Suppose that a prominent mountain range is oriented perpendicular to the prevailing wind direction (Fig. 16.12). As air is forced to rise along the *windward* slopes (facing the wind), it expands and cools, increasing its relative humidity. With sufficient cooling, clouds and precipitation develop. On the mountain's *leeward* slopes (downwind side), however, air descends and warms, reducing its relative humidity and the likelihood of clouds and precipitation. Hence, mountain ranges induce two contrasting climatic zones: moist on the windward slopes and dry on the leeward slopes. Dry conditions may stretch many hundreds of kilometers to the lee of prominent mountain ranges; this region is known as a **rain shadow.**

Where surface winds converge, air ascends. In one example (Fig. 16.13), horizontal winds blowing over a smooth lake surface slow down and converge when they cross the rougher surface of the land, where there is more frictional resistance. Again, ascending air means expansional cooling, increasing relative humidity, and perhaps clouds and precipitation. This is the mechanism that generates clouds and lake-effect snows downwind from the Great Lakes.

CLOUDS

A **cloud** is a visible mass of tiny water droplets and/or ice crystals suspended in the atmosphere. This section considers cloud formation, cloud classification, and the principal cloud types.

EARTHFACT

The Rainiest Place on Earth

The rainiest place on the planet is reported to be on the windward slopes of Mount Waialeale on the northeast coast of the island of Kauai in the Hawaiian chain. A rain gauge at 1,569 m (5,142 ft) above sea level receives an average annual rainfall of 1,199 cm (39.3 ft). Persistent northeast trade winds and orographic lifting of warm, humid air are responsible for this excessive precipitation. On the leeward slopes of this same volcanic mountain, average annual rainfall is less than 50 cm (20 in).

A close rival for Waialeale's distinction as the planet's wettest place is Cherrapunji in Assam, India. Situated in the foothills of the Himalayas, Cherrapunji holds the world's record for maximum monthly rainfall (930 cm or about 366 in in July 1861) and maximum annual rainfall (2,646 cm or about 1,042 in from August 1860 to August 1861).

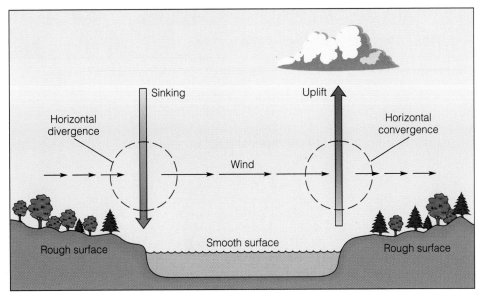

■ FIGURE 16.13 Horizonal winds slow and converge as they blow from a smooth lake surface to a rough land surface. Convergence of winds induces ascent of air, cloud development, and perhaps rain or snow.

Cloud Development

Laboratory studies demonstrate that when air is free of dust and other aerosols, condensation (or deposition) of water vapor requires *supersaturation,* or a relative humidity greater than 100%. With clean air, supersaturation is the only way that cloud droplets and ice crystals can grow from much smaller droplets and crystals via condensation and deposition.

In the real atmosphere, only slight supersaturation at most is required for cloud development because the atmosphere contains numerous suspended **nuclei,** tiny aerosols that provide relatively large surface areas on which condensation or deposition can take place and thereby produce cloud droplets and ice crystals directly. Nuclei are the products of forest fires, volcanic eruptions, wind erosion of the soil, salt-water spray, and emissions of industrial chimneys. One cubic centimeter of air typically contains about 10,000 nuclei.

In addition, the atmosphere has an abundant supply of *hygroscopic nuclei,* aerosols that have a chemical attraction for water molecules. Condensation begins on hygroscopic nuclei at relative humidities under 100%. Because nuclei are abundant and because many of them are hygroscopic, clouds are likely to develop whenever the relative humidity approaches 100%.

Nuclei are of two types: *cloud condensation nuclei (CCN)* and *ice-forming nuclei.* CCNs promote condensation at air temperatures both above and below freezing. In the atmosphere, water droplets condense and remain liquid even when the cloud temperature is well below 0°C (32°F). These are *supercooled* water droplets. Ice-forming nuclei promote freezing of water droplets or deposition of water vapor; they are much less abundant than CCNs and are active only at air temperatures well below freezing.

Cloud Classification

Several processes operating in the atmosphere give rise to a variety of forms of clouds. Clouds are classified on the basis of (1) temperature, (2) general appearance, and (3) altitude of their base.

Clouds at temperatures above freezing are composed of water droplets exclusively and are *warm clouds.* Clouds at temperatures below freezing are composed of ice crystals or supercooled water droplets or a combination of both; such clouds are *cold clouds.* Cirrus, stratus, and cumulus clouds are distinguished by their appearance. Cirrus are fibrous, stratus are layered, and cumulus occur as heaps or puffs.

Based on the altitude of their base, clouds are grouped into four families: high clouds, middle clouds, low clouds, and clouds with vertical development. Cloud base altitudes vary seasonally and with latitude, so the values presented in ○ Table 16.3 are intended only as general guidelines.

Gentle ascent of air (typically less than 5 cm per second or 1 mi per hour) over broad areas (overrunning) produces high, middle, and low clouds that spread laterally and form layers. These clouds are known as **stratiform clouds.** Much more vigorous ascent (sometimes in excess of 30 m per second or 70 mi per hour) is responsible for clouds having considerable vertical development. These **cumuliform clouds** are heaped or puffy in appearance and generally cover smaller areas than stratiform clouds.

Principal Cloud Types

The base of *high clouds* typically is at altitudes above 7,000 m (23,000 ft). Air temperatures are so low at these altitudes (under −25°C or −13°F) that these cold clouds are composed almost exclusively of ice crystals, giving them a fibrous

TABLE 16.3 Cloud Classification

GENUS	ALTITUDE OF CLOUD BASE ABOVE GROUND (km)	SHAPE AND APPEARANCE
High clouds		
Cirrus (Ci)	7–18	Delicate streaks or patches
Cirrostratus (Cs)	7–18	Transparent thin white sheet or veil
Cirrocumulus (Cc)	7–18	Layer of small white puffs or ripples
Middle clouds		
Altostratus (As)	2–7	Uniform white or gray sheet or layer
Altocumulus (Ac)	2–7	White or gray puffs or waves in patches or layers
Low clouds		
Stratocumulus (Sc)	0–2	Patches or layers of large rolls or merged puffs
Stratus (St)	0–2	Uniform gray layer
Nimbostratus (Ns)	0–4	Uniform gray layer from which precipitation is falling
Clouds with vertical development		
Cumulus (Cu)	0–3	Detached heaps or puffs with sharp outlines and flat bases, and slight or moderate vertical extent
Cumulonimbus (Cb)	0–3	Large puffy clouds of great vertical extent with smooth or flattened tops, frequently anvil shaped, from which showers fall, with thunder

SOURCE: M. Neiburger, J. G. Edinger, and W. D. Bonner. *Understanding our Atmospheric Environment.* New York: W. H. Freeman, 1973, p. 11. Copyright © 1973. All rights reserved.

(a)

(b)

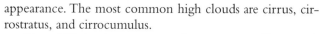
◉ **FIGURE 16.14** High, thin clouds: (*a*) cirrus, (*b*) cirrostratus, and (*c*) cirrocumulus.

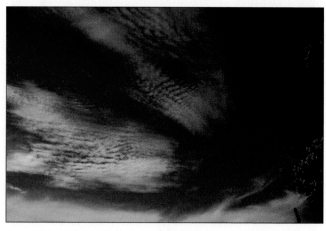

(c)

appearance. The most common high clouds are cirrus, cirrostratus, and cirrocumulus.

Cirrus (Ci) clouds occur as nearly transparent silky strands (◉ Fig. 16.14a). Strands are streaks of falling ice crystals blown by strong winds at high altitudes. Like cirrus clouds, *cirrostratus (Cs)* clouds are nearly transparent, so the Sun or Moon readily shines through them. They form a thin, white sheet that may cover the entire sky (Fig. 16.14b). *Cirrocumulus (Cc)* clouds appear as small, white, rounded patches arranged in a wavelike or mackerel pattern (Fig. 16.14c). All high clouds are too thin to prevent objects on the ground from casting shadows during daylight hours.

Middle clouds have their bases at altitudes between 2,000 and 7,000 m (6,500 and 23,000 ft). Temperatures of these clouds generally range between 0 and −25°C (32 and −13°F). They are composed of supercooled water droplets or a mixture of supercooled water droplets and ice crystals. *Altostratus (As)* clouds occur as uniformly gray or white layers that totally or partially cover the sky (■ Fig. 16.15a). These clouds are usually so thick that they very nearly block the Sun. *Altocumulus (Ac)* clouds consist of roll-like patches or puffs that often form waves or parallel bands (Fig. 16.15b).

Low clouds have bases ranging from Earth's surface (fog) up to perhaps 2,000 m (6,500 ft). Temperatures of these clouds are usually above −5°C (23°F); they are composed mostly of water droplets. *Stratocumulus (Sc)* clouds consist of large, irregular puffs or rolls often separated by areas of cloud-free sky (■ Fig. 16.16a). Rarely do stratocumulus clouds yield precipitation. *Stratus (St)* clouds appear as a uniformly gray layer stretching from horizon to horizon (Fig. 16.16b). Stratus in contact with Earth's surface are fog. Usually, only drizzle falls from stratus clouds, but moderate or heavy amounts of rain or snow may fall from the much thicker *nimbostratus (Ns)* clouds. Nimbostratus are darker and have a more ragged base than stratus.

Air ascending in convection currents can give rise to cumulus, cumulus congestus, and cumulonimbus clouds. *Cumulus (Cu)* clouds resemble puffs of cotton floating in the sky (■ Fig. 16.17a). Solar heating drives convection, so cumulus cloud development often follows the daily variation of solar radiation. On a day that dawns clear, cumulus clouds

■ **FIGURE 16.15** Middle clouds: (*a*) altostratus and (*b*) altocumulus.

(a)

(b)

■ **FIGURE 16.16** Low clouds: (*a*) stratocumulus and (*b*) stratus.

(a)

(b)

(a)

(b)

FIGURE 16.17 These clouds show vertical development and are produced in the updrafts of convective currents: (a) cumulus, (b) cumulus congestus, and (c) cumulonimbus.

(c)

may begin forming by midmorning, after the Sun has warmed the ground and spurred convection. Cumulus sky cover typically is most extensive by early to midafternoon, the warmest time of day. As sunset approaches, convection weakens, and cumulus clouds vaporize.

The stability of the troposphere determines to what altitude cumuliform clouds build and whether cumulus clouds develop into more ominous thunderstorm clouds. Stable air aloft inhibits vertical motion, and cumulus clouds show little vertical growth. Flat cloud tops indicate stable air aloft, and the weather is likely to remain fair. Unstable air aloft enhances vertical motion, and the tops of cumulus clouds surge upward. If the ambient air is unstable to great altitudes, the entire cloud mass takes on a cauliflower appearance as a *cumulus congestus* cloud (Fig. 16.17b) and then a *cumulonimbus* (thunderstorm) cloud (Fig. 16.17c).

Where convection is suppressed, so too is the development of cumuliform clouds. Cold surfaces chill and stabilize the overlying air and inhibit convection; hence, cumulus clouds do not readily form over snow-covered or cold-water surfaces.

 PRECIPITATION

Precipitation is water in the solid or liquid phase that falls from clouds to Earth's surface. Cloud formation is no guarantee of precipitation. Nimbostratus and cumulonimbus clouds produce the bulk of rain and snow, but less than 10% of all clouds yield any precipitation. Growth of cloud droplets or ice crystals is the key to precipitation formation, and adequate growth takes place only under special circumstances.

The water droplets and/or ice crystals composing clouds are so small that they will remain suspended in the atmosphere indefinitely unless they vaporize or grow considerably larger. Updrafts usually keep cloud particles from leaving the base of a cloud and falling to Earth's surface as rain or snow. Even if droplets or ice crystals fall from a cloud, they travel only a short distance before vaporizing in the clear (unsaturated) air beneath the cloud.

The speed of a falling cloud droplet or ice crystal in calm air is governed by two opposing forces: (1) gravity, which accelerates the particle downward, and (2) the resistance offered by the air through which the particle descends. As the particle accelerates downward, it meets with increasing air resistance whereas gravity remains essentially constant. The resisting force eventually balances the force of gravity, and the particle drifts downward at a constant speed known as the **terminal velocity.**

For a particle to remain suspended in air, updrafts must match its terminal velocity. Terminal velocity generally increases with the size of the particle. Hence, the larger the particle, the more vigorous the updraft must be to keep the particle in suspension. Cloud droplets and ice crystals are so small (most having diameters of 10 to 20 micrometers) that even weak updrafts readily counter their very low terminal velocities (typically only 0.3 to 1.2 cm per second).

For precipitation to form, cloud particles must grow large enough that they counter updrafts and survive the fall to Earth's surface without completely vaporizing. Considerable growth is necessary; about 1 million cloud droplets are needed to form a single raindrop (about 2 mm in diameter). The principal means whereby cloud particles grow large enough to precipitate is through the Bergeron process.

Bergeron Process

Perhaps 90% of precipitation on Earth originates through the Bergeron process. Named for the Scandinavian meteorologist Tor Bergeron, who first described it in 1933, the **Bergeron process** requires the coexistence of water vapor, ice crystals, and supercooled water droplets in *cold clouds.* Supercooled water droplets far outnumber ice crystals initially because cloud condensation nuclei (CCNs) are much more common than ice-forming nuclei.

Within the cloud, ice crystals grow rapidly at the expense of supercooled water droplets because the saturation vapor pressure is greater over supercooled water than over ice (see Table 16.1). At subfreezing temperatures, water molecules escape more readily from a liquid water surface than from the surface of ice; water molecules are more tightly bonded in the solid phase than in the liquid phase. Hence, in a cold cloud containing a mixture of ice crystals and supercooled water droplets, a vapor pressure that is *saturated* for water droplets is *supersaturated* for ice crystals.

Suppose that the vapor pressure is 2.86 mb in a cloud at a temperature of −10°C (14°F). From Table 16.1, this vapor pressure implies a relative humidity of 100% (saturation) for the air surrounding the water droplets and a relative humidity of 110% (supersaturation) for the air surrounding the ice crystals. In response to supersaturation within the cloud, water vapor deposits on ice crystals and ice crystals grow. Deposition removes water vapor from the cloud so that the relative humidity of the air surrounding the water droplets dips below 100%, and the droplets vaporize. Thus, through the Bergeron process, ice crystals grow at the expense of supercooled water droplets.

As the ice crystals grow, their terminal velocities increase, and they collide and coalesce with smaller, slower-moving supercooled water droplets and ice crystals (◻ Fig. 16.18). In this way, the ice crystals grow still larger and eventually fall from the cloud base. If air temperatures are subfreezing

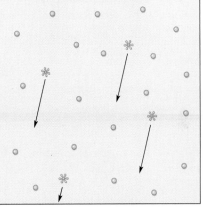

○ Supercooled   ✳ Ice crystals
water droplets

◻ **FIGURE 16.18** The Bergeron precipitation process by which ice crystals grow at the expense of supercooled water droplets in cold clouds. Descending crystals undergo further growth by colliding with smaller ice crystals and supercooled water droplets in their path.

Rainmaking

During a prolonged drought, farmers and ranchers may turn to artificial rainmaking to try to save their crops and livestock. Modern rainmaking involves cloud seeding, a technology based on an understanding of natural precipitation formation. Does cloud seeding work? If cloud seeding is successful, how much enhancement of rainfall or snowfall is likely? Are there drawbacks to cloud seeding?

Cloud seeding usually refers to efforts to enhance precipitation by stimulating the Bergeron process. As described elsewhere in this chapter, the Bergeron process occurs in cold clouds composed of a mixture of supercooled water droplets and ice crystals. In such clouds, ice crystals grow at the expense of supercooled water droplets.

With cloud seeding, nucleating (seeding) agents are introduced into cold clouds that are deficient in ice crystals. One common seeding agent is silver iodide (AgI), a substance with crystal properties similar to those of ice. Silver iodide crystals are freezing nuclei that are active at cloud temperatures of −4°C (25°F) or lower. Another seeding agent is dry ice, solid carbon dioxide at a temperature of about −80°C (−112°F). Pellets of dry ice chill surrounding supercooled water droplets, causing them to freeze to ice crystals that grow into snowflakes via the Bergeron process. In addition, latent heat released when supercooled water droplets freeze stimulates cloud growth by raising the temperature (and lowering the density) of cloud air.

Usually, a small aircraft equipped with a flare or hopper is used for cloud seeding. Silver iodide crystals are emitted from the flare, or dry ice pellets are dispensed from the hopper. Alternatively, ground-based generators emit a plume of silver iodide crystals that may drift into clouds and seed them.

When precipitation occurs after clouds were seeded, a question remains: Was the seeding responsible, or would the precipitation have fallen anyway without the seeding? Based on rigorously designed field experiments and sophisticated statistical analysis, scientists have determined that cloud seeding is sometimes but not always successful. When it is successful, at best it enhances precipitation by 15 to 20%.

Southern California's increasing demand for water has been the stimulus for a long-term cloud-seeding project over the windward western slopes of the Sierra Nevada. From January through March, aircraft seed orographic clouds with either silver iodide or dry ice. The objective of the *Sierra Cooperative Pilot Project (SCPP)* is to thicken the winter snowpack in the mountains, thereby increasing the spring runoff in the American River basin and the water supply for irrigation and domestic needs.

A proposal for large-scale cloud seeding often encounters strong opposition from some interest groups. A common argument is that successful cloud seeding merely redistributes a fixed supply of precipitation, so that more precipitation in one area may mean a compensating reduction in another. It is possible, for example, that cloud seeding that benefits ranchers on the high plains of eastern Colorado also deprives grain farmers of rain downwind in Kansas. Furthermore, successful cloud seeding is not good news for everybody in the area affected. Towns and cities dependent on snowpacks for their water supply welcome enhanced mountain snowfall, but too much mountain snow may close highways, heighten the danger of avalanches, and keep grazing lands under a blanket of snow longer into the spring. For these reasons, cloud seeding is controversial and has led to legal wrangles between adjoining counties, states, and provinces.

---

at least most of the way, ice crystals reach Earth's surface as snowflakes. If air temperatures below the cloud are above freezing, the snowflakes melt and fall as raindrops.

An understanding of the natural mechanisms of precipitation formation is the basis for efforts aimed at enhancing rain and snowfall. Precipitation enhancement is described in ▢ Perspective 16.2.

Forms of Precipitation

Besides the familiar rain and snow, precipitation also occurs as drizzle, freezing rain, ice pellets (sleet), and hail.

Drizzle consists of small water drops 0.2 to 0.5 mm in diameter. Drizzle drops are relatively small because they originate in stratus clouds that are so low and thin that the opportunity for cloud droplets to grow by colliding and coalescing with one another is limited.

Raindrops have diameters in the range of 1 to 6 mm. Nimbostratus and cumulonimbus clouds are the principal source of rain, and the bulk of rain begins as snowflakes or hailstones that melt as they fall through air that is above 0°C (32°F).

Freezing rain (or *freezing drizzle*) develops when rain falls from relatively mild air aloft into a shallow layer of subfreezing air next to Earth's surface. Drops become supercooled and freeze immediately on contact with cold surfaces. Freezing rain or drizzle forms a coating of ice that sometimes becomes heavy enough to disrupt traffic, snap power lines, and bring down tree limbs (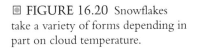 Fig. 16.19).

Snow consists of ice crystals in the form of flakes. Although it is said that no two snowflakes are identical, all snowflakes have six-sided symmetry. Depending on the water vapor concentration and temperature, snowflakes take the form of plates, stars or dendrites, columns, or needles (Fig. 16.20). Air is trapped between the accumulating flakes, so snow has an average density about one-tenth that of liquid water. Hence, 10 cm of snowfall on average melt down to 1 cm of liquid water.

Ice pellets, usually called *sleet,* are frozen raindrops 5 mm or less in diameter. They develop in much the same way as freezing rain except that the surface layer of cold air is thicker. Hence, raindrops freeze *prior* to striking the ground. As with freezing rain, accumulations of ice pellets cause hazardous walking and driving conditions.

Hail consists of rounded or jagged lumps of ice, often with internal concentric layering that resembles the structure of an onion (Fig. 16.21). Hail develops within a cumulonimbus (thunderstorm) cloud as vigorous updrafts transport ice pellets into the middle and upper reaches of the cloud. Along the way, ice pellets grow by intercepting supercooled water droplets. Eventually, the growing ice pellet becomes too heavy to be supported by the updraft and falls through the cloud, exiting the cloud base. Typically, the

■ FIGURE 16.19 The branches of this white birch were bent over by the weight of accumulating freezing rain.

■ FIGURE 16.20 Snowflakes take a variety of forms depending in part on cloud temperature.

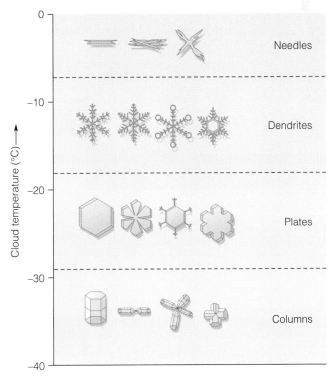

FIGURE 16.21 Hailstones are lumps of ice that fall from cumulonimbus (thunderstorm) clouds.

falling ice pellet enters air that is above freezing and begins to melt. If the pellet is large enough to start with, however, it survives the journey to the ground as a hailstone.

DEW, FROST, AND FOG

Dew and frost form when air next to the ground is chilled to saturation by radiational cooling at night. An object on Earth's surface (an automobile windshield, for example) emits infrared radiation to the atmosphere and eventually off to space, thereby cooling the object. Meanwhile, the atmosphere emits infrared radiation back to Earth's surface where some is absorbed by the object, thereby warming the object. On a clear night, the object emits more radiation than it receives from the atmosphere. Hence, the surface of the object becomes cooler than the air adjacent to it, and heat is conducted from the air to the object. With sufficient chilling, the air in immediate contact with the object cools enough to become saturated. If the air temperature remains above the freezing point, water vapor condenses on the object as **dew** (■ Fig. 16.22a); if the air temperature drops below freezing, water vapor deposits as **frost** (Fig. 16.22b).

(a)

FIGURE 16.22 On relatively cold surfaces, (*a*) condensation of water vapor produces dew, and (*b*) deposition of water vapor produces frost.

(b)

Dewpoint is the temperature to which air must be cooled, at constant pressure, to become saturated. In order for water vapor to condense as dew on the surface of an object, the temperature of that surface must drop *below* the dewpoint. When radiational cooling at constant pressure produces saturation at subfreezing temperatures, the temperature is called the *frost point*. Water vapor deposits as frost on the surface of an object if the temperature of that surface drops *below* the frost point.

Fog is a visibility-restricting suspension of tiny water droplets or ice crystals in contact with Earth's surface. Simply put, fog is a cloud on the ground. By international convention, fog is defined as a suspension that restricts visibility to 1,000 m (3,250 ft) or less; otherwise the suspension is called *mist*. Fog develops when air becomes saturated through radiational cooling, contact with a relatively cold surface, addition of water vapor, or expansional cooling (○ Table 16.4).

With a clear night sky, light winds, and an air mass that is humid near the ground and relatively dry aloft, radiational cooling may produce *radiation fog* (▣ Fig. 16.23). High humidity in the lowest portion of an air mass is usually due to evaporation of water from a moist surface. Hence, radiation fogs are most likely to form in marshy areas or where the soil has been saturated by recent rainfall or snowmelt. Typically, radiation fog lasts for only a few hours after sunrise. The fog gradually thins and disperses as saturated air at low levels mixes with drier air above the fog bank; the relative humidity drops and the fog droplets vaporize.

▣ FIGURE 16.23 Radiation fog is the consequence of nighttime radiational cooling of a layer of air that is relatively humid near the ground and dry aloft.

TABLE 16.4 Classification of Fog

FOG TYPE	PROCESS OF FORMATION
Radiation	Radiational cooling of humid air at night
Advection	Chilling of mild air as it passes over a relatively cool surface
Steam	Vaporization of warm water into cold air
Upslope	Expansional cooling of air ascending a hill slope

■ FIGURE 16.24 Sea fog forms when warm air moves over relatively cold seawater.

Mild, humid air blowing over a colder surface may be chilled to saturation in its lower layers. Fog formed in this way is known as *advection fog*. Sea fog often develops when warmer air passes over colder water (■ Fig. 16.24). The same fog-forming mechanism operates in spring in the northern states, when mild southerly winds pass over relatively cold, snow-covered ground.

Steam fog is produced when cold air comes in contact with warm water. It develops in early winter when extremely cold, dry air flows over an unfrozen body of water. The advecting air becomes warmer and more humid, and fog forms. Heating from below also destabilizes the air, causing the fog to appear as rising streamers that resemble smoke or steam. Steam fog over the North Atlantic is known as *Arctic sea smoke.*

Fog also forms on hillsides or mountain slopes as the result of upslope flow and expansional cooling of humid air. Fog formed in this way is called *upslope fog.*

CHAPTER SUMMARY

1. Vapor pressure is a direct measure of the water vapor concentration of air; relative humidity indicates how close air is to saturation.

2. Saturation vapor pressure represents a balance in the flow of water molecules entering and leaving the vapor phase. The saturation vapor pressure increases with rising temperature because temperature governs the rate of water vaporization.

3. When the actual vapor pressure equals the saturation vapor pressure, the relative humidity is 100%. For unsaturated (clear) air, the relative humidity varies inversely with air temperature. Hence, on a day with no warm or cold air advection, the relative humidity is lowest when air temperature is highest.

4. The relative humidity of unsaturated air increases when the air is cooled (decreasing the saturation vapor pressure) or when water vapor is added to the air (raising the vapor pressure).

5. Expansional cooling and compressional warming within the atmosphere are adiabatic processes; that is, no net heat exchange takes place between an air parcel and its environment when the parcel ascends or descends within the atmosphere.

6. Atmospheric stability is determined by comparing the temperature (or density) of an air parcel moving vertically within the atmosphere with the temperature (or density) of the surrounding (ambient) air. Stable air inhibits vertical motion; unstable air enhances vertical motion.

7. Expansional cooling and cloud development occur via the ascent of air in convective currents, along fronts or mountain slopes, or when surface winds converge.

8. Clouds, the visible products of condensation or deposition within the atmosphere, are composed of minute water droplets, ice crystals, or a combination of the two.

9. As the relative humidity approaches 100%, condensation and deposition occur on nuclei, tiny solid and liquid aerosols. Cloud condensation nuclei (CCNs) are much more abundant than ice-forming nuclei, and most ice-forming nuclei are active at temperatures well below the freezing point.

10. Many condensation nuclei are hygroscopic; that is, they have a chemical attraction for water molecules and promote condensation at relative humidities under 100%.

11. High, middle, and low clouds are caused by relatively gentle uplift (overrunning) of air over a broad region. Hence, these clouds are layered, that is, stratiform. Clouds having significant vertical development are produced by more vigorous uplift and are heaped or puffy in appearance, that is, cumuliform.

12. Because of relatively low terminal velocities, tiny droplets and ice crystals that compose clouds remain suspended in the atmosphere indefinitely unless they vaporize or undergo significant growth.

13. The Bergeron process is a mechanism whereby cloud particles grow large enough to fall to Earth's surface as precipitation. This process accounts for the bulk of precipitation that falls on Earth.

14. The Bergeron process requires the coexistence of ice crystals, supercooled water droplets, and water vapor in cold clouds. The saturation vapor pressure surrounding a supercooled water droplet is higher than that surrounding an ice crystal. Hence, air that is saturated for droplets is supersaturated for ice crystals, and ice crystals grow at the expense of water droplets.

15. Rain and snow fall from thicker clouds (nimbostratus or cumulonimbus), and drizzle falls from thinner low clouds (stratus).

16. Ice pellets (sleet) are raindrops that freeze prior to reaching the ground whereas freezing rain consists of supercooled raindrops that freeze on contact with subfreezing surfaces. Hail is a frozen form of precipitation produced in thunderstorms.

17. Dew and frost are products of nighttime radiational cooling of Earth's surface, which, in turn, chills the air in immediate contact with the surface to saturation. If the temperature of that surface drops below the dewpoint (or frost point), water vapor condenses as dew (or deposits as frost).

18. At a relative humidity of 100%, the dewpoint, wet-bulb temperature, and ambient air temperature are the same.

19. Fog is a visibility-restricting suspension of tiny water droplets or ice crystals in an air layer that is in contact with Earth's surface. Fog is a cloud at ground level.

20. Based on its mode of origin, fog is classified as radiation fog, advection fog, steam fog, or upslope fog.

IMPORTANT TERMS

adiabatic process	cumuliform clouds	frost	relative humidity
atmospheric stability	dew	nuclei	saturation vapor pressure
Bergeron process	expansional cooling	orographic lifting	stratiform clouds
cloud	fog	precipitation	terminal velocity
compressional warming	front	rain shadow	vapor pressure

REVIEW QUESTIONS

1. Describe what happens at the molecular scale when water evaporates.

2. All other factors being equal, as the temperature of the surface of a water body rises, the rate of evaporation:
   a. ___ increases; b. ___ does not change; c. ___ decreases.

3. Distinguish between air pressure and vapor pressure. How do they compare in magnitude at Earth's surface?

4. How do variations in air temperature influence the saturation vapor pressure and the relative humidity?

5. During what atmospheric condition is the vapor pressure equal to the saturation vapor pressure?

6. With no air mass advection, daily air temperature is usually _____ and relative humidity is usually _____ just after sunrise.

   a. ___ highest . . . lowest; b. ___ lowest . . . highest;
   c. ___ highest . . . highest; d. ___ lowest . . . lowest

7. If the vapor pressure is 3 mb and the saturation vapor pressure is 12 mb, the relative humidity is ___ percent.
   a. ___ 100; b. ___ 25; c. ___ 250; d. ___ 40; e. ___ 4

8. What is the relative humidity if the vapor pressure is 8 mb and the saturation vapor pressure is 8 mb?

9. A clear parcel of air will cool by ___ Celsius degrees if it is lifted 1500 m.
   a. ___ 10;   b. ___ 15;   c. ___ 12.5;   d. ___ 5;
   e. ___ 6.5

10. Expansional cooling and compressional warming of air are adiabatic processes. Explain this statement.

11. Explain why ascending parcels of saturated (cloudy) air cool more slowly than ascending parcels of unsaturated (clear) air.

12. How does the stability of ambient air influence the up and down movement of air parcels?

13. Determine whether the following soundings are stable, unstable, or neutral for both saturated (cloudy) and unsaturated (clear) air parcels:
    a. $+9°C/1,000$ m; b. $−6°C/1,000$ m; c. $−8°C/1,000$ m; d. $−10°F/1,000$ ft; e. $+2°F/1,000$ ft; f. $−14°C/1,000$ m.

14. Both clear air and clouds are associated with convection within the atmosphere. Explain why.

15. _____ air favors the development of convective clouds and precipitation.
    a. ___ Stable; b. ___ Unstable; c. ___ Neutral

16. How does topography influence the development of clouds?

17. What causes a rain shadow?

18. What is the significance of *hygroscopic* nuclei?

19. Distinguish between cloud condensation nuclei and ice-forming nuclei.

20. How might the composition of a cumulonimbus cloud vary with altitude?

21. Why do fair-weather cumulus clouds tend to vaporize toward sunset?

22. How does stability of the middle and upper troposphere determine whether a cumulus cloud builds into a cumulonimbus cloud?

23. If the temperature of the base of a cumulus cloud is 0°C (32°F) and the air temperature at the ground is 20°C (68°F), determine the approximate altitude of the cloud base above the ground.

24. How does an extensive snow cover influence the development of cumulus clouds?

25. On a bright sunny day, are cumulus clouds more likely to develop over a lake or over the adjacent land? Justify your response.

26. Describe the Bergeron process of precipitation formation.

27. Why is the saturation vapor pressure surrounding supercooled water droplets greater than that surrounding ice crystals at the same temperature?

28. Under what atmospheric conditions does freezing rain develop?

29. Why is rainfall likely to be heavier on top of a mountain than in a nearby valley?

30. What atmospheric conditions favor formation of radiation fog?

POINTS TO PONDER

1. How does the vapor pressure in the stratosphere compare with the vapor pressure in the troposphere?

2. Explain why when lifted, cold, cloudy air does not cool as rapidly as warm, cloudy air.

3. Urban-industrial areas are sources of abundant hygroscopic nuclei. What might this imply about the weather downwind of a large city?

4. Speculate on some ways whereby radiation fogs at an airport might be dispersed.

ADDITIONAL READINGS

Bohren, C. F. 1990. All that glistens isn't dew. *Weatherwise* 43, no. 5: 284–87.

Day, J. A., and V. J. Schaefer. 1991. *Peterson's first guide to clouds and weather.* Boston: Houghton Mifflin.

Gedzelman, S. D., and E. Lewis. 1990. Warm snowstorms: A forecaster's dilemma. *Weatherwise* 43, no. 5: 265–70.

Scorer, R., and A. Verkaik. 1989. *Spacious skies.* London: David & Charles.

Snow, J. T., and S. B. Harley. 1988. Basic meteorological observations for schools: Rainfall. *Bulletin of the American Meteorological Society* 69, no. 5: 497–507.

For current updates and exercises, log on to http://www.wadsworth.com/geo.

UNIVERSITY OF WISCONSIN SPACE SCIENCE AND ENGINEERING CENTER
http://ssec.wisc.edu/index/html

A number of useful weather satellite products are mentioned at this site.

WEATHER UNDERGROUND
http://sdm@madlab.sprl.umich.edu

This site contains a variety of current weather information and forecasts and includes information on environmental issues and special weather topics.

FLORIDA EXPLORES!
http://thunder.met.fsu.edu/explores/explores.html

Instructional materials related to satellites and satellite meteorology can be found at this site.

CHAPTER
17

Atmospheric Circulation

Wind is air in motion relative to the rotating Earth.

Prologue

On March 12, 1993, a major storm tracked out of the Gulf of Mexico and headed up the eastern seaboard. By the time the storm pushed out over the North Atlantic three days later, residents of coastal states from Alabama to Maine had felt the wrath of one of the most ferocious winter storms of the century. The storm's path enabled its circulation to tap unusually large amounts of water vapor from the warm ocean waters just off shore. Consequently, the storm produced paralyzing accumulations of snow over the highly populated region north and west of its track (☐ Fig. 17.1). Snowfall totals in many places exceeded 30 cm (12 in); 142 cm (56 in) fell at Mt. LeConte, Tennessee, 109 cm (43 in) at

Syracuse, New York, and 61 cm (24 in) at Mountain City, Georgia.

For the first time in history, snow forced the closing of every major East Coast airport. Thousands of people were isolated in the Appalachian Mountains; at one time more than 3 million customers had lost electrical power; hundreds of roofs collapsed under the weight of accumulated snow; and about 200 homes on the Outer Banks of North Carolina were severely damaged by storm waves. Wind gusts, in places exceeding hurricane force, blew the snow into huge road-clogging drifts. Meanwhile, in Florida the storm spawned 27 tornadoes.

All told, this so-called *superstorm* claimed about 270 lives, more than three times the combined death toll of Hurricanes Hugo and Andrew, and caused property damage in excess of $3 billion.

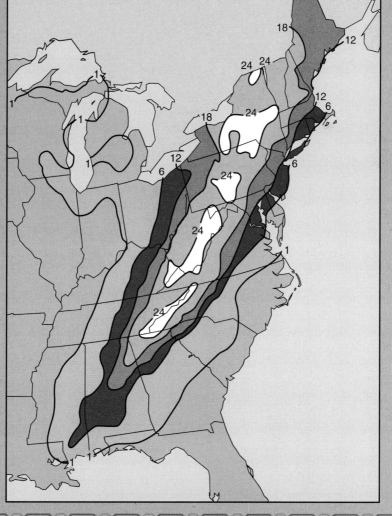

☐ FIGURE 17.1 Total snowfall (in inches) during the superstorm of March 12–15, 1993.

INTRODUCTION

Weather systems are responsible for a wide variety of atmospheric conditions, sometimes extreme as in the case of the superstorm just described, and sometimes tranquil. Some weather systems such as thunderstorms are relatively small and short-lived; others influence atmospheric conditions over thousands of square kilometers and persist for a week or longer. The type of weather produced by a system depends on the circulation of air within that system.

This chapter begins by describing the individual forces that initiate and shape atmospheric circulation and then explains how these forces are combined in highs (anticyclones) and lows (cyclones), the principal weather makers of the midlatitudes. This discussion, in turn, forms the basis for a description of the principal features of the planetary-scale circulation—the winds and pressure systems that operate around the globe.

WINDS: THE FORCES

Atmospheric circulation is experienced as **wind,** the motion of air relative to the rotating Earth. Forces that govern the wind arise from (1) air pressure gradients, (2) a centripetal force, (3) the Coriolis effect, (4) friction, and (5) gravity.

Pressure Gradient Force

A *gradient* in air pressure refers to a change in air pressure from one location to another. Air pressure gradients are caused by spatial differences in air temperature (principally) and/or humidity (see Chapter 14). Hence, a horizontal air pressure gradient exists between a cold air mass and a warm air mass.

Meteorologists routinely plot air pressure readings on a surface weather map. As explained in Chapter 14, these readings must first be adjusted to sea level to eliminate the influence of weather station elevation. Then lines, called *isobars,* are drawn connecting locations having the same air pressure. By convention, isobars are drawn at 4 mb intervals (e.g., 1016 mb, 1020 mb, 1024 mb).

An isobaric analysis reveals horizontal air pressure gradients and centers of high and low pressure. Where isobars are closely spaced (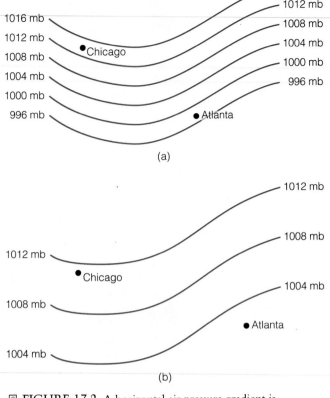 Fig. 17.2), air pressure changes rapidly with distance, and the pressure gradient is relatively steep. Where isobars are more widely spaced (Fig. 17.2), air

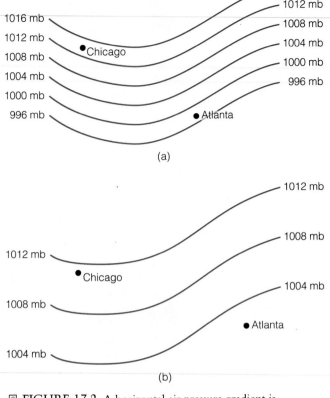

FIGURE 17.2 A horizontal air pressure gradient is (a) relatively steep where isobars are closely spaced and (b) relatively weak where isobars are widely spaced.

pressure changes more gradually with distance, and the pressure gradient is relatively weak.

The influence of an air pressure gradient on the wind can be illustrated by an analogy. Suppose a bathtub is partially filled with water, as in 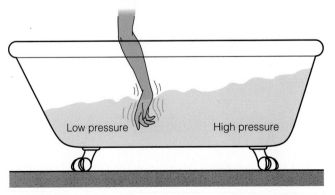 Figure 17.3. Sloshing the water from one end of the tub to the other creates a water pressure gradient along the bottom of the tub. At any instant in time, the water pressure on the tub bottom is higher where the water is deep and lower where the water is shallow. Thus, a horizontal water pressure gradient develops along the tub bottom. If the sloshing stops, the water level quickly returns to horizontal as water moves from the end of the tub where the water pressure is higher to the end where the water pressure is lower. The water depth and pressure become uniform, and the water pressure gradient along the tub bottom goes to zero.

Similarly, in response to an air pressure gradient, the wind blows from where air pressure is relatively high to-

FIGURE 17.3 In response to a water pressure gradient acting on the bottom of a tub, water moves from the end of the tub where the water pressure is greater to the end where the water pressure is less.

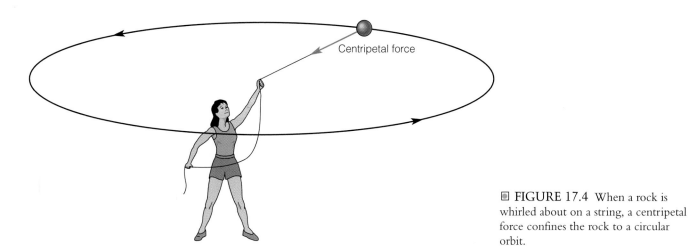

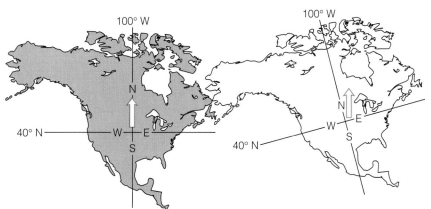

FIGURE 17.4 When a rock is whirled about on a string, a centripetal force confines the rock to a circular orbit.

ward where air pressure is relatively low. The wind is strong where the air pressure gradient is steep and light where the air pressure gradient is weak. The force that accelerates air in response to an air pressure gradient is the **pressure gradient force.** The pressure gradient force is always directed perpendicular to isobars from high to low pressure.

Centripetal Force

A centripetal force operates whenever an object follows a curved path. Consider an illustration. A rock is tied to a string and twirled so that the rock follows a circular orbit (Fig. 17.4). Cut the string and the rock flies off in a straight line. The rock behaves according to *Newton's first law of motion;* that is, an object remains in straight-line, unaccelerated motion unless acted upon by a *net* force. Prior to being cut, the string exerts a net force on the rock that confines it to a circular orbit. This net force, directed in toward the center of the orbit, is known as the **centripetal** (*center-seeking*) **force.** Cutting the string eliminates the centripetal force, and the rock follows a straight path.

A centripetal force operates whenever the wind follows a curved path. The centripetal force is responsible only for continuously changing the wind direction (curved rather than straight); it does not change the wind speed.

Coriolis Effect

A particular storm viewed from a distant fixed point in space appears to track directly from south to north. To an observer on Earth's surface, however, the storm appears to curve toward the northeast. Strangely, both descriptions of the storm's track are correct!

Both descriptions are correct because the two observers used different frames of reference as they followed the storm's movements. The Earthbound observer's frame of reference is the familiar north–south/east–west system. Although it is not obvious to the observer, the north–south/east–west coordinate system actually rotates as Earth rotates (Fig. 17.5). Thus, Earth and the coordinate system rotate under the storm as it travels over the surface. From space, the storm's movements are tracked with respect to a fixed nonrotating coordinate system. Hence, the apparent difference in the storm track (curved versus straight) arises from a difference in the frame of reference (rotating versus nonrotating).

Switching to a nonrotating frame of reference gives rise to a net force responsible for curved motion. This deflective force, due to Earth's rotation, is known as the **Coriolis effect.** It is named for Gaspard Gustav de Coriolis, who first described the phenomenon mathematically in 1835.

Wind is measured with respect to the north–south/east–west coordinate system. (By convention, wind direction is

FIGURE 17.5 A north–south/east–west frame of reference rotates as Earth rotates on its axis. This movement is the origin of the Coriolis effect.

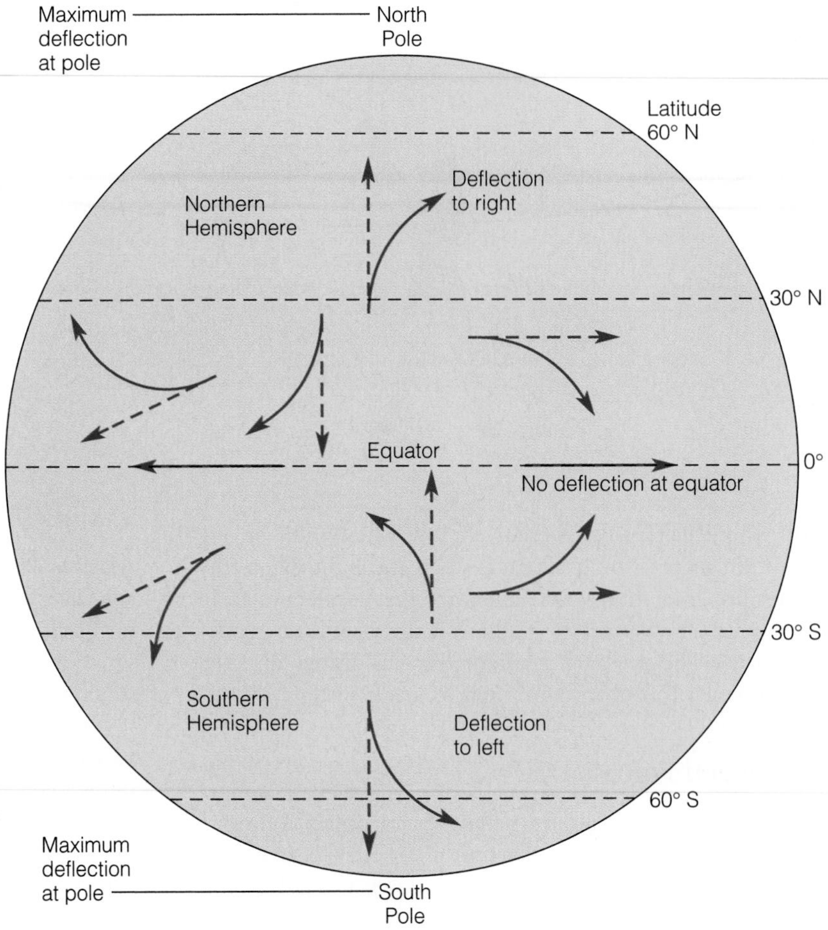

FIGURE 17.6 The Coriolis effect shifts the wind to the right of its initial direction in the Northern Hemisphere and to the left in the Southern Hemisphere. The Coriolis deflection is maximum at the poles and zero at the equator.

named for the direction *from which* the wind blows; thus, a north wind blows from north to south.) Hence, the Coriolis effect plays a role in large-scale atmospheric circulation. In the Northern Hemisphere, the Coriolis effect deflects the wind to the right of its initial direction. In the Southern Hemisphere, the Coriolis effect deflects the wind to the left (Fig. 17.6).

The reversal of the Coriolis deflection between the Northern and Southern Hemispheres is related to the difference in an observer's sense of Earth's rotation in the two hemispheres. The planet rotates counterclockwise when viewed from above the North Pole and clockwise when viewed from above the South Pole. This reversal in rotation direction translates into a reversal in Coriolis deflection between the two hemispheres.

The magnitude of the Coriolis deflection depends on latitude; it is zero at the equator and reaches a maximum at the poles. Imagine the daily rotation of towers located at different latitudes. Over the course of 24 hours, the towers at the North and South Poles make one complete rotation. Meanwhile, a tower at the equator moves end-over-end with no rotation. No rotation implies no Coriolis deflection. At any latitude between the equator and the poles, the towers rotate but not as much as those at the poles.

The Coriolis deflection also increases with increasing wind speed. In the same period of time, strong winds cover greater distances and experience greater deflection (due to Earth's rotation) than do weaker winds. For all practical purposes, the Coriolis effect is important only in winds within large-scale weather systems, such as those larger than ordinary thunderstorms.

Friction

Friction is the resistance an object encounters as it moves in contact with other objects. Although friction is usually associated with solids, it also affects fluids, both liquids and gases. As the horizontal wind blows in contact with Earth's surface, it is slowed by friction.

The rougher the surface, the more the wind is slowed. A forest offers more frictional resistance to the wind than does the smoother surface of a grassland. Friction diminishes rapidly with increasing altitude where the wind no longer comes in contact with obstacles such as shrubs, trees, and buildings that are responsible for surface roughness. Hence, horizontal winds strengthen with increasing altitude. Above an average altitude of about 1,000 m (3,300 ft), friction is a minor force that merely smooths the flow of air. The portion of the atmosphere where friction affects the wind is known as the **friction layer.**

Gravity

What we call **gravity** at Earth's surface is actually the net effect of two forces: (1) *gravitation,* or the force of attraction between Earth and all other objects, and (2) a weaker centripetal force attributed to all objects on Earth because of the planet's rotation. The gravity force at the surface accelerates any object downward at the rate of 9.8 m (32.1 ft) per second each second, provided air resistance can be ignored.

Gravity is always directed downward and perpendicular to Earth's surface. Hence, unlike the Coriolis effect and friction, gravity does not modify the horizontal wind. Gravity influences the vertical movement of air and is responsible for the downhill drainage of cold, dense air.

 # WINDS: COMBINING FORCES

Although it is convenient to examine the forces that govern atmospheric circulation separately, in reality those forces are not independent of one another. Their interaction controls the speed and direction of the wind. This section covers the interactions of forces that result in (1) the geostrophic wind, (2) the gradient wind, and (3) horizontal winds at Earth's surface.

Geostrophic Wind

The **geostrophic wind** is a horizontal wind that blows at constant speed in a straight line at altitudes above the friction layer. It is the product of a balance between the pressure gradient force and the Coriolis effect. According to Newton's first law of motion, a balance of forces means that an object either remains at rest or moves in a straight line with no change in speed.

As shown in Figure 17.7, the horizontal pressure gradient force (P_H) accelerates the wind perpendicular to isobars, from high pressure toward low pressure. Soon, however, the wind comes under the influence of the Coriolis effect (C) and is gradually deflected toward the right of its initial direction in this Northern Hemisphere example. Eventually, the two forces achieve a balance so that the geostrophic wind blows parallel to the isobars with the highest air pressure to the right of its direction.

 EARTHFACT

The Coriolis Effect and the Draining of Water

Water usually rotates as it drains from a sink or a bathtub. A popular misconception is that the direction of this rotation (clockwise or counterclockwise) is influenced by Earth's rotation (that is, by the Coriolis effect) and that the flow is consistently in one direction in one hemisphere and in the opposite direction in the other hemisphere. Actually, at the very small scale of water motion in a sink or bathtub, the Coriolis effect has negligible influence. Rotation during drainage may be either clockwise or counterclockwise. The direction most likely is due to the shape of the basin or some slight rotational motion of the water just before the drain is opened. Similarly, the Coriolis effect is too weak to influence the circulation in small-scale weather systems such as tornadoes and dust devils.

Gradient Wind

The **gradient wind** is similar to the geostrophic wind except that it describes a curved rather than a straight path. The gradient wind is also a horizontal wind that parallels isobars above the friction layer. A centripetal force constrains the wind to a curved path and has no influence on wind speed. The gradient wind is the product of the interaction of the horizontal pressure gradient force, the Coriolis effect, and a centripetal force.

Above the friction layer, the gradient wind blows about a high pressure center, called an **anticyclone** (or *high*), and a low pressure center, called a **cyclone** (or *low*). Consider, for example, an idealized anticyclone in which isobars form a bull's-eye pattern of concentric circles about the center of

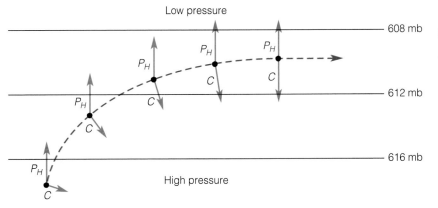

FIGURE 17.7 Forces operating in the geostrophic wind, a horizontal wind that blows in a straight line parallel to isobars above the atmosphere's friction layer.

C = Coriolis effect
P_H = Horizontal pressure gradient force
- - ➤ Geostrophic wind

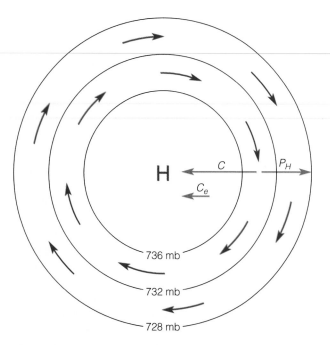

P_H = Horizontal pressure gradient force
C = Coriolis effect
C_e = Centripetal force
⟵ Gradient wind

■ **FIGURE 17.8** Horizontal winds in a Northern Hemisphere anticyclone (high) above the friction layer. Viewed from above, the winds blow clockwise and parallel to the isobars.

P_H = Horizontal pressure gradient force
C = Coriolis effect
C_e = Centripetal force
⟶ Gradient wind

■ **FIGURE 17.9** Horizontal winds in a Northern Hemisphere cyclone (low) above the friction layer. Viewed from above, the winds blow counterclockwise and parallel to the isobars.

the system (■ Fig. 17.8). The horizontal pressure gradient force (P_H) is directed radially outward, away from the center of the high. The Coriolis effect (C) is directed inward. The Coriolis effect is slightly greater than the pressure gradient force, and the difference between the two forces is equal to the centripetal force (C_e). Hence, in a Northern Hemisphere anticyclone above the friction layer, the wind blows clockwise and parallel to the isobars.

Consider also an idealized cyclone in which isobars form a bull's-eye pattern of concentric circles about the center of the system (■ Fig. 17.9). The horizontal pressure gradient force (P_H) is directed radially inward toward the cyclone center, and the Coriolis effect (C) is directed outward. The pressure gradient force is slightly greater than the Coriolis effect, and the difference between the two forces is equal to the centripetal force (C_e). Hence, in a Northern Hemisphere cyclone above the friction layer, the wind blows counterclockwise and parallel to the isobars.

Surface Winds

The roughness of Earth's surface influences horizontal winds blowing within the friction layer. Frictional resistance slows the wind and, in combination with the other forces, shifts the wind across isobars and toward low pressure. The angle between the wind direction and the isobars depends on the

surface roughness and varies from about 10° over smooth surfaces to 40° or more over rough terrain. In the Northern Hemisphere, surface winds blow clockwise and outward in an anticyclone (■ Fig. 17.10a) and counterclockwise and inward in a cyclone (Fig. 17.10b).

Reversal of the Coriolis deflection in the Southern Hemisphere means that cyclonic and anticyclonic circulations are opposite their Northern Hemisphere counterparts. In the Southern Hemisphere, at altitudes above the friction layer, winds blow parallel to isobars and clockwise in a cyclone and counterclockwise in an anticyclone. Horizontal surface winds spiral clockwise and inward in a cyclone and counterclockwise and outward in an anticyclone.

WEATHER OF HIGHS AND LOWS

The weather associated with anticyclones and cyclones is determined by the three-dimensional circulation that characterizes these pressure systems. Near Earth's surface horizontal winds diverge away from the center of a high (■ Fig. 17.11a). Air that slowly descends from above replaces air that diverges at the surface. Aloft, air converges toward the high center and replaces the descending air. Recall from Chapter 16 that descending air is compressionally warmed and its rel-

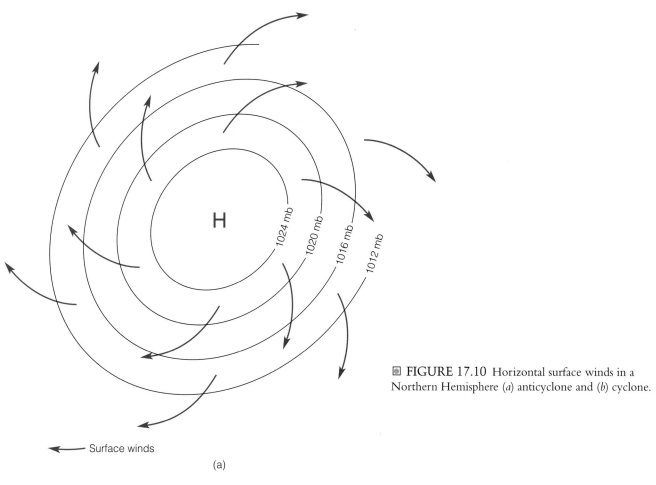

1024 mb

1020 mb

1016 mb

1012 mb

H

⬛ FIGURE 17.10 Horizontal surface winds in a
Northern Hemisphere (*a*) anticyclone and (*b*) cyclone.

⟵ Surface winds

(a)

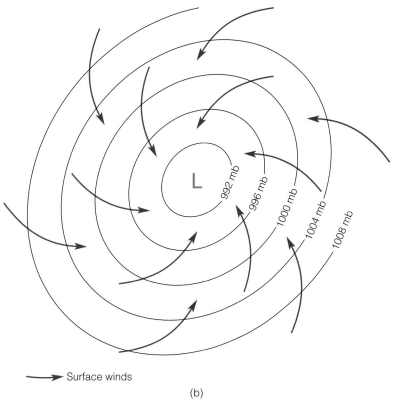

992 mb

996 mb

1000 mb

1004 mb

1008 mb

L

⟶ Surface winds

(b)

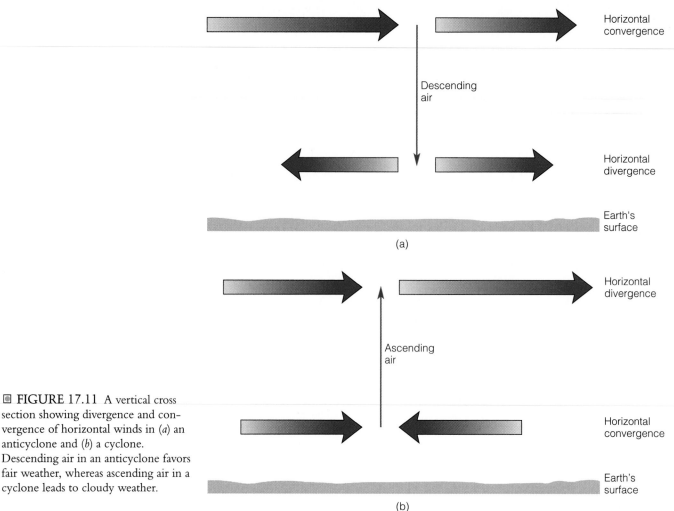

FIGURE 17.11 A vertical cross section showing divergence and convergence of horizontal winds in (*a*) an anticyclone and (*b*) a cyclone. Descending air in an anticyclone favors fair weather, whereas ascending air in a cyclone leads to cloudy weather.

ative humidity drops. Thus, anticyclones tend to be *fair weather* systems.

In addition, the horizontal air pressure gradient is typically weak (widely spaced isobars) over a broad area surrounding the center of an anticyclone. A weak horizontal pressure gradient means light or calm winds. If the night sky is clear, intense radiational cooling chills the ground and the overlying air, so dew, frost, or radiation fog may form (see Chapter 16).

In a Northern Hemisphere cyclone, surface winds blow counterclockwise and inward, converging toward the center of the system (Fig. 17.11b). Air ascends near the center and diverges aloft. Ascending air expands and cools, and its relative humidity increases. With sufficient cooling, the relative humidity approaches 100%, and clouds and precipitation develop. Hence, cyclones are typically *stormy weather* systems.

 WIND MEASUREMENT

Meteorologists monitor both wind direction and speed. A *wind vane,* such as the one in ▣ Figure 17.12a, consists of a free-swinging horizontal rod equipped with a flat vertical plate at one end and a counterweight (arrow) at the other end. The arrow points in the direction *from which* the wind blows. Some wind vanes are read visually; more sophisticated models are linked electronically to a dial or digital display calibrated in points of the compass or in degrees (90 for east, 180 for south, 270 for west, and 360 for north). A *cup anemometer* monitors horizontal wind speed (Fig. 17.12b) and works on the same principle as a bicycle speed computer.

A wind vane/anemometer system should be mounted on a tower so that the instruments are about 10 m (33 ft) above the ground. The instrument site should be well away from buildings that might shelter the instruments or any obstacle that might channel (and thus accelerate) the wind. Rooftop sites should be avoided because the wind accelerates over them.

 SCALES OF WEATHER SYSTEMS

For convenience of study, atmospheric circulation is subdivided into discrete weather systems operating at different spatial and temporal scales (◎ Table 17.1). Wind belts encircling the globe (e.g., westerlies, trade winds) are *planetary-scale systems. Synoptic-scale systems* are continental or oceanic

(a)

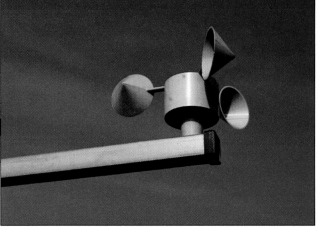

(b)

■ FIGURE 17.12 (*a*) A wind vane monitors horizontal wind direction, and (*b*) a cup anemometer monitors wind speed.

in extent; anticyclones (highs) and cyclones (lows) are examples. *Mesoscale systems* are regional and include, for example, thunderstorms and sea breezes. A weather system covering a very small area—a tornado, for example—operates at the smallest spatial subdivision of atmospheric circulation, the *microscale*.

Circulation systems differ in life expectancy as well as spatial scale. Patterns in the planetary-scale wind belts (discussed later in this chapter) may persist for weeks or even months. Synoptic-scale systems usually complete their life cycles within several days to a week or so. Mesoscale systems usually last for hours to perhaps a day whereas microscale systems persist for minutes or less.

In meso- and microscale systems, vertical (up and down) wind speeds are often comparable in magnitude to horizontal wind speeds. At the planetary and synoptic scales, however, horizontal winds are considerably stronger than vertical winds. Furthermore, the Coriolis effect is negligibly small in microscale systems and increases in importance with increasing spatial scale.

◉ PLANETARY-SCALE CIRCULATION

Atmospheric circulation operating at the planetary scale governs the development and movement of all smaller-scale weather systems. Hence, an examination of weather systems

TABLE 17.1 Scales of Atmospheric Circulation

CIRCULATION	SPATIAL SCALE	TEMPORAL SCALE
Planetary scale	10,000–40,000 km	Weeks to months
Synoptic scale	100–10,000 km	Days to weeks
Mesoscale	1–100 km	Hours to days
Microscale	1 m–1 km	Seconds to hours

appropriately begins with global wind belts and pressure systems with special emphasis on the characteristics of the mid-latitude westerlies. Weather systems operating at smaller scales are covered in Chapter 18.

Pressure Systems and Wind Belts

Several large high and low pressure systems appear on maps of average sea-level air pressure for January and July (■ Fig. 17.13). These are the *semipermanent pressure systems.* Although these anticyclones and cyclones are persistent features of the atmosphere's planetary-scale circulation, their strength and specific locations change seasonally; hence, they are termed *semipermanent.* Pressure systems include subtropical anticyclones and subpolar lows.

Subtropical anticyclones are centered, on average, near 30° N and S, over the North and South Atlantic, the North and South Pacific, and the Indian Ocean. These highs are massive systems that reach from the ocean surface to the tropopause. They have a major influence on the weather and climate of both tropical and middle latitudes.

Areas under the eastern flank of a subtropical anticyclone are dry whereas areas under the western flank are moist. From the center of a subtropical high, an extensive region of stable subsiding air stretches outward over its eastern flank. Compressional warming of descending air produces low relative humidities and sunny skies. The world's major tropical deserts, including the Sahara of North Africa, are situated under the eastern portion of subtropical anticyclones. On the western flank of a subtropical high, however, subsidence is less, and the air is not as stable, so episodes of cloudy, wet weather are more frequent. Hence, the American Southwest (on the eastern side of the subtropical *Hawaiian high*) is much drier than the American Southeast (on the western side of the subtropical *Bermuda-Azores high*).

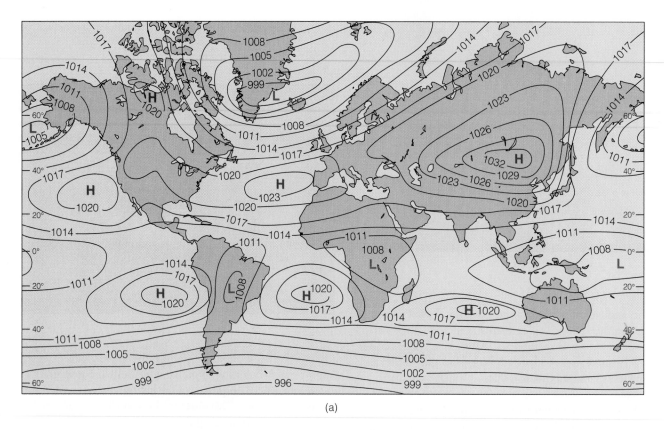

(a)

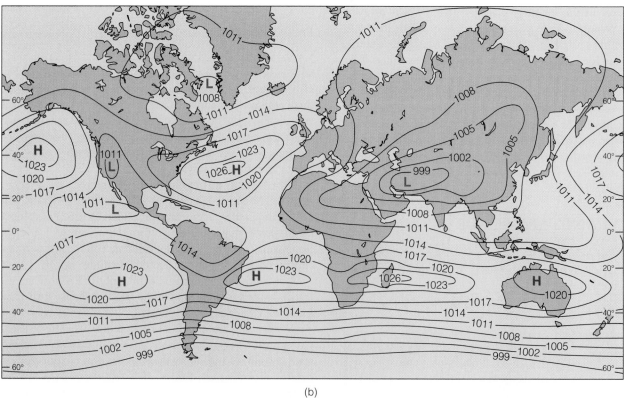

(b)

◉ FIGURE 17.13 Average sea-level air pressure for (*a*) January and (*b*) July.

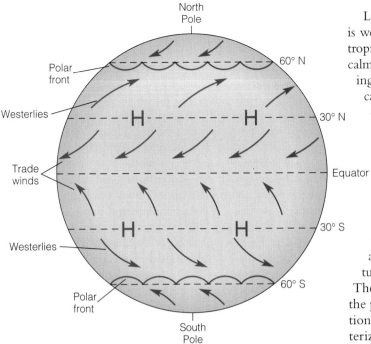

FIGURE 17.14 The trade winds blow out of the equatorward flanks of the subtropical anticyclones, and the midlatitude westerlies blow out of the poleward flanks of the subtropical anticyclones.

Like all anticyclones, the horizontal air pressure gradient is weak over a broad area surrounding the center of a subtropical high. Hence, surface winds are very light or even calm over vast stretches of the subtropical oceans. Ships sailing from Spain to the New World often were caught in calms for weeks at a time, forcing crews to jettison their cargo of horses when supplies of food and water ran low. For this reason, mariners refer to this area of calm winds over tropical seas as the *horse latitudes*.

Surface winds blowing outward from centers of subtropical highs constitute the westerlies and trade winds (◼ Fig. 17.14). In the Northern Hemisphere, north of the horse latitudes, prevailing surface winds blow from southwest to northeast. These are the highly variable **westerlies** of the middle latitudes (roughly in a belt between 30° and 60° N). South of the horse latitudes, prevailing surface winds blow from the northeast. These are the **trade winds,** the most persistent winds on the planet; in some regions, they blow from the same direction as much as 75% of the time. Analogous winds characterize the Southern Hemisphere's planetary-scale circulation, although reversal of the Coriolis deflection leads to southeast trade winds on the northern flanks of the Southern Hemisphere subtropical highs and northwest winds on the southern flanks.

Trade winds of the two hemispheres converge along the **intertropical convergence zone (ITCZ),** a discontinuous belt of thunderstorms paralleling the equator (◼ Fig. 17.15). On average, the ITCZ is located near the latitude

◼ **FIGURE 17.15** The intertropical convergence zone (ITCZ) is marked by a discontinuous line of thunderstorms as shown in this visible satellite image.

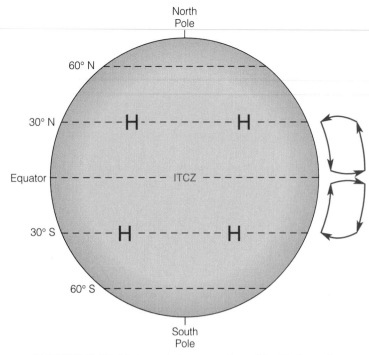

FIGURE 17.16 A vertical cross section of the Hadley cell circulation in tropical latitudes of both hemispheres.

In the Northern Hemisphere, surface westerlies override polar northeasterlies along the **polar front** (see Fig. 17.14). Recall that a *front* is a narrow zone of transition between air masses that differ in temperature and/or humidity. The polar front is not continuous around the globe and is well-defined in some areas and not in others. Where the polar front is well-defined (a relatively great temperature contrast across the front), synoptic-scale cyclones, cloudiness, and precipitation are likely to develop (see Chapter 18).

The planetary-scale winds aloft—that is, in the middle and upper troposphere—differ from those at the surface. Air subsides in subtropical anticyclones, sweeps toward the equator as the surface trade winds, and then ascends in the ITCZ. Aloft, winds blow poleward, away from the ITCZ and into the subtropical highs. Because of the Coriolis effect, these upper-level winds shift toward the right in the Northern Hemisphere (southwest winds), and toward the left in the Southern Hemisphere (northwest winds). In vertical profile, this circulation resembles a huge convective cell and is known as the **Hadley cell** (▣ Fig. 17.16). Hadley cells are located in the tropics on either side of the ITCZ.

In the middle and upper troposphere, prevailing winds of middle latitudes blow from west to east in a wavelike pattern of ridges and troughs (▣ Fig. 17.17). These are the upper-air westerlies. They are so important for midlatitude weather that they are treated in a separate section later in this chapter.

At high latitudes, air subsides and diverges at the surface away from shallow, cold anticyclones. In the Northern Hemisphere, these highs are well developed only in winter over the snow-covered continental interiors. In the Southern Hemisphere, cold highs persist over the Antarctic ice sheet all year. Aloft, polar winds weave from west to east.

Seasonal Shifts

Seasonal changes in the planetary-scale circulation affect regional climates. Pressure systems, the polar front, wind belts,

where Earth's average surface temperature is highest, the so-called *heat equator,* just north of the geographical equator.

Poleward of the subtropical anticyclones, surface westerlies flow into low pressure systems. In the Northern Hemisphere, two **subpolar lows** are centered near 60° N: the *Aleutian low* over the North Pacific and the *Icelandic low* over the North Atlantic. Midlatitude westerlies converge with polar northeasterlies within the subpolar lows. In the Southern Hemisphere, by contrast, the midlatitude northwesterlies and the polar southeasterlies converge along a belt of low pressure bordering the Antarctic continent.

FIGURE 17.17 In the middle and upper troposhere, the planetary-scale westerlies weave from west to east in a wavelike pattern of ridges and troughs.

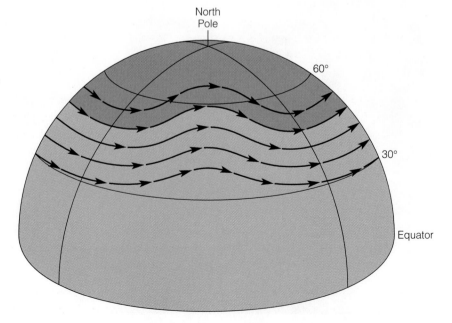

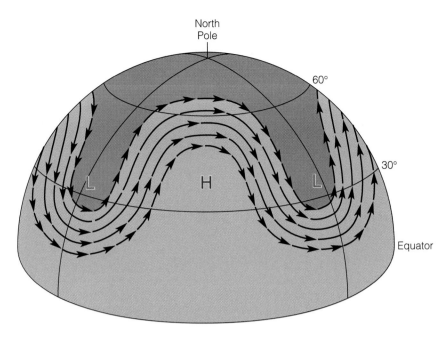

North
Pole

60°

30°

Equator

FIGURE 17.18 A looping meridional flow pattern in the midlatitude westerlies. Such a circulation pattern favors large contrasts in temperature across the United States and southern Canada.

and the ITCZ all follow the Sun, shifting poleward in spring and equatorward in autumn. Reversal of seasons between the two hemispheres means that the planetary-scale systems of both hemispheres move north and south in tandem. Furthermore, the subtropical anticyclones exert higher surface pressures in summer than in winter. The Icelandic low intensifies in winter and weakens in summer, and though the Aleutian low is well developed in winter, it disappears in summer.

Surface air pressures over the continents also change with the seasons primarily because of the contrast in thermal inertia between land and sea. Recall from Chapter 14 that in response to the same solar heating, the sea surface exhibits smaller temperature variations through the year than does the land surface. Hence, the continents are dominated by cold anticyclones in winter and relatively low pressure in summer. In winter, cold anticyclones develop over northwestern North America and over the Asian interior, where the most prominent of these systems is the massive *Siberian high*. In summer, belts of low pressure form across North Africa and from the Arabian peninsula eastward to Southeast Asia. Warm cyclones also develop in summer over arid and semiarid regions of the Americas.

Seasonal shifts in the atmosphere's planetary-scale circulation leave their mark on precipitation regimes over broad areas of the globe. The northward shift of the ITCZ triggers summer monsoon rains in Central America, sub-Saharan Africa, India, and Southeast Asia. As a subtropical anticyclone shifts north and south with the Sun, its dry eastern flank influences some localities in winter and others in summer and is responsible for pronounced rainy seasons and dry seasons. In sub-Saharan Africa, summers are wet and winters are dry; in southern California, summers are dry and winters are wet.

Billions of people depend on monsoon rains for their freshwater supply, but those rains are not always reliable. For

example, residents of sub-Saharan Africa have endured multidecadal drought (see ▢ Perspective 17.1).

◆ THE WEAVING WESTERLIES

The westerlies merit special attention because they govern the weather of much of North America. They provide upper-air support for developing cyclones and steer storms and air masses. Aloft, westerlies flow around the hemisphere in a wave pattern of ridges and troughs. These are **Rossby waves,** named for Carl G. Rossby, the Swedish-American meteorologist who discovered them in the late 1930s. Winds blow clockwise (anticyclonic) in ridges and counterclockwise (cyclonic) in troughs. Hence, westerly waves can be described in terms of wavelength (distance between successive troughs or, equivalently, successive ridges), amplitude (north-south extent), and the number of waves circling the hemisphere. All three characteristics change with time, and as a consequence, the weather also changes. In addition, ridges and troughs slowly progress from west to east.

At any time between two and five Rossby waves weave around the Northern Hemisphere. The westerlies are more vigorous in winter than summer and feature fewer waves having greater lengths and amplitudes. Ultimately, this seasonal contrast is due to the greater north-south air temperature contrast in winter.

The *weaving westerlies* consist of a north-south component of motion superimposed on an overall west-to-east flow. The north-south (meridional) component is responsible for north-south exchange of air masses and poleward heat transport (see Chapter 15). In the Northern Hemisphere, southwest winds carry warm air masses northward, and northwest winds transport cold air masses southward. Cold air is thus exchanged for warm air, and heat is transported poleward.

Perspective 17.1

Focus on the Environment: Drought in Sub-Saharan Africa

Distinct wet and dry seasons characterize monsoon climates. But the duration of the rainy season and the total amount of rainfall vary from one place to another and from one year to the next. Perhaps nowhere on Earth does the variability of monsoon rains cause more human misery than in sub-Saharan Africa. A combination of low average annual rainfall, considerable year-to-year variation in rainfall, and long-term drought has brought considerable hardship to the people of this region.

Sub-Saharan Africa, much of which is called the *Sahel*, is a transition zone between the Sahara to the north and the humid savanna to the south. The North African monsoon climate zone stretches from the Atlantic Ocean to the Red Sea and includes all or part of Mauritania, Senegal, Mali, Burkina Faso, Niger, Chad, and Ethiopia (☐ Fig. 1), which are among the poorest nations on Earth (○ Table 1).

The length of the rainy season and the average annual rainfall increase from north to south across sub-Saharan Africa. At the northern edge (about 16° N), the rainy season may not begin until June and then last only a month or two; mean annual rainfall is typically under 100 mm (about 4 in). At the southern edge (about 10° N), the rainy season may be underway as early as April and persist for five or six months; average annual rainfall exceeds 500 mm (about 20 in).

North-south shifts of the intertropical convergence zone (ITCZ) govern the timing and amount of rainfall in sub-Saharan Africa. As the ITCZ follows the Sun, its northward shift in spring triggers rainfall, and its southward shift in fall

☐ **FIGURE 1** Sub-Saharan Africa has a monsoon climate (wet summers and dry winters) and is subject to prolonged drought.

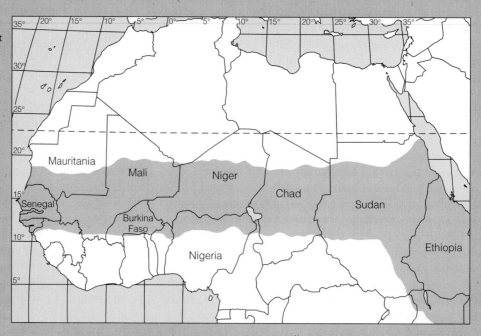

Sometimes, the westerlies blow almost directly from west to east with only a weak meridional component (Fig. 17.17). With this so-called *zonal flow pattern,* cold air masses remain to the north, and warm air masses stay to the south. Zonal flow floods much of the United States and southern Canada with *Pacific air,* an air mass that has its origins over the Pacific Ocean. Pacific air loses much of its moisture as it passes over the western mountain ranges and warms as it descends onto the Great Plains. Consequently, much of eastern and central North America is favored by generally uneventful mild, dry weather.

At other times, the westerlies exhibit considerable amplitude, blowing in an undulating pattern of deep troughs and sharp ridges (☐ Fig. 17.18). This so-called *meridional flow pattern* steers cold air masses southeastward and warm air masses northeastward so that significant temperature contrasts develop over the United States and southern Canada. Where contrasting air masses meet, fronts form and warm brings the rainy season to a close. During the rainy season, surface winds from the south and southwest transport humid air from over both the Atlantic and Indian Oceans. During the dry season, surface winds blow from the dry north and northeast.

In a summer when the ITCZ does not shift as far north as usual, arrives late, or shifts southward early, rainfall is below average. A succession of such summers means drought. In this century, the people of the Sahel have endured three major droughts, the most recent of which began in 1969 and persisted into the 1990s, albeit with a few brief interruptions. Rainfall during the 1961–1990 period was 20 to 40% lower than it was during the prior three decades, making this the most extreme drought on record anywhere.

A key element in unraveling the cause of drought in the Sahel is the research of Professor Sharon Nicholson and her colleagues at the Florida State University. They reconstructed the climate of the Sahel for the past few centuries based on analysis of lake-level fluctuations and historical accounts of changes in landscape. They argue that droughts lasting one to two decades are characteristic of the region's climate. What is unique about the most recent drought is its persistence.

Nicholson and others contend that the principal reason for drought in sub-Saharan Africa is changes in atmospheric circulation that are likely linked to patterns in sea-surface temperatures (SSTs). In fact, atmospheric scientists have discovered a strong correlation between drought in the Sahel and SSTs of the eastern tropical Atlantic. When SSTs are higher than normal south of the equator and southwest of West Africa (e.g., the Gulf of Guinea), rainfall is below the long-term average in the Sahel. Apparently, SST patterns affect shifts of the ITCZ.

In 1984, the worst year of the most recent drought, televised news reports brought home to people around the world the devastating human impact of the Sahelian drought. Horrible scenes of starving children, emaciated livestock, and withered crops spurred a massive international relief effort. The people of the region are particularly vulnerable to drought because most of them depend on agriculture for their livelihood. Prolonged drought forced them off lands that even in the best of times are marginal for the survival of crops and livestock. People forced from their land by deteriorating conditions migrated to urban areas in search of food and work, and many ended up in refugee camps.

| | TABLE 1 | Profile of the Nations of the African Sahel | | | |
NATION	POPULATION (millions)	PER CAPITA GNP ($)	LIFE EXPECTANCY (years)	MALNOURISHED CHILDREN (%)	LITERACY RATE (%)
Mauritania	2.083	493	46.0	65	34.0
Senegal	7.529	731	47.3	28	38.3
Mali	9.510	254	44.0	34	32.0
Burkino Faso	9.248	347	47.0	NA	18.2
Niger	7.986	296	44.5	38	28.4
Chad	5.692	213	45.5	NA	29.8
Ethiopia	51.379	120	45.0	43	NA

SOURCE: World Resources Institute.

air overrides cold air. The stage is thereby set for the development of cyclones and associated clouds and precipitation.

Figures 17.17 and 17.18 represent opposite ends of a range of possible Rossby wave patterns, featuring varying degrees of zonal (west-to-east) and meridional (north-to-south) components. The westerly wave pattern continually shifts back and forth between zonal and meridional flow. There is no regular zonal/meridional cycle, and the transition between wave patterns is usually abrupt, often taking place within a single day. These abrupt changes challenge weather forecasters because sudden shifts in the upper-air westerlies may steer a storm toward or away from a locality. Furthermore, a meridional flow pattern is usually more persistent than a zonal flow pattern. Whereas a zonal flow pattern might last for several days, a meridional flow pattern can persist for a few months or longer.

Occasionally, undulations of the weaving westerlies become so extreme that huge, rotating masses of air separate

The Windiest Place on Earth

On April 12, 1934, an anemometer at the 1,910 m (6,262 ft) summit of Mount Washington, New Hampshire, registered a peak wind gust of 373 km (231 mi) per hour, the highest wind speed recorded anywhere (■ Fig. 1). (No doubt stronger winds accompany violent tornadoes, but so far they have not been confirmed by instruments.) On that same day, the wind averaged 303 km (188 mi) per hour over a five-minute span. At 57 km (35 mi) per hour, the average annual wind speed on Mount Washington is the highest of any U.S. location. Average monthly wind speed ranges from about 40 km (25 mi) per hour in July to about 73 km (45 mi) per hour in January. Winds in excess of hurricane force (119 km or 74 mi per hour) are common in winter.

Elevation and exposure account for the windiness of mountaintops. As noted elsewhere in this chapter, the westerlies strengthen with altitude in response to the north-south air temperature gradient. Hence, mountaintops are particularly windy places. Furthermore, as winds flow up and over a mountain, they are squeezed between the summit and the overlying air layers. Constriction accelerates the wind. The prevailing westerlies are normally strong over New England, and the constricted flow over Mount Washington accelerates the winds to extremes.

■ FIGURE 1 The strongest winds ever recorded were measured at the 1,910 m (6,262 ft) summit of Mount Washington, New Hampshire.

from the main west-to-east flow (■ Fig. 17.19). This circulation pattern is analogous to whirlpools that develop in swiftly flowing mountain streams. Cutoff masses of air rotate in either a cyclonic or an anticyclonic direction. Because they block the usual west-to-east progression of weather systems, a *cutoff low* or *cutoff high* is known as a *blocking system*. Blocking systems usually last for several weeks or longer and thus are likely to lead to weather extremes such as drought or persistent heat or cold.

■ FIGURE 17.19 A blocking pattern in the midlatitude westerlies that may lead to extremes in weather.

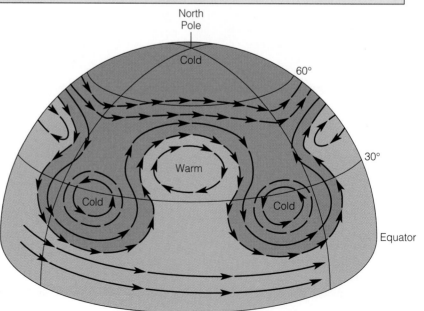

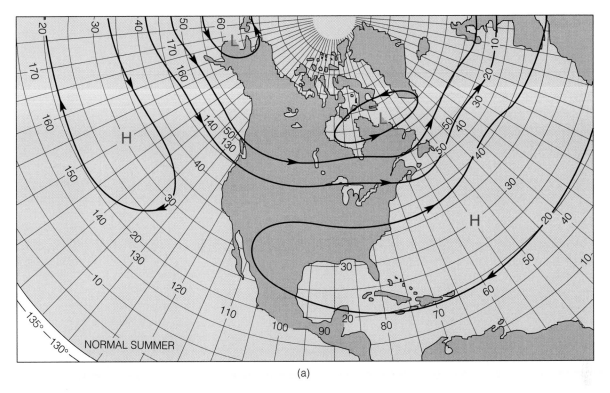

(a)

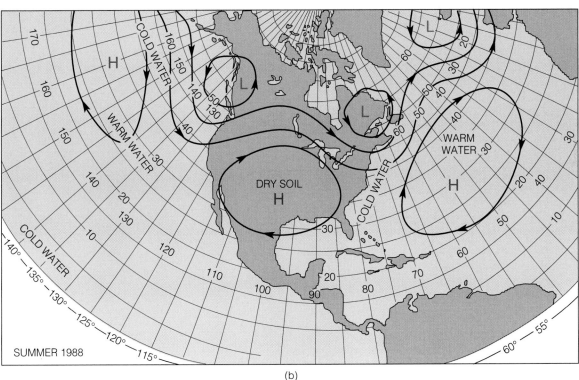

(b)

◻ FIGURE 17.20 (a) The long-term average summer circulation over North America and (b) the prevailing circulation during the summer of 1988. A pesistent blocking high over the center of the nation resulted in one of the severest droughts on record.

The severe drought that affected the central United States during the summer of 1988 was associated with a blocking weather pattern (◻ Fig. 17.20). Between early May and mid-August, a huge warm anticyclone stalled over the nation's midsection and diverted moisture-bearing weather

systems into central Canada, well north of their normal paths. For the Corn Belt, May through June was the driest since 1895.

Westerly circulation patterns responsible for weather ultimately may be linked to unusually high or low sea-surface

El Niño

Originally, El Niño was the name given by Peruvian fishermen to a period of unusually warm waters and poor fishing off the northwest coast of South America. The name *El Niño* was chosen because the event often coincided with the Christmas season. (*El Niño* means "the child" in Spanish, in particular the Christ child.) Usually, warm water episodes along the Peruvian coast last a month or two before water temperatures and the fishery return to normal. Occasionally, however, El Niño persists for a year or two and is accompanied by changes in sea-surface temperatures (SSTs) over vast stretches of the tropical Pacific, major shifts in planetary-scale atmospheric and oceanic circulations, and collapse of important fisheries. Today, scientists usually reserve the term *El Niño* for these long-lasting major ocean/atmosphere events.

El Niño is an interaction between sea and air. Normally, the southeast trade winds of the Southern Hemisphere drive warm surface waters westward, away from the northwest coast of South America. These warm surface waters are replaced by cold, nutrient-rich waters that well up from depths of 200 to 1,000 m (about 650 to 3,300 ft), a process known as *upwelling* (Fig. 1).

Although the zone of upwelling is relatively narrow (typically less than 200 km or 124 mi wide), the abundance of nutrients spurs the growth of plankton populations, which, in turn, support a highly productive fishery. In fact, the Peruvian anchovy harvest at one time was one of the world's largest fisheries. Prior to depletion by overfishing, the annual anchovy catch sometimes topped 10 million metric tons.

In 1966 Jacob Bjerknes, an atmospheric scientist at the University of California at Los Angeles, demonstrated a relationship between El Niño and the southern oscillation. The *southern oscillation* is a seesaw variation in air pressure between the eastern and western tropical Pacific. Surface air pressure over Indonesia and northern Australia varies inversely with surface air pressure over the tropical east Pacific. The horizontal pressure gradient thus changes as air pressure to the west rises and air pressure to the east falls (and vice versa).

Bjerknes found that El Niño begins when the east-west air pressure gradient across the tropical Pacific begins to weaken. Atmospheric scientists use the acronym *ENSO* to refer to this link between El Niño and the southern oscillation.

The usual contrast between relatively high air pressure over the eastern tropical Pacific and relatively low air pressure over the western tropical Pacific drives the trade winds. With the onset of El Niño, however, air pressure falls over the eastern tropical Pacific and rises over the western tropical Pacific. The air pressure gradient across the tropical Pacific weakens, and the trade winds slacken. During a particularly intense El Niño, trade winds west of the International Dateline (180° W longitude) reverse direction and blow from west to east.

In response to these changes in atmospheric circulation over the tropical Pacific, ocean currents and SSTs change. Warm surface waters, which are normally driven by the trade winds toward the western tropical Pacific, now drift slowly eastward and are deflected northward and southward by the continental landmasses, sometimes as far north as British Columbia, Canada, and as far south as central Chile. The warm water's eastward drift is so slow that it may take several

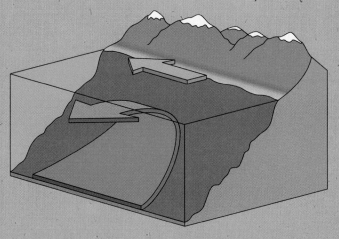

FIGURE 1 Upwelling of cold, nutrient-rich waters off the northwest coast of South America.

temperatures over the tropical Pacific. This possible linkage is discussed in Perspective 17.2.

Jet Stream

A relatively narrow corridor of very strong winds within the overall atmospheric circulation is known as a *jet stream*. A jet stream is located directly over the polar front near the tropopause. This **polar front jet stream** is part of the planetary westerlies and follows their meandering path with speeds frequently topping 160 km (100 mi) per hour. What causes the polar front jet stream, and what role does it play in midlatitude weather?

The association of a jet stream with the polar front stems from the effect of a horizontal temperature gradient on the horizontal air pressure gradient. Air pressure drops more

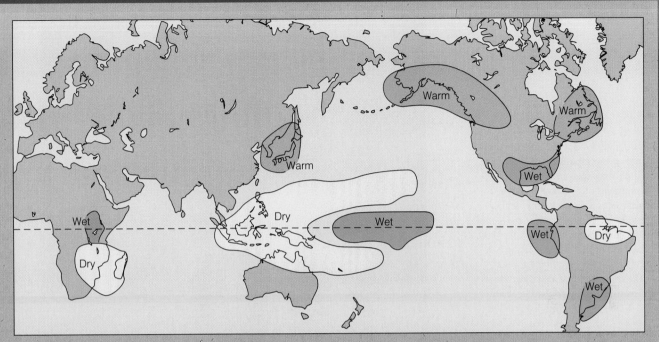

□ **FIGURE 2** Weather anomalies associated with a typical El Niño during the Northern Hemisphere winter.

months to reach the west coasts of North and South America.

Meanwhile, changes take place in the *thermocline,* the zone of rapid vertical change in temperature between relatively warm surface water and cold deep water. The tropical Pacific thermocline rises in the west and drops in the east, cutting off upwelling off the coast of Ecuador and Peru. Deprived of nutrients, phytoplankton production declines, and the fish harvest plummets.

Lower-than-normal SSTs in the western tropical Pacific and higher-than-normal SSTs in the eastern tropical Pacific coupled with the change in trade wind circulation give rise to anomalous weather patterns in the tropics and subtropics. For example, prevailing winds over Indonesia are onshore, so rainfall there normally is abundant. But during El Niño, prevailing winds blow offshore, and the weather of Indonesia is dry. El Niño–related drought also grips India and eastern Australia. Meanwhile, warmer-than-usual surface waters off the west coast of South America spur convection and heavy

rainfall along the normally arid coastal plain, causing the desert to bloom.

El Niño is also accompanied by weather extremes in the midlatitudes, especially in winter (□ Fig. 2). Changes in the planetary-scale circulation associated with SST anomalies shift the polar front jet stream north of its average winter latitude. This shift alters storm tracks so that winters are unusually wet over the Gulf Coast states and unusually mild over western Canada and portions of the northern United States.

El Niño can be expected about once every 3 to 7 years. Nine El Niño events have occurred over the past 40 years, the most recent of which was underway by late 1994 and persisted through early 1995. Prior ones took place in 1991–1993 and 1986–1987. The most intense El Niño of this century (so far) occurred in 1982–1983. So important is El Niño in year-to-year climate variability that signs of a developing El Niño are now routinely incorporated into long-range seasonal weather forecasts and regional agricultural strategies.

rapidly with altitude in a cold air mass than in a warm air mass. Hence, at the same altitude, air pressure on the warm side of the polar front is higher than it is on the cold side of the front (▣ Fig. 17.21). In response to that horizontal air pressure gradient, the wind initially blows across the front from the warm side toward the cold side. Quickly, however, the Coriolis effect deflects the wind to the right so that it blows parallel to the polar front with the cold air to the left of its direction.

With increasing altitude, the horizontal air pressure gradient steepens so that winds over the polar front also strengthen with altitude. Above the tropopause, within the stratosphere, however, the north-south temperature gradient reverses, the horizontal pressure gradient weakens with altitude, and winds also weaken. This means that winds are strongest near the tropopause, forming the polar front jet stream.

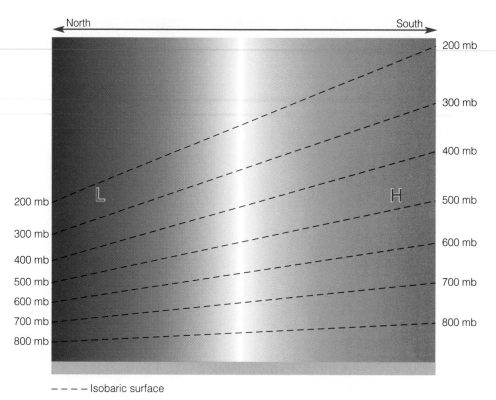

North   South

200 mb
300 mb
400 mb
500 mb
600 mb
700 mb
800 mb

L   H

200 mb
300 mb
400 mb
500 mb
600 mb
700 mb
800 mb

FIGURE 17.21 Because air pressure falls more rapidly with altitude in a cold air mass than in a warm air mass, surfaces of constant pressure slope downward from warm air toward cold air. Hence, aloft at the same altitude, air pressure is higher in warm air than in cold air. The resulting horizontal air pressure gradient accelerates the movement of air across the front from warm air toward cold air.

– – – – Isobaric surface

The contrast in air temperature across the polar front affects the speed of the polar front jet stream. In some regions, the temperature contrast across the front is relatively great, and the front is sharply defined whereas in other regions there is little temperature change and the front is diffuse and ill defined. Where the polar front is well-defined, jet stream winds accelerate. Such a segment of the jet stream, in which the wind may accelerate by 100 km (62 mi) per hour or more, is known as a **jet streak.** The strongest jet streaks develop in winter along the east coasts of North America and Asia where the land-to-sea temperature contrast is unusually great. There, jet streak wind speeds on some occasions have topped 350 km (215 mi) per hour.

FIGURE 17.22 Patterns of horizontal divergence and convergence associated with a jet streak, a segment of the polar front jet stream in which winds are exceptionally strong. Cyclones (lows) are most likely to develop under the left front quadrant of the jet streak. The dashed lines are *isotachs,* lines joining locations having the same wind speed.

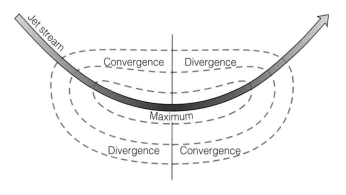

The polar front jet stream strengthens in winter, when the north-south temperature gradient is relatively great, and weakens in summer, when that gradient is less. Like other features of the planetary-scale circulation, the polar front jet stream also shifts seasonally. The average summer location of the polar front jet stream is across southern Canada, and its average winter position is over the southern United States. These locations are long-term averages; the jet stream can cover a considerable range of latitude from week to week, and even from one day to the next. As a rule, when the polar front jet stream is south of your location, the weather is relatively cold, and when it is north of you, the weather is relatively warm.

Upper-Air Support for Cyclones

Air circulation in cyclones is characterized by diverging horizontal winds aloft (see Fig. 17.11b). Horizontal divergence in the upper troposphere is required for a cyclone to develop and progress through its life cycle (see Chapter 18). The horizontal divergence aloft stems from a combination of (1) a polar front jet streak and (2) a trough in the westerlies. Without this so-called *upper-air support,* an incipient storm would fail to develop or would weaken rapidly.

Horizontal winds in a jet streak change speed and direction, and these changes generate a pattern of horizontal divergence and convergence aloft. In Figure 17.22, a jet streak (viewed from above) is divided into four quadrants. Horizontal divergence occurs in the left front and right rear quadrants, and horizontal convergence occurs in the right front and left rear quadrants. The strongest divergence oc-

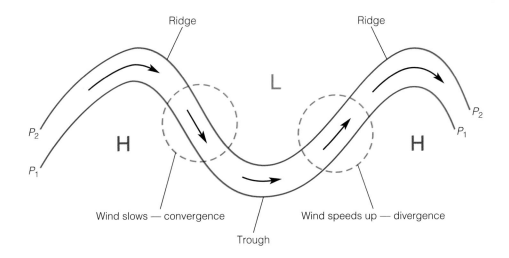

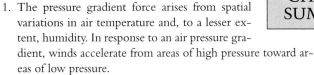

FIGURE 17.23 Patterns of horizontal divergence and convergence induced by a wave in the prevailing westerlies. A cyclone is most likely to develop to the east of a trough and west of a ridge. The solid lines are isobars.

curs in the left front quadrant, so a cyclone is most likely to develop under this sector of a jet streak.

Changes in horizontal wind speeds within westerly waves also produce divergence and convergence aloft. For the same air pressure gradient, geostrophic and gradient wind speeds are not equal. Anticyclonic gradient winds are stronger than geostrophic winds, and cyclonic gradient winds are weaker than geostrophic winds. Hence, as shown in Figure

17.23, westerly winds strengthen in a ridge and weaken in a trough. For this reason, winds diverge to the east of a trough (and west of a ridge). Westerly waves thus favor storm development in the same way as a polar front jet streak, that is, by inducing horizontal divergence aloft. Conditions aloft are most favorable for cyclone development when a jet streak appears on the east side of a trough.

CHAPTER SUMMARY

1. The pressure gradient force arises from spatial variations in air temperature and, to a lesser extent, humidity. In response to an air pressure gradient, winds accelerate from areas of high pressure toward areas of low pressure.

2. The centripetal force operates whenever the wind follows a curved path and is responsible for a change in the direction of the wind, but not a change in its speed.

3. The Coriolis effect arises from Earth's rotation on its axis. It deflects the wind to the right of its initial direction in the Northern Hemisphere and to the left in the Southern Hemisphere. The Coriolis effect is zero at the equator and increases with increasing latitude, reaching a maximum at the poles.

4. Friction slows horizontal winds blowing within about 1,000 m (3,300 ft) of Earth's surface, the so-called friction layer.

5. Gravity accelerates all objects downward, perpendicular to Earth's surface. Gravity is important only in up and down motion of air (e.g., cold-air drainage).

6. The geostrophic wind is an unaccelerated, horizontal wind that blows parallel to straight isobars at altitudes above the friction layer. The horizontal pressure gradient force balances the Coriolis effect in the geostrophic wind.

7. The gradient wind is a horizontal wind that parallels curved isobars at altitudes above the friction layer. The centripetal force, the horizontal pressure gradient force, and the Coriolis

effect operate in gradient winds. In the Northern Hemisphere, above the friction layer, winds blow parallel to isobars and clockwise around anticyclones and counterclockwise around cyclones.

8. In large-scale circulation (weather) systems, friction combines with the Coriolis effect to shift the wind across isobars and toward low pressure.

9. In the Northern Hemisphere viewed from above, surface winds blow clockwise and outward in anticyclones and counterclockwise and inward in cyclones.

10. In an anticyclone, horizontal divergence of surface winds means descending air, so a high is a fair weather system. In a cyclone, horizontal convergence of surface winds means ascending air, so a low is a stormy weather system.

11. Based on their spatial and temporal scales, atmospheric circulation systems are distinguished as planetary, synoptic, mesoscale, and microscale.

12. The principal features of the planetary-scale circulation are the intertropical convergence zone (ITCZ), trade winds, semipermanent subtropical anticyclones, westerlies, subpolar lows, polar front, polar easterlies, and polar front jet stream. These features shift poleward during spring and toward the equator during autumn.

13. Trade winds blow out of the equatorward portion of the subtropical anticyclones, and the westerlies blow out of the poleward portion.

14. The eastern flank of a subtropical anticyclone is characterized by subsiding stable air and dry conditions, whereas the western flank is less stable and wetter.

15. Aloft, in the tropics, trade winds reverse direction, thus completing the Hadley cell circulation. At higher latitudes, upper-air horizontal winds weave from west to east as Rossby long waves.

16. Contrasts in Earth's surface temperatures in winter favor relatively high pressure over the continents and low pressure over the ocean. In summer, this pattern reverses with low pressure prevailing over the continents and high pressure over the ocean.

17. Patterns in the upper-air westerlies vary in wavelength, amplitude, and number of waves. At one end of the spectrum, westerlies are mostly zonal; that is, they blow west-to-east with little amplitude. At the other end, westerlies are strongly meridional; that is, they blow west-to-east with considerable amplitude.

18. Shifts between zonal and meridional flow patterns in the westerlies affect north-south air mass exchange, poleward heat transport, and storm development and movement.

19. The polar front jet stream is a narrow corridor of very strong winds within the westerlies; it is located near the tropopause and above the polar front. A jet streak is a segment of the polar front jet stream where winds are particularly strong.

20. Cyclone development is most likely under the left front quadrant of a jet streak and to the east of an upper-level trough.

IMPORTANT TERMS

anticyclone
centripetal force
Coriolis effect
cyclone
friction
friction layer

geostrophic wind
gradient wind
gravity
Hadley cell
intertropical convergence
    zone (ITCZ)

jet streak
polar front
polar front jet stream
pressure gradient force
Rossby waves

subpolar lows
subtropical anticyclones
trade winds
westerlies
wind

REVIEW QUESTIONS

1. Compare the magnitude of vertical winds with that of horizontal winds in weather systems.
2. What causes horizontal gradients in air pressure?
3. What is the relationship between horizontal air pressure gradients and wind speed and direction?
4. Why does a centripetal force operate whenever the wind follows a curved path?
5. The Coriolis effect:
   a. ___ shifts large-scale winds to the right of their initial direction in the Southern Hemisphere; b. ___ is zero at the North Pole; c. ___ operates in both the geostrophic wind and gradient wind; d. ___ would occur even if the planet did not rotate; e. ___ does not affect the tropical trade winds.
6. How does the horizontal wind speed change with altitude within the friction layer?
7. What is the relationship between the roughness of Earth's surface and frictional resistance?
8. Provide an example of how gravity influences the circulation of air.
9. In the geostrophic wind, the horizontal pressure gradient force is balanced by:
   a. ___ gravity; b. ___ friction; c. ___ a centripetal force; d. ___ the Coriolis effect.

10. What is the function of centripetal forces in cyclones and anticyclones?
11. How does the roughness of Earth's surface affect the horizontal wind direction?
12. Viewed from above, horizontal surface winds in a Northern Hemisphere cyclone blow:
    a. ___ clockwise and outward; b. ___ clockwise and inward; c. ___ counterclockwise and outward; d. ___ counterclockwise and inward.
13. Describe the winds in an anticyclone (a) within the friction layer and (b) above the friction layer.
14. Why is fair weather usually associated with anticyclones and cloudy weather with cyclones?
15. Horizontal air pressure gradients are usually weak over a broad area surrounding the center of an anticyclone. What does this imply about the weather at the center of an anticyclone?
16. In what way are subtropical anticyclones *semipermanent*?
17. On the eastern flank of a subtropical anticyclone:
    a. ___ vertical winds are directed upward; b. ___ the atmosphere is very unstable; c. ___ the weather is cloudy and rainy much of the time; d. ___ the weather is mostly fair and dry; e. ___ are lots of thunderstorms.
18. What is the relationship between the trade winds and the subtropical anticyclones?

19. What is the relationship between the westerlies and the subtropical anticyclones?
20. Describe the linkage among the intertropical convergence zone (ITCZ), trade winds, and Hadley cells.
21. The intertropical convergence zone (ITCZ): a. ___ is a discontinuous band of thunderstorms; b. ___ occurs at high latitudes; c. ___ remains stationary all year long; c. ___ seldom is associated with rainfall; d. ___ is caused by the polar front jet stream; e. ___ has little or no impact on tropical climates.
22. How is a Hadley cell like a huge convective current?
23. Through the course of the year, the principal components of the planetary-scale circulation follow the Sun. Elaborate on this statement.

24. Distinguish between zonal and meridional flow patterns in the westerlies.
25. Explain why the weather is more extreme when the westerly wave pattern is strongly meridional.
26. What type of weather is associated with a zonal flow pattern over the United States?
27. What is a jet stream?
28. Describe the seasonal changes in the locations of the polar front jet stream.
29. How does a jet streak contribute to the development of a cyclone?
30. How does an upper-air trough in the westerlies contribute to the development of a cyclone?

POINTS TO PONDER

1. The gradient wind is not the consequence of balanced forces. Explain this statement.
2. Along a coast line, convective (cumulus) clouds are more likely to occur with onshore winds than with offshore winds. Explain why.
3. Speculate on why there is a connection between sea-surface temperatures and the circulation of the atmosphere.

4. Some climate models predict that global warming will be greater at high latitudes than at tropical latitudes. If this were to actually happen, the average north-south temperature gradient would decrease. How might this change in north-south temperature gradient affect the strength of the planetary-scale westerlies?

ADDITIONAL READINGS

Blackadar, A. 1986. Simple motions on the rotating Earth. *Weatherwise* 39: 99–103.
Higbie, J. 1980. Simplified approach to Coriolis effects. *The Physics Teacher* 18: 459–60.

Knox, P. N. 1992. A current catastrophe: El Niño. *Earth* 1, no. 5: 30–37.
Snow, J. T., et al. 1989. Basic meteorological observations for schools: Surface winds. *Bulletin of the American Meteorological Society* 70: 493–508.

WORLD WIDE WEB SITES

For current updates and exercises, log on to http://www.wadsworth.com/geo.

THE OHIO STATE UNIVERSITY ATMOSPHERIC SCIENCE PROGRAM
http://asp1.sbs.ohio-state.edu/
A major weather data center, this site has current weather maps, forecasts, and some aviation and marine weather information.

UNIVERSITY OF HAWAII METEOROLOGY
http://lumahai.soest.hawaii.edu/
This site contains current weather maps for tropical meteorology and the Pacific basin.

CHAPTER
18

Weather Systems

The atmosphere is dynamic so that weather varies from place to place and with time.

Prologue

A persistent atmospheric circulation pattern during the summer of 1993 was responsible for record flooding in the midwestern United States and drought over the Southeast (☐ Fig. 18.1). A cold upper-air trough stalled over the Pacific Northwest and northern Rockies, bringing unseasonably cool weather to that region. Meanwhile, the North Atlantic subtropical high shifted west of its usual location, causing the worst drought in seven years in the Carolinas and Virginia. Between the northwestern trough and the southeastern high, the principal storm track and an unseasonably strong jet stream weaved over the Midwest. As this circulation pattern persisted through June, July, and part of August, a nearly continual procession of weather systems brought heavy thunderstorms and record rainfall to the Missouri and upper Mississippi River valleys. In Wisconsin, Iowa, and Illinois, the June-July period was the wettest on record (☐ Fig. 18.2). Meanwhile, over the Southeast, subsiding, stable air inhibited thunderstorm development, so hot, dry conditions prevailed.

In the upper Mississippi River valley, the wet autumn of 1992 was followed by heavy winter snowfall and abundant spring snowmelt. Unusually wet conditions as the summer began set the stage for the most costly flooding in U.S. history. Heavy summer rains saturated soils and triggered excessive runoff and all-time record river crests. On the Mississippi River, the worst flooding was between Minneapolis, Minnesota, and Cairo, Illinois, and on the Missouri, between Omaha, Nebraska, and St. Louis, Missouri (☐ Fig. 18.3). At one point, an area the size of the state of Indiana was under water. Flooding was unprecedented in persistence as well as magnitude, with some localities experiencing more than one record river crest. At St. Louis, the Mississippi River remained above its previous record crest for three weeks.

The floods had extensive societal and economic impacts. Populous cities such as Des Moines, Iowa, and St. Joseph, Missouri, lost much of their freshwater supply; barge traffic was halted on the upper Mississippi and a portion of the Missouri Rivers from late June to early August; at least 50,000 homes were damaged or destroyed; and more than 4 million hectares (10 million acres) of cropland were inundated. All told, damage topped $15 billion, about one-third of which was due to crop losses. The death toll from the flooding was 48.

☐ FIGURE 18.1 This map shows the major features of the persistent weather pattern responsible for flooding rains in the central United States and drought in the Southeast during the summer of 1993.

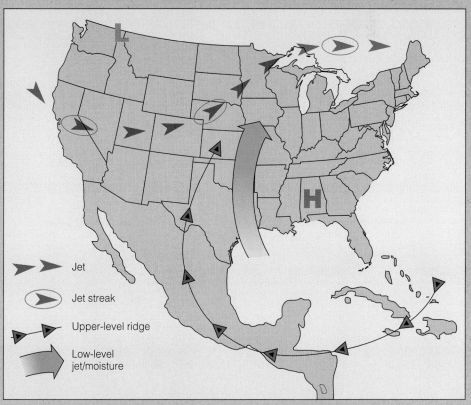

Jet

Jet streak

Upper-level ridge

Low-level jet/moisture

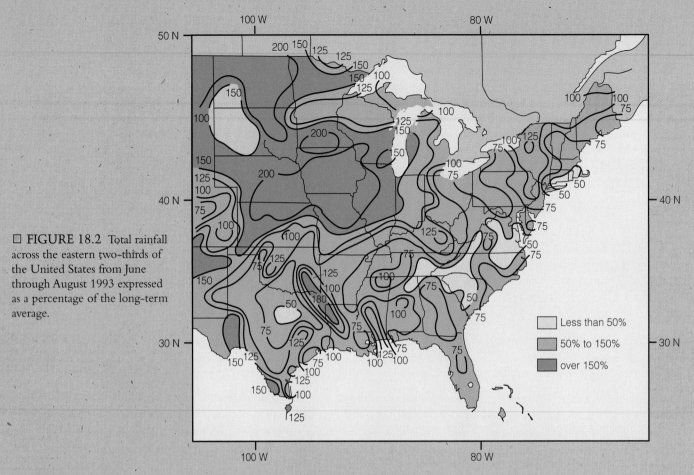

☐ FIGURE 18.2 Total rainfall across the eastern two-thirds of the United States from June through August 1993 expressed as a percentage of the long-term average.

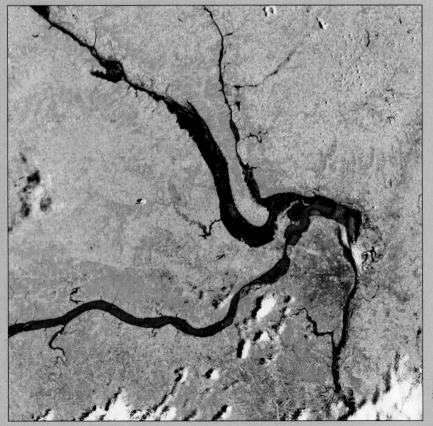

While some areas in the Midwest were receiving more than twice the average summer rainfall, parts of the Southeast had less than half of their long-term average rainfall (Fig. 18.2). As a whole, the region experienced its second driest July on record, and from Alabama and Georgia north to Tennessee and Virginia, July of 1993 was the hottest on record. The combination of drought and heat stress caused severe crop damage, especially in South Carolina where more than 95% of the corn crop and 70% of the soybean crop were lost. Total crop loss was estimated at $1 billion.

☐ FIGURE 18.3 A visible satellite image of the flood in the Midwest in 1993.

INTRODUCTION

The day-to-day variability that characterizes the weather of the midlatitudes is due to a ceaseless succession of weather systems. This chapter examines the characteristics and life cycles of the principal weather systems of the midlatitudes: air masses, fronts, cyclones (lows), anticyclones (highs), thunderstorms, tornadoes, and hurricanes. Special attention is paid to how these systems are interrelated and how the planetary-scale circulation governs their development and movements.

 AIR MASSES

An **air mass** is a huge volume of air, covering thousands of square kilometers, that is relatively uniform horizontally in temperature and humidity. The properties of an air mass depend on its *source region,* or the type of surface over which it develops. A source region, such as a great expanse of desert or ocean water, is relatively homogeneous over a broad area. An air mass must reside over its source region for several days to weeks to become uniform in temperature and humidity.

Air masses are either cold (polar, abbreviated as *P*) or warm (tropical or *T*), and either dry (continental or *c*) or humid (maritime or *m*). Air masses that form over snow-covered ground at high latitudes are relatively cold, and those that develop over low latitudes are relatively warm. Air masses that form over land are relatively dry, and those that develop over the ocean are humid. Hence, the four basic types of air masses are cold and dry continental polar (*cP*), cold and humid maritime polar (*mP*), warm and dry continental tropical (*cT*), and warm and humid maritime tropical (*mT*). Arctic (*A*) air, a fifth type, is distinguished from continental polar air by its lower temperatures. The principal source regions of North American air masses are plotted in Figure 18.4.

Continental tropical air is hot and dry because it develops primarily in summer over the deserts of Mexico and the southwestern United States. *Maritime tropical air* is warm and humid, and its source regions are tropical and subtropical seas. Source regions for *maritime polar air* are the cold ocean waters of the North Pacific and North Atlantic. *Dry continental polar air* develops over the northern interior of North America and changes temperature with the seasons. In winter, *cP* air is very cold because its source region is often snow covered, solar radiation is weak, nights are long, and radia-

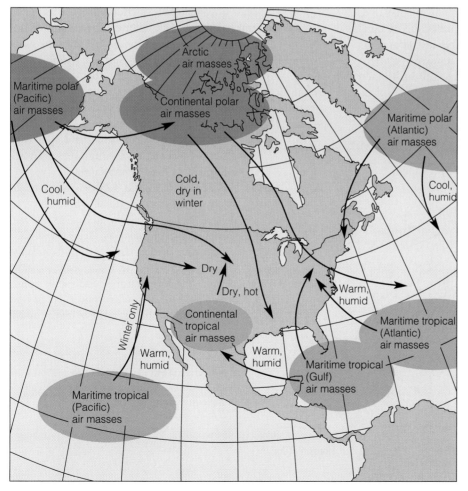

■ FIGURE 18.4 Air mass source regions of North America.

tional cooling is extreme. In summer, *cP* air is relatively mild because its source region is snow-free, days are long, and solar radiation is intense. *Arctic air* forms over the snow- or ice-covered regions of Siberia, the Arctic Basin, Greenland, and North America north of about 60° N in the same way as *cP* air does but in source regions that receive very little solar radiation in winter. Hence, arctic air masses are exceptionally cold and dry and are responsible for bone-numbing winter cold waves.

As an air mass moves away from its source region, it modifies; that is, its temperature and/or humidity change. *Air mass modification* occurs when the air mass exchanges heat and/or moisture with the surface over which it travels. Air masses are also modified by radiational heating or cooling, or large-scale ascent or descent.

In winter, as *cP* air surges southeastward from Canada over snow-free regions of the United States, its temperature modifies quite rapidly. Beyond its source region, polar air is usually colder than the ground over which it travels. The Sun heats the bare ground, and the warmer ground heats the overlying air mass, destabilizing it and triggering convection currents that transport heat vertically throughout the air mass. Although low temperatures might dip below −18°C (0°F) in the Northern Plains, they may not drop much below the freezing point by the time the *cP* air mass reaches the southern United States.

Modification is not as rapid for a tropical air mass. Outside its source region, tropical air is often warmer than the ground over which it travels. The air mass cools by contact with the ground, and cooling stabilizes the air and inhibits convection. Hence, a summer heat wave retains its high temperatures as it journeys from the Gulf of Mexico into southern Canada.

FRONTS

A **front** is a narrow zone of transition between air masses that differ in temperature or humidity or both. In most cases, the greatest contrast is in temperature, so a distinction is made between *cold fronts* and *warm fronts*.

As noted in Chapter 16, warmer air ascends along or just ahead of a front. If uplift is sufficient, expansional cooling raises the relative humidity to near 100% so that clouds and precipitation develop. Depending on the specific type of front, clouds and precipitation may be confined to a narrow band or stretch over a broad region. This section discusses the characteristics of the four basic types of fronts: stationary, warm, cold, and occluded.

A *stationary front* does not move. In most cases, surface winds on either side of the front blow parallel to it. As is the case for all fronts, a stationary front slopes back from Earth's surface toward colder, denser air. Rain and snow fall mostly on the cold side of a stationary front and are usually light but persistent.

If a stationary front begins to move so that the warm air advances while the cold air retreats, the front becomes a *warm front* (◻ Fig. 18.5). As a warm front approaches, overrunning causes stratiform clouds to gradually lower and thicken (◻ Fig. 18.6). Wispy cirrus clouds sometimes appear more than 1,000 km (620 mi) ahead of a surface warm front. Slowly, the clouds spread laterally, forming thin sheets of cirrostratus that turn the sky a milky white. Cirrostratus clouds eventually give way to altostratus clouds. Just after the altostratus thicken enough to block the Sun (or Moon), light rain or snow often begins. Steady, light-to-moderate precipitation falls from low, gray nimbostratus clouds and persists until the warm front passes at the surface, a period of perhaps 24 hours. Copious amounts of rain may fall ahead of the surface warm front, and if surface temperatures are low enough for snow, accumulations may be substantial.

Just ahead of the surface warm front, steady precipitation usually gives way to drizzle falling from low stratus clouds and sometimes to *prefrontal fog*. After the surface warm front finally passes through a locality, the fog vaporizes, skies at least partially clear, and temperatures and dew points rise. The zone of overrunning has moved on, and the locality is in the warm air mass.

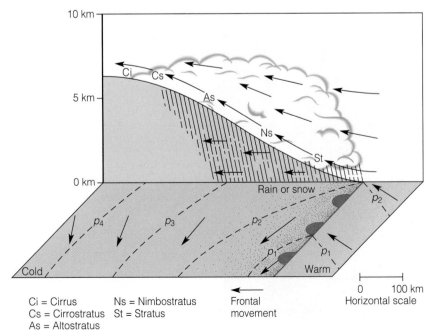

◻ **FIGURE 18.5** A vertical cross section of a warm front. Dashed lines are isobars with air pressure increasing from P_1 to P_4.

Ci = Cirrus   Ns = Nimbostratus
Cs = Cirrostratus   St = Stratus
As = Altostratus

Frontal movement

0   100 km
Horizontal scale

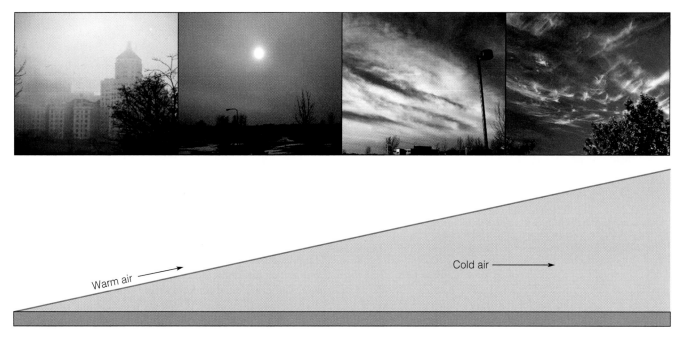

FIGURE 18.6 As a warm front approaches, stratiform clouds gradually lower and thicken. Cirrus clouds are followed by cirrostratus and then altostratus.

If a stationary front begins to move so that the cold air advances while the warm air retreats, the front becomes a *cold front*. The roughness of Earth's surface steepens a cold front into its characteristic nose-shaped profile (◫ Fig. 18.7). Because the frontal slope is relatively steep, the ascent of warm air (and clouds and precipitation) is confined to a narrow zone at or near the front's leading edge. Precipitation is likely to be showery. If the warm air is unstable, uplift is more vigorous and thunderstorms (cumulonimbus clouds) may form.

A cold front typically travels about twice as fast as a warm front and eventually may catch up and merge with a warm front, thereby forming an *occluded front*. In most North American cases, the air behind the advancing cold front is colder than the cool air ahead of the warm front. Hence, the cold air slides under and lifts the warm air, the cool air, and the warm front (◫ Fig. 18.8). Weather ahead of an occluded front is similar to that in advance of a warm front; the actual frontal passage may be accompanied by showers like those typical of a cold front.

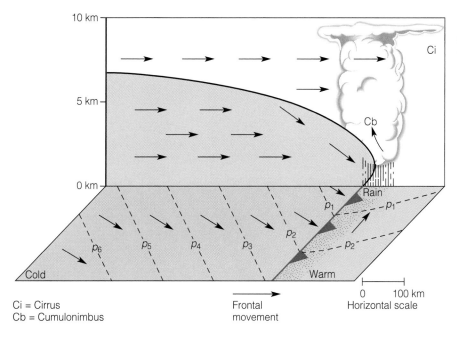

Ci = Cirrus
Cb = Cumulonimbus

FIGURE 18.7 A vertical cross section of a cold front. Dashed lines are isobars with air pressure increasing from P_1 to P_6.

MIDLATITUDE CYCLONES

Frontal weather is associated with an extratropical **cyclone,** the principal weather-maker of midlatitudes. Viewed from above, surface winds in a Northern Hemisphere cyclone blow counterclockwise and inward. Converging surface winds bring together contrasting air masses, fronts form, and the ascent of air along those fronts triggers cloudiness and precipitation.

Cyclogenesis, the birth of a cyclone, usually takes place along the polar front under an area of strong horizontal divergence aloft. As noted in Chapter 17, the strongest divergence aloft (the greatest upper-air support) is to the east of a trough in the westerlies and under the left-front quadrant of a jet streak. If divergence aloft removes more mass from a column of air than is brought in by converging surface winds, air pressure at the bottom of the column drops, a horizontal pressure gradient develops, the Coriolis effect kicks in, and a cyclonic circulation begins. The westerlies aloft then steer the cyclone as it progresses through its life cycle (◉ Fig. 18.9).

◉ **FIGURE 18.8** A vertical cross section of an occluded front.

Ac = Altocumulus   As = Altostratus
Cb = Cumulonimbus   Cs = Cirrostratus

◉ **FIGURE 18.9** The life cycle of a midlatitude cyclone (low). (*a*) As the air pressure drops, surface winds converge and the fronts begin to advance. (*b*) West of the low center, the front advances southeastward as a cold front, and east of the low center, the front advances northward as a warm front. (*c*) An occluded front begins to form, and (*d*) the warm sector detaches from the cyclone as the system's central pressure rises and its circulation weakens. The dashed lines represent horizontal winds in the mid and upper troposphere.

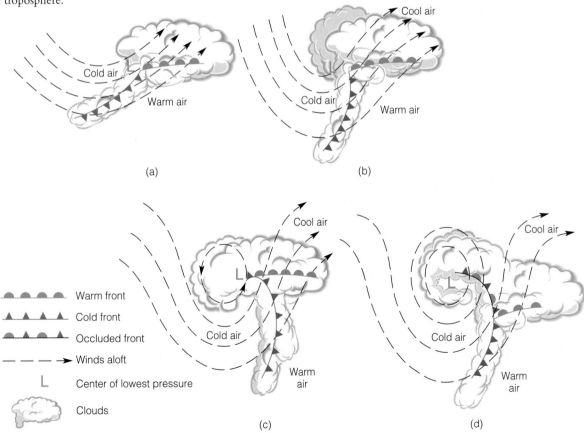

Just prior to cyclogenesis, the polar front is stationary. As the air pressure drops, surface winds converge and the front advances (Fig. 18.9a). West of the low center, the front pushes southeastward as a cold front, and east of the low center, the front moves northward as a warm front (Fig. 18.9b).

If upper-air support continues, the cyclone's central air pressure drops further, the associated horizontal pressure gradient steepens, and the winds strengthen. Because a cold front travels faster than a warm front, the area of the storm's warm sector (between the warm and cold fronts) shrinks, and an occluded front begins to form (Fig. 18.9c). With occlusion, the warm sector detaches from the cyclone (Fig. 18.9d), the system's central pressure rises, and its circulation weakens. The storm gradually loses its identity and may be accompanied by only cloudy skies and drizzle.

Cyclone Weather

Consider a typical winter cyclone that tracks over central Illinois. Although the storm's central pressure is still dropping, its circulation, clouds, and precipitation cover a broad area (▣ Fig. 18.10). The weather associated with a cyclone varies with the sector of the storm.

The coldest air is to the northwest of the low center where strong northwest winds transport continental polar or arctic air southeastward. West of the low center, precipitation tapers off to showers, and skies tend to clear as the storm center moves toward the northeast. The leading edge of the cold air mass (cold front) is south of the storm and is ac-

companied by a narrow band of showers and thunderstorms. The southwest sector has generally clear skies.

The mildest air is in the southeast (warm) sector of the storm where south and southeast winds transport maritime tropical air northward from over the Gulf of Mexico. Skies are generally partly cloudy, dewpoints are relatively high, and scattered showers and thunderstorms occur. Cloudiness increases, and showers become more frequent in advance of the eastward-moving cold front.

To the north and northeast of the cyclone center is a zone of extensive overrunning as warm, humid air surges over a wedge of cool air maintained at the surface by east and northeast winds. Skies are cloudy and precipitation is steady and substantial.

Cyclone Tracks

The main cyclone tracks over the lower 48 states of the United States are shown in ▣ Figure 18.11. Most lows track toward the northeast with the Icelandic low of the North Atlantic being their ultimate destination. Although many storms appear to originate just east of the Rocky Mountains, their life cycles actually begin over the Pacific Ocean. Cyclones moving through the western mountains temporarily lose their identity but redevelop on the Great Plains just east of the Front Range of the Rockies.

Cyclones that form in the south generally produce more precipitation than those that develop to the north because southern storms are closer to source regions of moisture-rich, maritime tropical air. Cyclones tracking from west to

▣ FIGURE 18.10 Weather surrounding a mature cyclone centered over Illinois: (*a*) air mass advection; (*b*) precipitation.

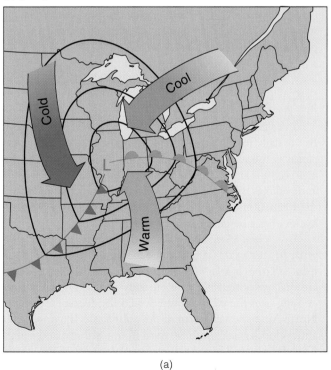

(a)

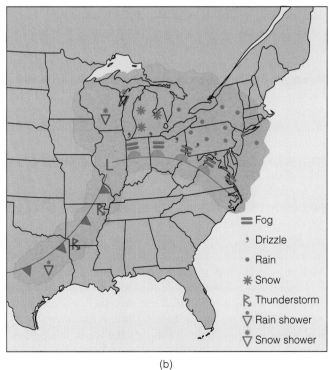

Fog
Drizzle
Rain
Snow
Thunderstorm
Rain shower
Snow shower

(b)

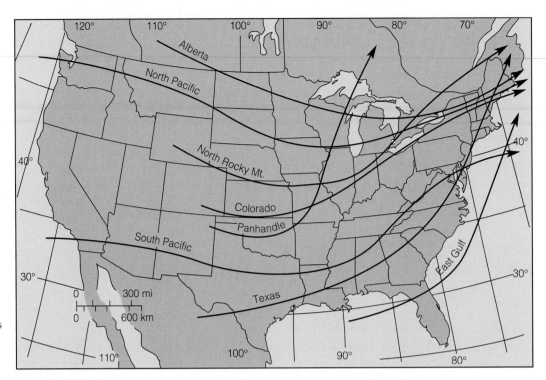

FIGURE 18.11
Principal tracks of cyclones over the lower 48 states of the United States.

east across southern Canada and the northern United States (*Alberta-track cyclones*) typically yield only light amounts of rain or snow, whereas cyclones that redevelop east of the Colorado Rockies (*Colorado-track cyclones*) and Atlantic and Gulf Coast storms often produce substantial accumulations of precipitation.

Cyclone tracks vary seasonally. In summer, when the mean position of the polar front jet stream is across southern Canada, few well-organized cyclones develop over the United States, and the Alberta storm track shifts northward across central Canada. In winter, however, when the average position of the polar front jet stream shifts southward, well-developed cyclones are more common over the United States.

A midlatitude cyclone has a warm side and a cold side. In Figure 18.10, the coldest air is northwest of the low center (where surface winds are from the northwest), and the warmest air is southeast of the low center (where surface winds are from the south). Hence, as a cyclone traverses the continent, the weather to the left (cold side) of the storm track differs markedly from the weather to the right (warm side) of the track.

Suppose that a Colorado-type winter cyclone tracks northeastward toward the Great Lakes region (Fig. 18.12). The storm follows either track A, which takes it west of Minneapolis, or track B, which takes it east of Minneapolis. The storm's effect on the weather in Minneapolis depends on which track is taken.

If the storm follows track A, Minneapolis residents experience the warm side of the storm. Several hours of steady rain (perhaps with snow or freezing rain at the onset) eventually give way to drizzle and fog. As the warm front passes through the city, skies partially clear, and winds shift from east to southeast to south, advecting warm and humid air at the

surface. The air temperature and dewpoint rise. The clearing is short-lived, however, as scattered showers and thunderstorms signal the approach of colder air. As the cold front passes through the city, winds blow first from the southwest, then the west, and finally the northwest. Skies clear again, and the air temperature and dewpoint gradually fall.

By contrast, if the storm takes track B, Minneapolis residents experience the cold side of the storm, and no fronts pass over the city. East and northeast winds drive steady snow or rain for 12 hours or longer. Then the winds gradually shift to the north, precipitation tapers off to snow flurries or showers, and air temperatures begin to drop. Finally, winds become northwesterly, clouds break up, and air temperatures continue falling.

ANTICYCLONES

In the daily march of weather systems in midlatitudes, a cyclone is followed by an anticyclone. An **anticyclone** (or *high*) is, in many ways, the opposite of a cyclone (or *low*). In anticyclones, subsiding air and diverging surface winds favor the formation of a uniform air mass and fair skies. An anticyclone may be either cold or warm.

A *cold anticyclone* coincides with a dome of continental polar (*cP*) or arctic (*A*) air and, depending on the specific air mass, is either a *polar high* or an *arctic high*. Cold anticyclones are shallow systems that are products of extreme radiational cooling over the often snow-covered continental interior of North America well north of the polar front. In winter, cold anticyclones track southeastward from northern and central Canada into the United States. Associated with these systems are the lowest air temperatures of the season (⬛ Fig. 18.13).

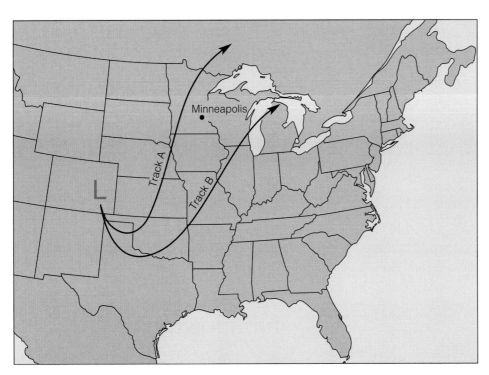

■ FIGURE 18.12 A Colorado-type cyclone follows either track A west of Minneapolis or track B east of that city. The specific track the storm takes has a major impact on the weather in Minneapolis.

■ FIGURE 18.13 A cold anticyclone (high) centered over eastern Montana on December 24, 1983. The solid lines are isobars. This system brought record cold as far south as southern Florida.

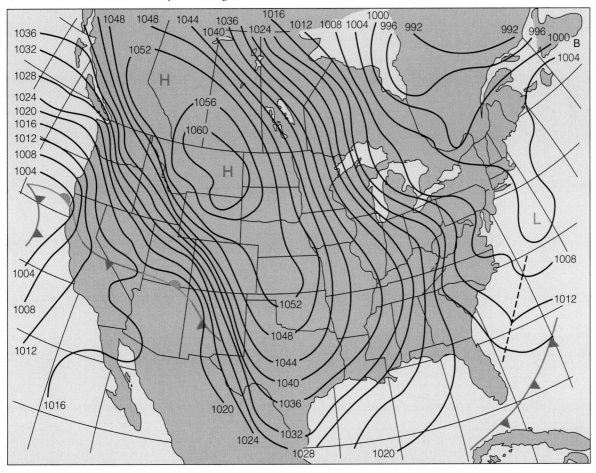

FIGURE 18.14 Cumuliform clouds billowing upward to form a thunderstorm (cumulonimbus) cloud.

A *warm anticyclone* forms south of the polar front and features extensive areas of subsiding warm, dry air. The system's circulation extends from Earth's surface up to the tropopause. Semipermanent subtropical anticyclones, described in Chapter 17, are examples of warm highs, but other warm anticyclones occur over the interior of North America, especially in summer.

The horizontal air pressure gradient strengthens with distance away from the central region of an anticyclone, and so does the wind. Stronger winds are accompanied by air mass advection. Typically, well to the east of the high center, northerly and northwesterly winds transport cold air southward, whereas to the west of the high center, southerly winds transport warm air northward. Hence, as a coastal storm pulls away from New England and a cold anticyclone follows in its wake, northwesterly winds mean falling temperatures as continental polar or arctic air invades the region.

 THUNDERSTORMS

A **thunderstorm** is a localized (mesoscale) weather system that is accompanied by lightning and thunder. It is the product of vigorous convection that reaches high into the troposphere, perhaps to the tropopause or higher. Ascending air currents are made visible by billowing cauliflower-shaped clouds, as shown in Figure 18.14. A thunderstorm consists of one or more convective cells, and each cell progresses through a three-stage life cycle: cumulus, mature, and dissipating.

During the *cumulus stage,* cumulus clouds build both vertically and laterally (Fig. 18.15a). Within perhaps 15 minutes, the tops of these clouds surge to altitudes of 8,000 to 10,000 m (26,000 to 33,000 ft). Meanwhile, neighboring cumulus merge so that by the end of the cumulus stage, the

FIGURE 18.15 The life cycle of a thunderstorm cell.

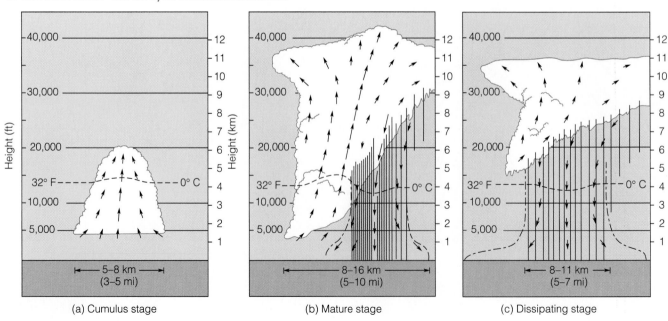

(a) Cumulus stage   (b) Mature stage   (c) Dissipating stage

FIGURE 18.16 The distinctive anvil top of a cumulonimbus (thunderstorm) cloud.

storm's lateral dimension may be 10 to 15 km (6 to 9 mi). A strong *updraft* keeps water droplets and ice crystals suspended in the upper reaches of the cloud, so no precipitation falls during the cumulus stage.

The cumulus stage yields to the *mature stage* when precipitation begins (Fig. 18.15b). Eventually, the weight of water droplets and ice crystals becomes so great that they are no longer supported by the updraft. Rain, ice pellets, and snow descend through the cloud and drag the adjacent air downward, creating a strong *downdraft* alongside the updraft. Typically, the mature stage lasts from 15 to 30 minutes.

A thunderstorm cell attains its maximum intensity toward the end of the mature stage. Precipitation is heaviest, lightning is most frequent, and hail, strong surface winds, and even tornadoes may develop. Tops of cumulonimbus clouds can exceed an altitude of 18,000 m (60,000 ft), and strong winds at such altitudes distort the cloud top into a characteristic anvil shape (Fig. 18.16).

The updraft weakens as precipitation falls against it. During the thunderstorm cell's *dissipating stage* (Fig. 18.15c), subsiding air eventually replaces the updraft throughout the system. The subsiding air is warmed by compression, the relative humidity drops, precipitation tapers off and ends, and the clouds gradually vaporize.

Classification

Thunderstorms are classified on the basis of the number, organization, and intensity of their constituent cells. *Single-cell thunderstorms* are usually relatively weak weather systems that appear to pop up randomly within a warm, humid air mass. Actually, single-cell thunderstorms do not occur randomly—they almost always develop along some boundary within an air mass. The boundary often is the edge of a bubble of cool, dense air that is the remnant of another thunderstorm cell that has dissipated. At least in middle latitudes, solar-driven convection alone usually is not sufficient to generate even a weak single-cell thunderstorm.

Typically, a thunderstorm cell completes its life cycle in 30 minutes or so, but sometimes lightning, thunder, and heavy rain persist for hours. The stormy weather hangs on because most thunderstorms are *multicellular,* and each cell is at a different stage of its life cycle. A succession of many cells is thus responsible for prolonged episodes of thunderstorm weather. Two important types of multicellular thunderstorms are the squall line and the mesoscale convective complex.

A **squall line** is an elongated cluster (or band) of thunderstorm cells that is most likely to develop in the warm sector of a mature cyclone ahead of and parallel to the cold front (Fig. 18.17). Squall-line thunderstorm cells are usually more intense than isolated single-cell thunderstorms. A nearly continuous gust front appears ahead of a squall line. A *gust front* marks the leading edge of cool, gusty air that de-

FIGURE 18.17 A squall line usually develops in the warm sector of a mature midlatitude cyclone, ahead of and parallel to a well-defined cold front.

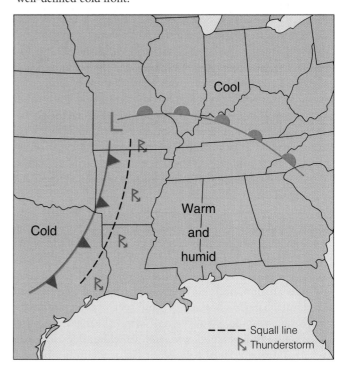

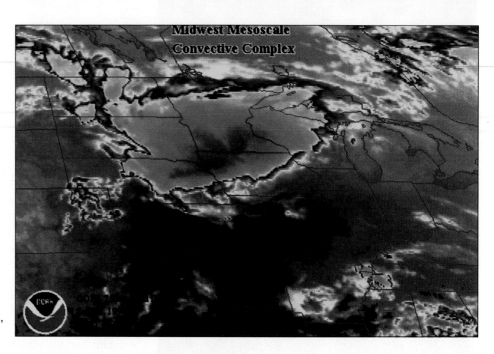

■ FIGURE 18.18 A color-enhanced, infrared satellite image of a mesoscale convective complex (MCC), a large cluster of many thunderstorm

scends from the base of a cumulonimbus cloud and flows horizontally along Earth's surface.

A **mesoscale convective complex (MCC)** is a roughly circular cluster of many interacting thunderstorm cells covering an area that may be a thousand times larger than that of a single isolated thunderstorm cell (■ Fig. 18.18). It is not unusual for an MCC to cover an area the size of the state of Kansas. MCCs are primarily warm-season (March through September) weather systems that develop at night over the eastern two-thirds of the United States, where more than 50 may be expected in a single season. An MCC is not associated with fronts. New cells form and old cells die continually, so the system may persist for 12 to 24 hours. Due to the longevity of an MCC and its typically slow movement, rainfall is widespread and substantial.

Cells in both squall lines and MCCs have the potential of producing severe weather, including hail, flash flooding, and tornadoes. But the most severe weather is likely to accompany a **supercell thunderstorm,** a relatively long-lived, large, intense system. It consists of a single cell having an exceptionally strong updraft, in some cases estimated at 240 to 280 km (150 to 175 mi) per hour. A second distinguishing characteristic of a supercell is the tendency of the updraft to develop a rotational circulation, known as a *mesocyclone,* that may evolve into a tornado.

By convention, a **severe thunderstorm** is one that is accompanied by locally damaging winds, frequent lightning, or large hail. As a general rule, the greater the altitude of the top of a thunderstorm, the more likely the system will be severe. The key ingredient in the development of a severe thunderstorm appears to be the magnitude of *vertical wind shear,* the change in horizontal wind speed or direction with increasing altitude. Weak vertical wind shear favors short-lived updrafts, low cloud tops, and weak thunderstorm cells whereas strong vertical wind shear favors vigorous updrafts,

great vertical cloud development, and severe thunderstorm cells.

Geographical and Temporal Distribution

The key to the spatial and temporal variability of thunderstorms is the basic mechanism of their formation. All thunderstorms require (1) humid air in the low to mid troposphere, (2) instability, and (3) a source of uplift.

Most thunderstorms develop within masses of warm, humid air—that is, maritime tropical (*mT*) air—that become destabilized. Maritime tropical air is initially stable but becomes unstable when lifted to the condensation level. Most thunderstorms develop when maritime tropical air is lifted (1) along fronts, (2) up mountain slopes, or (3) as the consequence of converging surface winds.

Central Florida has the highest frequency of thunderstorms in North America (■ Fig. 18.19). Residents of the interior of the Florida peninsula can expect 90 thunderstorm-days a year on average. (A *thunderstorm-day* is defined as a day when thunder is heard.) The Florida thunderstorm maximum is due to the convergence of sea breezes that develop along both the east and west coasts of the peninsula (■ Fig. 18.20). A *sea breeze* is an onshore flow of relatively cool air that develops during the day in response to differential heating of land versus sea. In response to the same intensity of solar radiation, a land surface heats up more than a sea surface (see Chapter 14). Relatively high air pressure over the sea and low pressure over the land give rise to a circulation that at low levels is directed from sea to land. The convergence of sea breezes over interior Florida induces the ascent of maritime tropical air and the formation of cumulonimbus clouds.

In general, thunderstorm development is inhibited when air masses reside or travel over relatively cold surfaces. Cold

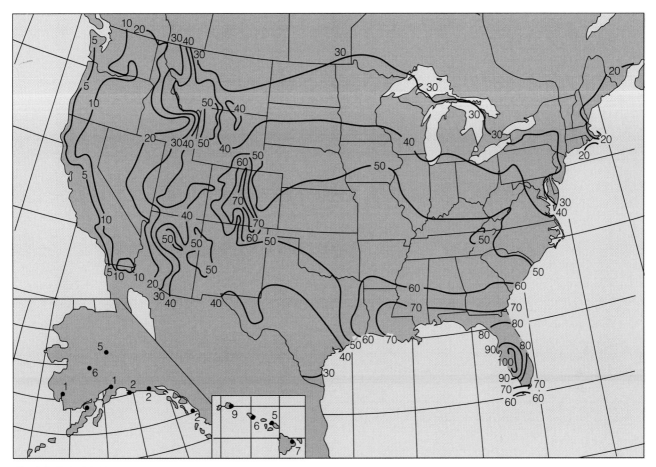

■ FIGURE 18.19 The frequency of thunderstorms over the United States expressed as the average annual number of thunderstorm-days.

surfaces chill and stabilize the overlying air, thereby suppressing convection. Hence, thunderstorms are rare over the snow-covered terrain of middle and high latitudes in winter. Thunderstorms are also unusual over coastal areas that are downwind from relatively cold ocean waters. In coastal California, for example, prevailing onshore winds advect a shallow layer of maritime polar air inland from off the relatively cold California current. The cool *mP* air inhibits deep convection and thunderstorm development. Thus, the average annual number of thunderstorm-days is only 2 at San Francisco, 6 at Los Angeles (Civic Center), and 5 at San Diego.

Severe Weather Pattern

Most severe thunderstorms break out over the Great Plains and are associated with mature cyclones. Severe thunderstorm cells usually form along a squall line within the cyclone's warm sector, ahead of and parallel to a well-defined cold front. Severe cells can produce large hail or heavy rain. Most tornadoes that form are weak, although some squall-line cells evolve into *supercells* that can spawn violent tornadoes.

■ FIGURE 18.20 The convergence of sea breezes over central Florida is the reason for that region's relatively high thunderstorm frequency.

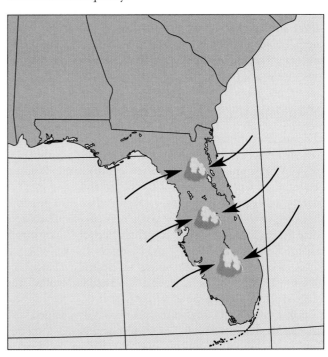

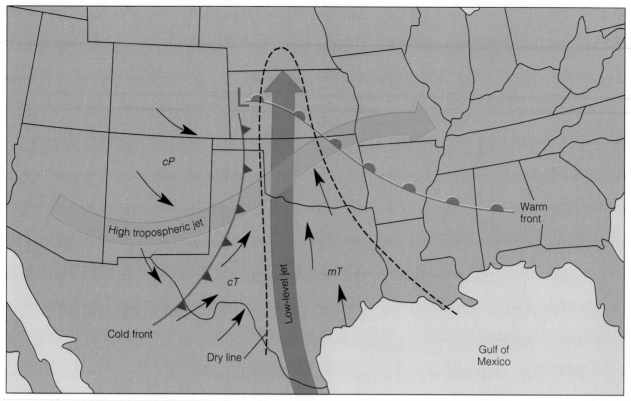

FIGURE 18.21 A weather pattern that favors the development of severe thunderstorms.

Figure 18.21 shows a weather pattern that favors the development of severe thunderstorms. A mature cyclone is centered over western Kansas with a cold front trailing southward over west Texas and a warm front stretching southeastward from the low center. The polar front jet stream and a low-level jet of maritime tropical air also appear on the map and cross to the southeast of the cyclone center. The low-level jet is a surge of exceptionally warm and humid air within about 3,000 m (10,000 ft) of Earth's surface. The most likely location for severe thunderstorms to develop is ahead of the cold front and near where the two jets intersect.

Thunderstorm Hazards

Thunderstorm hazards include lightning, downbursts, torrential rains, hail, and tornadoes. **Lightning** is a brilliant flash of light produced by an electrical discharge of about 100 million volts (Fig. 18.22). The potential for an electrical discharge exists whenever a charge difference develops between two objects. An electrically neutral object becomes negatively charged when it gains electrons (negatively charged subatomic particles) and positively charged when it loses electrons. When differences in electrical charge develop within a cloud or between a cloud and the ground, the stage is set for lightning.

On a clear day, Earth's surface is negatively charged, and the upper troposphere is positively charged. As a cumulonimbus cloud develops, however, this charge distribution changes. Within the cloud, charges separate so that the upper portion and a much smaller region near the cloud base become positively charged (Fig. 18.23). In between, a pancake-shaped zone of negative charge forms that is a few hundred meters thick and several kilometers in diameter. At the same time, the developing cumulonimbus induces a positive charge on the ground directly under the cloud.

Air is a very good electrical insulator; thus, as a thunderstorm forms and electrical charges separate and build, a potential soon develops for an electrical discharge. When the thunderstorm enters its mature stage, the electrical resistance of air breaks down, and electrons flow, thereby neutralizing the electrical charges. Lightning may forge a path between oppositely charged regions of a cloud or between a cloud and the ground. Electricity flows at nearly 50,000 km (31,000 mi) per second.

Where lightning occurs, so does thunder, although sometimes lightning can be seen in the distance and no thunder is heard. Lightning heats the air along a narrow conducting path to temperatures that may top 25,000°C (45,000°F). Such intense heating expands the air violently and produces a sound wave that is heard as **thunder.**

Both severe and nonsevere thunderstorms can produce a **downburst,** an exceptionally strong downdraft that exits the cloud base and, upon striking Earth's surface, diverges horizontally as a surge of potentially destructive winds. This flow can be simulated by aiming a garden hose nozzle downward so that a stream of water strikes the ground at an angle and bursts outward. Downbursts occur with or without rain.

⊡ FIGURE 18.22 Cloud-to-ground lightning.

They are hazardous to aviation (⊡ Fig. 18.24), blow down trees, and wreck buildings. Sometimes downburst damage is erroneously attributed to an unseen tornado.

A stationary or slow-moving intense thunderstorm can produce torrential rains that trigger a potentially disastrous **flash flood,** a sudden overflow of a river channel or other drainage way. Typically, a thunderstorm is stationary or slow moving because the system is embedded in weak steering winds aloft and/or there is a persistent flow of humid air up a mountain slope. Alternately, flood-producing rains may occur when a succession of thunderstorm cells mature over the same geographical area.

Flash flooding is especially hazardous in mountainous terrain, and motorists and campers are well advised to head for higher ground in the event of a flood warning. Even where the topography is relatively flat, prolonged periods of heavy

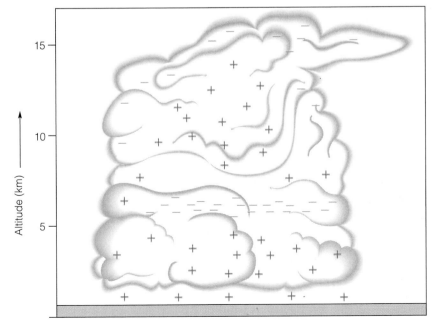

⊡ FIGURE 18.23 The separation of electrical charges in a thunderstorm (cumulonimbus) cloud.

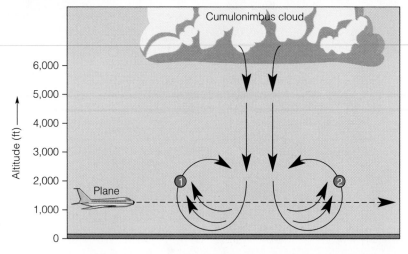

■ FIGURE 18.24 An aircraft encounters a downburst, a vigorous downdraft associated with a thunderstorm. The aircraft meets a strong headwind at point 1 and a strong tailwind at point 2. The result can be a dangerous loss of lift.

rain can greatly exceed the infiltration capacity of the ground. Excess water runs off to creeks, streams, or sewers or collects in other low-lying areas. If a drainage system cannot accommodate the sudden input of huge quantities of water, flash flooding is the consequence (■ Fig. 18.25).

■ FIGURE 18.25 In urban areas, flash flooding can trap motorists who venture onto low-lying streets and underpasses.

Hail is precipitation in the form of balls or lumps of ice, usually called *hailstones.* Hail falls from intense thunderstorm cells characterized by strong updrafts, great vertical development, and an abundant supply of supercooled water droplets. Hailstones range from the size of a pea to a golfball or even larger. The largest hailstone on record in the United States was about the size of a softball and was collected at Coffeyville, Kansas, on September 3, 1970. It weighed 758 g (1.67 lb).

A hailstone forms when a strong updraft transports an ice pellet through portions of a cumulonimbus cloud containing varying amounts of supercooled water droplets. The ice pellet then descends slowly through the cloud, or it may follow a more complex up-and-down motion as it is caught alternately in updrafts and downdrafts. In the process, the ice pellet grows by addition of freezing water droplets. In general, the stronger the updraft, the larger the ice pellet grows. Eventually, when it becomes too large and heavy for the updrafts to support it, the ice pellet descends and falls through the cloud base. If the ice does not melt completely during its journey through the above-freezing air beneath the cloud, it reaches the ground as a hailstone.

Hailstones are sometimes large enough to dent automobiles, but the most costly damage is to crops (■ Fig. 18.26). Hail usually falls during the growing season and can wipe out the fruits of a farmer's year of labor in a matter of minutes.

❖ TORNADOES

A **tornado** is a whirling column of air in contact with the ground that is made visible by clouds and/or dust and debris that are drawn into the system. Almost all tornadoes are associated with a thunderstorm. Some weak tornadoes spin off from a gust front, and some occur with a hurricane. Perhaps the most striking characteristic of a tornado is its *funnel cloud,* a localized lowering of the cloud base into a tapered column of water droplets (■ Fig. 18.27). A funnel cloud may not reach the ground as a tornado. If there is no corresponding whirl of dust or debris on the ground, the

The Distance to a Thunderstorm

Because light waves travel about a million times faster than sound waves, a lightning flash is seen almost instantly whereas the accompanying thunder is heard later. As a thunderstorm cell approaches a locality, the time interval between the lightning flash and the thunder becomes shorter. Thunder takes about 3 seconds to travel 1 km (5 seconds to travel 1 mi). Thus, if 9 seconds elapse between a lightning flash and thunder, the thunderstorm cell is about 3 km (1.8 mi) away. By a succession of such measurements it is possible to determine whether a thunderstorm cell is moving toward or away from your locality.

■ FIGURE 18.26 Hail can be devastating to crops.

system is reported as a *funnel aloft*. Sometimes a whirl of dust or debris appears on the ground even before a funnel cloud appears.

The path of a weak tornado typically is less than 1.5 km (1 mi) long and 100 m (330 ft) wide, and the system has a life expectancy of only several minutes. Wind speeds are under 180 km (110 mi) per hour. At the other extreme, an intense tornado can cause damage along a path more than 160 km (100 mi) long and hundreds of meters wide and can last a few hours. Estimated wind speeds in violent tornadoes range up to 500 km (300 mi) per hour.

Most tornadoes are associated with intense thunderstorm cells (supercells). Tornadoes and their parent cells most often track from southwest to northeast. Tornado trajectories are frequently erratic, however, with many tornadoes exhibiting a hopscotch pattern of destruction as they alternately touch

■ FIGURE 18.27 A funnel cloud and a swirl of dust and debris lifted from the ground are signs of a tornado.

down and lift off the ground. Average forward speed is around 48 km (30 mi) per hour, although there are reports of tornadoes racing along at 100 km (60 mi) per hour or higher.

An exceptionally steep horizontal air pressure gradient is responsible for a tornado's vigorous circulation. The air pressure between the exterior and interior of a tornado, a horizontal distance of perhaps 100 m (330 ft), may fall as much as 10%, the same as the normal drop in air pressure between sea level and an altitude of 1,000 m (3,300 ft). The inwardly directed pressure gradient force provides the centripetal force that keeps the air column rotating about a vertical axis.

A tornado is such a small system that the Coriolis effect is negligible. The winds rotate in either a clockwise or a counterclockwise direction, although the latter dominates among Northern Hemisphere tornadoes. This counterclockwise bias apparently is inherited from the parent thunderstorm, a much larger system in which the Coriolis effect is more pronounced.

Tornadoes and Supercell Thunderstorms

Potentially the most intense tornadoes are spawned by supercell thunderstorms in which a strong vertical shear in the horizontal wind interacts with the powerful updraft. Horizontal winds strengthen with altitude and shift from southeast at the surface to southwest or west aloft. This shear causes the air initially to rotate about a horizontal axis in a roll-like motion. The updraft tilts this tube of rotating air to a vertical orientation so that the updraft spins in a counterclockwise direction (viewed from above), forming a mesocyclone 3 to 10 km (2 to 6 mi) in diameter.

Often a mesocyclone is accompanied by an ominous-looking wall cloud. A *wall cloud* is a local lowering of the thunderstorm cloud base into a nearly circular mass about 3 km (2 mi) across. Rain-moistened air drawn upward into the mesocyclone expands and cools so that water vapor condenses and the cloud base builds downward.

Most mesocyclones and wall clouds do not evolve into a tornado. In those that do, the mesocyclone circulation builds both upward and downward within the supercell. The mesocyclone narrows and spirals downward toward Earth's surface. Relatively low air pressure in the mesocyclone causes water vapor to condense into a funnel cloud. As the spinning column of air narrows, its circulation strengthens, sometimes to exceptionally high speeds. This increase in speed is analogous to what happens to an ice skater performing a spin: as she pulls her arms closer to her body, her spin rate increases.

At about the same time that the tornadic circulation begins descending toward Earth's surface, a downdraft develops near the rear edge of the supercell. The downdraft strikes Earth's surface within minutes of the tornado touchdown and begins to wrap around the tornado and mesocyclone. Eventually, the downdraft completely surrounds the tornado and mesocyclone and cuts off the inflow of warm, humid air. The tornado weakens, and its funnel shrinks, tilts, and develops a ropelike appearance before completely dissipating.

Geographical and Temporal Distribution

The central United States is one of only a few places in the world where weather conditions favor development of tornadic thunderstorms; interior Australia is another. Although tornadoes have been reported in all 50 states and throughout southern Canada, most occur in **tornado alley,** a corridor from eastern Texas and the Texas Panhandle northward through Oklahoma, Kansas, and portions of Nebraska (▣ Fig. 18.28).

Each year, Americans can anticipate between 700 and 1,100 tornadoes. Between 1961 and 1990, the annual average was 803. Slightly more than half of all tornadoes develop during the warmest hours of the day (10 A.M. to 6 P.M.), and almost three-quarters of tornadoes in the United States occur from March to July. The months of peak tornado activity are April (13%), May (22%), and June (21%).

Tornado occurrences generally progress northward during the spring. In effect, the center of maximum tornado frequency follows the Sun, as do the midlatitude jet stream, principal storm tracks, and northward incursions of maritime tropical air. In late winter, maximum tornado frequency, on average, is along the Gulf Coast. By May, the maximum frequency shifts from the southeastern states to the southern Great Plains, and by June, the highest tornado incidence is usually over the northern plains and the Prairie Provinces east of the Rockies.

Tornado Hazards

Tornadoes threaten people and property because of their (1) extremely high winds, (2) strong updraft, (3) subsidiary vortices, and (4) abrupt drop in air pressure. Winds that may reach hundreds of kilometers per hour blow down trees, power poles, buildings, and other structures. Flying debris causes much of the death and injury associated with tornadoes as broken glass, splintered lumber, and even vehicles become lethal projectiles. In violent tornadoes, the updraft near the center of the storm's funnel may top 160 km (100 mi) per hour and is sometimes strong enough to lift a house off its foundation. Some tornadoes are composed of two or more subsidiary vortices that orbit about each other or about a common center within a tornado. These relatively massive, multivortex tornadoes are usually the most destructive of all tornadoes.

Structural damage to a building is caused either by winds slamming debris against the walls or by very strong currents of air that stream over the roof causing it to lift. As the roof is lifted, the walls collapse or are blown in, and the building disintegrates (▣ Fig. 18.29).

T. Theodore Fujita of the University of Chicago has devised a six-point intensity scale for tornadoes (◐ Table 18.1). Called the **F-scale,** it is based on rotational wind speed estimated from property damage and categorizes tornadoes as *weak* (F0, F1), *strong* (F2, F3), or *violent* (F4, F5). An F0 tornado causes only minor property damage, perhaps removing some shingles and breaking a few windows. Also, shallow-

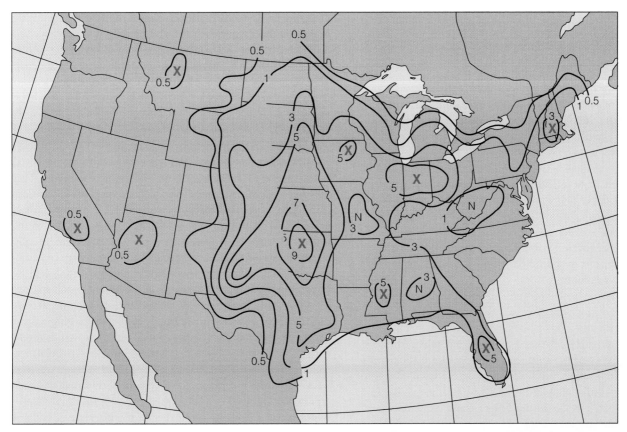

■ FIGURE 18.28 The frequency of tornadoes across the United States, expressed as the number of tornadoes occurring per year within a circle with a radius of 91 km (56 mi). X and N indicate the relative maximum and local minimum, respectively.

rooted trees may be blown over. F1 and F2 tornadoes can cause moderate to considerable property damage and even take lives. An F1 tornado can shift mobile homes off their foundations and push moving automobiles off the road, and an F2 tornado can rip roofs from frame houses and demolish mobile homes. An F3 tornado can partially destroy even well-constructed buildings and lift autos off the ground. At the violent end of the scale, destruction is devastating with the potential for many fatalities. An F4 tornado can toss automobiles about like toys. In an F5 tornado, sturdy frame houses are lifted and totally destroyed.

TABLE 18.1 The Fujita Scale of Wind Speed and Property Damage Potential

SCALE	WIND SPEED km/hr (mi/hr)	DAMAGE POTENTIAL
F0	68–118 (40–73)	Light
F1	119–181 (74–112)	Moderate
F2	182–253 (113–157)	Considerable
F3	254–332 (158–206)	Severe
F4	333–419 (207–260)	Devastating
F5	420–513 (261–318)	Incredible

■ FIGURE 18.29 Tornadoes can cause considerable structural damage to buildings.

Tornado Look-Alikes

Experienced watchers of the sky are well aware that a number of features resemble funnel clouds or tornadoes. These include waterspouts, virga, scud clouds, and dust devils. Even distant plumes of smoke are sometimes mistaken for tornadoes.

A *waterspout* is a whirlwind that occurs over the ocean or a large inland lake. It is so named because a rotating mass of water appears to stream out of the base of the system's parent cloud. Usually, a waterspout is weaker and smaller than a tornado and has a shorter life. The rare intense waterspout may well be a tornado that originated over land. Boaters are advised to steer clear of waterspouts.

Both curtains of *virga* (rain or snow that vaporizes before reaching the ground) and *scud clouds* (low clouds with a ragged appearance) sometimes assume a cylindrical or funnel-shaped profile that causes them to be mistaken for a tornado or funnel cloud (▢ Fig. 1). Watching these features for a minute or two, however, reveals that they lack organized rotation about a vertical axis, clearly distinguishing them from tornadoes or funnel clouds.

Wherever the soil is dry, solar heating may give rise to a swirling mass of dust, known as a *dust devil.* This system looks like a tornado, but is not attached to clouds and causes little if any property damage.

▢ **FIGURE 1** Virga is a shaft of falling rain or snow, much of which vaporizes before reaching the ground. Its funnel-shaped outline may be mistaken for a tornado.

Fortunately, F5 tornadoes are rare. Of the 800 or so tornadoes that occur in the United States each year on average, probably only one will be rated an F5. The American Meteorological Society reports that in a typical year, 79% of all tornadoes are weak, 20% are strong, and only about 1% are violent. The few violent systems are responsible for most fatalities, however. Between 1961 and 1990, the average annual mortality from tornadoes was 82 with 42% occurring in April and 17% in May.

Meteorologists rely on weather radar to warn the public of an approaching tornadic thunderstorm. ▢ Perspective 18.1 on page 436 discusses this topic in more detail.

HURRICANES

A **hurricane** is a tropical cyclone with maximum sustained wind speeds of 119 km (74 mi) per hour or higher. This weather system originates over tropical seas, usually in late summer or early fall. Hurricanes develop in a uniform mass of very warm, humid air, so they have no associated fronts or frontal weather. Air pressure is distributed symmetrically about the system center, so isobars form closely spaced concentric circles. Typically, the central pressure is considerably lower and the horizontal air pressure gradient much steeper in a hurricane than in a typical midlatitude cyclone. In addition, a hurricane is usually much smaller, averaging a third of the diameter of a cyclone. Hurricane-force winds rarely extend more than about 120 km (75 mi) from the storm center.

A hurricane is a warm, low pressure system that weakens rapidly with altitude; in the upper troposphere, anticyclonic winds blow above the hurricane. At the hurricane center is the *eye,* an area of almost cloudless skies, subsiding air, and light winds (less than 25 km per hour) (▢ Fig. 18.30). The eye's diameter usually ranges from 20 to 65 km (12 to 40 mi) but decreases as the hurricane intensifies and winds strengthen.

At a hurricane's typical forward speed, its eye may take up to an hour to pass over a locality. Hence, people may be deceived into thinking the storm has ended when clearing skies and slackening winds follow a hurricane's initial blow. They may well be experiencing the hurricane's eye—heavy rains and strong winds will soon resume but from a different direction.

Bordering the eye of a mature hurricane is the *eye wall,* a ring of cumulonimbus clouds that produce heavy rains and very strong winds. Potentially, the most destructive sector of a hurricane is near the eye on the side where the wind is in the same direction as the storm's forward motion. On that side, hurricane winds combine with steering winds to produce the storm's most powerful surface winds. Cloud bands accompanied by hurricane-force winds and heavy showers spiral outward from the eye wall.

Geographical and Temporal Distribution

Two conditions are necessary for a hurricane to develop: (1) warm surface waters and (2) sufficient Coriolis effect. The threshold for hurricane formation is a sea-surface temperature (SST) of at least 26.5°C (80°F) through a depth of 60 m (200 ft) or more. Such exceptionally warm water sustains the hurricane circulation through the latent heat released when water that evaporates from the ocean surface condenses within the storm. Temperature governs the rate of vaporization of water, so the higher the SST, the greater the supply of latent heat (energy) in the storm system. As a hurricane moves over colder water or land, however, it loses its warm-water energy source, and the system weakens.

The second requirement for hurricane formation is a Coriolis effect that is strong enough to initiate a cyclonic circulation. As noted in Chapter 17, the Coriolis effect decreases toward lower latitudes and becomes zero at the equator. The minimum latitude for hurricane formation is about 4°.

The combination of relatively high SSTs and sufficiently strong Coriolis effect occurs only over certain portions of the world's oceans, identified in ▣ Figure 18.31. Most hurricanes form in the latitude belt between 5° and 20°. Only hurricanes spawned over the tropical Atlantic, Caribbean Sea, and Gulf of Mexico pose a serious threat to coastal North America. During an average hurricane season, six hurricanes and four tropical storms (precursors to hurricanes) form over these waters. On average, two hurricanes strike the U.S. coast each year.

The requirement of relatively high SSTs also makes hurricanes distinctly seasonal. Because of the great thermal inertia of ocean water, SSTs reach a seasonal maximum long after the time of peak solar radiation. Consequently, most Atlantic hurricanes develop in late summer and early autumn; the official hurricane season runs from June 1 to November 30.

Life Cycle

The appearance of an organized cluster of thunderstorms over tropical seas is the first sign that a hurricane may be forming. This region of convective activity is labeled a *tropical disturbance* if a center of low pressure is detected at the surface. If conditions favorable to hurricane development

▣ FIGURE 18.30 A photograph of Hurricane Florence over the North Atlantic taken by astronauts aboard a space shuttle in November 1994. The cloud-free eye is visible at the center of the system.

▣ FIGURE 18.31 A global map showing hurricane breeding grounds and the average tracks of hurricanes.

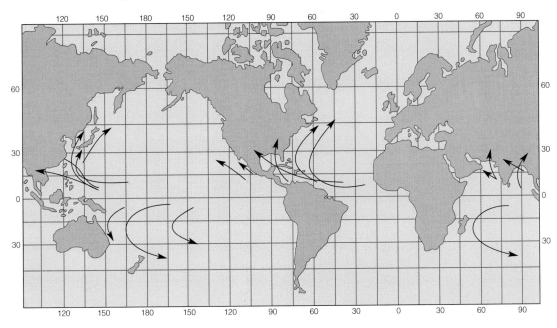

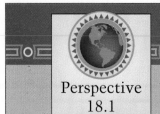
Weather Radar

Weather radar is an indispensable tool for detecting and tracking severe weather systems and providing the public with advance warning of the approach of hazardous weather. Thunderstorm cells are so small that they may not be detected visually by the widely spaced network of weather stations. Weather radar, on the other hand, scans a wide area continuously and can locate small, isolated patches of precipitation. Weather radar can detect a tornado as it develops within its parent cumulonimbus cloud, the spiral bands of a hurricane before they sweep onshore, downbursts that could be hazardous to aircraft, and rainfall rates and cumulative rainfall totals that forewarn of flooding.

National Weather Service weather radar operates continually in either a reflectivity or a velocity mode. In the reflectivity mode, the radar signal determines the location and movement of areas of precipitation, and in the velocity mode, the radar monitors the circulation within a weather system.

Weather radar continuously emits short pulses of microwave energy with wavelengths of 10 cm (☐ Fig. 1). These radar signals are reflected (or, more precisely, scattered) by rain, snow, or hail but not by the very small droplets or ice crystals that compose clouds. Thus, weather radar detects (*sees*) precipitation but not the parent clouds. Precipitation reflects (or scatters) some of the radar signal back to a receiving unit. From there, the return signal is electronically processed and displayed as a *radar echo* (electrical pulse) on a cathode ray tube similar to a television screen (☐ Fig. 2). The product is a map of radar echoes that represents the precipitation pattern surrounding the radar unit. Because the speed of the radar pulse is known, the time interval between the emission and reception of the radar signal can be calibrated to give the distance to the precipitation.

The strength (intensity) of a radar echo depends on the reflectivity of the targeted precipitation and is greatest for hailstones. The concentration of raindrops in the path of the radar beam also influences echo intensity. Hence, echo intensity is used as an index of rainfall rate and as a means of identifying severe thunderstorm cells, which often contain large hail. Based on continuous radar monitoring of rainfall rate, a special computer algorithm generates a map of rainfall totals over a specific period of time. Such maps help with flood forecasting.

Operated in the reflectivity mode, weather radar cannot detect a tornado directly, but in some cases, a hook-shaped echo appears on the radar screen (☐ Fig. 3). A *hook echo* is produced when rainfall is drawn around the mesocyclone within a severe thunderstorm cell and usually appears on the southeast side of the system.

In the velocity mode, weather radar operates on the same principle as the police radar that monitors traffic flow. The principle is called the Doppler effect after Johann Christian Doppler, the Austrian physicist who first explained the phenomenon in 1842. For this reason, a radar with velocity-detection capability is often referred to as a *Doppler radar*.

The *Doppler effect* refers to a shift in the frequency of sound waves or electromagnetic waves emanating from a moving source. For example, the pitch (frequency) of a train whistle is higher as a train approaches and drops off as the train pulls away. A

☐ FIGURE 1 The 10-meter diameter antenna for a weather radar is housed inside this cylindrical radome.

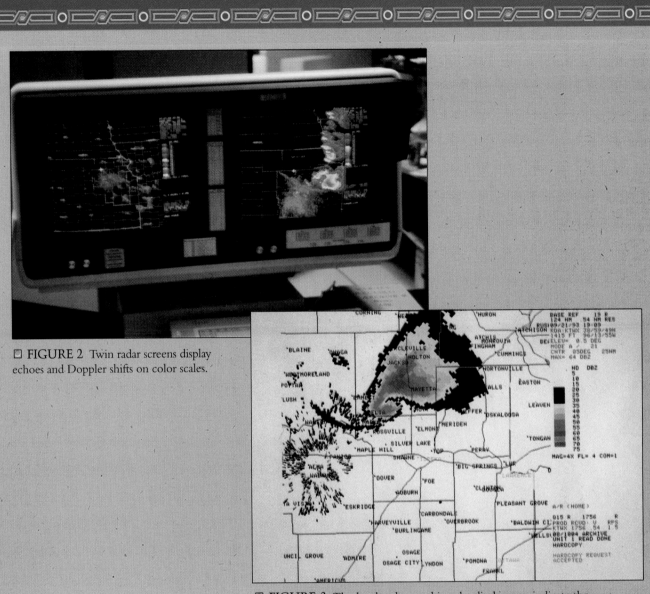

☐ FIGURE 2 Twin radar screens display echoes and Doppler shifts on color scales.

☐ FIGURE 3 The hook echo on this radar display may indicate the presence of a tornadic circulation.

Doppler weather radar monitors the speed of precipitation particles as they move radially away from or toward the radar antenna (☐ Fig. 4). With that movement, the frequency of the radar signal shifts slightly between emission and the return signal (echo). This frequency shift (the Doppler effect) is calibrated in terms of the motion of the particle. If the particle motion is perpendicular to the radar beam, however, no frequency shift occurs. Thus, Doppler radar does not detect the component of air motion at right angles to the radar beam.

The principal advantage of Doppler radar is that it monitors the circulation *within* a weather system rather than just the general movement of an area of precipitation. Hence, a Doppler radar sees inside clouds and severe thunderstorms and detects mesocyclones, developing tornadoes, gust fronts, and strong wind shears associated with downbursts. Doppler radar displays are color coded with greens and blues (cold colors) indicating motion toward the radar and reds and yellows (warm colors) indicating motion away from the radar.

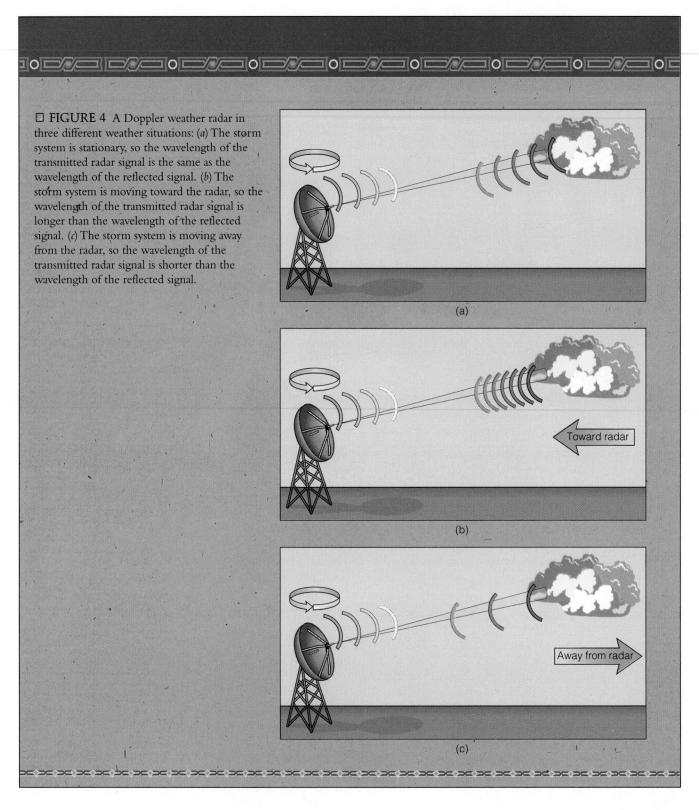

□ FIGURE 4 A Doppler weather radar in three different weather situations: (*a*) The storm system is stationary, so the wavelength of the transmitted radar signal is the same as the wavelength of the reflected signal. (*b*) The storm system is moving toward the radar, so the wavelength of the transmitted radar signal is longer than the wavelength of the reflected signal. (*c*) The storm system is moving away from the radar, so the wavelength of the transmitted radar signal is shorter than the wavelength of the reflected signal.

(a)

(b)

Toward radar

(c)

Away from radar

persist, a cyclonic circulation forms, and the central air pressure begins to fall. Water vapor condenses within the storm, releasing latent heat, and the heated air rises. Expansional cooling of the rising air triggers more condensation, release of even more latent heat, and a further increase in buoyancy. Rising temperatures, coupled with an anticyclonic outflow of air aloft, cause a sharp drop in air pressure, which, in turn,

induces convergence of air at the surface. The consequent uplift surrounding the developing eye leads to additional condensation and release of latent heat.

Thus, the tropical disturbance intensifies and winds strengthen. When sustained wind speeds top 37 km (23 mi) per hour, the developing storm is called a *tropical depression*. When wind speeds reach 63 km (39 mi) per hour, the sys-

■ **FIGURE 18.32** Extensive property damage in southern Florida caused by Hurricane Andrew in August 1992.

tem is classified as a *tropical storm* and assigned a name. (Male and female names alternate.) If winds hit 119 km (74 mi) per hour, the storm is officially designated a hurricane. As a hurricane decays, the storm is downgraded using this same classification scheme.

Hurricanes that threaten the Atlantic and Gulf Coasts of the United States usually drift very slowly westward with the trade winds across the tropical North Atlantic and into the Caribbean. At this stage, it is not unusual for the storm to travel at a mere 10 to 20 km (6 to 12 mi) per hour and take a week to cross the Atlantic. Once over the western Atlantic, however, the storm usually speeds up and begins to curve first northward and eventually northeastward, as it is caught in the midlatitude westerlies. Where this change in course takes place determines whether the hurricane enters the Gulf of Mexico, moves up the eastern seaboard, or curves back out to sea.

When a hurricane reaches a latitude of about 30° N, it may begin to acquire extratropical (midlatitude) characteristics, as colder air is drawn into the circulation and fronts develop. From then on, the storm follows a life cycle similar to that of any other midlatitude cyclone and often ends by occluding over the North Atlantic.

Hurricane Hazards

Hazards of hurricanes are (1) strong winds, (2) associated tornadoes, (3) heavy rains, and (4) storm surge. About 90% of hurricane-related fatalities are caused by coastal and inland flooding.

Strong wind gusts are responsible for much of the property damage associated with hurricanes (■ Fig. 18.32). Hurricane winds weaken rapidly once the storm makes landfall, so most wind damage is confined to within about 200 km (124 mi) of the coastline. The abrupt drop in wind speed once a hurricane makes landfall occurs for two reasons: (1) over land a hurricane is no longer in contact with its warm-water energy source, and (2) the rougher land surface shifts the winds toward the system's center, causing a rise in the central air pressure.

Once a hurricane makes landfall, tornadoes often develop. Typically, they occur to the northeast of the storm center, often outside the region of hurricane-force winds. Hurricanes also produce torrential rains with amounts typically in the range of 13 to 25 cm (5 to 10 in). Even if the storm tracks well inland, heavy rains often persist and may trigger flash flooding (■ Fig. 18.33 on page 442).

Potentially, the most devastating feature of a hurricane that strikes coastal areas is a **storm surge,** a flood of ocean water that accompanies the storm. Low air pressure in a hurricane causes sea level to rise (about 0.5 m for every 50 mb drop in air pressure), and hurricane-force winds drive ocean waters over low-lying coastal areas (■ Fig. 18.34).

A storm surge of 1 to 2 m (3 to 6.5 ft) can be expected with a weak hurricane, whereas the storm surge accompanying an intense hurricane may top 5 m (16.4 ft). In the most deadly natural disaster in U.S. history, an estimated 6,000 people perished, mostly by drowning, when a hurricane storm surge flooded low-lying Galveston, Texas, on September 8, 1900. Hurricane Camille, an exceptionally in-

The Hurricane Threat to the Southeastern United States

Hurricane Andrew took only three hours to cross southern Florida on the morning of August 24, 1992. In that short time, its winds, gusting in excess of 265 km (164 mi) per hour, contributed to the deaths of 25 people and left 180,000 homeless (□ Fig. 1). With property damage estimated at $25 billion, Hurricane Andrew was the most costly natural disaster in U.S. history. To make matters worse, Andrew continued to track west and northwestward across the Gulf of Mexico and two days later came ashore along the Louisiana coast where it claimed 4 more lives and caused $400 million in property damage.

Hurricanes Hugo in 1989 and Andrew in 1992 and an unusually active 1995 hurricane season have heightened public awareness of the hurricane threat to the Gulf and Atlantic Coasts. This recent episode of increased hurricane activity follows a two-decade lull that gave many residents of the coastal Southeast a false sense of security and encouraged population growth in areas that could be devastated by a major hurricane. Resort hotels, high-rise condominiums, and cottages were constructed perilously close to the shoreline (□ Fig. 2).

About 45 million permanent residents inhabit the hurricane-prone coasts of the United States, and that population continues to grow. The most rapid population growth is taking place from Texas through the Carolinas. Florida leads the nation in both population growth and hurricane potential. Compounding the problem are the weekend, seasonal, and holiday visitors who flock to seaside resorts. During vacation periods, the population in some areas swells 10- to 100-fold. Many of these resorts are in low-lying coastal areas or on exposed beaches that are subject to inundation by the waters of a hurricane storm surge (□ Fig. 3).

The hurricane threat is most serious for people living on or visiting the nearly 300 barrier islands that fringe portions of the Atlantic and Gulf Coasts. A *barrier island* is a long, narrow ribbon of sand separated from the mainland by a lagoon. Barrier islands protect coastal beaches and wetlands by absorbing the brunt of powerful storm-driven sea waves. Nevertheless, many barrier islands have been developed for resorts and cottages, and some coastal cities, including Miami Beach, are located on barrier islands.

Evacuation of people from barrier islands and other low-lying coastal areas is the most prudent strategy should a hurricane threaten (□ Fig. 4). Successful evacuation, however, requires sufficient advance warning of a hurricane's approach. Unfortunately, hurricanes are notorious for sudden changes in intensity, direction, and forward speed. This capricious behavior is especially troublesome for isolated localities and congested cities where highway systems may not have the capacity to handle great numbers of evacuees. Estimated evacuation time is 37 hours for the Florida Keys, for example, and 50 hours for New Orleans.

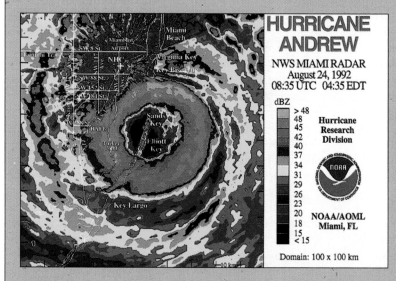

HURRICANE ANDREW

NWS MIAMI RADAR
August 24, 1992
08:35 UTC 04:35 EDT

dBZ

| > 48 |
48	Hurricane
45	Research
42	Division
40	
37	
34	
31	
29	
26	
23	
20	NOAA/AOML
18	Miami, FL
15	
< 15	

Domain: 100 x 100 km

□ FIGURE 1 A weather radar view of Hurricane Andrew as it made landfall south of Miami beach on August 24, 1992. Red indicates the area of heaviest rainfall.

tense storm with winds to 300 km (186 mi) per hour, produced a maximum storm surge of 7.3 m (24 ft) at Pass Christian, Mississippi, in August 1969.

Like tornadoes, hurricanes have an intensity scale, known as the **Saffir-Simpson Hurricane Intensity Scale** after its designers, H. S. Saffir, a consulting engineer, and R. H. Simpson, former director of the National Hurricane Center. The scale rates hurricanes from 1 (weak) to 5 (very intense). As ○ Table 18.2 on page 443 shows, each intensity category corresponds to (1) a range of central air pressure, (2) a range

☐ **FIGURE 2** A resort hotel constructed close to the shoreline is vulnerable to a hurricane storm surge.

☐ **FIGURE 3** These buildings were constructed to withstand hurricane storm surge. The walls of the first floor are designed to give way to flood waters without damaging the supporting beams.

☐ **FIGURE 4** Road signs direct residents of North Carolina's Outer Banks to an evacuation route in the event a hurricane threatens.

of wind speed, (3) storm surge potential, and (4) property damage potential. Of the 126 tropical storms or hurricanes that struck the United States Gulf or Atlantic Coasts between 1949 and 1990, 25 (19.8%) were classified as *major;* that is, they rated 3 or higher on the Saffir-Simpson scale.

One of today's major weather-related concerns is the hurricane threat to the southeastern United States. For details on this, see ☐ Perspective 18.2.

■ **FIGURE 18.33** A highway sign in central Virginia commemorates victims of flooding produced by the torrential rains of the remnants of Hurricane Camille in August 1969.

■ **FIGURE 18.34** A hurricane storm surge can cause considerable coastal flooding, shoreline erosion, and property damage.

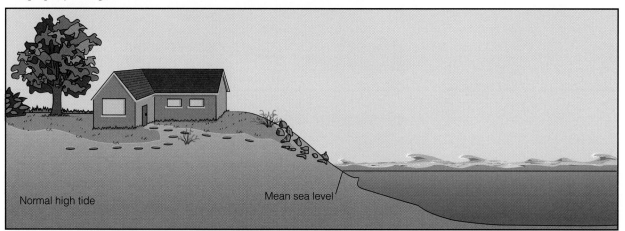

Normal high tide

Mean sea level

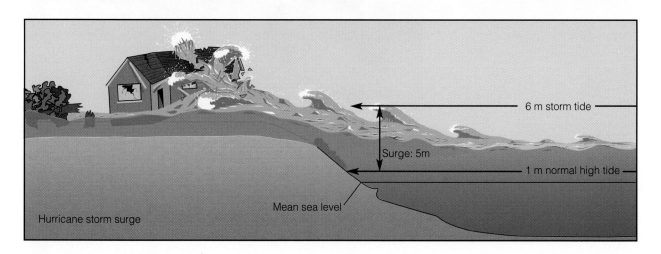

6 m storm tide

Surge: 5m

1 m normal high tide

Mean sea level

Hurricane storm surge

TABLE 18.2 Saffir-Simpson Hurricane Intensity Scale

Category	Central Pressure		Wind Speed		Storm Surge		Damage Potential
	mb	in.	km/hr	mi/hr	m	ft	
1	>979	(>28.91)	119–154	(74–95)	1–2	(4–5)	Minimal
2	965–979	(28.50–28.91)	155–178	(96–110)	2–3	(6–8)	Moderate
3	945–964	(27.91–28.47)	179–210	(111–130)	3–4	(9–12)	Extensive
4	920–944	(27.17–27.88)	211–250	(131–155)	4–6	(13–18)	Extreme
5	<920	(<27.17)	>250	(>155)	>6	(>18)	Catastrophic

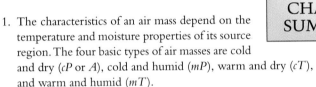

CHAPTER SUMMARY

1. The characteristics of an air mass depend on the temperature and moisture properties of its source region. The four basic types of air masses are cold and dry (*cP* or *A*), cold and humid (*mP*), warm and dry (*cT*), and warm and humid (*mT*).

2. A front is a narrow zone of transition between air masses that contrast in temperature and/or humidity. The four types of fronts are stationary, cold, warm, and occluded.

3. Upper-air support (horizontal divergence aloft) is required for formation of a cyclone. A midlatitude cyclone begins its life cycle along the polar front and intensifies as its central pressure drops. Winds strengthen and frontal weather develops. The storm finally occludes as the faster-moving cold front merges with the slower warm front. From then on, the system weakens.

4. On the cold (left) side of a cyclone track, winds shift counterclockwise with time, and there are no frontal passages. On the warm (right) side of a cyclone track, winds shift clockwise with time, and a warm front is followed by a cold front.

5. Cold anticyclones are shallow systems that are equivalent to domes of continental polar or arctic air. Warm anticyclones, such as the semipermanent subtropical highs, extend high into the troposphere and are accompanied by broad areas of subsiding, warm, dry air.

6. A thunderstorm is a mesoscale weather system produced by strong convection currents that surge to great altitudes in the troposphere. The life cycle of a thunderstorm cell is a three-stage sequence: cumulus, mature, and dissipating.

7. Thunderstorms occur as isolated cells, elongated clusters known as squall lines, nearly circular clusters known as mesoscale convective complexes (MCCs), and supercells.

8. Lightning is a brilliant flash of light produced by the discharge of electrons within a cloud or between a cloud and the ground.

9. Some thunderstorm cells produce downbursts, strong downdrafts that spread out at Earth's surface as potentially destructive winds.

10. Slow-moving thunderstorms or successive cells that mature over the same region may produce torrential rains and flash flooding. Flash flooding is a special hazard in mountainous terrain, where steep slopes channel excess runoff into narrow stream valleys.

11. Hail develops in intense thunderstorm cells characterized by strong updrafts, great vertical development, and an abundant supply of supercooled water droplets.

12. A tornado is a small mass of air that whirls rapidly about a nearly vertical axis and is made visible by clouds and dust and debris drawn into the system.

13. An exceptionally steep horizontal air pressure gradient between the very low pressure at the tornado center and the pressure at the outer edge is the force ultimately responsible for the violence of a tornado. A tornado is too small to be influenced by the Coriolis effect.

14. Weather conditions favorable for an outbreak of tornadoes progress northward (with the Sun) from the Gulf Coast in early spring to southern Canada by early summer.

15. Based on estimated rotational wind speed, tornadoes are rated as weak, strong, or violent on the F-scale. Most tornadoes are weak, but most fatalities are caused by rare violent tornadoes.

16. A hurricane originates over tropical oceans usually in late summer or early fall. It develops in a mass of warm, humid air, has no fronts, and is about one-third the diameter of an average midlatitude cyclone.

17. For tropical storms and hurricanes to develop, sea-surface temperatures must be 26.5°C (80°F) or higher through a depth of at least 60 m (200 ft), and the Coriolis effect must be of significant magnitude (latitude of at least 4°) to initiate a cyclonic circulation.

18. Once a hurricane makes landfall, it loses its warm-water energy source and experiences greater frictional resistance. Hence, its circulation weakens rapidly, so most wind damage is confined to the coast. Heavy rains often continue well inland, however, and can cause severe flooding.

19. A storm surge is a flood of ocean water driven by hurricane winds. The greatest potential for coastal flooding and erosion occurs when a storm surge coincides with high tide.

20. Based on intensity and potential for property damage, a hurricane is rated 1 (weak) to 5 (very intense) on the Saffir-Simpson Scale.

air mass
anticyclone
cyclone
downburst
flash flood
front

F-scale
hail
hurricane
lightning
mesoscale convective
   complex (MCC)

Saffir-Simpson Hurricane
   Intensity Scale
severe thunderstorm
squall line
storm surge

supercell thunderstorm
thunder
thunderstorm
tornado
tornado alley

1. What determines the humidity and temperature characteristics of an air mass?
2. Distinguish among the various air masses that regularly invade North America.
3. What causes air mass modification, and in what ways do air masses modify?
4. Explain why continental polar air modifies more rapidly than maritime tropical air. What is the significance of this difference for cold waves and heat waves?
5. Distinguish among stationary, warm, cold, and occluded fronts.
6. As a surface warm front approaches your locality, clouds lower and thicken in the following sequence: a. ___ nimbostratus, altostratus, cirrostratus; b. ___ cirrus, cirrostratus, altostratus, nimbostratus; c. ___ cumulonimbus, cumulus congestus, cumulus; d. ___ altostratus, stratus, cirrocumulus; e. ___ stratus, stratocumulus, altocumulus.
7. Explain why the pattern of surface winds about a cyclone favors the development of fronts.
8. The coldest air is located in the _____ sector of a mature, midlatitude cyclone.
   a. ___ northeast;   b. ___ southeast;   c. ___ southwest;
   d. ___ northwest
9. A midlatitude cyclone has a cold side and a warm side. Explain.
10. An Alberta-track cyclone usually produces less precipitation than a Colorado-track cyclone. Why?
11. What happens when a cyclone occludes?
12. What is the relationship between a cold anticyclone and a polar or arctic air mass?
13. Explain why the circulation in an anticyclone favors the development of frost or radiation fog.
14. Describe the characteristics of each stage in the life cycle of a thunderstorm cell.

15. The stage in the life cycle of a thunderstorm cell that is most likely to produce heavy rain, strong surface winds, and perhaps hail is:
   a. ___ cumulus; b. ___ mature; c. ___ dissipating.
16. Distinguish between a squall line and a mesoscale convective complex (MCC).
17. Describe how thunderstorms develop over the interior of the Florida peninsula.
18. Why are thunderstorms infrequent along the coast of southern California?
19. In what sector of a mature cyclone are thunderstorms most likely to develop?
20. Describe a weather pattern favorable for the development of severe thunderstorms in the Midwest.
21. What causes lightning? Why is it dangerous?
22. Speculate on why the damage caused by a downburst is sometimes mistaken for tornado damage.
23. Why is flash flooding a particular hazard in mountainous terrain? Why is it a potential problem in urban areas?
24. Distinguish between a tornado and a funnel cloud.
25. The principal force operating in a tornado is:
   a. ___ gravity; b. ___ friction; c. ___ the Coriolis effect;
   d. ___ the horizontal pressure gradient force.
26. Thunderstorms are most likely to produce a tornado in which of the following states?
   a. ___ Massachusetts;   b. ___ Kansas,   c. ___ Oregon;
   d. ___ Wyoming; e. ___ South Carolina.
27. What is the basis for the F-scale of tornado intensity? Where do most U.S. tornadoes rank on this scale?
28. Compare and contrast a hurricane with a mature, midlatitude cyclone.
29. What conditions are required for hurricane development?
30. What is the most destructive aspect of a hurricane approaching a low-lying coastal area?

1. A mature, midlatitude, winter cyclone follows a southwest to northeast track that takes the center of the system about 100 km south of Buffalo, New York. Describe the sequence of weather changes at Buffalo as the storm approaches and moves away from the city.
2. Why are tornadoes more frequent in spring than in autumn?
3. Surface winds in a hurricane are generally strongest on the side of the system where the steering winds are blowing in the same direction as the surface winds. Explain why.
4. How is it possible for a slow-moving, relatively weak nor'easter to cause more shoreline erosion than a fast-moving intense hurricane? Assume the hurricane makes landfall on the South Carolina coast.

American Meteorological Society. 1991. Tornado forecasting and warning. *Bulletin of the American Meteorological Society* 72: 1270–72.

Businger, S. 1991. Arctic hurricanes. *American Scientist* 79 (January-February): 18–33.

Cobb, H. 1989. The siege of New England. *Weatherwise* 42: 262–66.

Davies-Jones, R. 1995. Tornadoes. *Scientific American* 273, no. 2: 48–57.

Dolan, R., and H. Lins. 1987. Beaches and barrier islands. *Scientific American* 257, no. 1: 68–77.

Hughes, P. 1979. The great Galveston hurricane. *Weatherwise* 32: 148–56.

———. 1987. The blizzard of '88. *Weatherwise* 40, no. 6: 312–20.

Ludlum, D. M. 1988. The great hurricane of 1938. *Weatherwise* 41: 214–16.

Marshall, T. 1992. Dryline magic. *Weatherwise* 45, no. 2: 25–28.

Williams, J., et al. 1992/93. Hurricane Andrew in Florida. *Weatherwise* 45, no. 6: 7–17.

For current updates and exercises, log on to http://www.wadsworth.com/geo.

NORTHERN ILLINOIS UNIVERSITY (NIU) STORM CHASER HOME PAGE
http://taiga.geog.niu.edu/chaser.html

This site provides storm chasers with the latest National Weather Service reports on severe weather systems, serves as a forum for exchange of information, and includes book and film reviews and severe storms photo gallery.

FLORIDA STATE UNIVERSITY, DEPARTMENT OF METEOROLOGY HOME PAGE
http://thunder.met.fsu.edu/

Regional weather maps, satellite images, and radar summaries are included.

NATIONAL HURRICANE CENTER– TROPICAL PREDICTION CENTER
http://nhc-hp2.nhc.noaa.gov/index.html

This site offers a link to the center responsible for tracking and forecasting tropical storms and hurricanes.

FEDERAL EMERGENCY MANAGEMENT AGENCY (FEMA)
http://femapubl.fema.gov/homepage.html

The FEMA site provides information on weather-related natural disasters.

CHAPTER

19

The Earth-Moon System

This classic painting, done in the 1950s by Chesley Bonestell, a pioneer of space art, shows a molten early earth and a nearby, still-glowing moon. Although present ideas about the formation of the Earth-Moon system are quite different than in Bonestell's day, this painting may be fairly close to actual physical conditions just after the formation of Earth's moon. Bonestell lived to see humans walk on the Moon and spacecraft visit many of the worlds he had explored in his artistic imagination. In recognition of his work, an asteroid was named after him.

Prologue

In 55 B.C., Julius Caesar was preparing to invade Britain from the coast of what is now northern France when he got a rude surprise. He was familiar with sailing in the Mediterranean, but not in the open sea. In particular, he was not familiar with tides, which are virtually absent from the Mediterranean. He beached his ships one evening and awoke next morning to find that many of them had been carried off by the tide. His invasion had to be delayed until the ships were recovered. Other historic events have been affected by the tides as well.

On December 16, 1773, American colonists, frustrated by British tax policies, planned a dramatic protest. The merchant ships *Dartmouth, Endeavour,* and *Beaver* were in port with a cargo of tea, on which the colonists refused to pay tax. Negotiations with British authorities had broken down that afternoon, and the cargo was due to be seized at midnight if the tax was not paid. A group of colonists, dressed as Indians (a disguise that fooled no one), boarded the ship about 6 P.M. and threw the tea overboard in the famous Boston Tea Party. But what happened to the tea after that?

Ideally, the tea should have been thrown overboard just after high tide, so the ebbing tide could carry it out to sea. In reality, because the colonists had only a few hours to stage their protest, they could not plan for the tides. The Boston Tea Party took place not just at low tide, but at an exceptionally low tide at that. The incoming tide simply washed the tea ashore, and work parties with shovels cleaned it off the waterfront and threw it back into the harbor the next day (pollution was not a concern for most people in those days!).

The effect of tides on the Boston Tea Party is a humorous footnote to history. The effect of tides on the battle for Tarawa during World War II was anything but humorous (☐ Fig. 19.1). The atoll of Tarawa was one of

☐ **FIGURE 19.1** Tides had a near disastrous effect on the battle for Tarawa during World War II. (*a*) The location of Tarawa. (*b*) A map of Tarawa. Planners hoped to be able to take landing craft across the lagoon. (*c*) The planners intended the bombardment to begin at dawn, at low tide. As the tide rose, the invasion would begin. By noon, larger landing craft should be able to come ashore with reinforcements and supplies. (*d*) Unfortunately, the planners did not realize that the Moon's greater distance from Earth would result in unusually small tides. This graph shows the tide heights for November 1943 as calculated by a modern computer program.

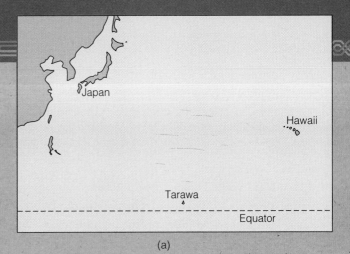

(a)

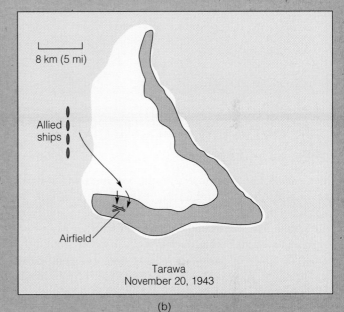

(b)

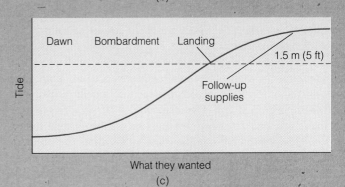

(c)

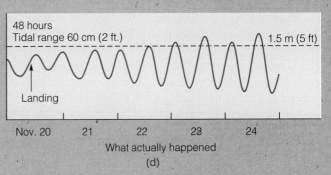

(d)

the easternmost Pacific islands held by the Japanese, and the assault was the first amphibious assault in the Pacific during World War II. Tide data for the island were sketchy but suggested a tidal range of about 2 m (7 ft). There were puzzling rumors of periods when the tides on Tarawa almost ceased, however, a condition local mariners called a "dodging tide."

The invasion was set for November 20, 1943, when tide conditions were expected to be favorable. The bombardment would begin at low tide in the early morning. As the tide rose and water levels in the lagoon reached 1.5 m (5 ft), landing craft would head ashore, and by noon, at high tide, heavier craft could come ashore bringing tanks and supplies.

Unfortunately, the rumors of almost-tideless periods at Tarawa were true. November 20 was near last-quarter moon. The gravitational pulls on the Moon and Sun were perpendicular to each other, resulting in a condition called *neap tide,* where the range of the tides is unusually small. Military planners knew about the risks of neap tide, of course, but did not realize the Moon was unusually far from Earth as well, weakening its tidal effects even more. Landing craft hit bottom hundreds of meters offshore, and the Marines had to wade ashore under heavy fire. Once ashore, they had to fight without assistance because supply ships could not come in. For 48 hours, the tidal range was only 60 cm (2 ft), and it did not increase to normal for four days. In the battle for Tarawa, 1,027 American soldiers were killed and 2,292 were wounded.

 ## OVERVIEW OF THE SOLAR SYSTEM

The *solar system* (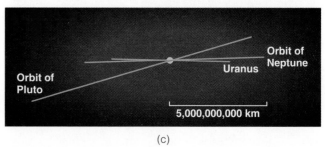 Fig. 19.2) is the system to which Earth belongs. At the hub of the solar system is the Sun, only an average star, but still holding over 99.9% of the total mass of the system. We will examine the Sun in Chapter 21.

The *planets* and other objects travel in **orbit** around the sun. The inner or *terrestrial* planets (Mercury, Venus, Earth and its moon, and Mars) are small, dense bodies made mostly of silicate minerals and metals (Fig. 19.3). The outer or *Jovian* planets (Jupiter, Saturn, Uranus, and Neptune) are larger and more massive than the inner planets, but with much lower densities. They probably consist of solid cores with enormous atmospheres. For this reason these planets are often called the *gas giants.* Most of the planets have **satellites** orbiting them. The term *satellite* refers to any body that orbits another. Most planets have natural satellites (like our Moon) orbiting them. The term *artificial satellite* refers to human-launched satellites. In the inner solar system rocky satellites dominate, but in the outer solar system, the satellites are made of rock and frozen gases, called *ices* by astronomers. The outermost planet, Pluto, is probably also made of ices and rock.

Small objects orbit the sun as well. Between Mars and Jupiter are the *minor planets* or asteroids, up to 1000 kilometers (600 miles) in diameter. Very small solid objects in space are called **meteoroids.** They orbit the sun according to the same laws as all the other planets. Meteoroids that glow from frictional heat as they enter Earth's atmosphere are called *meteors,* and those that reach Earth's surface are called *meteorites.* *Comets,* icy bodies that travel elongated orbits that take them far from the Sun, only become visible when they enter the inner solar system.

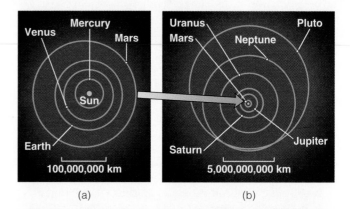

(a)   (b)

(c)

FIGURE 19.2 The solar system. (*a*) The orbits of the inner planets. (*b*) The orbits of the outer planets. Note how much farther apart the outer planets' orbits are than those of the inner planets. Also note that Pluto actually comes closer to the Sun than Neptune. (*c*) The orbits of the planets all lie nearly in the same plane, with the exception of Pluto. Because Pluto's orbit is tilted with respect to Neptune's, there is no chance of a collision between the two planets.

When comparing objects in the solar system, the distinction between planets, satellites, and even large asteroids is often not very useful. If Earth's moon or the large satellites of Jupiter were in separate orbits around the Sun, we would not

■ FIGURE 19.3 The relative sizes of the planets and the Sun. Jupiter is about 10 times the diameter of Earth, and the Sun is about 10 times the diameter of Jupiter.

hesitate to call them planets. The Moon and Mercury have some features in common even though one is a satellite and the other a planet. In this chapter (and in most books on planetary science), terms like *planetary body* and *planetary surface* are used to refer to any solid object in the solar system, regardless of whether it is a planet, satellite, or asteroid.

 ## INTRODUCTION

Visitors to our solar system from elsewhere would see a variety of types of planets, including two pairs they might well call *double planets*. Most planets like Jupiter dwarf their satellites, but Pluto and Charon on the outer edge of the solar system are so similar in size that it seems appropriate to think of them as a double planet. The other double planet in the solar system is Earth and its moon. By remarkable coincidence, our moon appears to be almost exactly the same size in the sky as the Sun. On those rare occasions when the Moon passes between the Sun and Earth, the close similarity in apparent size of the two objects results in eclipses that have no rival for variety and beauty anywhere else in the solar system. The Moon and Earth affect each other in complex ways, and the Moon preserves a record of the early days of the solar system that has been lost on Earth. We will begin our survey of the universe beyond Earth with a study of the Earth-Moon system.

 ## THE MOTIONS OF EARTH

Earth rotates and travels in its orbit slowly enough that we feel no motion. It took people many centuries to realize that Earth moved. In part, the problem lay in erroneous ideas about motion. Believers in a fixed Earth argued that we should feel motion. There were no smooth roads in those

days, and few, if any, people had ever experienced smooth motion, so it was natural to assume that all motion should be felt. They also argued that we should feel a wind if we were moving because they did not realize that Earth carries its atmosphere with it. With no understanding of gravity and no accurate formulas to calculate motion, many people assumed that a rotating Earth would fling objects into space. It is easy for us who have grown up with modern scientific ideas to forget how difficult these discoveries actually were.

As we see the sky from Earth, the heavens appear to rotate, with objects appearing to rise above the horizon in the east and set in the west. The heavens appear to rotate around points called the **celestial poles** (■ Fig. 19.4). Just as on Earth, we can draw an imaginary line in the sky midway between the celestial poles and call it the *celestial equator* (see Appendix G). At the North Pole, the north celestial pole is directly overhead, and the celestial equator is on the horizon. In the winter, when the Sun is below the horizon, the stars move in circles parallel to the horizon, neither rising nor setting. As we move south, the north celestial pole appears lower in the sky, and the celestial equator rises higher. There is a zone around the north celestial pole where some stars never set and a corresponding zone around the south celestial pole where some stars never rise (■ Fig. 19.5). At the equator, the celestial equator passes overhead, and both celestial poles are visible on opposite points on the horizon. All stars rise and set. The equator is the only place on Earth where every star in the sky can be seen at some time of the year. South of the equator, the south celestial pole is visible, and the north celestial pole is below the horizon. At the South Pole, the south celestial pole is overhead, the celestial equator is on the horizon, and again, objects in the sky move in circles parallel to the horizon. An observer at the South Pole and one at the North Pole see opposite halves of the sky. The celestial poles and equator are defined by Earth's rotation; an observer on Mars, say, would see the heavens appear to rotate around different celestial poles.

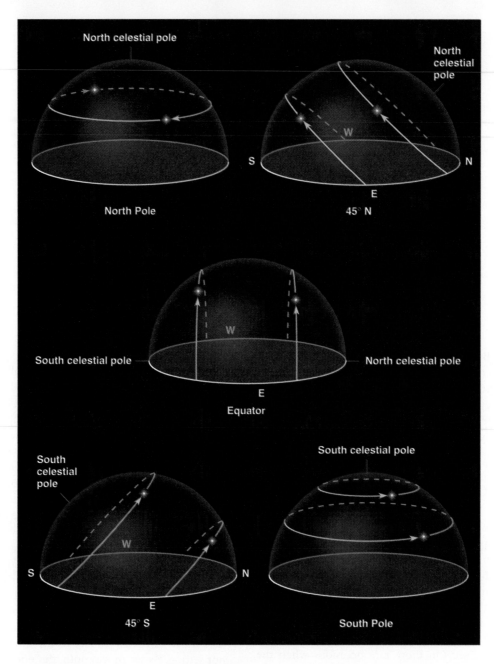

⊙ **FIGURE 19.4** The heavens appear to rotate around points called the *celestial poles.*

Like all the planets and satellites in our solar system, Earth and the Moon travel in **orbits.** Earth orbits the Sun once each year, and the Moon orbits Earth about once a month. As Earth moves around the Sun, the night side of Earth faces different parts of the sky, and different constellations become visible at different times of the year. Also, as Earth moves around the Sun, the Sun appears to trace a path against the stars called the **ecliptic** (⊙ Fig. 19.6). The ecliptic marks the plane of Earth's orbit. Because of the tilt of Earth's axis, or **axial tilt,** the ecliptic is tilted 23½° to the celestial equator. The Moon and other planets also travel in orbits nearly parallel to the plane of Earth's orbit, so the Moon and planets always appear very close to the ecliptic.

Because of Earth's axial tilt, first one hemisphere, then the other faces the Sun more directly as Earth travels in its orbit (⊙ Fig. 19.7). At noon on December 21, the Sun shines most directly on the Southern Hemisphere and obliquely on the Northern. The Sun appears to be at the southernmost point on the ecliptic, a moment called the *winter solstice.* It is high in the sky for Southern Hemisphere observers and low for those in the Northern Hemisphere. The Southern Hemisphere has its summer and the Northern Hemisphere has winter. By March 21, Earth has moved 90° in its orbit. The Sun appears on the celestial equator. On Earth, the Sun is overhead at noon on the equator. Day and night are both 12 hours long everywhere on Earth. We call this the *spring*

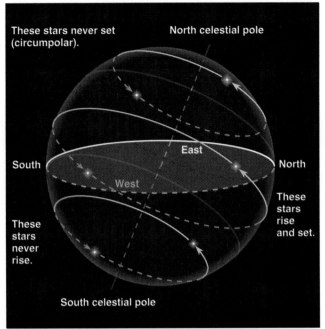

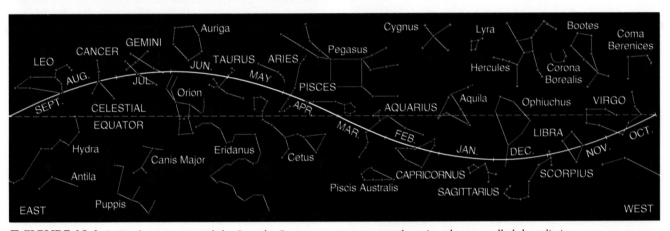

FIGURE 19.5 There is a zone around the north celestial pole where some stars never set and a corresponding zone around the south celestial pole where some stars never rise for Northern Hemisphere observers.

▣ **FIGURE 19.6** As Earth moves around the Sun, the Sun appears to trace a path against the stars called the *ecliptic*.

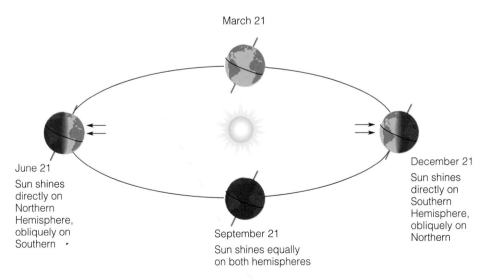

▣ **FIGURE 19.7** Because Earth's axis is tilted, first one hemisphere, then the other faces the Sun more directly as Earth travels in its orbit.

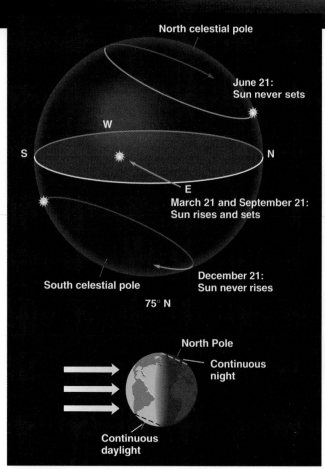

North celestial pole

June 21:
Sun never sets

March 21 and September 21:
Sun rises and sets

South celestial pole

December 21:
Sun never rises

75° N

North Pole

Continuous
night

Continuous
daylight

■ FIGURE 19.8 Above the Arctic Circle and below the Antarctic Circle, the Sun is sometimes a circumpolar star. We refer to continuous sunlight as the *midnight sun.*

Recall that everywhere except at the equator, there are zones around the celestial poles where some stars never rise and other stars never set. Stars that never set, as seen from a given latitude, are called *circumpolar.* Above the Arctic Circle and below the Antarctic Circle, these zones are large enough to include the Sun at certain times of the year (■ Fig. 19.8). On the Arctic Circle, the Sun never sets completely on June 21 and never rises completely on December 21 (the dates are reversed for the Antarctic Circle). Closer to the poles, the Sun is continuously above the horizon for a period in the summer, a condition called the *midnight sun.* There is a corresponding period in the winter when the Sun never rises. At the North Pole, the Sun rises at the spring equinox, reaches its highest altitude at the summer solstice, and sets at the fall equinox, not to reappear until the spring equinox. (The dates are reversed for the South Pole.) Only the poles have six months of continuous daylight or darkness. At other latitudes there are shorter periods of continuous daylight or

equinox (from the Latin for "equal night"). By June 21, Earth has moved 180° since the winter solstice. The Sun appears on the northernmost point of the ecliptic, shining most directly on the Northern Hemisphere and obliquely on the Southern. It is summer in the Northern Hemisphere and winter south of the equator. By September 21, Earth has moved another 90°. The Sun is once again on the celestial equator, and we have the *fall equinox.* The **equinoxes** are the dates when the Sun crosses the celestial equator. The **solstices** are the dates when the Sun is farthest north or south in the sky.

The Rarest Scheduled Event in History

Earth does not rotate a whole number of times as it orbits the Sun. Instead, it rotates 365.24219 times, or just a bit under 365¼ times. In four years, the calendar would be off by one day relative to the equinoxes, which is why we add an extra day every four years for leap year. This system was introduced by Julius Caesar over 2,000 years ago. But that leap year correction leaves an extra 0.00781 day unaccounted for. In A.D. 325, the Christian church defined the rules for determining the date of Easter. The date of Easter is based on the spring equinox, but by 1582, the discrepancy between the calendar and the seasons had grown to 10 days. Pope Gregory XIII appointed a commission to reform the calendar, which recommended that 1582 be shortened to bring the calendar back in line with the seasons. The missing days were simply dropped; October 4 was followed by October 15. (If you think many people were upset over paying a month's rent for a 21-day month, you're right!) The discrepancy amounts to an extra 3.1 days in 400 years, so the commission recommended that century years *not* be leap years unless they were divisible by 400. This system, called the Gregorian calendar, is still in use today. The year 1600 was a leap year, but 1700, 1800, and 1900 were not. The year 2000 *will* be a leap year, something that has not happened in 400 years. Leap Year Day in 2000 was scheduled before the Pilgrims landed. There is still a small discrepancy in the calendar that our remote descendants may wish to correct in several thousand years.

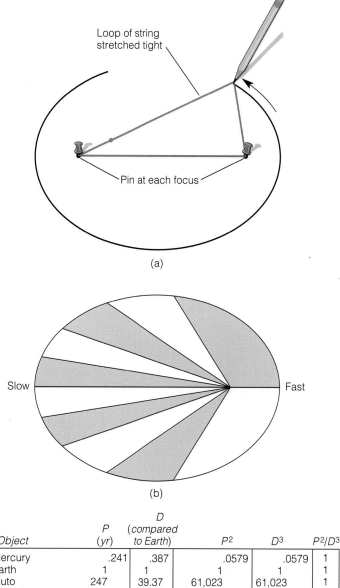

Object	P (yr)	D (compared to Earth)	P²	D³	P²/D³
Mercury	.241	.387	.0579	.0579	1
Earth	1	1	1	1	1
Pluto	247	39.37	61,023	61,023	1
Halley's comet	76.1	17.96	5,791	5,791	1
Comet Hale-Bopp	2,390	178.8	5,726,449	5,726,449	1

(c) A few examples of Kepler's third law

darkness, separated by periods of normal sunrises and sunsets.

Earth and the Moon, like all planets in the solar system, obey **Kepler's laws,** laws of planetary motion first discovered by Johannes Kepler in the seventeenth century (▣ Fig. 19.9):

- *Kepler's first law* is that planets and satellites travel in elliptical orbits, and their distance from their parent body varies. Earth's distance from the Sun varies from 147.1 to 152.1 million km (91.2 to 94.3 million mi). Somewhat surprisingly, Earth is nearest the Sun (*perihelion*) in January and farthest from the Sun (*aphelion*) in July. Thus, the variation in Earth's distance from the Sun has only a minor effect on climate compared to the tilt of Earth's axis.

- *Kepler's second law* holds that an imaginary line between a planet and the Sun (or a satellite and a planet) sweeps out equal areas in equal times. Thus, Earth travels around the Sun fastest at perihelion (30.3 km/sec) and slowest at aphelion (29.3 km/sec).

▣ FIGURE 19.9 Kepler's laws. (*a*) According to the first law, planets travel around the Sun in elliptical orbits, with the Sun at one focus. Shown here is one way to draw an ellipse. (*b*) According to the second law, a line joining the planet and the Sun sweeps out equal areas in equal times. (*c*) The third law holds that the square of an object's orbital period (*P*) and the cube of its distance (*D*) from its parent body are proportional.

- *Kepler's third law* holds that the farther a planet or satellite is from its parent body, the longer it takes to complete an orbit. First, the farther away an object is, the greater the circumference of its orbit. Also, the farther away an object is from its parent body, the weaker the gravitational pull on it. The exact relationship is that the *square* of the orbital period and the *cube* of the distance are proportional. A space probe orbiting the Sun at four times Earth's distance would take eight years to complete an orbit ($4 \times 4 \times 4 = 64 = 8 \times 8$).

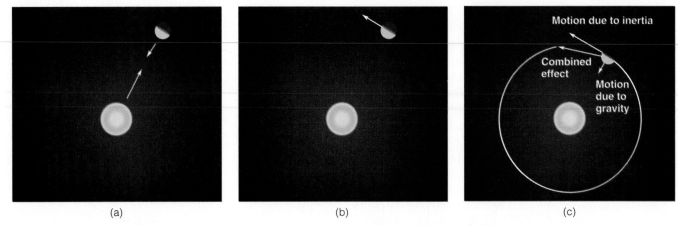

(a)   (b)   (c)

▣ **FIGURE 19.10** A planet orbits the Sun because of two opposing effects. (*a*) Left to itself, the force of gravity would pull the planet into the Sun. (*b*) On the other hand, the inertia of the planet would keep the planet moving in a straight line. (*c*) Gravity and inertia together cause the planet to travel in a fixed path, or orbit.

Half a century after Kepler discovered these laws, Isaac Newton discovered **gravity** and the fundamental laws of motion and **inertia** and showed that Kepler's laws were a consequence of these fundamental principles (▣ Fig. 19.10). Gravity alone would cause a planet to fall into the Sun. Inertia is the tendency of moving objects to remain in motion. Inertia alone would cause the planet to travel in a straight line. The combination of gravity and inertia causes planets to travel in elliptical orbits.

THE MOTIONS OF THE MOON AND ARTIFICIAL SATELLITES

Satellites travel in orbits around planets and obey Kepler's laws just as planets do. Earth and the other planets are simply satellites of the Sun. The Moon averages 388,400 km (240,808 mi) from Earth and moves in its orbit at just over 1 km (0.62 mi) per second. Just as Earth varies in distance from the Sun, the Moon varies in distance from Earth. At its nearest (**perigee**), it is 367,100 km (227,600 mi) away, and at its farthest (**apogee**), it is 409,700 km (254,000 mi) away. Since 1957, we have had the means of launching *artificial satellites* into orbit around Earth and the other planets. We do this by giving the satellite a high enough velocity that its inertia (tendency to travel in a straight path) just balances the gravitational force acting on it.

One practical application of Kepler's third law is of enormous commercial value. An artificial satellite orbiting just outside Earth's atmosphere takes 88 minutes (1.45 hours) to complete an orbit. The radius of its orbit is about 6,550 km (4,061 mi); in doing these calculations for planets, we calculate distances from the center of the planet. The Moon, at an average distance of 388,400 km (240,808 mi), takes about 27.9 days (669 hours) to complete one orbit. Somewhere between 6,550 and 388,400 km is an orbit with a period of 24 hours. Its radius is 42,239 km (26,188 mi)—35,868 km (22,238 mi) above Earth's surface. Such an orbit is called *geo-synchronous;* an artificial satellite in geosynchronous orbit keeps pace exactly with Earth's rotation (▣ Fig. 19.11). Geosynchronous weather satellites always look down on the same part of Earth. As seen from Earth, geosynchronous satellites always appear to be at the same spot in the sky. The satellite television antennas we see everywhere are all aimed at specific artificial satellites about 36,000 km (22,320 mi) in space. It is important to understand that geosynchronous satellites are *not* stationary; they are moving at 3 km (1.86 mi) per second, or 11,000 km (6,820 mi) per hour, just fast enough to keep pace with Earth's rotation.

THE MOON IN EARTH'S SKY

The Moon and all other planets are visible only by reflected sunlight. The Moon is bright because it is nearby and large in the sky, but actually its surface is quite dark, about as dark as an asphalt parking lot. As the Moon orbits Earth, we see varying portions of its surface lit by the Sun. We call the changing appearance of the Moon from night to night the **phases** of the Moon (▣ Fig. 19.12). When the Moon is between Earth and the Sun, we cannot see its illuminated side at all, and the Moon becomes invisible for a time. This phase is *new moon.* A day or so after new moon, the Moon becomes visible as a slender crescent, which grows from night to night. About a week after new moon, the Moon has moved one-quarter of the way around its orbit, and we see exactly half of the moon illuminated. This is *first-quarter Moon.* After first quarter, the Moon appears more than half illuminated, or *gibbous.* Finally, about two weeks after new moon, the Moon is opposite the Sun in the sky, and we see *full moon,* with the entire illuminated side of the Moon visible. The period from new moon to full moon, when the visible portion of the Moon grows from night to night, is called *waxing.* After full moon, the Moon is *waning,* as its visible portion shrinks from night to night. After going through a waning

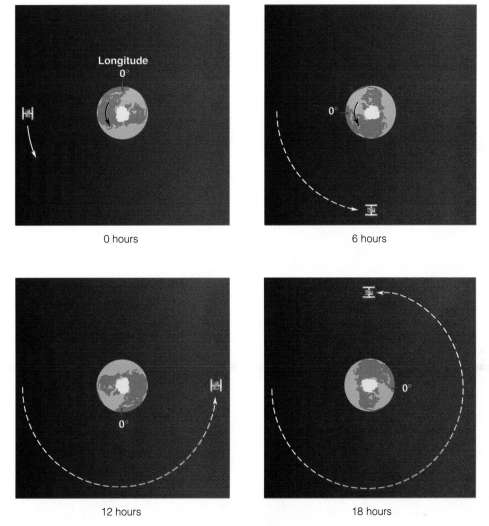

FIGURE 19.11 A satellite in geosynchronous orbit keeps pace exactly with Earth's rotation. The satellite moves faster than a point on Earth's surface because it travels a longer path.

gibbous period, *last-quarter Moon* occurs about a week after full moon, and we see half of the Moon illuminated. After last-quarter moon, the Moon appears as a waning crescent, until finally the Moon is once again between Earth and the Sun and the next cycle of phases begins. Although the Moon takes about 27 days to complete an orbit (perigee to perigee), the phases of the Moon take 29.5 days to repeat. The reason is that during the month Earth is also moving around the Sun, and it takes an additional two days for the Moon to catch up to Earth's orbital motion.

During crescent moon phase, the dark side of the Moon is not completely invisible. Earth is almost four times the diameter of the Moon and reflects a much larger percentage of sunlight into space. When the Moon is a crescent, Earth as seen from the Moon is almost full and more than 20 times brighter than a full moon. We can see the dark side of the Moon by reflected earthlight, or *earthshine* (■ Fig. 19.13). More poetically, earthshine is sometimes called *the new moon in the old moon's arms.*

Occasionally, the Moon actually passes in front of the Sun, and a **solar eclipse** occurs. Solar eclipses can happen only at new moon. At other times, the Moon passes through Earth's shadow, and a **lunar eclipse** occurs. Lunar eclipses can occur only at full moon. If the Moon's orbit coincided exactly with the plane of Earth's orbit, there would be eclipses at every new and full moon. The plane of the Moon's orbit is tilted by about 5°, however, and most of the time the Moon misses the Sun or Earth's shadow (■ Fig. 19.14). The Moon must be on the ecliptic for an eclipse to occur; in fact, that is the origin of the term *ecliptic*. About every six months, the plane of the Moon's orbit is closely enough aligned with the Sun that eclipses become possible. There can be anywhere from two to seven solar and lunar eclipses in a year.

If you hold your hand a few inches from a flat surface, you see that its shadow has two parts: a light outer zone, the *penumbra,* where only part of the light is blocked by your hand, and a dense inner portion, the *umbra,* where light is

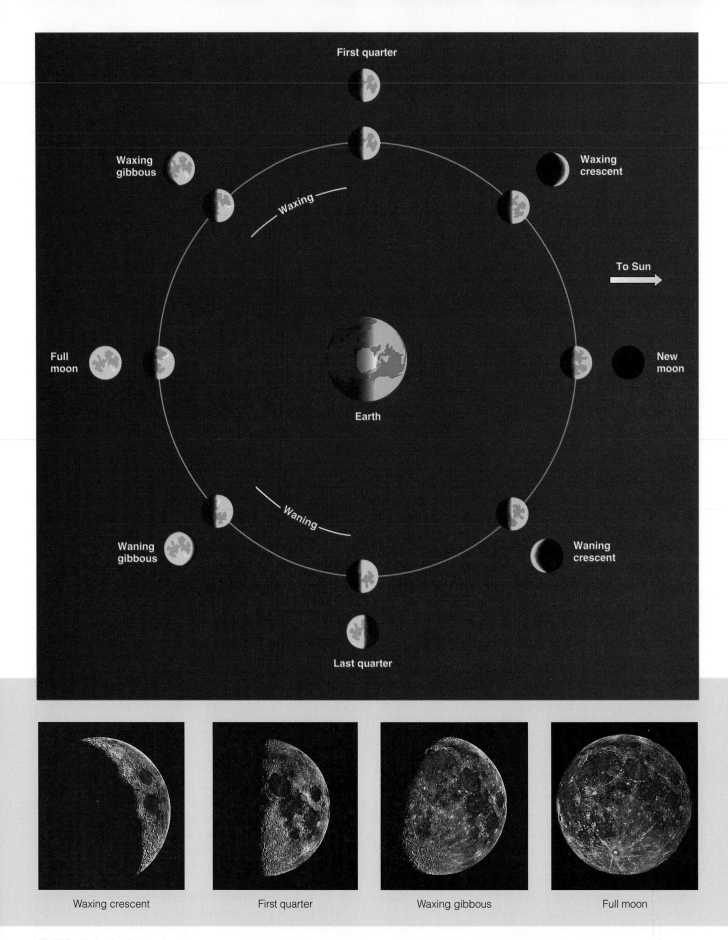

FIGURE 19.12 The phases of the Moon result from the Moon showing different amounts of its illuminated face as it orbits Earth.

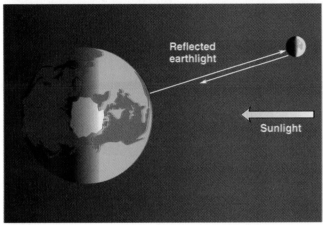

■ FIGURE 19.13 We can see the dark side of the crescent moon by reflected earthlight, or *earthshine*.

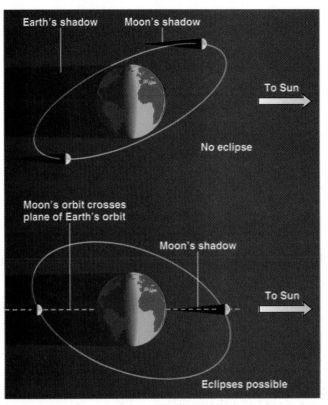

■ FIGURE 19.14 Most of the time, because of the tilt of its orbit, the Moon misses the Sun or Earth's shadow.

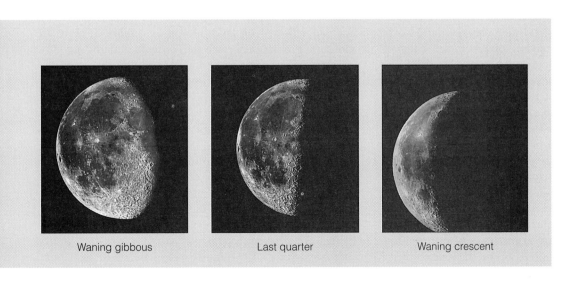

Waning gibbous   Last quarter   Waning crescent

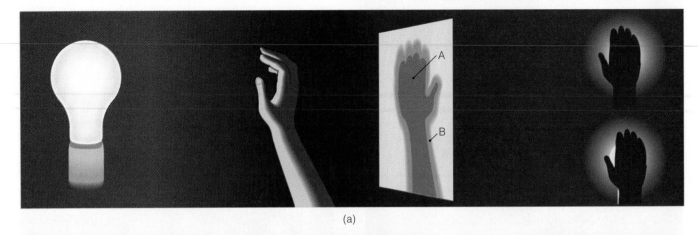

(a)

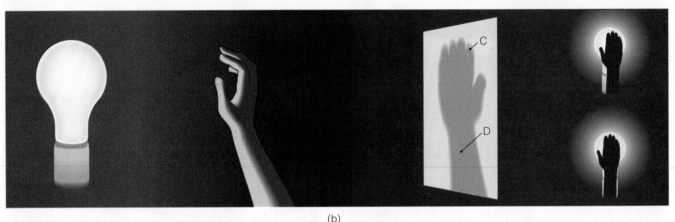

(b)

◼ **FIGURE 19.15** (*a*) If you hold your hand a few inches from a flat surface, you see that its shadow has two parts. The dark inner portion, the *umbra,* is where your hand blocks all light. In the lighter outer shadow, or *penumbra,* only some light is blocked. A fly at A would see a total eclipse of the lightbulb, but one at B would see only a partial eclipse. (*b*) The umbra of a shadow has limited length. If your hand is too far away, it is not big enough to cover the light and your shadow has no umbra. A fly at C would see a partial eclipse of the lightbulb. One at D would see your hand silhouetted against the light, an *annular eclipse.*

completely blocked (◼ Fig. 19.15). (The umbra is not completely dark because light reflected from other surfaces in the room still reaches it.) The shadows of Earth and the Moon also have an umbra and penumbra.

A solar eclipse is the Moon's shadow falling on the Earth (◼ Fig. 19.16). In the penumbra, observers see a *partial eclipse,* with the Moon obscuring only part of the Sun. Sometimes only the penumbra of the Moon's shadow strikes Earth (Fig. 19.16a). In most cases, the umbra also sweeps across Earth, and observers in a narrow zone see the Moon pass directly across the face of the Sun (Fig. 19.16b). As a solar eclipse progresses, the Moon first appears as a notch on the edge of the Sun. Most observers see only a partial eclipse; the notch grows in size for an hour or so, then decreases again. Unless the Moon covers more than 80% of the Sun, many people will not be aware an eclipse is happening at all. In the umbra, the Moon covers the Sun completely, creating a *total eclipse.* Very few spectacles in nature rival total eclipses of the Sun for beauty. As long as the smallest piece of the Sun is visible, it is bright daylight. In the last few seconds, the sunlight starts to fade rapidly; then *the sun goes out.*

When the last speck of Sun vanishes, the light goes from daylight to deep twilight in seconds (◼ Fig. 19.17). The sky becomes dark blue; Venus, other planets, and bright stars become visible. Around the horizon is a yellow glow from sunlight reflected off land outside the umbra. If there are clouds in the sky, they catch the scattered sunlight and glow brightly against the dark sky. The Moon is silhouetted as a perfectly round, black disk, surrounded by the Sun's pale outer atmosphere, or *corona* (Chapter 21). Total eclipses of the Sun last anywhere from a few seconds to a bit under eight minutes, although a duration of two or three minutes is about average.

In many eclipses, the Moon is too far away to cover the Sun completely. The umbra of the Moon is a narrow cone that does not quite touch Earth (Fig. 19.16c). Instead, observers see a narrow ring of sunlight surrounding the Moon. Such an eclipse is called *annular,* from the Latin word for "ring." An annular eclipse was visible across the United States on May 10, 1994. In a few cases, the tip of the Moon's umbra just nicks Earth. Observers who see the eclipse at sunrise or sunset see an annular eclipse, but those who see it

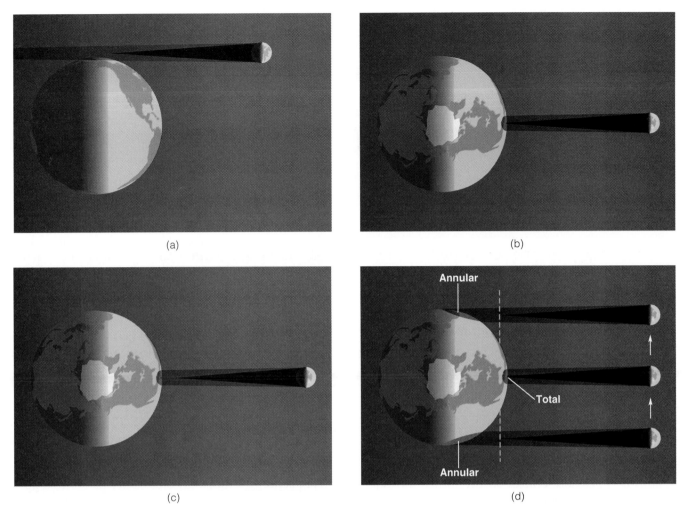

(a)

(b)

(c)

(d)

🔲 FIGURE 19.16 A solar eclipse occurs when the Moon's shadow falls on the Earth. (*a*) In some cases, only the penumbra touches Earth, and all observers see a partial eclipse. (*b*) When the umbra sweeps across Earth, viewers in a narrow zone see a total eclipse, while everyone else sees a partial eclipse. (*c*) When the Moon is too far away for the umbra to reach Earth, observers in a narrow zone see an annular eclipse, with the Moon surrounded by a narrow ring of sunlight. (*d*) Sometimes the umbra just nicks Earth. Observers at sunrise and sunset, who are 6,400 km (3,968 mi) farther away from the Moon, see an annular eclipse, while noontime observers see a very short total eclipse.

🔲 FIGURE 19.17 A total eclipse of the Sun is one of the most spectacular phenomena in nature. When the last speck of Sun vanishes, the light goes from daylight to deep twilight in seconds. The Sun goes out at midday. No matter how scientifically prepared you are, you understand for an instant why ancient peoples were terrified by eclipses. The white halo around the Moon is the Sun's corona (Chap. 21).

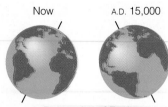

Precession of spinning top   Precession of Earth

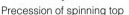 **FIGURE 19.19** The gravity of the Moon, Sun, and other bodies in the solar system try to tilt Earth's rotation axis and cause Earth to precess.

 FIGURE 19.18 Earth's umbra is big enough to cover the entire Moon. This multiple exposure shows the Moon's passage through the umbra during a total lunar eclipse. The telescope followed the apparent rotation of the stars while the Moon moved through the umbra. Note the red color of the eclipsed Moon.

at midday see a very short total eclipse (Fig. 19.16d). An eclipse of this sort was visible in the United States on May 30, 1984.

The Moon's umbra is barely able to reach Earth, but Earth's umbra is big enough to cover the entire Moon (Fig. 19.18). When the Moon is entirely covered by Earth's umbra, a *total lunar eclipse* occurs. Lunar eclipses are actually rarer than solar eclipses, but because Earth's shadow is cast on the Moon itself, everyone who can see the Moon can see the eclipse. Lunar eclipses are visible over more than half of Earth, compared to less than 20% of Earth's surface for partial solar eclipses and much less than 1% for total or annular eclipses. Sometimes the Moon passes through the edge of the umbra, and a *partial lunar eclipse* occurs. The Moon can also pass only through Earth's penumbra. Observers on the Moon would see the Sun partially eclipsed by Earth, but such *penumbral lunar eclipses* rarely darken the Moon enough to be noticeable to observers on Earth.

During total lunar eclipses, the Moon never becomes completely invisible. Sunlight refracted through Earth's atmosphere reaches the Moon, giving it a coppery color. An observer on the Moon would see Earth surrounded by a red-orange ring of refracted sunlight. (This is not an annular eclipse. Earth hides the Sun completely, but Earth's atmosphere is illuminated by the Sun.) The brightness of the Moon varies depending on how cloudy Earth's atmosphere is. Some of the darkest lunar eclipses occur when volcanic ash high in the atmosphere blocks sunlight. In some cases the Moon is so dark it is invisible to the unaided eye, though it is visible through a telescope or binoculars.

HOW EARTH AND THE MOON AFFECT EACH OTHER

Most people have seen the way a spinning top wobbles if it is not perfectly vertical. Any time a force (in this case, gravity) tries to change the axis of a spinning body (the top), it creates a twisting motion called **precession.** Similarly, the gravity of the Moon, Sun, and other bodies in the solar system try to tilt Earth's rotation axis and cause Earth to precess (Fig. 19.19). Earth's axis traces out a circle over a period of 25,770 years. As the axis precesses, the celestial pole appears to move in the sky. Currently, a bright star (Polaris) happens to be near the north celestial pole, but there is no corresponding star near the south celestial pole. For the next hundred years, the north celestial pole will move closer to Polaris, approaching closest in 2102, but will never hit it exactly. A thousand years ago, Polaris was not very close to the celestial pole, and a thousand years from now, it will not be very close. Other stars come and go as pole stars as precession continues.

Kepler's laws would hold exactly in an ideal solar system with a Sun and one planet. In our solar system, all the planets affect one another. The shape of Earth's orbit actually varies slightly over a cycle of about 100,000 years. The tilt of Earth's axis varies by about a degree over a period of about 41,000 years. Precession changes the orientation of Earth's axis over a cycle of about 26,000 years. Earth is at perihelion now during Northern Hemisphere winter; in 13,000 years it will be at perihelion during Northern Hemisphere summers. These periodic changes, called *Milankovitch cycles,* affect the amount of sunlight reaching Earth and can affect climate (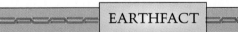 Fig. 19.20).

EARTHFACT

The World's Slowest Machines

During the Middle Ages and Renaissance, the showpiece of a city was often an elaborate mechanical clock in the city hall or cathedral. Many of these clocks included displays of the movements of the Sun, Moon, and stars, and a few even included gears to display precession. At one revolution in almost 26,000 years, these are the slowest machines ever built.

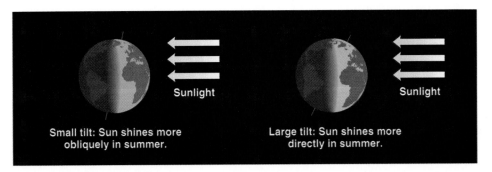

Small tilt: Sun shines more obliquely in summer.

Large tilt: Sun shines more directly in summer.

(a)

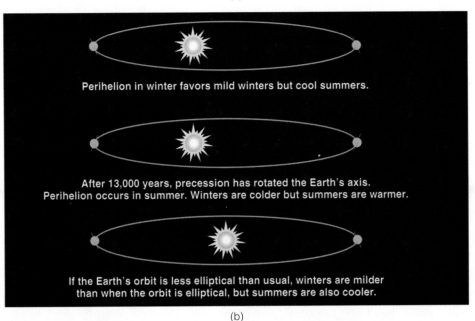

Perihelion in winter favors mild winters but cool summers.

After 13,000 years, precession has rotated the Earth's axis. Perihelion occurs in summer. Winters are colder but summers are warmer.

If the Earth's orbit is less elliptical than usual, winters are milder than when the orbit is elliptical, but summers are also cooler.

(b)

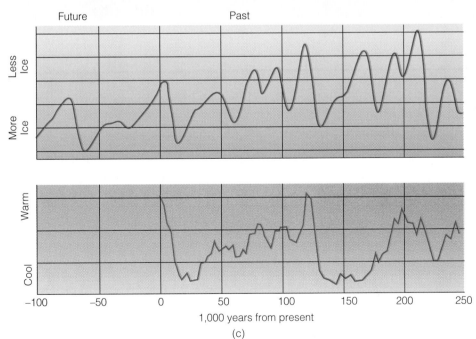

(c)

■ FIGURE 19.20 Milankovitch cycles. Changes in the shape of Earth's orbit, the tilt of Earth's axis, and the season of perihelion (determined by precession) result in periods of warmer or cooler climate. Cool summers favor ice ages. (*a*) Larger axial tilt favors warmer summers. (*b*) Perihelion in summer favors warmer summers, especially if the orbit is more elliptical than usual. (*c*) Calculated climatic effects of Milankovitch cycles compare well with the record of recent ice ages.

It doesn't matter how cold winter is, as long as summer is warm enough to melt the preceding winter's snow. Cool summers, rather than cold winters, favor ice ages. A large axis tilt results in cold winters, but also hot summers. Small axis tilts, on the other hand, favor cool summers. Aphelion in summer favors cool summers, especially if Earth's orbit is more elliptical than usual. The predicted climatic effects of Milankovitch cycles agree well with the actual record of recent ice ages, but Milankovitch cycles must have operated throughout Earth's history, even during ice-free periods. They certainly contribute to ice ages, but they cannot be the only cause.

 **TIDES**

Earth pulls more strongly on the near side of the moon than the far side, with interesting results. The Moon's radius is 1,737 km (1,077 mi). According to Kepler's third law, an object on the near side of the Moon should orbit Earth a little faster than an object on the far side—about 4 m per second (9 mph) faster (■ Fig. 19.21a). Because the Moon is a solid body, however, an object on the near side of the Moon is moving in a slightly smaller path than an object on the far side, and is moving more slowly—about 10 m per second

■ **FIGURE 19.21** Tidal bulges. (*a*) An object on the near side of the Moon should orbit Earth a little faster than an object on the far side—about 4 m per second (9 mph) faster. (*b*) Because the Moon moves as a solid body, an object on the near side travels a slightly shorter path than an object on the far side and moves about 10 m per second (22 mph) slower. (*c*) Because the near side of the Moon is moving slower than Kepler's third law requires, it bulges toward Earth. Because the far side is moving faster, it bulges away. (*d*) Earth and the moon actually orbit around the center of mass of the Earth-Moon system, a point about 1,650 km (1,023 mi) beneath Earth's surface. (*e*) The Moon's gravity attracts the oceans toward the Moon. The motion of Earth around the center of mass of the Earth-Moon system throws a bulge up on the opposite side of Earth from the Moon. The combination of the two effects creates two tidal bulges.

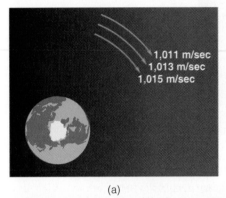

(a)

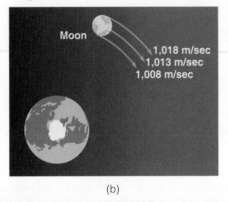

(b)

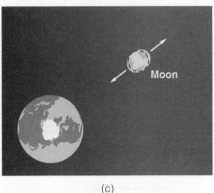

(c)

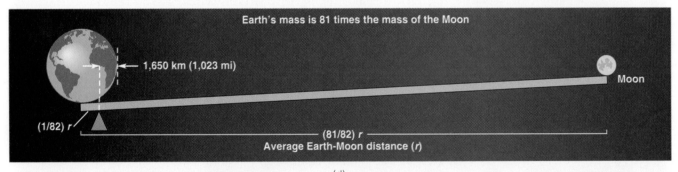

(d)

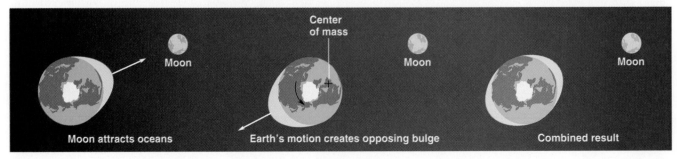

(e)

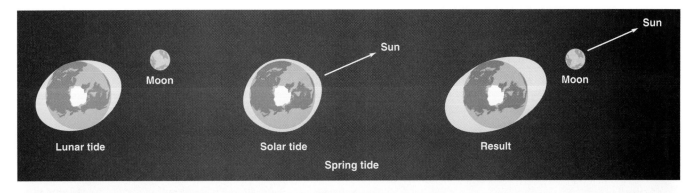

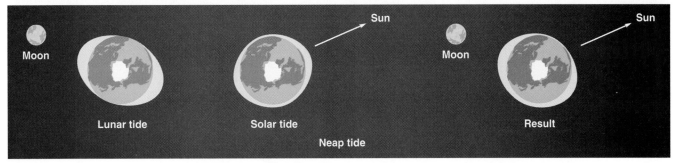

◻ FIGURE 19.22 At new or full moon, the tidal bulges created by the Sun and Moon are superimposed. At first and last quarter, the Sun's low tide partially cancels out the Moon's high tide and vice versa.

(22 mph) slower (Fig. 19.21b). Thus, the near side of the Moon is moving a bit slower than Kepler's laws require, and the far side a bit faster. Without the gravity of the Moon, an object on the near side would fall into a lower orbit, and one on the far side would fly off into a higher orbit. There is a tension here between two physical principles, and the tension is *real*—the Moon is pulled into an oval shape with its long axis pointing at Earth. There is a bulge facing Earth because the near side of the Moon is moving a bit slower than Kepler's laws require (Fig. 19.21c). Somewhat surprisingly, there is a bulge on the *opposite* side of the Moon as well, because the far side of the Moon is moving a bit faster than Kepler's laws require.

The effects of gravity acting with different force on opposite sides of a planet are called **tidal forces,** and they are very important. Because of tidal forces, Earth's rotation is slowing, the Moon is drifting farther from Earth, and the Moon always shows the same face to Earth. In the next chapter, we will see how the immense tidal forces created by Jupiter's gravity cause volcanoes on one of its moons and shattered a comet into pieces.

Just as Earth deforms the Moon into an oval shape, the Moon does the same thing to Earth. It's not exactly accurate to say the Moon orbits Earth. Both actually orbit around the *center of mass* of the Earth-Moon system, which is about 1,650 km or a bit over 1,000 miles beneath Earth's surface (Fig. 19.21d). The Moon's gravity attracts the oceans on the near side of Earth, but Earth's motion about the center of mass throws up a bulge on the opposite side as well (Fig. 19.21e). Thus, there are two tidal bulges, one facing Moon, the other opposite it. Because the Moon moves ap-

preciably during a day, and the tidal bulges follow the Moon, successive high tides are actually 12 hours and 50 minutes apart on the average. The Moon also deforms the solid earth by about 30 cm (1 ft). This *earth tide* can be detected with accurate measuring instruments.

The Sun also creates tides on Earth. The Sun is far more massive than the Moon, but much farther away, and its tidal effect is about half that of the Moon. At new or full moon, the tidal bulges created by the Sun and Moon are superimposed (◻ Fig. 19.22). High tides are higher than normal, and low tides are lower, a condition called *spring tide*. At first- or last-quarter moon, the Sun's high tide is superimposed on the Moon's low tide and vice versa. High tides are unusually low and low tides unusually high, a condition called *neap tide*. Neap tide was what created a near-disaster during the Battle of Tarawa (see the Prologue). If the Moon is near perigee, the lunar tidal range is greater. And if Earth is near perihelion, the solar tidal range is greater. At the Boston Tea Party, both factors combined with a spring tide to cause an unusually low tide. At Tarawa, the Moon was near apogee (small lunar tidal range), and Earth was near perihelion (large solar tidal range); these factors combined with neap tide to cause the tides almost to vanish.

If Earth were smooth and covered with deep seas, the tidal bulges would remain aligned with the Moon. In shallow seas, however, friction with the bottom interferes with tidal movements. The continents also interfere with the tides, creating a complex pattern (◻ Fig. 19.23). In most of the Atlantic Ocean, tides move from south to north. In the Pacific and Indian Oceans and parts of the North Atlantic, tides pivot around so-called *nodal points* like spokes of a

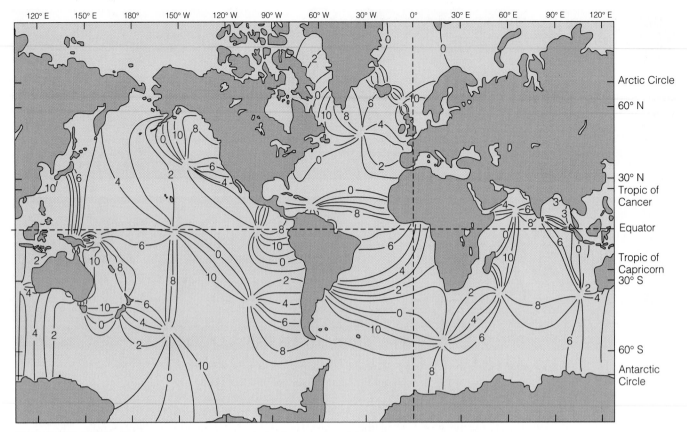

■ **FIGURE 19.23** The continents interfere with the free motion of the tides, creating a complex pattern. In this diagram, lines show the positions of high tide crests at hourly intervals.

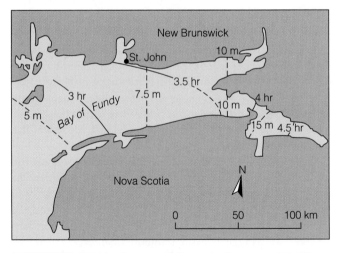

■ **FIGURE 19.24** The actual effects of tides are modified by local conditions. It takes an hour for the tides to travel up the Bay of Fundy, Canada (as shown by the solid lines), and the tidal range becomes greater as the tides move up the bay (as shown by the dashed lines).

wheel. Islands near nodal points, like Hawaii, have very little tide.

The actual effects of tides are modified by local conditions (■ Fig. 19.24). In enclosed bays, it may take hours for the tides to reach into the remotest parts of the bay. For this reason, tide prediction tables have to be created specifically for each desired location. Tides may be unusually high in en-closed bays where the incoming tide pushes water into the bay. In the Bay of Fundy, Canada, tidal ranges of 15 m (49 ft) occur. Weather conditions can also influence tidal effects. A spring tide combined with a strong wind from offshore is likely to result in coastal flooding.

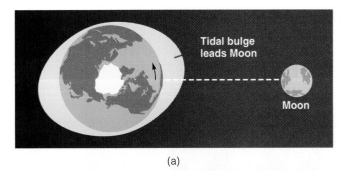

(a)

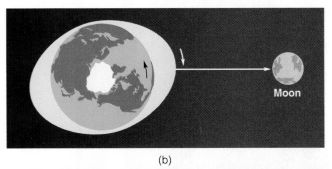

(b)

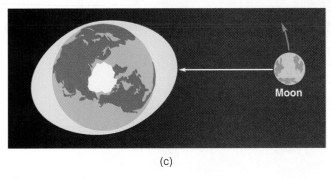

(c)

FIGURE 19.25 Tidal interactions of Earth and the Moon. (*a*) Friction drags tidal bulges ahead of the Moon. (*b*) The Moon attracts the near tidal bulge more strongly, braking Earth's rotation. (*c*) The near tidal bulge also attracts the Moon, accelerating it into a higher orbit.

The inability of tides to move freely affects the motions of Earth and the Moon in very important ways. Friction with the rotating Earth drags the tidal bulges along ahead of the Moon (Fig. 19.25a). The Moon's gravity pulls more strongly on the near bulge than the far one, thus braking Earth's rotation (Fig. 19.25b). Daily and annual growth rings in fossil marine organisms indicate that during the Cambrian Period, about 550 million years ago, Earth rotated about 400 times per year, or once in 22 hours. Earth's slowing has not been uniform over time, however. Periods of widespread shallow seas like the Ordovician or Cretaceous have been more effective at slowing the rotation than periods with few shallow seas like the present. Nevertheless, tidal slowing is observable, and extremely accurate timekeeping systems take it into account. Over a few thousand years, the total discrepancy can add up to many hours, as records of ancient eclipses show. Thus, there is no point in trying to create a calendar system more accurate than our present one.

By the time discrepancies in our calendar become significant, tidal slowing of Earth's rotation will have an equally great effect.

The Moon also deforms the solid earth and Earth deforms the solid moon. As Earth rotates, the bulge in the solid earth is constantly moving, and friction created as Earth deforms contributes to the slowing of Earth's rotation. Earth, which is 81 times as massive as the Moon, has had a much greater effect on the Moon's rotation. Earth has *tidally locked* the Moon's rotation so that it always shows the same face to Earth. (An observer on another planet would see all sides of the Moon as it orbits Earth. Thus, the Moon *does* rotate, but it rotates once a month.) Tidal forces have similarly locked the rotations of other planetary satellites as well.

Earth's tidal bulges also exert a gravitational pull on the Moon. The near bulge pulls more strongly on the Moon than the far bulge, pulling the Moon ahead in its orbit and throwing it into a higher orbit (Fig. 19.25c). Calculations indicate that the slowing of Earth's rotation and the increasing distance of the Moon will balance several billion years in the future when the Moon's period and Earth's rotation both take about 40 days. Earth's rotation will then also be tidally locked to the Moon.

THE MOON

Through telescopic observations, we can see two types of terrain on the Moon: dark, flat plains, which early astronomers thought were seas and called **maria** (singular, *mare*) from the Latin word for sea, and bright, heavily cratered **highlands** (Fig. 19.26). The dark maria make up the familiar features of the "man in the moon." The highlands are heavily cratered and older than the maria. In places we can still see buried craters within the maria, but only a few craters have formed later than the maria. From this evidence, we can infer that the Moon had an early history of very intense cratering, followed by the formation of the maria, and finally a much lower rate of cratering.

Apollo astronauts left seismometers on the Moon that monitored lunar seismicity for many years. On the return to Earth, the astronauts launched their cast-off booster stages on paths that carried the boosters to impacts on the far side of the Moon, so the deep interior of the Moon could be probed. A few impacts of meteors weighing up to a few tons have also been recorded. The seismic studies showed that the Moon has a highly fractured outer layer several kilometers thick, a crust perhaps 100 km (60 mi) thick, and a core about 740 km (450 mi) in diameter (Fig. 19.27). The lunar core is not very dense and lacks a sharp boundary. The Moon has no tectonic activity except for very tiny (magnitude 1 or less) "moonquakes" that occur at depths of about 1,000 km (600 mi). The moonquakes are caused by tidal forces exerted by the Earth.

The *Apollo* missions to the Moon brought back samples that could be dated by radiometric means and studied by

■ **FIGURE 19.26** We can see two types of terrain on the Moon: dark, flat plains, which early astronomers thought were seas and called *maria* (singular, *mare*) from the Latin word for sea, and bright, heavily cratered *highlands*. This false-color image taken by the *Galileo* spacecraft combines visible light, infrared, and ultraviolet into a single image. The maria show mostly as dark blue and occasionally orange, while the highlands show as red.

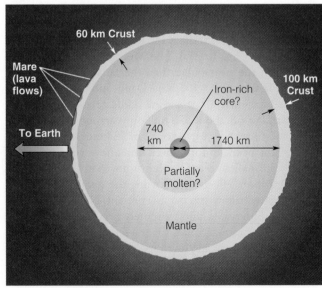

■ **FIGURE 19.27** The Moon has a highly fractured outer layer several kilometers thick, a crust of anorthosite perhaps 100 km (60 mi) thick, and a core about 740 km (450 mi) in diameter.

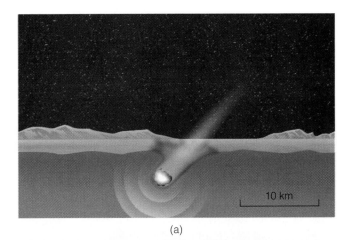

(a)

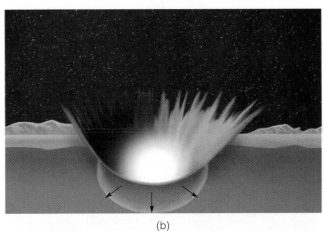

(b)

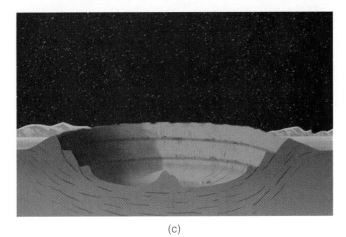

(c)

⬛ **FIGURE 19.28** How impact craters form. (*a*) The impacting object plows deep into the planet before coming to a stop. The kinetic energy of the meteor is converted into heat, and the meteor vaporizes. (*b*) The hot gases expand explosively, hurling material out of the crater to form an ejecta blanket and creating a tremendously powerful shock wave within the crust. (*c*) After the shock wave passes, the compressed crust may rebound to create a central peak. The events depicted here take place in seconds.

other techniques. The highland areas are largely made of a light, plagioclase-rich rock called *anorthosite*, which probably makes up most of the lunar crust. The highland rocks, which are about 4.5 billion years old, date back almost to the formation of the solar system. The maria are just what earthbound observers suspected: basalt plains. The maria are also very old—over 3 billion years. The circular shapes and mountain rims of many of the maria are the remnants of huge craters. The Moon's mountains are not at all like the mountains of Earth. They are the remnants of enormous craters, not products of plate tectonics. Clearly, to understand the Moon and other bodies in the solar system, we must understand impact, a process that is relatively uncommon on Earth.

IMPACT: A MAJOR GEOLOGIC PROCESS ON OTHER PLANETS

Our exploration of the planets has shown that meteor **impact,** though unfamiliar on Earth today, has been fundamental in the history of the solar system. Really large meteoroids strike Earth so rarely that none have hit in recorded history. The most recent great impact occurred about 20,000 years ago and formed Meteor Crater in Arizona.

A large meteoroid or small asteroid strikes a planet at many thousands of kilometers per hour. Any moving object has energy, called *kinetic energy,* which can be found from the following formula:

$$\text{kinetic energy} = \tfrac{1}{2} \, \text{mass} \times \text{velocity}^2$$

Small meteoroids up to a meter or so in diameter lose their energy through friction with the atmosphere. Most of them vaporize high above Earth and shine briefly as meteors or "shooting stars." Meteoroids up to a few meters across survive to reach the surface, but they are slowed down so much by the atmosphere that they do little damage when they hit. Large meteoroids, tens of meters across and larger, are not slowed significantly by an atmosphere. A meteoroid 10 m (30 ft) across, striking Earth at 20 km per second (45,000 mph), a typical speed, would have as much energy as a small atomic bomb. A meteoroid 1 km across (3,300 ft) striking Earth would release far more energy than is contained in all Earth's nuclear arsenals.

When a large meteoroid strikes (⬛ Fig. 19.28), all its kinetic energy transforms into heat. On worlds with no atmospheres, even microscopic meteoroids strike with enough energy to melt rock and form craters (⬛ Fig. 19.29). (You can duplicate the process of turning kinetic energy into heat on a small scale. Hold a nail against a hard surface and hit it hard several times with a hammer. It will become quite warm.) The heat released after a large impact vaporizes or melts the impacting meteor and a good deal of the surrounding rock. The vaporized rock expands violently and produces an enormous explosion, which excavates a **crater.**

Craters are always much larger than the impacting body that forms them—typically 20 to 30 times larger. The crater

The Tunguska Event

The largest meteoroid impact in recorded history took place in Siberia on June 30, 1908. The impact generated a shock wave recorded thousands of kilometers away and flattened trees for many kilometers around, but did not produce a crater (☐ Fig. 1).

The absence of a crater has spawned many bizarre theories: a collision with a mini–*black hole* or chunk of *antimatter,* or the explosion of a crippled alien spacecraft. The real explanation seems to be that the meteoroid was a relatively fragile rocky object that fragmented in the atmosphere several kilometers above the ground. Apart from heating due to friction, an entering meteoroid ex-

periences enormous stresses; entering Earth's atmosphere at many kilometers per second is like diving face first off a multistory building or driving a car into a bridge abutment at high speed. Had the meteoroid remained intact, it would have created a sizable crater, but when it fragmented, the fragments were small enough to vaporize instantly as meteors. Thus, all the kinetic energy of the meteor was released in seconds, not on the surface, but above it. Such a blast has all the characteristics of a large nuclear weapon except nuclear radiation.

Interestingly enough, the largest impact in recorded history to produce a crater also occurred in Siberia. On

☐ FIGURE 1 The largest historic impacts. (*a*) The Tunguska impact of 1908 flattened trees for many kilometers but left no crater because the incoming meteor vaporized a few kilometers above the surface. (*b*) Map showing the locations of the 1908 Tunguska and 1947 Sikhote-Alin impacts.

(a)

☐ FIGURE 19.29 On bodies with no atmosphere, even the tiniest meteors strike with enough energy to melt rock. In this picture a micrometeoroid has struck a lunar rock sample, melting some of it. The entire area of this picture is too small to see with the un-aided eye. An impact this tiny would not penetrate the outermost layer of a spacesuit and thus would not endanger an astronaut.

February 12, 1947, a large meteoroid struck eastern Siberia. This meteoroid also broke up in flight, but the fragments reached the surface and made over 100 craters up to 26 m (85 ft) across. Recently declassified military satellite data show that meteoroids with energy equiva-lent to small nuclear weapons (1,000 tons or more of high explosives) strike Earth and vaporize in the atmosphere about once a month on the average. As seen from the ground these meteoroids briefly rival the Sun in bright-ness.

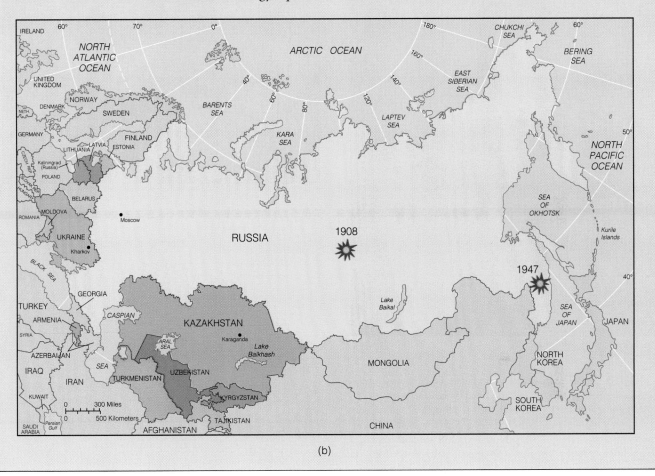

(b)

forms in seconds as the meteoroid stops and its kinetic energy turns into heat (Fig. 19.28a). The path of the meteor before impact has almost no effect on the shape of the crater, thus craters are round regardless of the angle of impact. Only the most grazing impacts produce oval craters. The impacting object is largely destroyed by the impact. Following the impact, great quantities of material are hurled out of the crater and land around it, producing an *ejecta blanket,* which is thickest near the crater rim (Fig. 19.28b). The ejecta blanket often is much lighter in color than the older rock on the surface, so new craters can be easily identified by a pattern of bright *rays* around them. Within the crust of the planet, the shock wave from the impact travels outward and subjects the rocks to enormous and very sudden forces. Rocks at the rim of the crater are turned upward and even overturned. After the shock wave passes, the compressed floor of the crater may rebound to raise a **central peak** (Fig. 19.28c).

On Earth, new craters are immediately attacked by erosion. A young crater like Meteor Crater (⬛ Fig. 19.30a) still has an obvious basin, upturned rim, and ejecta blanket. Older craters can be recognized only by geologic mapping and by signs of violent disruption of the rocks (Fig. 19.30b and c). On planets and other bodies without atmospheres, craters are scarcely modified by erosion. Small craters 1 to 10 km (0.6–6 mi) across are similar to Meteor Crater (⬛ Fig. 19.31a). Craters 10 to 100 km (6–62 mi) across may have unstable walls that slump downward along curving faults to produce terraces (Fig. 19.31b). Craters larger than about 50 km (31 mi) across usually have central peaks as well. Very large impacts, involving asteroids kilometers across, form **multiple-ring impact basins** hundreds of kilometers in

(a)

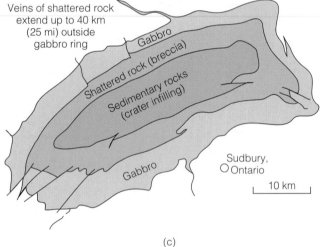

(b)

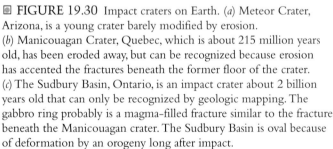

(c)

■ FIGURE 19.30 Impact craters on Earth. (*a*) Meteor Crater, Arizona, is a young crater barely modified by erosion. (*b*) Manicouagan Crater, Quebec, which is about 215 million years old, has been eroded away, but can be recognized because erosion has accented the fractures beneath the former floor of the crater. (*c*) The Sudbury Basin, Ontario, is an impact crater about 2 billion years old that can only be recognized by geologic mapping. The gabbro ring probably is a magma-filled fracture similar to the fracture beneath the Manicouagan crater. The Sudbury Basin is oval because of deformation by an orogeny long after impact.

diameter (Fig. 19.31c). These take the form of a series of concentric fault scarps. Often these fractures extend deep enough to form conduits for magma, as happened on the Moon. On the icy worlds of the outer solar system, impacts have also been common. Although these worlds are made of ice instead of rock, the impact craters on them are almost identical in form to those on rocky planets.

Craters are a valuable source of information on the geologic history of a planet (Fig. 19.30). First, the *principle of superposition* (Chapter 11) indicates that a crater must be younger than any crater it overlaps and older than any crater that overlaps it (■ Fig. 19.32a). Second, *crater degradation* gives us a rough idea of the age of craters (Fig. 19.32b). A

very young crater is bright and surrounded by bright rays. Older craters lack rays but are still sharply defined. Still older craters are overlapped by younger ones and have subdued features due to the effects of countless small meteor impacts. The very oldest craters may be all but unrecognizable. Finally, *crater saturation* can indicate how old a planetary surface is (Fig. 19.32c). The older a surface is, the more craters it has. None of these techniques can tell us the actual age of a crater because they all depend on rates of cratering, which may be different at different times or in different parts of the solar system, but we can use these methods to find relative ages and construct the sequence of events on a planetary surface.

(a)

(b)

(c)

◾ FIGURE 19.31 Typical crater forms. (*a*) Small craters are simple pits. Note how the walls have collapsed to leave mounds of debris on the crater floor. (*b*) Craters larger than about 50 km (31 mi) have terraced inner walls and often have central peaks. This crater is Tycho, one of the most prominent craters on the near side of the Moon. (*c*) Very large impacts create multiple-ring impact basins like Mare Orientale on the Moon, which is 1,000 km (620 mi) in diameter.

FORMATION OF THE EARTH-MOON SYSTEM

The *Apollo* data show that the crust of the Moon formed by about 4.5 billion years ago and was intensively cratered. Late in the formation of the Moon's crust, a few huge impacts created multiple-ring impact basins. A long interval of less-intense cratering followed. About 3 billion years ago, basalt magma welled to the surface along the faults around the multiple-ring basins and flooded the basins. Since then, the Moon has been quiet except for rare large impacts. The very intense early cratering gives strong support for the idea that

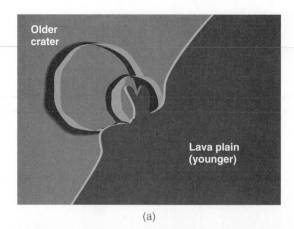

(a)

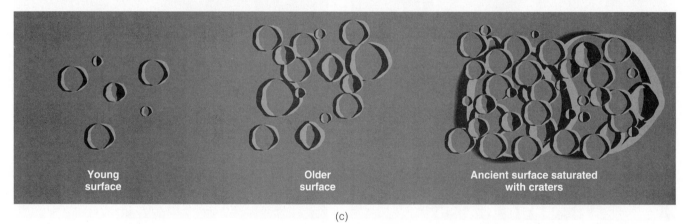

(b)

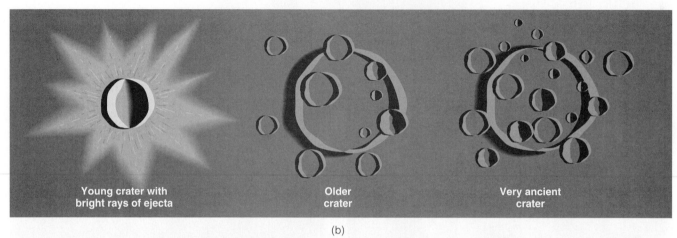

(c)

■ FIGURE 19.32 Craters as clues to solar system history. (*a*) According to the principle of superposition, younger craters overlap older ones. (*b*) Crater degradation provides clues to the age of a crater. Young craters are sharply defined and have bright rays whereas old craters are rounded and softened. (*c*) Crater saturation can indicate the age of a planetary surface. Young surfaces have few craters; old surfaces have many.

the planets formed by the accretion of smaller bodies. The craters we see on the Moon and other bodies are simply the scars left by the last few meteoroids to be swept up. In all likelihood, early Earth was just as battered, but erosion and crustal movements have obliterated the evidence (the oldest known rocks on Earth are only about 4.0 billion years old).

Theoretical studies of accretion suggest that planets should become very hot as they accrete. The more massive a planet is, the faster meteors fall into it and the greater the energy they release. Heat accumulates as impact debris buries the hot material left by earlier impacts. The anorthosite crust of the Moon testifies to a period when the outer surface was

Focus on the Environment: Impacts That Shatter Worlds

Large meteor impacts can have devastating environmental effects. A meteor only 10 m (30 ft) in diameter can release as much energy as a small atomic bomb. Meteors or small asteroids kilometers in diameter can release far more energy than all the world's nuclear weapons combined.

One such impact may have caused the extinction of the dinosaurs (see the Prologue for Chapter 1). The impact of an asteroid 5 km (3 mi) or so in diameter would excavate a crater perhaps 200 km (124 mi) across. Some of the environmental effects are obvious: atmospheric shock wave, ground shock wave, tsunami, and ejecta. Other effects are not so obvious: acid rain from nitrogen oxides produced by the heat of the impact, or heat radiated by ejects reentering Earth's atmosphere. As one scientist put it: "The problem isn't figuring out how mass extinctions could happen. The problem is figuring out how anything might survive."

Even bigger impacts happened early in the history of the solar system. An impact of a 50 km (31 mi) asteroid like those that made the lunar mare basins would vaporize so much rock that for a year or so Earth would have an "atmosphere" of vaporized rock. All land life would be destroyed, and the oceans boiled away. It may be that early Earth had oceans and life several times, only to be sterilized by giant impacts.

Even these events may have been dwarfed by collisions as the planets formed. Many computer simulations of planetary formation indicate that the solar system formed as a swarm of many small planets, as large as the Moon or Mars, and these in turn collided to form the present planets. An impact between early Earth and a Mars-sized object could account for many puzzling features of the Earth-Moon system. Such an impact would have melted most of Earth and the impacting body. Earth and the Moon would have been nearly as hot as the present surface of the Sun for thousands of years and would have had magma oceans hundreds of kilometers deep. Other giant impacts could account for other peculiarities of the solar system, such as Venus's slow backward rotation or the unusual axial tilt of Uranus. Mercury's large core can be explained if most of the planet's crust and mantle had been blasted away by a mega-impact.

---

molten, a so-called *magma ocean.* During this period, the heavy minerals in the magma settled, leaving the lighter feldspar behind (the magma is about the same density as solid feldspar grains, so feldspar grains would neither rise nor sink rapidly). The final stages of accretion of the Moon probably heated the surface so much that global melting occurred. Perhaps all the rocky planets of the inner solar system had a magma-ocean phase.

The formation of the Earth-Moon pair has long been a problem. One popular early theory was the *fission theory:* early Earth spun so rapidly that a piece flew off into orbit and became the Moon. Another theory, *co-creation,* held that the Moon formed in orbit around Earth, like the satellites of the outer planets. Both theories suffer from the fatal flaw that unlike most satellites in the solar system, the Moon does not orbit above Earth's equator but in the same plane as Earth's orbit and the orbits of the other planets. This fact suggests that the Moon formed as an independent planet. Also, the Moon is poor in *volatile* elements, or elements that evaporate easily in a vacuum. This observation, from the analysis of lunar samples, suggests that the Moon formed in a hotter part of the solar system than Earth. The *capture theory* proposed that the Moon formed as a separate planet, passed close to Earth, and was captured into its present orbit. The requirements for a successful capture are so stringent, however, that such a scenario is extremely unlikely.

In the last two decades, a new theory of the Moon's origin has gained favor. It avoids the problems that plagued all earlier theories. This theory, the *impact theory,* proposes that a Mars-sized object struck Earth a glancing blow, and that debris from the impact was flung off into orbit to re-accrete as the Moon (■ Fig. 19.33 and ■ 19.34). The heat released in such an impact would have been sufficient to melt much of Earth (see □ Perspective 19.1).

Earth and the Moon form a remarkable pair. Earth is one of the most active planets in the solar system, with a wide range of geologic processes shaping its surface and modifying its interior. The very activity that makes Earth such an interesting place erases its history in the process. Orbiting nearby is a sharply contrasting body, the Moon, that has very little activity and preserves the early record that Earth lacks. In the next chapter, we will apply some of the ideas about early solar system history learned from the Moon to the other objects in our solar system.

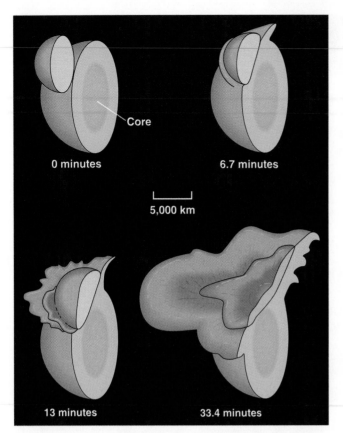

Core

0 minutes   6.7 minutes

5,000 km

13 minutes   33.4 minutes

▣ **FIGURE 19.33** These computer simulations show the result of an impact between a Mars–sized protoplanet and early Earth. Although the models look like two balls of clay colliding gently, the temperature of the expelled debris would rival the surface of the Sun. An unprotected human observing this event from a few Earth diameters away would be blinded and probably killed by the radiated heat.

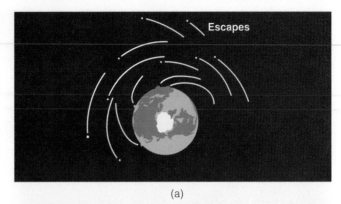

Escapes

(a)

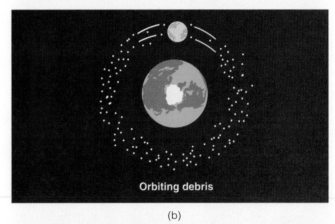

Orbiting debris

(b)

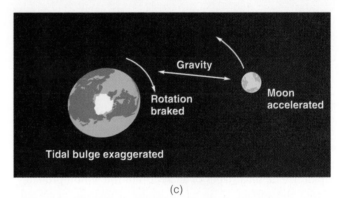

Gravity

Rotation braked

Moon accelerated

Tidal bulge exaggerated

(c)

▣ **FIGURE 19.34** Formation of the Moon after impact. (*a*) Most of the impact debris falls back to Earth. Some of the debris is propelled high enough to enter orbit; some is blasted free of Earth altogether and escapes into space. (*b*) The Moon accretes from the orbiting debris. Initially, it is fairly close to Earth. (*c*) Tidal interactions between Earth and the Moon slow the rapid rotation of early Earth and accelerate the Moon into a higher orbit.

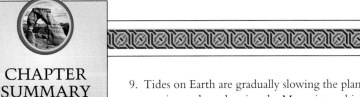
1. Planets orbit the Sun because of the combined effects of gravity and inertia. Gravity alone would pull a planet into the Sun; inertia alone would cause the planet to continue moving in a straight line. The combined effect of the two laws keeps the planet traveling in a fixed path, or orbit.

2. Three laws, called Kepler's laws, govern the motions of the planets: (a) Planets travel in elliptical orbits, with the Sun at one focus. (b) An imaginary line between the planet and the Sun sweeps out equal areas in equal times. (c) A planet's period of revolution and its distance from the Sun are related.

3. Satellites orbit planets in the same way that planets orbit the Sun and also obey Kepler's laws.

4. Earth's seasons are due to the tilt of its axis. Summer occurs when a hemisphere is tilted toward the Sun, winter when it is tilted away.

5. The phases of the Moon occur as the Moon orbits Earth and shows differing amounts of its illuminated face to Earth.

6. Solar eclipses occur when the Moon passes between the Sun and Earth, and occur only at new moon. Lunar eclipses occur only at full moon when the Moon passes through Earth's shadow. Most of the time eclipses do not occur because of the tilt of the Moon's orbit.

7. Tidal forces result from one celestial object's gravity attracting the near side of another object more than the far side. The Moon's tidal forces create ocean tides on Earth.

8. Both the Sun and the Moon exert tidal forces on Earth. At new and full moon, the effects are superimposed and spring tides occur. At first- and last-quarter moon, the effects partially cancel out and neap tides occur.

9. Tides on Earth are gradually slowing the planet's rotation and accelerating the Moon into a higher orbit. Earth's tidal forces have locked the Moon's rotation so that the Moon always shows the same face to Earth.

10. The gravitational attraction of the Moon and Sun causes the orientation of Earth's axis to change, or precess, in a cycle of about 26,000 years.

11. The Sun, the Moon, and the other planets cause slight changes in the tilt of Earth's axis and the shape of its orbit. These changes, called Milankovitch cycles, affect Earth's climate and may play a role in the timing of ice ages.

12. Meteor impact has been a very important process in shaping the planets of the solar system. Depending on the size of the impacting body, meteor craters can be simple pits, large craters with central peaks, or multiple-ring impact basins. The crater is always much larger than the impacting meteor.

13. Most of the craters we observe in the solar system formed during a period of very intense cratering early in the history of the solar system. On Earth, the early craters have been obliterated by erosion and tectonic activity.

14. The Moon's crust consists of a feldspar-rich rock called anorthosite. Dark plains on the Moon are basalt flows that fill large impact basins.

15. The Moon may have formed as the result of a giant impact of another planet with Earth. Material thrown out by the impact would have re-accreted in orbit to form the Moon. The impact would have destroyed the impacting body and melted much of Earth.

IMPORTANT TERMS

apogee	equinox	lunar eclipse	phases
axial tilt	gravity	maria	precession
celestial poles	highlands	meteoroids	satellite
central peak	impact	multiple-ring impact basin	solar eclipse
crater	inertia	orbit	solstice
ecliptic	Kepler's laws	perigee	tidal forces

REVIEW QUESTIONS

1. Why are craters usually round, regardless of the angle at which the impacting meteor strikes?
   a. _____ Most meteors are round; b. _____ the explosion creates a round crater; c. _____ erosion wears the crater into a circular shape.

2. Why do people see eclipses of the Moon more often than solar eclipses, even though solar eclipses happen more frequently?
   a. _____ Eclipses of the moon are visible from half the Earth at once; b. _____ the weather is more cloudy at new moon;

c. _____ eclipses of the Moon last longer; d. _____ people are more interested in viewing eclipses of the Moon.

3. Why are craters relatively uncommon on Earth, compared to the Moon?
   a. _____ Earth was struck less often; b. _____ craters on Earth have been largely eroded away; c. _____ Earth formed after the cratering had ceased.

4. The dark plains on the Moon are:
   a. _____ Sedimentary basins; b. _____ flood basalt plains; c. _____ anorthosite; d. _____ granitic batholiths.

5. Which of the following is the most widely accepted theory for the formation of the Moon?
   a. _____ capture theory; b. _____ impact theory; c. _____ fission theory; d. _____ co-creation theory.

6. Mountains on the Moon are:
   a. _____ formed by plate tectonics; b. _____ formed by erosion; c. _____ rims of ancient huge impact basins.

7. Describe Kepler's laws.

8. Describe how inertia and gravity combine to keep planets in their orbits.

9. What is kinetic energy? What happens to the kinetic energy of an impacting meteor? How does the size of the crater relate to the size of the meteor?

10. Describe the features associated with craters of different size ranges.

11. How can craters be used to infer the geologic history of a planet? Describe superposition, crater degradation, and crater saturation.

12. Describe partial, total, and annular solar eclipses.

13. Why does the Moon remain visible during a total lunar eclipse? What does the visibility of the Moon tell us about conditions on Earth?

14. Why is there a tidal bulge *opposite* the Moon?

15. How do phases of the Moon affect the tides?

16. For safety reasons, all manned missions to the Moon landed on the visible side. So how were scientists able to send seismic waves *through* the Moon to probe its interior?

17. Using a pocket calculator, complete this table and verify Kepler's third law. You may get small errors because of rounding.

NAME	PERIOD (yr)	DISTANCE (compared to Earth)	P × P	D × D × D
Mercury	0.2408	0.3871		
Venus	0.6152	0.7233		
Earth	1.0000	1.0000		
Mars	1.8807	1.5236		
Jupiter	11.8565	5.1996		
Saturn	29.4235	9.5308		
Uranus	83.7474	19.1417		
Neptune	163.7232	29.9277		
Pluto	248.081	39.4816		
Halley's comet	75.9957	17.9415		

POINTS TO PONDER

1. An irate buyer once complained to authorities that an Australian company had defrauded him because he had installed its solar collector exactly as instructed on the north side of his roof, and it didn't work. What went wrong?

2. Could an eclipse ever occur during a neap tide? Why or why not?

3. Can there ever be an annular eclipse of the Moon?

Benningfield, D. 1991. Mysteries of the Moon. *Astronomy* 19 (December): 50–55.

Grieve, R. A. F. 1990. Impact cratering on the Earth. *Scientific American* 262 (April): 66–73.

Hartmann, W. K. 1989. Birth of the Moon. *Natural History* (November): 68–77.

———. 1989. Piecing together Earth's early history. *Astronomy* 17 (June): 24–34.

Mechler, G. 1995. *The Sun and Moon.* New York: National Audubon Society Pocket Guides, Knopf.

Morrison, D. 1990. Target Earth: It will happen. *Sky and Telescope* 79 (March): 261–65.

Schultz, P. H., and J. K. Beatty. 1992. Teardrops on the pampas [aerial photographs revealing impact craters in central Argentina]. *Sky and Telescope* 83 (April): 387–92.

Taylor, G. J. 1994. Scientific legacy of Apollo. *Scientific American* 271 (July): 40–47.

WORLD WIDE WEB SITES

For current updates and exercises, log on to http://www.wadsworth.com/geo.

THE NASA HOME PAGE
http://www.nasa.gov/

This site is the jumping-off point for information on space activities. There are photo archives of spacecraft and mission patches as well as links to information on upcoming space missions. Extensive photo archives of images of Earth from space are also accessible.

CHAPTER
20

The Solar System

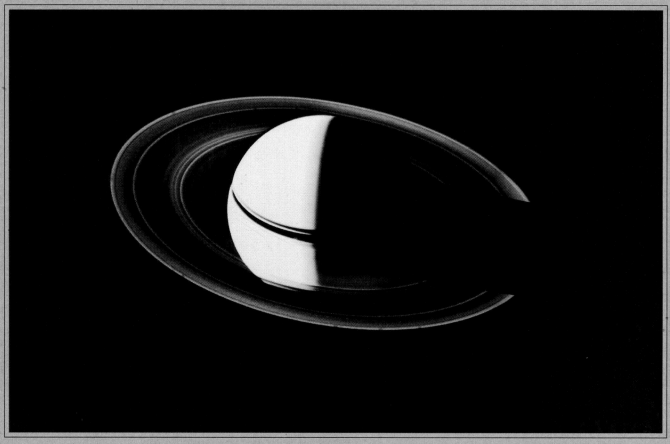

A crescent Saturn, seen by Voyager II in 1981. Planets appear as crescents only to observers beyond them, looking back toward the Sun. The shaded side of the rings is seen here. Sunlight filtering through thinner parts of the rings makes them shine brightly, but thicker parts of the rings are nearly opaque.

For the husband and wife asteroid-hunting team of Eugene and Carolyn Shoemaker and their colleague David Levy, March 23, 1993, began as just "one of those days." The previous day, they had discovered that someone had accidentally opened an expensive box of film, damaging much of it. After they had exposed only a few photographs, thin clouds moved in. With observing conditions poor, the observers decided to use the light-damaged film because the resulting pictures would not be of high quality anyway.

Upon developing the film, the astronomers found a bizarre object. Instead of the round fuzzy image of a typical comet, one picture showed a curious line, like nothing the astronomers had ever seen before. More detailed photographs later revealed a string of more than a dozen small comets (□ Fig. 20.1a). Evidently, a single comet had broken up only a short time earlier. Other cometary breakups had been observed, but none like this one. The comet, or swarm of mini-comets, was called Comet Shoemaker-Levy 9 (the ninth comet discovered by the team).

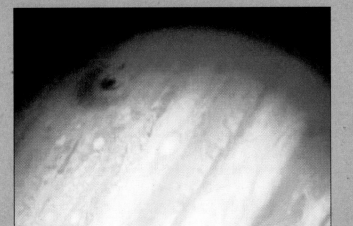

Comet P/Shoemaker-Levy 9 (1993e)
May 1994

Hubble Space Telescope · Wide Field Planetary Camera 2

(a)

(b)

(c)

□ **FIGURE 20.1** Comet Shoemaker-Levy 9 strikes Jupiter. (*a*) As the comet passed close by Jupiter, it was pulled apart by tidal stresses and broke into a string of over 20 small comets. (*b*) The impact created spectacular disturbances in the atmosphere of Jupiter. (*c*) The *Galileo* spacecraft captured this image of one of the comet impacts.

Subsequent calculations showed that the comet had missed Jupiter by a scant 21,000 km (13,000 mi) on July 8, 1992. Even more astonishing, the orbital calculations showed that the comet fragments would strike Jupiter beginning on July 16, 1994, and continuing for a week. The impacts would be by far the largest ever observed in human history.

Unfortunately, the impacts occurred just beyond the edge of Jupiter as seen from Earth, and only a few occurred when Jupiter was visible from North America. Fortunately, the *Galileo* spacecraft, en route to Jupiter, was in a position to observe the impacts and returned spectacular photographs despite its crippled main antenna (Fig. 20.1b). Had the impacts been visible from Earth, they would have produced bright flashes easily visible in telescopes. Even so, the effects of the impact were clearly visible on Earth. The impacts threw clouds of debris well beyond the edge of Jupiter that were brightly lit by the Sun. The impacts triggered spectacular changes in Jupiter's cloud system (Fig. 20.1c) that were visible even in small telescopes. The impacts created great dark splotches on Jupiter that eventually coalesced into a belt girdling the planet before they finally dissipated. Chains of craters on the Moon and other solar system bodies are now widely believed to record impacts of fragmented objects like Comet Shoemaker-Levy 9.

INTRODUCTION

The greatest period of exploration in human history is happening now. After Columbus's initial voyages to the Americas, it took Europeans centuries to explore the continents fully. In contrast, Mars, with an area equal to the entire land area of Earth, has been almost completely mapped since 1964, when the first close-up pictures were sent back by *Mariner IV.* Since then we have obtained close-up images of every planet except Pluto, all the large satellites of the solar system, several asteroids, and the nucleus of Halley's comet.

This explosion of information about the solar system has created a science that did not exist 30 years ago, *planetary geology.* For the first time, we can compare Earth to other planets and attempt to develop general theories about the formation and dynamics of planets.

GENERAL PROPERTIES OF PLANETS

Thanks to unmanned exploration of the solar system, we now have close-up images of dozens of solar system bodies. Recently, astronomers have begun to discover evidence for planets around other stars as well (Chapter 21). Many of the features of planets can be explained simply by the laws of physics. For example, why are planets round? Large planets and stars are round because of gravity. Because gravity attracts masses together, it tends to shape every large mass into a sphere. In the interior of a large planet, pressures are so great that rocks flow; bulges collapse under their own weight and hollows fill out until the planet is round (■ Fig. 20.2a). Tiny objects like the satellites of Mars lack enough gravity to pull them into a sphere, and these objects are notably

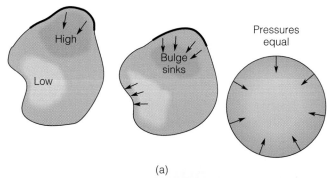

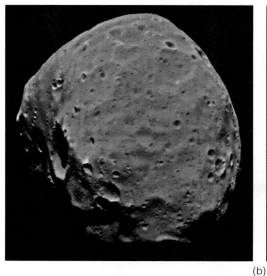

(a)

(b)

FIGURE 20.2 (*a*) Planets are round because their gravity pulls them into a spherical shape. (*b*) Small objects, like the tiny moons of Mars, are not massive enough for gravity to pull them into spheres (*left:* Phobos; *right:* Deimos). (*c*) Rotating planets bulge at the equator because the rotation of the planet partially opposes the pull of gravity.

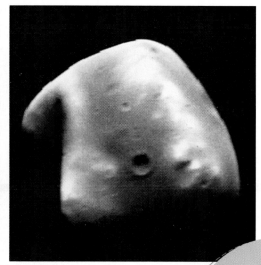

(c)

lumpy (Fig. 20.2b). The limiting size of nonspherical rocky bodies is a few hundred kilometers.

Fast-rotating planets are not perfectly spherical, but have *equatorial bulges* (Fig. 20.2c). Earth's equatorial bulge is only $\frac{1}{298}$ of its diameter because Earth is small, spins slowly, and is made of rock. Jupiter and Saturn have very pronounced bulges—about $\frac{1}{10}$ of their respective diameters—because they are much larger than Earth, spin faster, and are made mostly of gases. Gravity also causes planets to be denser at the center than on the surface. First, gravity crushes the atoms of the planet's interior together, making it denser. Second, the pressures inside the planet may be enough to collapse minerals into a new and closer atomic arrangement. Finally, gravity causes denser materials to sink into the center of a planet and allows lighter materials to rise to the surface. This is probably why Earth has a dense iron–nickel core.

The average density, or *bulk density,* of a planet can tell us a great deal about its makeup. The bulk density is the mass of the planet divided by its volume. We can calculate the volume of a planet from its dimensions, and we can find its mass from its gravitational effect on spacecraft or other bodies. Objects with densities around 1 gm/cm^3 are likely to be made of ice if they are small, or dense gases if they are large, like Jupiter or the Sun. Very large bodies with a bulk density less than 1 gm/cm^3, like Saturn, are likely to be made mostly

of gases. Saturn, with a bulk density of 0.7 gm/cm^3, is actually less dense than water. Densities between 1 and 3 gm/cm^3, like the large satellites of Jupiter, probably indicate a mixture of ice and rocky material, and densities around 3 gm/cm^3 (like our moon) indicate rock. Densities above 3 gm/cm^3 probably indicate that the planet has a dense core. Earth, with a bulk density of 5.5 gm/cm^3, is in this category.

Planets can have *atmospheres* only if they have enough gravity to hold them. The massive gas giants are large enough to retain all their gases and have enormous atmospheres. Venus and Earth are large enough to retain substantial atmospheres, but Mars retains only a thin atmosphere, and smaller bodies like the Moon retain none at all. Also, the hotter a planet is, the faster the atoms in its atmosphere

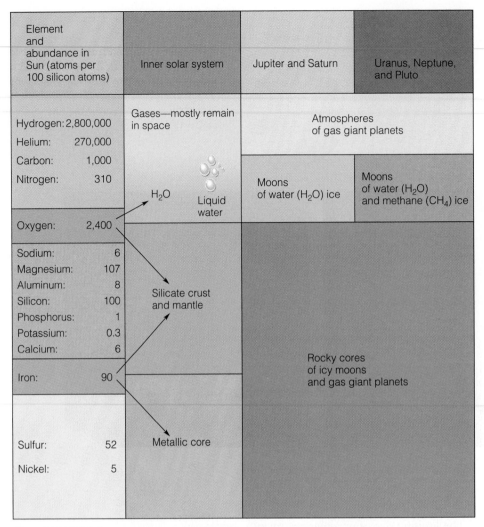

Element and abundance in Sun (atoms per 100 silicon atoms)	Inner solar system	Jupiter and Saturn	Uranus, Neptune, and Pluto
Hydrogen: 2,800,000 Helium: 270,000 Carbon: 1,000 Nitrogen: 310	Gases—mostly remain in space	Atmospheres of gas giant planets	
	H_2O Liquid water	Moons of water (H_2O) ice	Moons of water (H_2O) and methane (CH_4) ice
Oxygen: 2,400			
Sodium: 6 Magnesium: 107 Aluminum: 8 Silicon: 100 Phosphorus: 1 Potassium: 0.3 Calcium: 6	Silicate crust and mantle	Rocky cores of icy moons and gas giant planets	
Iron: 90			
Sulfur: 52 Nickel: 5	Metallic core		

◉ **FIGURE 20.3** The chemical makeup of the solar system. Close to the Sun, magnesium, iron, silicon, and oxygen dominate the makeup of the planets; light elements remain as gases. Between Mars and Jupiter, it was cold enough for water ice to form solid particles. At extreme distances from the Sun, methane and ammonia solids could also form.

move, and the more easily they escape. Mercury, which is close to the Sun, lacks an atmosphere, but Titan, the large and very cold satellite of Saturn, has a substantial atmosphere, even though Titan is less massive than Mercury.

Planets must have an atmosphere to have liquid on their surfaces. In a vacuum, all liquids eventually evaporate, so seas are impossible on a planet without an atmosphere. Even on a world like Mars, the atmosphere is so thin that liquids would evaporate. If a planet is too hot, liquids will evaporate; if the planet is too cold, liquids will freeze. For a planet to have seas, it must have a dense enough atmosphere and a narrow temperature range. Seas, lakes, and oceans are probably very rare in the universe. In our solar system, only on Earth and possibly Saturn's satellite Titan are conditions right for liquids on the surface. There may be liquid water beneath the surface of some of Jupiter's moons, sealed in by an icy crust.

FORMATION OF THE SOLAR SYSTEM

If the solar system formed from the same material as the Sun, we would expect the planets to have the same chemistry as the Sun (◉ Fig. 20.3), except for those elements that would not condense into solids. In the warm, inner part of the solar system, most of the lightest elements would form gases. The remaining elements would form solid grains that would undergo **accretion** into larger masses as the grains collided. The most abundant elements would be silicon, oxygen, magnesium, and iron. The composition of Earth as a whole, with its iron core and mantle of iron and magnesium silicates, matches this composition.

A class of meteorites called **chondrites** comes very close to the composition of the Sun, minus those elements that would form gases, and most planetary scientists consider

(b)

(c)

(d)

■ FIGURE 20.4 From chondrites to rocks. (*a*) A chondrite meteorite, which is probably leftover raw material from the formation of the solar system. (*b*) This rock, a probable sample of Earth's mantle, is rich in the same elements as chondritic material. (*c*) If chondritic magma erupts and cools quickly, *basalt* results. (*d*) If an originally chondritic magma cools slowly, light feldspar collects at the top to form *anorthosite.*

chondrites the raw material of the solar system (■ Fig. 20.4a). The age of meteorites, 4.6 billion years, is a principal piece of evidence for the age of the solar system and Earth (Chapter 11). The commonest minerals of chondritic meteorites are also the minerals of Earth's upper mantle: olivine and pyroxenes (Fig. 20.4b). If chondritic material melts and solidifies quickly, basalt forms (Fig. 20.4c) (Chapter 3). Basalt is probably abundant on other planets. The dark plains on the Moon are basalt, and shield volcanoes on Mars and Venus are also very likely basalt. If a large quantity of chondritic material melts and cools slowly, the ferromagnesian minerals will sink, leaving a feldspar-rich upper layer that cools to become *anorthosite* (Fig. 20.4d), the material that makes up the light-colored highlands of the Moon. With nothing more than the abundances of elements in the Sun and some very simple reasoning, we can account for much of the geology of the inner solar system.

Earth, however, has a crust rich in a material that has not been found anywhere else in the solar system so far: *granite.* Granite is a product of repeated partial melting and remelting of rocks. Granitic crust on another planet would be a clear sign the planet has a dynamic interior.

The outer solar system is cold enough for water to condense as ice and remain solid even in a vacuum. The atoms in gases also move more slowly there and are more easily retained by planets. Jupiter, Saturn, Uranus, and Neptune probably began by accreting solid cores that became massive enough for these planets to attract and hold huge gaseous envelopes as well. The satellites of the outer planets, and probably the planet Pluto as well, are largely made of water ice. In addition, comets are made mostly of water ice. Very far from the Sun, materials like methane and ammonia would also condense as solids, and these frozen gases may be present on the satellites of Uranus and Neptune, on Pluto, and in comets.

What's in a Name?

The past few decades of planetary exploration have revealed new lands larger in area than the entire Earth, full of craters, mountains, and valleys that need names. The tradition used in naming features on the planets has roots that go back to the 1600s. Early observers thought the dark plains on the Moon were seas, and gave them fanciful Latin names. We still use the term *mare* (sea) and the plural *maria* (MARR-i-a) even though we know there is no liquid water on the Moon. In 1651, Giovanni Riccioli started the tradition of naming craters on the Moon after great astronomers, a custom we still follow. (A large crater was later named after Riccioli himself.)

When Mars and Mercury were photographed, it became clear that there were far more craters than there were astronomers, great or otherwise, so the field was broadened first to great scientists in general, then to great figures in any intellectual field. On Mercury, therefore, we find Beethoven, Tolstoi, and Shakespeare commemorated in craters. A serious effort has been made to recognize figures from non-European cultures as well. Political figures are excluded; there is a Tchaikovsky, but no Lenin; a Longfellow, but no Jefferson. The very few exceptions are rulers who also had notable scientific accomplishments, like Julius Caesar, who is honored on the Moon for his calendar reforms.

☐ FIGURE 1 (*a*) Part of a map of Mars prepared by the U.S. Geological Survey. (*b*) A geologic map of the same area.

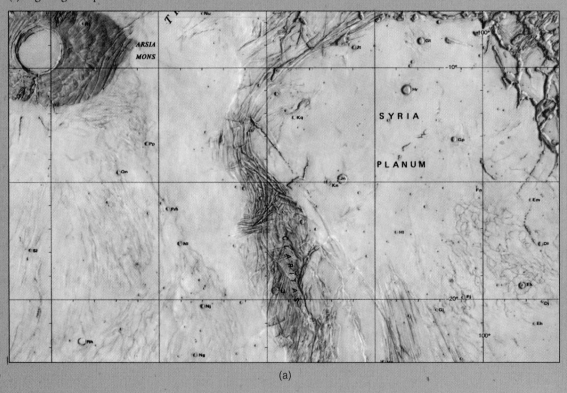

(a)

 ## SILICATE AND METAL PLANETS

Mercury, Venus, Earth, and Mars, the four planets of the inner solar system, as well as Earth's moon, are small rocky worlds rich in iron and magnesium silicates. Some of the asteroids may be made of the same materials. These worlds generally consist of an outer crust, a silicate mantle, and a core. Large planets like Earth and Venus have cores mostly of metallic iron; smaller bodies probably have iron-sulfur cores.

On some bodies, especially in the outer solar system, many naming systems follow themes or mythology. On volcanic *Io,* one of Jupiter's satellites, major geographic features are named from the world's various fire myths. On Venus, geographic features are named for female figures from history and mythology. On Saturn's satellite *Mimas,* craters are named for figures from the legend of King Arthur, and so on. Objects discovered in historic times always have some prominent feature named for their discoverer.

Objects other than craters have names based on a Latin geographic term, plus a given name. Some common terms include *Vallis* (valley), *Rupes* (cliff), *Terra* (large landmass), *Fossa* (trench or deep valley), *Linea* (line), *Regio* (region), *Mons* (plural *Montes,* mountain), *Dorsum* (ridge), and *Rima* (rill or small valley). For example, the great valley on Mars is called *Vallis Marineris* (valley of the *Mariner* spacecraft, which first photographed it), and the great shield volcano is *Olympus Mons* (Mount Olympus). Latin is used in naming because it is traditional and nonpolitical.

All this unparalleled productivity in naming is overseen by a committee of the International Astronomical Union. About 10,000 names have been adopted for solar system geographic features so far. The various spacefaring nations informally recognize this agency as the authority because it is international and politically neutral.

In the United States, mapping the planets is the job of the U.S. Geological Survey. Not only does this agency draft and distribute geographic base maps of the planets, but it prepares geologic maps of them as well (☐ Fig. 1).

(b)

In 1973 and 1974, *Mariner 10* photographed Mercury at close range (☐ Fig. 20.5). Mercury is Moon-like on the outside, but Earth-like on the inside. Mercury is heavily cratered, with a great multiple-ring impact basin, the Caloris Basin (Fig. 20.5a). (See ☐ Perspective 20.1 for an explanation of place-names on other planets.) The craters on Mercury are less abundant than those on the moon, however, and many craters are shallow and flat-bottomed (Fig. 20.5b). Evidently, early in its history, Mercury was flooded repeatedly by lava, which covered or filled many craters.

(a)

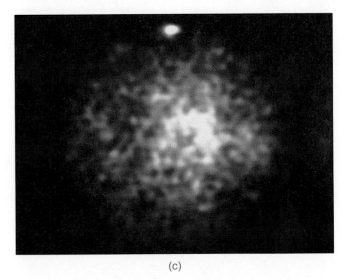

(c)

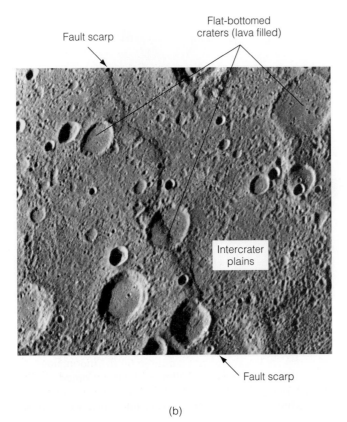

(b)

■ FIGURE 20.5 Mercury. (*a*) The Caloris Basin, a multiple–ring impact basin. (*b*) Mercury's intercrater plains and flat-floored craters appear to indicate repeated flooding of the surface by lava. Mercury also has many thrust fault scarps, indicating shrinkage of the crust. (*c*) Radar reflections from Mercury show strong reflectivity from the planet's north pole. The reflections (using 3.5-centimeter wavelength radar) can be explained if ice exists at the pole.

Some planetary geologists think Mercury's outer surface may have been largely molten during much of the early cratering, so that only the latest craters were preserved.

Mercury is almost as dense as Earth, with a density of about 5 gm/cm^3. To have such a great density, Mercury must have a very large core, perhaps two-thirds of its total diameter. This large core seems to explain Mercury's unexpectedly large magnetic field, which is about 1% as strong as Earth's. Mercury's surface is broken by a number of thrust fault scarps that indicate that early in the planet's history, its interior shrank slightly, perhaps by a few kilometers. There is no evidence for recent internal activity on Mercury.

Mercury's rotation is locked to the Sun in a *resonance:* Mercury rotates three times for every two orbits, so that on every other orbit, the same part of Mercury faces the Sun. This resonance is due to powerful *tidal forces* (Chapter 19) exerted by the Sun that have slowed Mercury's rotation.

Perhaps nothing could seem more unlikely than ice on Mercury, but radar images of Mercury reveal bright reflective spots at Mercury's poles that match the radar signature of ice (Fig. 20.5c). Mercury's rotation axis has almost no tilt. As seen from Mercury's poles, the Sun is always just on the horizon, so despite the planet's proximity to the Sun, the surface at the poles is always very cold because of the grazing illumination by the Sun. It is not known if the ice is on the surface or just beneath it, covered by a thin layer of protective debris. Some lies at the bottoms of polar craters, forever shielded from the Sun.

On paper, Venus is a twin to Earth, only a bit smaller and less massive. It probably is very similar internally to Earth, but its surface is radically different. Venus has a thick atmosphere—90 times as dense as Earth's atmosphere—that is composed mostly of carbon dioxide (Fig. 20.6a). Earth has nearly as much carbon dioxide as Venus, but on Earth, organisms have locked most of the carbon dioxide in carbonate rocks. On Venus, the carbon dioxide is still in the atmosphere, where it traps solar heat with ruthless efficiency. This *greenhouse effect* (Chapter 15) causes the surface temperature of Venus to be a searing 480°C (900°F). The surface of Venus is entirely hidden by clouds of sulfuric acid droplets (H_2SO_4). Venus's atmosphere contains traces of sulfur dioxide (SO_2), possibly released by volcanic activity. Venus rotates slowly (once every 243 days) in the opposite direction from all other planets except Uranus.

The U.S. *Pioneer Venus Orbiter* and *Magellan* spacecraft and Soviet *Venera* probes have produced detailed maps of the surface of Venus (Fig. 20.6b and c) by radar mapping. On a large scale, there are elevated plateaus. The largest, *Aphrodite Terra,* is about as large as South America; another large plateau, *Ishtar Terra,* is about the size of Australia. There are also smaller highlands, but overall the topography of Venus is much different from Earth. Venus lacks the sharp contrast between continents and ocean basins that we see on Earth. The large plateaus on Venus seem to be very large volcanic plateaus rather than continents of granitic crust. Venus has nothing resembling Earth's mid-ocean ridges, transform faults, or trenches. Yet detailed radar maps show great areas of rifted and folded crust (Fig. 20.6d), and there are signs of volcanic activity over much of the planet. Clearly, Venus is a very dynamic planet, but it does not work the way Earth does. There are some impact craters on Venus, but most of the planet's original craters have probably been buried by volcanic deposits. In fact, much of Venus's surface seems to have been buried by global volcanism about half a billion years ago. In some cases, craters appear to have been formed by clusters of meteorites, apparently because the impacting bodies fragmented in Venus's dense atmosphere.

Mars (Fig. 20.7), the nearest planet with a visible surface, has long been a favorite source of speculation. It has a thin carbon dioxide atmosphere, polar caps of ice and frozen carbon dioxide, and an axial tilt and rotation period much like Earth's. About half of Mars is high, densely cratered, ancient crust much like the lunar highlands. There are a few very large impact basins, somewhat modified by erosion and volcanic activity. On the other half of Mars, most craters have been covered or obliterated by erosion, windblown dust, and volcanic activity. Wind is the major erosive agent on Mars now. At times, the entire planet is obscured by great dust storms. Streaks of windblown dust downwind from craters and other obstructions are common, and in places there are great fields of dunes. Iron oxide colors much of the Martian surface reddish; the red color is visible even from Earth without a telescope, so that Mars is widely known as the "red planet."

The present Martian atmosphere is much too thin to permit liquid water, but Mars had liquid water in the past. On the flanks of some Martian plateaus are ancient river valleys and great channels (Fig. 20.7a). These valleys and channels do not dissect the Martian landscape into drainage basins the way terrestrial streams dissect Earth; instead, it appears that these streams flowed only briefly. Some channels seem to mark great catastrophic floods. Judging from the cratering of these ancient fluvial landscapes, the water ran on Mars about a billion years ago.

Much of Mars's water probably was, and is, in the form of *permafrost* (Chapter 6). There are areas of chaotic and broken landscape at the sources of some channels, as if the surface had been undermined by the melting of subsurface ice. Internal heat may have melted the permafrost in these areas. Impact may also have melted permafrost. Some craters on Mars have ejecta blankets consisting of sharp-edged lobes that look very much like mudflow deposits. These deposits probably formed when impact melted permafrost. The mixture of water and shattered rock was flung out of the crater as mud and flowed outward as a fast-moving mudflow (Fig. 20.7b). Finally, in some deep Martian valleys, there is evidence that water lasted long enough to lay down sedimentary deposits. For water to last that long, Mars must have been warmer and had a thicker atmosphere than at present.

Mars has a great tectonic plateau, called the *Tharsis Plateau,* capped by enormous shield volcanoes. The highest of these, *Olympus Mons* (Mount Olympus), is twice as large in all dimensions as the shield volcanoes of Hawaii: 500 km

■ **FIGURE 20.6** Venus. (*a*) From space, Venus shows only its cloud-covered atmosphere. (*b*) The surface of Venus, photographed by a Soviet *Venera* spacecraft. The spacecraft, partially visible in the picture, has the typical style of early Soviet spacecraft: rugged, simple, no frills, but effective. The hinged arm is a simple chemical analyzer, while the teeth on the spacecraft base are for measuring angles in the photograph and to help right the spacecraft if it tipped over on landing. (*c*) The U.S. *Pioneer Venus Orbiter* carried a radar altimeter that outlined the topography of Venus, shown in this map. (*d*) The U.S. *Magellan* and Russian *Venera* spacecraft have carried side-looking radars that have produced detailed maps of parts of Venus. The intricately folded area shown here is several hundred kilometers wide.

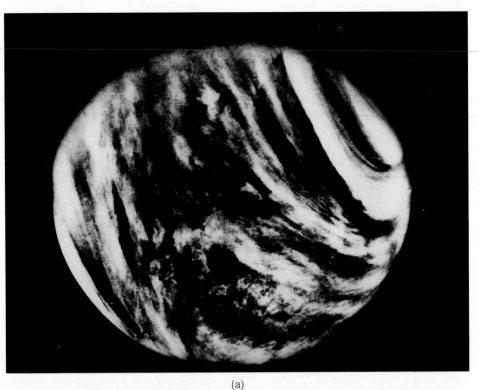

(a)

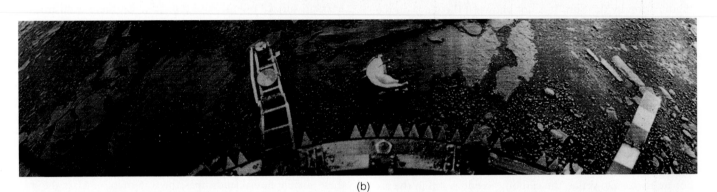

(b)

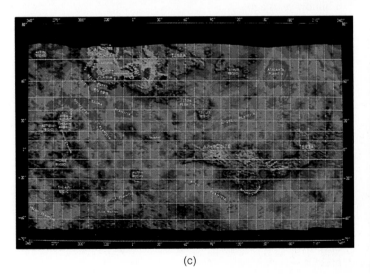

(c)

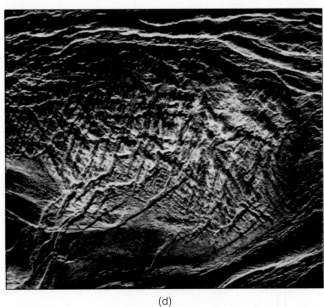

(d)

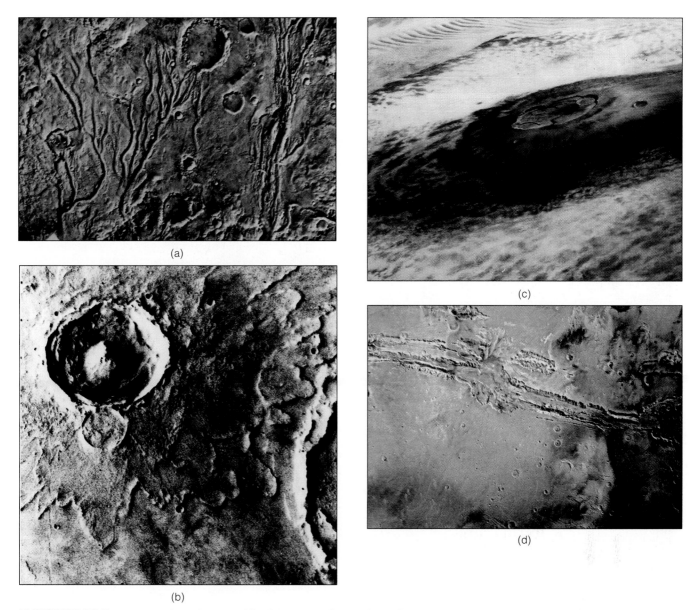

(a)

(b)

(c)

(d)

■ FIGURE 20.7 Mars. (*a*) Channel systems like this one are clear evidence for running water on Mars at one time. (*b*) A rampart crater. Its distinctive ejecta lobes probably formed when the impact melted permafrost, mixing the resulting water with ejecta to create vast mudflows. (*c*) The vast shield volcano Olympus Mons dwarfs any shield volcano on Earth. (*d*) The Vallis Marineris is a rift valley larger than any canyon on Earth.

(300 mi) across at the base and 25 km (80,000 ft) above the average elevation of the surface (Fig. 20.7c). We do not know if these volcanoes are still active or how often they erupt.

Splitting the Tharsis Plateau is a rift valley that dwarfs any on Earth: *Vallis Marineris* (Mariner Valley). This valley is 5,000 km long (3,000 mi), hundreds of kilometers wide, and 10 km (30,000 ft) deep (Fig. 20.7d). The Grand Canyon would be a minor ravine in the Vallis Marineris canyon system.

The Tharsis region shows that Mars has some internal activity, enough to produce volcanoes and rifts. Yet Mars lacks a global plate tectonic system. Evidently, the lithosphere on Mars is too thick, and mantle convection too weak, for the

crust to break up into many plates. The immensity of the Tharsis volcanoes is probably due to their building up on a stationary crust rather than on a moving plate. If the Pacific plate were stationary on Earth, the Hawaiian hot spot might have built a volcano comparable to Olympus Mons.

Explaining the climates of Venus, Earth, and Mars has been called "the Goldilocks problem": Venus is too hot, Mars is too cold, and Earth is just right. With slightly different histories, all three planets could have developed atmospheres capable of sustaining life. Recently, indeed, evidence of possible ancient life on Mars has been found (■ Perspective 20.2). There are indications that Mars has ancient sedimentary deposits, which are possible evidence of

Life on Mars?

On August 7, 1996, the National Aeronautics and Space Administration announced at a news conference that a team of researchers had found the strongest evidence yet for ancient life on Mars. The evidence was found in a meteorite, known by its catalog number ALH84001 (□ Fig. 1).

The meteorite, found in Antarctica, was one of a very rare class of meteorites called *SNC meteorites* (from the initials of three localities where this class of meteorites were first found). These SNC meteorites are mostly geologically young igneous rocks, very different in texture from most meteorites. Only about a dozen meteorites of this type are known. They were finally recognized as originating on Mars because gases trapped in the meteorites match the composition of the Martian atmosphere exactly. The evidence for life consists of distinctive organic molecules called *polycyclic aromatic hydrocarbons*, or PAHs, microscopic nodules of carbonate and sulfide minerals and tiny crystals of magnetite similar to those created by some types of terrestrial microorganisms, and very tiny spheres and tubes that have been interpreted as possible fossils of microorganisms (□ Fig. 2).

There have been false alarms in the past about detection of extraterrestrial life in meteorites. On a number of occasions, researchers have claimed to have found chemical evidence of extraterrestrial life, only to discover later that the chemicals were the result of terrestrial contamination. (The Antarctic meteorites are especially valuable precisely because they are found in an uncontaminated, nearly lifeless environment. They are handled with the same ultrastrict standards required for the handling of lunar rock samples.) Thus, all researchers involved in studies of possible Martian life stress that they have discovered only *possible* evidence for life. They all agree that none of the pieces of evidence alone is definite proof for life. Nevertheless, they feel that the simplest explanation for all the pieces of evidence together is biological activity. Skeptics say that there is not yet enough evidence to rule out nonbiological origins for the features observed in the meteorites. In addition, the supposed fossils are only about one-tenth as large as the smallest terrestrial bacteria and some critics believe it is not possible for organisms to be that small. They claim that any system of organic molecules capable of growing and reproducing would be too large to fit in such a tiny space. (Viruses can be smaller, but they are not considered living because they can reproduce only within some other living cell but not on their own.)

Additional studies are under way. Other SNC meteorites are being scrutinized closely for signs of Martian life, and the original meteorite is being examined still more closely. In October of 1996 another Martian meteorite, EETA 79001, was discovered to have organic compounds unusually rich in the carbon isotope carbon-12, an isotopic signature typical of biologically formed compounds. Finding fossil microorganisms preserved in the act of cell division is one of the most convincing lines of evidence that terrestrial microfossils are genuine fossils, and researchers hope to find similar evidence in a Martian meteorite. The issue may not be finally resolved, however, until samples are returned directly from Mars by future space missions.

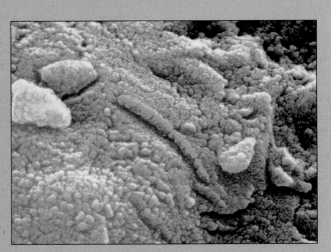

□ **FIGURE 2** Possible fossils of microorganisms. The suspected fossils are highlighted in turquoise.

□ **FIGURE 1** Meteorite ALH84001.

ancient early seas (and a thicker atmosphere), but Mars is not massive enough to retain a thick atmosphere for billions of years. Earth has nearly as much carbon dioxide as Venus, but most of it is chemically bound in carbonate rocks. Either Venus never had any process capable of regulating carbon dioxide, or the regulatory processes broke down.

An additional problem in explaining Earth's climate is the "faint early Sun problem." Stars brighten as they age (Chapter 21). Four billion years ago, the Sun was only about 70% as bright as today, yet Earth has been warm enough to have liquid water on its surface throughout its history. To have avoided freezing early in its history, Earth must have had a much thicker atmosphere with a much greater greenhouse effect.

Oceans are believed to have been crucial in Earth's history. Nowadays biological processes bind carbon dioxide into carbonate rocks on Earth, but inorganic processes can do the same thing, and they could have done so on the early Earth. But these processes all take place in liquid water. If Venus became too hot for liquid water early in its history, it could never have locked up carbon dioxide in the rocks as happens on Earth. Carbon dioxide would have accumulated in Venus's atmosphere, making the planet still hotter and resulting in its present climate.

 ## THE GAS GIANT PLANETS

The four *gas giant* planets (Jupiter, Saturn, Uranus, and Neptune) owe their huge sizes, as well as their collective name, to their great atmospheres (Fig. 20.8). They all have

▣ FIGURE 20.8 A composite photograph, showing the gas giants and Earth to the same scale. The photographs of the giant planets were all obtained by the *Voyager* spacecraft.

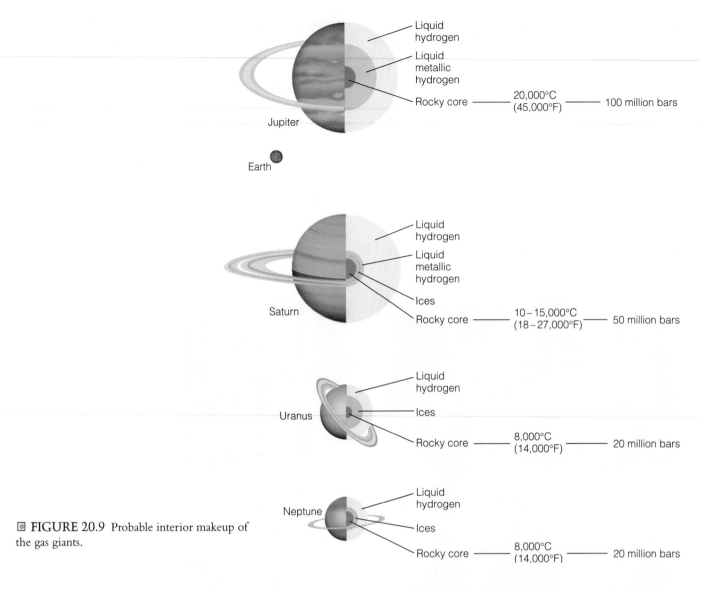

Liquid
hydrogen

Liquid
metallic
hydrogen

Rocky core —— 20,000°C —— 100 million bars
(45,000°F)

Jupiter

Earth

Liquid
hydrogen

Liquid
metallic
hydrogen

Ices

Saturn

Rocky core —— 10–15,000°C —— 50 million bars
(18–27,000°F)

Liquid
hydrogen

Ices

Uranus

Rocky core —— 8,000°C —— 20 million bars
(14,000°F)

Liquid
hydrogen

Neptune

Ices

Rocky core —— 8,000°C —— 20 million bars
(14,000°F)

◼ FIGURE 20.9 Probable interior makeup of
the gas giants.

compositions much like the Sun, being rich in hydrogen and helium, probably with rocky cores deep within (◻ Fig. 20.9). They are all cold at their visible surfaces, as we might expect given their distance from the Sun, but hot within. They also all have rings, which we will discuss separately in the next section.

When the planets accreted, the gas giants formed where it was cold enough for water vapor and other gases to freeze and be available for accretion, unlike the inner solar system where ices would not condense. In the cold outer solar system, it was also easier for small planets to retain gases than in the warmer, inner regions. Finally, when the early Sun began to shine by thermonuclear reactions, it probably emitted vigorous streams of atomic particles that swept the inner solar system free of gas, but perhaps did not affect the more distant parts of the solar system so strongly.

Jupiter has been described as a "star that failed." An object more than about 50 times as massive as Jupiter would have developed internal temperatures and pressures high enough to begin fusing hydrogen to helium and would be a star instead of a planet. Theoretical calculations show that Jupiter

is almost as large as a gas giant planet of its composition can be. A less massive planet would be smaller in diameter because it has less material, but a more massive planet would also be smaller because its gravity would compress its interior more.

Jupiter has no visible solid surface. What we see of Jupiter is a swirling mass of clouds. Temperature increases rapidly toward the planet's interior, so the clouds represent gases that condense at various temperature levels. The clouds are probably ammonia (NH_3), followed by a layer of ammonium hydrosulfide (NH_4HS), and finally water. We cannot see below the water cloud layer. Late in 1995, the *Galileo* mission dropped a probe into Jupiter's atmosphere. When it stopped transmitting about 160 km (100 mi) below the cloud tops, the pressure was about 20 times that of Earth's atmosphere, and the temperature was 140°C (284°F). Deep within Jupiter, where it is far too hot for most molecules to form, there is probably a domain of very hot, very dense gas—mostly hydrogen. Still deeper, the pressures are so great that the electrons are squeezed off hydrogen atoms, and the hydrogen becomes *liquid metallic hydrogen,* a form of hydrogen predicted in theory but not yet made in the laboratory. The heavier elements in Jupiter probably form a solid core, estimated at about twice Earth's diameter, but under temperatures and pressures we can scarcely imagine. The temperature at the center of Jupiter may be 20,000°C (45,000°F) and the pressure 100 million bars—30 times the pressure in Earth's core.

Jupiter has an immense magnetic field that is probably generated by currents in its liquid metallic hydrogen interior. Charged particles from the Sun are trapped and accelerated by this field and create a vast radiation belt around the planet. Passing through this radiation belt, the *Pioneer* and *Voyager* spacecraft absorbed radiation doses that nearly overpowered their instruments and would have killed a human in a short time. The satellites of Jupiter orbit in this radiation bath; exploring them in person will be dangerous.

The other gas giants are similar to Jupiter in many ways, but they are less massive and have lower pressures and interior temperatures. The gas giants also appear to become richer in heavier elements the farther they are from the Sun.

With less heat from the Sun and less internal heat of their own, the outermost atmospheres of the gas giants beyond Jupiter are hazier and less turbulent than Jupiter's, and their cloud belts less visible. Like Jupiter, Saturn is believed to have an interior with a deep layer of liquid metallic hydrogen. Uranus and Neptune are smaller and less massive than Saturn. They lack the immense pressures necessary to produce liquid metallic hydrogen. Instead, they probably consist of an outer layer of hydrogen and methane; a thick layer of water, ammonia, and methane; and finally a rocky core a bit larger than Earth.

Uranus and Neptune are cold enough for methane to condense and form the principal ingredient in the high-altitude haze on these planets. Although they are nearly the same size—about four times the diameter of Earth—Uranus and Neptune are quite different. Uranus presented a nearly featureless blue globe to *Voyager*, whereas Neptune has a dynamic atmosphere with cloud belts, layers of high-level clouds, and great storm systems.

The outer solar system is cold enough for ice to last billions of years, even in a vacuum, without melting or vaporizing. Here we find satellites and planets of ice. One of the surprises of planetary exploration has been the diversity and internal activity of these objects.

THE JUPITER SYSTEM

The four large satellites of Jupiter are called the *Galilean satellites* in honor of Galileo, who discovered them. From the innermost satellite outward, they are *Io, Europa, Ganymede,* and *Callisto.* Io and Europa are comparable in size to Earth's moon; Ganymede and Callisto are larger than the planet Mercury (▦ Fig. 20.10).

▦ FIGURE 20.10 A composite of *Voyager* photographs showing the Galilean satellites of Jupiter to the same scale. Clockwise from the upper right are Europa, Callisto, Ganymede, and Io.

What Makes Icy Moons Black?

The dark coating on Ganymede and Callisto occurs on many objects in the outer solar system. It is literally as black as soot and probably is made of the same material: carbon. Atomic particles and ultraviolet light from the Sun probably strip hydrogen atoms off hydrocarbons, leaving a carbon residue. Why it coats some moons and not others is a mystery.

Io, Jupiter's innermost large satellite, is a bit smaller than Earth's moon and about as far from Jupiter. *Voyager* discovered Io to be a world stranger than any science fiction author would dare to invent. It is covered with spectacular red, yellow, and white deposits, punctuated by black volcanic vents. Nine volcanic vents were active during the two *Voyager* encounters, making Io the most volcanically active body in the solar system. Io has almost no impact craters, probably because craters are rapidly buried by volcanic eruptions.

What could power volcanoes on a world as tiny as Io? Steam provides the pressure that powers volcanic eruptions on Earth, but Io has such weak gravity that it long ago lost all its water vapor. To drive the eruptions, we must find something that is abundant in the universe, volatile enough to vaporize easily and power eruptions, yet heavy enough for a body as small as Io to retain, and, preferably, capable of accounting for Io's strange surface. Sulfur fills the bill on all counts. The red and yellow areas are probably different crystalline forms of sulfur, while the white areas are probably sulfur dioxide frost.

The energy that powers Io's eruptions ultimately comes from Jupiter. The planet's immense gravitational force pulls more strongly on Io's near side than on its far side. Consequently, Io is stretched into an ellipsoid with its long axis pointing at Jupiter and always shows the same face to Jupiter, just as the Moon does to Earth. Every time Io passes another Galilean satellite, especially Europa, the other satellite tries to twist Io into a new orientation. This gravitational tug-of-war constantly kneads the interior of Io like a piece of putty and generates a great deal of heat.

Europa, the second large satellite out from Jupiter, is the smallest of the Galilean satellites but also the brightest. Close-up imagery shows that Europa has a brilliant white surface of ice almost devoid of craters; it is possibly the smoothest surface in the solar system. The surface is covered with a network of cracks. Europa probably has a thin icy crust with liquid water or icy slush beneath. Plastic flow of Europa's icy crust probably fills in large craters and accounts for its crater-free surface. The tidal tug-of-war between Jupiter, Io, and Europa probably heats Europa enough to account for its slushy interior.

Ganymede, the third Galilean satellite, is the largest satellite in the solar system, being larger than Saturn's Titan by just a few kilometers. Ganymede has a heavily cratered dark surface, but fresh craters are brilliant white, showing that its crust is icy. Ganymede possesses two distinct types of crust. The older, darker crust is smooth and fractured into large pieces, many of which can be reassembled just like the continents on Earth. These old blocks of crust are separated by bands of younger, lighter crust with a ribbed or grooved texture. The grooved terrain represents new ice that filled the cracks between the separating blocks. There is no indication of subduction or compression of the crust on Ganymede; the blocks of old crust probably drifted more like rafts of ice than like tectonic plates. One hypothesis is that Ganymede's crust broke apart and drifted atop an early liquid or slushy mantle. Another idea is that Ganymede once had a continuous crust, but as the interior froze, the ice expanded and split the crust apart. Ganymede has little or no internal activity now. Whatever formed the grooves is no longer operating, and the grooved terrain is heavily cratered.

Callisto, the outermost Galilean satellite, is large but dark. As on Ganymede, young craters expose brilliant white ice, which darkens with age. Callisto is heavily cratered, but seems never to have had internal activity. The most spectacular features on Callisto are two great multiple-ring impact basins.

THE SATURN SYSTEM

Like Jupiter, Saturn has many satellites. Apart from gigantic Titan, the satellites of Saturn are all small icy bodies, but with a remarkable variety of surfaces (■ Fig. 20.11).

Mimas and *Tethys* are heavily cratered, each with an enormous impact basin. On both, the impact must have been almost enough to shatter the satellite. Indeed, opposite the impact basin on Tethys is a great rift valley running much of the way around the satellite, as if the impact nearly split it in two. *Iapetus* is heavily cratered with a remarkable color pattern. Like Ganymede and Callisto, Iapetus has been coated with a dark material, but only on one side. Like all satellites of the gas giants, Iapetus keeps the same face toward Saturn all the time. One part of the satellite always faces forward as it travels through space. Iapetus has swept up a dark coating only on its leading face, and the trailing face is still bright.

Some satellites of Saturn have internal activity. *Rhea* and *Dione* are covered with wispy white streaks, probably frost that formed when water vented out of their interiors. Half of *Enceladus* is heavily cratered, while the other half is smooth and crisscrossed by fractures.

Titan is just a bit smaller than Ganymede and probably similar internally, but it is unique in the solar system in having a dense atmosphere. The atmosphere is mostly nitrogen with methane and other hydrocarbons (■ Fig. 20.12). Ultraviolet light from the Sun causes chemical reactions in these gases, making an orange photochemical haze. (On

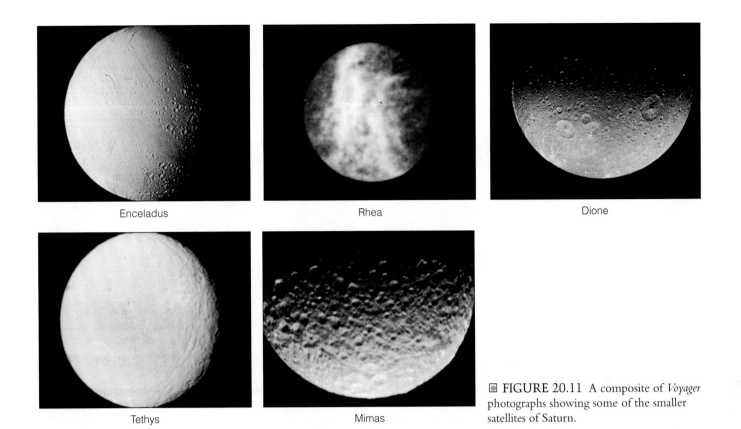

Enceladus

Rhea

Dione

Tethys

Mimas

■ FIGURE 20.11 A composite of *Voyager* photographs showing some of the smaller satellites of Saturn.

■ FIGURE 20.12 Titan. (*a*) Viewed with the Sun almost behind it, Titan appears as a thin crescent. Titan's atmosphere scatters sunlight far around the night side, outlining the satellite with a complete ring of light. (*b*) Titan shows only a featureless orange disk with a blue layer of high-altitude haze above it.

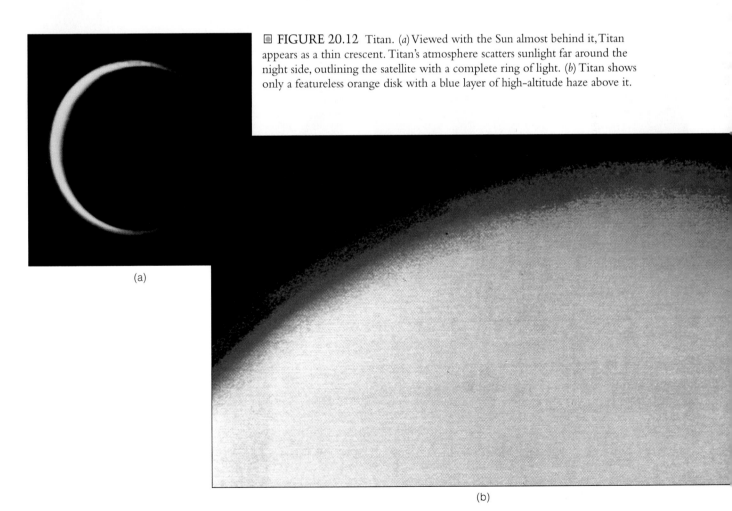

(a)

(b)

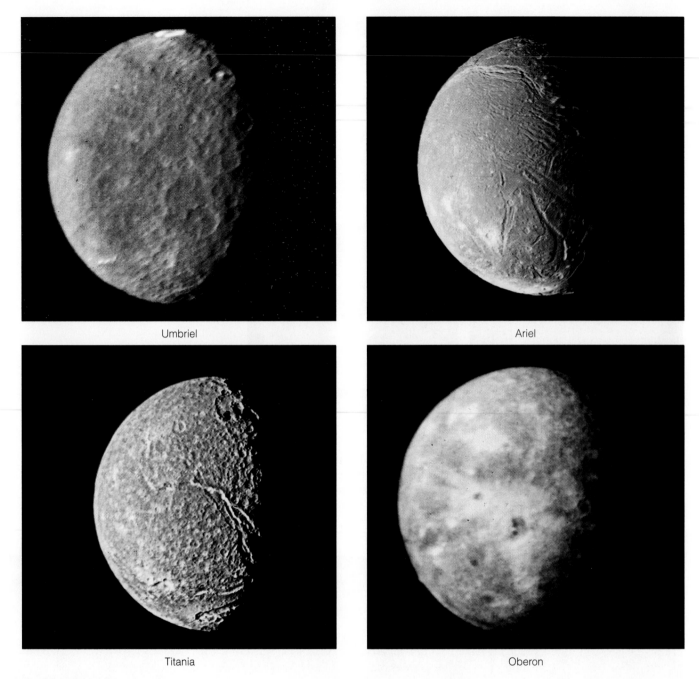

Umbriel

Ariel

Titania

Oberon

 FIGURE 20.13 A composite of *Voyager* photographs showing some of the satellites of Uranus.

Earth, it would be called smog; the cause is very different but the chemistry is probably much the same.) The haze, which extends up to 300 km (186 mi) above the surface, is thick enough to hide the surface entirely. Above the orange haze is a thin, bluish haze layer.

The pressure and temperature on Titan's surface are sufficient for some light hydrocarbons like methane (CH_4) to liquefy, so Titan may be one of the few worlds in the solar system with liquid on its surface. Titan is a prime target for future probes to the outer solar system.

THE SATELLITES OF URANUS, NEPTUNE, AND PLUTO

Generally, the inner satellites of Uranus are more varied and internally active than the outer satellites (■ Fig. 20.13). *Oberon*, the outermost satellite, is a cratered ice body partially covered with a dark material. *Titania*, the largest satellite, has a cratered surface crisscrossed with faults and fractures. *Umbriel* is heavily cratered, inactive, and covered with a very dark coating. Why Umbriel alone is so

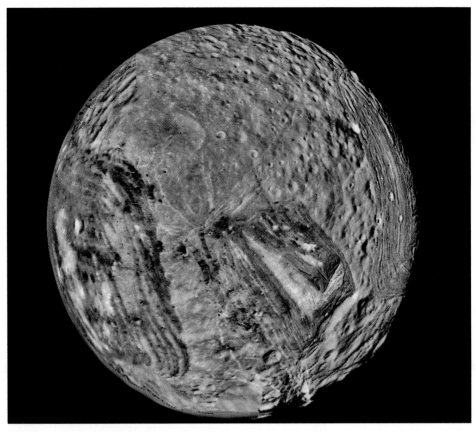

■ FIGURE 20.14 Miranda, Uranus's inner moon, shows a spectacular surface that may reflect a catastrophic early history.

dark is unknown. *Ariel,* the second-innermost large satellite, is almost entirely covered with rift valleys and fractured crust.

Miranda, the innermost satellite of Uranus, is as bizarre a place as the solar system has to offer (■ Fig. 20.14). Most striking are Miranda's strange surface markings, including a great rectangle covered with V-shaped patterns of grooves and two vast ovals of concentric grooves. Miranda also has a great rift valley with a sheer ice cliff 5 km (15,000 ft) high.

The likeliest explanation for Miranda's strange surface markings is the sinking of dense material into the interior and the rise of lighter ice to the surface. One hypothesis holds that long ago the satellite was fractured into several pieces by a great impact. The pieces would have orbited Uranus in very similar orbits and would gradually have re-accreted in a series of fairly gentle collisions. Miranda is massive enough for gravity to shape the satellite into a sphere and to pull denser pieces of material into the satellite's interior.

Neptune's largest satellites both have odd orbits. *Triton* (■ Fig. 20.15b) is traveling in a *retrograde* orbit, opposite Neptune's direction of rotation; it is the only large satellite in the solar system to have such an orbit. One of Neptune's other major satellites, *Nereid,* has one of the most eccentric orbits of any body in the solar system. At its most distant, Nereid is seven times farther from Neptune than at its closest.

Triton is the coldest body yet studied in the solar system. Its surface temperature of $-220°C$ ($-360°F$) is cold enough for nitrogen to freeze on its surface. Triton has a curious surface of small polygonal mounds and depressions, probably caused by upward flow of ices within the crust, much the way salt domes (Chapter 10) flow on Earth. Triton is also crisscrossed by networks of long fractures. Parts of the crust appear to have melted and refrozen to produce large flat-floored basins. Geyserlike eruptions, probably powered by nitrogen vaporized by heat from the distant Sun, occur on Triton. A thin atmosphere of nitrogen freezes during Triton's winters (40 years long), covering the polar regions with nitrogen frost.

The outpost of the solar system, *Pluto* was not known until 1930; its satellite *Charon* was not discovered until 1978. Pluto is the lone planet in the solar system not to have been observed by spacecraft. A proposed *Fast Pluto Flyby* mission would use Jupiter's gravity to accelerate a probe to Pluto in just a few years. Planning for this mission is under way for launch sometime after the year 2000. What we know of these bodies will remain sketchy for some time to come (■ Fig. 20.16), but Pluto is probably similar in many ways to Triton.

Pluto and Charon come closer than any other objects, even Earth and the Moon, to being a double planet system. At only about two-thirds the size of Earth's moon, Pluto is by far the smallest planet. The bulk density of Pluto and Charon, about 1.8 gm/cm^3, suggests that the two objects are ice with rocky cores. The two objects are different: Pluto has a reddish tint, probably produced by nitrogen compounds on its surface, whereas Charon is nearly a neutral gray. Pluto has the most eccentric and tilted orbit of any planet—until

2009, Pluto will actually be closer to the Sun than Neptune. Pluto and Neptune cannot collide because their orbits do not intersect. Perhaps Pluto is simply the largest object in an outer belt of ice and rock asteroids. More than 40 small asteroid-like objects orbiting beyond Neptune in the so-called *Kuiper Belt* have been discovered since 1992. Some travel farther from the sun than Pluto.

 # PLANETS WITH RINGS

All of the gas giant planets have rings. Ring systems are swarms of individual orbiting particles. From the way light filters through the rings, ring particles are known to vary in size from microscopic to chunks a meter or so across. Only Saturn has rings that are obvious through an Earth-based telescope. Jupiter's ring is very thin, consisting of microscopic particles. Uranus and Neptune have very narrow rings. Ring systems may be left over from the formation of the planet, like miniature asteroid belts, or they may possibly be formed by the breakup of larger satellites through tidal forces or collisions.

The orbits of ring particles cannot intersect sharply, because collisions between ring particles would destroy the particles or expel them from the ring. Therefore, rings can exist only if the orbits of the particles are almost perfectly circular and in the equatorial plane of the planet (▣ Fig. 20.17). Because rings must lie exactly in the equatorial plane

of the planet, they are also extremely thin. The rings of Saturn are 274,000 km (171,000 mi) in diameter but less than a kilometer thick.

Even though the orbits of ring particles are precisely orchestrated, the rings would still dissipate with time, because the satellites of the planet constantly disturb the orbits of the particles. The sharp edges of Saturn's rings and the incredibly narrow rings of Uranus seem to be stabilized by **shepherd moons** (▣ Fig. 20.18). Any particles that drift out of the ring are tugged by the tiny gravitational pull of the shepherd moon as they drift by and are steered back into the ring.

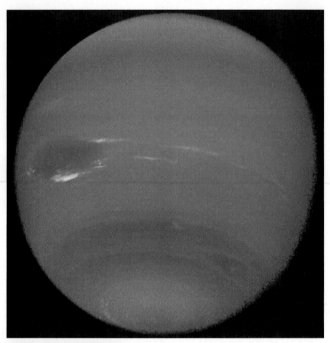

(a)

▣ FIGURE 20.15 (*a*) Unlike bland Uranus, Neptune has a dramatic and dynamic atmospheric circulation. (*b*) Triton has a surface marked by long fissures and a polar cap of nitrogen frost.

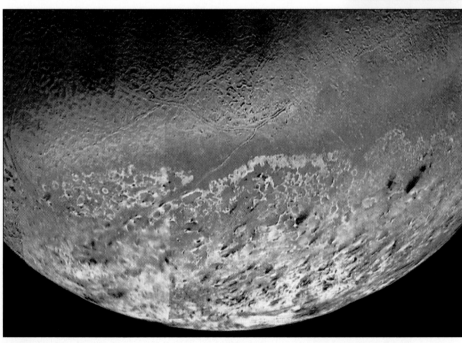

(b)

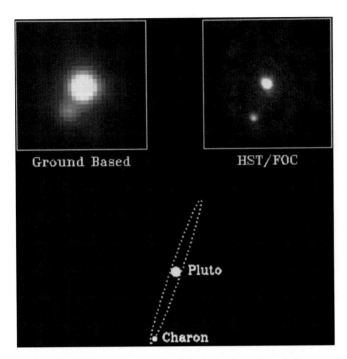

FIGURE 20.16 Pluto and Charon. This picture of Pluto (top) and Charon (bottom), obtained by the Hubble Space Telescope, is the best available image of these worlds.

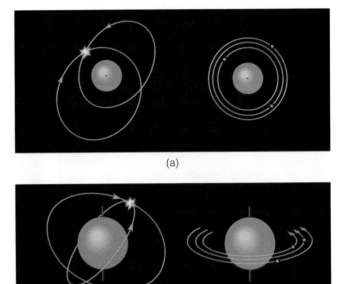

FIGURE 20.17 Ring systems and how they work. (*a*) Ring particles must travel in circular orbits, or they will collide. (*b*) Ring orbits must lie in the plane of the planet's equator, or ring particles will collide.

 SMALL SOLAR SYSTEM OBJECTS

Countless small objects are orbiting the Sun; they are minor in mass but very important scientifically because they include some of the only surviving samples of original solar system material. Orbiting between Mars and Jupiter are a host of small worlds, the *asteroids* or *minor planets* (◻ Fig. 20.19), which are probably remnants of early solar system material that never accreted into a larger planet. The largest, *Ceres,* is about 1,000 km (600 mi) in diameter, and perhaps 50 are larger than 100 km (60 mi). Asteroids are named after mythological figures, cities, countries, and people and are also designated by number in order of discovery. New asteroids are still being discovered and named; for example, asteroids 3350 through 3356 were named for the seven astronauts killed aboard the space shuttle *Challenger* in 1986. As of 1997, about 6,300 asteroids had been discovered. The total mass of the asteroids is less than Earth's moon. Asteroids are often pictured as dense swarms of tumbling rocks, but in reality they are tens of millions of kilometers apart. At least one asteroid, 243 *Ida,* has a satellite. Radar images of two other asteroids, 4769 *Castalia* and 4179 *Toutatis,* reveal them to be double asteroids close together, probably held in contact by their weak gravity.

Comets are probably icy equivalents of the asteroids, leftovers from the formation of the solar system. Comets travel very elongated orbits that take them far from the Sun and bring them close to it only rarely. Halley's comet, for exam-

ple, comes closer to the Sun than Venus, but travels beyond Neptune at its farthest. Far from the Sun, comets travel as small, icy asteroids, but as they approach the Sun, they begin to warm up. Their ices vaporize and form a thin cloud around the solid core, or **nucleus** (◻ Fig. 20.20a). This

FIGURE 20.18 Shepherd moons. The closer a moon or ring particle is to a planet, the faster it orbits. Shepherd moons on the inside of a ring overtake errant ring particles and give them a gravitational pull as the moon passes. The ring particles are accelerated into a higher orbit, back into the ring. Shepherd moons on the outside of a ring give errant ring particles a pull as the particles overtake the moon. The ring particles are decelerated into a lower orbit, back into the ring.

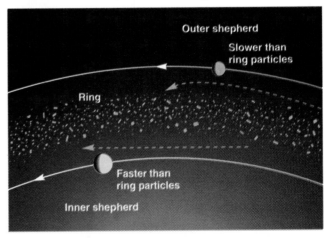

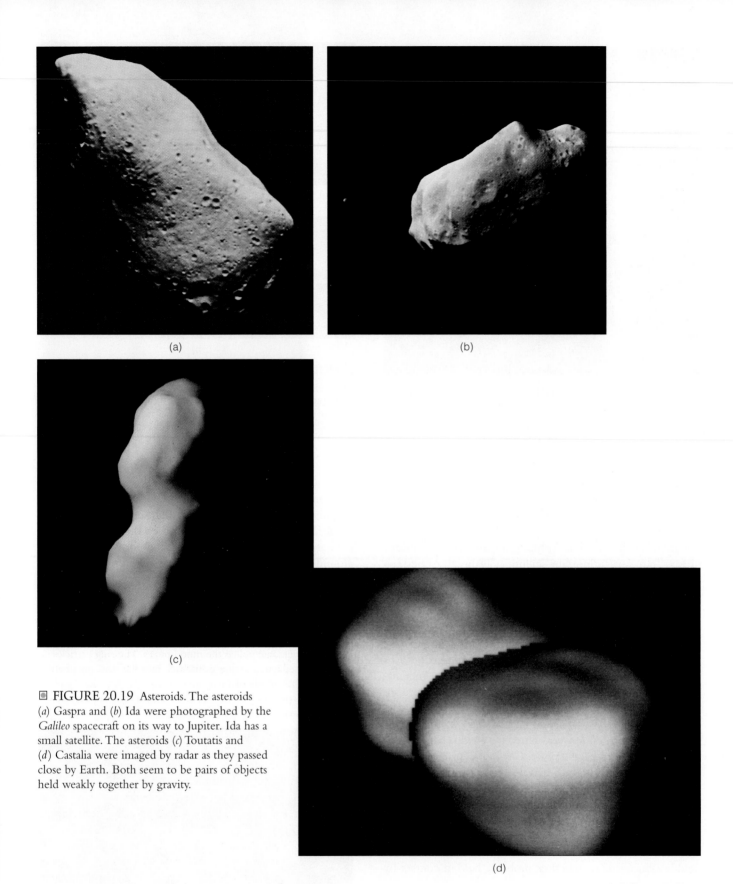

(a)

(b)

(c)

(d)

■ FIGURE 20.19 Asteroids. The asteroids
(*a*) Gaspra and (*b*) Ida were photographed by the
Galileo spacecraft on its way to Jupiter. Ida has a
small satellite. The asteroids (*c*) Toutatis and
(*d*) Castalia were imaged by radar as they passed
close by Earth. Both seem to be pairs of objects
held weakly together by gravity.

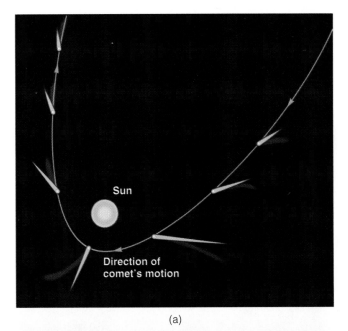

(a)

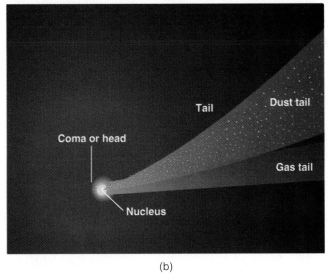

(b)

⬚ FIGURE 20.20 Comets. (*a*) Heating by the Sun causes the nucleus of a comet to evaporate, creating a thin cloud of gas that shines by reflected sunlight. We see this cloud as the head of the comet. Sunlight and streams of charged atomic particles from the Sun push some of the gas and dust away from the head of the comet, creating a tail. (*b*) There are actually two tails: a fainter, straight tail of gases and a bright, often curving tail of dust. The gas molecules, being light, are pushed rapidly away from the comet whereas the heavier dust particles move more slowly. Over a few days, the comet moves relative to the dust tail so that the tail does not always extend in a straight line from the comet.

cloud is the *coma,* or head, of the comet. If the comet comes still closer to the Sun, the evaporation of the nucleus speeds up and the coma brightens. Gas and dust are pushed away by the pressure of sunlight and particles streaming from the Sun to form the *tail* of the comet, which always points away from the Sun (Fig. 20.20b). The dust and gas shine by reflected sunlight, and the gases also emit some light of their own. The tail can be millions of miles long, but the gases in it are very thin, comparable to evaporating a teaspoon of water and dispersing the vapor through the Grand Canyon.

In 1986, Halley's comet became the first comet ever to be photographed at close range by spacecraft (⬚ Fig. 20.21). Its nucleus turned out to be a gourd-shaped icy mass about 16 km (10 mi) long, covered with a very dark surface coating. Bright jets of vapor streamed from the nucleus, probably from cracks in the ice (Fig. 20.21c). The nucleus of Halley's comet is surprisingly light, with a bulk density of about 0.2 gm/cm^3. The material of the nucleus is evidently not solid ice but more like crusty snow.

Each passage by the Sun causes a comet to lose mass, and every comet, even Halley's, will eventually evaporate or break up. During its 1986 visit, Halley's comet lost a meter or so of material from its surface. The comets we observe must be recent visitors to the inner solar system; otherwise they would long ago have disappeared. Small particles lost by comets continue to orbit the Sun. *Meteor showers* occur when Earth crosses such a swarm of orbiting particles. Some meteor showers travel the same path as known comets; others are probably the remnants of long-extinct comets.

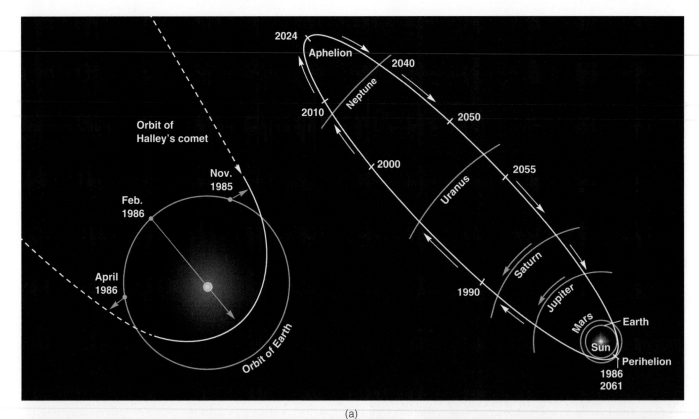

(a)

(b)

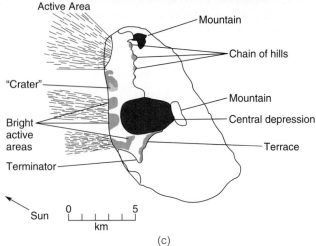

(c)

◾ **FIGURE 20.21** Halley's comet at its 1986 return. (*a*) Halley's comet passed on the far side of the Sun from Earth in 1986. The comet was far from Earth and dim and was best seen in the Southern Hemisphere. From the viewpoint of the general public, the apparition was disappointing. It is scant consolation that the apparition in 2061 should be very favorable. (*b*) This photo of Halley's comet was taken with a large telescope in 1986. (*c*) The first close-up view of the nucleus of a comet was taken by the *Giotto* spacecraft of the European Space Agency. The nucleus of Halley's comet is gourd-shaped and very dark. Escaping gas, mostly water vapor, is brightly lit by the Sun and silhouettes parts of the nucleus.

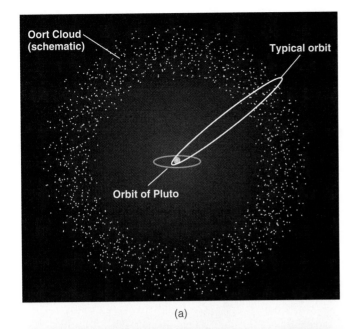

(a)

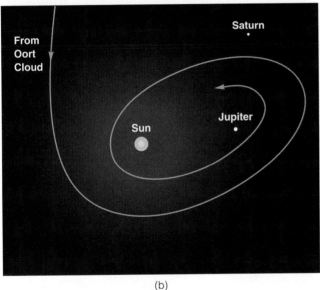

(b)

☐ FIGURE 20.22 The origin of comets. (*a*) Comets probably originated in the outer solar system and now orbit far from the Sun in the Oort Cloud. Close encounters with the outer planets probably hurled the comets into highly elliptical, long-period orbits. (*b*) On their infrequent returns to the outer solar system, Oort Cloud comets may pass close enough to one of the outer planets to have their orbits changed. In some cases, the comet can be "captured" into a short-period orbit. The comet will evaporate away after only a few thousand passes by the Sun.

Comets are thought to have originated in the outer solar system. Close encounters with the outer planets hurled the comets into extremely elliptical orbits with very long periods. Because of Kepler's second law, comets in such orbits spend the vast majority of their time near the far points of their orbits. Most comets are now thought to inhabit a vast swarm called the *Oort Cloud* extending to hundreds of times the distance of Pluto from the Sun (☐ Fig. 20.22). During

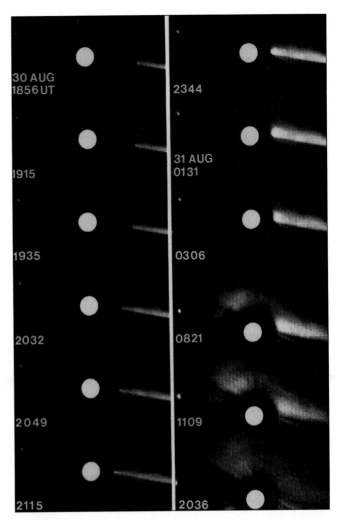

☐ FIGURE 20.23 This fascinating series of photographs was taken by an orbiting spacecraft. A bright comet approached the Sun from beyond, as seen from Earth. No observer on Earth saw the comet because it was almost directly in line with the Sun. A few hours later, the comet had vanished. It struck the Sun, evaporating instantly and leaving only a bright plume of gas visible for a few hours. The comet's effect on the Sun was about like a snowflake on a forest fire.

their rare visits close to the Sun, comets may experience close encounters with the planets and be "captured" into short-period orbits. A few comets actually hit the Sun (☐ Fig. 20.23), but their effect on the Sun is negligible. Once a comet is captured, it is only a matter of 10,000 to 100,000 years before all its volatile material is gone.

Meteors are small solid bodies that enter Earth's atmosphere. The very tiniest are slowed down gently high in the atmosphere and drift to the surface as fine dust. Larger meteors are heated to incandescence by friction as they enter the atmosphere. They flare briefly as "shooting stars" and vaporize 100 km (60 mi) or so above the surface. Meteors large enough to survive the passage through the atmosphere reach the surface as meteorites. Meteorites are divided into two main classes: **iron meteorites,** made mostly of a mixture of iron and nickel, and **stony meteorites,** made mostly of olivine and pyroxene (☐ Fig. 20.24). *Stony-iron* meteorites,

(a)

■ **FIGURE 20.24** Meteorites. (*a*) A chondrite, one variety of stony meteorite. Chondrites may be leftover raw materials from the formation of the solar system. (*b*) An iron meteorite, actually a nickel–iron alloy. Etching with a special chemical reveals the distinctive crystalline structure of the meteorite. Etching is a conclusive test for identifying iron meteorites, but the chemical mixture is dangerous and should not be used without expert supervision. (*c*) A tektite, probably rock melted by a large impact and streamlined as it flew, still molten, through the air.

(b)

(c)

as the name implies, are mixtures of the two materials. One important class of stony meteorites are the *chondrites,* which are composed of small pellets a few millimeters in diameter in a silicate matrix. The pellets may represent some of the grains that first accreted during the formation of the solar system. A few stony meteorites are almost identical to lunar rock samples and probably originated when impacts blasted material off the Moon. A few other meteorites are igneous rocks only about a billion years old and are believed to have been ejected from Mars. Related to meteorites are *tektites,* blobs of glass that probably formed when terrestrial or lunar rocks were melted by large impacts (Fig. 20.24c).

Bright meteors often look as if they fall only a short distance away. Actually, meteorites cool down and stop glowing long before they hit the surface. Unless the impact was actually seen or felt, or the meteor made an audible sound as it passed by, the meteor probably never reached the surface. If an impact is suspected, scientists collect eyewitness accounts from the area to pinpoint the location and then search for the meteorite.

CHAPTER SUMMARY

1. The solar system consists of nine major planets with their satellites, thousands of minor planets, and probably millions of comets.
2. The rocky planets of the inner solar system consist of magnesium and iron silicate minerals with metallic cores.
3. The Gas Giant planets of the outer solar system consist of enormous thick gaseous envelopes grading downward into progressively hotter and denser gases. They have roughly the same overall composition as the Sun. They may have small cores of heavier elements similar to rocky planets in composition.
4. The ice-rock satellites of the Gas Giants, as well as the planet Pluto and its satellite, consist of icy crusts with rocky cores. Most of the ice is frozen water, although ammonia, methane, and nitrogen exist in frozen form on the colder moons.
5. Temperature in the early solar system determined the composition of planets. The materials that condensed out of the initial cloud of dust and gas were the elements in the Sun minus those elements that mostly formed gases. In the inner solar system solid bodies rich in iron, magnesium, silicon, and oxygen formed. In the outer solar system solid water ice also formed.
6. Planets are round because gravity pulls all large objects into a spherical shape. Rapidly rotating planets bulge at the equator.
7. The bulk density of planets is a clue to their internal composition. Low-density objects are probably mostly ice if small, or dense gases if large. Medium-density objects are mostly rock. High densities indicate rocky bodies with dense cores.
8. Ring systems are swarms of orbiting particles. Small shepherd moons help maintain their sharp edges.
9. Comets originated in the outer solar system and were flung into extremely elongated orbits by encounters with the outer planets. This swarm of comets is the Oort Cloud. On their rare return visits to the outer solar system, they may be captured into shorter-period orbits.
10. As a comet approaches the Sun, it is heated and gives off dust and gas. The dust and gas shine by reflected sunlight to produce the glowing tail of the comet.

IMPORTANT TERMS

accretion	ices	minor planet	satellite
chondrites	iron meteorites	nucleus	shepherd moon
comet	Jovian planets	orbit	stony meteorite
gas giant	meteorite	planet	terrestrial planet

REVIEW QUESTIONS

1. The energy for the volcanoes on Io is provided by:
a. _____ decay of radioactive elements; b. _____ heating by meteor impacts; c. _____ solar radiation; d. _____ friction from gravitational and tidal forces.
2. Match each of the Galilean satellites of Jupiter with a significant surface feature:

a. Io   _____ Large multiple-ring impact basins
b. Europa   _____ Tidal heating and sulfur volcanoes
c. Ganymede _____ Very smooth surface, crisscrossed with fractures
d. Callisto _____ Large pieces of crust moved across surface

3. Saturn's satellite Titan is unusual because it:
   a. _____ is the largest satellite in the solar system; b. _____ has volcanoes; c. _____ has a thick atmosphere; d. _____ has no ice.

4. Which of the following is an accurate description of rings and the rules that govern their shape?
   a. _____ Rings are made of gas and can be any shape; b. _____ rings are separate particles and must have circular orbits above a planet's equator; c. _____ rings are separate particles and must have circular orbits, but can orbit over a planet at any angle; d. _____ rings are separate particles and can have orbits of any shape.

5. Shepherd moons help maintain the shapes of rings by:
   a. _____ keeping particles in the rings through gravity; b. _____ sweeping up any stray particles; c. _____ wearing away to replace lost ring particles; d. _____ shielding the rings from meteor impacts.

6. Why are planets round?

7. Why do most planets lack seas and oceans?

8. Why are iron, magnesium, silicon, and oxygen the principal ingredients of the terrestrial planets?

9. Why can Mercury be described as Moon-like on the outside but Earth-like on the inside?

10. If Mercury is so close to the Sun, how can it have ice at its poles?

11. How are Venus and Earth similar? How are they different?

12. Why is Venus hotter than Mercury even though it is farther from the Sun?

13. What are some of the features that indicate Mars once had liquid water?

14. Describe the major volcanic and tectonic features on Mars. How do they compare in size to similar features on Earth?

15. Describe the internal structure of Jupiter.

16. What is so unusual about Uranus's satellite Miranda? How do astronomers explain these features?

17. Describe the surface of Triton.

18. How do Uranus and Neptune differ?

19. What is the difference between a meteor and a meteorite? Why do most meteors fail to reach Earth's surface?

20. Describe what happens to a comet as it passes close to the Sun. What direction does the tail point?

POINTS TO PONDER

1. Other planets rotate at different rates than Earth. How might a future colony on some other planet keep time? (There is no single right answer.)

2. If the nucleus of Halley's Comet is 10 kilometers thick, and it loses a meter or so of material on every visit to the inner solar system once every 76 years, how much longer is it likely to last? How does that compare to the interval between ice ages or the length of the Tertiary Period? How many comets like Halley's could have evaporated since the formation of the solar system?

ADDITIONAL READINGS

Cooper, H. S. 1994. *Venus, the evening star.* Baltimore: Johns Hopkins University Press.

Littman, M. 1990. *Planets beyond.* Cambridge, Mass.: Sky Publishing.

Mechler, G. 1995. *Planets and their moons.* New York: National Audubon Society Pocket Guides, Knopf.

———. 1995. *The Sun and Moon.* New York: National Audubon Society Pocket Guides, Knopf.

Miller, R., and W. K. Hartmann. 1993. *The grand tour: A traveller's guide to the solar system.* Workman Publishing.

Morrison, D. 1993. *Exploring planetary worlds.* San Francisco: Scientific American Library, Freeman.

Neal, V., C. S. Lewis, and F. H. Winter. 1995. *Spaceflight.* New York: Macmillan.

Rogers, J. H. 1995. *The giant planet Jupiter.* Cambridge: Cambridge University Press.

The following periodicals are available by subscription, but are also widely available in bookstores and public libraries:

Astronomy (Kalmabach Publishing, 1027 N. Seventh Street, Milwaukee, WI 53233) is a well-illustrated, popular-level monthly.

Sky and Telescope (Sky Publishing Corp., 49 Bay State Road, Cambridge, MA 02238-9102) is somewhat advanced, but unexcelled for reporting the latest news and concepts in astronomy.

Scientific American (Scientific American, Inc., 415 Madison Avenue, New York, NY 10017) has articles on astronomy 5–10 times per year. These articles are surveys of important advances and are good sources for overall understanding of particular subjects.

WORLD WIDE WEB SITES

For current updates and exercises, log on to http://www.wadsworth.com/geo.

SKY ON-LINE
http://www.skypub.com/

Operated by Sky Publishing, publishers of *Sky and Telescope* magazine, this site offers information on current phenomena in the sky, including comet and asteroid positions, eclipses, and interesting planetary events. There are also many links to other astronomical sites.

NASA PHOTO ARCHIVE
http://www.hq.nasa.gov/office/pao/library/photo.html

This site is the entry point to NASA's vast photo archives. Planetary photographs and atlases are just a few of the things available.

CHAPTER

21

Stars and Galaxies

Remnants of a cosmic dust cloud hide bright young stars in this image taken by the Hubble Space Telescope.

Prologue

No question about the stars arouses more interest among non-scientists than the question of intelligent extraterrestrial life (□ Fig. 21.1). Beginning about 1960, a number of astronomers felt that improvements in knowledge about the formation of the solar system and the origin of life made it possible to make a serious estimate of the likelihood of extraterrestrial life and intelligence. Most such estimates are based on a formula devised by Frank Drake of Cornell University:

□ **FIGURE 21.1** Astronomers may not have detected extraterrestrial life, but many other people believe they have, mostly by sighting *unidentified flying objects*, or *UFOs*.

The two pictures here, one from California (*left*) and one from England (*right*), shed some light on the UFO question. They were both drawn about 1900 and show balloon-like craft with searchlights. These are exactly what people about 1900 would have expected a flying machine to look like. Clearly, the observers saw a bright light, probably a star or planet, and subconsciously added other details. (We can easily rule out other possibilities, such as aliens trying to escape notice by using Earth-type aircraft; if we have viewing devices that can see in total darkness, we can expect advanced aliens not to need searchlights.) There is no reason to believe modern UFO sightings are any different.

number of intelligent civilizations in our galaxy

= number of suitable stars in our galaxy

× number of ecologically suitable planets per star

× fraction of planets where life develops

× fraction of planets where intelligence develops

× fraction of planets that develop a technological society

× fraction of planetary lifetime that intelligence survives

Obviously, the farther down the list we go, the more speculative the numbers become. We can estimate the number of suitable stars in our galaxy at perhaps 100 billion, we have good reasons to suspect most average solitary stars have planets, and there is no reason to think that our solar system has unusually many or few planets. It is by no means unreasonable to estimate billions of planets with life.

When it comes to intelligent life, however, our guesses become less and less well informed. For more than half of its lifetime, Earth had only single-celled life. Life on land has existed for only 5 to 10% of Earth's history, and

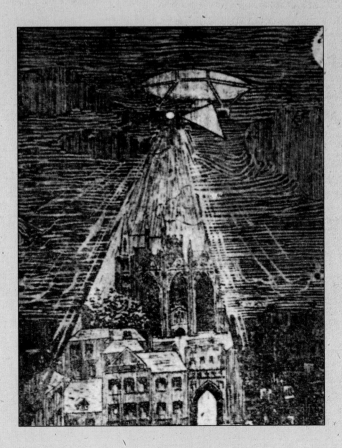

intelligent life for only a million years or so, or 0.02% of Earth's history. We have no idea how long intelligence will survive on Earth.

Recently, some scientists have begun to question whether the estimates of intelligence in the universe are too optimistic. They reason that if intelligence is as common as some scientists have claimed, at least one civilization should have developed a means of interstellar travel and colonized the galaxy. Since we see no signs of a galactic civilization, something must be wrong with the reasoning. Perhaps we are the first intelligence or one of a very few intelligences. Perhaps intelligent civilizations do not survive long, or do not have a drive to explore, or have strict rules against interfering with primitive planets. Perhaps interstellar travel on a large scale is simply too difficult. We just do not know.

Some people urge caution in making contact with extraterrestrials, but we have already revealed ourselves. Earth emits much more radio energy than the Sun, or indeed most stars, and has been doing so for half a cen-

tury. The signals from the infamous 1938 broadcast of "War of the Worlds" are now more than 50 light-years away, and there are thousands of stars within that radius of Earth. Our own radio telescopes would have no trouble detecting Earth at such a distance.

Extraterrestrials observing Earth's radio traffic would probably not be able to decipher message content (probably a good thing, considering some of the things we broadcast!), but they could still tell a lot. Radio signals would be strongest when the broadcast stations are just on the edge of Earth as seen from space. Our extraterrestrials could make a simple map of Earth's technological centers. By timing the signals, they could determine the length of our day and our speed of rotation. From this information they could determine Earth's diameter. By observing us over a year, they could determine the length of our year, the size of our orbit, and the mass of our Sun. Our own radio telescopes are entirely capable of gathering this sort of data for an Earth-like planet many light-years away.

 # INTRODUCTION

Of all the ideas in science, it would be hard to pick one greatest idea, but if we were to try, a good candidate would be this short sentence:

The Sun is a star.

Think of what this idea means. The lights we see in the night sky are suns. They must be at incredible distances to appear so faint. If we understand our own sun, we can understand the stars. Perhaps these distant suns have planets, and maybe even intelligent life.

The Sun, like most stars, shines because of nuclear reactions in its core that *fuse* hydrogen atoms into helium. Some older stars get their energy by creating heavier atoms. Nuclear reactions in stars are responsible for almost all atoms heavier than helium. If there is a second-greatest idea in science, it may well be that the atoms that make up our world—and us—formed in stars. One of the tragic ironies of modern life is that, just when we are learning so much about the heavens, human activities make it impossible for many people to see the stars at all (□ Perspective 21.1).

We call enormous numbers "astronomical." Astronomers deal with quantities of distance, mass, and energy so vast that they tax our imaginations and mathematical skills. In this chapter we will make a very brief survey of the universe beyond the solar system, examining stars and galaxies and how they evolve. We will begin with the nearest star of all, the Sun.

 # THE SUN

In many ways the Sun is an ideal star to have nearby. Its steady energy output creates a stable environment for life on Earth. For the astronomer, the Sun is as typical an example of a star as one could hope for.

The Sun has three visible layers (□ Fig. 21.2). The bright disk of the Sun is the innermost layer, or *photosphere,* with a surface temperature of about 5,500°C (9,000°F). Above the photosphere is the *chromosphere,* which is about 8,000 km (5,000 mi) deep. Thin though the chromosphere is, in one way it is the most important layer of the Sun or any other star. The thin gases of the chromosphere absorb some of the light from the photosphere beneath it. As we shall see later in this chapter, the absorption of light by atoms in the chromosphere enables astronomers to determine the chemical composition of the Sun and other stars. The outermost visible layer of the Sun is the *corona,* which is visible during total eclipses of the Sun as a ghostly white halo reaching out 800,000 km (500,000 mi) or more until it fades into invisibility. The coronal gases are very thin but also very hot, reaching temperatures of more than 1,000,000°C because atomic particles in the corona are accelerated to very high speeds by the Sun's magnetic fields.

The Sun's visible surface shows a variety of features. **Sunspots** are great dark disturbances that can be many times larger than Earth. Observations of sunspots from day to day show that the Sun rotates. The Sun is gaseous, not solid, and

Focus on the Environment: Seeing the Stars

The night sky seen by Galileo and Copernicus is a stranger to most modern Americans. Only those who camp in remote places far from artificial light see the stars as our ancestors did. Under the best conditions, about 3,000 stars are visible to the unaided eye. In most suburban settings, a few hundred stars are visible; from the centers of large cities, only a few dozen, and sometimes none at all can be seen.

Human activities diminish the view of the heavens in many ways. It takes about half an hour for the human eye to become *dark adapted,* or fully sensitive to faint light. Bright light, like auto headlights or streetlights, interferes with dark adaptation. Artificial light is also responsible for *light pollution,* the bright sky over any large city. Observatories such as Mount Wilson and Lick Observatories in California and Kitt Peak Observatory in Arizona once enjoyed dark skies, but have seen the quality of the night sky sharply degraded in recent years by light from nearby cities. Some light pollution is inevitable, but most is simply waste: light from poorly shielded fixtures illuminates the sky instead of the streets and buildings it should be lighting. Some cities have passed ordinances to limit light pollution; emphasizing the economic waste of light pollution is one of the strongest arguments astronomers have found for gaining support in their battle against it.

Human interference affects more than just visible light. Astronomers learn almost as much from radio emissions from the universe as they do from visible light, but large portions of the radio spectrum are completely swamped by human-made radio emissions. The total energy ever received from outside the Milky Way by all Earth's radio telescopes would not light a flashlight bulb for a thousandth of a second; such faint emissions are no match for radio stations broadcasting at 50,000 watts. Some astronomically critical radio frequencies are safeguarded by international agreement, but there is constant pressure to open these frequencies to commercial and government applications. The increasing use of cellular phones, for example, is greatly increasing pressure to open new radio frequencies for use.

Another subtle form of pollution has only recently gained attention. Human activities release large quantities of *particulate matter* into the air: dust from construction, soot and sulfur particles from burning fuels, and so on. Particulate matter has reduced average visibility in the eastern United States from over 150 km (100 mi) to 30 km (20 mi). Particulate matter impairs the view in scenic areas, reduces the clarity of the sky, and increases light pollution by scattering and reflecting stray light in the sky.

For large observatories, it is essential to find the best viewing conditions: dark, clear skies far from cities, and steady air without turbulence. Mountain peaks near coastlines and on islands are the best locations known. In some cases, environmentalists have opposed the construction of observatories in remote areas, but astronomers and environmentalists have a common interest in wanting to prevent further development in such areas.

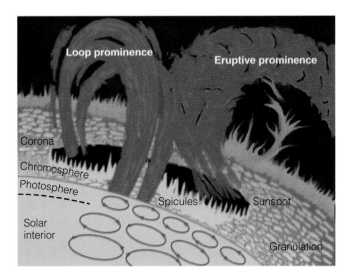

■ FIGURE 21.2 The visible surface of the Sun, showing sunspots, prominences, and other solar features. Granulation results from convection currents, the last stage in transporting energy from the solar interior. The cutaway diagrams at the lower left show convection cells that cause the granulation pattern and indicate the thickness of the photosphere and chromosphere.

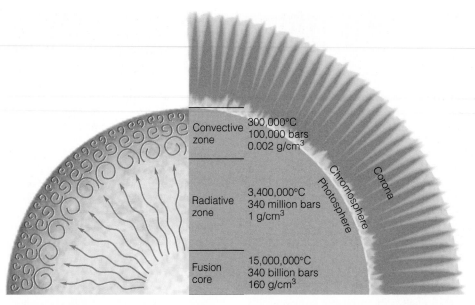

Convective zone
300,000°C
100,000 bars
0.002 g/cm^3

Radiative zone
3,400,000°C
340 million bars
1 g/cm^3

Fusion core
15,000,000°C
340 billion bars
160 g/cm^3

Chromosphere

Photosphere

Corona

■ FIGURE 21.3 The structure of the Sun. The Sun derives its energy from nuclear fusion in its small central core. Temperatures in the core are estimated at about 15,000,000°C. The tremendous pressure of 340 billion bars necessary to contain such hot gases compresses the gases until they weigh about 160 g/cm^3, or about 15 times denser than lead. Nevertheless, the gas is still gas. Deep in the Sun, energy escapes outward by radiation, but near the surface, convection occurs, and the heat is transported by rising currents of hot gas. It takes about 100,000 years for energy to get from the core to the surface of the Sun.

does not rotate uniformly. The equatorial regions rotate in about 28 days, but the middle latitudes take about 30 days. Small sunspots persist for a week or so, but large ones may last through several rotations of the Sun. Sunspots appear dark only because they are cooler and less luminous than the average surface of the Sun. In reality, sunspots are hotter than many stars and give off appreciable light of their own. *Flares,* or intense outbursts of electrically charged particles, are associated with sunspots. In the chromosphere, great surges of hot gas called *prominences* can rise hundreds of thousands of kilometers above the surface.

Sunspots and other solar disturbances rise and fall in frequency in a cycle of about 11 years. Prominences and flares emit streams of electrons and protons that reach Earth a day or so after eruption. When bursts of these particles strike Earth's magnetic pole regions, they cause *auroral displays* (Chapter 15). More important, they generate enormous electric currents in the upper atmosphere that disrupt communications and even damage electronic equipment. Fortunately, since the particles take about a day to reach Earth, it is possible to monitor the Sun and issue warnings when major disturbances occur.

Even though the Sun is just an average star, it puts out energy on a staggering scale. It would take all the power plants on Earth over a million years to produce as much energy as the Sun does in one second. The Sun derives its energy from **nuclear fusion,** in which four hydrogen atoms collide, in a series of steps, to form a helium nucleus. This process is much the same as takes place in a thermonuclear weapon, but on a vastly greater scale. The fusion occurs in the Sun's small central *core* where temperatures are estimated to be about 15,000,000°C (■ Fig. 21.3).

DISTANCES TO THE SUN AND STARS

We can determine the distances to objects in the solar system and to nearby stars by *triangulation* (■ Fig. 21.4). By observing an object from two stations a known distance apart, we can find the distance to the object. Objects in the inner solar system can be observed from opposite sides of Earth. Astronomers initially used this technique to calculate the distance to Venus using observations of the planet crossing in front of the Sun that were made on opposite sides of Earth (Fig. 21.4a). A similar technique was used to determine the distance to asteroids passing very close to Earth (Fig. 21.4b). Once one distance was obtained, Kepler's third law could be used to determine the relative spacing of the planets' orbits. Thus, with the actual distance to one object, the distance of every object in the solar system could be established. (Nowadays we can bounce radar signals off nearby planets and measure the distance directly.)

The stars are so far away that they look the same from any point on Earth, but we can make observations from opposite sides of Earth's orbit, which are 300 million km (186 million mi) apart (Fig. 21.4c). When we photograph the stars at six-month intervals, some of them appear to shift very slightly because of Earth's motion. This shift, or *parallax,* is extremely tiny, but it can be measured for nearby stars.

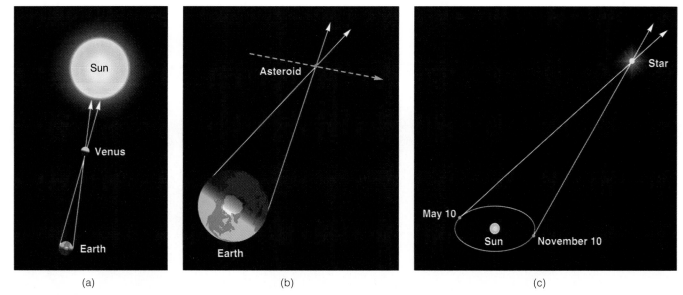

FIGURE 21.4 With triangulation, observations made on opposite sides of Earth (a) of Venus crossing, or transiting, the face of the Sun or (b) of asteroids passing very close to Earth could be used to calculate the distance to those objects. (c) The distance to a nearby star is determined by measuring the position of the star from opposite sides of Earth's orbit, 300 million km (186 million mi) apart.

The distances to the stars are so great that we need a new unit of distance, the **light-year,** to express them. The light-year is not a unit of time; it is the distance light travels in a year, or about 10 trillion km (6 trillion mi). The nearest star beyond the Sun, Alpha Centauri, is 4.3 light-years away. Measuring the parallax shift of Alpha Centauri is equivalent to viewing a quarter at a distance of 3.5 km (2.2 mi).

Films like *Star Wars* create the impression that traveling from star to star is only a bit more complex than driving down to the corner store for a loaf of bread. In fact, distances in the universe are so vast it is hard to comprehend them. Everyday speeds will not begin to cover the distances in the universe. For example, an automobile traveling a steady 90 km (55 mi) per hour would take about two and a half days to cross the United States, six months to travel to the Moon, and 54 years to cover the distance to Venus at its nearest. If we had started out from the Sun at that speed in A.D. 1, we would now be just beyond Saturn, only a third of the way out of the solar system.

Light travels at 298,000 km (186,000 mi) per second, so it could circle Earth seven times in one second and reach the Moon in 1.5 seconds. The Sun is about 8.5 minutes away at the speed of light. Light takes only about 5.5 hours to travel from the Sun to Pluto, but 4.3 years to reach the next star. Light takes 100,000 years to cross our galaxy, 2.3 million years to travel to the nearest galaxy like our own, and about 12 to 15 billion years to reach us from the edge of the visible universe.

The idea of covering interstellar distances at high speed is appealing, but highly problematical. To accelerate a hundred-ton spaceship (comparable to a space shuttle) to only one-tenth of the speed of light would require Earth's entire energy output for about two months. The energy needs increase much faster than speed; to reach nine-tenths of the speed of light would take Earth's energy output for roughly half a century. To reach the speed of light itself would require infinite energy, and nothing that we know in physics suggests that we can ever exceed the speed of light.

THE STARS AS SEEN FROM EARTH

Every culture has told stories about the stars and grouped the stars into **constellations** (Fig. 21.5). The 89 constellations now accepted in astronomy are based on traditional European and Near Eastern star lore, with a few seventeenth- and eighteenth-century additions. The constellations we see from Earth are patterns of stars more or less in the same direction as we see them from Earth. Most of the objects in a constellation are at widely different distances and are not physically connected with each other. For example, "in" the constellation Virgo, we might find the Moon (380,000 km [236,000 mi] away), Jupiter (780 million km [484 million mi]), the bright star Spica (220 light-years), and the Virgo Cluster of galaxies (40 million light-years). If we were to travel to a distant star, none of the constellations would be recognizable any more (Fig. 21.6).

Only a few of the brightest stars have proper names. Most of the proper star names now used in astronomy were invented by the Arabs. Stars are also named from their designations in star catalogs. One system, introduced in 1603 by Johannes Bayer, uses Greek letters, starting with alpha for the brightest star in a constellation and proceeding in order of

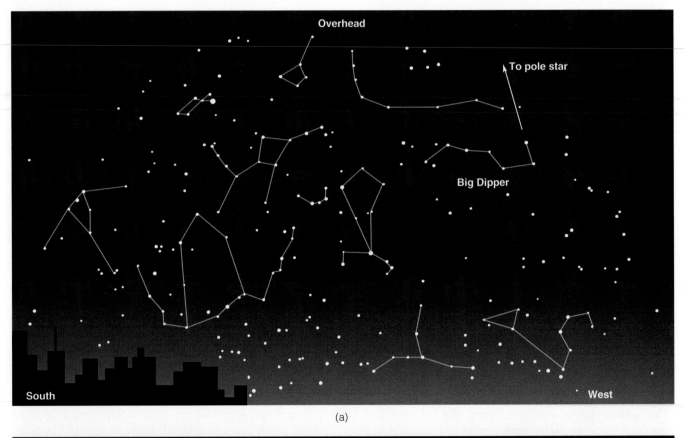

(a)

(b)

▣ **FIGURE 21.5** As these star maps of the same region of the sky show, constellation figures differ radically from one society to the next: (*a*) Western constellations; (*b*) traditional Chinese constellations. Both maps are looking south at 9 P.M. on June 21.

■ **FIGURE 21.6** The constellations we see from Earth are chance alignments of stars. The familiar Big Dipper would not look the same from anywhere else in the universe: (*a*) view of the Big Dipper from Earth; (*b*) view looking down from the north celestial pole; (*c*) view from a point (X) in space.

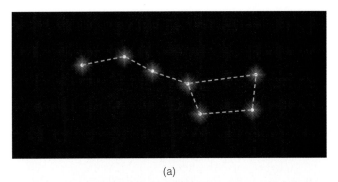

(a)

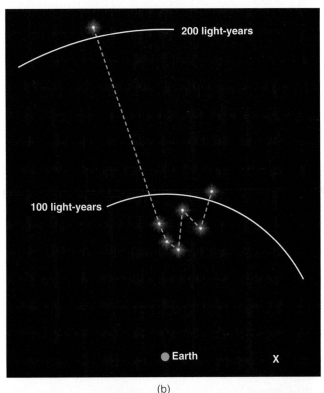

(b)

decreasing brightness. Thus, Sirius, the brightest star next to the Sun, is also Alpha Canis Majoris, the brightest star in the constellation Canis Major. The overwhelming majority of very faint stars have no designation unless they happen to be of unusual interest.

Stars vary in brightness for two reasons. First, they are at varying distances from us, so distant stars appear fainter. Second, stars actually do emit different amounts of light. For example, two bright stars in the summer sky, Altair (16 light-years away) and Deneb (about 1,600 light-years away) appear about equally bright, even though Deneb is 100 times farther away. Light intensity drops off in proportion to the square of the distance; thus, Deneb must be emitting about 100×100 or 10,000 times as much energy as Altair.

Astronomers describe brightness in terms of **magnitude,** carrying on a tradition begun by the ancient astronomer Hipparchus nearly 2,000 years ago. Faint objects have large magnitude numbers; bright objects have small numbers. Hipparchus assigned the 20 or so brightest stars in the sky to magnitude 1. Stars just visible to the unaided eye were magnitude 6. Astronomers have refined this system so that each step of one magnitude corresponds to an increase or decrease of about 2.5 times in brightness, and a five-magnitude step means a 100-fold increase or decrease. Very bright objects can have zero or even negative magnitudes. The nearest star beyond the Sun, Alpha Centauri, has a magnitude of zero. Sirius, the brightest star next to the Sun, has a magnitude of −1.4. Venus at its brightest has a magnitude of −4, the full moon about −12, the Sun about −27.

The brightness of stars as seen from Earth is called *apparent magnitude.* If we can determine the distances to the stars we can determine their true brightness compared to each other. Astronomers define the *absolute magnitude* of a star as the brightness it would have at a distance of 32.6 light-years.

(c)

The absolute magnitude of the Sun is 4.8; it would be entirely invisible to the unaided eye from 50 light-years away, a stone's throw in interstellar terms. This humbling fact suggests that interstellar navigation might just be a bit more challenging than many science fiction epics would have us believe.

Most stars are constant in brightness, but there are many types of **variable stars** whose brightness varies. *Eclipsing binaries* consist of a faint star orbiting a bright one. Normally, we see the combined brightness of both stars, but when the faint star hides the bright one, the brightness drops. By timing the disappearances of eclipsing binaries, astronomers obtain valuable information about the diameters of the stars. Other variable stars pulsate and vary in brightness, some regularly and rhythmically, others spasmodically and unpredictably. Other stars, called *novas,* experience explosive outbursts that may increase the brightness of the star by hundreds of times. Finally, some stars experience catastrophic explosions that may destroy the star. These stars, called **supernovas,** flare briefly to billions of times their normal brightness and are perhaps the most violent events we ever witness.

 ## STARLIGHT AND WHAT IT TELLS US

White light is actually a mixture of the colors of the rainbow. When the colors are spread out by a prism or other means, they form a sequence called the *spectrum* (red, orange, yellow, green, blue, and violet). The colors of visible light are actually only a tiny part of the *electromagnetic spectrum,* which also includes radio, infrared, ultraviolet, X-ray, and gamma radiation (Chapter 15).

A detailed examination of the Sun's spectrum reveals that it is crossed by thousands of fine dark lines. These lines are produced when atoms absorb or emit specific wavelengths of light. Each element has its own characteristic pattern of lines. Not only do different elements have different spectral lines, but the pattern produced by each element depends on the temperature and pressure in the star. Also, spectral lines

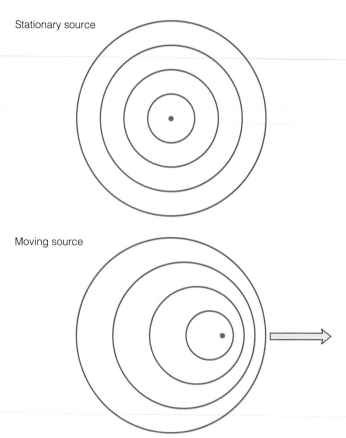

Stationary source

Moving source

☐ **FIGURE 21.7** The Doppler effect. Waves from a moving source are compressed to higher frequency and shorter wavelength as the source approaches, and stretched to lower frequency and longer wavelength as the source moves away.

are modified by electrical or magnetic fields. The study of spectra and the information they contain is called *spectroscopy.* Spectroscopy allows astronomers to learn much about the physical conditions in stars. Spectral lines were discovered more than 140 years ago, yet astronomers are still finding new ways to get information out of the spectra of stars.

Besides allowing astronomers to determine the chemical compositions of stars, spectroscopy provides another powerful tool: the **Doppler effect.** Most people have noticed that the pitch of an ambulance siren drops as the ambulance goes by. As the sound source moves toward us, the sound waves are compressed, making the pitch of the sound higher (☐ Fig. 21.7). As the sound source moves away from us, the sound waves are stretched out, lowering the pitch of the sound. In a similar way, light from an approaching star has its wavelengths shortened, or *blueshifted,* and light from a receding star has its wavelengths lengthened, or *redshifted.*

Light that is redshifted or blueshifted merely changes color. There is no way to tell from color alone that a star is moving. The key to the Doppler effect is that spectral lines also move (☐ Fig. 21.8). The change in position is easily measured on a photographic plate. The Doppler effect allows astronomers to determine if objects are moving toward or away from Earth. Using this information, we can detect

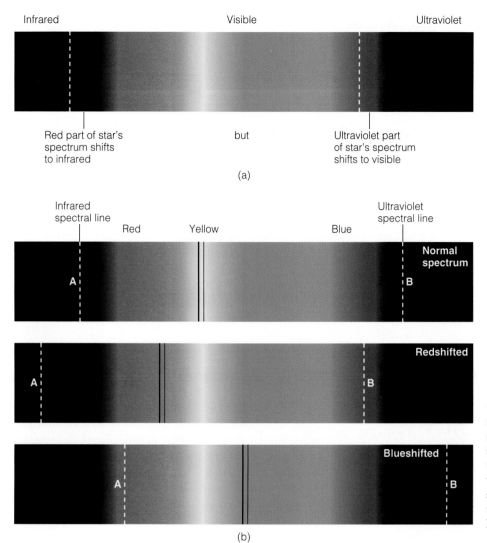

Infrared   Visible   Ultraviolet

Red part of star's spectrum shifts to infrared   but   Ultraviolet part of star's spectrum shifts to visible

(a)

Infrared spectral line   Red   Yellow   Blue   Ultraviolet spectral line

Normal spectrum

A   B

Redshifted

A   B

Blueshifted

A   B

(b)

■ FIGURE 21.8 Red shifting. (*a*) Contrary to many popular illustrations, redshifted stars do not look red; as the visible spectrum of a star is shifted to longer wavelengths, its ultraviolet spectrum is shifted into the visible range. All the colors of the spectrum are still present. (*b*) The key to the Doppler effect is that spectral lines change position.

the rotations of stars and galaxies and the movements of multiple stars that are too close together to see individually.

The stars are not fixed, but move measurably over the years. Most stars would take centuries to move the apparent width of the Moon, but they do move. Astronomers can measure the motion of the stars on photographs taken years apart. If we know the distance to a star, we can calculate its velocity across our line of sight. The Doppler effect enables us to determine the star's velocity along our line of sight. From these two velocities, we can determine the true velocity and direction of the star's movement in space.

While the stars are in motion, we ourselves are moving too. Just as a driver in a snowstorm sees the snowflakes appear to radiate away from his direction of motion, most of the nearby stars in the sky appear to be moving away from a certain region of the sky. This direction, in the constellation Hercules (straight overhead for U.S. observers in the early evening in summer), is the direction the solar system is moving at about 20 km (12 mi) per second, or one light-year in 16,000 years.

Even to the unaided eye, stars range in color from reddish through yellow to bluish white. When we determine distances to the stars, still more differences appear: some very nearby stars are extremely faint, while very distant stars are sometimes very luminous. Spectroscopy provides a means of making sense out of the variety of the stars. Astronomers divide stars into **spectral classes,** labeled with letters. The major classes of stars, from hottest (blue–white) to coolest (reddish), are now designated O, B, A, F, G, K, and M. For stars that fall between the main classes, subdivisions are used: an F5 star is halfway between classes F and G, for example. Our own Sun is classed as G2.

In 1912 Enjar Hertzsprung and Henry N. Russell plotted a graph of absolute magnitudes of stars against spectral class. They found that most stars plotted on a diagonal line, with the O stars brightest and the M stars faintest (■ Fig. 21.9). Such a diagram is called a **Hertzsprung-Russell diagram,** or *H-R diagram* for short. Astronomers refer to the main band of stars as the **main sequence.** The Sun is a main sequence star. Some stars are plotted well above the main

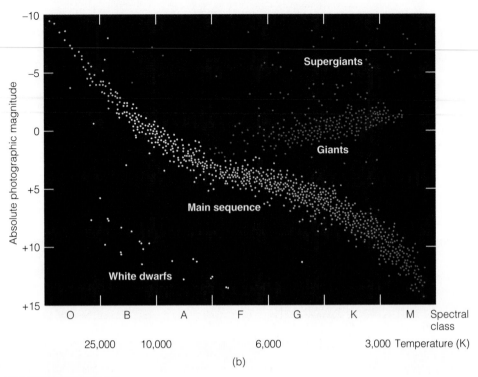

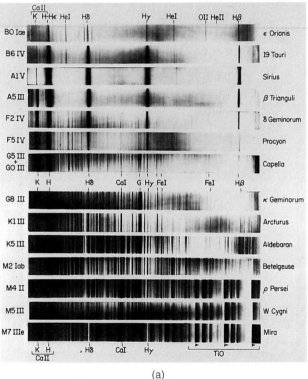

FIGURE 21.9 Spectral classification of stars. (*a*) Spectra of stars of different classes. Note how different spectral lines are prominent only in certain classes. The very hottest stars show atoms stripped of many of their electrons. In cool stars even a few sturdy molecules can exist. (*b*) The Hertzsprung-Russell diagram is a powerful tool for studying stars. Most stars fall on the main sequence. Giant and supergiant stars plot in the upper right of the diagram, while dwarfs plot below the main sequence. (Astronomers use the Kelvin temperature scale, which measures temperature from absolute zero. Kelvin temperatures are 273 degrees higher than centigrade temperatures.)

sequence, meaning they are more luminous than most stars. These stars are called *giants* and *supergiants*. Still other stars plot well below the main sequence, meaning that they emit light very feebly. These stars are called *dwarfs*.

The color of a star, or any other very hot object, is directly related to its temperature. An incandescent lightbulb filament and a red giant star emit light of nearly the same color because they are at about the same temperature (about 2,700°C or 4,890°F). Also, objects of the same temperature give off the same amount of light per unit area; if two stars of the same color and at the same distance differ in brightness, it is because the stars differ in size. Giant and dwarf stars are literally giants and dwarfs in size compared to normal stars. The Sun is 1.3 million km (800,000 mi) in diameter; giant stars can be hundreds of millions of kilometers in diameter, while dwarf stars may be only a few thousand kilometers in diameter.

 ## THE BIRTHS OF STARS

Stars begin to form when clouds of interstellar gas and dust start to contract through their own gravity. In interstellar space, far from other stars, a cloud of gas can be very thin and still be dense enough to begin contracting. Many astronomers believe violent explosions of older stars create shock waves in nearby dust and gas clouds that help start the contraction process; at the same time, the remnants of the exploded stars enrich the gas in heavy chemical elements. In all likelihood, a cloud will condense into many stars, which form a *star cluster.*

When part of an interstellar cloud contracts, most of its mass falls to the center, adding energy to the center and heating it up. The center of the cloud becomes a *protostar,* emitting mostly infrared radiation. Finally, the temperatures and pressures within the protostar reach the point where nuclear reactions begin, and the star "turns on." Matter in the surrounding disk may accrete to form planets or, if the condensations are massive enough, companion stars. A protostar must be at least 80 times as massive as Jupiter to become a true star.

STARS WITH COMPANIONS

Many, perhaps most stars, have companions, either other stars or planets. Some stars are **binary stars** or *multiple stars,* with several stars orbiting around one another. Our nearest neighbor, Alpha Centauri, is a triple star, with a *primary* star much like our Sun, a smaller and cooler *secondary* star orbiting about as far away as Uranus orbits our Sun, and a very faint companion many times farther away from the primary than Pluto is from the Sun. Binary stars are of great value to astronomers because analysis of their orbits enables astronomers to determine the masses of stars.

From what we know of the formation of the solar system, it seems very likely that solitary stars will have planets, formed from the material that did not condense into the central protostar. Perhaps widely separated multiple stars like Alpha Centauri have planets as well. A number of stars are known to have disks of solid matter orbiting them, probably solar systems in the making (see Fig. 1b on page 520). Locating other planetary systems is a very great challenge. One approach is to detect very tiny, regular variations in the positions of stars. As a planet orbits a star, the star and planet both orbit around their *center of mass;* if the star and planet were joined by a beam, the point where the beam would balance is the center of mass. Tiny motions of the star can also be detected using the Doppler effect. The motions are very tiny: it would be hard to detect the effect of Jupiter on the Sun from a nearby star, for example. Another approach is to block the light of the star and search for the reflected light from planets. All of these techniques are extremely difficult to carry out with existing technology, but are becoming more feasible as instruments improve. Beginning in 1993, evidence for other planetary systems has accumulated rapidly (◎ Table 21.1). Some of these newly discovered objects are so massive they may be so-called *brown dwarfs,* which are on the borderline between large planets and tiny stars.

TABLE 21.1 Probable Planets Around Other Stars (as of March 1997)

STAR NAME	DISTANCE FROM EARTH (light-years)	ORBITAL PERIOD	DISTANCE FROM PARENT STAR (millions of km)	MASS COMPARED TO EARTH	MASS COMPARED TO JUPITER
HD 114762	140	84 days	51	>2,860	>9
PSR 1257+12	1,000	23 days	28		>.000047
		65 days	54	>3.4	>.011
		98 days	70	>2.8	>.0088
		170 years	6,000	100	0.3
Gliese 229	18	200 years	6,600	6,300–19,000	20–60
51 Pegasi	40	4.4 days	7.5	>190	>0.6
70 Virginis	80	117 days	65	>2,050	>6.5
47 Ursae Majoris	46	3 years	315	>725	>2.3
Rho Cancri	46	15 days	17	>285	>0.9
Lalande 21185	8.2	30 years	1,600	350	1.1
		6 years	330	285	0.9
Tau Bootis	49	3.3 days	7	>1240	>3.9
Upsilon Andromedae	54	4.6 days	8.6	>190	>0.6
16 Cygni	70	2.2 years	255	476	1.5
PSR B1620-26	12,400	100	6,000	100	0.3

As of March 1997, a dozen other stars with very massive companions (possible brown dwarfs or very large planets) are known and as many others reported but unconfirmed. The list is expanding too fast to keep up with.

The Hubble Space Telescope

One of the oldest dreams of astronomers was placing a large telescope in space beyond the distortions imposed by Earth's atmosphere. This dream was realized in 1991 when the Hubble Space Telescope was launched. The telescope was named for Edwin Hubble, a pioneer in the study of distant galaxies. The dream quickly turned into a nightmare when it was discovered that the main telescope mirror had been ground to the wrong shape. Although the defect could be somewhat compensated by computer processing, a special space shuttle mission was required in 1993 to correct the problem.

Once repaired, the Hubble Space Telescope equaled or surpassed astronomers' initial expectations. Among the most spectacular images it returned were highly detailed images of regions of new star formation. In some images, disks of dark matter, probably similar to the cloud from which our solar system formed, can be seen surrounding newly formed stars. Other images show ragged clouds of dark dust and gas being slowly blown away by the pressure of radiation and atomic particles streaming away from new stars. These images provide a much closer look at the details of star formation than had ever been possible before; they have confirmed many of the ideas astronomers held concerning the formation of stars and solar systems (Fig. 1).

(a)

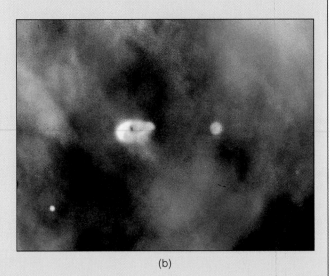

(b)

FIGURE 1 (*a*) This spectacular photo taken by the Hubble Space Telescope shows a region of young star formation. Radiation from newly formed stars is blowing away the dust and gas cloud that gave birth to the stars, but areas of thick material are more resistant and protect the cloud behind them, leaving the eerie dark pillars as remnants of the original cloud. (*b*) This close-up view of another region of young star formation shows several disks surrounding young stars. These disks are probably the early stages of the formation of a solar system. Our own solar system is believed to have formed from a similar disk.

 EVOLUTION OF STARS

Once stars begin to shine, they assume a position on the main sequence and tend to stay there, brightening somewhat as they age. Almost every aspect of the life of a main sequence star is determined by one fact: its mass. Very faint stars called *red dwarfs* emit light feebly and remain cool. More massive stars are hotter and brighter. Massive stars have more

fuel to sustain their output, but use it more rapidly than faint stars do. Red dwarf stars can last tens or hundreds of billions of years, doling out their fuel at a miserly rate. Our own Sun will shine as a main sequence star for perhaps 10 billion years, whereas a bright blue-white supergiant like Deneb or Rigel might last only a few million years.

When main sequence stars run out of available fuel, the nuclear reactions in the center of the star die out. Without

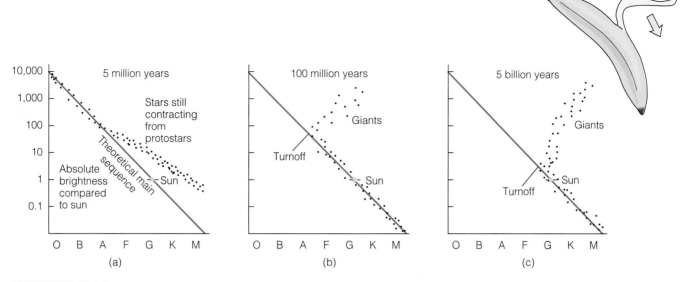

■ FIGURE 21.10 (a) In very young star clusters, all the stars are on the main sequence. In fact, the least massive stars may not even have reached the main sequence. (b) As the star cluster ages, the brighter stars become giants and leave the main sequence. The older the cluster, the farther down the main sequence the turnoff occurs. (c) As the H-R diagram for a star cluster evolves, it comes to resemble a partially peeled banana.

the intense radiation pressure produced by these nuclear reactions, the star begins to contract under its own gravity. As matter falls in toward the center of the star, it releases gravitational energy that heats the star until finally a new sequence of nuclear reactions may begin. The renewed energy output heats the outer part of the star, causing it to expand. At this point the star leaves the main sequence. The outer layers expand and cool, causing the star to redden. The star becomes a **red giant** or even a *supergiant.*

The most massive stars leave the main sequence first. We can see the evolution of stars clearly by plotting H-R diagrams for star clusters whose members all formed at the same time (■ Fig. 21.10). Very young clusters may even have remnants of their parent gas clouds still visible and contain nothing but main sequence stars and stars that have not yet reached the main sequence. In older clusters, some of the brightest stars are leaving the main sequence, and in very old clusters, most of the stars brighter than the Sun have left the main sequence. As a star cluster evolves, its H-R diagram comes to resemble a banana with a strip peeled from it; as the star cluster ages, the giant star branch becomes larger, and the branching point moves farther down the main sequence.

 THE FATES OF STARS

All objects in the universe exist because of a balance between gravity and some counteracting force. In Chapter 2, we saw how the fundamental forces of nature govern the structure and bonding of atoms. It is fitting that we return

to these same basic forces here when we examine matter on the largest scale.

Left to itself, gravity would pull all masses together to a central point. In planets, gravity is counteracted by the atomic bonding between atoms and the repulsion between the negatively charged electrons of neighboring atoms. In normal stars, gravity is counteracted by the thermal motion of the atoms in the hot gas of the star and the outward pressure exerted by radiation.

When the radiation pressure in a star falters, gravity begins pulling the gas of the star inward. As the gas falls inward, it gains energy and heats up the interior of the star. The gas also becomes more tightly compressed, and the pressure increases. At this point, one of three alternatives ensues depending on the mass of the star: the temperature and pressure inside the star rise until a new generation of nuclear reactions begins, the star continues to collapse until electromagnetic or nuclear forces stop the collapse, or the star collapses until gravity crushes its mass into a single point—a *black hole.* The more massive the star is, the more dramatic its end will be.

Small stars up to several times the mass of the Sun become red giants. The Sun will experience a fate typical of most small stars. In about 10 billion years, the Sun will have used up all the hydrogen in its core. As the Sun's hydrogen supply runs out and its energy output falters, it will begin to contract. As the Sun contracts, its internal temperature and pressure will increase until helium begins to fuse to make carbon. The new energy output will heat the outer gases of the Sun, causing them to expand and probably making Earth too hot for life. Eventually, the Sun may expand to envelop Earth itself.

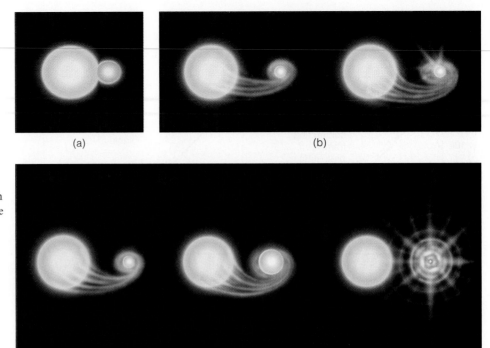

■ **FIGURE 21.11** Close binary stars can evolve in strange ways. (*a*) Some binary stars swell until the two stars come into contact with each other. (*b*) A nova occurs when a white dwarf companion draws matter from its primary star and accumulates a thick enough layer for nuclear reactions to begin on the surface of the star. (*c*) A Type I supernova occurs when a white dwarf accretes a large envelope of gas. When the pressure deep in the star becomes great enough, the star resumes nuclear fusion and blasts off its outer layers.

(a)

(b)

(c)

Many red giants pulsate and vary in brightness. Some red giants pulsate rhythmically, while others flare up in irregular bursts. Red giant stars also shed matter into space, some steadily, others violently. Eventually, all red giants lose their thin outer envelopes. The core runs out of nuclear fuel and collapses until a new counteracting force comes into play. The electron shells of the atoms are crushed out of existence, and the electrons wander between densely packed atomic nuclei. The forces between these electrons prevent the star from collapsing further. By this time, the star's diameter may be only about as large as Earth's, even though the star is as massive as the Sun. The matter of the star, called *degenerate matter*, is so dense that a teaspoonful would weigh many tons on Earth. In this final stage, the star is a **white dwarf.** If Earth survives the Sun's red giant stage, the final white dwarf will emit only a fraction of the present energy of the Sun, and Earth will be frozen solid.

Binary or multiple stars that are far apart evolve independently of one another. When multiple stars are close together, they can exchange material with one another and evolve in very complex ways (■ Fig. 21.11).

If the partner star is a white dwarf, some very violent events can happen. If a large amount of hydrogen accumulates on a white dwarf, nuclear fusion can occur on the surface of the star (Fig. 21.11b). The resulting outburst causes the star to brighten briefly by hundreds of times, becoming a *nova* (Latin for new). Even more dramatic outbursts are possible: the white dwarf can accumulate mass until its internal pressures become great enough for a new cycle of nuclear reactions to begin, for example, fusing helium to carbon. The renewed nuclear activity blows the white dwarf apart, creating a *Type I supernova* (Fig. 21.11c).

Very massive stars have an even more dramatic end: they become *Type II supernovas,* stars that explode and briefly outshine all the other stars in our galaxy combined (■ Fig. 21.12). Because old stars lose mass in the red giant stage, it is hard to predict exactly which stars will meet this most violent of fates.

Red giants with 20 or so times the mass of the Sun develop cores with great enough temperatures and pressures to form elements heavier than carbon: carbon fuses to form oxygen, oxygen fuses to silicon, silicon to iron (Fig. 21.12a). Eventually, the core becomes so massive that it cannot withstand the pressure of gravity. The core collapses until a new counteracting force halts the collapse, but the only forces capable of doing this are nuclear forces. Thus, the core of the star collapses until it becomes, in effect, a gigantic atomic nucleus: a *neutron star* (Fig. 21.12b). It may have more mass than the Sun but be only 15 km (10 mi) across. A teaspoonful of this matter would weigh thousands of tons on Earth.

Where there was once the hot, dense interior of the star, there is briefly a void. The newly formed neutron star core, more massive than the Sun but not much bigger than a mountain on Earth, has an incredible gravitational pull. In this enormous gravitational field, the matter of the star falls inward, reaching perhaps a tenth of the speed of light.

The results are impressive, to say the least. Several times the mass of our Sun crashes into the surface of a neutron star at up to a tenth of the speed of light. Nuclear reactions run rampant; enough energy and fast-moving nuclear particles are available to build nuclei at least as heavy as plutonium and probably far heavier. The neutron star core itself is almost incompressible even under these conditions, and the

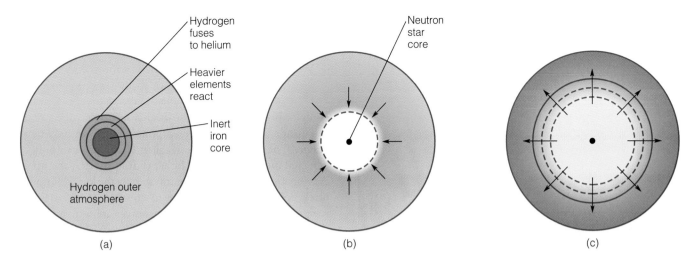

Hydrogen
fuses
to helium

Heavier
elements
react

Inert
iron
core

Hydrogen outer
atmosphere

(a)

Neutron
star
core

(b)

(c)

▣ FIGURE 21.12 A Type II supernova occurs when a massive red giant collapses. (*a*) The red giant has a core with concentric shells. In progressively deeper shells, heavier elements are being consumed. The innermost core consists of inert iron. (*b*) The core eventually becomes massive enough to collapse to a neutron star. The outer layers of the star fall inward onto the collapsed core. (*c*) The infalling material is compressed and heated to an almost unimaginable degree. Nuclear reactions run rampant. The impacting gas bounces off the neutron star core and outward as a shock wave that tears off the outer layers of the star, exposing the nuclear core to view.

impacting matter rebounds as a shock wave. This event, the *core bounce,* aided by energy from nuclear reactions, tears the star apart. In a few hours, the shock wave reaches the surface, tearing off the outer layers and exposing the hot interior of the star (Fig. 21.12c). The star brightens to billions of times its normal brightness. Most of the light is actually due to the decay of radioactive cobalt and nickel, created from iron in the core of the star.

Supernovas are commonly seen in distant galaxies. If our own galaxy has supernovas as often as other galaxies, one probably occurs every few years. Yet only a dozen or so supernovas in our own galaxy have been witnessed from Earth. The rest have been obscured by gas and dust clouds in our galaxy. The last visible supernovas in our galaxy happened in 1572 and 1604.

Almost four centuries of not-very-patient waiting by astronomers ended in 1987 with the first supernova visible to the unaided eye since 1604 (▣ Fig. 21.13). The supernova occurred not in our galaxy, but in the Large Magellanic Cloud, a small satellite galaxy to our own about 180,000 light-years away. Supernova 1987 confirmed the generally accepted picture of supernovas. All the phenomena astronomers expected to see after the outburst have occurred on schedule.

A supernova can be recognized for many thousands of years by the expanding shell of gas blasted off the star. Such a shell, detectable both in visible light and by its radio emissions, is called a *supernova remnant.* The remnant neutron star is often detectable as a pulsating radio source, or *pulsar.* Pulsars emit tremendous bursts of energy in radio, visible light, and X-ray wavelengths because of matter falling onto small regions of the neutron star (▣ Fig. 21.14). These pulses

▣ FIGURE 21.13 Supernova 1987 at its brightest. Supernova 1987 was the first bright supernova to be studied with modern methods.

are extremely regular, ranging in period from 0.001 second to a few seconds. The period of the pulses is the rotation period of the neutron star. As the parent star collapses, its rotation speeds up, just as a pirouetting skater speeds up by drawing in her arms.

Supernova remnants often appear to be associated with areas of new star formation, and it is very likely that a supernova triggers the formation of new stars. If the expanding blast wave from a supernova strikes a cloud of interstellar gas and dust, it can compress the cloud enough for parts of the cloud to start to contract gravitationally. The supernova

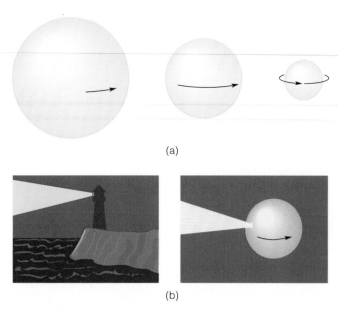

(a)

(b)

■ **FIGURE 21.14** Neutron stars and pulsars. (*a*) Just as an ice skater spins faster by pulling her arms in, a neutron star spins faster than its parent star. (*b*) Pulsars are probably much like lighthouses; we detect the pulses when the rotation of the star sweeps a beam of radiation over us. Pulsars emit radio waves, visible light, and X-rays.

debris will also mingle with the cloud, enriching it in heavy elements.

In normal stars, it is not possible to form large amounts of elements much heavier than iron. Some heavy elements form in red giant stars, but very heavy elements like lead, gold, and uranium form only in supernovas. The fact that we find these elements in the solar system is a sign that our Sun is perhaps a third-generation star. The Sun is only a third as old as the Milky Way Galaxy, and there was time for many cycles of stellar birth and death before the Sun formed. The

implications of this idea are profound. Most of the atoms in our world formed in a star billions of years ago and many light-years away.

If the collapsed core of a supernova is more than 1.4 times as massive as the Sun, it will not form a neutron star. No known force can halt the gravitational collapse. The star will contract until its gravity is so immense that not even light can escape. The star becomes a point, detectable only by its gravity. These bizarre objects, called *black holes,* can only be detected indirectly. A black hole orbiting another star might draw matter from the companion star. As the gas fell into the black hole, it would accelerate to enormous velocities and emit intense X rays. A few X-ray sources are so massive that they appear likely to be black holes (■ Fig. 21.15).

BEYOND THE NEAREST STARS

Most of the stars we see with the naked eye at night are within a few hundred light-years. A few are as far away as 2,000 light-years, only about 1/50 the diameter of our galaxy.

The **Cepheid variables** give us a yardstick to our own and other galaxies. Cepheid variables are yellow-white giant stars that pulsate and vary in brightness, but in a regular way. The brighter the absolute magnitude of a Cepheid, the larger the star is and the slower it pulsates. It is easy to measure the period of a Cepheid variable, and armed with this information, we can determine the absolute brightness of the star. We can then compare the absolute and apparent magnitudes of the star to determine its distance (see ■ Perspective 21.2 for a discussion of Henrietta Leavitt's pioneering work on Cepheid variables).

There are many kinds of objects in the sky besides stars: *star clusters, nebulae* or clouds of interstellar dust and gas, and **galaxies** (■ Fig. 21.16). Because most of these objects are

■ **FIGURE 21.15** Black holes cannot be detected directly, but some binary stars have supermassive components that are likely to be black holes. A black hole can draw material from its companion star. As material spirals in toward the black hole, it spins faster and faster, flattening into an accretion disk that emits great amounts of radio and X-ray radiation. Intense jets of gas are ejected along the axis of the accretion disk. This model accounts well for the features of a number of puzzling radio and X-ray sources in the sky.

(a)

(b)

(c)

■ FIGURE 21.16 Some deep-sky objects. (*a*) The Horsehead Nebula in Orion is a dark nebula, a nebula so dense that it obscures the bright nebula and stars beyond. The bright nebula shines because of nearby stars. A dark nebula may be dense enough to contract into new stars. (*b*) The Ring Nebula in Lyra (M57) is a planetary nebula, a bubble of gas given off by a former red giant. The white dwarf core of the star is at the center of the nebula. (*c*) The Pleiades in Taurus are a very young star cluster. Wisps of the gas cloud that gave birth to the cluster are still visible.

Women Discoverers of the Universe

Until recently, women who advanced far in science often did so in spite of opposition and hostility. Astronomy was no exception, yet women have probably made more fundamental discoveries in astronomy than in any other science.

In 1979, Linda Morabito, a navigation specialist for the *Voyager* mission, was examining images of Jupiter's innermost moon Io for faint background stars that could be used to determine the exact path of *Voyager*. What she found instead was a wisp of light that turned out to be a plume of gas emitted by one of Io's volcanoes. She had discovered the first active volcano beyond Earth.

In 1967, S. Jocelyn Bell (now Burnell), a graduate student at Cambridge University, had the tedious task of examining hundreds of feet of chart recordings from a radio telescope. She found puzzling "scruff," or very faint signals, on the records and determined that the emissions were coming from a particular point in the sky. Closer study showed that the object was emitting regular radio pulses every second or so. The signals were so regular and so rapid that for a time serious thought was given to the idea they might be artificial. Finally, Bell's mentor, Anthony Hewish, proposed that the signals came from a rotating neutron star. We now call such an object a *pulsar*.

Women have also been historians and popularizers of science. The *History of Astronomy* published by Agnes Clerke (1842–1907) in 1887 contained some of the first astronomical photographs ever published. This mode of publishing was so new the photos had to be pasted into the book by hand. At a time when biographies tended to be stuffy, her work was noted for its human and at times slightly irreverent depiction of great astronomers.

The area where women have made the greatest impact on astronomy is in our knowledge of the stars and the cosmic distance scale. Annie Jump Cannon (1863–1941) did much of the pioneering work in developing the modern system of spectral classification of stars. Cecilia Payne-Gaposchkin (1900–1979) contributed much to our understanding of the physical conditions in stars.

In 1912, Henrietta Leavitt (1868–1921) began a study of *Cepheid variables*. She analyzed the Cepheids in the Magellanic Clouds, small satellite galaxies of the Milky Way, because she could find many of them in a small area of the sky. To her surprise, Cepheids with the same period of light variation all turned out to have the same brightness. Since the Cepheids were all at about the same distance, the period of Cepheids must be directly related to their absolute magnitude. This discovery meant that we could find the absolute magnitude of a Cepheid just by timing its light variations, and by comparing the absolute and apparent magnitudes, we can determine the distance to the star.

The situation was confused for a while because there are actually two kinds of Cepheids, one about three times brighter than the other. Harvard astronomer Harlow Shapley and another woman astronomer, Henrietta Swope (1903–1981), unraveled this problem. In 1962, Swope published her determination of the distance to the Andromeda Galaxy: 2.2 million light-years. Leavitt and Swope did much of the work in giving us the Cepheid variables as a cosmic yardstick. By any standard, this must be one of the most important discoveries in the history of science.

---

far away and visible only in telescopes, astronomers refer to them as *deep-sky objects*. Like stars, most are identified by a catalog number.

Stars tend to form in batches and to remain together for long periods of time. Tightly bunched groupings of stars are called *star clusters*. Star clusters are very useful to astronomers because they contain batches of stars of the same age that are valuable for testing theories of stellar evolution. Some star clusters are close enough for us to see their individual stars without a telescope. Two clusters in the constellation Taurus are an example: the *Pleiades* (PLEE-a-deez) are a young, tightly packed cluster about 410 light-years away, still with remnants of its ancestral gas cloud (Fig. 21.16c), and the

Hyades (HY-a-deez) are an older, looser cluster about 130 light-years away.

There are many types of gas and dust clouds in space, collectively called *nebulae*. *Planetary nebulae* are spherical shells of gas given off in the final stages of a red giant's life cycle. They are called planetary because they have a disklike appearance in a telescope, like a planet (Fig. 21.16b). *Bright nebulae* shine because they reflect the light of bright, nearby stars, or because a nearby star bombards the nebula with enough energy to cause its atoms to emit their own light. *Dark nebulae* are seen only in silhouette, when they block the light of stars or nebulae behind them (Fig. 21.16a). Bright and dark nebulae are the places where new stars form.

Planets Around a Pulsar

No planet could survive a supernova explosion: if the Sun were to become a supernova a billion times brighter than its present brightness, the temperature at the distance of Pluto would be about 8,000°C (14,430°F), high enough to vaporize any material. Yet in 1992, the first conclusive evidence ever discovered for planets outside the solar system was found, and it indicated planets around a pulsar known as PSR 1957+12 (see Table 21.1).

The planets could not be original, but after a supernova explosion, debris orbiting the remnant of the star could re-accrete into a second generation of planets. Pulses emitted by pulsars are so regular they can literally be predicted within a millionth of a second months in advance. As PSR 1957+12 and its planets move around their center of mass, pulses alternately arrive early or late; even movements of a few kilometers by the pulsar can be detected. Timing of the pulsar signals indicates there are three planets: a Moon-sized planet about 28 million km (17 million mi) from the pulsar, a planet 3.4 times as massive as Earth at a distance of 54 million km (33.5 million mi), and a planet 2.8 times as massive as Earth at a distance of 70 million km (43 million mi). The outermost planet revolves around the pulsar in 98 days. Because of the extreme regularity of pulsar signals, it is actually much easier to detect planets around pulsars than around normal stars.

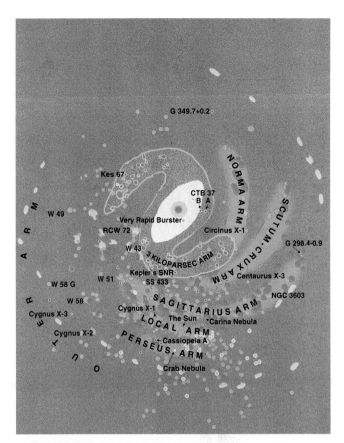

◙ FIGURE 21.17 As yet, there is no definite map of our own galaxy. This figure shows one attempt to map the Galaxy using radio emissions from distant gas clouds to map the regions we cannot see directly.

OUR GALAXY

Our galaxy, the **Milky Way Galaxy,** is a vast, disk-shaped aggregate of stars about 100,000 light-years across (◙ Fig. 21.17). The disk thickens at the center into a *hub* about 10,000 light-years in diameter. Our solar system lies on the outskirts of the Galaxy, about two-thirds of the way from the center to the edge. When we see the Milky Way, we are looking along the plane of the disk.

If we were to shrink the solar system, defined by the orbit of Pluto, to a quarter, the Sun would become a microscopic speck and Earth would be tinier still, traveling in an orbit about the size of the period at the end of this sentence. On this scale, the nearest star would still be 100 m (300 ft) away. Many people think the stars are not far beyond Pluto. Have a friend hold up a quarter 100 m away to get an idea of just how far beyond the solar system the stars really are. On this scale, the Galaxy would be 2,500 km (1,300 mi) across, comparable to the United States east of the Mississippi. In the central hub of the galaxy, comparable to Ohio, the stars would be dust specks a meter or so apart. Our solar system, somewhere around Green Bay, Wisconsin,

Bangor, Maine, or Cape Hatteras, North Carolina, is in a sparsely populated region where the dust specks would be 100 m or so apart. Two sand grains at opposite ends of a football stadium fill the stadium as densely with sand as space is filled with stars.

In the summer, when the Milky Way appears wide and bright, we are looking toward the center of the Galaxy (◙ Fig. 21.18). We cannot see the center in visible light because stars, dust, and gas are in the way, but we can detect its radio emissions. The center of the Galaxy lies in (actually far behind) the constellation Sagittarius. In the winter, when the Milky Way appears narrow and faint, we are looking toward the sparse outer edges of the disk. When we look toward the Big Dipper or the southern sky in autumn, we are looking out of the plane of the Galaxy into intergalactic space.

The key to mapping our galaxy is the study of other galaxies. Nearby galaxies have remarkable star clusters around them called **globular star clusters;** these great spherical swarms of 100,000 or more stars in a ball are perhaps 100 light-years across. The globular star clusters form a spherical halo around the center of each galaxy. In our own sky, however, the great majority of globular clusters are tightly bunched in one part of the sky. Clearly, if our own galaxy is like the millions of others we can detect, we must

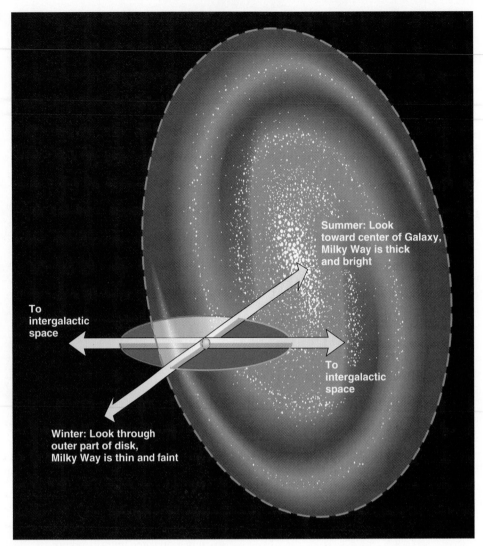

FIGURE 21.18 In the summer, we look toward the center of the Galaxy, and the Milky Way appears wide and bright. In the winter, we look toward the sparse outer edges of the disk, and the Milky Way appears narrow and faint. When we look toward the Big Dipper or the southern sky in autumn, we are looking out of the plane of the Galaxy into intergalactic space.

lie far off to one side of the halo of globular clusters (■ Fig. 21.19).

The final piece in the puzzle of our galaxy was the discovery of Cepheid variables and their use as a cosmic yardstick. Because the globular clusters contain Cepheids, it is possible to determine their distances. The center of the globular cluster halo, about 30,000 light-years away, is also the center of our galaxy.

 ## THE NATURE OF GALAXIES

To most people, "galaxy" means **spiral galaxy.** Spiral galaxies are enormous disks of stars like our own Milky Way, all held together by their mutual gravitational attraction and all orbiting around their common center of gravity, the central hub. As noted earlier, our Sun is moving relative to nearby stars. But the Sun and its entire stellar neighborhood are also moving around the center of the galaxy. The Sun takes about 220 million years to orbit the galaxy, traveling about one light-year in 1,300 years.

In the estimated 12 billion years since the universe formed, most stars have made 50–100 orbits around their galaxies, yet in most galaxies the spiral arms have only two or three turns, not 50 or 100. Thus, the spiral arms are much younger than the galaxies themselves. Young blue-white stars rich in heavy elements mark the spiral arms of galaxies (■ Fig. 21.20a and b). One theory holds that the rotation of the galaxy compresses interstellar gas to start the process of star formation. Another theory suggests that supernovas compress interstellar gas and trigger star formation and that

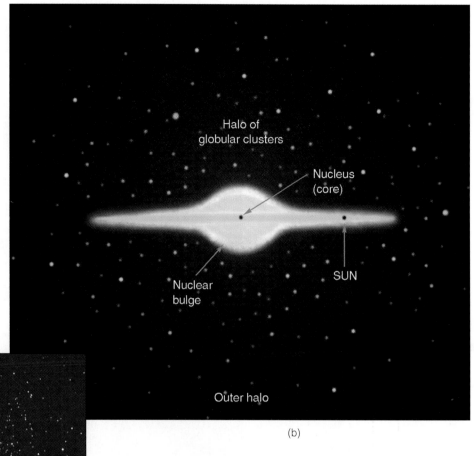

Halo of
globular clusters

Nucleus
(core)

Nuclear
bulge

SUN

Outer halo

(b)

 FIGURE 21.19 (*a*) The globular cluster 47 Tucanae is a vast swarm of perhaps 100,000 stars packed into a sphere a few hundred light-years across. Although the center of the cluster appears to be a solid ball of stars, the stars are actually a fraction of a light-year apart. (*b*) Globular clusters form spherical halos around other galaxies, but the globular clusters around our own galaxy are heavily concentrated in one part of the sky. From this pattern, we know we are well outside the center of the Galaxy.

(a)

THE REALM OF THE GALAXIES

Cepheid variables are of incalculable value in determining distances in the universe, but even with the Hubble Space Telescope the brightest Cepheids cannot be seen more than a hundred million light-years away. We can see Cepheids in many nearby galaxies, but not in very distant ones.

Astronomers call an object of known absolute brightness a *standard candle*. Cepheids are one standard candle, but there are others. For example, the brightest galaxies in large clusters of galaxies all appear to be about the same in absolute brightness. So do Type I and Type II supernovas. Using different standard candles allows astronomers to cross-check their distance estimates and assess the reliability of different methods.

There is one last distance estimator astronomers use. All over the universe, galaxies appear to be receding from us. As nearly as we can determine, the farther away galaxies are, the

rotation of the galaxy smears the young star groups into a spiral pattern. Probably, both mechanisms are at work (Fig. 21.20c and d).

Galaxies occur in other shapes besides spirals. *Elliptical galaxies* are the largest, some containing a trillion or more stars. These vast systems long ago used up most of their gas for star formation and consist entirely of old stars. Elliptical galaxies tend to be smooth and featureless, offering few clues to their origin or evolution. Other galaxies have no regular structure and often have bizarre forms. In some cases, it is clear that the galaxies have been disrupted by close encounters or collisions with other galaxies.

☑ **FIGURE 21.20** Young stars, old stars, and spiral arms. (*a*) This view of a spiral galaxy (M51) was taken in blue light. We can see that the spiral arms of the galaxy contain mostly hot, young stars. (*b*) This view of the same galaxy was taken in infrared light. The hot blue stars are invisible, and we see older red giants instead. These stars have made many orbits of the galaxy and no longer have a strong spiral pattern. (*c*) Spiral arms can form when the stars in a galaxy attract one another and influence one another's orbits. The process of star formation can begin when a spiral density wave of this sort sweeps up interstellar gas clouds and compresses them in much the same way that traffic bunches up on a busy highway. (*d*) Interactions between galaxies can also create spiral arms.

(a)

(b)

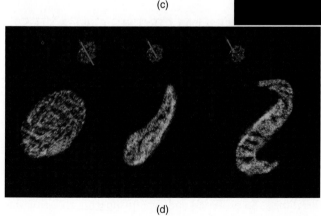

(c)

(d)

faster they are receding, at a rate of 50 to 100 km (31 to 62 mi) per second for every million light-years. For the most distant objects known, billions of light-years away, the only clue we have to their distance is the speed at which they are moving away from us.

Galaxies occur in clusters. Our own Milky Way is a member of a sparse group called the *Local Cluster,* which includes two other large spiral galaxies about 2.3 million light-years away and about 20 *dwarf galaxies,* with up to a few billion stars (☑ Fig. 21.21). Dwarf galaxies are so small and faint that only the nearest ones can be seen. Many dwarf galaxies appear to be satellites of larger galaxies. Our own galaxy has two small satellite galaxies, the *Magellanic Clouds.* These small galaxies, about 180,000 light-years away, are visible only from the Southern Hemisphere. Supernova 1987A occurred in one of them before humans had migrated to North America.

Other galactic clusters have hundreds or thousands of members. The nearest such grouping, the *Virgo Supercluster,* is about 40 million light-years away. Our Local Cluster appears to be an outlying fringe of this supercluster. On a

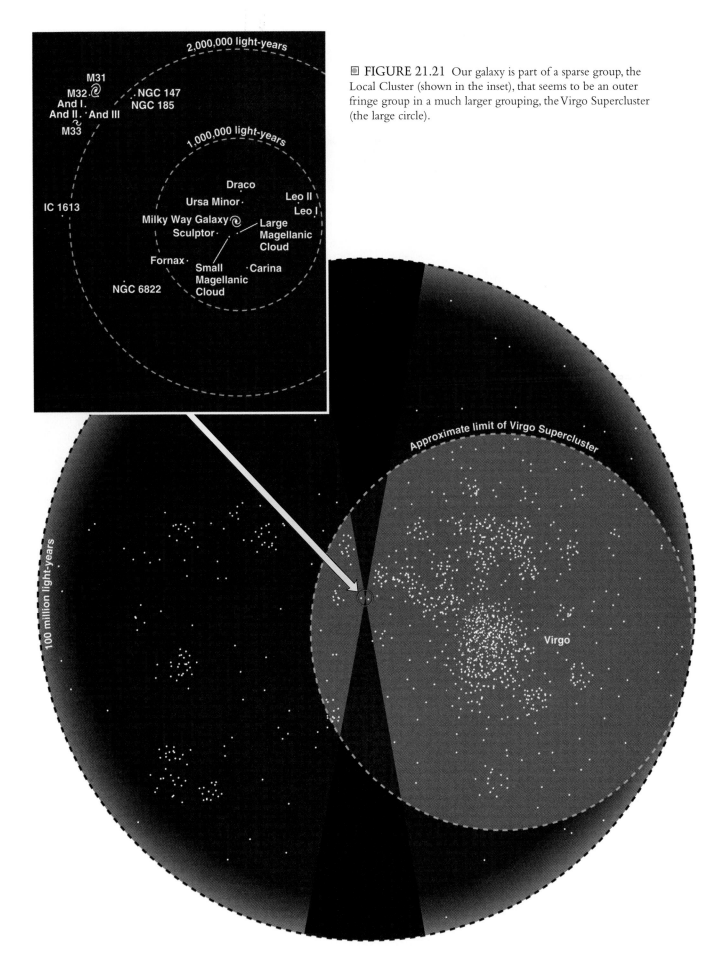

FIGURE 21.21 Our galaxy is part of a sparse group, the Local Cluster (shown in the inset), that seems to be an outer fringe group in a much larger grouping, the Virgo Supercluster (the large circle).

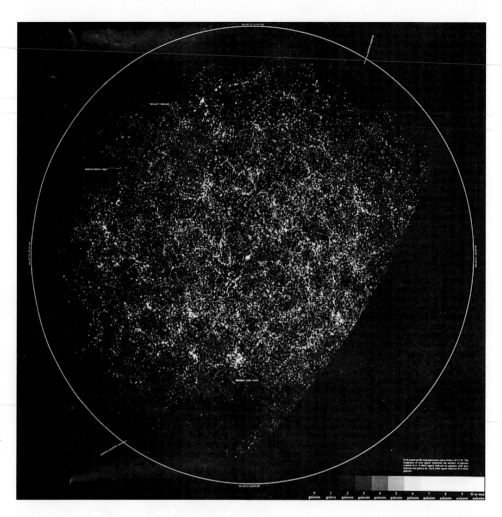

 FIGURE 21.22 This plot, based on a survey of a million galaxies, shows a patchy pattern in which streamers of galaxies surround vast voids.

very large scale, galaxies appear to be clustered into great interconnected filaments separated by voids hundreds of millions of light-years across (Fig. 21.22).

THE BIGGEST QUESTIONS OF ALL

The study of the origin, overall structure, and future of the universe is called *cosmology.* The present period may be the most revolutionary in cosmology's history. Here we can touch on only a few of the exciting developments in this field.

For half a century, astronomers have known that galaxies everywhere are rushing away from us and that the farther away they are, the faster they are receding. The apparent expansion of the universe does not mean we are at the center. All galaxies are receding from one another, and an observer on any galaxy would see an **expanding universe** just as we do. The likeliest explanation of the expanding universe is that the universe began in a **Big Bang,** a tremendous explosion that flung matter and energy outward. In 1965, radio astronomers detected faint *microwave background radiation* filling all of space—the echo of the initial explosion. The

best estimate of the age of the universe is about 12 to 15 billion years, the time it would require for galaxies, starting from a common point and receding at their observed speeds, to reach their present positions.

All the energy in the present universe was compressed into a tiny volume at the instant of the Big Bang. By using the principles of physics, cosmologists can estimate what conditions in the earliest universe were like. The younger the universe, the smaller it was, the more densely it was filled with energy, and the hotter it was. At about 0.001 second after the Big Bang, the universe had expanded and cooled enough for neutrons and protons to form out of still more elementary particles. At about three minutes, protons and neutrons could combine to form stable nuclei. By about 500,000 years, the universe was cool enough for atoms to form.

The fate of the universe depends on how much matter it holds, and that question is one of the most important in astronomy. From studies of the motions of stars in galaxies, and of the motions of galaxies within clusters of galaxies, it appears that at least 90% of the matter in the universe is not in the form of bright stars, but in a *nonluminous* form that emits little light or radiation of any kind. This matter is often called "missing," but it is not really missing, just very hard to detect directly. The nonluminous mass in the universe could take many possible forms. It could exist as clouds of cold gas between the galaxies or as large numbers of faint stars surrounding galaxies. Other possibilities are more exotic. Some theories of physics predict that rare but extremely massive subatomic particles might exist. Other theories predict that supposedly massless particles called neutrinos might actually have a very small mass. Although popular articles often refer to "missing mass," for astronomers the question is not "where is the missing mass?" but "which of many possible forms does the nonluminous matter take?"

If the universe has enough matter, its gravitational attraction will eventually halt the expansion of the universe, and matter will begin collapsing inward. Perhaps our universe will end in a "Big Crunch," followed by another Big Bang and a new universe. If there is not enough matter to halt the expansion, the universe will continue to expand indefinitely. In all probability, new discoveries in physics will require us to revise our theories about the origin and fate of the universe in ways we cannot foresee. We simply do not know what the final answer will be.

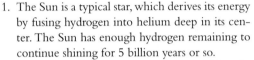

CHAPTER SUMMARY

1. The Sun is a typical star, which derives its energy by fusing hydrogen into helium deep in its center. The Sun has enough hydrogen remaining to continue shining for 5 billion years or so.

2. The visible portion of the Sun consists of three layers: the photosphere, the bright disk of the Sun; a thin overlying layer of gas called the chromosphere; and a tenuous outer envelope called the corona.

3. Disturbances on the Sun include dark, cool sunspots and great jets of gas called prominences.

4. The stars as we see them differ in color and brightness. Variations in color reflect differences in temperature, with the hottest stars being blue-white and the coolest stars red.

5. Differences in brightness are due to differences in distance and to some stars being more luminous than others.

6. Astronomers measure the brightness of stars in terms of magnitude. Apparent magnitude is the brightness of an object seen from Earth; absolute magnitude is the brightness the object would have at a standard distance.

7. Star patterns in the sky, or constellations, differ from culture to culture and are merely chance alignments of stars in space.

8. The brightest stars have traditional proper names, but more commonly stars are known by catalog number. Most faint stars have no formal name.

9. The stars move across the sky slowly but measurably.

10. Many stars consist of binary or multiple stars in orbit around one another.

11. The most powerful tool astronomers use in studying the stars is spectroscopy, the analysis of starlight spread out into a spectrum. Spectroscopy enables astronomers to determine a star's chemical composition, velocity, and rotation.

12. Stars form when dense clouds of interstellar gas and dust begin to contract under their own gravity. Eventually, the contracting cloud, or protostar, becomes hot and dense enough for nuclear reactions to begin within it.

13. Stars shine stably until their hydrogen runs out; the brighter the star, the shorter its stable lifetime. The star then begins to contract until a new cycle of nuclear reactions may begin. The outer envelope of the star is heated and swells, and the star becomes a red giant.

14. Many red giants shed their outer envelopes, leaving only the core of the star as a white dwarf. Others collapse and explode violently as supernovas, leaving a neutron star or black hole remnant.

15. Distances to the stars are measured in light-years, the distance light travels in one year.

16. Distances to nearby stars can be determined by triangulation, or the parallax method. Distances to more distant stars can be found by spectroscopic methods or by analyzing star movements in star clusters. For very distant objects, Cepheid variables are used to determine distance.

17. Stars are part of vast systems called galaxies. Our Milky Way Galaxy is a spiral galaxy about 100,000 light-years across. Besides stars, our galaxy includes star clusters and nebulae of many varieties. Other galaxies are elliptical in form.

18. Galaxies group into larger units, galactic clusters and superclusters.

19. The recession of distant galaxies and faint microwave radiation filling space indicate that the universe formed about 12–15 billion years ago in a gigantic explosion called the Big Bang.

Big Bang
binary star
Cepheid variable
constellation
Doppler effect
expanding universe

galaxy
globular star cluster
Hertzsprung-Russell (H-R)
    diagram
light-year
magnitude

main sequence
Milky Way Galaxy
nuclear fusion
red giant
spectral class

spiral galaxy
sunspot
supernova
variable stars
white dwarf

1. The bright visible disk of the Sun is the:
   a. _____ chromosphere; b. _____ photosphere; c. _____ corona.

2. The cool, dark disturbances seen on the surface of the Sun are:
   a. _____ prominences; b. _____ flares; c. _____ sunspots.

3. What is the difference between apparent and absolute magnitude?
   a. _____ apparent magnitude is the brightness we see, absolute magnitude is the brightness a star would have at a standard distance; b. _____ absolute magnitude is the brightness we see, apparent magnitude is the brightness a star would have at a standard distance; c. _____ apparent magnitude is always brighter than absolute magnitude; d. _____ absolute magnitude is always brighter than apparent magnitude.

4. True or False: A magnitude zero star emits no light.
   a. _____ true; b. _____ false.

5. The distance to nearby stars is determined by:
   a. _____ timing radar signals; b. _____ measuring their brightness; c. _____ timing their light variations; d. _____ triangulation.

6. Spectral lines are caused by:
   a. _____ light being absorbed by specific elements in stars; b. _____ light being blocked by fainter stars; c. _____ parts of the star being too cool to emit light; d. _____ interruptions of the star's energy output.

7. When light is shifted by the Doppler effect:
   a. _____ approaching objects become blue and receding objects become red; b. _____ approaching objects become red and receding objects become blue; c. _____ the color of the object does not change but spectral lines are shifted; d. _____ approaching objects become brighter, receding objects become fainter.

8. The principal reason why the stars have different colors is that:
   a. _____ their light shines through gas clouds of different colors; b. _____ they have different temperatures; c. _____ their light is shifted by the Doppler effect; d. _____ they are made of chemicals that burn with different colors.

9. The sequence of evolution of a typical star is best described as:
   a. _____ protostar, white dwarf, main sequence, red giant;
   b. _____ protostar, main sequence, red giant, white dwarf;
   c. _____ red giant, main sequence, protostar, white dwarf;
   d. _____ protostar, red giant, main sequence, white dwarf.

10. The mass of a star is so significant in its life history because:
    a. _____ mass determines whether the star can have planets;
    b. _____ mass governs the final fate of the star; c. _____ mass governs how quickly the star uses its nuclear fuel; d. _____ both b. and c.; e. _____ all of a., b., and c.

11. Which of the following is *not* a common cause of violent outbursts in stars?
    a. _____ Collisions between stars; b. _____ The core of an old massive star suddenly collapses; c. _____ A white dwarf accumulates matter from a nearby companion until nuclear fusion takes place on its surface; d. _____ A white dwarf accumulates matter from a nearby companion until it resumes nuclear fusion in its interior.

12. What is the Hertzsprung-Russell diagram? Why is it so important?
    a. _____ It is a plot of spectral type and brightness, used for tracing stellar evolution; b. _____ It is a plot of brightness and distance, used for mapping the galaxy; c. _____ It is a plot of sunspots over time, used for following solar activity; d. _____ It is a plot of galaxy redshift, used for determining how fast the universe is expanding.

13. Why are star clusters so valuable to astronomers?
    a. _____ The stars in them are all the same type; b. _____ The stars in them are all the same size; c. _____ The stars in them are all the same age; d. _____ The stars in them are all the same temperature.

14. Why are Cepheid variables important?
    a. _____ They are all the same age, so they can calibrate the ages of galaxies; b. _____ Their absolute magnitude is easily determined, so they can be used for finding distance to distant galaxies; c. _____ They are all the same size, so we can determine the sizes of stars; d. _____ They all have planets that might possibly have life.

15. How do globular clusters provide a clue to our place in the galaxy?
    a. _____ Globular clusters surround the center of the galaxy;
    b. _____ Our galaxy is like a globular cluster, only bigger;
    c. _____ Globular clusters outline the spiral arms; d. _____ Globular clusters mark the outer rim of the galaxy.

16. The stars most responsible for outlining the spiral arms of galaxies are:
    a. _____ hot, young, blue-white stars; b. _____ old, cool, red stars; c. _____ Cepheid variables; d. _____ white dwarfs.

17. Why is the Milky Way wide and bright in summer and thin and faint in winter?
    a. _____ The sky is clearer in the summer; b. _____ In summer, we look toward the center of the galaxy; c. _____ In winter,

we look toward the center of the galaxy; d. _____ In winter, we look out of the plane of the galaxy.

18. Why can't we see the hub of our galaxy directly?
a. _____ It is obscured by clouds of dust and gas; b. _____ We live in a galaxy with no hub; c. _____ It is too far away to see; d. _____ It is too faint to see.

19. All of these are evidence for the Big Bang except one. Which is not?
a. _____ microwave background radiation; b. _____ recession of distant galaxies; c. _____ the faintness of distant galaxies.

20. Henrietta Leavitt's great contribution to astronomy was:

a. _____ devising the spectral classification system; b. _____ discovering the key to distance scales in the universe; c. _____ discovering the Big Bang; d. _____ discovering pulsars.

21. What is light pollution? Why is it a problem?
a. _____ noxious materials given off by electric power plants that lower air quality; b. _____ stray light in the sky that interferes with astronomical observations; c. _____ mild pollution, as opposed to serious or heavy pollution; d. _____ production of light gases like methane that contribute to global warming.

POINTS TO PONDER

1. The nuclear reactions that power the Sun convert matter into energy at the rate of 4.3×10^9 kg/second. If the mass of the Sun is 2×10^{30} kg, how long could the Sun produce energy? (The actual figure is much less, but still a long time because only a small portion of the matter in a nuclear reaction converts to energy.)

2. Space paintings commonly show distant skies peppered with bright galaxies. Is this a realistic depiction? (Hint: You are already surrounded by galaxies. What does your night sky look like?)

3. If you have access to a telescope and somebody who knows the skies, observe a bright object like the Orion Nebula or the Andromeda Galaxy. Compare what you see with a photograph. Why do you suppose the photograph is so much more spectacular?

ADDITIONAL READINGS

Barnes, J., L. Hernquist, and F. Schweizer. 1991. Colliding galaxies. *Scientific American* 265 (August): 40–47.

Berman, R. 1995. *Secrets of the night sky.* New York: Morrow.

Black, D. C. 1991. Worlds around other stars. *Scientific American* 264 (January): 76–82.

Freedman, W. L. 1992. The expansion rate and size of the universe. *Scientific American* 267 (November): 54–60.

Hawking, S. 1993. *Black holes and baby universes and other essays.* New York: Bantam.

Kippenhahn, R. 1994. *Discovering the secrets of the Sun.* New York: Wiley.

Mechler, G. 1995. *Galaxies and other deep-sky objects.* New York: National Audubon Society Pocket Guides, Knopf.

Mechler, G., and M. Chartrand. 1995. *Constellations.* New York: National Audubon Society Pocket Guides, Knopf.

Miley, G. K., and K. C. Chambers. 1993. The most distant radio galaxies. *Scientific American* 268 (June): 54–61.

Moore, P. 1994. *Atlas of the universe.* New York: Rand McNally.

Powell, C. S. 1993. Inconstant cosmos. *Scientific American* 268 (May): 110–18.

———. 1993. Young suns: Telescope technology pulls the veil from infant stars. *Scientific American* 268 (March): 30.

Rees, M. J. 1990. Black holes in galactic centers. *Scientific American* 263 (November): 56–60.

Sagan, C. 1994. *Pale blue dot: A vision of the human future in space.* New York: Random House.

Silk, J. 1994. *A short history of the universe.* New York: Scientific American Library.

Soker, N. 1992. Planetary nebulae: These fluorescent clouds of gas represent the last gasp of dying, sunlike stars. *Scientific American* 266 (May): 78–85.

Stahler, S. W. 1991. The early life of stars. *Scientific American* 265 (July): 48–55.

Townes, C. H., and R. Genzel. 1990. What is happening at the center of our Galaxy? *Scientific American* 262 (April): 46–55.

van den Bergh, S., and J. E. Hesser. 1993. How the Milky Way formed. *Scientific American* 268 (January): 72–78.

van den Heuvel, E. P. J., and J. van Paradijs. X-ray binaries. *Scientific American* 269 (November): 64–71.

For current updates and exercises, log on to http://www.wadsworth.com/geo.

TODAY@NASA
http://www.hq.nasa.gov/office/pao/NewsRoom/today.html

This is a site that, among other things, releases the most recent and interesting photos from the Hubble Space Telescope.

NASA PHOTO ARCHIVE
http://www.hq.nasa.gov/office/pao/library/photo.html

The entry point to NASA's vast photo archives includes Hubble Space telescope images plus collections of many other pictures of deep space objects.

EXTRASOLAR PLANETS CATALOG
http://www.obspm.fr/departement/darc/planets/catalog.html

A French site with references (mostly highly technical), but it also has a table of current data on all known or suspected extrasolar planets.

APPENDIX
A

English-Metric Conversion Chart

ENGLISH UNIT	CONVERSION FACTOR	METRIC UNIT	CONVERSION FACTOR	ENGLISH UNIT
Length				
Inches (in.)	2.54	Centimeters (cm)	0.39	Inches (in.)
Feet (ft)	0.305	Meters (m)	3.28	Feet (ft)
Miles (mi)	1.61	Kilometers (km)	0.62	Miles (mi)
Area				
Square inches (in^2)	6.45	Square centimeters (cm^2)	0.16	Square inches (in^2)
Square feet (ft^2)	0.093	Square meters (m^2)	10.8	Square feet (ft^2)
Square miles (mi^2)	2.59	Square kilometers (km^2)	0.39	Square miles (mi^2)
Acre (ac)	0.404	Hectares (Ha)	2.47	Acres (ac)
Volume				
Cubic inches (in^3)	16.4	Cubic centimeters (cm^3)	0.061	Cubic inches (in^3)
Cubic feet (ft^3)	0.028	Cubic meters (m^3)	35.3	Cubic feet (ft^3)
Cubic feet (ft^3)	0.028	Cubic meters (m^3)	264	Gallons (gal.)
Cubic yards (yd^3)	0.8	Cubic meters (m^3)	1.3	Cubic yards (yd^3)
Gallons	3.79	Liters (l)	0.264	Gallons (gal.)
Cubic miles (mi^3)	4.17	Cubic kilometers (km^3)	0.24	Cubic miles (mi^3)
Flow				
Cubic feet/sec (cfs)	0.028	Cubic meters/sec (m^3/s)	35.3	Cubic feet/sec (cfs) (= 448.8 gal.min)
Gallons/min (gpm)	3.79	Liters/min (lpm)	0.264	Gallons/min (gpm) ($\times$ 1440-gallons/day,gpd)
Cubic feet/sec (cfs)	—	—	7.48	Gallons/sec (gps)
Weight				
Ounces (oz)	28.3	Grams (g)	0.035	Ounces (oz)
Pounds (lb)	0.45	Kilograms (kg)	2.20	Pounds (lb)
Short tons (st)	0.91	Metric tons (t)	1.10	Short tons (st)

Temperature $°C = 5/9(°F - 32)$.
To convert degrees Fahrenheit (°F) to degrees Celsius (°C):
Subtract 32 degrees from °F, then divide by 1.8.
To convert degrees Celsius (°C) to degrees Fahrenheit (°F):
Multiply °C by 1.8, then add 32 degrees.

Periodic Table
of the Elements

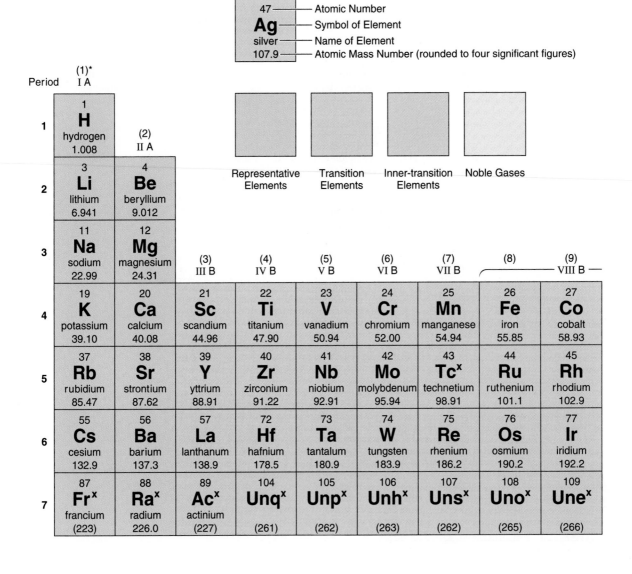

		47	— Atomic Number
		Ag	— Symbol of Element
		silver	— Name of Element
		107.9	— Atomic Mass Number (rounded to four significant figures)

Representative Elements

Transition Elements

Inner-transition Elements

Noble Gases

Period	(1)* I A	(2) II A	(3) III B	(4) IV B	(5) V B	(6) VI B	(7) VII B	(8) ⌒ VIII B ⌒	(9)
1	1 **H** hydrogen 1.008								
2	3 **Li** lithium 6.941	4 **Be** beryllium 9.012							
3	11 **Na** sodium 22.99	12 **Mg** magnesium 24.31							
4	19 **K** potassium 39.10	20 **Ca** calcium 40.08	21 **Sc** scandium 44.96	22 **Ti** titanium 47.90	23 **V** vanadium 50.94	24 **Cr** chromium 52.00	25 **Mn** manganese 54.94	26 **Fe** iron 55.85	27 **Co** cobalt 58.93
5	37 **Rb** rubidium 85.47	38 **Sr** strontium 87.62	39 **Y** yttrium 88.91	40 **Zr** zirconium 91.22	41 **Nb** niobium 92.91	42 **Mo** molybdenum 95.94	43 **Tc**[x] technetium 98.91	44 **Ru** ruthenium 101.1	45 **Rh** rhodium 102.9
6	55 **Cs** cesium 132.9	56 **Ba** barium 137.3	57 **La** lanthanum 138.9	72 **Hf** hafnium 178.5	73 **Ta** tantalum 180.9	74 **W** tungsten 183.9	75 **Re** rhenium 186.2	76 **Os** osmium 190.2	77 **Ir** iridium 192.2
7	87 **Fr**[x] francium (223)	88 **Ra**[x] radium 226.0	89 **Ac**[x] actinium (227)	104 **Unq**[x] (261)	105 **Unp**[x] (262)	106 **Unh**[x] (263)	107 **Uns**[x] (262)	108 **Uno**[x] (265)	109 **Une**[x] (266)

Lanthanides	58 **Ce** cerium 140.1	59 **Pr** praseodymium 140.9	60 **Nd** neodymium 144.2	61 **Pm**[x] promethium (147)	62 **Sm** samarium 150.4

Actinides	90 **Th**[x] thorium 232.0	91 **Pa**[x] protactinium 231.0	92 **U**[x] uranium 238.0	93 **Np**[x] neptunium 237.0	94 **Pu**[x] plutonium (244)

x: All isotopes are radioactive.

( ) Indicates mass number of isotope
    with longest known half-life.

*   Number in ( ) heading each column
    represents the group designation
    recommended by the American
    Chemical Society Committee
    on Nomenclature.

								(18) Noble Gases
								2 **He** helium 4.003
			(13) III A	(14) IV A	(15) V A	(16) VI A	(17) VII A	
			5 **B** boron 10.81	6 **C** carbon 12.01	7 **N** nitrogen 14.01	8 **O** oxygen 16.00	9 **F** fluorine 19.00	10 **Ne** neon 20.18
(10)	(11) I B	(12) II B	13 **Al** aluminum 26.98	14 **Si** silicon 28.09	15 **P** phosphorus 30.97	16 **S** sulfur 32.06	17 **Cl** chlorine 35.45	18 **Ar** argon 39.95
28 **Ni** nickel 58.71	29 **Cu** copper 63.55	30 **Zn** zinc 65.37	31 **Ga** gallium 69.72	32 **Ge** germanium 72.59	33 **As** arsenic 74.92	34 **Se** selenium 78.96	35 **Br** bromine 79.90	36 **Kr** krypton 83.80
46 **Pd** palladium 106.4	47 **Ag** silver 107.9	48 **Cd** cadmium 112.4	49 **In** indium 114.8	50 **Sn** tin 118.7	51 **Sb** antimony 121.8	52 **Te** tellurium 127.6	53 **I** iodine 126.9	54 **Xe** xenon 131.3
78 **Pt** platinum 195.1	79 **Au** gold 197.0	80 **Hg** mercury 200.6	81 **Tl** thallium 204.4	82 **Pb** lead 207.2	83 **Bi** bismuth 209.0	84 **Po**^{x} polonium (210)	85 **At**^{x} astatine (210)	86 **Rn**^{x} radon (222)

63 **Eu** europium 152.0	64 **Gd** gadolinium 157.3	65 **Tb** terbium 158.9	66 **Dy** dysprosium 162.5	67 **Ho** holmium 164.9	68 **Er** erbium 167.3	69 **Tm** thulium 168.9	70 **Yb** ytterbium 173.0	71 **Lu** lutetium 175.0
95 **Am**^{x} americium (243)	96 **Cm**^{x} curium (247)	97 **Bk**^{x} berkelium (247)	98 **Cf**^{x} californium (251)	99 **Es**^{x} einsteinium (254)	100 **Fm**^{x} fermium (257)	101 **Md**^{x} mendelevium (258)	102 **No**^{x} nobelium (255)	103 **Lr**^{x} lawrencium (256)

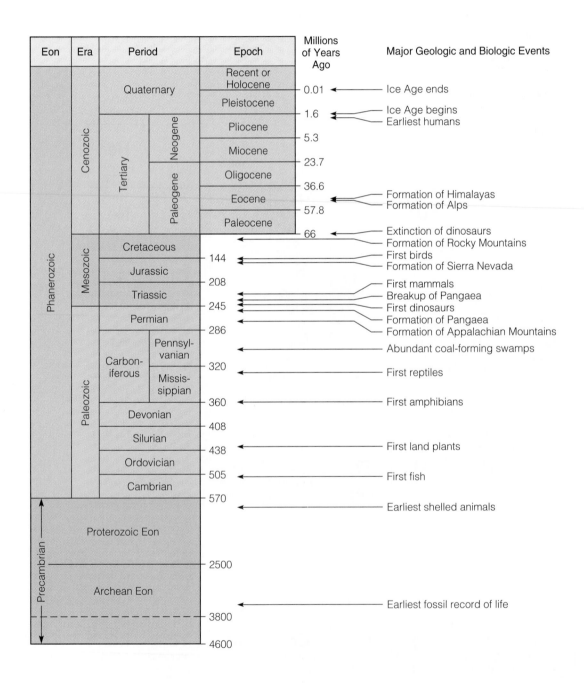

Eon	Era	Period		Epoch	Millions of Years Ago	Major Geologic and Biologic Events
Phanerozoic	Cenozoic	Quaternary		Recent or Holocene	0.01	Ice Age ends
				Pleistocene	1.6	Ice Age begins / Earliest humans
		Tertiary	Neogene	Pliocene	5.3	
				Miocene	23.7	
			Paleogene	Oligocene	36.6	
				Eocene	57.8	Formation of Himalayas / Formation of Alps
				Paleocene	66	Extinction of dinosaurs
	Mesozoic	Cretaceous				Formation of Rocky Mountains
		Jurassic			144	First birds / Formation of Sierra Nevada
		Triassic			208	First mammals / Breakup of Pangaea
					245	First dinosaurs
	Paleozoic	Permian			286	Formation of Pangaea / Formation of Appalachian Mountains
		Carboniferous	Pennsylvanian			Abundant coal-forming swamps
			Mississippian		320	First reptiles
		Devonian			360	First amphibians
		Silurian			408	
		Ordovician			438	First land plants
		Cambrian			505	First fish
					570	Earliest shelled animals
Precambrian		Proterozoic Eon			2500	
		Archean Eon			3800	Earliest fossil record of life
					4600	

D Principal Rock Types

IGNEOUS ROCKS	Generally hard, lack foliation or bedding.	
	Volcanic	Generally grains invisible to unaided eye, except for possible phenocrysts. Often contains cavities or vesicles (gas bubbles).
	Plutonic	Grains usually visible to unaided eye.
SEDIMENTARY ROCKS	Usually layered, often contain fossils.	
	Detrital	Consist of grains cemented together.
	Biochemical	Grains may or may not be visible to unaided eye; generally soft.
METAMORPHIC ROCKS	Often foliated, often recognizable by distinctive metamorphic minerals like garnet, staurolite, kyanite, or sillimanite. Frequently deformed.	
	Foliated	Platy structure or grain (foliation).
	Nonfoliated	Recognizable by crystalline texture, deformation, or metamorphic minerals.
VOLCANIC IGNEOUS ROCKS	Fine-grained volcanic rocks are difficult to identify exactly without laboratory analysis. Other types are defined by texture.	

MINERALS	DISTINCTIVE FEATURES	ROCK NAME
Quartz, potassium feldspar, plagioclase, biotite, or amphibole. Never olivine.	Often light in color (but can be dark), sometimes streaked or banded, sometimes has a porcelain-like appearance.	Rhyolite
No quartz, potassium feldspar, or olivine. Usually pyroxene, plagioclase, sometimes amphibole.	Very difficult to tell from basalt without laboratory analysis. Usually dark, often with vesicles.	Andesite
No quartz or potassium feldspar. Often olivine. Usually pyroxene and plagioclase.	Dark to black, often with vesicles.	Basalt
None visible. Can be any composition.	Very porous but does not float.	Scoria
None visible. Usually rhyolite composition.	Very porous, usually light color. Floats or sinks slowly.	Pumice
None visible. Usually rhyolite composition.	Glassy. Black or brown, conchoidal fracture.	Obsidian
Any composition.	Obvious fragments cemented together.	Volcanic breccia
Any composition.	Fine fragments cemented together, may be layered. Easy to mistake for sedimentary rock.	Tuff

PLUTONIC IGNEOUS ROCKS

Quartz, potassium feldspar, plagioclase, biotite, or amphibole, often muscovite. Never olivine.	Familiar building stone. Often pink potassium feldspars. Quartz is the diagnostic mineral.	Granite
No quartz, potassium feldspar, or olivine. Usually pyroxene, plagioclase, sometimes amphibole.	Very difficult to tell from gabbro without laboratory analysis.	Diorite
No quartz or potassium feldspar. Often olivine. Usually pyroxene and plagioclase.	Usually darker than diorite.	Gabbro
Olivine and pyroxene, minor plagioclase.	Very dark, dense.	Ultramafic rocks. Dunite: mostly olivine. Pyroxenite: mostly pyroxene.
Same minerals as granite.	Very coarse, may have rare minerals as well.	Pegmatite
Quartz, potassium feldspar.	Fine-grained, sugary texture, occurs as dikes in granitic rocks.	Aplite

DETRITAL OR CLASTIC SEDIMENTARY ROCKS

Rounded grains > 2 mm in diameter. Cemented gravel.	Conglomerate
Angular grains > 2 mm in diameter. Cemented broken rock fragments.	Sedimentary Breccia
Grains .06–2 mm in diameter. Cemented sand.	Sandstone
Sand grains include abundant feldspar.	Arkose
Grains < .06 mm in diameter. Cemented mud.	Mudrock
Mudrock with platy bedding, splits into thin sheets.	Shale
Mudrock without platy bedding, grains impart gritty feel.	Siltstone
Mudrock without platy bedding, grains so small rock feels smooth or slippery when wet.	Claystone

BIOCHEMICAL SEDIMENTARY ROCKS

Gray or buff. Hardness 3, fizzes vigorously when acid dropped on it.	Limestone
White, powdery, fizzes vigorously when acid dropped on it.	Chalk
Gray or buff. Hardness 3, fizzes weakly when acid dropped on it.	Dolostone
White, pink, or gray. Hardness 2, often coarsely crystalline.	Rock gypsum
White, pink, or gray. Hardness 2, often coarsely crystalline, salty taste.	Rock salt
Can be white and chalky or gray and hard. When very hard, has hardness 7. Does not fizz in acid. Often found in cavities in limestone and dolostone. Also occurs in bedded form, white, gray, or other colors. Consists of quartz in microscopic crystals.	Chert
Brown or gray, woody texture, crumbly, burns.	Lignite coal
Black, commonly with thin layers, streaks of glassy-looking black material common.	Bituminous coal

NONFOLIATED METAMORPHIC ROCKS

Distinctive Minerals	Texture, Appearance	Rock Name	Original Rock Type
Quartz	Very hard, does not split between grains.	Quartzite	Sandstone
Calcite	Sugary or coarsely crystalline, fizzes in acid.	Marble	Limestone
Organic compounds	Black, metallic sheen, burns.	Anthracite coal	Bituminous coal
Chlorite, epidote	Green or bluish.	Greenstone	Basalt or andesite

FOLIATED METAMORPHIC ROCKS

None visible	Gray, red, or green, fine platy foliation.	Slate	Mudrock
Micas	Silvery sheen, otherwise like slate.	Phyllite	Mudrock
Mica, garnet, staurolite	Prominent foliation, may have large crystals.	Schist	Mudrock
Garnet, kyanite, sillimanite	Minerals segregated into bands or streaks.	Gneiss	Many rock types
Amphiboles	Black or dark green with large splintery amphibole crystals.	Amphibolite	Basalt or andesite
Bluish amphiboles (glaucophane)	Schist with bluish gray appearance, found in subduction zone settings.	Blueschist	Mudrock

E Common Mineral Identification

If the mineral has these properties:	It could be a . . .
Metallic luster, dense, sulfur smell.	Sulfide
Good cleavage, fizzes in acid. In sedimentary or metamorphic rocks.	Carbonate
Good cleavage, light color and density, insoluble in water. In sedimentary rocks.	Sulfate
Good cleavage, light color and density, often soluble in water. In sedimentary rocks.	Halide
Hard, scratches glass.	Silicate
Pronounced platy cleavage. In igneous or metamorphic rocks.	Phyllosilicate
Hard, markedly elongate crystal form, good splintery cleavage. In igneous or metamorphic rocks.	Inosilicate

General rules for identifying minerals:
1. *Consider the setting.* Carbonates and sulfates rarely occur in igneous rocks, staurolite and garnet rarely anywhere else but metamorphic rocks.
2. *Bet on the familiar.* As it's sometimes put, "If you see hoofprints, think horses, not zebras." If there's a common mineral that matches most of the characteristics of your specimen, it's almost certain that you have a common mineral and not some rare alternative.

Minerals with generally metallic luster. Note: Many metallic materials collected by novice mineral collectors are actually artificial, not minerals.

NAME	FORMULA	COLOR	DENSITY	HARDNESS	CLEAVAGE	GEOLOGIC SETTING	REMARKS
Gold	Au	Yellow	15–19	2.5–3	None—malleable	Low-temperature metamorphosed volcanic rocks, often old volcanic island arc terranes. Also gains in stream deposits.	Visible gold is extremely rare. Gold is much denser than lead, flattens and does not break if hit with a hammer (malleable), and is unaffected by most acids. If a mineral lacks any of these properties, *it is not gold.*
Copper	Cu	Red-orange	8.9	2.5–3	None—malleable	Volcanic rocks and conglomerate in Michigan, granite and metamorphic rocks of copper deposits.	Found in northern Michigan and some western copper deposits.
Galena	PbS	Silvery	7.6	2.5	Excellent cubic	In carbonate rocks in U.S. midwest.	Density and cubic cleavage are distinctive.
Pyrite	FeS$_2$	Yellowish	5.0	6–6.5	None	Very common, all types of rock.	Fool's gold. Not as dense as gold and not malleable. Often in good crystals. Not as yellow as gold.
Chalcopyrite	CuFeS$_2$	Yellow, often with greenish cast	4.2	3.5–4	None	In carbonate rocks in U.S. midwest and low-temperature metamorphic rocks.	Usually yellower than pyrite, often with a slight yellow-green cast.
Magnetite	Fe$_3$O$_4$	Dark gray	5.2	6	None	Igneous and metamorphic rocks.	Magnetic. No other tests needed.

Minerals that can be either metallic or nonmetallic in luster.

NAME	FORMULA	COLOR (METALLIC)	COLOR (NON-METALLIC)	DENSITY	HARDNESS	CLEAVAGE	GEOLOGIC SETTING	REMARKS
Hematite	Fe$_2$O$_3$	Gray	Red or brown	5.2 in massive forms	1 in earthy forms, up to 6 in metallic forms	Rarely visible	Sedimentary and metamorphic rocks.	Metallic: gray, flaky texture (specular hematite). Nonmetallic: massive or earthy, sometimes oolitic (small rounded pellets). Always streak.
Sphalerite	ZnS	Brown or gray	Yellow to red	4.0	3.5–4	Good, very complex, in six directions	In carbonate rocks in U.S. midwest and low-temperature metamorphic rocks.	Streak always lighter than mineral. Usually with galena and chalcopyrite.
Graphite	C	Dark gray	Dark gray	2.2	1	Rarely visible	Metamorphic carbon-rich rocks.	Very soft, leaves mark on paper, greasy feel.

Minerals with nonmetallic luster and good cleavage.

NAME	FORMULA	COLOR	DENSITY	HARDNESS	CLEAVAGE	GEOLOGIC SETTING	REMARKS
Gypsum	$CaSO_4H_2O$	Clear, white, pink	2.3	2	Excellent, thin sheets	Sedimentary rocks	Softness and lack of solubility distinctive.
Halite	$NaCl$	Clear, white	2.2	2.5	Excellent cubic	Sedimentary rocks	Salty taste. No other test needed.
Fluorite	CaF	Clear, white, green, violet	3.2	4	Excellent, four directions	Carbonate sedimentary rocks	Colors are often distinctive, cleavage complex, but distinctive.
Calcite	$CaCO_3$	Clear, white	2.7	3	Excellent, cleaves into skewed box shapes	Carbonate sedimentary rocks	Fizzes vigorously in acid.
Dolomite	$CaMg(CO_3)_2$	Clear, white, light colors	2.85	3.5–4	Excellent, cleaves into skewed box shapes	Carbonate sedimentary rocks	Looks almost identical to calcite, but slightly harder, fizzes weakly. Cleavage planes often slightly curved.
Kyanite	Al_2SiO_5	Blue, white	3.6	5 along crystal, 7 across	Cleaves along length of crystal	High-temperature metamorphic rocks; gneisses	Color, occurrence, and the hardness variation are distinctive.
Sillimanite	Al_2SiO_5	White, clear	3.2	7	One good cleavage direction	High-temperature metamorphic rocks; gneisses	Common, but inconspicuous. Often fibrous in appearance.
Amphibole	A group name. Complex, but Si/O ratio is 4/11.	Black, green, brown; some can be light or white	3–3.5	5–6	Cleavage planes parallel to length of crystal at 56 and 124 degree angles	Igneous and metamorphic rocks	In granites, more likely to occur than pyroxene.
Pyroxene	A group name. Complex, but Si/O ratio is 1/3.	Black, green; some can be light or white	3–3.5	5–6	Cleavage planes parallel to length of crystal at near right angles	Mafic igneous rocks, metamorphic rocks	Dark minerals in igneous rocks with long crystals are most likely pyroxene.
Biotite	Complex. K, Fe, Mg, and Si/O ratio of 2/5.	Black, brown, light when weathered	3	2.5–3	Perfect cleavage into thin, elastic sheets	Igneous and low-temperature metamorphic rocks	Color and cleavage distinctive. Weathered varieties are often mistaken for gold but are far less dense.
Muscovite	Complex. K, Al, and Si/O ratio of 2/5.	Clear, white, gray, silvery	2.8	2–2.5	Perfect cleavage into thin, elastic sheets	Granitic and low-temperature metamorphic rocks	Color and cleavage distinctive. No other minerals are similar to the micas
Plagioclase	Mixture of *anorthite* $CaAl_2Sl_2O_8$ and *albite*, $NaAlSi_3O_8$	White, gray	2.6–2.7	6	Good, two directions	Igneous and some metamorphic rocks	Large crystals may have a strated appearance. Some dark gray varieties show color flashes and are popular ornamental stones.
Potassium feldspar	$KAlSi_3O_8$	Pink, white	2.6	6	Good, two directions	Granitic and metamorphic rocks	Pink color is common but not universal.

Minerals with nonmetallic luster and no cleavage. (Some minerals have cleavage, but it is rarely visible because they rarely occur in large crystals.)

NAME	FORMULA	COLOR	DENSITY	HARDNESS	GEOLOGIC SETTING	REMARKS
Corundum	Al_2O_3	Gray, rarely clear and colored	4.0	9	Silica-poor igneous and metamorphic rocks. Never with quartz.	Hardness is distinctive.
Limonite	Group name for iron oxides and hydroxides	Yellow, brown, black	2–4.5	1–5	Sedimentary rocks, soils.	The main coloring agent in soils. Can be yellow, soft, and earthy or hard, black, and massive.
Bauxite	Group name for aluminum hydroxides	White, red, brown	2–2.5	1–3	Ancient highly weathered soils.	Often has pellet-like structure with nodules a few mm in diameter.
Apatite	$Ca_5(PO_4)_3(OH)$	Yellow, green, brown, others	3.2	5	Igneous and metamorphic rocks.	Can be confused with many other minerals. Hardness and occurrence are the best guides.
Malachite	$Cu_2(CO_3)_2(OH)_2$	Bright green	4.0	3.5–4	Copper deposits.	Color distinctive. Often banded. Fizzes in acid.
Azurite	$Cu_3(CO_3)_2(OH)_2$	Intense blue	3.8	3.5–4	Copper deposits.	Color distinctive. Radiating crystals. Fizzes in acid.
Olivine	$(Fe, Mg)_2 SiO$ $NaAlSi_3O_8$	Green	3.3–4.4	7	Silica poor igneous rocks. Never with quartz.	Occurrence, color, lack of cleavage distinctive.
Garnet	Fe, Mg, Ca, Al with Si/O ratio 1/4	Brown or red	3.5–4.3	7.5	Metamorphic rocks of medium to high grade.	Often in good equidimensional crystals. Color and occurrence distinctive.
Staurolite	$Fe_2Al_9O_6(SiO_4)_4(OH)_2$	Brown	3.7	7–7.5	Metamorphic rocks of medium to high grade.	Rectangular crystals frequently intergrown in cross shape. Color, crystal form, and occurrence are distinctive.
Andalusite	Al_2SiO_5	White, gray, brown	3.2	7.5	Low-pressure and low-temperature metamorphic rocks.	Most often as rectangular crystals with a dark X marking visible on the ends.
Epidote	Ca, Al, with Si/O ratio 2/7	Pea green	3.4	6–7	Granitic rocks.	Color and occurrence in granite rocks distinctive. Good cleavage but usually in fine-grained masses and veins.

NAME	FORMULA	COLOR	DENSITY	HARDNESS	GEOLOGIC SETTING	REMARKS
Tourmaline	Complex, with boron and Si_6O_{18} rings	Black, sometimes pink or green	3.1	7–7.5	Granitic rocks.	Most often as deep black accessory mineral in granite. Crystals have triangular cross-section.
Chlorite	Mg, Fe, with Si/O ratio 2/5	Green	2.6–2.9	2–2.5	Low-grade metamorphic rocks.	Crystals rarely visible. The principal coloring agent in low-grade schists and slates.
Serpentine	$Mg_3Si_2O_5(OH)_4$	Usually dark green or blue-green	2.3–2.7	2–5	Ophiolites, low-grade metamorphic rocks.	Crystals rarely visible. Sometimes occurs as the fibrous form asbestos.
Talc	$Mg_3Si_4O_{10}(OH)_2$	White or light green	2.7–2.8	1	Low-grade metamorphic rocks.	Softness and greasy feel distinctive.
Opal	SiO_2 plus water	Light colors	1.9–2.2	5–6	Weathered rocks.	A mineraloid. Usually has a waxy look, conchoidal fracture. Rarely shows a play of color.
Chert	SiO_2	White, gray, light colors, but can be any color	2–2.6	7	Sedimentary rocks.	Can be soft and chalky or massive. Microcrystalline quartz. Very common.
Quartz	SiO_2	Clear, white, light colors, can by any color	2.65	7	All rock types.	The most common mineral in the Earth's surface. If it scratches glass, suspect quartz.

APPENDIX
F Topographic Maps

Figure F1a shows part of a reservoir with a small island. The island is the top of a submerged hill. In (b) through (e), the water in the reservoir is lowered in steps of ten meters. As the water drops, more and more land is exposed. At each step, the shoreline marks a line of constant elevation. In each figure, the former shorelines are also marked. In the final figure (f) the reservoir is completely drained and the old shorelines make up a topographic map of the area.

Elevation is shown on topographic maps with contour lines, lines of equal elevation like the shorelines above. Contour lines follow two basic rules:

1. They never cross (why not?, answer follows).

2. They never end, though they may be cut off at the edge of the map. A contour just above sea level may wrap all the way around a continent before closing, but it does eventually close. In some places, contour lines are omitted to make the map more readable, but the actual line of constant elevation on the ground is continuous.

Answer to question: If contour lines crossed, the same point on the ground would simultaneously be at two different elevations. In theory, an overhanging cliff or natural arch could result in crossing contours, but in practice such features are almost never large enough to show on most maps.

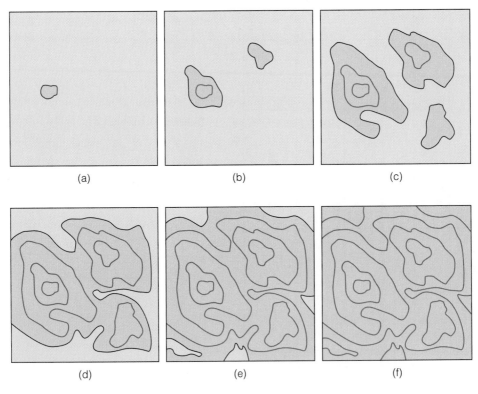

■ FIGURE F1

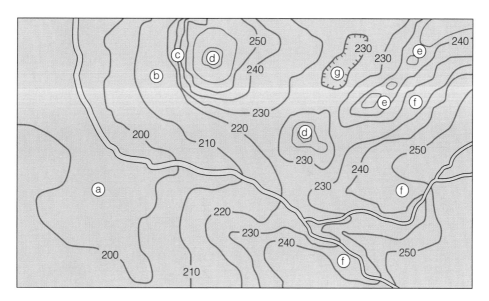

▣ FIGURE F2

Earth scientists can tell a lot about an area's geology just by reading topographic maps and interpreting the land-forms. Figure F2 shows some common landforms on topographic maps.

a. Flat areas have no topographic contours.
b. Areas of gentle slope have widely spaced contours.
c. Areas of steep slope have closely spaced contours.
d. Hills are usually marked by sets of concentric contours increasing in elevation toward the center.

e. Ridges are like hills but there may be a set of summits in a row, or a single, long-closed contour.
f. Valleys are marked by contours decreasing in elevation toward the center. Often there is a stream in the bottom. The contours make a V pointing upstream as they cross the valley bottom.
g. Closed depressions are rare landforms. They are usually shown by tick marks pointing downhill.

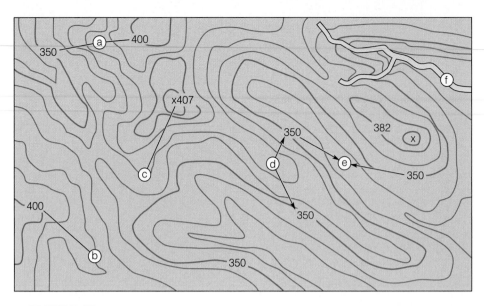

In areas of complex topography (Fig. F3) it may not be immediately obvious what a feature is. Nevertheless, by reading the map carefully, you can figure out the topography. Don't let the complexity intimidate you; proceed slowly and carefully.

If you don't know the contour interval, find two labeled contours and count the intervals between them (a). The contour interval is the difference between the labeled contours divided by the number of intervals. On this map, the interval is ten meters.

If you're not sure what a contour is, find a contour that is labeled and count up or down, using the contour interval of the map, to find the elevation. Find the nearest label and follow the contour as close to your location as you can. Contour (b) is at 380 meters.

Some elevation points like hilltops are also marked on topographic maps. You can count up or down from these points in the same way. Contour (c) is at 370 meters.

Is feature (d) a valley, or is (e)? Find contours of known elevation to determine the slope on either side of the feature. The land slopes on both sides away from (d), so (d) must be a ridge. It slopes toward (e) on both sides, so (e) must be a valley.

Is (f) a ridge or a valley? Streams only flow in valleys, so it must be a valley.

G Coordinates in Earth and Sky

Location on the Earth

A cross section through the center of Earth is a circle, and just as on a circle, we measure location on Earth in terms of degrees. Earth's axis meets the surface at the north and south geographic poles. Midway between the poles we draw an imaginary line, the equator, that divides Earth in half. If we draw a line from the surface of Earth to the center, the angle that line makes with the equatorial plane is called the latitude of the point (Figure G1). Latitude is measured north or south from the equator. The equator has latitude zero degrees; the poles have latitude 90 degrees. Lines of equal latitude on Earth are circles called parallels of latitude.

Location east and west on Earth is called longitude. Unlike latitude, there is no natural marker on Earth from which to reckon longitude. By international agreement, zero longitude runs from pole to pole through Greenwich Observatory, England. If we look down on Earth from above one pole, we can draw a line from any point to the pole. The angle this line makes with a similar line through Greenwich Observatory is the longitude of the point (Figure G2). Longitude is measured east and west from 0 to 180 degrees. Lines of equal longitude run from pole to pole and are called meridians. Because the meridians converge at the poles, a degree of longitude varies from 69 miles (111 km) at the equator to zero at the poles. (The metric system was set up so that the distance from the equator to the pole should be exactly 10,000 km. So the circumference of Earth is easy to remember: 40,000 km.)

The location of any point on Earth can be described in terms of its latitude and longitude (Fig. G3). For very accurate locations, degrees are subdivided into 60 minutes and minutes into 60 seconds. One degree of latitude (or longitude at the equator) is about 69 miles (111 km), one minute about 1.1 mile (1.8 km), and one second about 100 feet (30 m).

With computer-generated maps, the convention in computing is that north latitude is considered positive and south negative, so that latitude ranges from −90 degrees at the south pole to 90 degrees at the north pole. West longitude is defined as negative. Longitude ranges from −180 degrees to 180 degrees. New Orleans has a longitude of −90 degrees, Calcutta a longitude of about +90 degrees.

Questions:

1. If you drilled a hole from the United States through the center of Earth, would you come out in China? Would you come out in land or sea?
2. Where is "noplace" (0 latitude, 0 longitude)? Is it on land or sea?
3. What was so unusual about the New Zealand team at the 1992 Barcelona Olympics?

Answers:

1. You'd come out in the Indian Ocean. Apart from a few islands, there is no land on the opposite side of the Earth anywhere in the continental United States.
2. It's in the Atlantic west of Africa.
3. They were almost exactly on the opposite side of the Earth from their home country.

Location in the Sky

As Earth rotates, the heavens appear to rotate around points called the celestial poles. We can divide the heavens in half by an imaginary line equidistant from the poles, called the *celestial equator.* The equivalent of latitude in the sky is called declination and it is measured on the sky exactly the same way latitude is measured on Earth (Fig. G4).

The apparent path of the Sun against the stars (actually the plane of Earth's orbit) is tilted with respect to the celestial equator. The two crossing points are called the *equinoxes,* and the two points farthest north and south are called the *solstices* (Fig. G5). The spring or vernal equinox, the location of the Sun on March 21, is the zero point for celestial longitude. The equivalent of longitude on the sky is called right ascension (Fig. G6). Unlike longitude on Earth, right ascension is measured in hours, a reminder of the way people used the stars for timekeeping before there were clocks. The sky is divided into 24 hours of right ascension, each hour corresponding to 15 degrees. Right ascension, in keeping with its timekeeping origins, is subdivided into minutes and seconds, which obviously are not the same as angular minutes and seconds. A minute of right ascension on the celestial equator is 1/4 degree or 15 angular minutes. The sun is at right ascension zero on March 21, at 6 hours on June 21,

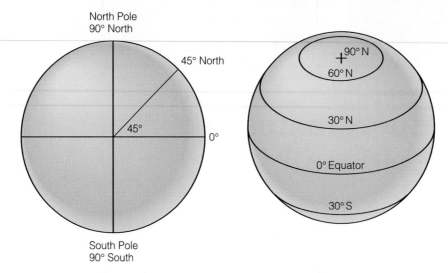

■ **FIGURE G1**

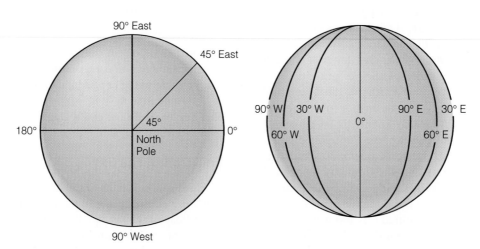

■ **FIGURE G2**

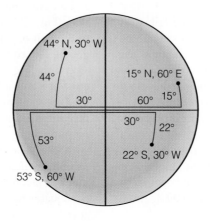

■ **FIGURE G3**

12 hours on September 21, and 18 hours on December 21. We see the stars of those parts of the sky at midnight six months before or after those dates; we see zero hours right ascension in the sky during the fall.

Directions on star charts reverse east and west compared to maps of Earth. On Earth, we are on the outside of a sphere looking in. When we look at the sky, we can imagine we are on the inside of a sphere looking out, hence the reversal of east and west (Fig. G7). In the sky, north is always the direction toward the north celestial pole; south is the direction toward the south celestial pole. West is 90 degrees clockwise from the north-south direction; east is 90 degrees counterclockwise. In general, sky directions do not line up exactly with ground directions (Fig. G8). There's nothing odd about this. On Earth, north is always toward the North Pole. Two people on opposite sides of the pole facing each other are both facing north. But not many people visit the poles of Earth, so we rarely experience such unusual situations. We can see half of the sky at once, however, so we must learn to understand directions on a sphere.

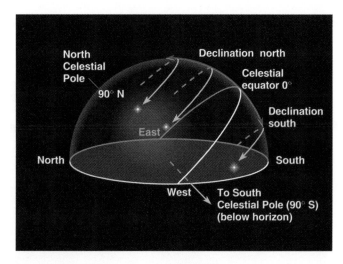

☐ FIGURE G4

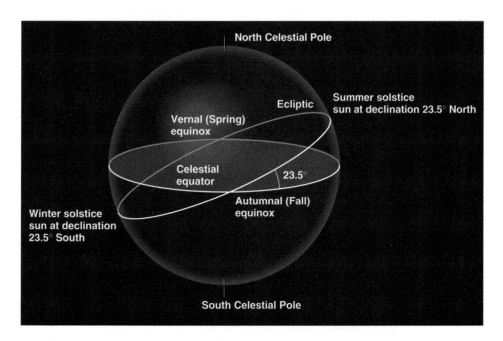

☐ FIGURE G5

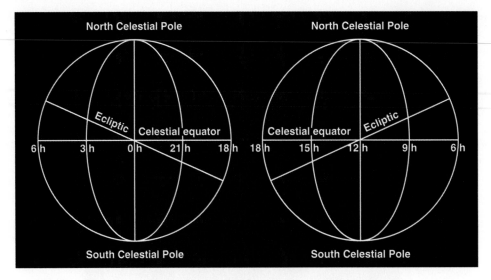

■ FIGURE G6

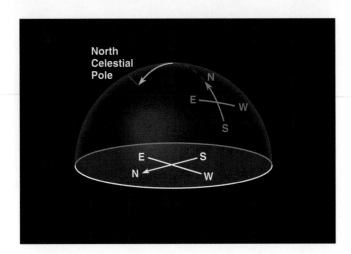

■ FIGURE G7

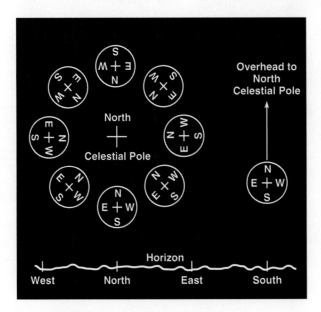

■ FIGURE G8

APPENDIX

H Earth Data

DIMENSIONS		
Mean radius:	6,371 km	(3,960 mi)
Equatorial radius:	6,378 km	(3,964 mi)
Polar radius:	6,357 km	(3,951 mi)
Polar circumference:	40,009 km	(24,866 mi)
Equatorial circumference:	40,077 km	(24,908 mi)

ROCK VOLUMES		
Total volume:	1.08×10^{12} km^3	(2.6×10^{11} mi^3)
Volume of continental crust:	6.21×10^{8} km^3	(1.49×10^{8} mi^3)
Volume of oceanic crust:	2.66×10^{8} km^3	(0.64×10^{8} mi^3)
Volume of mantle:	8.99×10^{5} km^3	(2.16×10^{5} mi^3)
Volume of core:	1.75×10^{5} km^3	(4.31×10^{4} mi^3)

DENSITIES		
Mean density of Earth:		(5.5 gm/cm^3)
Mean density of the crust:		(2.8 gm/cm^3)
Density of the continental crust:		(2.7 gm/cm^3)
Density of the oceanic crust:		(3.0 gm/cm^3)
Density of the mantle:		(4.5 gm/cm^3)
Density of the core:		(10.7 gm/cm^3)

AREAS		
Total area:	5.10×10^{8} km^2	(1.97×10^{8} mi^2)
Ocean area:	3.61×10^{8} km^2 70.6% of total	(1.39×10^{8} mi^2)
Land area:	1.49×10^{8} km^2 29.3% of total	(5.75×10^{7} mi^2)
Asia:	4.66×10^{8} km^2	(1.80×10^{7} mi^2)
Africa:	2.98×10^{7} km^2	(1.15×10^{7} mi^2)
North America:	2.1×10^{7} km^2	(8.2×10^{6} mi^2)
South America:	1.9×10^{7} km^2	(7.6×10^{6} mi^2)
Europe:	1.1×10^{7} km^2	(4.2×10^{6} mi^2)
Australia:	7.8×10^{6} km^2	(3.0×10^{6} mi^2)

WATER VOLUMES		
Total volume:	1.41×10^{9} km^3	(3.39×10^{8} mi^3)
Volume of oceans and seas:	1.37×10^{9} km^3	(3.3×10^{8} mi^3)
Volume of glaciers:	2.5×10^{7} km^3	(7×10^{6} mi^3)
Volume of groundwater:	8.4×10^{6} km^3	(2×10^{6} mi^3)
Volume of lakes:	1.25×10^{5} km^3	(3×10^{4} mi^3)
Volume of rivers:	1.25×10^{3} km^3	(3×10^{2} mi^3)

RELIEF		
Mean height of land above sea level:	875 m	(2,871 ft)
Mean depth of ocean:	3,800 m	(12,467 ft)

APPENDIX I

Köppen-Geiger Climatic Classification System

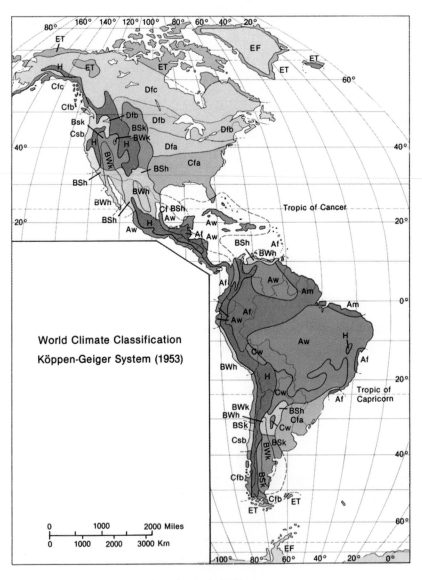

World Climate Classification
Köppen-Geiger System (1953)

Af: Tropical Rainforest climate
Am: Tropical Monsoon climate
Aw: Tropical Savanna climate
BSh: Tropical Steppe climate
BSk: Mid Latitude Steppe climate
BWh: Tropical Desert climate
BWk: Mid Latitude Desert climate
Cfa: Humid Subtropical climate

Cfb, Cfc: Marine climate
Csa, Csb: Mediterranean climate
Cwa, Cwb: Subtropical Monsoon climate
Dfa, Dwa: Humid Continental–Warm Summer climate
Dfb, Dwb: Humid Continental–Cool Summer climate
Dfc, Dwc, Dfd, Dwd: Taiga (or Subarctic) climate
EF: Polar (or Icecap) climate
ET: Tundra climate

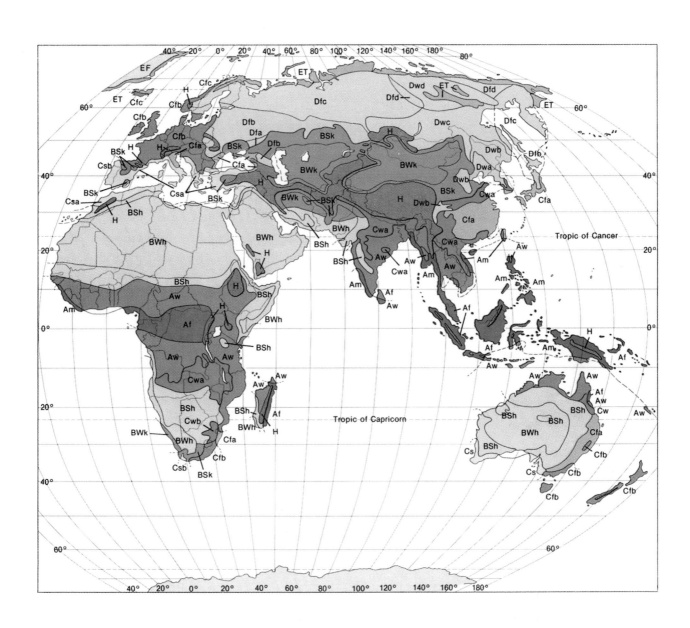

World Weather Extremes

Maximum Temperature
58°C (136°F), El Azizia, Libya, September 13, 1922

Minimum Temperature
−89°C (−129°F), Vostok, Antarctica, July 21, 1983

Maximum Precipitation in 24 hours
1880 mm (73.62 in.), Cilaos, La Reunion, Indian Ocean, March 15–16, 1952

Maximum Precipitation in one month
9300 mm (366.14 in.), Cherrapunji, Assam, India, July 1981

Maximum Precipitation in one year
26,461 mm (1041.78 in.), Cherrapunji, Assam, India, August 1860 to August 1861

Minimum Precipitation in one year
0.0 mm (0.0 in.), Arica, Chile, October 1903 to December 1917

Maximum Snowfall in one month (United States)
991 cm (390 in.), Tamarack, California, January 1911

Maximum Snowfall in one year (United States)
2850 cm (1122 in.), Paradise Ranger Station, Mt. Rainier, Washington, 1971–72

Maximum Air Pressure
1083.8 mb (32.005 in.), Agata, Siberia, December 31, 1968

Minimum Air Pressure
870 mb (25.69 in.), Typhoon Tip, Pacific Ocean, October 12, 1979

Largest Recorded Earthquakes

	DATE	LOCATION	RICHTER MAGNITUDE	SEISMIC MOMENT MAGNITUDE
1	May 22, 1960	Chile	8.5	9.5
2	March 28, 1964	Alaska	8.4	9.2
3	March 9, 1957	Aleutians	8.1	9.1
4	November 4, 1952	Kamchatka	8.2	9.0
5	January 31, 1906	Ecuador	8.2	8.8
6	February 4, 1965	Aleutians	8.2	8.7
7	November 11, 1922	Chile	8.3	8.5
8	March 2, 1933	Japan	8.5	8.4
9	August 15, 1950	India–China	8.6	
10	December 16, 1920	North China	8.6	

SOURCE: K. Abe, "Magnitudes and Moments of Earthquakes," in *Global Earth Physics, A Handbook of Physical Constants,* American Geophysical Union Reference Shelf, Volume 1, pp. 206–213.

Largest Known Holocene Volcanic Eruptions

	YEAR	VOLCANO	LOCATION	EJECTA*	REMARKS
1	4650 BC	Mount Mazama	Oregon	10+	Formed Crater Lake
2	1470 BC	Santorini	Greece	10+	Destroyed Minoan civilization
3	186	Taupo	New Zealand	80+	Pyroclastic flows travelled 100 km
4	260	Ilopango	El Salvador	10+	
5	536	Rabaul	New Guinea	10+	Global climatic effects
6	850	Hekla	Iceland	10+	
7	1783	Laki	Iceland	1	Largest historic fissure flow
8	1815	Tambora	Indonesia	150	Global climatic effects
9	1883	Krakatau	Indonesia	20+	Blast heard 5,000 km away
10	1912	Katmai	Alaska	10+	

*Ejecta refers to cubic kilometers of ash erupted. All figures are estimates. Since 10,000 BC at least seven other eruptions are known to have vented more than 10 cubic km of ash. Laki, 1783, is also included as the largest historic lava flow. Dates of the first six events are from radiocarbon dates and are approximate.

SOURCE: T. Simkin et al., 1984, *Volcanoes of the World,* Stroudsburg, PA, Hutchinson Ross, p. 232.

K

Star Charts

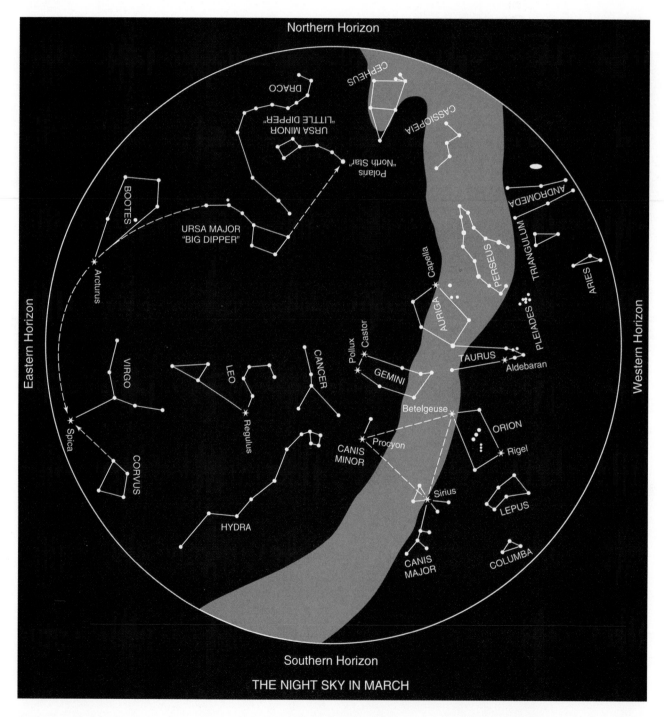

To use: Hold chart vertically and turn it so the direction you are facing shows at the bottom.

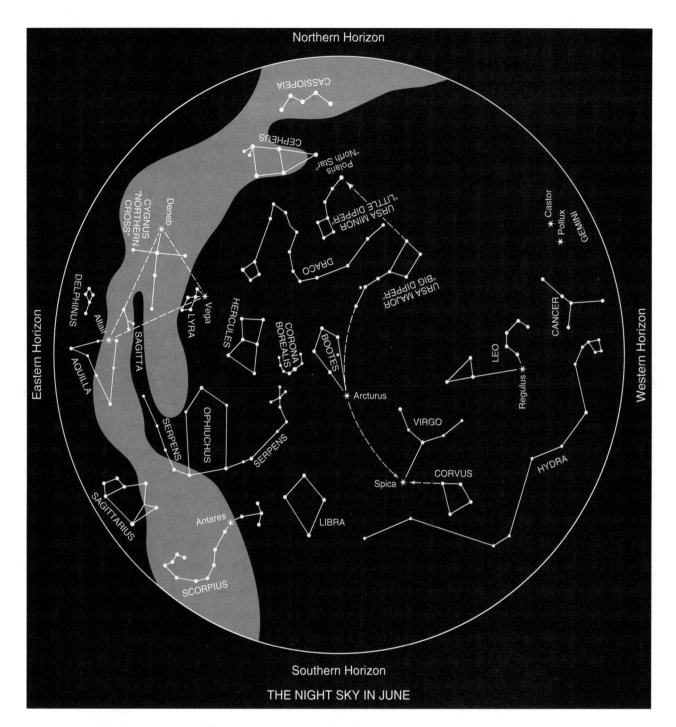

THE NIGHT SKY IN JUNE

To use: Hold chart vertically and turn it so the direction you are facing shows at the bottom.

Northern Horizon

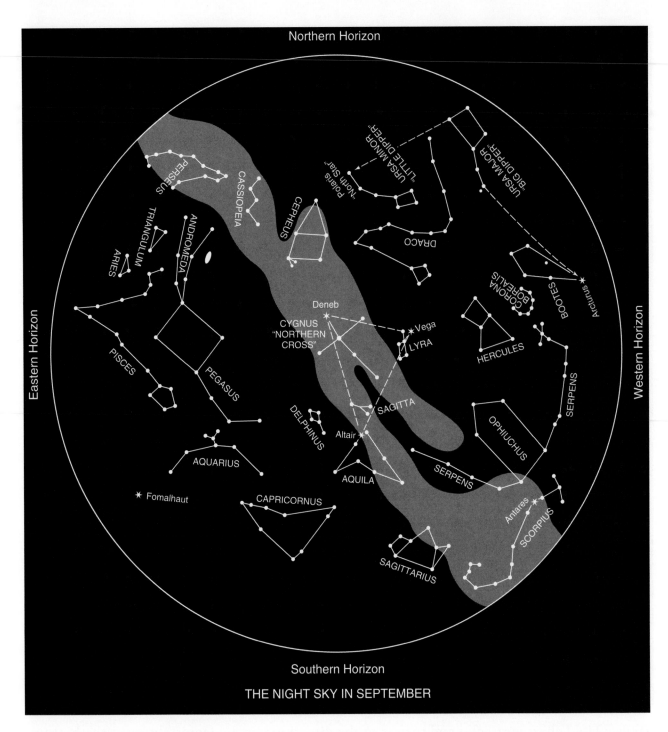

THE NIGHT SKY IN SEPTEMBER

Eastern Horizon

Western Horizon

Southern Horizon

To use: Hold chart vertically and turn it so the direction you are facing shows at the bottom.

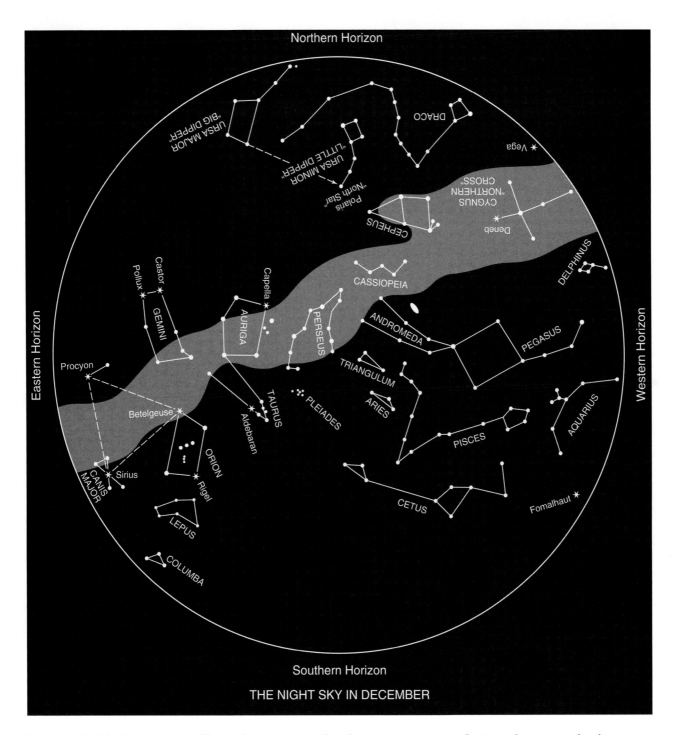

The Night Sky in December

To use: Hold chart vertically and turn it so the direction you are facing shows at the bottom.

APPENDIX

L   Planetary Data

			Physical Data				
PLANET	EQUATORIAL (km)	DIAMETER* (Earth = 1)	OBLATENESS	MASS** (Earth = 1)	DENSITY (gm/cm³)	SURFACE GRAVITY (Earth = 1)	ESCAPE VELOCITY (km/s)
Mercury	4,878	0.382	0	0.055	5.43	0.38	4.25
Venus	12,104	0.95	0	0.82	5.24	0.90	10.36
Earth	12,756	1	0.0034	1	5.52	1	11.18
Mars	6,794	0.53	0.009	0.107	3.93	0.38	5.02
Jupiter	142,800	11.2	0.064	318	1.32	2.69	59.6
Saturn	120,660	9.5	0.102	95.1	0.70	1.19	35.6
Uranus	51,000	4.0	0.024	14.5	1.2	0.93	21.1
Neptune	49,500	3.9	0.027	17.2	1.76	1.22	24.6
Pluto	2,290	0.2	?	0.002	2	0.05	1

*Equatorial diameter of Earth = 12,756 km.
**Mass of Earth = 5.9742×10^{24} kg.

	Axis Tilt and Rotation Period	
PLANET	EQUATORIAL TILT TO ORBITAL PLANE (degrees)	SIDEREAL ROTATION PERIOD* (days or hours)
Mercury	0	58.65 d
Venus	178	−243 d
Earth	23.44	23.934 h
Mars	25.20	24.623 h
Jupiter	3.12	9.842 h
Saturn	26.73	10.65 h
Uranus	97.86	−17.2 h
Neptune	28.8	16.05 h
Pluto	119	−6.39 d

*The "−" signs indicate retrograde rotation.

564

	Orbit Data				
PLANET	SEMIMAJOR (10^6 km)	AXIS (AU)	ORBITAL PERIOD	ORBITAL ECCENTRICITY	INCLINATION TO ECLIPTIC (degrees)
Mercury	57.9	0.387	87.97 d	0.2056	7.0
Venus	108.2	0.723	224.7 d	0.0068	3.39
Earth	149.6	1	365.26 d	0.0167	0
Mars	227.9	1.524	1.881 y	0.0934	1.85
Jupiter	778.3	5.203	11.86 y	0.0485	1.3
Saturn	1427	9.539	29.46 y	0.0556	2.49
Uranus	2870	19.19	84.01 y	0.0472	0.77
Neptune	4497	30.06	164.79 y	0.0086	1.77
Pluto	5900	39.44	248.5 y	0.250	17.2

M   The Brightest Stars

STAR	POPULAR NAME	APPARENT MAGNITUDE	APPARENT BRIGHTNESS COMPARED*	ABSOLUTE MAGNITUDE	ABSOLUTE LUMINOSITY COMPARED**	DISTANCE (LY)	SPECTRAL TYPE
	Sun	−26.8	1.4×10^{10}	4.83	1	0.000015	G2
α Cma A	Sirius	−1.47	1	1.4	24	8.7	A1
α Car	Canopus	−0.72	0.50	−3.1	1,500	98	F0
α Boo	Arcturus	−0.06	0.27	−0.3	110	36	K2
α Cen	Rigel Kentaurus	0.01	0.26	4.4	1.5	4.2	G2
α Lyr	Vega	0.04	0.25	0.5	54	26.5	A0
α Aur	Capella	0.05	0.25	−0.6	150	45	G8
β Ori A	Rigel	0.14	0.23	−7.1	59,000	900	B8
α Cmi A	Procyon	0.37	0.18	2.7	7.1	11.4	F5
α Ori	Betelgeuse	0.41	0.18	−5.6	15,000	520	M2
α Eri	Achernar	0.51	0.16	−2.3	710	118	B3
β Cen AB	Hadar	0.63	0.14	−5.2	10,000	490	B1
α Aql	Altair	0.77	0.13	2.2	11	16.5	A7
α Tau A	Aldebaran	0.86	0.12	−0.7	160	68	K5
α Vir	Spica	0.91	0.11	−3.3	1,800	220	B1
α Sco A	Antares	0.92	0.11	−5.1	9,400	520	M1
α PsA	Fomalhaut	1.15	0.090	2.0	14	22.6	A3
β Gem	Pollux	1.16	0.089	1.0	34	35	K0
α Cyg	Deneb	1.26	0.081	−7.1	59,000	1,600	A2
β Cru	(Beta Crucis)	1.28	0.079	−4.6	5,900	490	B0.5
α Leo A	Regulus	1.36	0.074	−0.7	160	84	B7

*Compared to Sirius, the brightest star other than the Sun.
**Compared to the Sun.

APPENDIX

N   The Nearest Stars

STAR	APPARENT MAGNITUDE	APPARENT BRIGHTNESS COMPARED*	ABSOLUTE MAGNITUDE	ABSOLUTE LUMINOSITY COMPARED**	DISTANCE (LY)	SPECTRAL TYPE
Sun	−26.8	1.4×10^{10}	4.83	1	0.0000015	G2
Proxima Centauri	11.5	0.0000065	15.5	0.000054	4.2	M5
α Centauri A	0.01	0.26	4.4	1.5	4.2	G2
α Centauri B	1.5	0.065	5.8	0.41	4.2	K5
Barnard's star	9.5	0.000041	13.2	0.00045	5.9	M5
Wolf 359	13.5	0.0000010	16.8	0.000016	7.6	M6
Lalande 21185	7.5	0.00026	10.4	0.0059	8.1	M2
Luyten 726-8A	12.5	0.0000026	15.4	0.000059	8.2	M5
Sirius A	1.47	1	1.4	24	8.7	A1
Sirius B	7.2	0.00034	11.5	0.0021	8.7	White dwarf
Ross 154	10.6	0.000015	13.3	0.00041	9.4	M4
Ross 248	12.2	0.0000034	14.8	0.00010	10.3	M5
ε Eridani	3.7	0.0086	6.1	0.31	10.7	K2
Ross 128	11.1	0.0000094	13.5	0.00034	10.8	M4
Luyten 789-6	12.2	0.0000034	14.6	0.00012	11.0	M6
61 Cygni A	5.2	0.0021	7.6	0.078	11.2	K5
61 Cygni B	6.0	0.0010	8.4	0.037	11.2	K7
ε Indi	4.7	0.0034	7.0	0.14	11.2	K5
τ Ceti	3.5	0.010	5.7	0.45	11.4	G5
Lacaille 9352	7.4	0.00028	9.6	0.012	11.4	M1
Procyon A	0.4	0.18	2.7	7.1	11.4	F5
Procyon G	10.8	0.000012	13.1	0.00049	11.4	White dwarf

*Compared to Sirius, the brightest star other than the Sun.
**Compared to the Sun.

ANSWERS TO ODD-NUMBERED REVIEW QUESTIONS

Chapter 1

1. c   3. d   5. b   7. c
9. Crust: granite (continents) and basalt (oceans). Mantle: magnesium and iron silicate minerals. Core: mostly iron and nickel. Outer core is liquid; inner core is solid.
11. Earth-universe, atmosphere, hydrosphere, biosphere, solid Earth. The evaporation of ocean water to fall as rain and snow is an interaction between the atmosphere and hydrosphere.

Chapter 2

1. b   3. c   5. b   7. b   9. b   11. e
13. They have an orderly internal atomic structure.
15. A silicate mineral contains the SiO_4 unit. Ferromagnesian silicates contain abundant iron and magnesium; nonferromagnesian silicates do not.
17. Mineral hardness refers to resistance to scratching and is controlled by the strength of atomic bonds.
19. Value and amount of the mineral, distance to market, cost of wages, taxes, and equipment.
21. Construction: sand and gravel. Nonmetallic: diamonds, gypsum, salt, fluorite, many others. Energy: coal, petroleum, natural gas, uranium, geothermal energy.
23. Metallic bonding makes metals good conductors of electricity and easily deformable.

Chapter 3

1. d   3. d   5. a   7. c   9. c   11. e   13. c
15. They have higher silica content.
17. Lithification is the conversion of sediment to rock. Compaction is enough to lithify clay, but coarser sediment is usually cemented.
19. Slate, a low-grade rock, has not been altered as much in mineral composition and texture as schist, a high-grade rock.
21. The calcite is originally extracted by marine organisms. Replacement of calcium by magnesium after deposition changes much limestone to dolostone.
23. Evaporites form when dissolved minerals remain after a body of water evaporates. Rock salt and rock gypsum are examples.

Chapter 4

1. e   3. d   5. a   7. b   9. b   11. e   13. a
15. Mechanical weathering is the disintegration of rocks by physical processes rather than chemical reactions. It speeds up chemical weathering by increasing the surface area exposed to chemical attack.
17. Acid solutions contain abundant hydrogen ions, which can replace other positively charged ions in minerals.
19. Parent material is the most important factor in the makeup of young soils. Climate influences soil formation because abundant water and warm temperatures speed up weathering.
21. The amount of rainfall determines how well lubricated potential failure surfaces are. Saturated soils are more prone to movement than dry soils. Permafrost leads to saturated surface soil and solifluction. Freezing and thawing or wetting and drying cycles both result in increased soil creep.
23. A rockslide is the movement of a mass of rock along a bedding plane; a slump is the movement of unconsolidated material. Slumps occur above excavations because the excavation weakens the support for the material above the excavation.

Chapter 5

1. c   3. b   5. c   7. c   9. b   11. d   13. e
15. Water evaporates from the oceans, falls as rain or snow on the land, and returns to the oceans by runoff or the atmosphere by evaporation and transpiration by plants.
17. The larger the perimeter of the channel cross section and the rougher the bed, the more friction impedes flow.
19. The stream would cut deeper.
21. It takes time for water to percolate through the rocks. As long as more rain falls before the water table can become level, the water table will mimic topography.
23. If the water is under pressure, but not enough pressure to reach the surface, it must still be pumped.
25. Leaching of contaminants from the surface, injection into deep waste disposal wells, and salt-water intrusion due to excessive pumping.

Chapter 6

1. c   3. c   5. e   7. c   9. c   11. d   13. b
15. Rock debris embedded in the ice abrades. Flow of the ice over fractured rock pulls loose pieces away (plucking).
17. Terminal moraines form from material that is carried to the terminus of a glacier. Recessional moraines are much like terminal moraines that mark successive positions of the terminus as the ice retreats. Lateral moraines form from material carried along the sides of the glacier.
19. Medial moraines form when the lateral moraines of two glaciers merge. Each medial moraine marks the merger of two glaciers.
21. Wind erodes by transporting loose material and abrading hard material. Wind is most effective at moving silt and fine sand and otherwise not very effective.
23. Mechanical weathering dominates in deserts because there is little moisture for chemical weathering.

Chapter 7

1. b   3. d   5. b   7. b   9. c
11. Fitting continental margins and magnetic stripes, paleomagnetism, paleoclimate indicators.
13. Glossopteris is a fossil tree. It and associated land plant and animal fossils occur on all the Gondwanaland continents and this distribution is best explained if the continents were once joined together.
15. Thermal convection is the transport of heat by the rising of hot material and the sinking of cool material. It is believed to be the driving force for plate tectonics.
17. Divergent: new oceanic lithosphere forms; as plates pull apart, new magma from the mantle fills the gaps. Transform: plates slide past each other with little volcanic activity. Convergent: an oceanic plate descends into the mantle to be recycled. Crustal deformation and igneous activity occur on the overriding plate.

Chapter 8

1. d   3. b   5. e   7. b   9. a
11. Most (over 90%) earthquakes occur at plate boundaries.
13. Tsunami are probably created when submarine landslides displace large volumes of sea water. Tsunami are destructive because they can be very large and strike distant locations with no warning.
15. Since fluids do not transmit S-waves, the fact that the S-wave shadow zone covers the entire opposite side of the Earth from an earthquake shows that the core is fluid.
17. To observe seismic waves of different frequency and direction.
19. By studying samples of deep rocks brought to the surface by volcanic and tectonic processes, and by comparing the seismic properties of rocks in the laboratory to the properties of the actual Earth.

Chapter 9

1. d   3. b   5. d   7. d   9. d   11. d
13. Gases, volcanic ash and other pyroclastic materials, mud, lava flows.
15. The magma expands rapidly like a can of warm, shaken soda. These flows are extremely hot and fast moving.
17. Crater Lake formed when a violent eruption expelled much of the magma chamber of a volcano and the cone collapsed. The summit caldera of Mauna Loa or Kilauea formed when magma drained away by eruption and the summit of the volcano gently subsided.
19. A dormant volcano has not erupted in a long time but has the potential for future eruptions. An extinct volcano has ceased activity entirely. The terms cannot be precisely defined because some volcanoes have erupted after being quiet for thousands of years.

Chapter 10

1. c   3. e   5. d   7. d
9. Hydrostatic pressure is equal in all directions; directed pressure differs in magnitude with direction. Deep in the Earth, pressure is nearly hydrostatic.
11. Geologic maps allow geologists to summarize rock relationships over large areas. Rock type, structures like folds and faults, contacts or rock boundaries and strike and dip all are plotted on geologic maps.
13. Ocean-ocean, continent-ocean, continent-continent.
15. Student-performed exercise.
17. Gravity, magnetic, electrical and seismic surveys.

Chapter 11

1. d   3. a   5. d   7. d   9. e   11. a   13. b
15. Hutton: formulated the principle of cross-cutting relationships. Lyell: formulated the principle of uniformitarianism. Steno: formulated the principle of superposition.
17. The big problem was avoiding water loss. Early land plants and animals remained in wet environments and reproduced in water.
19. Moon rocks and meteorites are used to estimate the age of the Earth. Rocks of that age are very unlikely to be found on Earth because of weathering, erosion, and metamorphism.
21. Radioactive decay is the change of one element to another by emission of particles from atomic nuclei. Since decay happens at constant rates, measurement of radioactive elements and the products formed from them allows us to determine the ages of rocks.
23. Carbon-14 depends on the continual production of carbon-14 by solar atomic particles, and its incorporation into living things. It cannot be used to date non-living or very old materials.

25. Relative dating determines the order of events but not their ages in years. Absolute dating determines ages in years, but just as in history, ages can be somewhat approximate.

Chapter 12

1. b   3. d   5. c   7. d   9. b   11. b   13. d

15. Minerals that are soluble in the superheated vent water are insoluble in cold sea water and precipitate quickly.

17. Abyssal plains are very old oceanic crust lacking in seamounts and covered by thick pelagic sediment. They are found far away from spreading plate margins, especially in the Atlantic.

19. Oceanic ridges are sites of crustal extension; land mountains are sites of compression.

21. The EEZ is the zone claimed by the United States and other nations within which they have exclusive right to sea-floor resources. Oil and natural gas are the most important resources in most EEZs, although manganese nodules and sea-floor hydrothermal metal deposits are potentially minable.

Chapter 13

1. c   3. e   5. c   7. a   9. c   11. d   13. c

15. Polar: cold, salty water sinks and deep water rises to replace it. Equatorial: gyres diverge and deep water rises to fill the gap. Coastal: surface water flows away from a coast and deep water rises to replace it.

17. Deep water waves are not affected by the bottom; shallow water waves are.

19. A longshore current is generated when waves consistently approach a shoreline from a certain direction. Water returning to the sea from the shore is pushed along by the next wave.

21. See Figure 13.15.

23. One bulge is raised by the Moon's gravity, the other by motion around the Earth-Moon center of mass.

25. Warm but very salty water from the Mediterranean sinks and flows out along the bottom of the Strait of Gibraltar whereas less salty Atlantic water flows in at the surface.

Chapter 14

1. It has been said that "climate is what you expect and weather is what you get." Climate is the long-term average of weather conditions plus extremes. Actual weather conditions can depart drastically from the average.

3. d

5. Natural: volcanic, wind-blown dust, sea salt. Human-caused: construction, burning of fossil fuels.

7. Heat is the total amount of thermal energy a substance contains; temperature is the average energy of its atoms. A thin gas can be very hot but contain little heat; water can be lukewarm but contain a great deal of heat.

9. It is the temperature at which all thermal motion of atoms ceases.

11. a

13. Radiation, direct transport as electromagnetic radiation, like heat from the Sun. Conduction, from atom to atom, like the warming of a metal rod with one end in a flame. Convection, bodily movement of warm material, like the boiling of water in a pot.

15. Air is a poor conductor of heat. Movement of warm air is a more efficient way to transport heat.

17. The higher the specific heat of a material, the greater thermal inertia a given mass will have.

19. Air pressure is the weight of overlying air per unit area. It decreases rapidly with altitude.

21. No; it tapers gradually into space.

23. Humid air exerts lower pressure because water vapor is less dense than air.

25. Falling air pressure is generally associated with unstable conditions and stormy weather; high pressure with stable conditions and fair weather.

27. More compact and more rugged, and does not use a toxic substance.

29. Most auroras occur near the magnetic poles.

Chapter 15

1. e

3. Wien: the hotter an object the shorter the wavelength of the radiation it emits. Stefan-Boltzmann: the total energy an object emits is proportional to the fourth power of its temperature. Application: the Sun (hot) emits enormous energy in visible light; the Earth (cool) emits weakly in the infrared.

5. They are the limits of the zone where the Sun can appear directly overhead.

7. Reflection is the bouncing of radiation off a boundary in a definite direction, scattering is the bouncing of radiation in all directions. Reflection is a special case of scattering.

9. Clouds have a much higher albedo.

11. b

13. Carbon dioxide and to a lesser extent methane and nitrous oxide.

15. Radiational heating exceeds cooling at Earth's surface; cooling exceeds heating in the atmosphere.

17. a

19. Latent heat is released when water droplets condense.

21. d

23. Heating dominates at low latitudes, cooling at high latitudes.

25. Variations in air temperature are due to inequalities in heating of the Earth's surface plus air mass advection.

27. Clouds can lower air temperature by blocking sunlight but also raise it by trapping infrared radiation.

29. No. Advection implies movement of air, hence wind.

Chapter 16

1. Water molecules acquire enough energy to overcome the attraction of neighboring molecules.

3. Air pressure is the weight of overlying air. Vapor pressure is the fraction of air pressure due to water vapor. Air pressure is much greater.

5. The two pressures are equal if relative humidity is 100%.

7. b

9. b

11. Latent heat is released as water vapor condenses.

13. Only (d) and (f) are unstable for clear air; (a), (b), (c), and (e) are stable for clear air. For cloudy air, (b) is neutral; (c), (d), and (f) are unstable; (a) is stable.

15. b

17. Orographic cooling and condensation on the windward side of a mountain range, followed by sinking and warming on the lee side.

19. Cloud condensation nuclei are active at all temperatures; ice-forming nuclei are much less abundant and active only at very low temperatures.

21. Because the convection that sustains them diminishes.

23. About two kilometers (20 degrees C/10 degrees C per km lapse rate).

25. Over land because it heats more rapidly.

27. Because water molecules escape more easily from liquid water than from ice.

29. Because of orographic cooling and condensation; also closer to cloud base.

Chapter 17

1. Horizontal winds are much greater.

3. The steeper the pressure gradient, the higher the wind. Wind flows from higher to lower pressure.

5. c

7. The rougher the surface, the greater the resistance.

9. d

11. It causes the wind to blow across isobars toward lower pressure.

13. In the friction layer they spiral outward in a clockwise direction. Above the friction layer they follow the isobars in a clockwise direction.

15. Light or calm winds.

17. d

19. Westerlies spin off the higher-latitude flanks of the anticyclones.

21. a

23. The Intertropical Convergence Zone moves back and forth, remaining in the summer hemisphere. The polar front expands and strengthens in the winter hemisphere and contracts and weakens in the summer hemisphere.

25. There is greater temperature contrast between converging air masses.

27. A narrow band of high-speed wind in the upper atmosphere.

29. Where the air speeds up, divergence, hence low pressure, occurs.

Chapter 18

1. Whether it forms over land or ocean, in polar or tropical latitudes.

3. Exchange of heat or moisture with the surface. Polar air masses warm as they move to lower latitudes over warmer ground; tropical air masses cool, but more slowly, as they move to higher latitudes.

5. Warm front: advancing warm air overrides cooler air. Clouds drop gradually in altitude. Cold front: advancing cold air pushes warm air off the ground; vertical clouds, abrupt onset, sometimes violent weather. Occluded front: converging cool air masses push intervening warm air off the ground; combines characteristics of warm and cool fronts. Stationary fronts do not move.

7. Warm air is drawn in on the low-latitude side and cool air on the poleward side; eventually a sharp boundary between air masses of different temperatures results.

9. The leading side of a cyclone is warm because low-latitude air is being drawn in; the trailing side is cool because high-latitude air is being drawn in.

11. Converging air masses wrap around the low so tightly that warm air is forced off the surface. The air pressure begins to rise and the cyclone weakens and eventually dissipates.

13. Clear skies in the high pressure system result in efficient radiational cooling at night.

15. b

17. Convergence of sea breezes from both coasts, plus rising of heated air from the land.

19. Along the cold front on the leading edge of the warm sector.

21. Lightning results from discharges between areas of electrical charge imbalance in thunderstorms. It can electrocute or burn victims.

23. Steep valleys and rapid runoff funnel water from a large area into a single channel. The rapid runoff from paved surfaces causes flash flooding in urban areas.

25. d

27. Wind speed estimated from damage. Most U.S. tornadoes are F0 or F1.

29. Warm surface water and sufficient Coriolis effect. Hurricanes never occur at the Equator, where the Coriolis effect is zero. The South Atlantic, of all oceans, has no hurricanes because the sea surface is not warm enough.

Chapter 19

1. b   3. a   5. b

7. Planets travel in elliptical orbits with the Sun at one focus; a line from the planet to the Sun sweeps out equal areas in equal times; the period of a planet, squared, is proportional to its average distance, cubed.

9. Kinetic energy is energy due to motion. The kinetic energy of an impacting meteoroid is largely converted to heat. The resulting crater is 20–30 times the diameter of the meteoroid.

11. Superposition allows us to determine which craters formed earlier, degradation allows us to distinguish young craters from older ones, and saturation allows us to estimate the ages of cratered surfaces.

13. Sunlight refracted through the atmosphere illuminates the eclipsed Moon. If the atmosphere is unusually cloudy or dust-laden, the eclipsed Moon will be unusually dark.

15. Tide ranges tend to be greatest at new and full Moon, least at first and last quarter.

17. Student-performed exercise.

Chapter 20

1. d   3. c   5. a

7. They are either too hot or too cold for liquids to exist. If the planet has no atmosphere, liquids will evaporate.

9. Its surface is cratered and lava-covered like the Moon, but its density, large core, and magnetic field are more Earth-like than Moon-like.

11. Venus is nearly the same size and mass as Earth, but lacks plate tectonics and has a much denser atmosphere.

13. Water-cut valleys are the most persuasive evidence.

15. Jupiter's visible clouds are the top of a thick atmosphere. Deep within, Jupiter probably consists of liquid metallic hydrogen. It may have a small solid core.

17. Although nearly the same size and mass, Neptune has a much more active atmosphere than Uranus. Also Uranus has a far greater axial tilt than Neptune.

19. A meteor is a glowing interplanetary body heated by friction in Earth's atmosphere. If it survives to reach the surface, it is termed a meteorite. Most meteors vaporize before reaching the surface.

Chapter 21

1. b   3. a   5. d   7. c   9. b   11. a   13. c   15. a   17. b   19. c   21. b

GLOSSARY

absolute dating The process of assigning actual ages to geologic events. Various radioactive decay dating techniques yield absolute ages.

absolute zero The temperature at which theoretically all molecular activity ceases—although there is some atomic-level activity; 0 K.

absorption Conversion of electromagnetic radiation to heat energy.

abyssal plain The flat surface of the sea floor, covering vast areas beyond the continental rises of passive continental margins.

accretion Formation of a large object by incorporation of small objects. Refers to the formation of the planets by sweeping up small objects, and the formation of continents by the addition of terranes.

accretionary prism The mass of deformed sediment that accumulates on a subduction zone as plate convergence scrapes up sea-floor sediment.

active continental margin A continental margin that develops at the leading edge of a continental plate where oceanic lithosphere is subducted.

adiabatic process In the atmosphere, changes in air temperature due to expansion (cooling) or compression (warming) only. There is no net heat exchange between ascending or descending air parcels and the surrounding (ambient) air.

aerosols Tiny solid or liquid particles of various composition that are suspended in the atmosphere.

aftershock An earthquake following a main shock resulting from adjustments along a fault. Aftershocks are common after a large earthquake, but most are smaller than the main shock.

air mass A huge volume of air covering thousands of square kilometers that is relatively uniform horizontally in temperature and water vapor concentration.

air mass advection Horizontal movement of air or air masses from one place to another.

air pressure The cumulative force exerted on any surface by the molecules composing air; usually described as the weight of a column of air per unit surface area.

albedo The portion of the radiation striking a surface that is reflected by that surface; usually expressed as a percentage.

alluvial fan A cone-shaped deposit of alluvium; generally deposited where a stream flows from mountains onto an adjacent lowland.

angular unconformity An unconformity below which older rocks dip at a different angle (usually steeper) than the overlying younger rocks. See also *disconformity* and *nonconformity*.

anticline An up-arched fold in which the oldest exposed rocks coincide with the fold axis, and all strata dip away from the axis.

anticyclone A dome of air that exerts relatively high surface pressure compared with the surrounding air; same as a *high*. In the Northern Hemisphere, surface winds in an anticyclone blow clockwise and outward when viewed from above.

aphotic zone The dark region in the ocean below the depth of sunlight penetration. See also *photic zone*.

apogee The point in a satellite's orbit that is farthest from Earth.

aquiclude Any layer that prevents the movement of groundwater.

aquifer A permeable layer that allows the movement of groundwater.

arête A narrow, serrated ridge separating two glacial valleys or adjacent cirques.

artesian system A system in which groundwater is confined and builds up high hydrostatic (fluid) pressure.

asthenosphere The part of the upper mantle over which lithospheric plates move. A zone of ductile mantle rocks between about 200 and 400 km deep.

astronomy The study of bodies beyond Earth.

atmosphere A thin envelope of gases, suspended solid and liquid particles, and clouds that encircles the planet.

atmospheric stability Property of the atmosphere that either enhances (unstable) or suppresses (stable) upward and downward motion of air; depends on the vertical temperature profile of ambient air and whether ascending or descending air is saturated (cloudy) or unsaturated (clear).

atom The smallest unit of matter that retains the characteristics of an element.

atomic number The number of protons in the nucleus of an atom.

atomic weight The total mass of protons and neutrons in the nucleus of an atom.

aurora A luminous phenomenon in the night sky consisting of overlapping curtains of light in the ionosphere at high latitudes; caused by emission of light by atoms and molecules that are excited by beams of electrons generated by a complex interaction between the solar wind and Earth's magnetosphere.

axial plane The plane along the center of a fold.

axial tilt The tilt of a planet's rotation axis relative to the plane of its orbit.

barrier island An elongate sand body oriented parallel to a shoreline, but separated from the shoreline by a lagoon.

basal slip A type of glacial movement that occurs when a glacier slides over the underlying surface.

base level The lowest limit to which a stream can erode.

batholith A discordant, irregularly shaped pluton composed chiefly of granitic rocks.

baymouth bar A spit that has grown until it completely cuts off a bay from the open sea.

beach A deposit of unconsolidated sediment extending landward from low tide to a change in topography or where permanent vegetation begins.

Bergeron process Precipitation formation in cold clouds whereby ice crystals grow at the expense of supercooled water droplets.

Big Bang A model for the evolution of the universe in which the universe begins in a dense, hot state and expands and cools to a less dense state.

binary star A pair of stars that orbit around their common center of mass, much the way planets orbit the Sun.

biosphere The domain of living organisms on Earth.

blackbody A perfect radiator or absorber; an object that absorbs all the radiation striking it and emits the maximum possible radiation at all wavelengths.

bonding The process whereby atoms are joined to other atoms usually by sharing or exchanging electrons.

Bowen's series A composite series of silicate minerals that accounts for the observed mineral compositions of igneous rocks. It has two branches: ferromagnesian minerals and plagioclase feldspars. The final minerals to form are potassium feldspar and quartz.

braided stream A stream possessing an intricate network of dividing and rejoining channels. Braiding occurs when sediment transported by the stream is deposited within channels as sand and gravel bars.

breaker A wave that oversteepens as it enters shallow water until the crest plunges forward.

brittle Strain that results in the breakage of a material.

caldera A large, steep-sided circular to oval volcanic depression formed by summit collapse resulting from partial draining of an underlying magma chamber or from eruption of material that formerly supported the volcano.

cave A naturally formed subsurface opening that is generally connected to the surface and is large enough for a person to enter.

celestial poles The points in the sky around which the heavens appear to rotate. They are overhead as seen from the north and south geographic poles and on the horizon as seen from Earth's equator.

central peak The mountain in the center of an impact crater formed when the densely compressed floor rebounds after the initial impact.

centripetal force An inward-directed force that confines an object to a curved path; the resultant of other forces.

Cepheid variable A yellow-white giant star that pulsates regularly. Useful because they can be used to determine the distance to distant star clusters and galaxies.

chemical formula A way of symbolizing the composition of chemical compounds by use of symbols to represent chemical elements and numbers to represent their proportions.

chemical sedimentary rock Rock formed from materials dissolved during chemical weathering.

chemical weathering The process whereby rock materials are decomposed by chemical alteration of the parent material.

chondrites Meteorites with a pellet-like structure that have metallic elements in about the same ratio as the Sun; believed to be leftover samples of the raw materials of the original solar system.

cinder cone A small volcano composed chiefly of pyroclastic materials.

cirque A steep-walled, bowl-shaped depression at the upper end of a glacial valley. Formed by erosion by a valley glacier.

cleavage Tendency of a mineral to split along smooth planes that are planes of weakness between atoms in the mineral.

climate Weather of some locality averaged over some time period plus extremes in weather observed during the same period or during the entire period of record.

comet An icy body orbiting the Sun usually in a highly elongated orbit. When it approaches the Sun, its material evaporates to form a visible tail.

complex movement A combination of different types of mass movements in which one type is not dominant; most complex movements involve sliding and flowing.

compressional warming A temperature rise that accompanies a pressure increase on a parcel (unit mass) of air, as when air parcels descend within the atmosphere.

conduction Transfer of energy (electrical, heat) within and through a conductor by means of internal particle or molecular activity.

cone of depression The lowering of the water table around a well in the shape of a cone; results when water is removed from an aquifer faster than it can be replenished.

constellation A pattern of stars as seen in Earth's sky; the component stars are at different distances and usually not physically related in any way.

contact metamorphism Metamorphism of the rock adjacent to an intrusion due to the heat from the intrusion.

continental drift The theory that the continents were once joined into a single landmass that broke apart with the various fragments moving with respect to one another.

continental glacier A large glacier covering a vast area (at least 50,000 km^2) and unconfined by topography. Also called an ice sheet.

continental margin The area separating the part of a continent above sea level from the deep-sea floor.

continental rise The gently sloping area of the sea floor beyond the base of the continental slope.

continental shelf The area between the shoreline and continental slope where the sea floor slopes gently seaward.

continental slope The relatively steep area between the shelf-slope break (at an average depth of 135 m) and the more gently sloping continental rise or oceanic trench.

convection Vertical air circulation in which cool dense air sinks and forces warm less dense air to rise.

convergent plate boundary The boundary between two plates that are moving toward one another. See also *subduction zone*.

core The innermost part of the Earth below the mantle at about 2,900 km; divided into an outer liquid core and an inner solid core.

Coriolis effect A deflective force arising from the rotation of Earth on its axis; affects primarily large-scale atmospheric and oceanic circulation. Winds and ocean currents are deflected to the right of their initial direction in the Northern Hemisphere and to the left in the Southern Hemisphere.

correlation Demonstration of the age equivalence of rock units in different areas.

crater A circular or oval depression formed by explosion. The crater at the summit of a volcano results from the violent expulsion of gases, pyroclastic materials, and lava whereas impact craters result from meteor strikes.

craton The stable interior of a continent; includes exposed shields as well as areas of ancient crust covered by sedimentary rocks.

creep The imperceptible downslope movement of soil or rock; it is the slowest type of flow.

crust The outermost part of Earth overlying the mantle. Continental crust is 30 to 70 km thick and largely made of granitic rocks; oceanic crust averages about 5 km thick and is largely basalt.

crystal A geometric solid resulting from the orderly addition of atoms to a mineral. Materials with an orderly atomic arrangement are described as *crystalline* but do not always have the form of crystals.

cumuliform clouds   Clouds that exhibit significant vertical development; usually produced by updrafts in convection cells.

cyclone   A weather system characterized by relatively low surface air pressure compared with its surrounding area; same as a *low*. Viewed from above, surface winds blow counterclockwise and inward in the Northern Hemisphere.

daughter element   A stable element resulting from the radioactive decay of a parent element. See also *parent element*.

debris avalanche   A complex type of mass wasting common in steep mountains; typically starts as a rock fall.

debris flow   A mass wasting process; much like a mudflow but more viscous and containing larger particles.

deflation   The removal of loose surface sediment by the wind.

delta   An alluvial deposit formed where a stream discharges into a lake or ocean.

density   Mass per unit volume; usually expressed as grams per cubic centimeter.

depositional environment   Any area where sediment is deposited such as on a stream's floodplain or on a beach.

desert   Any area that receives less than 25 cm of rain per year and has a high evaporation rate.

detrital sedimentary rock   Sedimentary rock consisting of the solid fragments (detritus) of preexisting rocks (e.g., sandstone and conglomerate).

dew   Tiny droplets of water formed by condensation of water vapor on a relatively cold surface.

dike   A tabular or sheet-like discordant pluton.

dip   A measure of the maximum angular deviation of an inclined plane from the horizontal.

disconformity   An unconformity above and below which the rock layers are parallel.

divergent plate boundary   The zone between plates that are pulling apart, most often at a midoceanic ridge.

divide   A topographically high area that separates adjacent drainage basins.

Doppler effect   The change in wavelength of radiation due to the approach or recession of an object. Radiation from an approaching object has its wavelength shortened; radiation from a receding object has its wavelength lengthened.

downburst   A strong and potentially destructive thunderstorm downdraft that spreads horizontally after striking the Earth's surface.

downwelling   The slow transfer of ocean surface water to depth.

drainage basin   The surface area drained by a stream and its tributaries.

drainage pattern   The regional arrangement of stream channels in a drainage system.

drumlin   An elongate hill of till formed by the movement of a continental glacier.

ductile   Strain in which a material deforms without breaking.

dune   A mound or ridge of wind-deposited sand.

Earth science   The family of sciences related to the study of Earth and its systems.

earthflow   A type of mass wasting process; involves downslope flowage of water-saturated soil.

earthquake   Vibration of the Earth caused by the sudden release of energy, usually as a result of displacement of rocks along faults.

Earth-universe system   The system consisting of the Earth and its interconnections with the external universe.

ecliptic   The apparent path of the Sun in the sky; marks the plane of the Earth's orbit. The Moon and other planets travel very near the plane of the ecliptic.

elastic rebound theory   A theory that explains the sudden release of energy when rocks are deformed by movement on a fault.

electromagnetic spectrum   Range of radiation types arranged by wavelength, frequency, and energy level.

electron   A negatively charged particle of very little mass that orbits the nucleus of an atom.

element   A substance composed of all the same atoms; it cannot be changed into another element by ordinary chemical means.

emergent coast   A coast where the land has risen with respect to sea level.

end moraine   A pile of rubble deposited at the terminus of a glacier.

epicenter   The point on the Earth's surface directly above the focus of an earthquake.

equinoxes   The first days of spring and autumn when day and night are of equal length at all latitudes (except at the poles). On those days, the noon sun is directly over the equator (solar altitude of 90 degrees).

erosion   The removal of weathered materials from their source area.

esker   A long, sinuous ridge of sediment formed by deposition by running water in a tunnel beneath stagnant ice.

Exclusive Economic Zone   An area extending 371 km seaward from the coast of the United States and its territories in which the United States claims all sovereign rights.

exfoliation dome   A large rounded dome of rock resulting from the process of exfoliation.

exotic terrain   A segment of crust attached to a continent by plate convergence.

expanding universe   The term for the recession of galaxies in all directions away from our own. See also *Big Bang*.

expansional cooling   A temperature drop that accompanies a pressure reduction on a parcel (unit mass) of air, as when air parcels ascend within the atmosphere.

fault   A fracture in rock along which movement has occurred parallel to the fracture surface.

felsic magma   Refers to magma containing more than 65% silica and considerable sodium, potassium, and aluminum, but little calcium, iron, and magnesium. See also *intermediate magma* and *mafic magma*.

fetch   The distance the wind blows over a continuous water surface.

flash flood   A sudden rise in river or stream levels causing flooding.

floodplain   A low-lying, relatively flat area adjacent to a stream that is partly or completely covered by water when the stream overflows its banks.

focus   The place within the Earth where an earthquake originates and energy is released.

fog   A cloud in contact with the Earth's surface that reduces visibility to less than 1.0 km (0.62 mi).

foliation   A platy or planar structure in metamorphic rocks that forms as a result of deformation; it is usually parallel to the axial planes of folds.

foreland   The less-deformed interior portion of an orogenic belt, with little igneous activity and mild or no metamorphism.

fossil   Remains or traces of prehistoric organisms preserved in rocks of Earth's crust.

friction   The resistance an object encounters as it comes into contact with other objects; in the atmosphere, friction is important within about 1 km (0.62 mi) of the Earth's surface.

front A narrow zone of transition between air masses that differ in temperature, humidity, or both; fronts may be cold, warm, stationary, or occluded.

frost Ice crystals formed by deposition of water vapor on a relatively cold surface.

frost wedging The opening and widening of cracks by the repeated freezing and thawing of water.

F-scale Tornado intensity scale, developed by T. T. Fujita, that rates tornadoes from F0 to F5 on the basis of rotational wind speed estimated from property damage.

galaxy A large gravitationally connected mass of billions of stars, usually with a spiral or elliptical form.

gas giant A large planet having a thick atmosphere and small solid core; the planets Jupiter, Saturn, Uranus, and Neptune.

geologic map A map displaying locations of rock boundaries, geologic structures, and orientations of rock layers.

geologic time The time from Earth's origin 4.6 billion years ago to the present.

geologic time scale A chart with the designation for the earliest interval of geologic time at the bottom, followed upward by designations for more recent intervals of time.

geology The science concerned with the study of Earth; includes studies of Earth materials (minerals and rocks), surface and internal processes, and Earth history.

geostrophic wind Unaccelerated horizontal wind that blows in a straight line parallel to isobars above the friction layer; results from a balance between the horizontal pressure gradient force and the Coriolis effect.

geothermal gradient The temperature increases with depth in the Earth; it averages about 25 Celsius degrees per km near the surface.

geyser A hot spring that intermittently ejects hot water and steam.

glacial budget The balance between accumulation and wastage in a glacier.

glacial drift A collective term for all sediment deposited by glacial activity, including till deposited directly by glacial ice and outwash deposited by streams derived from melting ice.

glacier A mass of ice on land that moves by plastic flow and basal slip.

globular star cluster A large spherical cluster of up to a million stars that surround galaxies.

***Glossopteris* flora** A Late Paleozoic association of plants found only on Southern Hemisphere continents and India.

graded stream A stream possessing an equilibrium profile in which a delicate balance exists between gradient, discharge, flow velocity, channel characteristics, and sediment load such that neither significant erosion nor deposition occurs within the channel.

gravity The force that holds all objects on the Earth's surface; the net effect of gravitation and the centripetal force due to the Earth's rotation.

greenhouse effect Terrestrial infrared radiation is absorbed and radiated primarily by water vapor and, to a lesser extent, by carbon dioxide and other trace atmospheric gases, thereby slowing the escape of heat to space from the Earth-atmosphere system and significantly elevating the average temperature at the Earth's surface.

groundwater The underground water stored in the pore spaces in rocks, sediment, or soil.

guide fossil Any fossil that can be used to determine the relative geologic ages of rocks and to correlate rocks of the same age in different areas.

Gulf Stream An oceanic current that flows northward along the east coast of North America; part of the larger North Atlantic gyre.

guyot A flat-topped seamount of volcanic origin rising more than 1 km above the sea floor.

gyre A system of oceanic currents rotating clockwise in the Northern Hemisphere and counterclockwise in the Southern Hemisphere.

Hadley cell Vertical air circulation in tropical and subtropical latitudes of both hemispheres; resembles a huge convective cell with rising air near the equator and sinking air in subtropical anticyclones.

hail Precipitation in the form of rounded or jagged chunks of ice; often characterized by internal concentric layering. Hail is associated with intense thunderstorm cells that have strong up drafts and relatively high moisture content.

half-life The time required for one-half of the original number of atoms of a radioactive element to decay to a stable daughter product, e.g., the half-life of potassium-40 is 1.3 billion years.

hanging valley A tributary glacial valley whose floor is at a higher level than that of the main glacial valley.

hardness Resistance of a mineral to scratching.

heat The total kinetic energy of the atoms or molecules composing a substance.

Hertzsprung-Russell (H-R) diagram A plot of star brightness against temperature that reveals patterns of stellar evolution.

highlands (lunar) Ancient high-standing crust of the Moon, made largely of a feldspar-rich rock called anorthosite.

horn A steep-walled, pyramidal peak formed by the headward erosion of cirques.

hot spring A spring in which the water temperature is warmer than the temperature of the human body (37°C).

hurricane Intense warm-core oceanic cyclone that originates in tropical latitudes with sustained winds of 119 km (74 mi) per hour or higher.

hydrologic cycle The continuous recycling of water from the oceans, through the atmosphere, to the continents, and back to the oceans.

hydrosphere The domain of water on Earth.

hypothesis A provisional explanation for observations; subject to continual testing and modification. If well supported by evidence, hypotheses are then generally called *theories*.

ices Materials of low melting point that form as solids in the outer solar system.

igneous arc The zone of maximum igneous activity in an orogenic belt, characterized by volcanism, intrusions, and usually high-temperature metamorphism.

igneous rock Any rock formed by cooling and crystallization of magma or lava, or by the accumulation and consolidation of pyroclastic materials.

impact The collision of an object with a planet, and the processes that result from the collision.

inertia The tendency of bodies to continue moving once in motion, or to remain still if at rest.

inselberg An isolated steep-sided erosional remnant rising above a desert plain.

intensity A subjective measure of the amount of damage done by an earthquake as well as people's reaction to it.

intermediate magma A magma having a silica content between 53 and 64% and an overall composition intermediate between felsic and mafic magmas.

intertropical convergence zone (ITCZ) Discontinuous belt of thunderstorm cells that parallels the equator; marks the convergence of trade winds of the Northern and Southern Hemisphere.

intrusion A large mass of solidified magma within the crust of the Earth.

ion An electrically charged atom produced by adding or removing electrons from its outermost electron shell.

ionic bonding A chemical bond resulting from the attraction of positively and negatively charged ions.

ionosphere Region of the upper atmosphere between 80 and 900 km (50 and 600 mi) altitude that contains a relatively high concentration of ions, electrically charged particles.

iron meteorites A group of meteorites composed mostly of iron and nickel.

island arc A curved chain of volcanic islands parallel to a deep-sea trench where oceanic lithosphere is subducted causing volcanism and producing volcanic islands.

isostasy The effect of buoyancy on Earth's crust so that light and thick crust rises and thin and dense crust sinks.

jet streak An area of accelerated air flow within a jet stream.

jet stream A narrow, fast-moving current of air in the upper atmosphere.

joint A fracture in rock along which no movement has occurred parallel to the fracture surface.

Jovian planet Any of the four planets that resemble Jupiter (Jupiter, Saturn, Uranus, and Neptune). All are large and have low mean densities, indicating they are composed mostly of lightweight gases and ices. See also *gas giant*.

kame Conical hill of stratified drift originally deposited in a depression on a glacier's surface.

karst topography A topography with numerous caves, springs, sinkholes, solution valleys, and disappearing streams developed by groundwater erosion.

Kepler's laws The three laws of planetary motion discovered by Johannes Kepler; planets travel in elliptical orbits, a line joining the planet and the Sun sweeps out equal areas in equal intervals of time, and the square of a planet's period is proportional to the cube of its average distance from the Sun.

kettle A circular to oval depression formed when a once buried or partially buried block of glacial ice melts.

latent heating Transport of heat from one place to another within the Earth-atmosphere system as a consequence of phase changes of water. Heat is supplied for evaporation and sublimation of water at the Earth's surface, and heat is released with condensation and deposition (cloud formation) within the atmosphere.

lateral moraine The sediment deposited as a long ridge of till along the margin of a valley glacier.

laterite A red soil rich in iron or aluminum or both that forms in the tropics by intense chemical weathering.

lava Magma at the Earth's surface; the molten rock material that flows from a volcano or a fissure.

light year The unit of distance used in astronomy; the distance light travels in a year, about 10 trillion km.

lightning A flash of light produced by an electrical discharge in response to the buildup of electrical potential between cloud and ground or between different portions of the same cloud.

lithification The process of converting sediment into sedimentary rock.

lithosphere The outer, rigid part of Earth consisting of the upper mantle, oceanic crust, and continental crust.

loess Windblown silt and clay deposits derived from deserts, Pleistocene glacial outwash, and floodplains of stream in semiarid regions.

longshore current A current between the breaker zone and the beach flowing parallel to the shoreline and produced by wave refraction.

longshore drift The movement of sediment along a shoreline by longshore currents.

low-velocity zone A region in the mantle where seismic waves travel slowly because the rocks are close to the melting point (though still solid); corresponds closely to the asthenosphere.

lunar eclipse The passage of the Moon through the Earth's shadow.

luster The appearance of a mineral in reflected light; the most important distinction is between metallic and nonmetallic minerals.

mafic magma A silica-poor magma containing between 45 and 55% silica and proportionately more calcium, iron, and magnesium than intermediate and felsic magmas.

magma Molten rock material generated within the Earth.

magnitude The total amount of energy released by an earthquake at its source.

main sequence The diagonal band on the Hertzsprung-Russell diagram that includes most ordinary stars.

mantle The thick layer between Earth's crust and core.

maria Dark lava plains on the Moon. Singular form is *mare*.

marine terrace A wave-cut platform now elevated above sea level.

mass wasting The downslope movement of rock, sediment, or soil under the influence of gravity.

meandering stream A stream possessing a single, sinuous channel with broadly looping curves.

mechanical weathering The disaggregation of rock materials by physical forces yielding smaller pieces that retain the chemical composition of the parent material.

mesoscale convective complex (MCC) A nearly circular organized cluster of many interacting thunderstorm cells covering an area of many thousands of square kilometers.

metamorphic facies A group of metamorphic rocks characterized by mineral assemblages formed under the same broad temperature-pressure conditions.

metamorphic rock Any rock that has been altered by heat, pressure, or chemically active fluids or a combination of these agents of metamorphism.

meteorite A mass of matter of extraterrestrial origin that has fallen to Earth.

meteorology Study of the Earth's atmosphere, atmospheric processes, and weather.

Milankovitch cycles Regular variations in the Earth's rotation and orbit that may influence climate and the onset of glacial episodes.

Milky Way galaxy Our own galaxy, a spiral galaxy about 100,000 light years in diameter. We are about 30,000 light years from its center.

mineral A naturally occurring, inorganic, crystalline solid having characteristic physical properties and a narrowly defined chemical composition.

minor planet An object smaller than a planet in orbit around the Sun; this term is preferred by astronomers over the synonym *asteroid*.

Modified Mercalli Intensity Scale A scale having values ranging from I to XII that is used to characterize earthquake intensity based on damage.

Mohorovicic discontinuity (Moho) The boundary between Earth's crust and mantle.

mudflow A mass wasting process; a flow consisting of mostly clay- and silt-sized particles and more than 30% water.

multiple-ring impact basin The structure formed by the largest meteoroid impacts; consists of concentric circular fractures hundreds of kilometers in diameter.

nearshore sediment budget The balance between additions and losses of sediment in the nearshore zone.

neutron A particle of neutral electric charge in the nucleus of an atom.

nonconformity An unconformity in which sedimentary rocks above an erosion surface overlie igneous or metamorphic rocks.

normal fault A dip–slip fault on which the rocks above the fault have moved down relative to the rocks below, resulting in extension of the crust.

nuclear fusion The joining of light atomic nuclei, resulting in the release of energy. The principal energy source in the stars.

nuclei Tiny solid or liquid particles of matter on which condensation or deposition of water vapor takes place; necessary for cloud development.

nucleus The central part of an atom consisting of one or more protons and neutrons.

nuée ardente A mobile dense cloud of hot pyroclastic materials and gases ejected from a volcano. A synonym for *pyroclastic flow*.

oceanic ridge A submarine mountain system found in all of the oceans; it is composed of volcanic rock (mostly basalt) and displays features produced by tension.

oceanic trench A long, narrow depression in the sea floor where subduction occurs.

oceanography Study of Earth's oceans and their physical, chemical, geologic, and biological properties.

ophiolite A sequence of igneous rocks thought to represent a fragment of oceanic lithosphere; composed of peridotite overlain successively by gabbro, sheeted basalt dikes, and submarine basalt lavas.

orbit The path one heavenly body follows under the influence of another body's gravitational attraction.

orogeny The process of forming mountains, especially by folding and thrust faulting; an episode of mountain building. Usually marked by intense igneous activity, deformation, and metamorphism.

orographic lifting The forced rising of air up the slopes of a hill or mountain.

outgassing Release of gases to the atmosphere from hot, molten rock during volcanic activity; thought to be the origin of most atmospheric gases.

oxbow lake A cutoff meander filled with water.

ozone shield Absorption of high intensity solar ultraviolet radiation during the natural formation and destruction of ozone (O_3) within the stratosphere; protects organisms at the Earth's surface.

paleomagnetism The study of remanent magnetism in rocks so that the intensity and direction of the Earth's past magnetic field can be determined.

Pangaea The name proposed by Alfred Wegener for a supercontinent consisting of all the Earth's landmasses that existed at the end of the Paleozoic Era.

parent element An unstable element that changes by radioactive decay into a stable daughter element. See also *daughter element*.

parent material The material that is mechanically and chemically weathered to yield sediment and soil.

passive continental margin The trailing edge of a continental plate consisting of a broad continental shelf and a continental slope and rise, commonly with an abyssal plain adjacent to the rise. Passive continental margins lack volcanism and intense seismic activity.

pediment An erosion surface of low relief gently sloping away from a mountain base.

pedocal A soil of arid and semiarid regions with a thin A horizon and a calcium carbonate-rich B horizon.

perigee The closest point in a satellite's orbit to Earth.

permeability A material's capacity for transmitting fluids.

photic zone The sunlit layer of water in the oceans where plants photosynthesize.

photosynthesis The process whereby plants use sunlight, water, and carbon dioxide to manufacture their food; a byproduct is oxygen (O_2).

planet A large, nonluminous satellite of a star, typically more than a few thousand kilometers in diameter.

planetary albedo The fraction of solar radiation that is scattered or reflected back into space by the Earth-atmosphere system.

plastic flow The flow that occurs in response to pressure and causes permanent deformation as in glacial ice.

plate tectonic theory The theory that large segments of the lithosphere move relative to one another.

playa lake A broad, shallow, temporary lake that forms in an arid region and quickly evaporates to dryness.

pluton An intrusive igneous body that forms when magma cools and crystallizes within the Earth's crust (e.g., batholith and sill).

plutonic rock Rock that crystallizes from magma intruded into or formed in place within the Earth's crust.

point bar The sedimentary body deposited on the gently sloping side of a meander loop.

polar front Narrow transition zone between cold polar easterlies and mild midlatitude westerlies.

polar front jet stream A corridor of relatively strong westerlies situated in the upper troposphere and directly over the polar front.

polar wandering The apparent change in position of Earth's magnetic and geographic poles over time; in reality the poles are fixed but the continents move.

poleward heat transport Flow of heat from tropical to middle and high latitudes in response to temperature gradients brought about by latitudinal imbalances in radiational heating and cooling. Poleward heat transport is accomplished by air mass exchange, migrating storms, and ocean currents.

porosity The percentage of a material's total volume that is pore space.

precession The slow oscillation of Earth's axis over a cycle of 26,000 years due to the gravitation of the Sun and Moon acting on Earth's equatorial bulge.

precipitation Water in solid or liquid form that falls to the Earth's surface from clouds.

pressure gradient force A force operating in the atmosphere that accelerates air parcels away from regions of high pressure toward regions of low pressure in response to an air pressure gradient.

pressure release A mechanical weathering process in which rocks formed deep within the Earth, due to release of pressure, expand upon being exposed at the surface.

principle of cross-cutting relationships A principle used to determine the relative ages of events; holds that an igneous intrusion or fault must be younger than the rocks that it intrudes or cuts.

principle of fossil succession A principle holding that fossils, and especially assemblages of fossils, succeed one another through time in a regular and determinable order.

principle of inclusions A principle that holds that inclusions, or fragments, in a rock unit are older than the rock unit itself, e.g., granite fragments in a sandstone are older than the sandstone.

principle of original horizontality A principle that holds that sediment layers are deposited horizontally or very nearly so.

principle of superposition A principle that holds that younger layers of strata are deposited on top of older strata.

principle of uniformitarianism A principle that holds that we can interpret past events by understanding present-day processes; based on the assumption that natural laws have not changed through time.

proton A positively charged particle found in the nucleus of an atom.

P-wave A compressional, or push-pull, wave; the fastest seismic wave and one that can travel through solids, liquids, and gases; also known as a *primary wave*.

P-wave shadow zone The area between 103° and 143° from an earthquake focus where little P-wave energy is recorded.

pyroclastic material Fragmental material such as ash explosively ejected from a volcano.

radiation A form of energy or energy transport via electromagnetic waves; capable of traveling through a vacuum.

radioactivity The spontaneous decay of an atom to an atom of a different element.

radiosonde A small balloon-borne instrument package equipped with a radio transmitter that measures vertical profiles of temperature, pressure, and humidity in the atmosphere.

rain shadow A region situated downwind of a high mountain barrier and characterized by descending air and a relatively dry climate.

rainshadow desert A desert found on the leeward side of a mountain range; forms because moist marine air moving inland yields precipitation on the windward side of the mountain range and the air descending on the leeward side is much warmer and drier.

recessional moraine A type of end moraine formed when a glacier's terminus retreats, then stabilizes and deposits till.

red giant A large cool star, often hundreds of millions of kilometers in diameter, with a very tenuous outer envelope of gas.

reef A moundlike, wave-resistant structure composed of the skeletons of organisms.

reflection (atmosphere) The process whereby a portion of the radiation that is incident on a surface is turned back by that surface.

reflection (earthquake) The return toward the source of some of a seismic wave's energy when it encounters a boundary separating materials of different density or elasticity within the Earth.

refraction The change in direction and velocity of a seismic wave when it travels from one material into another of different density and elasticity.

regional metamorphism Metamorphism that occurs over a large area resulting from high temperature and pressure, and the action of chemically active fluids within the Earth's crust.

regolith The layer of unconsolidated rock and mineral fragments covers much of the Earth's surface.

relative dating The process of determining the age of an event relative to other events; involves placing geologic events in their correct chronological order, but involves no consideration of when the events occurred in terms of number of years ago.

relative humidity A measure of how close air is to saturation at a specific temperature; expressed as a percentage.

Richter Magnitude Scale An open-ended scale that measures the amount of energy released during an earthquake.

rock cycle A sequence of processes through which Earth materials may pass as they are transformed from one rock type to another.

rockfall A common type of extremely rapid mass wasting involving the movement of solid rock debris down steep slopes.

rockslide A type of rapid mass wasting in which rocks move downslope along a more or less planar surface.

Rossby waves Series of long-wavelength troughs and ridges that characterize the planetary-scale westerlies; also called *long waves*.

Saffir-Simpson Hurricane Intensity Scale Hurricane intensity scale based on central pressure, and specifying a range of wind speed, height of storm surge, and damage potential; 1 is minimal, 5 is most intense.

salinity A measure of the dissolved solids in seawater. Most commonly expressed in parts per thousand (commonly abreviated ‰).

salt crystal growth A mechanical weathering process in which rocks are disaggregated by the growth of salt crystals in crevices and pores.

satellite Any smaller object that travels in an orbit around a larger object, like the Moon orbiting Earth.

saturation vapor pressure The maximum vapor pressure in a sample of air at a specific temperature.

scattering Process whereby small particles disperse radiation in all directions.

scientific method A logical, orderly form of inquiry that involves gathering data, formulating and testing hypotheses, and proposing theories.

sea-floor spreading The theory that the sea floor moves away from spreading ridges and is eventually consumed at subduction zones.

seamount A structure of volcanic origin rising more than 1 km above the sea floor.

second law of thermodynamics All systems tend toward disorder.

sedimentary rock Any rock composed of sediment (e.g., sandstone and limestone).

sedimentary structure Any structure in sedimentary rock such as cross bedding, mud cracks, and animal burrows that formed at the time of deposition or shortly thereafter.

seismology The study of earthquakes.

sensible heating Transport of heat from one location or object to another via conduction and convection.

severe thunderstorm Convective weather systems accompanied by locally damaging surface winds, frequent lightning, or large hail.

shear strength The resisting forces helping to maintain slope stability.

shepherd moon A small satellite orbiting at the edge of a planetary ring that maintains the sharp edges of the ring.

shield volcano A large, dome-shaped volcano with a low rounded profile built up mostly of overlapping basaltic lava flows (e.g., Mauna Loa and Kilauea on the island of Hawaii).

shoreline The line of intersection between the sea or lake and the land.

silica A compound of silicon and oxygen atoms with the formula SiO_2.

silicate A mineral containing silica (e.g., quartz [SiO_2] and potassium feldspar [$KAlSi_3O_8$]).

silica tetrahedron The basic building block of all silicate minerals. It consists of one silicon atom and four oxygen atoms.

sill A tabular or sheetlike concordant pluton.

sinkhole A depression in the ground that forms in karst regions by the solution of the underlying carbonate rocks or by the collapse of a cave roof.

slump A type of mass wasting that occurs along a curved surface of failure and results in the backward rotation of the slump block.

soil Regolith consisting of weathered material, water, air, and humus that can support plants.

soil degradation   Any process leading to a loss of soil productivity; may involve erosion, chemical pollution, and compaction.

soil horizon   A distinct soil layer that differs from other soil layers in texture, structure, composition, and color.

solar altitude   The angle of the sun (90 degrees or less) above the horizon.

solar eclipse   The passage of the Moon in front of the Sun.

solid Earth system   The rocky part of the Earth's surface and interior, as opposed to the atmosphere, biosphere, and hydrosphere.

solifluction   A type of mass wasting involving the slow downslope movement of water-saturated surface materials.

solstice   When the sun is at its maximum poleward locations relative to the Earth (latitudes of 23 degrees 30 minutes North and South); first days of summer and winter.

specific heat   Amount of heat required to raise the temperature of 1 gram of a substance by 1 Celsius degree.

spectral class   A series of classes describing the temperatures and physical conditions of stars, ranging from O (hottest) through B, A, F, G (our Sun) and K to M (coolest).

spheroidal weathering   A type of chemical weathering in which corners and sharp edges of angular rocks weather more rapidly than flat surfaces, thus yielding spherical shapes.

spiral galaxy   A disk-shaped mass of billions of stars, marked by spiral bands of young stars.

spit   A continuation of a beach forming a point of land that projects into a body of water, commonly a bay.

spreading ridge   A midocean ridge where sea-floor spreading occurs.

spring   A place where groundwater flows or seeps out of the ground. Springs occur where the water table intersects the ground surface.

squall line   A narrow band of intense thunderstorms occurring parallel to and ahead of a fast-moving, well-defined cold front.

Stefan-Boltzmann law   The total energy radiated by a blackbody at all wavelengths is directly proportional to the fourth power of the absolute temperature (in Kelvins) of the body.

stony meteorites   A group of meteorites composed of iron and magnesium silicate minerals; comprise about 93% of all meteorites.

storm surge   A rise in sea level along a shore caused primarily by strong onshore winds and, to a lesser extent, low air pressure associated with a storm (often a hurricane); may cause considerable coastal erosion and flooding.

strain   Deformation caused by stress.

stratiform clouds   Layered clouds, such as altostratus or stratus, usually produced by air mass overrunning.

stratosphere   The atmosphere's thermal subdivision situated between the troposphere and mesosphere; primary site of ozone formation. Within the stratosphere, air temperature is constant in the lower portion and increases with altitude in the middle and upper reaches.

stratovolcano   A conical mountain composed of pyroclastic layers, lava flows typically of intermediate composition and mudflows; also called *composite volcano*.

stress   Force per unit area.

strike   The direction of a line formed by the intersection of the horizontal plane with an inclined plane, such as a rock layer.

subduction   The process whereby the leading edge of one plate descends beneath the margin of another plate.

subduction zone   A long, narrow zone at a convergent plate boundary where an oceanic plate descends beneath another plate (e.g., the subduction of the Nazca plate beneath the South American plate).

submarine canyon   A steep-sided canyon below sea level in the continental shelf and slope.

submarine fan   A cone-shaped sedimentary deposit that accumulates on the continental slope and rise.

submergent coast   A coast along which sea level rises with respect to the land or the land subsides.

subpolar lows   High-latitude, semipermanent cyclones marking the convergence of planetary-scale surface southwesterlies of midlatitudes with surface northeasterlies of polar latitudes; Icelandic and Aleutian lows are examples.

subtropical anticyclones   Semipermanent warm, high pressure systems centered over subtropical latitudes of the Atlantic, Pacific, and Indian oceans.

sunspot   A disturbance on the surface of the Sun, darker because it is cooler than the surrounding area.

supercell thunderstorm   A relatively long-lived, large, and intense cell characterized by an exceptionally strong updraft; may produce a tornado.

supernova   Catastrophic collapse and explosion of a massive star.

surface wave   A seismic wave that travels along the outer surface of the Earth or on rock boundaries within the Earth.

S-wave   A shear wave that moves material perpendicular to the direction of travel, thereby producing shear stresses in the material it moves through; also known as a *secondary wave*.

S-wave shadow zone   Those areas more than 103° from an earthquake focus where no S-waves are recorded.

syncline   A down-arched fold in which the youngest exposed rocks coincide with the fold axis, and all strata dip inward toward the axis.

system   A collection of related parts that interact in organized ways.

temperature   A measure of the average kinetic energy of the individual atoms or molecules composing a substance.

terminal velocity   Constant downward-directed speed of a particle within a fluid due to a balance between gravity and fluid resistance.

terrestrial planet   Any of the four innermost planets of the solar system (Mercury, Venus, Earth, and Mars). All are small and have high mean densities, indicating that they are composed of rock and metallic elements. See also *Jovian planet*.

theory   An explanation for some natural phenomenon that has a large body of supporting evidence; to be considered scientific, a theory must be testable (e.g., plate tectonic theory).

thermal convection   A type of circulation during which warm material rises, moves laterally, cools and sinks, and is reheated and reenters the cycle. Applies to any materials, including the formation of cumulus clouds in the atmosphere and the flow of mantle material in Earth's interior.

thermal expansion and contraction   A type of mechanical weathering in which the volume of rock changes in response to heating and cooling.

thermohaline circulation   Deep-ocean circulation resulting from differences in density of adjacent water masses.

thrust fault   A type of reverse fault with a fault plane dipping less than 45°.

thunder   Sound accompanying lightning; produced by violent expansion of air due to intense heating by a lightning discharge.

thunderstorm   A small-scale weather system produced by strong convection currents that reach to great altitudes within the troposphere. Consists of cumulonimbus clouds accompanied by lightning and thunder and, often, locally heavy rainfall (or snowfall) and gusty surface winds.

tidal forces Effects that result from the gravity of one heavenly body pulling more strongly on the near side of another object than the far side.

tide The regular fluctuation in the sea's surface in response to the gravitational attraction of the Moon and Sun.

tombolo A type of spit that extends out into the sea or a lake and connects an island to the mainland.

tornado A small mass of air that whirls rapidly about an almost vertical axis. The tornado is made visible by clouds, and by dust and debris drawn into the system.

Tornado Alley Region of maximum tornado frequency in North America; a corridor stretching from central Texas northward into Oklahoma, Kansas, and Nebraska, and eastward into central Illinois and Indiana.

trade winds Prevailing planetary-scale surface winds in tropical latitudes; blow from the northeast in the Northern Hemisphere and from the southeast in the Southern Hemisphere.

transcurrent fault A fault involving horizontal movement so that blocks on opposite sides of a fault plane slide sideways past one another.

transform fault A type of fault that changes one type of motion between plates into another type of motion.

troposphere Lowest thermal subdivision of the atmosphere in which air temperature normally drops with altitude; site of most weather.

tsunami A destructive sea wave that is usually produced by an earthquake but can also be caused by submarine landslides or volcanic eruptions.

turbidity current A sediment-water mixture denser than normal seawater that flows downslope to the deep-sea floor.

unconformity An erosion surface separating younger strata from older rocks.

upwelling The slow circulation of ocean water from depth to the surface.

U-shaped glacial trough A valley with very steep or vertical walls and a broad, rather flat floor. Formed by the movement of a glacier through a stream valley.

valley glacier A glacier confined to a mountain valley or to an interconnected system of mountain valleys.

vapor pressure That portion of the total air pressure exerted by the water vapor component of air.

viscosity A fluid's resistance to flow.

volcano A conical mountain formed around a vent as a result of the eruption of lava and pyroclastic materials.

volcanic ash Powdered rock erupted by a volcano.

volcanic (extrusive igneous) rock Igneous rock formed when magma is extruded onto the Earth's surface where it cools and crystallizes, or when pyroclastic materials become consolidated.

water table The surface separating the zone of aeration from the underlying zone of saturation.

wave Oscillation on a water surface; most waves are generated by wind blowing over water.

wave base A depth of about one-half wave length, where the diameter of the orbits of water particles in waves is essentially zero; the depth below which water is not affected by surface waves.

wave-cut platform A beveled surface that slopes gently in a seaward direction; formed by erosion and landward retreat of a sea cliff.

wave refraction The bending of waves so that they more nearly parallel the shoreline.

weather The state of the atmosphere at some place and time described in terms of such variables as temperature, cloudiness, and precipitation.

weathering The physical and chemical breakdown of materials of the Earth's crust by interaction with the atmosphere and biosphere.

westerlies Winds that encircle midlatitudes from west to east in a wavelike pattern of ridges and troughs.

white dwarf A very dense and small but very hot star where nuclear reactions have ceased.

Wien's displacement law The higher the temperature of a radiating object, the shorter is the wavelength of maximum radiation intensity; applies to *blackbodies.*

wind Air in motion relative to the Earth's surface.

zone of aeration The zone above the water table that contains both water and air within the pore spaces of the rock, sediment, or soil.

zone of saturation The zone below the water table in which all pore spaces are filled with groundwater.

ILLUSTRATION CREDITS

Chapter 1

1.12: Victor Royer.

Chapter 3

3.21: Reprinted with permission from AGI Data Sheet 35, 4, *AGI Data Sheets,* 3d ed., 1989, American Geological Institute.

Chapter 4

4.3a: From A. Cox and R. R. Doell, "Review of Paleomagnetism." *GSA Bulletin* 71 (1960): 758, Figure 33.

Chapter 5

Perspective 5.2, Fig. 1: Modified from *U.S. News and World Report* (18 March 1991): 72–73.

Chapter 7

7.3: From E. Bullard, J. E. Everett, and A. G. Smith, "The Fit of the Continents Around the Atlantic." *Philosophical Transactions of the Royal Society of London* 258 (1965). Reproduced with permission of the Royal Society, J. E. Everett, and A. G. Smith.

7.5: Modified from E. H. Colbert, *Wandering Lands and Animals* (1973): 72, Figure 31.

7.8: Reprinted with permission from Cox and Doell, "Review of Paleomagnetism," *GSA,* v. 71, 1960, p. 758 (Fig. 33), Geological Society of America.

7.9: From A. Cox, "Geomagnetic Reversals." *Science* 163 (17 January 1969): 240, Figure 4. Copyright 1969 by the AAAS.

7.23a, b, c: Reprinted with permission from Dietz and Holden, "Reconstruction of Pangaea: Breakup and Dispersion of Continents, Permian to Present," *Journal of Geophysical Research,* 1970, v. 75, no. 26, pp. 4939–4956. © 1970 by the American Geophysical Union.

Chapter 8

Perspective 8.1, Fig. 1: From *Earthquakes* by Bruce A. Bolt. Copyright © 1988 by W. H. Freeman and Co. Reprinted by permission.

8.4: Data from National Oceanic and Atmospheric Administration.

8.8: From *Nuclear Explosions and Earthquakes: The Parted Veil* by B. A. Bolt. Copyright © 1976 by W. H. Freeman and Co. Reprinted by permission.

8.9a: Data from C. F. Richter, *Elementary Seismology.* 1958. W. H. Freeman and Co.

8.11: From M. L. Blair and W. W. Spangle, *U.S. Geological Survey Professional paper 941-B.* 1979.

8.18: From M. L. Blair and W. W. Spangle, *U.S. Geological Survey Professional paper 941-B.* 1979.

8.21: From G. C. Brown and A. E. Musset, *The Inaccessible Earth* (London: Chapman & Hall, 1981): 17 and 124, Figures 12.7a and 7.11. Reprinted by permission of Chapman & Hall.

Chapter 10

10.8: USGS *Earthquakes and Volcanoes,* v. 25, no. 1, 1994. Adapted from a cross-section by T. L. Davis and J. Namson, 1994.

Chapter 11

11.7: From *The Story of the Great Geologists* by Carroll Lane Fenton and Mildred Adams Fenton. Used by permission of Doubleday, a division of Bantam Doubleday Dell Publishing Group, Inc.

11.9: From *Geologic Time.* 1981. U.S. Geological Survey.

11.10: From A. R. Palmer, "The Decade of North American Geology, 1983 Geologic Time Scale," *Geology* (Boulder, Colorado: Geological Society of America, 1983): 504. Reprinted by permission of the Geological Society of America.

Chapter 12

Perspective 12.2, Fig. 1: From Phyllis Young Forsyth, *Atlantis: The Making of a Myth* (Montreal: McGill-Queen's University Press): 13, Figure 2.

12.11: Modified from Bruce C. Heezen and Charles D. Hollister, *The Face of the Deep* (New York: Oxford University Press, 1971): 297, Figure 8.15. Copyright © 1971 by Bruce Heezen and Charles Hollister. Used by permission of Oxford University Press, Inc.

12.13: From Alyn and Alison Duzbury, *An Introduction to the World's Oceans.* Copyright © 1984 Addison-Wesley Publishing Company, Inc., Reading, Massachusetts. Reprinted by permission of Wm. C. Brown Publishers, Dubuque, Iowa.
12.15b: Adapted from R. D. Ballard and T. H. Van Andel, *Geological Society of America Bulletin* 88 (1977): 507–530.
12.22: From *U.S. Geological Survey.*

Chapter 13

Perspective 13.2, Fig. 1: From *U.S. Geological Survey Circular* 1075.
13.4a, c: Dennis Tasa.
13.6: Adapted from G. Wurst, *Journal of Geo-Physical Research* 66 (1961): 3261–3271.
13.18: From *U.S. Geological Survey Circular* 1075.

Chapter 14

Perspective 14.2, Fig. 1: NOAA, Climate Prediction Center.
14.12: NOAA data.

Chapter 15

Perspective 15.2, Fig. 1: From University of East Anglia/British Meteorological Office.
15.11: From "Ozone: What is it, and why do we care about it?" *NASA Facts* (Greenbelt, MD: Goddard Space Flight Center, 1993).
15.15: From R. G. Fleagle and J. Businger, *An Introduction to Atmospheric Physics.* New York: Academic Press, 1980, p. 232.
15.16: NOAA and Scripps Institution of Oceanography.
15.20: From W. S. Broecker, Columbia University's Lamont-Doherty Earth Observatory.

Chapter 17

17.1: Nese, J. M. et al., *A World of Weather, Fundamentals of Meteorology,* 1996, Dubuque, IA: Kendall/Hunt Publishing Company, p. 12.

17.5: Modified from F. K. Lutgens and E. J. Tarbuck, *The Atmosphere: An Introduction to Meteorology,* 4th ed., 1989, p. 169. Adapted by permission of Prentice Hall, Inc.

Chapter 18

18.1: NOAA/National Weather Service.
Perspective 18.1, Fig. 4: Student Activities in meteorology, NOAA/Environmental Research Laboratories/Forecast Systems Laboratory Pub., Act No. 1, Version 2.
18.2: NOAA, Climate Analysis Center.
18.5: From Moran, J. M., and M. D. Morgan, 1994. *Meteorology: The Atmosphere and the Science of Weather,* New York: Macmillan College Publishing, p. 247.
18.7: From Moran, J. M., and M. D. Morgan, 1994. *Meteorology: The Atmosphere and the Science of Weather,* New York: Macmillan College Publishing, p. 249.
18.8: From Moran, J. M., and M. D. Morgan, 1994. *Meteorology: The Atmosphere and the Science of Weather,* New York: Macmillan College Publishing, p. 250.
18.13: NOAA/National Weather Service.
18.15: NOAA/National Weather Service, Advanced Spotters' Field Guide.
18.19: NOAA.
18.24: Nese, J. M., et al., 1195. *A World of Weather: Fundamentals of Meteorology,* Dubuque, IA. Kendall/Hunt Publishing Company, p. 269.
18.28: NOAA/National Weather Service.
18.31: NOAA/National Weather Service.

Chapter 19

19.33: Adapted from Hartmann, *Natural History,* Nov. 1989, pp. 68–77.

PHOTOGRAPH CREDITS

Chapter 1

Opener: Photo courtesy of NASA.
1.1: Photo courtesy of D.J. Nicholas, U.S. Geological Survey.
1.2: Photo courtesy of D.J. Nicholas, U.S. Geological Survey.
1.5: Photo courtesy of U.S. Geological Survey.
1.8: Dr. Jean Lorre/SPL/Science Source/Photo Researchers Inc.
1.10: Photo courtesy of National Park Service.
1.11a: Photo courtesy of Sue Monroe.
1.11b: Photo courtesy of Sue Monroe.

Chapter 2

Opener: Photo courtesy of D. Penland, National Museum of Natural History, Smithsonian Institute.
2.1a: Jerry Jacka Photography.
2.1b: Photo courtesy of Shane McClure and Robert Kane, Gemological Institute of America.
2.2: Photo courtesy of Steve Dutch.
2.7a: Photo courtesy of James S. Monroe.
2.7b: Runk/Schoenberger/Grant Heilman Photography Inc.
2.7c: Photo courtesy of Sue Monroe.
2.8: Photo courtesy of Sue Monroe.
2.9: A.J. Copley/Visuals Unlimited.
2.11a: Photo courtesy of D. Penland, National Museum of Natural History, Smithsonian Institute.
2.11b: Wards Natural Science Establishment Inc.
2.14a: Photo courtesy of Sue Monroe.
2.14b: Photo courtesy of Sue Monroe.

Chapter 3

Opener: Photo courtesy of R.V. Dietrich.
3.3: Photo courtesy of Sue Monroe.
3.4a–d: Photo courtesy of Sue Monroe.
3.5a,b: Photo courtesy of Sue Monroe.
3.5c: Photo courtesy of David J. Matty.
3.7a,b: Photo courtesy of Sue Monroe.
3.8: Photo courtesy of Sue Monroe.
3.10a–d: Photo courtesy of Sue Monroe.
3.11a: Photo courtesy of Sue Monroe.
3.11b: Photo courtesy of R.V. Dietrich.

3.12a–d: Photo courtesy of Sue Monroe.
3.14c: Photo courtesy of Sue Monroe.
3.15a,b: Photo courtesy of Sue Monroe.
Perspective 3.1, Fig. 1: Photo courtesy of Sue Monroe.
Perspective 3.1, Fig. 2: Photo courtesy of R.L. Elderind/ U.S. Geological Survey.
3.18a,b: Photo courtesy of Sue Monroe.
3.19a: Photo courtesy of Sue Monroe.
3.20: Photo courtesy of Sue Monroe.
3.22: Photo courtesy of Sue Monroe.
3.23a: Photo courtesy of Montana Historical Society.
3.23b: Steve McCutcheon/Visuals Unlimited.

Chapter 4

4.1: Photo courtesy of U.S. Geological Survey.
4.3: Photo courtesy of University of Colorado.
4.4a: Photo courtesy of W.D. Lowry.
4.8b: Photo courtesy of John S. Shelton.
4.8c: Walt Anderson/Visuals Unlimited.
Perspective 4.1, Fig. 1: Photo courtesy of Kansas State Historical Society.
4.11: Science VU/Visuals Unlimited.
4.14b: T. Spencer/Colorific.
4.16: Photo courtesy of Sue Monroe.
4.17b: Photo courtesy of John S. Shelton.
4.18b: Photo courtesy of Precision Graphics.
4.19: Stephen R. Lower/GeoPhoto Publication Company.
4.21b: Photo courtesy of James S. Monroe.
Perspective 4.2, Fig. 1: UPI/Bettmann.
4.24b: Dell R. Foutz/Visuals Unlimited.

Chapter 5

Opener: Photo courtesy of the Kentucky Department of Parks.
5.1: Photo courtesy of Michael Lawton.
5.6: Photo courtesy of John S. Shelton.
5.7b: Photo courtesy of John S. Shelton.
5.11: Photo courtesy of John S. Shelton.
5.16: Photo courtesy of John S. Shelton.
5.23: Photo courtesy of John S. Shelton.
5.26: Daniel W. Gotshall/Visuals Unlimited.
5.28: Photo courtesy of U.S. Geological Survey.

Chapter 6

Opener: Photo courtesy of Michael J. Hambrey.
6.1: Offentliche Kunstammlung Basel, Kupferstichkabinett.
6.3: Engineering Mechanics, Virginia Polytechnic Institute and State University.
6.8: Photo courtesy of Sue Monroe.
6.9: Photo courtesy of Swiss National Tourist Office.
6.10: Photo courtesy of R.V. Dietrich.
6.12: Photo courtesy of Engineering Mechanics, Virginia Polytechnic Institution & State University.
Perspective 6.1, Fig. 2: Photo courtesy of U.S. Geological Survey.
6.19: Photo courtesy of Marion A. Whitney.
Perspective 6.2, Fig. 1: Steve McMurray/Magnum Photos.
6.20c: Photo courtesy of David J. Matty.
6.20d: Martin G. Miller/Visuals Unlimited.
6.23: Photo courtesy of James S. Monroe.
6.24: Photo courtesy of Alan L. and Linda D. Mayo, GeoPhoto Publishing Company.
6.25: Photo courtesy of James S. Monroe.
6.26: Photo courtesy of James S. Monroe.

Chapter 7

Opener: Photo courtesy of NASA.
7.4: Photo courtesy of Scott Katz.

Chapter 8

Opener: Tony Freeman/Photo Edit.
8.1b: Lee Stone/Sygma.
8.1c: Ted Soqui/Sygma.
8.1d: Don Boomer/Time Inc.
8.1e: Photo courtesy of Rick Rickman.
8.5: Bunyo Ishikawa/Sygma.
8.6: Photo courtesy of Dennis Fox.
8.7: Photo courtesy of J.K. Hillers, U.S. Geological Survey.
8.13b: Photo courtesy of U.S. Geological Survey.
8.14: Photo courtesy of M. Celebi/U.S. Geological Survey.
8.15: Cindy Andrews/Gamma Liaison Inc.
8.16: Photo courtesy of Martin E. Klimek, Marin Independent Journal.
8.17: Photo courtesy of Bishop Museum.
8.19: Photo courtesy of ChinaStock Photo Agency.

Chapter 9

Opener: Bille Eldred/National Geographic Image Collection.
Perspective 9.1, Fig. 1: D.R. Crandell/U.S. Geological Survey.
Perspective 9.1, Fig. 2a–c: Gary Rosenquist/Earth Images.
Perspective 9.1, Fig. 3: U.S. Geological Survey.
9.5: Photo courtesy of Steve Dutch.
9.6: Photo courtesy of Ward's Natural Science Establishment, Inc.
9.7: Photo courtesy of Steve Dutch.
9.10c: Photo courtesy of NOAA.
9.11: Photo courtesy of U.S. Geological Survey.
9.12: J. Langevin/Sygma.
9.13: Anthony Suau/Gamma Liaison.
9.14c: Yomiuri/AP Wide World Photos.
9.16: Photo courtesy of John S. Shelton.
9.17b: Photo courtesy of NASA.
9.17d: Photo courtesy of U.S. Geological Survey.
9.18: D.R. Crandell.
9.20a: Photo courtesy of U.S. Geological Survey.
9.20b: Photo courtesy of U.S. Geological Survey.
9.20c: Photo courtesy of Steve Dutch.
9.24: Photo courtesy of Steve Dutch.

Chapter 10

Opener: Photo courtesy of Peter Kresan.
10.1: Nicholas Devoree III/Photographers Aspen, Inc.
10.2: Dianne Roberts/National Geographic Image Collection.
10.7a: Photo courtesy of B. Bradley, Department of Geology University of Colorado.
10.7b: Doug Kokell/Visuals Unlimited.
10.9a: Photo courtesy of John S. Shelton.
10.14b: Catherine Ursillo/Photo Researchers, Inc.
10.14c: Photo courtesy of Steve Dutch.
10.18b: Photri.

Chapter 11

Opener: Photo courtesy of James S. Monroe.
11.2: Photo courtesy of R.V. Dietrich.
11.3a,b: Photo courtesy of David J. Matty.
11.4a: Photo courtesy of James S. Monroe.
11.6b: Photo courtesy of James S. Monroe.
11.6d: Photo courtesy of Dorothy L. Stouth.
11.6f: Photo courtesy of James S. Monroe.
11.14: V. Khristoforov/Sovfoto.
11.15a: Photo courtesy of James S. Monroe.
11.15b–e: Photo courtesy of Sue Monroe.
11.16a,b: Photo courtesy of Neville Pledge, South Australian Museum.
11.17: Photo courtesy of Field Museum of Natural History Chicago.
11.20: Photo courtesy of Smithsonian Institution.
Perspective 11.2, Fig. 1: Photo courtesy of University of Nebraska State Museum.
11.24: Photo courtesy of Field Museum of Natural History Chicago.

Chapter 12

Opener: Douglas Faulknet, Science Source/Photo Researchers, Inc.
12.2: The Granger Collection.
12.5c: Photo courtesy of James S. Monroe.
12.8: World Ocean Floor Map by Bruce C. Heezen and Marie Tharp, 1977. Copyright Marie Tharp 1977.
Perspective 12.2, Fig. 2: Painting by Lloyd K. Townsend, copyright National Geographic Society.

12.19: Photo courtesy of Dr. Bruce Heezen, Lamont-Doherty Geologic Observatory, Columbia University, Courtesy Scripps Institution of Oceanography, University of California San Diego.
12.20: Photo courtesy of Carl Roessler.

Chapter 13

Opener: Photo courtesy of James S. Monroe.
13.1: Photo courtesy of Rosenberg Library, Galveston, Texas.
13.2: Photo courtesy of Rosenberg Library, Galveston, Texas.
13.8a,b: Photo courtesy of Karl Kuhn.
13.11a: Photo courtesy of Precision Graphics.
13.12: Photo courtesy of John S. Shelton.
13.13a: Photo courtesy of James S. Monroe.
13.13b: Photo courtesy of Michael Slear.
13.14b: Photo courtesy of John S. Shelton.
13.16b: Photo courtesy of James S. Monroe.
13.17b: Photo courtesy of James S. Monroe.
Perspective 13.2, Fig. 3a,b: Photo courtesy of U.S. Army Corps of Engineers.
13.21b: Photo courtesy of John S. Shelton.
13.22b: Photo courtesy of Nick Harvey.
13.22c: Photo courtesy of Suzanne and Nick Geary/Tony Stone World Wide.
13.23: Photo courtesy of GEOPIC/Earth Satellite Corporation.
13.24: Photo courtesy of Jerry Westby.
13.25: Photo courtesy of James S. Monroe.

Chapter 14

Opener: Photo courtesy of J.M. Moran.
14.1: Photo courtesy of J.M. Moran.
14.2: Photo Researchers, Inc.
14.3: Photo courtesy of NOAA, National Geophysical Data Center, Boulder, CO.
Perspective 14.1, Fig. 1: Photo courtesy of J.M. Moran.
14.5: Photo courtesy of J.M. Moran.
14.6: Photo courtesy of J.M. Moran.
14.7a,b: Photo courtesy of J.M. Moran.
14.9: Photo courtesy of J.M. Moran.
14.14a: Photo courtesy of J.M. Moran.
14.19: Photo Researchers, Inc.
14.21: Photo Researchers, Inc.

Chapter 15

Opener: Photo courtesy of J.M. Moran.
15.10: Photo courtesy of J.M. Moran.
15.12: Photo Researchers, Inc.
Perspective 15.1, Fig. 1: Photo Researchers, Inc.
15.14: Photo courtesy of J.M. Moran.

Chapter 16

Opener: Photo courtesy of J.M. Moran.
16.1: Photo Researchers, Inc.

16.3: Photo courtesy of J.M. Moran.
16.9: Photo courtesy of J.M. Moran.
16.14a–c: Photo courtesy of J.M. Moran.
16.15a,b: Photo courtesy of J.M. Moran.
16.16a,b: Photo courtesy of J.M. Moran.
16.17a–c: Photo courtesy of J.M. Moran.
16.19: Photo courtesy of J.M. Moran.
16.21: Photo courtesy of J.M. Moran.
16.22a: Photo Researchers, Inc.
16.22b: Photo courtesy of J.M. Moran.
16.23: Photo courtesy of J.M. Moran.
16.24: Photo courtesy of J.M. Moran.

Chapter 17

Opener: Photo courtesy of J.M. Moran.
17.12a: Photo courtesy of J.M. Moran.
17.12b: Photo Researchers, Inc.
17.15: Photo Researchers, Inc.
Perspective 17.1, Fig. 1a,b: Photo courtesy of J.M. Moran.

Chapter 18

18.3: Photo Researchers, Inc.
18.6a: Photo courtesy of J.M. Moran.
18.14: Photo courtesy of J.M. Moran.
18.16: Photo courtesy of J.M. Moran.
18.18: Photo courtesy of NOAA.
18.22: Keith Kent, Science Library/Photo Researchers, Inc.
18.25: Catherine Ursillo/Photo Researchers, Inc.
18.26: National Center for Atmospheric Research/University Corporation for Atmospheric Research/National Science Foundation.
18.27: Howard Bluestein/Photo Researchers, Inc.
18.29: Walter Stricklin/Atlanta Journal-Constitution/Gamma Liaison.
Earthfact, Fig. 1: Photo courtesy of J.M. Moran.
18.30: Photo courtesy of NASA.
Perspective 18.1, Fig. 1: Photo courtesy of J.M. Moran.
Perspective 18.1, Fig. 2: Photo courtesy of J.M. Moran.
Perspective 18.1, Fig. 3: Photo courtesy of NOAA/National Weather Service.
18.32: Mark M. Lawrence/The Stock Market.
Perspective 18.2, Fig. 1: Photo courtesy of NOAA/National Weather Service.
Perspective 18.2, Fig. 2: Photo courtesy of J.M. Moran.
Perspective 18.2, Fig. 3: Photo courtesy of J.M. Moran.
Perspective 18.2, Fig. 4: Photo courtesy of J.M. Moran.
18.33: Photo courtesy of J.M. Moran.

Chapter 19

Opener: Photo courtesy Herb Orth, Life Magazine/Time Warner Inc., painting by Chesley Bonestell, The Estate of Chesley Bonestell.
19.8: Ken Graham/Panoramic Images.
19.12a–d: Photo courtesy of Lick Observatory.

19.13a: Lorenz Denney/Science Photo Library/Photo Researchers, Inc.

19.14a–c: Photo courtesy of Lick Observatory.

19.17: Photo courtesy of NASA.

19.18: Photo courtesy of Jim Rouse, of the 8-16-89 lunar eclipse.

19.26: Photo courtesy of NASA/Galileo Project.

Earthfact, Fig. 1: SovPhoto.

19.29: Photo courtesy Geotimes, 1976, May 25–27.

19.30a: Francois Gohier/Photo Researchers, Inc.

19.30b: Photo courtesy of NASA.

19.31a: Photo courtesy of NASA.

19.31b: Photo courtesy of NASA.

19.31c: Photo courtesy of NASA.

Chapter 20

Opener: NASA.

20.1a: Photri.

20.1b: Science Library/Photo Research.

20.1c: Photo courtesy of NASA.

20.2a: Finley Holiday Film Corp./NASA/JPL.

20.2b: Photo courtesy of NASA/JPL.

20.4a: Photo courtesy of Fotosmith.

20.4b–d: Photo courtesy of Steven Dutch.

Perspective 20.1, Fig. 1: Photo courtesy of U.S. Geological Survey.

20.5a: Photo courtesy of NASA.

20.5b: Photo courtesy of NASA.

20.5c: Photo courtesy of JPL/California Institute of Technology/NASA.

20.6a: Finley Holiday Film Corp./NASA/JPL.

20.6b: Photo courtesy of NASA/JPL.

20.6c: Photo courtesy of NASA/JPL.

20.6d: Finley Holiday Film Corp./NASA/JPL.

20.7a: A.S.P./Science Source/Photo Researchers, Inc.

20.7b: Photo courtesy of NASA.

20.7c: Finley Holiday Film Corp./NASA/JPL.

20.7d: Finley Holiday Film Corp./NASA/U.S. Geological Survey.

Perspective 20.2, Fig. 1: Finley Holiday Film Corp./NASA.

Perspective 20.2, Fig. 2: NASA/Science Photo Library/Photo Researchers, Inc.

20.8: NASA/JPL.

20.10: Photri.

20.11: Finley Holiday Film Corp./NASA/JPL.

20.12a: Photo courtesy of NASA.

20.12b: Photo courtesy of NASA.

20.13a–d: Finley Holiday Film Corp./NASA/JPL.

20.14: Finley Holiday Film Corp./NASA/JPL.

20.15a: Photo courtesy of NASA.

20.15b: Photo courtesy of NASA.

20.16: Photo courtesy of NASA.

20.19a–d: Photo courtesy of NASA.

20.21b: Tony Stone Images.

20.23: Photo courtesy of Naval Research Laboratory.

20.24a: Photo courtesy of Fotosmith.

20.24b: Photo courtesy of Fotosmith.

20.24c: Science Library/Photo Researchers, Inc.

Chapter 21

Opener: Finley Holiday Film Corp./NASA/JPL.

Earthfact, Fig. 1a,b: Finley Holiday Film Corp./NASA/JPL.

21.13: Photo courtesy of Chris Floyd.

21.15: University of Cambridge, Mullard Radio Astronomy Observatory, with compliments of Professor Anthony Hewish.

21.16a: Royal Observatory, Edinburgh, and Anglo-Australian Telescope Board.

21.16b: Photo courtesy of National Optical Astronomy Observatories.

21.16c: Royal Observatory, Edinburgh, and Anglo-Australian Telescope Board.

21.17: J. Baum & N. Henbest/Science Photo Library/Photo Researchers.

21.19a: Photo courtesy of National Optical Astronomy Observatories.

21.20a,b: Photo courtesy of National Optical Astronomy Observatories.

21.22: Photo courtesy of R.J.E. Peebles.

INDEX